# Comprehensible biochemistry

# Comprehensible biochemistry

**Michael Yudkin**
Lecturer in Biochemistry, University of Oxford
Tutor in Biochemistry, University College, Oxford

**Robin Offord**
Lecturer in Molecular Biophysics, University of Oxford
Tutor in Biochemistry, Christ Church, Oxford

**Longman**

ISBN 0 582 44251 6

Library of Congress Catalog Card Number: 73-85212

A United States version of this book
entitled **Biochemistry** is published
by Houghton Mifflin Company, Boston

Printed in Great Britain by
William Clowes & Sons, Limited
London, Beccles and Colchester

# Contents

# Acknowledgements

We are glad to have this opportunity of expressing our thanks to Dr. H. B. F. Dixon, Dr. K. G. H. Dyke, Dr. B. C. Loughman, Professor J. Mandelstam, Dr. A. Miller, Dr. J. D. Priddle, Mr. D. Webster, Dr. D. H. Williamson and Dr. J. Woodhead-Galloway, each of whom has read and criticized some part of the draft of this book. Their comments have greatly improved our manuscript, and have saved us from committing ourselves in print to several errors.

In some cases we have not accepted the suggestions that the reviewers made, and we alone, therefore, are responsible for any errors and misinterpretations that may remain in the text.

Cambridge University Press have allowed us to use some material from our earlier *Guidebook to Biochemistry* (1971).

We are greatly indebted to Mr. Robert Welham of Longman Group Limited for his help at all stages during the planning, writing and production of this book.

We wish to thank all those people and organizations that have kindly granted us permission to reproduce copyright illustrations. Details of the figures concerned are listed below:

Fig. 3.6 from Pauling, L., *The Nature of the Chemical Bond*, (copyright 1939 and 1940 by Cornell University, 3rd edn. copyright 1960 by Cornell University);

Figs. 4.5, 4.7, 5.8 from Dickerson, R. and Geis, I., *Structure and Action of Proteins*, 29 and 41, Harper and Row (1970);

Figs. 4.6, 4.10, 5.9, 5.15, 6.4, 6.8, 6.21, 12.3, 12.6, 17.1, 19.1, 26.2 from Yudkin, M. D. and Offord, R. E., *Guidebook to Biochemistry*, 23, 24, 31, 40, 43, 48, 46, 77, 81, 113, 123, 168, Cambridge University Press (1971);

Fig. 4.18 supplied by Blake, C. C. F. and Swan, I.D.A., Laboratory of Molecular Biophysics, University of Oxford;

Fig. 4.19 from Watenpaugh, K. D. et al., *Cold Spring Harbor Symposium XXXVI*, Fig. 6;

Fig. 7.7 from Setlow, R. and Pollard, E., *Molecular Biophysics*, 147, Addison-Wesley (1962) (by permission of Fuller, W., Kings College, London);

Fig. 7.8 from Davidson, J. T., *The Biochemistry of the Nucleic Acids*, 139, Methuen and Co. Ltd. (1969);

Fig. 7.13a from Holley, R. W., *Scientific American* **214**, 30 (1966), (copyright 1966 by Scientific American Inc. All rights reserved);

Fig. 7.13b from Madison, J. T., *Annu. Rev. Biochem.*, **37**, 140 (1968);

Fig. 7.13c from Sanger, F., *Biochem. J.*, **124**, 833 (1970);

Fig. 7.15a from Breedis, C., Berwick, L. and Anderson, T. F., *Virology*, **17**, 84–94 (1962), (supplied by Anderson, T. F. Institute for Cancer Research, Philadelphia);

Fig. 7.15b from Kellenberger, E., *Path. Microbiol.*, **28**, 540–560 (1965), (supplied by Kellenberger, E., Biozertum der Universität, Basel);

Fig. 7.16 from Setlow, R. and Pollard, E., *op. cit.*, 152, (by permission of Holmes, K. C. and Caspar, D. L.);

Fig. 7.17 from Offord, R. E., *J. Mol. Biol.*, **17**, 370 (1966);

Fig. 8.24 from Kilbourn, B. T., *J. Mol. Biol.*, **30**, 559 (1967);

Fig. 10.1 supplied by Poole, J. C. F., Sir William Dunn School of Pathology, University of Oxford;

Figs. 10.2, 10.3 from duPraw, E., *Cell and Molecular Biology*, 447 and 130, Academic Press Inc., New York (1968);

Figs. 10.4, 13.1 supplied by Brangeon, J. and Juniper, B. E., Botany Department, University of Oxford;

Figs. 10.5, 10.6 supplied by Kay, D., Sir William Dunn School of Pathology, University of Oxford;

Fig. 11.1 from Ross, B. D., *Perfusion Techniques in Biochemistry*, 149, The Clarendon Press, Oxford (1972), (adapted from Miller *et al.*, *Experimental Medicine*, 94);

Fig. 18.2a from Lowy, J. and Hanson, E. J., *J. Mol. Biol.*, **11**, 293 (1965);

Fig. 18.3 from Gibbons, I. and Grimstone, A., *J. Biophys. Biochem. Cyt.*, **7**, 697 (1970);

Figs. 24.3, 24.4 from Yanofsky, C., *J. Am. Med. Ass.*, **218**, 1029 (1971);

Fig. 25.1 from Meselson, M. and Stahl, F., *Proc. Nat. Acad. Sci. USA*, **44**, 675, (supplied by Meselson, M., Biological Laboratory, Harvard University);

Figs. 27.2, 27.4, 27.5 from Dayhoff, M. O. (ed.), *Atlas of Protein Sequence and Structure 1972*, National Biomedical Research Foundation, Washington;

Fig. 27.7 based on data from Black, J. A. and Dixon, G. H., *Nature*, **218**, 736 (1968);

Figs. 29.2, 29.3, 29.4 from Gerhart, J. C. and Pardee, A. B., *J. Biol. Chem.*, **237**, 893 and 894 (1962);

Figs. A2.19a,b from Sanger, F., Brownlee, G. and Barrell, B., *J. Mol. Biol.*, **13**, 373–398 (1965).

We have taken the opportunity provided by the reprinting of this Edition to correct some errors. We have also rewritten a number of short sections of the text.

We are grateful to Dr R. N. Campagne, Professor R. T. Ross, Mr P. R. Butler, Dr M. Davies, Professor A. L. Underwood and Dr G. W. Crosbie for sending us a number of valuable criticisms.

# Foreword

The authors set themselves the admirable aim of selecting from the wealth of biochemical information those aspects which can be related to the life of the cell and the organism as a whole, and of emphasizing unifying principles. They have gone a long way in achieving this goal. In clear, simple and attractive language they present a coherent and thus comprehensible picture, rather than amassing unconnected pieces of information. It is a further advantage of the book that it describes key experiments by which important information has come to light.

January 1975                                                          Hans Krebs

# Preface

There are many textbooks of biochemistry at present available, of which some are quite up-to-date and several are more comprehensive than ours. Why, then, have we thought it worth writing a new textbook?

In the past fifteen or twenty years biochemistry has become a fashionable subject for research, and it now comprises a huge body of information. The very scope of this knowledge presents serious problems to the student. Which are the facts that it is important for him to learn and which are less important? By what means can he create order in his mind from the jumble that is presented to him? How can he place a particular aspect of biochemistry in its biological context and assess its significance to the living state?

In writing this book we have tried to keep these problems at the front of our minds and have therefore aimed at making our treatment of biochemistry comprehensible rather than comprehensive. We have adopted several means towards this end. In the first place we have intentionally omitted some parts of the subject which are comparatively specialized and seem to us not to illustrate the most important principles. Although everyone will have his own opinion on which topics should be omitted, we hope that our choice will seem reasonable for a book of this size. Secondly we have tried to emphasize the biological significance of the biochemical facts that we discuss. Thus in describing haemoglobin, for instance, we consider what *physiological* effects follow from a change in the *molecular* structure of the protein; and we treat the pentose phosphate pathway (to give a quite different example) not so much as a subsidiary means of degrading glucose but rather as the principal means of reducing NADP. Thirdly we have tried to unify the different aspects of biochemistry, an endeavour that we regard as particularly important at a time when the subject is becoming increasingly fragmented. Thus although Part 1 is devoted largely to the structure of macromolecules we discuss, as far as possible, structure and function together; although Ch. 24 is devoted to genetics we discuss genetics and metabolism together; and so on. There are certain features that can be detected throughout the biochemical systems that we consider – to name a few, the reliance of living systems on specific macromolecules, the importance of non-covalent interactions in macromolecular structure and synthesis, and the significance of reactions that are displaced from equilibrium for the integration of metabolism. We refer to these characteristics repeatedly in different contexts, and we make a special feature of providing extensive cross-references in the book. It may at first seem odd to refer forwards (to a page that you have not yet come to) as well as backwards, but it will seem much less odd – indeed it will, we hope, prove invaluable – on a second reading when you already have some picture in your mind of the whole compass of the book.

Our intentions, then, have been: to highlight those biochemical structures, reactions and pathways that exemplify important principles, and to give less attention to the others; to put the facts that we discuss in their biological context and stress their biological importance, and to integrate the subject by tracing the characteristic features of living systems through every aspect of biochemistry.

If we have succeeded in these aims, this book will be suitable not only for honours students of biochemistry but also for students that need biochemistry as a subsidiary subject for any kind of medical, biological or agricultural science. In order to maintain the balance of the book we have not concentrated exclusively on the needs of any one of these groups, but we hope that the general approach we have adopted (and particularly our concern with the biological relevance of biochemical phenomena) will be of benefit to all of them.

This is not a very elementary book, and to those with no background whatever in biochemistry, or with comparatively little chemistry, its first appearance may be daunting. Such readers may find it useful to approach the book through one of the introductory textbooks that are available, such as Jevons, *The Biochemical Approach to Life*; Bartley, Birt and Banks, *The Biochemistry of the Tissues*; or our own *Guidebook to Biochemistry*.

MDY
REO

# Conventions and Abbreviations

## CHEMICAL COMPOUNDS

Biochemical reactions commonly take place at or about pH 7. This poses the problem of how one should write the structural formulae of compounds that take part in acid-base equilibria. Such compounds will be wholly dissociated, partly dissociated or undissociated at pH 7 depending on the $pK$ of their particular dissociation equilibrium.

To avoid confusion we have tried wherever possible to write the structures of the molecules we discuss in the un-ionized form, irrespective of their true $pK$ and actual state of ionization at pH 7. Thus we show in equation (14.11), p. 265 the synthesis of lactic acid when in fact what is produced is mainly lactate ion,

$$CH_3.CHOH.C\overset{-}{O}O^-,$$

balanced by a solvated proton, $H_3O^+$. We depart from this rule in the few cases where it would hinder rather than help comprehension.

In the case of the hydrogen carriers NAD and NADP we ignore ionization completely (see pp. 29 and 234) and write NAD and NADP for the oxidized form and $NADH_2$ and $NADPH_2$ for the reduced form. These abbreviations are still allowed by international convention, although the symbols $NAD^+$, $NADP^+$, NADH and NADPH are now preferred. We have not used this convention because we believe that the alternative one makes the events in, for example, Ch. 12 easier to follow.

We have followed the frequently adopted practice of neglecting the ionization of the phosphate group by writing Ⓟ for phosphate in an organic compound and $P_i$ for the inorganic phosphate ion. Similarly, organic pyrophosphate is written (P—P) and the inorganic ion $PP_i$. Thus the reaction

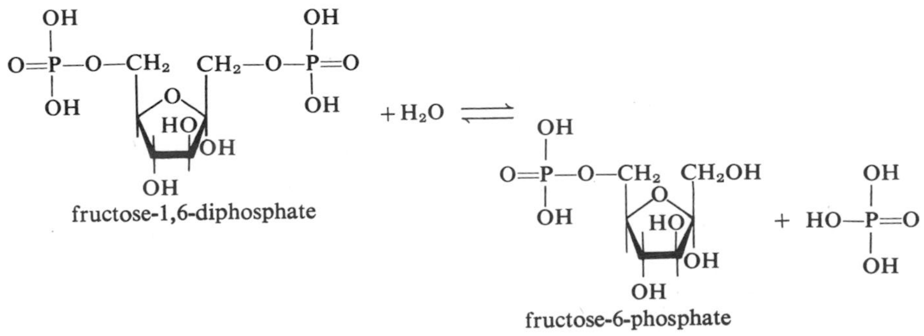

fructose-1,6-diphosphate $+H_2O \rightleftharpoons$ fructose-6-phosphate $+$ $HO—P=O$

(in which the possible ionizations of the phosphate groups are already neglected) is written as

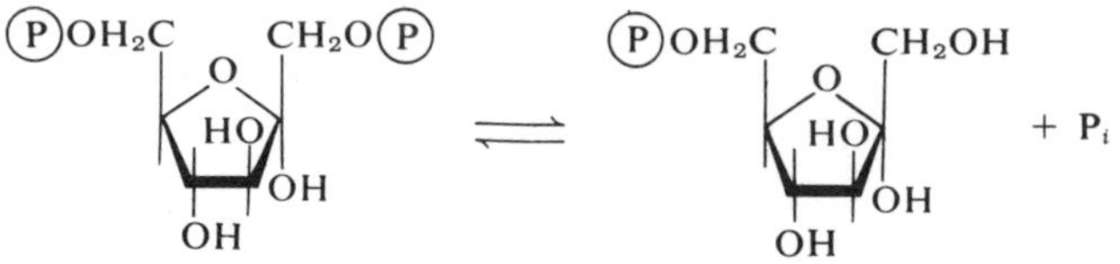

$$\begin{array}{ccc}
\text{\textcircled{P}OH}_2\text{C} \quad \text{CH}_2\text{O} \text{\textcircled{P}} & & \text{\textcircled{P}OH}_2\text{C} \quad \text{CH}_2\text{OH} \\
\rightleftharpoons & & + \text{ P}_i
\end{array}$$

Apart from the increased rapidity with which reactions can be written, this convention has the advantage that we need no longer write $H_3PO_4$ as a substrate or a product and give the impression that so many biochemical reactions use or generate a strong acid.

As a consequence of neglecting ionization equilibria, equations do not always balance as regards H— and —OH groups. Thus, if we were not to neglect ionization the equation on p. 281 should be

$$CH_3.CO.COO^- + CO_2 + ATP \rightleftharpoons \underset{\underset{CH_2.COO^-}{|}}{CO.COO^-} + ADP + P_i$$

pyruvate ion                           oxaloacetate ion

rather than

$$CH_3.CO.COOH + CO_2 + ATP \rightleftharpoons \underset{\underset{CH_2.COOH}{|}}{CO.COOH} + ADP + P_i$$

as written there. It will be seen that in the latter form there is a proton missing from the left-hand side. We do not put it in the equation in case we should give the impression that it takes part in the mechanism of the main reaction.

Where an overall reaction scheme involves coupled reactions (see p. 23) we have sometimes found it convenient to use the Baldwin notation

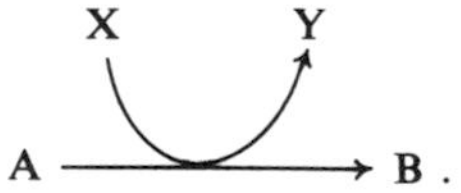

This notation is not meant to imply that the reaction scheme is mechanistically irreversible. It would be possible, but clumsy, to write the arrows for the return reaction.

## UNITS

We have departed from the SI convention in two ways: we have continued to use the Ångström unit (1 Å = $10^{-10}$ metre, i.e. 100 picometres) in discussing the dimensions of biological molecules, and the calorie (1 calorie $\simeq$ 4.2 joules) in discussing the thermodynamics of chemical and physico-chemical processes.

We believe that this is justified. The original measurements were all made in non-SI units and all discussion of them to date has been in terms of non-SI units; this will probably continue to be the case for some time. We feel that to convert to SI would be to interpose an unnecessary barrier between the reader and the subject as it actually has been and is being practised.

It is easy to use the conversion factors – see for example p. 523.

# Chapter 1 Introduction

Compared with such venerable studies as physics, chemistry or natural history, biochemistry is a young science. Its entire history is encompassed within this century – most of it, indeed, within the past forty years. For all its youth, biochemistry has already passed through several stages that we can recognize.

The first stage was dominated by attempts to identify the molecules that are characteristic of living systems; this phase of research explored life from the viewpoint of the organic chemist and was thus concerned more with analysis of the static situation than with the dynamics of the processes that occur in living organisms. Its results were dramatic, leading to an understanding of the structure of most of the small molecules that occur in cells – not only those small molecules that are found free but also those that are combined with one another to make up very large molecules such as proteins and nucleic acids.

The success of these studies, although striking, was limited in two respects. First, it became clear that many of the methods that had proved fruitful in solving the structures of the small molecules found in biological systems were inapplicable to studies of macromolecules; for want of suitable methods, investigations of the macromolecules had to be left in abeyance until the development of such techniques as chromatography, specific cleavage of macromolecules, and X-ray crystallography (see Chs 4 and 7 and Appendix 2). Second, having established the structure of most of the small biological molecules, biochemists became increasingly interested in how these were degraded and synthesized, and hence in how the energy of degradative reactions could be used to drive syntheses. This interest led to the second stage of biochemical history, that of intermediary metabolism. Pathways were established for the breakdown of carbohydrates, fats, proteins and nucleic acids, and later for the synthesis of sugars, fatty acids, amino acids and nucleotides (see Part 2). The link between breakdown and synthesis was established with the discovery of phosphate-bond energy (see Ch. 2).

By 1960 most of the major pathways had been worked out, and biochemical research was making advances on several different fronts which the study of intermediary metabolism had exposed. In one direction there was the problem of the structure of macromolecules and their interactions with one another and with small molecules (see Part 1). In another direction there was the question of reproduction, of how organisms came to resemble their parents not only macroscopically but also in their biochemical characteristics (see Ch. 24). A further type of problem was the means by which macromolecules were synthesized, and how the specificity of their structure (by now known to be one of their crucial characteristics) was maintained (see Part 3). Yet another field was regulation: how are the hundreds of pathways and thousands of reactions of intermediary metabolism integrated, and what controls the synthesis of the enzymes that catalyse them? (See Part 4.)

Investigations of some of these problems still dominate biochemistry today. Although we know a fair amount about the relation between heredity and biochemical make-up, and about the synthesis of some types of macromolecule, the other questions are being pursued at an ever more sophisticated level: the study of macromolecular structure now includes studies of assemblies of macromolecules in, for example, membranes (see Ch. 7) and ribosomes (see Ch. 25), and the study of control of protein synthesis is now moving towards studies of differentiation and the cancerous state.

The rapid advance of biochemistry poses problems not only for the student but also for the authors of a textbook. There is a serious problem of balance between the different aspects of the subject. A similar problem exists for most subjects, but it is perhaps more difficult in biochemistry than in some others to solve it satisfactorily: the normal problem of assessing the relative importance of different aspects of the subject is compounded in biochemistry by an historical problem: how much space should be given to the well-established parts of biochemistry, compared with those that are still in the making? One danger is evidently to dwell on those facts that are so firmly established as to be beyond contradiction and to eschew all others, a procedure that makes for a dull and staid book. The opposite danger is to emphasize the thrill of the fashionable frontier and to disregard the solid background of accomplished work, a procedure that makes for a flashy and ephemeral book. If we have not succeeded in solving these problems we are at least aware of them.

What has, in any case, been achieved by the work of biochemists is this: we now understand *in principle* how a living cell maintains the structure of its specific molecules and reproduces itself. In the remainder of this chapter we shall outline with extreme brevity how these things are accomplished.

The molecules that are most characteristic of living cells are large and complicated, but they are composed by joining together in various permutations a large number of small units. In two kinds of macromolecule the structures are maintained with precision: these are the proteins, in which there are twenty types of small unit (amino acids), and the nucleic acids, in any one molecule of which there are (with a few exceptions) four types of unit (nucleotides). The precision of ordering the amino acids to form proteins, and the nucleotides to make nucleic acids, can be regarded as absolute. Proteins have not only a defined order of amino acids but also a defined three-dimensional shape: this enables them to perform a large number of functions within the cell of which perhaps the most striking is to act as highly specific catalysts (enzymes) of biochemical reactions. The defined order of nucleotides in nucleic acids enables them to carry information; they are thus able to act as the hereditary material of the cell. Two identical copies of this material are passed to the daughter cells on division and these are used to specify the structure of the proteins.

In order to maintain their own structure, which is far more organized than that of the environment, and to grow, cells need energy. This energy is obtained in green plants from the radiant energy of the sun and in animals by degrading foodstuffs, chiefly carbohydrates and fats. Both in plants and in animals the energy is used for the synthesis of amino acids, nucleotides, sugars, etc. (the constituents of macromolecules) and also for the synthesis of the macromolecules from these precursors. Intervening between the processes that supply energy and those that utilize energy is a special compound called ATP. The synthesis of ATP requires energy (thus ATP is made by the processes that provide energy) and its degradation releases energy (thus ATP is broken down by the processes that need energy). This coupling of the synthesis of ATP to processes that release energy, and of the degradation of ATP to processes that require energy, is one of the striking characteristics of living matter.

In addition to enabling the cell to synthesize ATP, the degradation of foodstuffs (particularly carbohydrates and fats) provides a large number of carbon-containing compounds which are intermediates between the foodstuffs themselves and their final products of degradation. Many of these compounds serve as precursors in the synthesis of amino acids, nucleotides and other compounds, most of which are building blocks for macromolecules. The pathways of degradation of foodstuffs and of synthesis of these building blocks form a highly complex and intricate network of reactions, often referred to as intermediary metabolism. Each reaction usually accomplishes only a small chemical change, and each is catalysed by an enzyme.

The cell achieves great economy in its use of foods, and in its synthesis of building blocks, by a system of highly specific regulatory mechanisms. These affect both the rate at which a particular enzyme is synthesized and the rate at which a given number of molecules of enzyme catalyses a particular reaction. By means of these systems of regulation the cell is able to integrate its many biochemical pathways so as to avoid waste of energy and of raw material.

This very brief account of the principles of biochemical organization is of necessity greatly over-simplified. It has been our intention in introducing this book to give a highly condensed abstract, so that the discussion in later chapters can be seen in the context of the general principles we have mentioned. You will find that the different parts of the book are introduced separately in rather more detail than we have given here. Thus Ch. 3 is a general introduction to the treatment of macromolecules in the chapters that follow it; Ch. 9 gives a brief summary of the intermediary metabolism that is discussed in Part 2; Ch. 24 introduces the principles of molecular genetics that are worked out in Part 3, and the beginning of Ch. 28 explains the need for and gives an outline of cellular regulation. In order to write about biochemistry it is necessary to separate out the different aspects of the study and focus attention on one at a time, but it is our hope that the readers of this book will be left with an integrated rather than a disjointed view of the chemical activities of the cell.

# Chapter 2 Kinetics and thermodynamics in biochemistry

This chapter has two main aims:

1. To introduce the basic ideas of physical chemistry in a way that can be understood by those with little training in mathematics or the physical sciences.

2. To describe the specifically biochemical applications of physical chemical principles.

If you have had courses in physical chemistry you will probably wish to omit pp. 7–14. Probably you will also not need to follow in detail all of the qualitative, non-mathematical arguments that occur in the subsequent pages of this chapter. Instead you will want to look through the chapter in order to see how the physical chemistry with which you are already familiar is relevant to the biological systems.

Subsequent chapters will refer from time to time to the arguments given in Chapter 2. We hope that you will at least make sure that you are acquainted with the form of these arguments, and that you know where to find the details if you need them, before going on to the rest of the book.

# Chapter 2 Kinetics and thermodynamics in biochemistry

This book is about the description of biological events in chemical terms. Most chapters will be concerned with how some specific objective is achieved. For example, we shall study the remarkable way in which the structure of a macromolecule is precisely designed to enable it to carry out its function; we will give many examples of the way in which an intricate series of chemical reactions is directed so that it gives rise to some essential product, and we will show how all this complexity in design and function is handed down intact from generation to generation.

These are discussions of the 'how' of living processes, necessarily concentrating on the structures and mechanisms involved. For a more complete understanding, we shall also have to look at biology through the medium of physical chemistry. We shall want to know if a process can be operated without an external supply of energy, how far it will go to completion and how fast it can be operated. These questions – on the one hand 'how much energy' is involved and 'how far' a process will go, and on the other hand 'how fast' – are the proper concern of thermodynamics and kinetics respectively. The rest of this chapter is devoted to developing some of the more useful ideas that come under these headings. Thermodynamics and kinetics possess the only vocabulary in which we can discuss, in basic terms, any of the events with which we are to deal. We shall see that thermodynamic ideas are particularly useful when we come to explain the way in which the cell deploys its energy, while a knowledge of chemical kinetics is essential to understanding the means by which it can act so rapidly (Ch. 6).

## THERMODYNAMICS – 'HOW FAR, IF AT ALL?'

It has long been recognized that the extent to which a chemical reaction may proceed depends on two main factors.

### Affinity

Perhaps the most fundamental factor is the intrinsic *affinity* between the reactants. Thus, sulphuric acid will successfully displace carbonic acid ($H_2CO_3$, i.e. $CO_2 + H_2O$) or hydrochloric acid from their sodium salts, hydrochloric will displace carbonic acid but not sulphuric, and carbonic acid cannot displace either of the others.

We might thus draw up a table of strengths of the affinity of the acids for sodium ion. Sulphuric acid would be at the top, carbonic acid would be at the bottom and hydrochloric acid would be in the intermediate position. Such tables existed as long as 250 years ago.

A mechanical analogy may be useful here: water will flow from A to B or C (Fig. 2.1), from B to C, but not from C to either B or A, nor from B to A. This

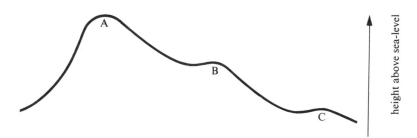

**Fig. 2.1**

linking of the concept of chemical affinity and the potential energy of classical mechanics is quite apt, as we shall see.

We need not look at present for reasons for the order in which chemical affinities are found to lie. Affinities depend on the complexities of the electronic structures of the molecules and such reasons are often as difficult to find as the reasons for one hill being higher than another. All we must remember is that in designing the chemical activities of the cell, nature, acting through evolution by natural selection, must take account of differing affinities. If a desired process calls for reactions that cannot be operated because of unfavourable affinities, some special measures must be taken to circumvent the problem without violating these fundamental principles. We are undertaking a study of chemical thermodynamics principally to equip ourselves to be able to trace the details of this essential accommodation to the principles of nature.

## Concentration

One way of reversing an unfavourable order of affinities is to control *concentration*, as was first noticed in a non-living system. The observation that led to the discovery was as follows. The reaction:

$$Na_2CO_3 + CaCl_2 \rightleftharpoons CaCO_3 + NaCl \tag{2.1}$$

proceeds from left to right and $CaCO_3$ (chalk) is precipitated. However, it was discovered that at the edges of salt lakes, with the very high concentration of NaCl in the evaporating brine, the reaction was reversed so that chalk or limestone rocks dissolved and sodium carbonate was precipitated.

This observation led to the idea of the reversibility of chemical reactions (and explains why the symbol ⇌ is used in writing chemical equations).

An example of a biological reaction that is readily reversed is the isomerization of glucose-6-phosphate to fructose-6-phosphate:

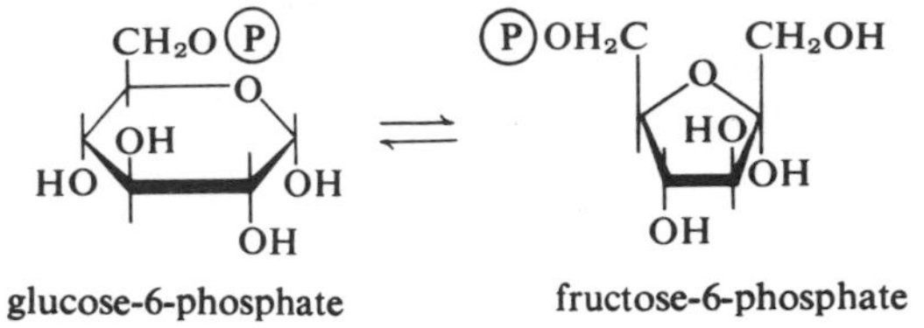

$$\text{glucose-6-phosphate} \quad\rightleftharpoons\quad \text{fructose-6-phosphate} \tag{2.2}$$

glucose-6-phosphate        fructose-6-phosphate

The chemical affinities are better satisfied by the arrangement of atoms in the glucose-6-phosphate than by that in the fructose-6-phosphate and so the reaction tends to proceed from right to left. However, we shall see later that if the concentration of the glucose-6-phosphate is raised the reaction will be forced back against the affinities and some fructose-6-phosphate will be formed.

There is a limit beyond which concentrations cannot be increased – the limit being the maximum solubility of the reactant in question. The alternative to *pushing* the reaction over by increasing a concentration is to *pull* the reaction over by decreasing one.

We will discuss many examples of the manipulation of concentrations of all types of reactions later on, particularly in terms of the energy that is required, or can be obtained, in such an operation (p. 17). For the moment we should note that this is the first of many times that we shall be able to point to the superiority of living systems over non-living ones. The pulling over of reactions in non-living systems is almost exclusively restricted to spontaneous precipitation of some poorly soluble product (or evolution of a gas), which keeps the concentration low. Living matter, on the other hand, possesses highly sophisticated systems (chs 8 & 18) for shuttling reactants away from the site of one reaction (thus keeping the concentration low and pulling the reaction over) and toward the site of a second (and thus bringing about a high concentration and pushing it over).

You will notice that we can preserve the mechanical analogy of Fig. 2.1: what we are doing by piling up a reactant is equivalent to building a water tower on B, from which, when it is full enough, water will flow to A. Alternatively, in lowering a concentration, we are digging a well at A from which, if it is deep enough, we may take water that comes from B.

Some of this may be expressed mathematically in the *Law of mass action*. In a reversible reaction which is at equilibrium

$$A + B \rightleftharpoons C + D$$

$$\frac{[C] \times [D]}{[A] \times [B]} = K \tag{2.3}$$

where the square brackets mean 'concentration of' and $K$ is the so-called equilibrium constant. We can see that, since $K$ is a constant, an increase in [C] must be matched by an increase in [A] or [B]. This is the outcome that is suggested by the discussion in the preceding few paragraphs.

## Simple equilibria of particular biological importance

We shall discuss many equilibrium processes in this book, but there are two types that deserve special mention. These are equilibria involving hydrogen ions and equilibria involving the binding of one molecule to another by non-covalent forces – that is, without chemical reaction.

*Hydrogen-ion equilibria – the concept of* pH *and* pK

The following definitions are crucial:

*An acid is a substance able to donate protons* $(H^+)$.
*A base is a substance able to accept protons.*

Consider the ionization of acetic acid and of methylamine.*

$$CH_3COOH \rightleftharpoons CH_3COO^- + H^+ \qquad (2.4)$$

and

$$CH_3NH_2 + H^+ \rightleftharpoons CH_3NH_3^+ \qquad (2.5)$$

Acetic acid earns its name by donating a proton to the solution and methylamine qualifies as a base by accepting a proton from solution.

These reactions are reversible and will have equilibrium constants:

$$K_{(eqn\,2.4)} = \frac{[CH_3COO^-][H^+]}{[CH_3COOH]} \qquad (2.6)$$

$$K_{(eqn\,2.5)} = \frac{[CH_3NH_3^+]}{[CH_3NH_2][H^+]}. \qquad (2.7)$$

Note that these particular equilibrium constants are not *dimensionless* constants. $K_{(eqn\,2.4)}$ has the dimension of 'molar' while $K_{(eqn\,2.5)}$ has the dimension molar$^{-1}$.

The values for $K_{(eqn\,2.4)}$ and $K_{(eqn\,2.5)}$ are known. They are $2.24 \times 10^{-5}$ M and $4.1 \times 10^{10}$ M$^{-1}$ respectively. These figures enable us to calculate the ratio between the concentrations of the charged and uncharged forms of these molecules. From equations (2.6) and (2.7) respectively,

$$\frac{[CH_3COO^-]}{[CH_3COOH]} \cdot [H^+] = 2.24 \times 10^{-5} \text{ M} \qquad (2.8)$$

$$\frac{[CH_3NH_3^+]}{[CH_3NH_2]} \cdot \frac{1}{[H^+]} = 4.1 \times 10^{10} \text{ M}^{-1}. \qquad (2.9)$$

* Both acetic acid and methylamine were known very early on as biological products – the former in vinegar, the latter in decaying tissues, particularly those of fish. However, we have chosen them not for this reason but because the functional groups —COOH and —NH$_2$ are important in biology (see especially Chs 4, 5, 6, 7 and 8), while the group —CH$_3$ is the one least likely to introduce unnecessary complications.

It is very useful to be able to carry out these calculations. We shall come across many cases in later chapters in which a biological process depends on an ionizable group being entirely in one form or another and we shall want to determine what conditions will be required to bring this about. Equations (2.8) and (2.9) show that *the ratio between the concentrations of the charged and uncharged forms is governed by* $[H^+]$. (The concentration of $H^+$ is normally expressed on a logarithmic scale. One reason for this is the very large range of concentrations of $H^+$ that may obtain. The concentration usually varies between $10^{-1}$ M and $10^{-14}$ M in the majority of non-living systems but, as we shall see, biological systems exhibit a more restricted range. This logarithmic scale is called the pH scale where, as most of you will know already,

$$pH = \log \frac{1}{[H^+]} \cdot )$$

(2.10)

Figures 2.2 and 2.3 show, as a function of pH, the ratio between the charged and uncharged form of acetic acid and methylamine respectively, calculated from equations (2.8), (2.9) and (2.10). After what we have learned about the effect of changes in concentration on the position of equilibrium we should not be

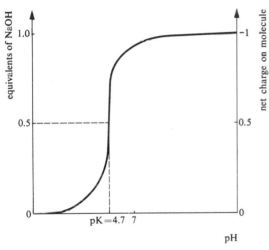

**Fig. 2.2**   The titration of acetic acid          **Fig. 2.3**   The titration of methylamine

surprised to find that, at low pH (high concentration of protons) the protonated forms $CH_3COOH$ and $CH_3NH_3^+$ predominate. At high pH (low concentration of protons) the unprotonated forms $CH_3COO^-$ and $CH_3NH_2$ predominate. (You will find that it is very helpful to construct sentences like this for yourself whenever you have to work out which form of an ionizable group will predominate under a given set of conditions.)

If we return from the situation at the extremes of pH we will find that the midpoint of the curve has some interesting properties. It is the point of *inflection* of the curve; that is, the point at which the ratio between the two forms changes most sharply for a given change (up or down) in pH. At this point, the concentration of the charged form equals the concentration of the uncharged form.

Therefore, at this pH, $[H^+] = K_{(eqn\ 2.4)}$ (for the acetic acid ionization), or $1/[H^+] = K_{(eqn\ 2.5)}$ (for the methylamine ionization).

In the case of acetic acid, therefore, the pH of the midpoint is given by $\log 1/[H^+] = \log 1/K_{(eqn\ 2.4)}$.

The expression $\log 1/K$ is called $pK$ by analogy with pH. The numerical value is 4.65 in this case ($\log 1/(2.24 \times 10^{-5})$).

We shall make use of $pK$ later on as an indication of the point of maximum sensitivity to pH. It also forms a useful index of the *strength* of an acid. The stronger an acid the more ready it is to donate protons and thus the more protons (lower pH) will be required to reverse the ionization. This means that the $pK$ of a stronger acid will be lower (e.g. that of formic acid is 3.7). By a similar argument the $pK$ of a weaker acid will be greater (e.g. that of propionic acid is 4.9).

If we move 1 pH unit from the $pK$, what effect do we observe on the extent of ionization? If we *reduce* the pH by 1 pH unit, we increase $[H^+]$ by a factor of 10 ($\log \frac{1}{10} = -1$). If we *increase* $[H^+]$ tenfold, then clearly, from equation (2.4), we reduce the ratio between the charged and uncharged forms from 1 to 0.1. Similarly the ratio is *increased* from 1 to 10 by moving 1 pH unit in the other direction. Each further pH unit that we move brings a tenfold change in the ratio. We can obtain ratios as shown in Table 2.1.

**Table 2.1**

| pH | $pK - 3$ | $pK - 2$ | $pK - 1$ | $pK$ | $pK + 1$ | $pK + 2$ | $pK + 3$ |
|---|---|---|---|---|---|---|---|
| Ratio | 1:1000 | 1:100 | 1:10 | 1:1 | 10:1 | 100:1 | 1000:1 |
| % charged | 0.1 | 1 | 9 | 50 | 91 | 99 | 99.9 |

The table can be converted to the pH scale in any case in which the $pK$ is known. Thus for acetic acid it is centred on a $pK$ of 4.65.

A very similar operation is possible with the $pK$ of methylamine. For uniformity with the treatment above we can rewrite equation (2.7) as

$$\frac{1}{K_{(eqn\ 2.5)}} = \frac{[CH_3NH_2]\,[H^+]}{[CH_3NH_3{}^+]}. \tag{2.11}$$

$1/K_{(eqn\ 2.5)}$ is the equilibrium constant $K$ of the donation of a proton to the solution by $CH_3NH_3{}^+$. (If $CH_3NH_3{}^+$ can act as a proton donor it is an acid, by the definition that we give on p. 10. It is in fact known as the *conjugate acid* of the base $CH_3NH_2$. Similarly acetate ion $CH_3COO^-$ is the *conjugate base* of acetic acid (we will find the concept of conjugate base and acid essential in Ch. 4, p. 65). In cases of difficulty refer to the definition of acid and base that was given at the beginning of this section.)

$1/K_{(eqn\ 2.5)}$ is therefore the equilibrium constant of the acidic dissociation of $CH_3NH_3{}^+$. The numerical value is $1/(4.1 \times 10^{10}) = 2.4 \times 10^{-11}$ M. The $pK$ of the acid dissociation is therefore 10.6. $pK = \log K_{(eqn\ 2.11)}$ and the same table applies as before but centred on a value of 10.6.

The curves in Figs 2.2 and 2.3 are of more than theoretical interest. They are titration curves for these substances. Methylamine, if present entirely in the uncharged form, will require one equivalent of protons to convert it to the charged form. In titrations used in volumetric analysis these protons are normally provided by an aqueous solution of strong acid. A strong acid is one which is always in the fully ionized state in the range of pH in which we wish to use it. Hydrochloric acid is suitable, and the vertical axis of Fig. 2.3 indicates the number of equivalents of HCl needed to produce any given state of ionization in the methylamine.

Similarly, a strong base is required to titrate $CH_3COOH$ by detaching the proton. An aqueous solution of NaOH is suitable and the appropriate scale has been used in Fig. 2.2.

## The ionization of water

Water can itself ionize by the equation

$$H_2O \rightleftharpoons H^+ + OH^- \qquad (2.12)$$

which has an equilibrium constant, $K$

$$K_{(eqn\ 2.12)} = \frac{[H^+][OH^-]}{[H_2O]}. \qquad (2.13)$$

(In fact $H^+$ is *solvated* by a molecule of $H_2O$ to $H_3O^+$. This makes no difference to the outcome and may be neglected in the interests of simplicity.)

Obviously, when water ionizes, the concentration of the non-ionized form $H_2O$ falls. However, the concentration of water is 55 M – the molecular weight is 18 and there are 1000 grams ($= 55 \times 18$) per litre. This concentration far exceeds any feasible concentration of $H^+$ or $OH^-$ and therefore, although $H_2O$ is lost in forming $H^+$ and $OH^-$, its concentration decreases only imperceptibly. Thus, as a first approximation, $[H_2O]$ is constant.

We may therefore replace $K_{(eqn\ 2.12)}$ for water with $K_{(eqn\ 2.12)} \times [H_2O]$ which equals, from equation (2.13), $[H^+][OH^-]$. This new constant, the *ionic product* of water, $K_w$, has had its value determined at $10^{-14}$ M$^2$.

(This example illustrates the rule that, in equilibria involving $H_2O$ as a *reactant*, its concentration is conventionally set at 1 M, rather than 55.5 M. All equilibrium constants involving water as a reactant therefore need adjusting by a factor of 55.5 M. We have followed this convention, since it is universal. It does not lead to incorrect results, so long as the appropriate factor is borne in mind (see also p. 34).)

The ionic product enables us to calculate the concentration of $H^+$ at neutrality, that is when $[H^+] = [OH^-]$. Substituting:

$$[H^+]^2 \text{ at neutrality} = 10^{-14} \text{ M}^2$$

$$[H^+] = 10^{-7} \text{ molar},$$

that is the pH of a neutral solution is 7.

We mentioned earlier that the range of pH in biological systems is rather limited. The pH of blood is 7.35, and this value is close to that found within cells. However, we must emphasize that the meaning of pH must be explored with some care in any particular case since cellular fluids are not free, dilute solutions. They are highly concentrated solutions of large and small molecules and the effective concentration of water will be well below 55.5 M.

Apart from such curiosities as those bacteria that produce $H_2SO_4$ and generate an external pH for themselves of $<1$ (they can, with the assistance of other species of bacteria, extract the sulphur for the $H_2SO_4$ from concrete and can eventually eat holes in masonry), the furthest departures from a pH of 7 in living organisms are in the extra-cellular fluids of digestion. For instance the peptic secretion of the gut is equivalent to, roughly, $10^{-2}$ M HCl. We have seen that HCl is fully ionized in aqueous solutions and there will therefore be values of $[H^+]$ and $[Cl^-]$ at $10^{-2}$ M each. $[H^+]$ of $10^{-2}$ M is equivalent to a pH of 2. The alkaline section of the gut (the duodenum for instance) is at a less extreme value – about 8.6.

The last few pages have served to emphasize several useful points. The main ones were:

(1) The balance in equilibria can, as suggested in the previous section, be disturbed by changing concentration. This has been exemplified by the effect of changing $[H^+]$ on the amount of the charged form of molecules.
(2) The usefulness of the concept of p$K$ – it is both a measure of the strength of the acid and indicates the point at which the state of ionization is most sensitive to pH.

## Non-covalent binding

By the time that you reach the end of Part 2 of this book you will probably have come to feel that non-covalent binding between molecules (particularly that between, on the one hand, a very large molecule and, on the other, a very small molecule) is one of the key factors in the maintenance of the living state. The list of cases that you will by then have encountered includes the binding of hormones and vitamins to their sites of action, the binding of an antigen to an antibody, the binding of oxygen to haemoglobin, the binding of substrates to enzymes and the binding between genes and the products derived from them. We shall explore the relationship between such binding and biological activity in particular detail for the enzymes, but shall at least refer to the role of the binding in the biological activity of the remaining examples. We therefore need some general aids to the description of such systems, one of which is the *binding constant*.

If we consider the process to be of the form

$$\text{free substance} + \text{binding agent} \rightleftharpoons \text{bound complex}$$

we can write

$$K_{(binding)} = \frac{[\text{bound complex}]}{[\text{free substance}][\text{binding agent}]}$$

where $K_{(binding)}$ is the association constant of the binding process. We could equally well write

$$K_{(dissociation)} = \frac{[\text{free substance}][\text{binding agent}]}{[\text{bound complex}]}$$

where $K_{(dissociation)} (= 1/K_{(binding)})$ is the *dissociation* constant of the binding process. As before, these constants will rarely be dimensionless.

We shall see in the chapters of Part 2 that binding takes place at specific sites on macromolecules. These sites are constructed so as to have a high efficiency of binding (high $K_{(binding)}$; i.e. low $K_{(dissociation)}$). Typical values for $K_{(dissociation)}$ may be as low as $10^{-3}$ or $10^{-4}$ M. Exceptional values may be even lower (so called transition state analogues (Ch. 6) may reach $10^{-6}$; the '*lac* repressor' involved in controlling the expression of the *lac* genes (Ch. 28) may reach $10^{-10}$ M).

A simple calculation shows that even a $K_{(dissociation)}$ of $10^{-4}$ M has a dramatic effect in restricting location within the solution of the substance that is bound. Consider a case in which the binding agent is present at a concentration of 1 mM and the initial concentration of the substance to be bound is 1 μM at $K_{(dissociation)} = 10^{-4}$ M. Then

$$10^{-4} = \frac{[\text{free substance}]}{[\text{bound complex}]} \cdot 10^{-3}.$$

(The concentration of the binding agent is still written as $10^{-3}$ M since it can fall by no more than one-thousandth of $10^{-3}$ even if the total initial amount of the free substance is bound in forming the complex.)

Therefore the ratio free:bound is $10^{-1}$, that is 90 per cent of the material is bound. This represents a considerable ability to abstract molecules from free solution and concentrate them in the close neighbourhood of another type of molecule. In the case that we have just given, if the binding agent were not there, 1 μmole would be found anywhere within a volume of 1 litre. On adding the binding substance, 90 per cent of this amount (0.9 μmole) will be found, in effect, within the volume of space occupied by 1 mmole of binding material.

From the definition of Avogadro's number, 1 mole of any material contains $6 \times 10^{23}$ molecules. Therefore 1 mmole has $6 \times 10^{20}$ molecules. If the binding substance is a protein it may have molecular dimensions of the order of $20 \times 20 \times 20$ Å, that is a volume of $(20 \times 10^{-8})^3$ cm$^3$ = $8 \times 10^{-21}$ cm$^3$. Therefore the volume actually occupied by the $6 \times 10^{20}$ molecules is $6 \times 10^{20} \times 8 \times 10^{-21} = 48 \times 10^{-1}$ cm$^3$ = 4.8 cm$^3$. Therefore 0.9 μmole is now somewhere within a certain 4.8 cm$^3$ of the original 1000 cm$^3$ of solution. That is, the effective concentration is raised from 0.9 μM to $(0.9 \times 1000)/4.8 = 187$ μM – an approximately 200-fold increase. If we are able to define the binding site on the protein more closely, the effective volume within which the bound substance is to be found decreases still further and an even higher enhancement of concentration may be claimed. This is a rough calculation, but we feel that it has shown the potential importance that a low dissociation constant of binding may have. We propose to substantiate this conclusion by more specific examples in later chapters.

Having surveyed these specific types of equilibria, we shall now return to the general discussion. We shall consider the energetic aspects of equilibrium processes which are of great assistance in understanding the mode of operation of living systems.

## The relationship between equilibrium and available energy

It is everyday experience that all processes can be divided into two categories, those that tend to occur by themselves and those that must be driven by the expenditure of energy. Water running downhill is an example of the first class, a vehicle moving uphill is an example of the second. Furthermore we know that it is processes of the first class that yield the energy to drive those of the second (a water-wheel may be used to pull a trolley uphill.) This division applies just as much at the molecular level; it is of great importance in understanding the design of living systems. In particular, by analogy with the mechanical examples cited, *it is possible to couple the chemical energy yielded by a process of the first type to drive a chemical or physical process of the second type.*

Living matter depends on this possibility. Organisms exploit it to bring about syntheses that would not otherwise be feasible, and also in order to couple chemical energy with mechanical work. (Examples of coupling between chemical and mechanical processes are muscular contraction and the forced transport of solutes into areas of higher concentration – see Ch. 18.) Just as the wholesale exploitation of energy coupling is a characteristic of life at the molecular level, so the consequences of that coupling – the ability to grow, to move and to organize – are the characteristics by which we recognize life at the macroscopic level.

To assess the importance of any particular process in the economy of the living state, we need to know in which category it lies and to have some idea of the amount of energy that could be obtained from it or that would be needed for it. There exists a useful thermodynamic quantity to help us. Consider a process $A + B \rightleftharpoons C + D$ that is taking place at constant temperature and pressure – the conditions under which living matter normally operates. The maximum useful energy produced or the minimum that is consumed is known as $\Delta G$, the change in the so-called Gibbs free energy of the reacting system. By convention, if $\Delta G$ has a negative value when going from $A + B$ to $C + D$, energy may be obtained from the process and thus the process $A + B \rightarrow C + D$ is of the first, energy-yielding, category. If the value is positive then $A + B \rightarrow C + D$ is of the second, energy-requiring, category.

Clearly, if $\Delta G$ is negative for $A + B \rightarrow C + D$ then it will be positive for $C + D \rightarrow A + B$. This convention is not so illogical as it may seem: if we obtain energy (negative $\Delta G$) from the process the system itself must lose it.

Thus the sign of $\Delta G$ tells us in which category the process lies. Our second requirement, to know how much energy is involved, is met by its magnitude. The greater the negative magnitude of $\Delta G$, the more energy is to be had. The greater the positive magnitude, the more is needed.

It must not be thought that if $\Delta G$ is negative the process will always and instantaneously occur. To return to the water analogy, consider a reservoir a certain height above the surrounding countryside. There is clearly a strong tendency for the water to run down; whether it actually does so or not, and the rate if it does, depends on factors unrelated to the strength of the tendency. If the dam is effective the water will never run down; if there is an infinitesimal hole it may do so but at an imperceptible rate; if the dam is destroyed it will do so with great violence.

Thus returning to chemical reactions we can speak of a reaction being energetically favoured (negative $\Delta G$) but kinetically hindered (see p. 34 ff). The violent combustion of our bodies in air has a large negative $\Delta G$. We survive because at room temperature there is *kinetic* hindrance of the process slowing it down to an infinitesimal rate – this is the so-called 'potential barrier', with which we shall deal in great detail later on (p. 36 ff).

The value of $\Delta G$ is obviously closely connected with the concept of intrinsic affinity but must also depend on the other important factor that we mentioned, namely the *quantities* of the components present.

The relationship that incorporates all these factors is

$$\Delta G = -RT \log_e K + RT \log_e \frac{[C][D]}{[A][B]} \qquad (2.14)$$

where $R$ is the gas constant, $T$ the absolute temperature and the other symbols have the same meaning as in equation (2.3). The derivation of this equation can be found in any textbook of physical chemistry, and you should consult one if you think that it would be helpful. The following argument might assist those without mathematical training to see that the equation is, at any rate, intuitively reasonable. Consider the reversible reaction $A + B \rightleftharpoons C + D$, in which $A + B \rightarrow C + D$ is energetically favoured over $C + D \rightarrow A + B$. If we start with equal concentrations of the reactants, the overall reaction will proceed from left to right and the concentration of C and D will build up, while that of A and B will decrease. The $\Delta G$ of $C + D \rightarrow A + B$ will increase in negative value as the concentrations of C and D increase and the $\Delta G$ of $A + B \rightarrow C + D$ will fall in negative value as the concentrations of A and B fall. Eventually the concentrations will reach a point at which the $\Delta G$'s for the forward and back reactions are equal. At this point no further change in the overall concentrations of the reactants will occur.

What we have described, of course, is the approach to equilibrium of the reversible reaction $A + B \rightleftharpoons C + D$. Therefore, it should not surprise us to find a relationship between $\Delta G$ and the equilibrium constant, $K$, since the latter quantity simply expresses the ratio at equilibrium between the concentrations of the forward and backward reactants. When the reaction reaches equilibrium the expression $\frac{[C][D]}{[A][B]}$ becomes equal to $K$, the equilibrium constant and equation (2.14) reduces to

$$\Delta G = -RT \log_e K + RT \log_e K = 0$$

In other words, when the reaction reaches equilibrium, there is no further change in free energy, as we of course know.

A knowledge of the way in which $\Delta G$ depends on concentration is useful in itself, as we shall see when we consider the energetics of the process by which biological membranes control the concentrations of solutes in living systems (Chs. 8 and 18). Moreover, we shall make considerable use of the relationship that follows from it:

$$\Delta G^\circ = -RT \log_e K \qquad (2.15)$$

where $\Delta G^\circ$ is the *standard* change in the Gibbs free energy of the forward reaction at 25°C. ($\Delta G^\circ$ of the backward reaction has the same value but the opposite sign.) 'Standard' means that the effective concentration of all reactants is taken as molar, so that the term $\left( + RT \log_e \dfrac{[C]\,[D]}{[A]\,[B]} \right)$ in equation (2.14) disappears.

For numerical calculations this expression reduces to

$$\Delta G^\circ = -1364 \log_{10} K \text{ cal/mol} \qquad (2.16)$$

These equations are of considerable value, since we often use them to calculate the extent to which a reaction will proceed, as well as the overall direction (given by the sign of $\Delta G$) and the possible yield of chemical energy. Since $\Delta G^\circ$ and $K$ are virtually interchangeable, we can calculate one from the other at once.

The concept of $\Delta G^\circ$ is also very useful when we wish to assess the strength of the bonds that hold macromolecules together (Ch. 3) and also when applied to those two types of equilibria, pH and binding, that were dealt with specially in the preceding section.

For example we can see that the free energy of ionization of $CH_3COOH$ is $-1364 \log_{10} K_{(\text{eqn }2.4)} = 6340$ cal/mole, while that of the ionization of $CH_3NH_2$ is $-1364 \log_{10} K_{(\text{eqn }2.5)} = -14\,500$ cal/mole.

Similarly the strength of the interaction needed to produce a dissociation constant of $10^{-4}$ is $\Delta G^\circ = 4 \times 1364 \simeq 5500$ cal/mole.

Before going on we should observe that too rigid a use of the standard state produces a difficulty when dealing with biological systems. A molarity of 1 would, for the hydrogen ion, be equivalent to a pH of 0 (p. 11). This is clearly unrealistic since the state of ionization of many biological compounds changes on going from pH 7 (the approximate pH of most biological fluids) to pH 0 and the reactants become, in effect, different compounds. The energy calculations would then be greatly confused by the contribution of the free energy of ionization which, as we have just seen, can be quite significant. All values of $\Delta G^\circ$ are therefore corrected to those obtaining at pH 7 and the notation becomes $\Delta G^{\circ\prime}$.

Values of $\Delta G^{\circ\prime}$ are not known for all biochemical reactions. Even where values are known many have been determined in free solution and not under conditions realistically reproducing those in the cell.

There is also the problem that, so long as processes are doing work or being used to bring about a net synthesis of product they are only *tending* toward equilibrium rather than having attained it. That is, the product of a reaction is drawn off from the system so fast that the concentration of product never reaches the level suggested by $K = [\text{product}]/[\text{substrate}]$. We shall see later that the attainment of equilibrium literally means that the process is over. The treatment of thermodynamics that we have given here is based on the assumption that equilibrium is attained, but, where necessary, corrections drawn from the theory of *non-equilibrium thermodynamics* can be applied.

In spite of these problems the values that are available probably do not misrepresent the position too seriously. They have proved so useful that we may expect further and more accurate values of $\Delta G^{\circ\prime}$ to become available as time goes on. Meanwhile, we *must* make do with what we have, since there is no other way of dealing with many of the problems we shall be facing in later chapters. In particular, we must use $\Delta G^{\circ\prime}$ to deal with the question of the coupling of energy-yielding processes to energy-demanding ones.

## Free energy in metabolic reactions

To illustrate the use of the ideas contained in the previous section, let us consider some biochemical reactions drawn from later chapters. First a reaction to which we referred on p. 9, in which $\Delta G^{\circ\prime}$ is fairly near zero.

$$\text{glucose-6-phosphate} \rightleftharpoons \text{fructose-6-phosphate}$$

$$\Delta G^{\circ\prime} = +500 \text{ cal/mole}$$

(2.2)

This is a reaction of some importance in both the synthesis and degradation of carbohydrates (pp. 260 and 333). From equation (2.16)

$$\log_{10} K = -\frac{500}{1364} = -0.367 = \bar{1}.633$$

$$K = \text{antilog } \bar{1}.633 = 0.43.$$

Thus, at equilibrium at pH 7, the concentration of fructose-6-phosphate will be close to that of glucose-6-phosphate. If there is 1 μmole of glucose-6-phosphate present there will be 0.43 μmole of fructose-6-phosphate at equilibrium.

There are important consequences of the fact that the equilibrium constant is near to one. Biochemical reactions exist to bring about the net formation of a compound which may be required either for itself or as the starting material of a further process (Ch. 9). Reactions with equilibrium constants near unity can, in distinction to some that we shall discuss below, bring about a useful net formation of product in either direction. If one were to add 14.3 μmoles of fructose-6-phosphate to the system, a net formation of 10 μmoles of glucose-6-phosphate would result. If one were to add 14.3 μmoles of glucose-6-phosphate there would be a net synthesis of 4.3 μmoles of fructose-6-phosphate, a reasonable amount, even though the reaction in this direction is slightly disfavoured in terms of $\Delta G^{\circ\prime}$. The slightly elevated concentration of glucose-6-phosphate is sufficient to convert an unfavourable $\Delta G^{\circ\prime}$ to a favourable $\Delta G$ (equation (2.14)). The ease with which this can be done is the reason why, on p. 9, we selected this equation as an example of a reaction that may be easily influenced by changes in concentration of reactants. Thus the reaction may be used in degradation of carbohydrate, which requires it to proceed from left to right (p. 260), and in synthesis, which requires it to proceed from right to left (p. 333).

Let us turn to another reaction in carbohydrate metabolism, the equilibrium between dihydroxyacetone phosphate and glyceraldehyde-3-phosphate.

$$
\begin{array}{ccc}
\text{CH}_2\text{OH} & & \text{CHO} \\
| & & | \\
\text{C}=\text{O} & \rightleftharpoons & \text{CHOH} \\
| & & | \\
\text{CH}_2\text{O}\,(\text{P}) & & \text{CH}_2\text{O}\,(\text{P}) \\
\text{dihydroxyacetone} & & \text{glyceraldehyde-} \\
\text{phosphate} & & \text{3-phosphate}
\end{array}
$$

(2.17)

$$\Delta G^{\circ\prime} = +1800 \text{ cal/mole}$$

The equilibrium constant is

$$\text{antilog } -\frac{1800}{1364} = \text{antilog } \bar{2}.68, \text{ i.e. about } \frac{1}{21}.$$

At equilibrium there will be 21 μmoles of dihydroxyacetone phosphate to one of glyceraldehyde-3-phosphate. Now we shall see (p. 260) that carbohydrate breakdown requires the reaction from left to right. Although more extreme examples are to come, the 'wrong' product seems even here to be very much favoured by the equilibrium constant. How can the process be used for a significant net formation of glyceraldehyde-3-phosphate and thus a significant breakdown of carbohydrate? The answer is that an enzyme system exists which is ready to take such glyceraldehyde-3-phosphate as there is and convert it very rapidly to the next product in the chain of reactions leading to carbohydrate breakdown. In an attempt to restore equilibrium more glyceraldehyde-3-phosphate will be formed to replace it. So long as the enzyme system continues to tap off the glyceraldehyde-3-phosphate as it is formed, a useful net formation of product will occur. Thus, virtually any quantity of dihydroxyacetone phosphate can be transformed to glyceraldehyde-3-phosphate.

We must stress that the overall equilibrium constant of the process which removes glyceraldehyde-3-phosphate should favour removal rather than synthesis. Putting this in free energy terms, the $\Delta G^{\circ\prime}$ of the removal process must be sufficiently negative overall to overcome the positive $\Delta G^{\circ\prime}$ of reaction (2.17).

It may be helpful to consider this analogy. Imagine a rope with buckets at each end. If the rope is passed over a pulley, the downward movement of one bucket results in the upward movement of the other. One bucket can be raised any distance, so long as the other can be lowered sufficiently.

There is another most important example of the use of this tapping-off process to control the direction of net formation of product. This is given on p. 284, in which one of the two products of a reaction is destroyed so that the reaction is pulled over and the extent of the synthesis of the other product is maximized. Closely analogous examples occur on pp. 431 and 444.

Now for a reaction with a very high positive value of $\Delta G^{\circ\prime}$.

$$\text{ribulose-5-phosphate} + \text{phosphate} \rightleftharpoons \text{ribulose-1,5-diphosphate} \qquad (2.18)$$

$$\Delta G^{\circ\prime} = +2400 \text{ cal/mole}, \therefore K = 1.74 \times 10^{-2}$$

Ribulose-1,5-diphosphate is to be used for the synthesis of compounds such as carbohydrates. (It is, among other things, an early product of photosynthesis – see Ch. 19.) This reaction cannot readily achieve a net flow to the right by tapping off the diphosphate to some other reaction which results in a lower energy state, as was done with the removal of glyceraldehyde-3-phosphate in equation (2.17). This is because the available free energy of the products that we wish to make from this compound is so high. Here, using the bucket and pulley analogy, we cannot lower the bucket far enough because it has a fixed, high destination.

Furthermore, the value of the equilibrium constant makes it very difficult to achieve a useful net formation of the diphosphate by adding reactants on the left of the equation as can be done in the case of equation (2.1) (p. 8). We have already pointed out that we cannot exceed the concentration of a saturated solution.* How then are this reaction and the many others like it, equally

---

* Also, $H_2O$ has been omitted from the right hand of equation (2.18) for the reasons stated on p. xii. $[H_2O]$ will be large and will contribute to the fact that $\Delta G$, as well as $\Delta G^{\circ\prime}$, does not favour the forward reaction.

vital to living matter and equally unfavourable on energetic grounds, to be brought about? The answer is, as we saw on p. 16, that processes may be *coupled*. The problem is solved by coupling the processes of high positive $\Delta G$ to those of high negative $\Delta G$.

Before we can consider the system used for coupling, we must round off this survey of the energetics of biochemical reactions by quoting an example of a reaction with a high negative $\Delta G$, the equilibrium between phospho-*enol*pyruvate and pyruvate (see p. 264).

$$
\begin{array}{c}
CH_2 \\
\parallel \\
CO\,\text{(P)} + H_2O \rightleftharpoons \\
\mid \\
COOH \\
\text{phospho-\emph{enol}pyruvate}
\end{array}
\qquad
\begin{array}{c}
CH_3 \\
\mid \\
C{=}O + P_i \\
\mid \\
COOH \\
\text{pyruvate}
\end{array}
\qquad (2.19)
$$

$$\Delta G^{\circ\prime} = -13\,200 \text{ cal/mole}$$

This is a step in the degradation of carbohydrate. It is clear that there is at least the possibility of a useful energy yield in this process which might be coupled to those of high positive $\Delta G$. This possibility, as we shall see later, is made a reality. For the moment we note that the equilibrium constant is

$$\text{antilog} \frac{13\,200}{1364} = 4.5 \times 10^9$$

in favour of the forward (left to right) reaction.

*Operation of reactions in either direction*

We can use this example to help to explain the misuse so often made in biochemistry of the terms 'reversible' and 'irreversible' as applied to metabolic reactions. These terms are not to be taken to refer to the reversibility or otherwise of the *mechanism* of the reaction. They refer only to the ease or otherwise of achieving net formation of product in both directions, which turns out to depend on two factors.

We have explained the importance, in deciding whether or not the useful operation in both directions will be easy, of the equilibrium constant. If the equilibrium constant is very unfavourable to net reaction in one direction, unacceptably high concentrations of reactant would be needed to overcome it (see p. 284 and, for an additional, very important, example, p. 334).

We must now mention an additional factor that affects the ease with which useful operation in both directions can be achieved. This derives from the fact that, as we have mentioned (p. 18), processes are generally *displaced* from equilibrium. That is, in a process A $\rightleftharpoons$ B the concentration of B is never allowed to rise to the value given by

$$K = \frac{[B]}{[A]} \; .$$

If [B] is held to a lower value [B]′ by the rapid removal of B from the site of reaction we write

$$\Gamma = \frac{[B]'}{[A]}$$

or, to allow for any similar perturbations in [A] that there might be,

$$\Gamma = \frac{[B]'}{[A]'} .$$
(2.20)

$\Gamma$ is called the *mass-action* ratio. In contrast to $K$ it is not a constant but depends on the extent to which the system is kept from reaching equilibrium under a given set of conditions. (When the system is at equilibrium, $\Gamma$ rises to equality with $K$.) $K/\Gamma$ is in fact a measure of the *degree of displacement* of the system from equilibrium. If $K/\Gamma \gg 1$ or $\ll 1$ the system is greatly displaced, if $K/\Gamma = 1$ the system is at equilibrium.

[B]' and [A]' are known as the *steady-state* values of the concentrations of B and A. This name expresses the fact that, under a given set of conditions, the concentration levels are in *balance* between two tendencies. There is, on the one hand, the attempt of the chemical reaction to supply the reactant and raise the concentration to the equilibrium value. On the other hand there is the effect of the process that is removing reactant and driving the level away from its equilibrium value.

When we wish to assess the feasibility of achieving a net synthesis of A by raising [B], as has been the case in the examples quoted in the preceding few pages, we must now remember that we have further to go in raising [B] than if the reaction were proceeding under equilibrium conditions.

There are many biological reactions in which [B] is depressed to only an unimportant extent and we may then neglect the non-equilibrium nature of living processes in our discussion. In some others, however, [B] is depressed by several orders of magnitude and this will have a serious effect on the ease with which we can bring about a net synthesis of A. This additional difficulty can, like the simple effect of an unfavourable $K$, be overcome by *coupling* to an energy-producing process. We shall include this point in the discussion of coupling that follows and deal with the individual reactions where they occur in Parts 3 and 4 of this book.

Whether or not one of these procedures, manipulation of concentration or coupling, can enable us to operate a given process for useful net synthesis in both directions, there is always *some* reversal, however insignificant, of every reaction. We shall therefore use double arrows for all the reactions in this book.

## Energy sources and energy coupling

A chemical reaction is in a sense solely concerned with electrons. If the electrons that maintain the structure of chemical compounds shift permanently from one configuration to another, we say that a chemical reaction has taken place. If the available free energy of the first configuration is higher than that of the second (more negative $\Delta G$) then the reaction could be made to yield energy. If the reverse is true it will require energy. If we imagine ourselves designing a living organism we must search for a source of energy to couple to its essential energy-requiring processes. We are thus really looking for electrons in a state of high potential energy (high negative $\Delta G$) and a sink of lower potential energy (less negative $\Delta G$) into which we can put them.

With a few trivial exceptions there is only one *primary* source of electrons at a high potential, able to fall to a lower, which is suitable for coupling to the energy need of living organisms.

This is the photosynthetic process (Ch. 13), carried out in a solid-state device found in plant cells which is known as the chloroplast. It consists of an assembly of proteins (Ch. 4), prosthetic groups (p. 78) and lipids. A quantum of light is absorbed and its energy ($\Delta G^{\circ\prime}$ of about 30 000 cal/gram equivalent ($6 \times 10^{23}$) of red quanta) is given up in promoting an electron to a higher potential energy state. The electron then falls back to its original state. In a way which is understood only in outline (Ch. 13), the energy of the fall is coupled to the synthesis of specialized molecules known as 'high-energy' compounds. These compounds, which are discussed below, can undergo energy-yielding reactions, the energy of which can be coupled to energy-requiring reactions. They can be regarded as a *secondary* source of free energy. By the use of this small group of high-energy compounds a large number of molecules can be made, the synthesis of which would otherwise be energetically disfavoured to a greater or lesser extent – for example carbohydrate and fat (Chs 19 and 20). These products then comprise a *tertiary* source of electrons at a useful potential energy level. The plants themselves use such photosynthetically produced compounds as food reserves (p. 173); animals obtain these compounds either by consuming vegetable matter directly or, if they are carnivorous, via the food chain.

Given that we have, in the form of carbohydrates, fats and so on, a source of electrons at a high configurational energy, we must now look for a sink at a lower energy state into which they may fall, so that the energy of the fall may be coupled to endergonic processes. Simple processes like

$$Fe^{2+} \rightleftharpoons Fe^{3+} + e^-$$

remind us that to donate an electron is to be oxidized. Although other molecules exist that are capable of acting as acceptor, molecular oxygen has an almost unrivalled ability to act as an electron acceptor. It owes this ability to certain peculiarities of the way in which its electron shells are filled.

Oxidative metabolism is, therefore, the most important source of energy. Oxidation reactions of the type

$$AH_2 + \tfrac{1}{2}O_2 \rightleftharpoons A + H_2O \tag{2.21}$$

can easily yield a $\Delta G^{\circ\prime}$ of more than $-50 000$ cal/mole.* This would be more than enough to drive the endergonic processes of the cell, reaction (2.18) for example. Cells that utilize oxidative reactions to produce energy also have a solid-state device, this time called the mitochondrion (Ch. 10), to trap this energy in a process called oxidative phosphorylation. The components of the mitochondrion are, as far as the different functions permit, similar to those of the chloroplast. The 'high-energy' compounds produced are either identical or closely related.

It is now time to describe the high-energy compounds and the way in which they are used in energy coupling.

---

* You may be accustomed to treating oxidation reactions differently from all others and expressing the energy levels in terms of standard electrode potentials. These potentials are directly convertible to $\Delta G^{\circ\prime}$ (a change in standard potential of $+1$ volt for 1 gram equivalent of electrons at pH 7 corresponds to a $\Delta G^{\circ\prime}$ of 23 000 cal/mole). In order to stress the unity of all energy-yielding reactions the one scale, $\Delta G^{\circ\prime}$, will be used throughout this book.

Both the photosynthetic and the oxidative processes produce the molecule adenosine triphosphate (ATP). The structure is given in Fig. 2.4. It is a property of the phosphoric-anhydride bond (marked with a star in the figure) that it has a considerable free energy of hydrolysis.

$$R-O-\overset{\overset{\displaystyle O}{\|}}{\underset{\underset{\displaystyle O^-}{|}}{P}}-O-\overset{\overset{\displaystyle O}{\|}}{\underset{\underset{\displaystyle O^-}{|}}{P}}-O^- \rightleftharpoons R-O-\overset{\overset{\displaystyle O}{\|}}{\underset{\underset{\displaystyle O^-}{|}}{P}}-O^- + \text{phosphate}$$

$$\Delta G^{\circ\prime} = -7000 \text{ cal/mole}$$

It is difficult to arrive at values for $\Delta G^{\circ\prime}$ of the two reactions

$$\text{ATP} \rightleftharpoons \text{ADP} + P_i \tag{2.22}$$

$$\text{ATP} \rightleftharpoons \text{AMP} + PP_i \tag{2.23}$$

that are completely satisfactory. Each $\Delta G^{\circ}$ is affected by a number of factors, the precise effect of which is difficult to determine. To take a single example, all the reactants and products in these two equations have a strong tendency to form complexes with $Mg^{2+}$, an ion which occurs in appreciable concentrations in physiological fluids. Since the free energy of hydrolysis of the $Mg^{2+}$ complexes will be very different from the free energy for the parent compounds, some correction ought to be made to take this factor into account. Unfortunately, it is not easy to make the correction with any precision. Because of this and other similar difficulties the values that we use in the book ($-7600$ cal/mole for equation (2.22) and $-7400$ cal/mole for equation (2.23)) are to be regarded as approximations only.* They are unlikely to be so far wrong as to invalidate the qualitative conclusions that we base on them.

During the photosynthetic or oxidative synthesis of ATP (*photosynthetic phosphorylation* (p. 248) and *oxidative phosphorylation* (p. 233)) the energy of the falling electron is somehow used to drive reactions such as (2.22) from right to left. The resulting ATP can now move to a site in the cell at which, for example, reaction (2.18) is to proceed.

We can therefore combine reactions (2.18) and (2.22), and obtain:

$$\text{ribulose-5-phosphate} + \text{ATP} \rightleftharpoons \text{ribulose-1,5-diphosphate} + \text{ADP}. \tag{2.24}$$

$$\Delta G^{\circ\prime} = 2400_{(\text{eqn } 2.18)} - 7600_{(\text{eqn } 2.22)} = -5200 \text{ cal/mole}$$

The equilibrium constant is now antilog $5200/1364 = 6.5 \times 10^3$ in favour of the *formation* of ribulose-1,5-diphosphate. In contrast to the situation when equation (2.18) is taken alone, the synthesis of the diphosphate now proceeds almost to completion.

* See R. Alberty, *J. Biol. chem.* **224**, 3290 (1969) for further information on this point.

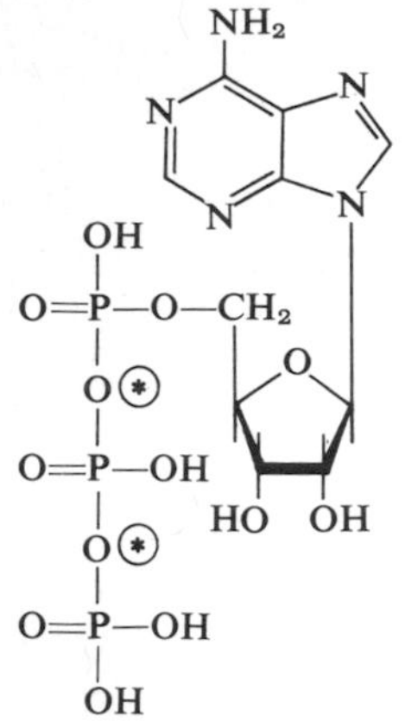

**Fig. 2.4** The chemical structure of ATP

Coupling of this sort, with high energy compounds as intermediate carriers of the energy, takes place in very many biochemical reactions.* A further example will show both how ATP is used for energy coupling and also a problem that arises when the reversal of a coupled reaction is required.

### The reversal of reactions involving ATP

In carbohydrate breakdown (Ch. 14) fructose-6-phosphate has to be converted to fructose-1,6-diphosphate against a $\Delta G^{\circ\prime}$ of $+3400$ cal/mole. It is helped to do so by coupling with the highly negative $\Delta G^{\circ\prime}$ of hydrolysis of ATP.

$$\text{fructose-6-phosphate} + \text{ATP} \rightleftharpoons \text{fructose-1,6-diphosphate} + \text{ADP}$$

$$(2.25)$$

The overall $\Delta G^{\circ\prime}$ is $-4200$ cal/mole.

The conversion of fructose-6-phosphate to fructose-1,6-diphosphate has to be reversed in carbohydrate synthesis (Ch. 19). It cannot be reversed by any practicable manipulation of [ADP] since the equilibrium constant against reversal is about $10^3$, from the value of $\Delta G^{\circ\prime}$ and equation (2.16); there is the additional problem that this is one of those processes that are found in the cell to be significantly displaced from equilibrium (see p. 18). (In any case the cell attempts to keep the ratio [ATP]/[ADP] constant at all costs, for reasons to do with the control of metabolic rate (see Ch. 29).)

Therefore when net reversal is required, another enzyme is used that catalyses, not the transfer of the esterified phosphate to and from adenosine phosphates, but the conversion to and from *inorganic phosphate*.

$$\text{fructose-1,6-diphosphate} \rightleftharpoons \text{fructose-6-phosphate} + \text{P}_i$$

$$\Delta G^{\circ\prime} = -3400 \text{ cal/mole} \qquad (2.26)$$

The equilibrium constant greatly favours net formation of fructose-6-phosphate.

There are many such biochemical processes which are coupled so that they can be driven in the direction that is energetically unfavourable but which are catalysed in the opposite direction by a *different enzyme*. The reversal of such reactions does not then involve the reversal of the process to which they were coupled in the forward direction.

When we encounter a reaction that is displaced from equilibrium to a significant extent ($K/\Gamma \gg 1$, p. 22) we must take the fact into account when assessing the possibility of using coupling to achieve a net synthesis in the energetically disfavoured direction. We do this by adding (or subtracting, if the displacement from equilibrium is already in the direction that we want) a correction to $\Delta G^{\circ\prime}$.

It can be shown to follow from equation (2.14) that this correction is

$$\Delta G_{(\text{displacement})} = -1364 \log_{10} K/\Gamma \quad \text{cal/mole.} \qquad (2.27)$$

(Note that this relationship confirms that the correction vanishes when equilibrium is attained since, when this is so, $\Gamma = K$, i.e. $K/\Gamma = 1 : \log 1 = 0$.)

---

* The product of the energy-requiring reaction is not always a phosphorylated compound even when the reaction is being driven by a high-energy phosphate (e.g. the synthesis of glutamine, p. 363). We must also point out that coupling with the breakdown of high-energy compounds drives mechanical endergonic processes as well as chemical ones (pp. 300ff).

You can see that it is only when $K/\Gamma$ becomes a very large number indeed that $\Delta G_{\text{(displacement)}}$ becomes more significant than the $\Delta G^{\circ\prime}$ values of the important energy-producing reactions. Large values ($10^3$ or more) of $K/\Gamma$ do exist for reactions of large or small $\Delta G^{\circ\prime}$ and they serve to focus our attention on certain reactions that are used for the *control* of the extent to which metabolic pathways operate (Ch. 29).

You may now realize that we had some time ago an example of what is in a sense an energy coupling. The tapping-off of glyceraldehyde-3-phosphate to a lower energy state (p. 20) employs a process of negative $\Delta G$ to drive reaction (2.17) which has a positive $\Delta G^{\circ\prime}$. The process of negative $\Delta G$ is the series of reactions (Chs 14 and 15) that accomplish the stepwise oxidation of the triose to three molecules of carbon dioxide.

*Other high-energy compounds*

ATP, although the principal medium of energy exchange, is not the only high-energy compound. Table 2.2 lists a number of others.

**Table 2.2** High-energy* compounds

| Type | $\Delta G^{\circ\prime}$ of hydrolysis (cal/mole) | Common examples | See page |
|------|-----------|-----------------|----------|
| Pyrophosphates 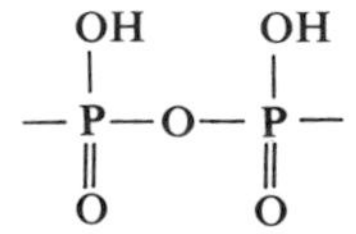 $$\begin{array}{cc} OH & OH \\ | & | \\ -P-O-P- \\ \| & \| \\ O & O \end{array}$$ | approx. $-7000$ | ATP, ADP; other nucleoside di- and triphosphates | 259, 284, 277, 334, and many others |
| Acyl phosphates 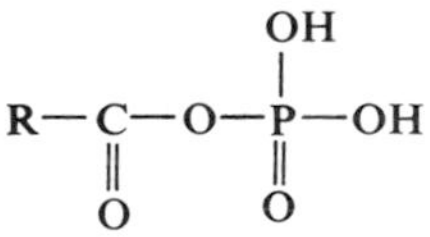 $$\begin{array}{c} OH \\ | \\ R-C-O-P-OH \\ \| \quad \| \\ O \quad O \end{array}$$ | approx. $-10\,000$ | Glycerate-1,3-diphosphate | 262 |
| Enol phosphates 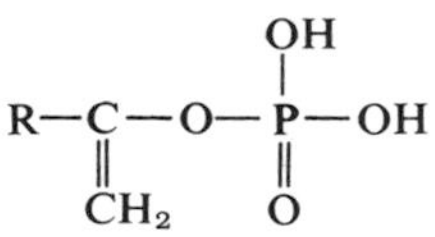 $$\begin{array}{c} OH \\ | \\ R-C-O-P-OH \\ \| \quad \| \\ CH_2 \quad O \end{array}$$ | approx. $-13\,000$ | Phospho-*enol*pyruvate | 264 |
| Thioesters 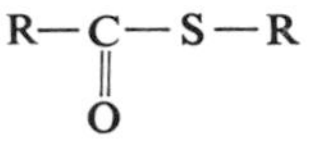 $$\begin{array}{c} R-C-S-R \\ \| \\ O \end{array}$$ | approx. $-8000$ | Acetyl CoA, some enzyme-substrate complexes | 273, 262 |
| Guanidine phosphates 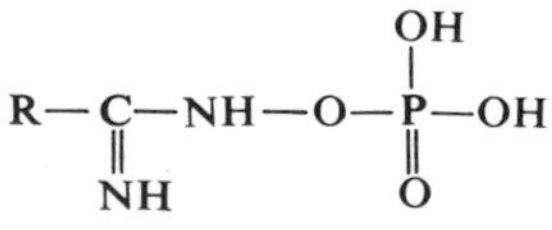 $$\begin{array}{c} OH \\ | \\ R-C-NH-O-P-OH \\ \| \qquad\quad \| \\ NH \qquad\quad O \end{array}$$ | approx. $-7000$ | Creatine phosphate $$\begin{array}{c} R \text{ is } CH_3 \\ | \\ N- \\ | \\ CH_2 \\ | \\ COOH \end{array}$$ | 300 |

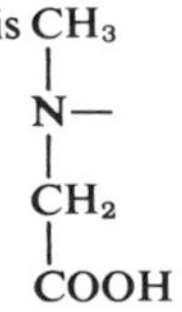

* 'High energy' is a despised term in some quarters and does indeed have its drawbacks. However, it is a useful shorthand expression and as long as it is used with care it is worthy to be retained.

The $\Delta G^{\circ\prime}$ values may be contrasted with those of 'low energy' compounds, for example the phosphate esters of various sugar compounds (such as glucose, fructose, and in the first of the phosphate groups of ATP itself, ribose) which are 'low energy', having $\Delta G^{\circ\prime}$ values for hydrolysis of 2000 to 5000 cal/mole.

Phospho-*enol*pyruvate (eqn 2.6) can now be seen as an example of a high energy compound. It is not used directly to drive reactions but indirectly, through the manufacture of ATP.

$$\text{phospho-}\textit{enol}\text{pyruvate} + \text{ADP} \rightleftharpoons \text{pyruvate} + \text{ATP}$$

The overall energy change is

$$\Delta G^{\circ\prime} = -13\,200 + 7600 = -5600 \text{ cal/mole.}$$

Thus there is a net formation of ATP.

This example also serves to point out that there are means of producing ATP other than photosynthetic or oxidative phosphorylation. This third type of process, in which the phosphorylation arises as part of the mechanism of a metabolic reaction, is called *substrate-level phosphorylation*. Although substrate-level phosphorylations are not of such general significance as the other two types, important examples do exist (pp. 262 and 278).

*Structural features of high-energy compounds*

It is possible to point to the structural features of the molecules in Table 2.2 that confer upon the appropriate bonds their necessary tendency toward breakage (which because of the relation between $\Delta G^\circ$ and equilibrium constant, means high energy of bond formation). In the case of the pyrophosphate or phosphoric anhydride type, there are two principal features. The first is that the p$K$ values of the phosphate groups are such that ATP, for example, exists at physiological pH mainly in the tetra-anion form.

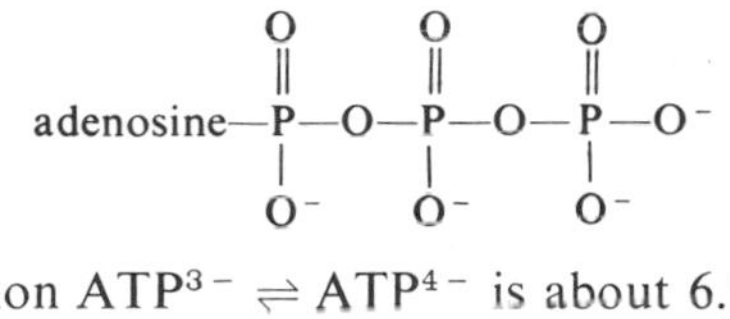

(The p$K$ of the ionization $\text{ATP}^{3-} \rightleftharpoons \text{ATP}^{4-}$ is about 6.5: calculate for yourself the percentage of the $\text{ATP}^{4-}$ form at physiological pH (7.3, say)).

There is thus a great deal of electrostatic repulsion between the negatively-charged groups and this makes an important contribution to the tendency of the molecule to fly apart.

Then there is the question of *resonance*. You will observe that there are, at either side of the weakened bond, two groups that will tend to attract electrons away from the central grouping.

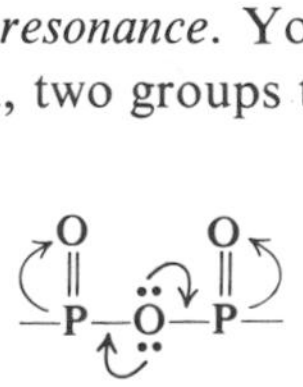

This competing resonance, which is absent in the products, contributes greatly to the weakening of the bond. Explanations in terms of resonance that are rather similar may be given for the next type of structure shown in the table, the acylphosphates. This clearly resembles the pyrophosphate in the essential structural features. Phospho-*enol*-pyruvate is an exceptionally powerful high-energy compound, and is now coming into use as a phosphate-donating reagent in classical organic synthesis. The phosphate group in this case forces the molecule to remain in the enol form even though the keto form ($R-C=O$) is much more

$$\overset{|}{CH_3}$$

stable. However, when the phosphate group is removed the enol structure is enabled to relax to the keto form, and this fact provides much of the driving force for the hydrolysis. There is also the fact that the carbon atom carrying the phospho-*enol* group is $sp_2$ hybridized, which is an intrinsically weaker form of bonding than the $sp_3$ adopted by carbon atoms carrying ordinary phosphate esters. Thioesters are weaker than carboxyl esters because of the inability, relative to oxygen, of sulphur to allow its non-bonding electrons to resonate with those in the carbonyl group.

The guanidine phosphates owe their high-energy status to the prevention, by the phosphate group, of the rather exceptional degree of resonance of the guanidinium ion. The guanidinium ion is therefore much more stable than the phosphorylated compound and there is consequently a large negative free energy of hydrolysis.

These are brief outlines of the type of arguments that may be deployed to explain one of the most important features of this class of molecule: the thermodynamic instability. We say 'one of', but is not the high $\Delta G^{\circ\prime}$ of hydrolysis the *only* point of importance? The answer is no. In addition to the question of the degree of thermodynamic instability, there is the almost unrelated, but equally important, question of the degree of kinetic *stability*. We mean by this that, if there were not some factor slowing down the rate at which these molecules may follow their strong inbuilt tendency to fall apart, they would be of no use, since they could not last long enough to travel from their site of synthesis to their site of action. In fact each has structural features that slow down the disruption of the molecule; thus ATP will exist in free solution for some days. Some last for much shorter periods of time (glycerate-1,3-diphosphate, for example), some for longer (creatine phosphate). All, in spite of the high $\Delta G^{\circ\prime}$ values, last long enough to carry out their functions. Clearly, when considering the fitness of a molecule for its role as a high-energy compound, we must bear the kinetic factors in mind just as much as the static, thermodynamic factors.

### Coupling of oxidative and reductive reactions

It is possible to couple oxidative and reductive reactions without the intervention of ATP, or of any of the other high energy compounds that we have so far considered.

For example, in fat metabolism (Chs. 16 and 20) certain derivatives of β-oxo and β-hydroxy acids must be interconverted. This reaction can be written formally as,

$$CH_3(CH_2)_nCO.CH_2COOR^* + 2H \rightleftharpoons CH_3(CH_2)_nCHOH.CH_2COOR^*$$
$$\Delta G^{\circ\prime} \simeq -10\,000 \text{ cal/mole} \tag{2.28}$$

* The nature of, and need for, the R groups are explained fully in Chs. 16 and 20.

The hydrogen is donated from the specialized, hydrogen-carrying molecule nicotinamide adenine dinucleotide phosphate (NADP). (The structure of this molecule is shown in Table 4.2, p. 79.)

The donation of hydrogen may be written as

$$NADPH_2 \rightleftharpoons NADP + 2H \qquad (2.29)$$
$$\text{(reduced form)} \qquad \text{(oxidized form)}$$
$$\Delta G^{\circ\prime} = +4100 \text{ cal/mole}$$

and so the overall, coupled process is

$$CH_3(CH_2)_nCO.CH_2.COOR + NADPH_2 \rightleftharpoons$$
$$CH_3(CH_2)_nCHOH.CH_2COOR + NADP \quad (2.30)$$

$\Delta G^{\circ\prime}$ for (2.30) is about $-6000$ and the reaction readily proceeds from left to right.

When reaction (2.28) is operated from right to left, a hydrogen *acceptor* is required. NADP could in theory be used for this purpose, but in fact it is only occasionally so used in living systems. The closely similar molecular nicotinamide adenine dinucleotide (NAD), which differs only very slightly in structure from NADP (see Table 4.2), is normally used instead.

The structures of NAD and NADP are, in fact, so similar that the energetics of reactions that involve them would not be significantly different if one replaced the other. We can therefore re-write equation (2.30) to represent the use of NAD as an hydrogen acceptor:

$$CH_3(CH_2)_nCHOH.CH_2COOR + NAD \rightleftharpoons$$
$$CH_3(CH_2)_nCO.CH_2COOR + NADH_2 \quad (2.30a)$$
$$\Delta G^{\circ\prime} = +6000 \text{ cal/mole}$$

How is reaction (2.30a) driven from left to right in the face of this unfavourable $\Delta G^{\circ\prime}$? As is usual when an energetically unfavourable reaction has to be operated, it is *coupled* to one that is energetically favourable. The reaction that is used for coupling in the present case is

$$NADH_2 + \tfrac{1}{2}O_2 \rightleftharpoons NAD + H_2O \qquad (2.31)$$
$$\Delta G^{\circ\prime} = -52\,000 \text{ cal/mole}$$

(Reaction (2.31), which is a multi-step process, is discussed at length in Ch. 12.)

If we combine reactions (2.30a) and (2.31) we obtain a specific example of the general oxidative reaction we gave before as reaction (2.21):

$$CH_3(CH_2)_nCHOH.CH_2.COOR + \tfrac{1}{2}O_2 \rightleftharpoons$$
$$CH_3(CH_2)_nCO.CH_2.COOR + H_2O \quad (2.32)$$
$$\Delta G^{\circ\prime} = -46\,000 \text{ cal/mole}$$

There is now no difficulty at all in operating the process from left to right. Indeed, there is free energy to spare and some of it is used to bring about the phosphorylation of 3 molecules of ADP to ATP. We have mentioned this *oxidative phosphorylation* of ADP before, and will return to it in Ch. 12.

In contrast to NADH, NADPH does not generally enter oxidative phosphorylation, but is almost always used solely as a hydrogen donor. As we shall see in part 3, it is produced during the oxidation of only a few substrates and provides a convenient means of interrupting the oxidation of $AH_2$ at an early stage and using the reducing power elsewhere.

Equation (2.30) is a typical example of the general rule that the reducing power of $NADP_{(reduced)}$, in distinction to that of $NAD_{(reduced)}$, is coupled to synthetic reactions (pp. 241, 328, 343 and 344). We may note that reduced NADP is the other immediate product, beside ATP, of the fall of the electron in photosynthesis. This example further emphasizes the role of reduced NADP in synthesis since reduced NADP and ATP are used directly in the synthesis of carbohydrate in the plant (see Ch. 19).

We can summarize the preceding few pages by saying:

(1) The change in the Gibbs free energy, $\Delta G$, is a valuable measure of the availability of, or need for, energy in a reaction.

(2) $\Delta G$ is related to the equilibrium constant and this makes it possible to use the energy term to calculate the direction of net reaction.

(3) Living matter operates energetically disfavoured reactions by coupling them with energetically favoured ones. The fall of a light-energized electron back to its ground state and the fall of electrons to molecular oxygen are the principal among the fundamental energy-yielding processes. These processes produce 'high-energy' compounds. The high-energy compounds are used to couple the fundamental energy sources to the energy-requiring processes of the cell.

We have made a great deal of use of the concept of $\Delta G$ and it is time to return to look at its thermodynamic basis more closely.

## The thermodynamic significance of $\Delta G$

The water analogy that we have used on several occasions has served us well, but we must now move beyond it. The first reason is that water flowing downhill is an irreversible process, while we have confined ourselves to reversible reactions. Second, the concept of energy in chemical reactions goes beyond the mechanical energy (or heat, to which it is strictly equivalent) that is produced. It may be surprising to learn that $\Delta G$ is not a measure of heat alone, since it is expressed in units that involve calories, but further discussion may clear the matter up.

Many chemical reactions do of course produce heat, and warm-blooded animals make use of some of these to keep their body temperatures above that of their surroundings, with all the advantages of enhanced activity and stability of internal environment that this brings. Experimentally we may measure the heat output in such reactions by means of a *calorimeter*. This device is simple in principle, consisting of a reaction chamber insulated from the surroundings and a means of measuring the temperature changes that occur. Modern versions are extremely sensitive, and can measure, for example, the heat output of a small number of bacteria, or the heat changes accompanying the combination of a few micromoles of reactants.

What is found when a calorimeter is used is that some reactions that proceed substantially to completion are accompanied not by an output of heat but by an *uptake* of heat and a cooling of the system. Similarly, some physico-chemical processes, such as the dissolving of ammonium sulphate or urea in water, occur quite readily, but with a substantial cooling of the system.

Let us write the heat uptake as $\Delta H$ and use a similar sign convention to that for $\Delta G$. Thus a negative $\Delta H$ will indicate the production of heat (loss of heat by the system) and a positive $\Delta H$ will indicate the absorption of heat (gain of heat by the system). $\Delta H$ normally includes any mechanical work done on or by the system as well as heat energy, and $\Delta H$ is then known as the *enthalpy* change rather than just the heat change. The units may still involve calories, because of the complete equivalence of heat and mechanical energy. $\Delta H°$ will be the standard enthalpy change, defined in the same way as $\Delta G°$.

What we have said above – that $\Delta H$ is not the complete measure of the free energy change – may be written

$$\Delta G° \neq \Delta H°$$

or, to make things look less enigmatic,

$$\Delta G° = \Delta H° \text{ and some other factor.}$$

To underline the need for this other factor we shall consider some processes that proceed without an external energy source and which, therefore, have a negative $\Delta G°$. If $\Delta G°$ were equal to $\Delta H°$, $\Delta H°$ would be negative in all such cases, that is to say heat would be evolved. In fact we have already noted that $\Delta H°$ is sometimes positive.

*A process with negative $\Delta G°$ and positive $\Delta H°$*

You will be aware that many heavy metals, lead, for example, are highly poisonous. A principal reason for this is that the metal ion binds to proteins and, apart from preventing the groups to which it binds from participating in biological activity, it upsets the balance of forces (Ch. 3) that holds them in their proper, functioning configuration. The result is that the normal, tight structure (e.g. Fig. 4.10, p. 75) unfolds to a much more open type. This occurs very readily, but calorimeters show a great deal of heat energy is absorbed (typical values are in the region of 60 000 cal/mole). Therefore, the 'other factor' must have a considerable negative value, to give an overall negative sign to $\Delta G$ and to make the reaction

$$\text{Protein} + Pb^{2+} \rightleftharpoons \text{(protein–Pb complex)}$$

go predominantly from left to right.

*A process with negative $\Delta G°$ and a $\Delta H°$ of zero*

An elastic band (a biological product) will snap back from a stretched configuration to its resting shape without an external energy source to drive it. Rubber (stretched) $\rightleftharpoons$ rubber (unstretched) goes predominantly from left to right. $\Delta G°$ is therefore negative. Measurement shows that $\Delta H°$ is approximately zero since the heat change is almost exactly equal and opposite in effect to the mechanical work done. How then does the negative $\Delta G°$ arise?

The answer is in all cases related to the *degree of order in the system*. When an elastic band is stretched, the long chains of molecules of which it is composed are themselves stretched out into neat, parallel lines. When it is released, they assume a random, jumbled configuration. It is a natural tendency of all matter to pass from a highly ordered state to a disordered one. As time passes, intricate buildings crumble into shapeless mounds, while the reverse does not happen. A pack of cards, initially arranged in suits and in numerical order within the suits, will pass to a random arrangement on shuffling, while the chances of a reversal to the initial order by the same means are extremely low. With no enthalpy change to be overcome, an ordered array of chains in the elastic band rapidly tumbles into a random mass of coils, once the restraining force has been withdrawn. Similarly, in our example of the effect of lead ions on protein, the complex architecture of the protein molecule, once one of its stabilizing forces is interfered with, will strongly tend to fall into a less ordered state. This tendency provides the driving force necessary to overcome an unfavourable enthalpy change.

We know, further, that to impose order on a chaotic, random situation takes energy. It is so when erecting a building, when sorting cards into order, and when assembling molecules with a complex internal structure.

For mathematical purposes, it can be shown that the change of order in a system may be expressed in terms of the change in *entropy*, $S$. The precise form of the relationship between entropy and degree of order is

$$S = R \log_e Z \qquad (2.33)$$

where $Z$ is a direct measure, calculated from geometrical factors alone, of the change in the degree of order between the initial and final states.

The way in which $Z$ is calculated may be found in any textbook of statistical thermodynamics (unfortunately, the calculation becomes unmanageable for structures of any complexity and we are forced back to experimentally determined values). However, all we need to note is that $S$, which we will use rather than $Z$ for day-to-day purposes, is also related to degree of order. The more ordered a system is, the *lower* $S$ will be. It may seem surprising at first that in the spontaneous change from the ordered to the disordered state entropy *increases* and therefore that the total entropy of the universe will go on increasing until all spontaneous changes have been accomplished. This sort of statement often worries the student, who rightly suspects any process in which it seems that we get something for nothing. However, we should take comfort in the fact that it is due merely to one of those annoying sign conventions in which thermodynamics abounds. The founders of the subject could just as well have given $S$ the opposite sign, and then entropy would have *decreased* during the spontaneous changes of which we have written and everyone would have been content.

It is fortunately possible to incorporate the contributions made by $\Delta H$ and $\Delta S$ to $\Delta G$ in one expression. The expression, which will be of great value to us later on, is:

$$\Delta G^\circ = \Delta H^\circ - T \Delta S^\circ \qquad (2.34)$$

where $T$ is the absolute temperature.

We can now see that a process that needs no energy source (negative $\Delta G^\circ$) may proceed because of a negative $\Delta H^\circ$ (the system loses heat or mechanical energy); or because of a positive $\Delta S^\circ$ (the system falls from a more-ordered to a less-ordered state); or because of both; or, as often happens, because the contribution of one outweighs an opposite contribution from the other.

Similarly, if we find a process that has to be driven (positive $\Delta G°$), this will be because of a positive $\Delta H°$ (the system has to gain heat or mechanical energy) or because of a negative $\Delta S°$ (the system has to be arranged into a more ordered state) and so on.

The molecular approach to biology is unusually rich in examples in which the relationship between the order-derived term, entropy, and the progress of a reaction may be traced in reasonable detail. An example has been given above (the toxic effect of heavy metals) and others will be given in later chapters (pp. 77 and 120 ff). One further example that we shall mention here concerns the binding of a molecule to a surface on a macromolecule that is specially shaped to receive it. This, as we said before, might be the binding between an enzyme and the molecule that it has to help through a reaction, p. 114; between an antigen and an antibody, p. 101; or between a hormone and its receptor site, p. 100. In all these cases water molecules (which usually surround the reactants) will probably have been immobilized on the surface of the protein until displaced by the incoming molecule. It can be calculated that the entropy change involved in the transition from water molecules in a fixed orientation to the free, randomly-moving and tumbling state, can be large enough to assist such binding processes, upon which life depends.

Finally, entropy and life are often spoken of as being linked in another way. It is sometimes said that living matter is unique in that it alone possesses continuously decreasing entropy. This statement is partly true, but it causes a lot of confusion and it may be helpful to clear it up here. It is certainly true that an organism that grows, or even one that merely maintains itself, is *organising* matter and is thus decreasing entropy. One could feed a pig on sausages and the biological machinery of the animal will take the less organized matter of the sausage and turn it into infinitely more complex tissue. (The reverse transformation, pig to sausage, is much easier and can be accomplished by a very much simpler, mechanical, device such as a mincer. This example may help to show why, to many people, the study of biological devices is more interesting than the study of mechanical ones.)

There is, however, nothing mysterious about this. The entropy in the universe as a whole continues to increase. In order to *decrease* the entropy in that volume of space occupied by the tissues of the animal, payment must be made. If disordered material is to be made into ordered material, then energy (enthalpy) has to be consumed and converted according to the relationship $\Delta G = \Delta H - T\Delta S$, as we saw in the previous section of this chapter. In fact, the rise and maintenance of life on earth has required a continuous input of energy from the sun, and so the universe as a whole is required to supply energy. This, in the long term, means a net increase in entropy of the *whole* system – biological and non-biological – considered together.

## CHEMICAL KINETICS – 'HOW FAST?'

Now that we know how to treat the question of the direction in which a reaction is likely to proceed, we are equipped to deal with many biochemical problems.

However, we must return to face the question of *how fast* the reaction will be. This really includes the question of whether it will occur at any useful rate at all since some reaction rates may be so slow as to produce no appreciable product over geological periods of time, even if they are thermodynamically favoured. Other reactions, no more favoured than the first in terms of $\Delta G$, may be essentially complete in $10^{-9}$ seconds.

What accounts for these spectacular differences in rate, and why does the free energy term, i.e. the equilibrium constant, gives us no help in predicting rates? The rest of this chapter is designed primarily to answer these questions, and also to give sufficient understanding of the kinetic factors to help us to appreciate in particular the remarkable design of the enzymes – the agents that increase the rates of biochemical reactions to an extraordinary extent and thereby contribute so greatly to the successful operation of living processes.

We began the first part of this chapter by describing the elementary types of observations that led to the general formulations that we required, and we shall do the same here. The earliest major study of kinetics was carried out over 100 years ago on a biological product, sucrose (p. 328). The compound consists of the monosaccharides glucose and fructose joined by a glycosidic link (p. 46). This linkage may be broken by acid-catalysed hydrolysis in the test tube.

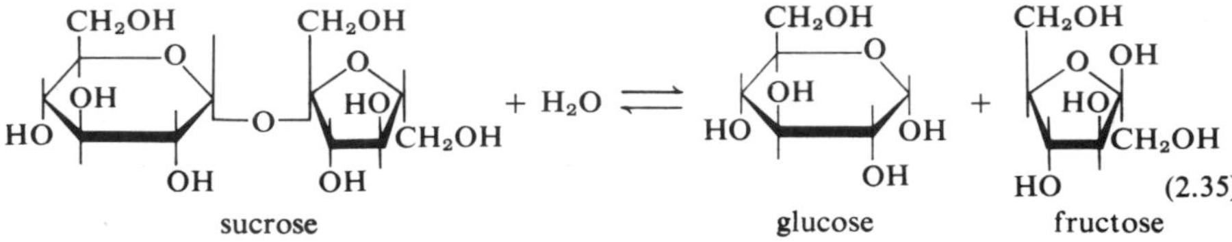

and the progress of the reaction may be followed by observing the change in optical rotation. It was found that the rate at which the reaction progressed from left to right varied with time and was proportional to the concentration of the sucrose that remained at any instant.

That is

$$\text{Rate from left to right} = k_{+1} \times \frac{\text{concentration of sucrose}}{\text{(in moles/litre)}} \qquad (2.36)$$

$k_{+1}$ is called the *rate constant* of the forward reaction. (In fact we now recognize that the equation ought to have one more term, since $H_2O$ appears in the chemical equation. Thus, still using the convention that [  ] means 'concentration in moles per litre', the true rate equation is

$$\text{Rate} = k_{+1} \, [\text{sucrose}][H_2O]. \qquad (2.36a)$$

The concentration of water is very large. Any concentration of sucrose that can be achieved in practice will be much less than this, and the drop in concentration of water caused by the consumption of one mole in the reaction will, as a consequence, be negligible. If the concentration of water is constant, it may be incorporated in the rate constant and thus change its numerical value. This is usually done as was the case for equilibrium constant (p. 13) and we return to the original expression

$$\text{Rate} = k_{+1} \, [\text{sucrose}]$$

with the value of $k_{+1}$ 55.5 times larger than the true figure.)

All subsequent work has confirmed that rates of reaction depend on the products of concentration of the reactants (thus it follows that if two molecules of a particular reactant are involved, the rate will be proportional to the square of the concentration of that reactant, and so on). In the hydrolysis of sucrose the concentrations of products, glucose and fructose, begin to build up. Thus, since we are dealing with a process that is in theory reversible, we may write:

$$\text{Rate from right to left} = k_{-1}\,[\text{glucose}][\text{fructose}].$$

In the same way as the rate from left to right falls off from the start of the reaction with the depletion of sucrose, so the rate from right to left increases as glucose and fructose begin to accumulate. (In order to keep the example simple we are neglecting the possibility of side reactions.) Eventually, the system will come to equilibrium, when the rate from right to left equals that from left to right. At this point

$$k_{+1}\,[\text{sucrose}] = k_{-1}\,[\text{glucose}][\text{fructose}].$$

Therefore:

$$\frac{k_{+1}}{k_{-1}} = \frac{[\text{glucose}][\text{fructose}]}{[\text{sucrose}]}$$

The right-hand half of this expression is equal to the equilibrium constant of the reaction, by definition (p. 9). And so

$$\frac{k_{+1}}{k_{-1}} = K. \tag{2.37}$$

We saw on p. 22 that, when the system was displaced from equilibrium, $\Gamma$, the mass action ratio, $= [B]'/[A]'$. Now, since $K = k_{+1}/k_{-1}$, forward velocity $= k_{+1}[A]'$ and reverse velocity $= k_{-1}[B]'$, $K/\Gamma$ (the degree of displacement from equilibrium, p. 22) is

$$\frac{K}{\Gamma} = \frac{k_{+1}/k_{-1}}{[B]'/[A]'} = \frac{k_{+1}[A]'}{k_{-1}[B]'} = \frac{\text{forward velocity}}{\text{reverse velocity}}.$$

We shall see (p. 509 ff) that equation (2.37) is a useful one, but it does not on its own help us to break through the barrier between the kinetic and thermodynamic approaches to chemical reaction. If we wish to calculate $k_{+1}$ from $K$, we have first to know the value of $k_{-1}$ and vice versa.

There is one question that gives the key to the nature of the rate constant, and indeed of the whole problem of chemical kinetics. The question is: 'If there is any thermodynamic tendency at all for the reaction to occur, why doesn't it do so at once?' The answer comes from the examination of the *theory of reaction intermediates*.

You will probably be familiar with the type of diagram shown in Fig. 2.6, but for those who are not we shall approach it by means of Fig. 2.5. In this figure, the vertical axis is the $-\Delta G°$ scale, as before, and the horizontal axis is the so-called 'reaction co-ordinate' in which successive stages in the event are indicated by increasing positive displacements along the $x$ axis.

Figure 2.5 indicates the situation where, in a reaction A $\rightleftharpoons$ B, A is at a higher

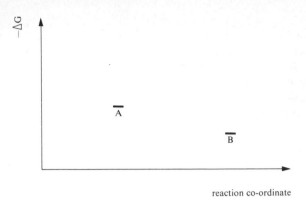

reaction co-ordinate

**Fig. 2.5**

state of available free energy $(-\Delta G^\circ)$ than B. The equilibrium constant, $K$, as we have seen, is given by

$$\Delta G^\circ_A - \Delta G^\circ_B = 1364 \log_{10} K$$

i.e.

$$K = 10^{\frac{\Delta G^\circ_A - \Delta G^\circ_B}{1364}}$$

and any great discrepancy between the $\Delta G^\circ$ values should mean that the change A → B effectively goes to completion.

However, we should note that in very many such pairs, the compound of high available free energy has as much of a distinct identity as the one with lower energy. It also has a name, it can be synthesized (or purchased) and stored, and it will stay in being for long enough to enable us to determine its properties. We must therefore propose that there is a barrier (Fig. 2.6) of even higher available free energy between A and B, to explain the failure of A to change to B completely and at once. If we call the peak of the barrier I, the equilibrium A ⇌ I will be almost totally in favour of A, and A will thus be stable. For some reactions the barrier is very low indeed. We will meet at least one case later, in the reaction

$$\alpha\text{-glucose} \rightleftharpoons \beta\text{-glucose} \quad (\text{p. 170}).$$

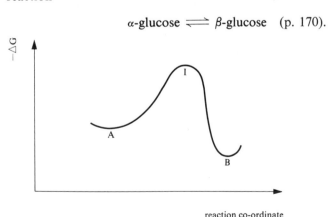

reaction co-ordinate

**Fig. 2.6**  The energy barrier

This case makes the necessary points very clearly. Neither form of glucose can be kept from approaching an equilibrium mixture for sufficiently long for the material to be stored. One cannot for example, purchase $\alpha$- and $\beta$-glucose separately, but only as the equilibrium mixture. Nonetheless, if the two are prepared separately they remain unchanged for long enough for some properties (optical rotation, for example) to be determined (see Fig. 8.5 p. 170).

What we are saying, in effect, is that if the barrier is high then the rate constant will be low and the reaction will be slow. Alternatively, if the barrier is low the rate constant will be high and the reaction will be a rapid one. The relative heights of A and B control the equilibrium constant (eqn 2.16) but clearly do not enter into the present issue.

We must look a little more deeply into the nature of the energy barrier so as to see if we can relate its height more quantitatively to the rate constant. First we may note that the state I may correspond to the temporary *activated complex*, which modern chemical theory requires as an intermediate state while bonds in A are rearranged to produce the new grouping of bonds and atoms that will become B. For example, in the triose phosphate isomerase reaction, a vital step in the degradation and synthesis of carbohydrates (p. 19 above, pp. 261 and 324) the course of the reaction is postulated to be

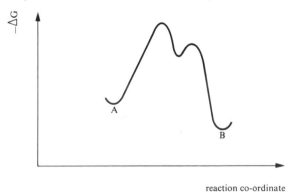

$$ \text{(2.38)} $$

| dihydroxy-acetone phosphate | 'ene-diolate' | glyceraldehyde phosphate |

It would be reasonable to suppose that the strictly temporary *transition state* (in this case known as an enediolate intermediate), with all its displaced charges and odd bonding, is more unstable than either the initial or final states. 'More unstable', in this context, means 'having a higher available free energy', so that we place it above both A and B, which is just what the argument on p. 36 requires. The true activated complex, as we have defined it above, need not be the highest point in all reactions, since some step on the way from A to I or from I to B may be even less stable (see Figs 2.7 and 2.8).

Fig. 2.7

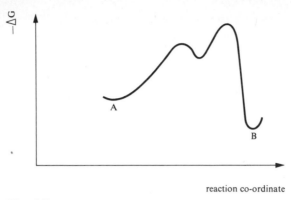

reaction co-ordinate

**Fig. 2.8**

However, what is clear is that if the reaction is to proceed, there is a peak to be scaled. How is this done, since it appears to involve the chemical equivalent of water running uphill?

The answer is that the position of A, I and B in reaction schemes such as Figs 2.5 to 2.8 are not static. Random thermal fluctuations cause a continual variation in the intrinsic energies of molecules. Energies of translation (movement in space), rotation (tumbling) and vibration (of bonds in the molecule) are continually modified by interaction with other molecules. Thus the positions of A, I and B that have been shown so far are *averages* and the energies of individual molecules of A, I and B will be distributed around these mean values (Fig. 2.9). Such a distribution can be described by a mathematical equation and we may derive an expression from this equation that relates the kinetic constant $K_{+1}$ to the height of the barrier $\Delta G^{\ddagger}$.

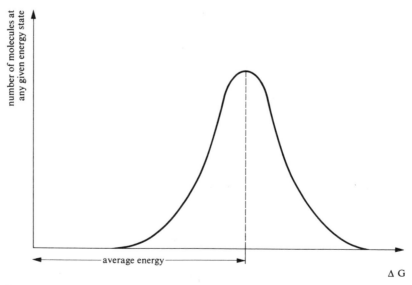

**Fig. 2.9** The distribution of energies about a mean value. (The positions of A, I and B in Figs 2.5, 2.6, 2.7 and 2.8 are average energy values of the type shown here)

The distribution equation is of the type proposed by *Boltzmann*

$$N_e = N_0 \exp\left(-\frac{\Delta G}{RT}\right) \tag{2.39}$$

where $N_0$ is the number of molecules in any given state of free energy, and $N_e$ is the number of molecules in an energy state $\Delta G$ above it. $T$ is the absolute temperature, and if $\Delta G$ is in calories per mole, then $R$, the gas constant, is 1.98 cal deg$^{-1}$ mole$^{-1}$. The rate constant, $k_{+1}$, is proportional to the number of molecules that are able to surmount the barrier and this number, from the expression above, is proportional to $\exp\left(-\Delta G^\ddagger/RT\right)$. Therefore, we may write

$$k_{+1} = \exp\left(-\frac{\Delta G^\ddagger}{RT}\right) \times \text{a constant factor.}$$

It can be shown that the constant factor is $RT/Nh$, where $R$ and $T$ have the same meaning as before. $N$ is the number of molecules per gram equivalent ($6.02 \times 10^{23}$ for any substance) and $h$ is Planck's constant. Thus, the equation that we have been working so long to obtain is, finally,

$$k_{+1} = \frac{RT}{Nh} \exp\left(-\frac{\Delta G^\ddagger}{RT}\right). \tag{2.40}$$

Since $\Delta G^\ddagger = \Delta H^\ddagger - T\Delta S^\ddagger$ by analogy with eqn (2.34), it is possible, and sometimes advantageous (see p. 114 ff) to write the expression as

$$k_{+1} = \frac{RT}{Nh} \exp\left(-\frac{\Delta H^\ddagger - T\Delta S^\ddagger}{RT}\right) = \frac{RT}{Nh} \exp\left(\frac{\Delta S^\ddagger}{R}\right) \times \exp\left(-\frac{\Delta H^\ddagger}{RT}\right). \tag{2.40a}$$

## The uses of the distribution and kinetic equations

We shall use equations (2.39) and (2.40) several times in this book, but shall use them immediately for only two purposes. The first is to point to the way in which enzymes operate in accelerating the rates of reactions (this is among the most important applications of physico-chemical concepts to biology and is more fully developed in Ch. 6); the second is to calculate the effect of fluctuations in temperature on the rate of chemical reactions (which, again, has relevance to the mode of action of enzymes, see p. 132 ff).

To turn to the first application, what we shall have to explain when we come to consider the enzymes is their immense ability to accelerate chemical processes – the kinetic constant, as we shall see in Ch. 6, may be increased by a factor of $10^{12}$. The explanation of this stems from the fact that $\Delta G^\ddagger$ for many of the chemical transformations that are accelerated by enzymes may easily be as much as, say, 28 000 cal/mole. The fraction of molecules able to overcome the barrier at blood heat, 310°K ($= 273 + 37$), is

$$\exp\left(-\frac{28\,000}{1.98 \times 310}\right) = \exp(-46) \text{ or } 6 \times 10^{-19}$$

– a very small proportion. $k_{+1}$ will therefore be very low.

If we could lower the barrier by even as small a factor as two the use of eqn (2.39) shows that the number of molecules now able to get across will increase by a factor of $e^{23}$ or about $8 \times 10^9$. $k_{+1}$ will therefore increase by this factor. Thus quite dramatic increases in rate are possible in return for relatively small changes in the height of the potential barrier.

This is the principal way in which enzymes, like all catalysts, work. Modern techniques have told us so much of their structure that we can point, in terms of detailed interactions between specific atoms, to the ways (there are a surprising number) in which they lower the barrier. This will take up a great deal of Ch. 6.

The effect of temperature on the rate of chemical reaction may be easily predicted from equation (2.40). Consider a chemical reaction taking place first at a temperature of $T°$ and then at $T + 10°$. From equation (2.40a), the ratio of the forward velocities at the two temperatures:

$$\frac{v_{+1}^{T+10}}{v_{+1}^{T}}$$

(this ratio, known as the '$Q_{10}$' is given by

$$Q_{10} = \frac{k_{+1}^{T+10}}{k_{+1}^{T}} = \frac{R(T + 10)/Nh}{RT/Nh} \times \frac{\exp(\Delta S^{\ddagger}/R)}{\exp(\Delta S^{\ddagger}/R)} \times \frac{\exp[-\Delta H^{\ddagger}/R(T + 10)]}{\exp(-\Delta H^{\ddagger}/RT)}.$$

Dividing through, we obtain

$$Q_{10} = \frac{T + 10}{T} \times \frac{\exp[-\Delta H^{\ddagger}/R(T + 10)]}{\exp(-\Delta H^{\ddagger}/RT)}$$

which easily simplifies to

$$Q_{10} = \frac{T + 10}{T} \exp\left(+ \frac{10\,\Delta H^{\ddagger}}{RT(T + 10)}\right). \tag{2.41}$$

Since $T \gg 10$ at all biological temperatures,

$Q_{10} \simeq \exp(+ 10\,\Delta H^{\ddagger}/RT^2)$ or, in the region of $310°K$,

$$Q_{10} \simeq \exp(+\Delta H^{\ddagger}/19\,000).$$

Note that the entropy term, being independent of temperature, is eliminated and $Q_{10}$ depends solely on $\Delta H^{\ddagger}$.

Most of the simple reactions with which we are familiar have $\Delta H^{\ddagger}$ values in the region of 10 000 to 20 000 cal/mole. (It is this fact that gives them an easily measurable rate and has enabled us to study them.) Since $e \simeq 2.7$, $Q_{10}$ is often found to lie between 2 and 3.

Table 2.3 shows some values for non-enzymic reactions.

**Table 2.3**

| Reaction | $Q_{10}$ | $\Delta H^{\ddagger}$ |
|---|---|---|
| Hydrolysis of sucrose catalysed by acid (p. 34) | 4.0 | 26 000 |
| Hydrolysis of casein (p. 91) catalysed by acid | 3.0 | 20 600 |
| Decomposition of $H_2O_2$ (uncatalysed) | 2.6 | 18 000 |
| Decomposition of $H_2O_2$ (catalysed by Pt) | 1.9 | 12 000 |
| Hydrolysis of urea catalysed by acid | 3.6 | 24 100 |

When you come to Ch. 6 you will be able to compare this table with the one on p. 133 which lists the same quantities, $Q_{10}$ and $\Delta H^{\ddagger}$, for these reactions when they are catalysed by enzymes. Since, as we have said above, enzymes lower $\Delta G^{\ddagger}$ you ought not to be surprised to find that they lower $\Delta H^{\ddagger}$. You will see that $Q_{10}$ is, consequentially, reduced.

There is one process for which the $Q_{10}$ is spectacularly greater than 2–3. This is the denaturation of proteins, for which a typical value of $\Delta H^{\ddagger}$ might be 57 000 cal/mole. $Q_{10}$ is an exponential function of $\Delta H^{\ddagger}$ and will in this case be about 20, a much greater value than the $Q_{10}$ for reactions that have values of $\Delta H^{\ddagger}$ closer to those shown in the table. The thermodynamics of the denaturation of proteins is discussed fully in Ch. 4 and, with particular reference to enzymes, in Ch. 6.

This brings to a conclusion our survey of the place of physical chemistry in biology. We have seen that both equilibrium thermodynamics and kinetics have a contribution to make. Our treatment has necessarily been in rather concentrated form, but we shall continue to exemplify the principles that we have set out in the appropriate places throughout this book.

# Part 1 Macromolecules

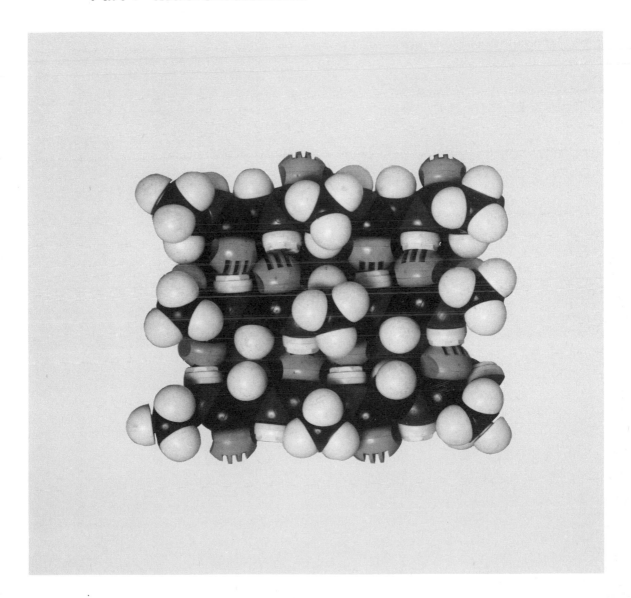

# Chapter 3   Introduction to macromolecules

Living matter is distinguished by its reliance on the special properties of certain classes of extremely large molecule. Of these classes the proteins, the nucleic acids and the polysaccharides are particularly prominent. All of these share certain principles of construction, although these are at first sight obscured by the differences that exist as a necessary consequence of the great diversity of the functions they undertake.

## The common features of macromolecules

The principal common feature is that all three are chain polymers formed by *condensation*, that is the combination of smaller molecules with the elimination of a molecule of water. In each case the smaller molecules are drawn from a related series. Proteins are composed of amino acids, of which the general formula is

$$R.CHNH_2.COOH$$

(see Ch. 4), polymerized by condensation between their amino and carboxyl groups. The resulting bond between the amino acids (which are now called amino-acid residues) is known as the peptide bond. The end of the chain bearing the free amino group is called the amino terminus, that bearing the carboxyl group is called the carboxyl terminus (see Fig. 3.1).

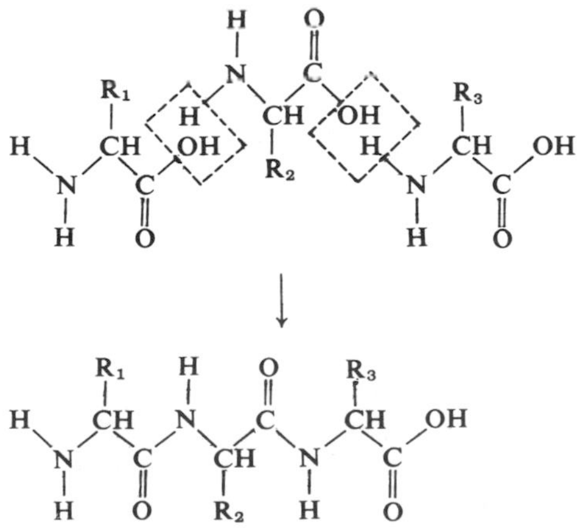

**Fig. 3.1**   The formal chemistry of the formation of the peptide bond. The amino terminus of the peptide chain is on the left, the carboxyl terminus is on the right

Similarly nucleic acids consist of nucleotides. These are of the general formula:

(Heterocyclic organic base)—pentose sugar—phosphate

(see Ch. 8) and they are joined by condensation between the phosphate group and an —OH group on the pentose of an adjacent nucleotide. The linkage between the nucleotide residues is called a phosphodiester bond – so named because each phosphate group carries two ester linkages. One end of the chain has a pentose in which the 3′ position takes no part in the bonding and the other has a pentose in which the 5′ position takes no part. These ends are called the 3′ and 5′ ends respectively (Fig. 3.2).

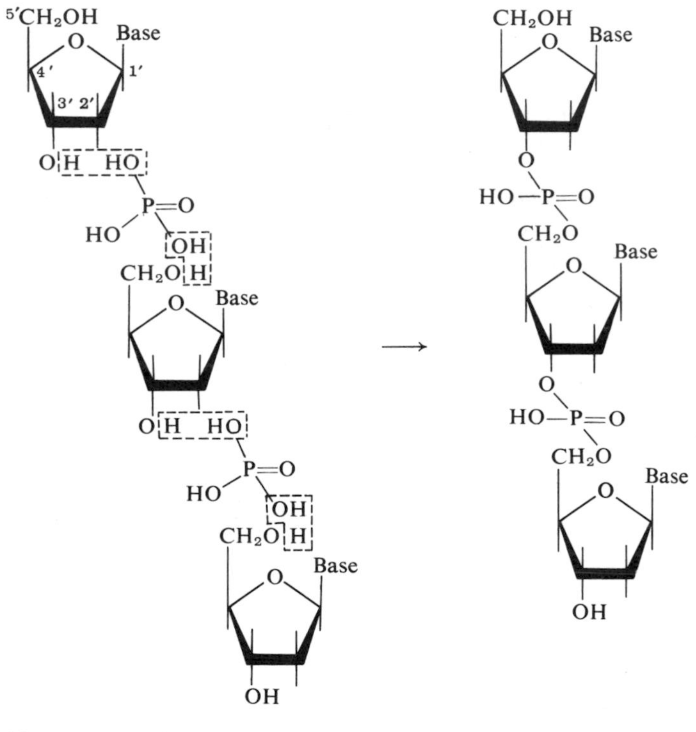

**Fig. 3.2** The formation of the phosphodiester bond. The numbering system used in the pentose ring is indicated in the top left-hand nucleotide. The 3′ end of the chain is at the bottom of the nucleotide chain, the 5′ end is at the top

Polysaccharides consist of sugars (Ch. 8) condensed through their —OH groups. The resulting bond between sugar residues is called the glycosidic linkage. In a glycosidic linkage the keto or aldo carbon atom of one of the sugars is always involved. The other —OH group may belong to any of the carbon atoms of the second sugar. Because of the need for a keto or aldo carbon atom in every glycosidic bond only one end of the chain will have a sugar with this position free; the other end of the chain will always have a sugar with the keto (or aldo) position 1 combined. Since the uncombined C1 position has reducing properties these ends are called the reducing and non-reducing ends respectively (see Fig. 3.3).

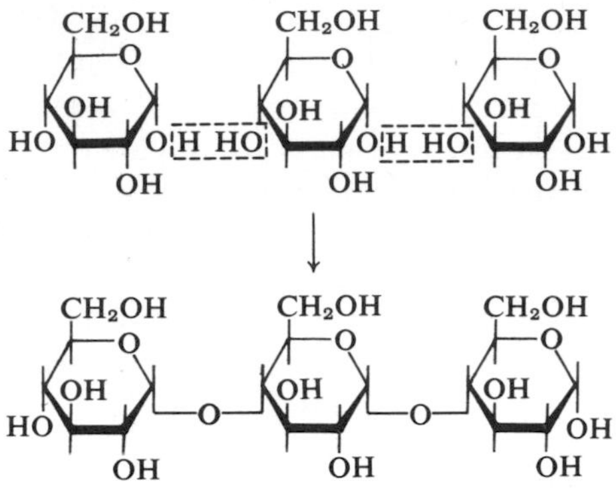

**Fig. 3.3** The formation of the glycosidic bond. The non-reducing end is on the left of the chain, the reducing end is on the right

You will notice that in each of the three cases an unbranched polymer has been formed. However, opportunities for cross-linking exist in all three. In the proteins the amino acids may possess groups in the side chains which can undergo cross-linking reactions. In fact this occurs only between a few such groups (see p. 71). (In proteins it hardly ever occurs by formation of a peptide-type bond between side-chain —COOH and —NH$_2$ groups.) At points at which it does occur there can clearly be branching or joining of chains. In the nucleic acids the pentose may possess more than the two —OH groups shown (the sugar here is deoxyribose; ribose, which is also used in nucleic acids, has an —OH at position 2′ (see Fig. 3.2)). The theoretical possibility therefore exists of branched or joined chains, but they seem never to occur. Polysaccharides, on the other hand, are frequently branched, particularly when the molecule has as its main function the storage of sugars as food reserves. This sort of function is not critically dependent on structure and so the variation in structure from molecule to molecule caused by more or less random cross-linking of the chains can be tolerated. Bonds other than the glycosidic link are occasionally used for cross-linking (see p. 180).

### *The differences between the classes of macromolecule*

It is when one looks at the nature of the homologous series involved, and the way in which selections are made from them to build up the molecule, that differences between the three types of polymer start to appear. The proteins have at their disposal twenty different amino acids (Table 4.1) with a wide range of types of side chain. The difference when we turn to consideration of the nucleic acids is striking. Here, once the choice of the pentose has been made (Fig. 7.2, p. 136) the entire molecule is usually made up by drawing on only four types of nucleotide (Fig. 7.1). Certain specialized nucleic acids do exist which use a much greater variety of nucleotides (see Fig. 7.2), but these extra nucleotides are in most cases derived from the more common ones by chemical modification.

The difference between the proteins and nucleic acids reflects the different demands made on them. As we shall see below, proteins have to carry out a wide range of functions, from the mechanical to the catalytic. Since the same amino acids are involved in every case, it is necessary that there should be a large range of these, so that sufficient structural permutations are possible. Nucleic acids, on the other hand, have only one main type of function, the storage and transfer of information (see chs 7, 24 and 26). Here the permutation of just four types of smaller unit suffices.

The position with the polysaccharides is determined by the fact that the types of functions usually undertaken here, mainly acting as structural material and for food storage (ch. 8), do not call for any great subtlety in combination of units. Relatively few sugars are involved in the formation of the common polysaccharides. The most striking feature is that in many cases a given polysaccharide will draw on only one or occasionally two types of sugar molecule. Glucose (Fig. 8.2) is very commonly used, but it will be seen from Fig. 8.2 and Table 8.1 that there is quite a large number of other sugars, some of which are used rather rarely, in only a few polymers. Thus, unlike the proteins and nucleic acids, the polysaccharides achieve their full range of properties not by permutation but by changes in the nature of the residues.

*The control of the order of the constituent residues*

We see, therefore, that all these types of macromolecule are constructed by building up from sets of smaller molecules. For any given protein, nucleic acid or polysaccharide, how close a control does the biosynthetic machinery exert over the order in which their constituent residues are incorporated into the polymer?

The answer is surprising and was indeed at one time thought so improbable as to be unworthy of serious consideration. It appears that in the proteins and nucleic acids there is next to complete control. Barring accidents, a protein containing many hundreds of amino acids or a nucleic acid containing many thousands of nucleotides will be turned out by the synthetic machinery of the cell time and time again, without alteration. One species of protein or nucleic acid is absolutely distinguished from any other by its amino-acid or nucleotide sequence. This tight control over the sequence is essential because the functions of these macromolecules are sharply dependent on structure and even a small change in properties can be fatal to the delicate balance of physical and chemical events in living material. The manner by which control is exerted is described on p. 427 ff. (proteins) and p. 439 ff. (nucleic acids).

The question of sequence does not arise in those polysaccharides that consist of a single sugar. Where more than one is used there is sometimes a measure of control of the sequence, so that if there are two they may alternate along the chain. In other molecules where the desired function is less critically dependent on structure there is even less control.

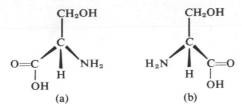

Fig. 3.4   (a) An L-amino acid, (b) A D-amino acid.

*The macromolecules and molecular asymmetry*

A further feature held in common by the three classes of macromolecule is related to *molecular asymmetry*.

Pairs of structures exist which, although they have the same atoms joined by bonds of the same length and set at the same angles, are dissimilar. The solution to this apparent paradox is found in the concept of molecular asymmetry. A full discussion of asymmetry in chemical structure will be found in textbooks of chemistry. Here, we will note that structures like the amino acid serine (Fig. 3.4) or glucose (Fig. 3.5) exist as mirror-image pairs which, in spite of all other simi-larities, lack complete identity in the same way that is true of the left and right human hands. These asymmetric pairs are known as *enantiomers*. The common feature of virtually all of them is that they contain a carbon atom bearing four different substituents.

As is widely known, when polarized light is passed through a solution of one of the two enantiomeric forms of a molecule, the so-called *plane of polarization* is rotated. It is rotated in the opposite direction by a solution of the other enanti-omer. (Enantiomers are also known as optical isomers.) As a result of this, polari-metry is one of the few non-biological methods (such is the relatively greater fre-quency of molecular asymmetry in biological systems) of distinguishing between enantiomers. However, *there is no easy way* to predict the direction of rotation of the plane of polarization – to the left or to the right – from any simple feature of the structure. We can only say that, under any given conditions, they will be equal and opposite for the two enantiomers. In fact the magnitude and even direction of rotation may be changed by altering such parameters as temperature and the wavelength of the polarized light.

With polarized light now seen in its proper place as a secondary phenomenon we can all the more appreciate that the importance of molecular asymmetry is purely geometrical. There is no difference between the enantiomers in their inter-nal geometry: the lengths of and angles between all bonds are the same. There-fore all the *intrinsic* properties of the enantiomers are the same. Differences are

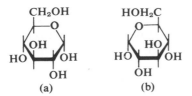

Fig. 3.5   (a) A D-sugar, (b) an L-sugar

only possible when they interact with another optically active structure.* Anyone who has tried to put on his or her shoes in a hurry in darkness will have some idea of the ability of asymmetric structures to fit in to one another, or, if a wrong pair is chosen, *not* to fit, with possibly disastrous consequences. The shoes are asymmetric structures existing in left and right handed varieties, which are tailored to fit feet, which are also asymmetric and exist in analogous varieties. The failure of a foot to fit adequately into the wrong shoe is an exact analogy with the difficulties that are encountered at the molecular level when asymmetric structures are involved. We shall see (p. 118) that a major difference between the chemistry of living and non-living matter is that biochemical reactions rarely take place in free solution, but instead on the *surface* of some larger structure. It is the intricate convolution of these surfaces that makes possible the rapid and totally efficient discrimination between closely related substances. Living processes depend on this ability, which extends to discrimination between enantiomers. This is one of the most difficult operations in conventional chemistry, but is found to be accomplished with ease in living tissue. In fact, one of the best known *in vitro* methods of separating enantiomers – fractional crystallization with alkaloids – is simply an appeal to the properties of a biologically derived class of asymmetric substances. In this method, the substance that is to be resolved into enantiomers (X, say) is combined with one of the enantiomers of an asymmetric compound that has already been resolved (A, say). The two possible compounds that result from this, (X+right-handed-A) and (X+left-handed-A), have *two* centres of symmetry. There will now be a difference in interatomic distances and angles and a separation can be effected on the basis of the differences of properties that these confer. The only problem is – where is the resolved sample of A to come from? The almost universal answer is 'from nature', where the resolution has already been accomplished by the biological methods that we shall describe.

The same convolution of structure that makes possible so fine a degree of discrimination demands the controlled use of optical isomers in its own construction. If in a protein one amino-acid residue is required to come into contact with another, each *must* be the correct isomer. The other isomer would turn away by 108° (the angle between the valency bonds of a saturated carbon atom) with fatal consequences for the ability of the protein to function and the organism to survive.

For this reason there is found in the majority of constituents of living matter only one of the possible configurations of the amino acids, that known as L (see Ch. 4). Similarly, natural sugars are almost all D (Ch. 8). Why these particular configurations were chosen in preference to their opposites is not clear. Chemical and most physical properties depend on size, the atoms involved and bond lengths. These will all be the same for both enantiomers. However, once the choice was made (by whatever means) it had to be adhered to for ever. The 'wrong' isomer is, at least potentially, a most effective poison. If it came to be incorporated into the macromolecules of the organism that ingested it, the struc-

---

* It is possible for the second optically active grouping to occur *in the same molecule* as the first. The different forms of compounds with two asymmetric centres are called *diastereo-isomers*. We shall meet biological compounds exhibiting this phenomenon later (pp. 64 and 166ff).

ture would be disrupted and life would cease. (In fact the liver possesses an enormously powerful system for the destruction of D-amino acids. It is believed by some that this system is a protection against the danger of the accidental incorporation of a D-amino acid.) Proteins, nucleic acids (by virtue of the D-pentoses that they contain) and polysaccharides all show molecular asymmetry. Further details, including the way in which the phenomenon was studied, are to be found in the appropriate chapters.

The analogies between the three types of macromolecule seem all the closer if one takes the lipids (Ch. 8) as a contrary example. These molecules are not individually as large as macromolecules, though they are larger on the whole than the residues of the three classes that we have covered. They are not necessarily optically active. They do form assemblies which are of very large molecular weight but which are joined, not by loss of water to form a covalent bond, but by non-covalent forces (see below). This gives them an entirely different character and we shall not discuss them further for the present except to point out that concepts such as the specificity of sequence do not apply.

Subsequent chapters will describe something of what is known about the structure of these types of macromolecule, the way in which structure influences function, and the way in which the organism ensures that the correct structure is obtained during synthesis of the macromolecules. Those chapters will seek to show that the remarkable properties which distinguish living matter from non-living are very largely a result of the properties of macromolecules and that these derive in their turn from recognizable features of the structure. The development of life had to await the discovery of ways of ensuring that the structures with beneficial properties could be repeated and safeguarded against deleterious changes from generation to generation.

## Non-covalent interactions

The covalent bonds involved in the formation of the macromolecules have been mentioned and we must now consider the non-covalent ones. As will be seen in the appropriate chapters these are just as important to the structure and function of the finished product as the covalent bonds. In fact, a look at the role of these non-covalent forces gives some clue as to why biological macromolecules became necessary in the first place.

Little variation is possible in the strength, direction or other properties of any given covalent bond. They are more or less incompressible and if we try to extend them, very little change occurs in the length until they rupture completely. A considerable amount of energy (of the order of 100 000 cal/mole) is required to do even this. It is, similarly, quite difficult to distort the angles between covalent bonds emanating from an atom. We therefore have the picture of the possibilities offered by covalent bonding as being fairly tightly limited to bonds of a fixed length disposed at a fixed number of rigidly controlled angles. This is not to say that the number of permutations that covalent interactions allow is small. They are in fact quite sufficient for the needs of non-living systems. However, the immense variety of biochemical reactions and the need for their precise control and integration easily exhaust the versatility of the covalent chemical interactions. Fortunately, strengths and geometrical properties of non-covalent bonds depend far more on environment and can be varied by the expenditure of far less energy, and as a consequence they possess the extra versatility that living systems need.

However, it is precisely this variability that makes it difficult for the non-covalent bonds to be maintained in a useful form when only small molecules are involved, moving in free, dilute solution. When, on the other hand, the inter-acting elements are anchored to one another in macromolecules the situation changes dramatically. Combinations of non-covalent forces can now be produced that are useful, powerful and capable of precise and infinite variation.* We shall see in a moment that, in terms of bond strength and range of action, there is a great variety of non-covalent interactions.

These combinations of forces and the local areas of high concentration of reactants that they make possible (pp. 15 and 121) are the real foundations of the differences between biological and non-biological chemistry and thus of the differences between the living and the non-living state.

This point is rarely stressed in textbooks. Some feeling for its validity can be obtained from a glance at those macromolecular products of the living organism in which the rule is suspended and covalent bonds predominate. In such products there are a large number of covalent linkages between different parts of the molecule: they include horn (p. 97), the exoskeleton of insects (p. 97) and hard (lignified) wood (p. 177). These are organic materials that we intuitively regard as the 'dead' parts. They change little on death of the organism or separation from it. This intuition is simply a recognition of the suppression of the effects of weaker but more responsive non-covalent interactions by the stronger, inflexible, covalent bonds.

## Types of non-covalent bonds

The *ionic bond* is formed between ionized groups of opposite sign.

The force of interaction $F$ between the electric charges $Q_1$, $Q_2$ placed a distance $r$ apart in a medium of dielectric constant $\epsilon$ is given by the familiar inverse-square law,

$$F = \frac{Q_1 Q_2}{\epsilon r^2}. \tag{3.1}$$

Dielectric constants vary but that of water is, by virtue of its molecular properties (see below) exceptionally high. The result of this is that the force between charges is reduced by nearly two orders of magnitude from what it would have been *in vacuo*. The energy (force × distance moved against the force) required to separate two attracting changes is correspondingly decreased. (Purely mechanical energy such as this is clearly enthalpy, $\Delta H$ (p. 31), but we will be quoting $\Delta G$ values for the energies required to break the bonds. Is this inconsistent? The

* It is even possible to exploit the different positions of rotation of groups about saturated covalent bonds. This is not usually possible in small molecules because, unless (as does sometimes happen) there is some exceptional hindrance to rotation, there is so little energy required to go from one position to another that they do not exist as separate entities. Macromolecules, however, with their complex shape and variety of forces, are easily able to stabilize useful states of rotation. This too contributes to the superiority of macromolecules over small molecules. We shall see at least one case (Fig. 612, p. 121) in which a small molecule shows (rather imperfectly) one of the exceptional properties of a macromolecule as a consequence of a restricted rotational state.

answer is that equations such as (3.1) are compatible with the uses to which we will put them for the following reason. If there is an entropy term in the $\Delta G$ (due to a certain degree of orientation of a dipole, for example) then it will most probably depend directly on the value of $F$ (which is directly related to $\Delta H$) and vary in proportion with it. Therefore the various estimates for $F$ that we shall derive will apply both to $\Delta G$ and $\Delta H$ providing we restrict the scope of our use of them to obtain approximate relationships to distance and dielectric constant. *In vacuo* this energy would be of the order of 50 000 cal/mole, comparable to that of a covalent bond which may be 100 000 cal/mole. It is for this reason that the ionic bond is spoken of as the strongest of the non-covalent bonds. However, using equation (3.1) we see that, in water, the energy required to break an ionic bond drops below 1000 cal/mole. This value approaches that of the strength of bond which can just be broken by random thermal agitation: by using the Boltzmann distribution (equation (2.39), p. 39) this value can be shown to be about 600–1000 cal/mole.*

We shall see (p. 57) that those parts of macromolecules with charges are often at the surface. They are thus exposed to the screening influence of water and therefore play a less important role in intra-molecular bonding than was believed, before the tendency of charges to be at the surface was realized. Nonetheless they can be found in roles of some significance (see p. 106), and their strength can be enhanced if water can be wholly or partly excluded from the region in which they operate by other parts of the macromolecule (p. 118).

*Permanent dipoles*

*The hydrogen bond.* When hydrogen is attached to an electronegative atom a distribution of electrons results in which the hydrogen atom will have less than its share (—OH and —$NH_2$ are typical of such groupings in biological systems). As a result of this unequal distribution the hydrogen atom will have a slight positive charge $\delta^+$ and the other atom a slight negative charge $\delta^-$. This pair of charges, known as a *dipole*, is capable of forming electrostatic interactions with other charges and dipoles. In particular, there are in biological systems a number of other groups besides those containing hydrogen, in which inequalities of charge distribution occur, with the consequent formation of electric dipoles. $\diagup C{=}O$ is an example. The electrostatic interaction between a hydrogen-containing dipole and another dipole is called a hydrogen bond. It is possible that such an arrangement:

$$\diagup \overset{\delta_1^- \, \delta_1^+}{NH} \qquad \overset{\delta_2^- \, \delta_2^+}{O{=}C} \diagdown$$

has also a small amount of covalent character (sharing of electrons) but this hypothesis is difficult to test.

Groups capable of forming hydrogen bonds are unusually abundent in macromolecules – doubtless this is no coincidence. It is quite rare in macromolecules for a hydrogen-bonding group not to have its potential for forming the bond realized.

* This is because $RT$ at normal temperatures $\simeq 600$ cal/mole. From equation (2.39), therefore, $e^{-1}$ ($\simeq 0.37$) of the molecules will have sufficient energy to break the bond. This is a sufficiently large fraction for it to be possible for all the molecules to do so in a very short period of time (cf. the effect of a barrier that is 28 000 cal/mole, p. 39).

Adequate calculations of the parameters of this interaction may be made using electrostatic theory alone. First, equation (3.1) may be used to show that the force at a distant point along the line of the dipole is proportional to $1/\epsilon r^3$. (This, incidentally, is fairly obviously the law controlling the interaction between a dipole and a single, separated charge.)

The $1/r^3$ relationship can then be used to compute the effect of the force so calculated on another dipole. It is found that this force is proportional to $1/\epsilon r^4$. (The details of this quite simple two-part calculation will be found in any elementary textbook covering electrostatics.)

This, then, is the relation controlling the dipole–dipole interaction of the type found in the hydrogen bond. It is clearly a very short-range force and susceptible to electrostatic screening. The maximum potential force *in vacuo* is about 5000 cal/mole, so that the screening of water will bring it well below the limit of stability at room temperature. Why then is the hydrogen bond so universally regarded as one of the most important interactions controlling the behaviour of biological molecules?

In answer to this it must first be said that the importance of the bond has probably been somewhat exaggerated. As we shall see later it is chiefly found stabilizing structures that have already come into being as the result of other forces. However, the principal answer is that it is, as we have seen, such a short-range force that it can only have any useful strength when the dipoles are in physical contact. In fact in the hydrogen bond pictured above the distance between the N and the O is 2.8–3.0 Å. This is actually *less* than the sum of the normal radii of the atoms and shows that the hydrogen bond has squashed the atoms closer together than they would normally come. Water is, therefore, quite unable to interpose once the bond is formed and it can bring about no screening. However, water is present around the site of synthesis of macromolecules and the actual net contribution of each hydrogen bond to the stability of the macromolecule is the difference between the strength *in vacuo* and the energy that is required to clear the water molecules out of the way as the structure folds up. This then will probably be quite a small contribution individually, but each macromolecule has a large number of hydrogen bonds and they might well amount to a sizeable force in total.

The hydrogen bond has traditionally been ascribed a unique significance in biology because its weakness allows it to be formed and broken easily. However, we shall have seen by the end of this chapter that this bond is just an intermediate member of a whole range of forces, going from very strong to very weak. All of these may be called upon to meet particular requirements of biological systems.

We should note that permanent dipoles can induce a dipole in nearby groups. There will be an attractive force between the original dipole and the dipole that it induces.

## Transient dipoles

Certain other dipole forces which are even shorter range complete the series of electrostatic interactions. They are frequently known by the generic name *Van der Waals* forces.

The charge clouds that surround atoms and maintain chemical bonds are never still. There is a continuous slight fluctuation as, in classical terms, the electrons travel their orbits. The results of these temporary inequalities in the distribution of charge is the formation of *transient dipoles*. These transient dipoles can interact with charged groups and with the permanent type of dipole that we have discussed already. In addition they can induce a dipole in a neighbouring group in the same way that the permanent dipoles can.

The interactions between transient dipoles and the dipoles that they induce are the most important contribution to the attractive forces between neutral atoms. They are known as *London*, *Heitler* or *dispersion* forces.

Calculations concerning these forces are rather more complex, but the outcome is that all three types (transient dipole–permanent dipole, transient dipole–induced dipole and permanent dipole–induced dipole) are governed by $1/\epsilon r^7$. They are thus individually very short-range forces indeed, and subject to screening by water. The same point applies here as did for the hydrogen bond – that is, when the interacting elements are brought sufficiently close by other means, these forces can then add a significant measure of stability. It is found that the transient dipole–induced dipole interaction is by far the strongest of the three types with binding energies of the order of magnitude of 1000 cal/mole when the elements are packed tightly together.

These dispersion forces may therefore be among the most important in deciding the tertiary structure of macromolecules. This is because almost all of the thousands of atoms in the molecule are involved, rather than just those groupings with the particular capacity for, say, an ionic or hydrogen bond. Moreover, although the individual forces are of excessively short range, it can be shown by integration that the resultant force of a very large number of such dipoles (such as is found in a macromolecule) falls off as $1/\epsilon r$, *more* slowly even than the inverse square law. Consequently, the resultant force is one of the longest-range interactions known. Such resultants may well have an important role in the early phase of the folding-up process of a macromolecule – they may help bring the various parts of the molecule close together, so that the shorter-range forces may take over and determine the final details of the structure.

Ionic bonds, hydrogen bonds and some of those bonds to be described in a moment can be identified simply by inspection of the structure of a macromolecule (see pp. 75 and 117). It is easy to see when positive and negative charges are close together and easy to identify, say, the characteristic $>N-H\cdots O=C<$ with the atoms crowded together. It is much less easy to identify the weak dipole forces by direct inspection. They therefore receive much less attention when the relationship between structure and function is discussed. We believe that, while this is understandable, it should not obscure the role that these forces play.

## *Other types of force*

Turning from straightforward electrostatic interactions we shall see that *resonance* influences the structure of macromolecules. One especially important example appears on p. 71 and this is possibly not the only example of the phenomenon in structural biochemistry. Another type of attractive force occurs between electron donors and acceptors. These are known as *charge-transfer forces* and may assist in the stabilization of the double-helix structure of DNA (see p. 140).

There is, however, a non-electrostatic phenomenon of much wider significance. This is the so-called *hydrophobic interaction*. As two oil drops coalesce when they touch in water, so adjacent hydrophobic structures (non-polar molecules, paraffins or the side chains of certain amino acids for example) find that the closer they are together, the more stable the arrangement becomes. The underlying theory of this common-sense conclusion depends on the lattice structure of liquid water.

To explain this briefly, it is widely known that the symbol $H_2O$ does not fully represent the situation encountered at (and in fact, somewhat above) the temperatures that are appropriate for living organisms. It can be seen from the discussion of the hydrogen bond that water could hydrogen-bond to itself (see Fig. 3.6) to form a crystal-like lattice. This is the situation in ice, where the random thermal motion of the water molecules is low enough to allow all possible

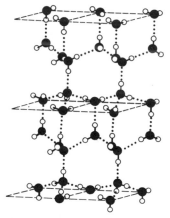

**Fig. 3.6**   The lattice structure of water

interactions to form and to give a solid: above 0°C thermal agitation begins to break up the lattice but at, say, 37°C, there are still many intermolecular bonds left and water should still be written $(H_2O)_n$. On average $n \simeq 30$; we must imagine aggregates of approximately this size continually losing odd members to other aggregates as well as gaining them.

If we now consider a macromolecule coming into existence in an aqueous environment, it is possible to predict what will happen to the water-repelling groups that it possesses. (There are, again by no coincidence, many types of such groups found in macromolecules.) Events will be governed by the fact that the water-repelling group, by definition, cannot join in the lattice structure of water. Two such groups are shown in Fig. 3.7, first separated and then placed close together. A smaller total surface area is presented to the solution when they are close together and the water orders itself about the combined structure.

The situation of the groups in juxtaposition has been compared with that of a body surrounded by an ice cap, for that is just what the lattice is. It seems clear that the energy needed to separate these two groups (which is the same thing as the energy of interaction or bond strength) is greater than it would have been if

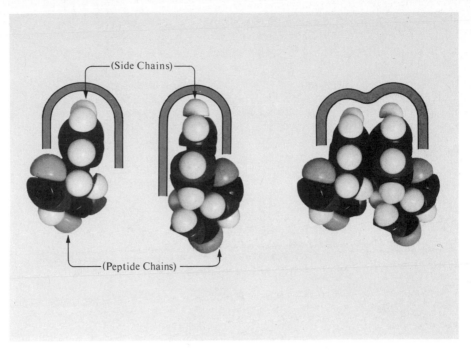

**Fig. 3.7** The hydrophobic interaction (see text) The dark tone is used throughout this book to indicate hydrophobic structures. Similarly, a light tone is used to indicate hydrophilic structures (e.g. Fig. 3.8).

the lattice were not present. In addition the two groups are now free to interact by dispersion and other forces as was described above (see also Ch. 6). Thus the simple picture is of the strength of the interaction being due to the groups being 'frozen' together, with the consequent necessity to 'break the ice' to get them apart.

(There is a controversy over whether the energy in breaking the surrounding water structure is purely enthalpic or entropic (see a textbook of thermodynamics and Ch. 2). Certainly, when the structure is broken, water molecules that were previously fixed in an ordered structure are free to move and to rotate. This increase in the freedom of movement of the water molecules is directly equivalent to an increase in entropy (see p. 33). However, either model (or any compromise combination of the two) derives from the same basic physical structure.)

A hydrophobic interaction between two atoms may require 1000–2000 cal/mole to break. This is as strong as or stronger than the other bonds when water is present and hydrophobic interactions are of prime importance to the structure and function of macromolecules (see pp. 75, 147, 172 and 185).

Hydrophobic interactions play a major part in maintaining the structure of both proteins (Ch. 4) and nucleic acids (Ch. 7). Both types of macromolecule have elements that favour hydrophobic bonding and elements that are hydrophilic. The hydrophobic elements will naturally lie, as far as possible, in the centre of the structure, away from the solvent water. The hydrophilic elements will, on the contrary, lie on the surface of the molecule where they may interact with the water.

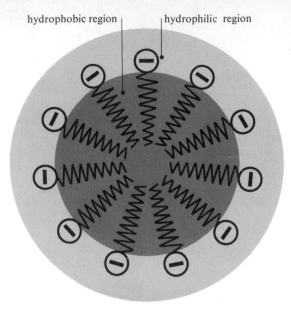

hydrophobic region | hydrophilic region

**Fig. 3.8**  A detergent micelle (see text)

The analogies with a detergent *micelle* (Fig. 3.8) are clear. The function of a detergent is to render the hydrophobic component of dirt soluble so that it may be washed out. Detergent molecules, like the macromolecules, have both hydrophobic and hydrophilic regions. The hydrophobic parts associate with each other and with the dirt. The hydrophilic parts interact with the water in precisely the same way as do the corresponding parts of macromolecules and, as in that case, bring about the desired solubility in water. This explains the overall similarity of Fig. 3.8 with the generalized picture of a protein (Fig. 4.9), a nucleic acid (Fig. 7.9) and a lipid assembly (Fig. 8.22).

As well as simply maintaining the structure, hydrophobic regions provide chemical environments quite dissimilar to those found in free aqueous solution. The unusual reactivity of certain key groups, in the catalytic site of some enzymes, for example (p. 118), is a consequence of their being in such environments.

Thus, in biochemistry, water should not be seen simply as a passive solvent, though that is part of its function, nor yet just as a source of the protons and $OH^-$ ions needed for biochemical reactions, though that too is a part of its function (pp. 122 and 255), but as actively and uniquely concerned in maintaining the character and stability of the essential elements of living matter.

## Summary

1. Macromolecules are of the greatest importance to the maintenance of living structures and living processes. The different types of macromolecule have important principles of construction in common.
2. Covalent bonds are of course necessary for the construction of macromolecules. However, even the great versatility of covalent bonds is insufficient to provide the range of properties required to sustain life.
3. There are many types of non-covalent bonds. These have a wide range of

properties which, when deployed in macromolecules of suitable structure, add the extra order of versatility required. In particular there is a wide choice of strength and range – examples are known of forces which depend on distance according to $1/r$, $1/r^2$, $1/r^3$, $1/r^4$ and $1/r^7$.

4. The hydrogen bond, dispersion forces and the hydrophobic interaction are of particular importance.

5. Few non-aqueous solvents could provide an effective analogue of the hydrophobic interaction. This is one reason for the unique place of water in living systems.

# Chapter 4　**Protein structure**

We pointed out in the last chapter that the proteins differ from one another in the selection and arrangement of their constituent amino acids. Succeeding chapters will show something of the types of function undertaken by proteins. Their remarkable range is achieved either solely by permutation of amino acids or at most with the addition of one or two other small molecules that are not amino acids (see p. 78). We shall now examine the amino acids used in proteins from the point of view of the types of forces mentioned in the last chapter, and see what properties each brings with it to help in the task of producing a functioning molecule.

Proteins are of considerable size, having usually hundreds of amino acids and, consequently, molecular weights of the order of 10 000–100 000. They might therefore be expected to be sprawling, ill-defined structures. In fact the majority of proteins are compact, highly convoluted molecules with the position of each atom relative to the others determined with great precision. An error in position of a constituent part of as little as the diameter of one atom may be sufficient to inactivate a protein. Thus we have to consider more than just the few residues which contribute directly to the function of the protein by mediating in the interaction between the protein and other molecules (see next chapter). This is because many other residues will make as vital a contribution indirectly by maintaining the required precision in the structure of the protein itself. Both types of contribution involve the covalent and non-covalent interactions mentioned in the last chapter, and a major part of this chapter will be devoted to examining the way in which these interactions are employed in the stabilization of the structure. Their application to the interaction of proteins with other substances is best left to the discussions of protein function in Chapters 5 and 6.

## The amino acids – their general properties

Table 4.1 shows those amino acids that are normally found in proteins. (There are many other naturally occurring amino acids (i.e. molecules having an amino and a carboxyl group), but they do not appear in proteins. Such non-protein amino acids are found singly or combined in molecules of relatively small size in natural fluids. Some are intermediates in metabolic pathways.)

We will now consider two general properties of the amino acids, their stereochemistry and their ionization, before going on to deal with the way in which the particular properties of individual groups are used to control the structure and function of proteins.

*Stereochemistry*

The stereochemical properties of the amino acids depend on the fact that all of them, with the exception of proline, have an amino and carboxyl group joined to the same carbon atom (called the α-carbon). Thus the α-carbon atom is an asymmetric centre for all amino acids since it has the necessary four dissimilar substituents. The only optically inactive amino acid is glycine $CH_2NH_2.COOH$, for obvious reasons.

Proline has a —COOH group but carries its amino group in a saturated ring. This is a secondary, cyclic amino group, and these are conventionally known as imino groups. (Thus proline and its derivative, 3-hydroxy proline are more properly known as imino acids.) Since the ring has saturated valencies, the α-carbon atom has a tetrahedral configuration and the four different substituents necessary for optical activity.

The configuration about the asymmetric carbon atom is always that in Fig. 3.4a as opposed to that in Fig. 3.4b (p. 49), irrespective of which R-group is involved.

If we look down the bond joining the H-atom to the α-carbon in Fig. 3.4a we note that it is possible to read the **COOH**, **R** groups and **NH₂** groups in clockwise order (this is true of proline also, see Fig. 4.1). This property of L-amino acids is known as the CORN law and distinguishes the normal biological configuration (known as L) from the other (known as D). We have seen (p. 50) why it is necessary that only one form of each amino acid is likely to be tolerated. It is not necessary to assume that it had to be L for all, once it had been shown to be so for one – it could quite well have been D for some amino acids and L for others.

How then was it proved that all of the naturally occurring amino acids are in fact of the same configuration? The answer is that chemical transformations were used to turn samples of one amino acid into another (these reactions were restricted to atoms other than the α-carbon atom for fear of altering the configuration during the transformation). A semi-synthetic amino acid produced in this way (alanine, say) could be confidently expected to have the same configuration as that of the natural form of the amino acid from which it is produced (e.g. serine). It only remained to compare the direction of rotation of polarized light

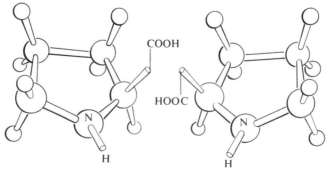

**Fig. 4.1**   The stereochemistry of proline (a) L-proline, (b) D-proline

**Table 4. 1** The side chains of the amino acids commonly found in proteins.

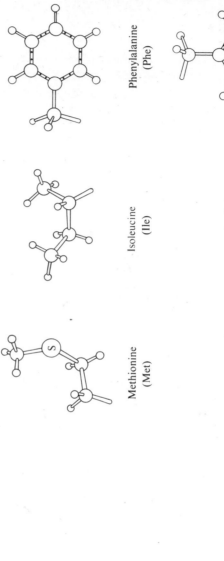

HYDROPHOBIC

No of carbon atoms in side chain

| 0 | 1 | 2 | 3 | 4 | Cyclic |
|---|---|---|---|---|---|

Glycine (Gly)

Alanine (Ala)

Valine (Val)

Methionine (Met)

Leucine (Leu)

Isoleucine (Ile)

Main chain

Proline (Pro)

Tryptophan (Trp)

Phenylalanine (Phe)

Tyrosine (Tyr)

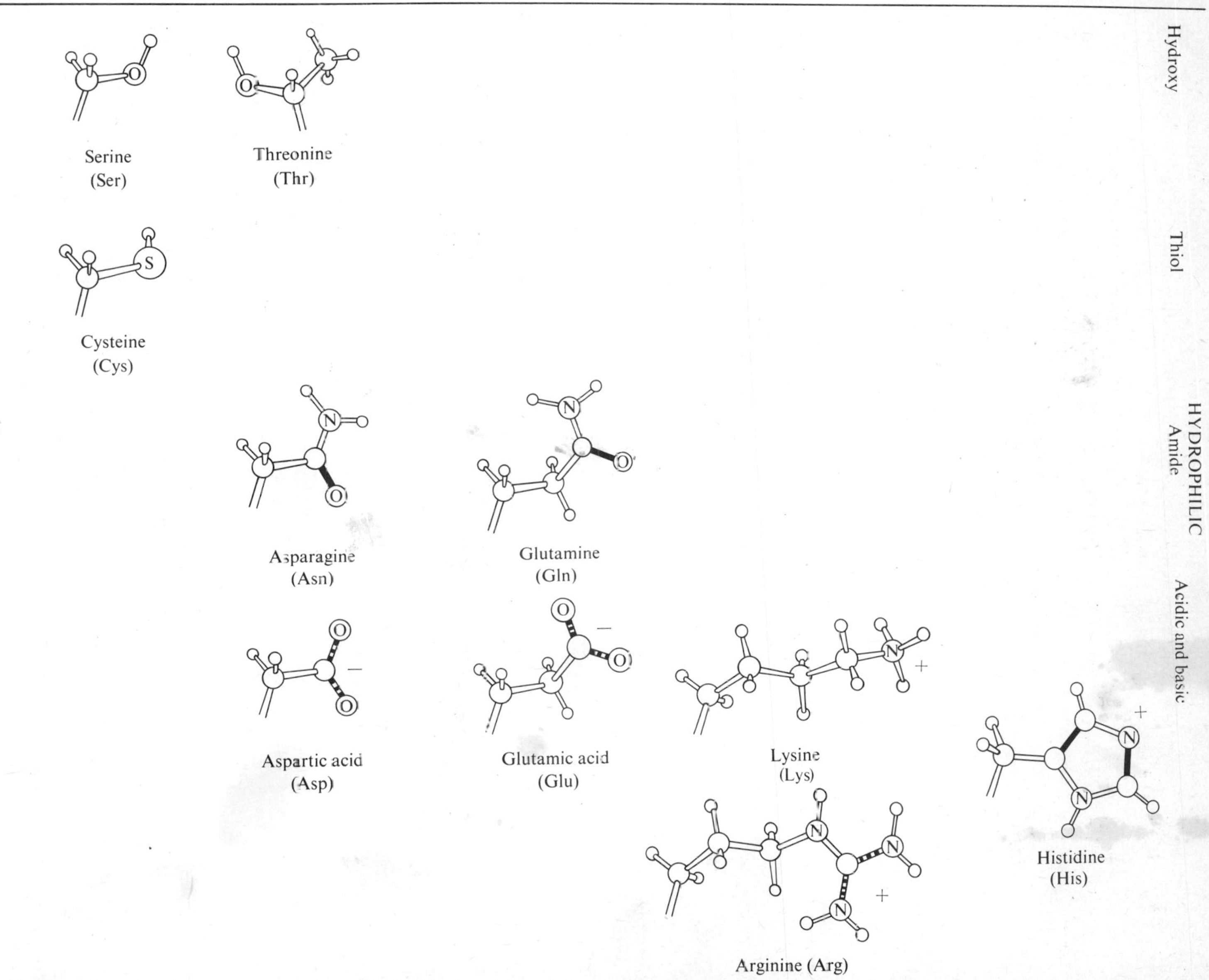

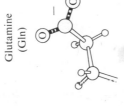

HYDROPHILIC
Hydroxy
Thiol
Amide
Acidic and basic
Serine (Ser)
Threonine (Thr)
Cysteine (Cys)
Asparagine (Asn)
Glutamine (Gln)
Aspartic acid (Asp)
Glutamic acid (Glu)
Lysine (Lys)
Arginine (Arg)
Histidine (His)

The three-letter symbols are those used in uniting amino-acid sequences. Double bonds are shown as solid bars.

of the semi-synthetic alanine with that of the natural form of alanine. If they were the same (as is actually the case) then the configuration of natural alanine is the same as that of natural serine.*

All the amino acids were shown in such a way to have the same configuration. However, before X-ray crystallography was developed, there was no way of telling which of the two possibilities it was. An arbitrary choice was made, in much the same way as the direction of flow of electricity was arbitrarily taken to be from positive to negative. Unlike the latter choice, which turned out to be incorrect, that for D and L was, happily, confirmed. The L configuration, once arbitrarily deemed to be that in Fig. 3.4a, is now known with certainty to be so.

Two of the amino acids, as will be seen from Table 4.1, have a second centre of asymmetry in the side chain. These are threonine and isoleucine. We saw on p. 50 that when there are two centres of asymmetry in a molecule the properties of the different forms need not be identical. In the present case such dissimilarities as exist are very slight. Nonetheless there are again both a natural form (called 'threo') and an unnatural form (called 'allo') of these compounds. In spite of the great similarity between the properties of threo and allo forms, discrimination operates in the same way as with the D and L forms. Of the four possible forms, L-threo, L-allo, D-threo and D-allo, only the first is found in proteins.

*Ionization*

The ionization of groups in the amino acids is of importance in understanding the relation between structure and function in the proteins, as we shall see below. It will be helpful if we examine first the behaviour of the $\alpha$-$NH_2$ and -COOH groups. Let us consider the ionization of free glycine as an example. It is best studied by means of a titration curve (Fig. 4.2; read pp. 11–13, Ch. 2).

We can clearly see, when we compare this graph with the titration curves in Ch. 2, that there are two ionizing groups that are undergoing titration and that they have p$K$ values of 2.4 and 9.6, respectively.

Glycine has an $\alpha$-amino group which undergoes the ionization

$$- NH_2 + H^+ \rightleftharpoons - NH_3^+$$

and is therefore a base ($\equiv$ proton acceptor).

It also has an $\alpha$-carboxyl group, which undergoes the ionization

$$- COOH \rightleftharpoons - COO^- + H^+$$

and is therefore an acid ($\equiv$ proton donor).

It is tempting to conclude that we are titrating the acidic group with NaOH and the basic group with HCl and therefore that the p$K$ of the —COOH group is 9.6 and that of the —$NH_2$ groups is 2.4. However, the titration curves of acetic acid and methylamine (Figs 2.2 and 2.3, p. 11) prepared us to expect quite different p$K$ values for these groups. It is certainly true that in glycine the two groups are

* Actually, solutions of L-alanine rotate the plane of polarized light to the right and solutions of L-serine rotate to the left. This example bears out what we said in Ch. 3, that there is no obvious relationship between absolute configuration and the direction of rotation. It was because of the possibility of confusion with D and L that the old notations for rotation to the right (*d*) and to the left (*l*) have been dropped in favour of (+) and (−) respectively.

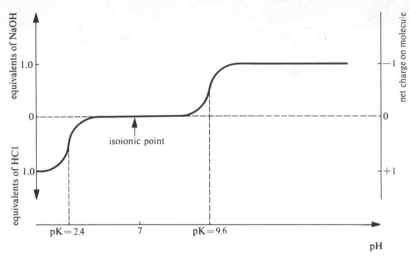

**Fig. 4.2** The titration of glycine (see text)

in close proximity to one another and will interact by electrostatic force. However, such an interaction could not be strong enough to change, for example, the pK of the $-NH_3^+$ from 10.6 (methylamine) to 2.4 as suggested here. We can prove this point if we recall that pK values are equilibrium constants expressed in another form ($pK = \log 1/K$). The standard free energy of the interaction that would be necessary to bring about the change in pK is given, from equation (2.16), p. 18, by

$$\Delta G° = -1364 \, (\log K_{(observed)} - \log K_{(methylamine)})$$
$$= -1364 \times -8.2$$
$$= 12\,000 \text{ cal/mole}.$$

This figure, and the analogous one that we must assume if the pK of the $COO^-$ ionization is 9.6, is improbably high for electrostatic interaction occurring in an aqueous environment (Ch. 3). The inductive effect of the different substituent on the ionizing group (e.g. $-CH_2.COOH$ rather than $-CH_3$) will also be much smaller than this figure.

We are left to assume that the pK of 2.4 refers to the ionization of the $\alpha$-COOH group and the pK of 9.6 refers to that of the $-NH_2$ group. Thus in going from neutrality to pH 1 we are titrating $-COO^-$ by adding protons: that is we are titrating the conjugate *base* (p. 12) of the acidic COOH group and it is quite reasonable to use an acid to do so. Similarly we are titrating $-NH_3^+$, the conjugate *acid* (p. 12) of $-NH_2$, when we move from pH 7 to pH 10, and we therefore require alkali.

The assignment of the pK of 9.6 to the $-NH_3^+ \rightleftharpoons -NH_2 + H^+$ ionization is confirmed by the effect of formaldehyde on the titration. The pK of 9.6 is replaced by one of about 7, while the pK of 2.4 is unchanged. This is so because formaldehyde modifies amino groups:

$$-NH_2 + HCHO \rightleftharpoons -N\overset{\displaystyle H}{\underset{\displaystyle CH_2OH}{\diagdown}} \tag{4.1}$$

and so changes the value of the $-NH_2$ pK but has no effect on carboxyl groups.

The changes in p$K$ from those seen in methylamine and acetic acid are now much more reasonable and the much smaller $\Delta G°$ required (which you can calculate for yourself) can be accounted for by electrostatic and inductive effects without difficulty.

As a result of this titration, it appears that glycine has both its ionizable groups fully charged at neutral pH (the exact point at which the charges are balanced, the *isoionic point*, is about pH 6). This electrical neutrality by balance (a *zwitterion* structure) is obviously of a different type and confers different properties to those of truly electrically neutral substances like methanol, $CH_3OH$.

## Specific properties of the amino acids

The amino acids in the table are arranged according to the characteristic of the side chain that appears to be the most important in determining its contribution to protein structure and activity. You will notice that in each case there is a gradation of properties across the range. It is more important that you should grasp these general points than commit the detailed structural formulae to memory without understanding what they imply.

### Hydrophobic ('apolar') amino acids

These are the amino acids that have side chains most able to form hydrophobic interactions. Clearly, as the size of chain increases so does the potential strength of the hydrophobic interaction. Thus a range exists from glycine (hardly hydrophobic) to the longer-chain aliphatics and the aromatics (capable of many hydrophobic contacts). Once the hydrophobic interaction has brought the residues together in a polypeptide, the individual van der Waals and other short-range forces may come into play. It is clear that in addition to changes in size the various shapes add to the range of possible contacts – compare leucine and iso-leucine, for example.

The gap that occurs between alanine and valine is as significant as many of the structures that are actually shown. There is no doubt that biological systems are capable of making the compound that could occur here (L-α-aminobutyric acid) since it is one of the naturally occurring, non-protein amino acids mentioned on p. 60. Doubtless it would be of additional value in enlarging the number of types of hydrophobic contact, but we must assume that this extra versatility is not

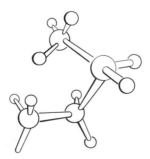

**Fig. 4.3**   Nor-leucine

worth the price of having to elaborate an additonal control mechanism to ensure that it, like the others, would always go into its correct place in the protein (p. 441).

Another amino acid that might have been used is nor-leucine (Fig. 4.3). In this case, however, any role that it might have played can be undertaken by methionine which, as you can see by comparison of the two structures, resembles it very closely. (So close is the resemblance, that certain micro-organisms can be deceived when fed with nor-leucine and will incorporate it into the positions within the protein appropriate to methionine. This is an example of mimicry at the molecular level, to which we will return later, p. 80.)

Two of the members of this class have groups in their side chains that are capable of polar interactions in addition to hydrophobic ones of the rest of the side chain. The indole $>$N of tryptophan can take part in hydrogen bonds, as can the —OH group of tyrosine. This —OH group can also ionize to —O$^-$, though at a pH (the p$K$ is about 10) which is above that normally encountered in living matter. These examples serve to show that amino-acid side chains may show two types of character in different parts of the structure. We shall see that other types of amino acid show dual characteristics, as do sugars (p. 172), nucleotides (p. 140) and lipids (p. 184).

### Proline and hydroxy-proline

The peptide bond formed between a proline —COOH and an —NH$_2$ group of a succeeding amino acid is normal, but the bond between its imino group and the preceding —COOH will have special properties due to the lack of a proton. Among these properties we may note that the proline N cannot join in hydrogen bonds and so the presence of proline disrupts such structures as the $\alpha$-helix (Fig. 4.5) and $\beta$-pleated sheet (Fig. 4.6). Many normal protein-digesting enzymes (proteases) will not attack the imino peptide bond and special imino-peptidases are required.

3-Hydroxyproline is mainly found in collagen (p. 96). It is probably hydroxylated after it is incorporated into the peptide chain.

### Hydrophilic (polar) amino acids

*The hydroxy- and amide type.* The functional groups in these, —OH and (amide) $>$C=O and —NH$_2$, are well adapted to the formation of hydrogen bonds with each other, with other amino acids, and with non-protein molecules. These groups are carried on side chains of varying size and shape. (We must not forget that the peptide back-bone has amide $>$C=O and $>$NH groups and that these can enter into similar interactions, see p. 72.)

Amino acids of this type are of assistance in maintaining the solubility of proteins since they can form hydrogen bonds with water and so strengthen the solute–solvent interactions on which the solubility of any substance depends.

In addition, the —OH group of serine, which normally has the same relative lack of reactivity as the —OH groups of ethanol, can be rendered much more reactive by the extraordinary chemical and physical environments that the convolutions of protein structures can produce. In particular, it can be rendered much more prone to acylation and to subsequent deacylation. Such activation frequently enables serine to act as the principal catalytic residue in the active sites of enzymes (p. 119). The —SH group of cysteine (the sulphur analogue of serine), which is intrinsically more reactive than the —OH group, can also find itself in an activating environment and act in a similar way in the catalytic sites of enzymes (p. 262).

*The charged amino acids* contribute to the forces existing in a protein by virtue of their ionizable groups. Any amino acid can contribute if it is the *N-terminal* (free α-NH$_2$ group) or *C-terminal* (free α-COOH group) residue of a poly-peptide. In addition there are the charged amino acids proper – aspartic and glutamic acids, with negatively charged groups, and arginine, lysine and histidine with positively charged groups. Two other side chains, those of cysteine and tyrosine (see below) contribute charges more rarely. Once again titration curves may be used to study the ionizations. It must be stressed, however, that the in-fluence of neighbouring groups in a protein can provide an interaction energy of 3000–4000 cal/mole and thus p$K$ values can well vary by 3 from the most usually encountered value. We discuss at length in Ch. 6 a case in which such a variation is of biological importance.

The principal acidic group is —COOH. Titration curves show that, at the carboxyl terminus, the p$K$ is normally around 3.5. (The value differs from that in a free amino acid because the α-NH$_2$ is at the other end of the polypeptide chain and can be many Angstroms away, whereas in an amino acid the two groups are in close proximity and are able to interact. A nearby positive charge stabilizes the —COO$^-$ group by formation of an ion pair. More protons (i.e. a lower pH) are therefore required to drive the equilibrium towards the —COOH form.) The p$K$ values of the —COOH of the side chains of aspartic and glutamic acids are, for the free amino acids, 3.5 and 4.7 respectively. Thus, leaving aside the groups with p$K$ values that are shifted to a really exceptional extent, all the —COOH groups of a protein will be fully charged at neutral pH.

Cysteine and tyrosine can ionize:

$$-\text{SH} \rightleftharpoons -\text{S}^- + \text{H}^+ \tag{4.2}$$

$$-\underset{\phantom{x}}{\bigcirc}-\text{OH} \rightleftharpoons -\underset{\phantom{x}}{\bigcirc}-\text{O}^- + \text{H}^+ \tag{4.3}$$

but the p$K$ values (about 9 and 10, respectively) mean that unless neighbouring groups produce an unusual environment these groups are normally uncharged at pH 7.

Although there is only one main type of acidic group there are three main types of basic group. The —NH$_2$ group is found at the amino terminus with p$K$ values around 8 (similar values are found for the imino group of proline) and on the side chain of lysine with p$K$ values around 10. The guanido group of arginine has a symmetrical structure in which there are unusually good possibilities for re-sonance (Table 4.1 and p. 28). As a consequence the positively charged form is exceptionally well stabilized and it does not lose its proton until [H$^+$] is very low, i.e. pH is very high. The p$K$ is in fact between 12 and 13.

The imidazole group of histidine (Table 4.1) is the weakest of the three bases and usually has a p$K$ between 5.5 and 7.5.

$$HC\!\!=\!\!\!C\!\!-\!\!\!\begin{array}{c} \\ N \quad NH \\ \underset{H}{C} \end{array} + H^+ \rightleftharpoons HC\!\!=\!\!\!C\!\!-\!\!\!\begin{array}{c} \\ HN^+ \quad NH \\ \underset{H}{C} \end{array} \qquad (4.4)$$

It is thus the only group that we have mentioned that has its average p$K$ in the range of pH of most biological systems. It is therefore easily protonated and deprotonated at the pH of living matter and might reasonably be expected to be used, because of this property, in many situations. Whenever a biological property, such as enzyme activity, can be titrated as a function of pH, and a p$K$ of about neutrality is obtained (p. 132), it is easy to suspect that a histidine side chain has been identified in a vital role. This correlation may be correct, but if p$K$ values can vary in particular environments by up to 3 pH units as was suggested on p. 68, then from the values that we have given the groups involved could be

$$-COOH, \ -NH_2, \ -SH \ \text{or} \ \text{—}\hexagon\text{—OH}.$$

The charged amino-acid side chains, then, differ in their shapes, in the sign of the charge that they carry, and in the strength (p$K$) of the ionization. We shall see shortly how they can contribute to the properties of proteins by forming electrostatic interactions with groups of opposite sign, and by altering reactivities by virtue of the electrostatic fields that they possess.

Finally we may mention that these residues may sometimes be involved in the stabilization of structure by interactions other than salt bridges. Their $>$N—H and $>$C$=$O groups may also form hydrogen bonds. In addition there are opportunities for hydrophobic interactions, especially in the aliphatic side chains of lysine and, to a lesser extent, of arginine.

*Cysteine*

The analogies that this amino acid shows to serine in structure and behaviour have already been mentioned. (These analogies cannot be taken too far since, in an ingenious experiment, the serine —CH$_2$OH of the active site of an enzyme was changed to —CH$_2$SH with a complete loss of activity.) We have seen that cysteine is capable of ionizing with a p$K$ of over 8, but there is no evidence that ionic bonds formed by cysteine are of any wide significance. A much more important point is that cysteine residues of proteins may form disulphide bridges by oxidative conversion of two residues of cysteine (pronounced *systayeen*) into a cystine (pronounced *systeen*) residue (see Fig. 4.4). This seems to be used principally in proteins that have to function in the more severe conditions that exist outside cells (e.g. in the gut, blood-stream, etc.) but, even here, many proteins are without disulphide bridges. A more specialized instance of the use of cysteine for covalent binding is the formation of thioether bonds in the attachment of haem (Table 4.2) to cytochrome *c* (p. 236).

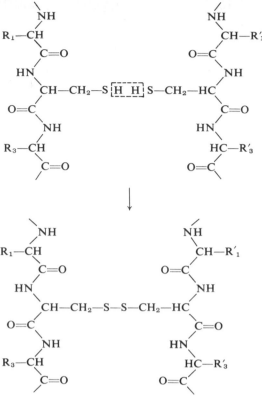

**Fig. 4.4** The formation of a disulphide bridge

## The interactions that control protein structure

We examined the range of available forces in the last chapter and have in this one examined the range of amino-acid side chains. We shall now be able to see how well the latter are adapted to the task of deploying the former. Before doing so we must define a few terms.

In considering the structure of proteins, the sequence of incorporation of amino acids into the polypeptide chain, the so-called *primary structure*, is most important. Although a protein of 300 residues has about $10^{400}$ possible permutations, only one is produced (e.g. Figs 5.10 or 7.15). (The analytical methods by which the sequence may be determined are described in a special section at the end of this chapter; the biological mechanism by which the sequence is maintained is described in Ch. 24.)

The properties of a given protein will depend on a particular juxtaposition of side chains, with the consequent interplay of their different types of characteristics. This juxtaposition is not merely a question of the primary structure, that is it is not merely one-dimensional, along the chain. The chain exists in three dimensions and it is its convolution in a precisely determined shape, the *tertiary structure*, that allows the interplay of characteristics of side chains at its fullest. (The experimental methods by which the tertiary structure may be determined are discussed at the end of this chapter. The means by which the protein maintains its tertiary structure are described in the sections immediately following this one.)

Proteins are often divided into the categories *globular* and *fibrous*. An absolutely spherical shape is not required for a protein to be classed as globular, but when the ratio of length to width (known as the *axial ratio*) exceeds approximately 5:1 it is classified as fibrous.

## Covalent forces

The *peptide bond* (p. 45) is clearly vital to the integrity of the structure of the molecule and is the basis of primary structure.

In addition, certain properties of the peptide bond which depend on the phenomenon of *resonance stabilization* have a bearing on three-dimensional structure. The resonance stabilization results from the fact that it is possible for there to be a considerable amount of de-localization in the electrons forming the $\diagup C = O$ and $\diagup C - N \diagdown$ linkages of the peptide bond. The result is a $\pi$-type molecular orbital with electron density above and below the plane defined by O – C – N. The consequence of this is that the bonds leading to all three atoms lie in the same plane – a simple result of the rules of orbital behaviour. Conversely, once the overlap occurs the atoms are locked in the planar configuration and a considerable amount of energy (21 000 cal/mole) is required to free them from it and to permit a 90° rotation of groups about the C—N bond. (There is no corresponding restriction of conformation in the phospho-diester and glycosidic bonds (Ch. 3).)

The *disulphide bridge* is the second most important covalent interaction. It is much involved in the cross-linking between different parts of the polypeptide chain and between different polypeptide chains (e.g. Fig. 5.10). It results, as we have seen, from the oxidation of two cysteine residues to form cystine (Fig. 4.4). Disulphide bridges are frequently found in proteins as a general aid to the stabilization of three-dimensional structure ('*tertiary structure*') and they are also used where special mechanical properties are required (e.g. see p. 97).

(In highly exceptional circumstances, peptide-type bonds, formed between side chain —NH$_2$ and —COOH groups, may be used to give an extra measure of stability (e.g. p. 99). More often, if the disulphide bridge is not to be used, certain other types of covalent bond are formed (p. 97).)

Although disulphide bridges are very strong, when compared to the strength of non-covalent bonds they are very short-range since, as with all covalent bonds, even a slight extension breaks them completely. They therefore only stabilize the tertiary structure when it has reached something approximating to its final form: in proteins that possess disulphide bridges the cysteine residues must be brought together by other forces before the bridge can be formed and its influence felt.

## Non-covalent forces

The *ionic bond* can occur by the interaction between, on the one hand, the positive charges on histidine, lysine, arginine and the $\alpha$-amino group and, on the other, the negative charges of aspartic acid, glutamic acid and the $\alpha$-carboxyl

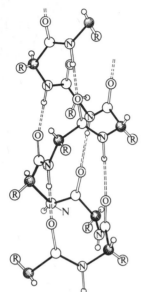

**Fig. 4.5**  The α-helix

group. Intramolecular ionic bonds are rather infrequently used in the stabilization of protein structure, but when they are so used, it is often with great effect (see Fig. 5.15). This view may be contrasted with that widely held a few years ago when the contribution of the hydrophobic and dispersion forces was insufficiently appreciated. The great strength of the ionic bond in a vacuum (p. 52) and the large number of groups that could form them led to an overestimation of their role in the stabilization of protein structure. In fact ionized groups are more frequently found stabilizing interactions between proteins and other molecules (see pp. 117 and 123). (However, ionized groups interact with the solution and these interactions form part of the balance of hydrophobic and hydrophilic forces.)

The *hydrogen bond* is found principally between the side chains of the hydrophilic group of amino acids (Table 4.1). We must also note that the peptide backbone itself is capable of a more active role in stabilizing the structure than simply holding the amino acids together. It has $>$C$=$O and $>$NH groups regularly disposed along its length and these groups are eminently capable of hydrogen bonding to each other. As the groups are regularly spaced, it is not surprising that hydrogen bonding between them can give rise to regular structures. Features of this type in which amino-acid residues have a regular repeating relationship to one another in space are described as *secondary structure*, and two forms, the α-helix and β-pleated sheet (Figs 4.5 and 4.6), are found in most proteins. These two structures are apparently dominated by the hydrogen bonds, but, just as important, they represent an extremely dense packing together of the amino acids. Thus many more close contacts between atoms are promoted than would occur in other regularly bonded structures that could be formed by the peptide chain but (perhaps for this reason) are not. The γ-helix (Fig. 4.7) is an example of a highly hydrogen-bonded structure that is insufficiently stable to be used in proteins because of the lack of interatomic contacts.

The reason why proteins are not entirely composed of elements with regular secondary structure is the disruptive effect of other, over-riding interactions between certain of the side chains (see below). However, the α-helix and the β-pleated sheet are quite common and are used as reinforcing members (struts and plates) in many proteins (Fig. 4.8). In a few proteins they form a predominant part of the structure (see p. 93). The properties of such proteins will then depend to a significant extent on the properties of the regions of ordered secondary structure and these in their turn depend on the properties, in particular the

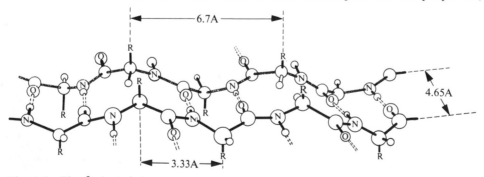

**Fig. 4.6**  The β-pleated sheet

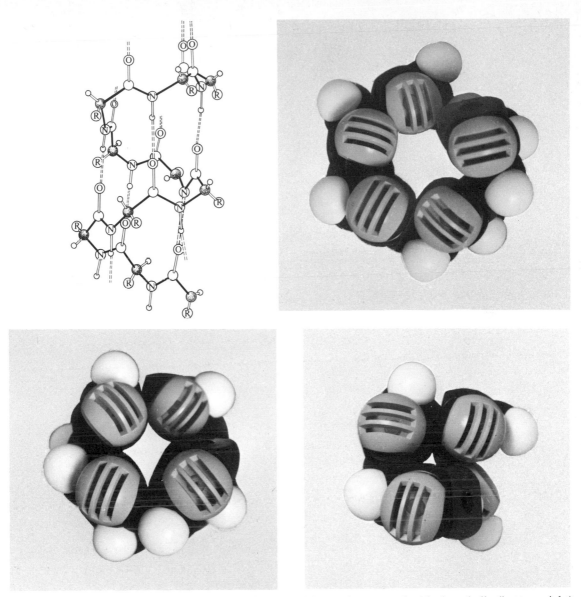

**Fig. 4.7** The γ-helix (left, top and bottom) compared with the α-helix (bottom right) and the π-helix. Note the sizes of the central holes.

elasticity, of the hydrogen bond (see p. 93). Hydrogen bonds are also of great importance in the interaction of proteins with other molecules (see pp. 117 and 119).

*Other dipole forces* must certainly be present in proteins, since so many elements of the structure produce dipoles. They are thought to be quite important, especially in stabilizing the *quaternary structure*, in the association between protein molecules to form functioning units of a higher order of complexity (p. 76). Because of the large number of relatively feeble interactions that are involved little is known in detail of their contribution.

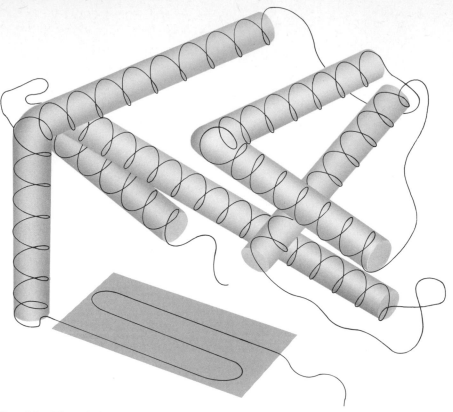

**Fig. 4.8** The reinforcement of protein structure by $\alpha$-helical and $\beta$-pleated sheet regions

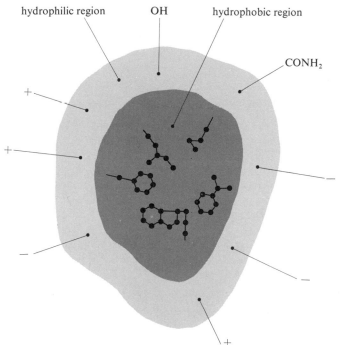

**Fig. 4.9** An idealised picture of a protein showing the division into hydrophobic and hydrophilic regions (compare Figs 3.8, 7.9 and 8.22)

The *hydrophobic* and short-range *dispersion* interactions are among the most powerful in stabilizing the structure; they are also prominent among the forces involved in protein interaction with other molecules (e.g. p. 104). Now that X-ray crystallography (see below) is making it possible to look at the tertiary structure of proteins it can be seen just how important these are (Figs 4.9, 4.10 and 4.11). The majority of the hydrophobic elements cluster together at the centre and only a few of them are exposed to the aqueous solution. The hydrophilic elements, on the other hand, are almost all exposed.

Fig. 4.10 shows, in a case where the structure is known with precision, a few of the amino-acid side chains, selected to illustrate the contribution of the various types of forces to the stabilization of the structure.

Some proteins consist of aggregates of protein subunits. This so-called *quaternary* structure should not be thought of in terms of a random aggregate. The number and type of subunits involved and the geometry of the structure is pre-

**Fig. 4.10** A computer-drawn diagram (with perspective) showing some of the types of interaction stabilizing the conformation of the enzyme lysozyme (see also Fig. 6.8). Selected parts are shown of the polypeptide backbone of the protein and a very few of the side chains. Note the disulphide bridge (the cystine residue: one of four in the protein); the ionic bond between the —NH₃⁺ of the lysine and the —COO⁻ of the carboxyl terminus of the protein (the only ionic bond in the protein); the close approaches between hydrophobic residues (just a few of those occurring in the complete structure). The clearest of the hydrophobic interactions is probably the 'sandwich' of a methionine side chain between two tryptophans. The atoms are not drawn to their full size for clarity; if they were it would be seen that many of the side chains are virtually touching. (This figure was prepared with the assistance of Prof. A. C. T. North.)

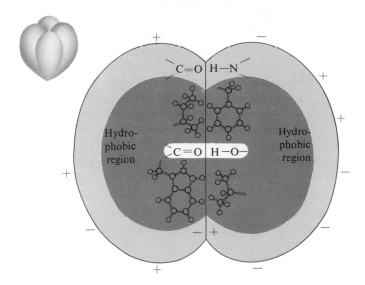

**Fig. 4.11**  Idealized section through a tetrameric arrangement of protein molecules (inset) showing types of interaction at an interface between two of the monomers.

cisely controlled (see Fig. 6.19, p. 127). Some proteins rely on the extra possibilities conferred by a quaternary structure for the full expression of their activity (see pp. 106, 128 and 313 ff). Figure 4.11 shows that quaternary structure is stabilized once again by the full range of non-covalent forces.

Proteins are synthesized as long chains (see pp. 446 ff) and are believed to begin to fold up and move toward their final configuration even before they are complete. We imagine that the longer-range forces ($1/r$ and $1/r^2$) are particularly important in the early stages. As the various parts of the chain approach each other water will begin to be excluded. Once the hydrophobic interactions are approximately correct there must be a final phase of settling down by minor adjustments. These would bring the cysteine residues into the correct position for bridge formation and the hydrogen bonds into line, and then maximize the number of very short-range forces.

## Denaturation

Proteins, as found in living tissue, are usually freely soluble materials with some definite chemical or mechanical function. When harshly treated (see below) they lose the ability to carry out their function and often become far less soluble. They are then said to be denatured. Curdled milk and boiled egg-white are examples of denatured protein.

The explanation for the changes in properties on denaturation is that the tertiary structure has been disrupted. The term is used to distinguish the relatively mild attack needed to disrupt the tertiary structure from the harsher conditions that bring about a general breakdown of many covalent bonds. Thus we are looking at the reversal of the folding-up process described above. Any agent leading to the weakening of any of the interactions maintaining the tertiary structure will be a denaturant. If the tertiary structure is unfolded, residues that are intended to join in hydrophobic and other interactions within the molecule may find themselves on the surface. Here they are very likely to interact with similar residues now exposed on the surface of adjacent molecules. The resulting intermolecular network will obviously be less soluble than the assembly of native molecules, since, in the native state, the residues of the micelle surface are very largely those which are more prone to interact with water than with each other (p. 57). The second feature of denaturation, the loss of biological activity, is also a consequence of unfolding, since, as has already been stated, the functional ability of the protein is closely dependent on the correct tertiary structure.

With our knowledge of the forces maintaining tertiary structure we can explain the action of individual denaturants. A few breaks in the peptide chain will disrupt the structure and agents that can effect these will denature (for examples, see pp. 91 and 98). Reducing or oxidizing agents denature largely by the cleavage of the disulphide bridge; extremes of pH denature by charging or discharging ionizable groups; agents that interfere with hydrogen bonding will denature; detergents denature by disrupting the hydrophobic interactions; and so on. An increase in temperature leads, by increasing the random thermal motion of the constituents, to an increased likelihood of all bonds being broken.

The thermodynamics of thermal denaturation are remarkable. Careful measurement of the principal thermodynamic functions gives values of which the following, obtained for the denaturation of the enzyme trypsin at 47°C, are typical. (The kinetics of this process are such that it is possible to observe it occurring in either direction. This is not always true of protein denaturation.)

$$\text{Trypsin (native)} \rightleftharpoons \text{trypsin (denatured)} \tag{4.5}$$

$$\Delta G^\circ = -560 \text{ cal/mole}$$
$$\Delta H^\circ = +67\ 600 \text{ cal/mole}$$
$$\Delta S^\circ = 213 \text{ cal/deg/mole}$$

The equilibrium constant for the reaction is, from equation (2.16), p. 18, about 2.5. This is not a particularly large value and from the relation $\Delta G^\circ = \Delta H^\circ - T\Delta S$ we can see that at even 10°C lower $\Delta G^\circ$ is $+1570$ cal/mole and the equilibrium constant in favour of the native state is about 10. (We should not confuse this effect of temperature on the *position* of equilibrium with that on the *rate* at which the equilibrium is achieved. For the latter, we require $\Delta H^\ddagger$, the enthalpy of activation, see pp. 40 and 133.)

This moderate stability, changing on a rise of temperature to a moderate instability, is the result of a balance between a very large, positive $\Delta H^\circ$ value (which tends to *prevent* denaturation occurring) and a very large, positive $\Delta S^\circ$ value (which tends to *make* it occur). The large value of $\Delta H^\circ$ is easily explained by the large number of bonds that have to be broken, and thus the large amount of mechanical work that has to be done to pull the interacting elements apart. The large value of $\Delta S^\circ$ is at first sight less easy to explain since we are not creating a large number of new molecules which, by their freedom to move and rotate, usually account for an increase in entropy. In the present case, of course, we know that the high value of $\Delta S^\circ$ comes from the loss of the highly ordered intramolecular structure of the protein.

The $\Delta G^\circ$ value indicates that the tertiary structure is only just stable at the working temperature of the protein. Should we be surprised at this?

The tertiary structure should not be too stable, since single amino-acid changes could not then so readily cause significant alterations in function. (We shall see that the mutations that provide the raw material for evolution usually involve single amino-acid changes (Ch. 27).) Such changes will usually be for the worse, but then the individual carrying them will tend to be eliminated by natural selection. From the point of view of evolutionary improvement of the species, it is worth tolerating this wastage to be able to take advantage of the favourable changes, infrequent though these are. The fact that proteins, and thus ourselves, exist on a knife-edge of stability (for a quantitative example see p. 107) may now be seen in part as the price we must pay for the enhanced possibility of evolutionary improvement.

## The formation of tertiary structure

There are two views on the way in which this highly ordered structure is brought into being. One is that the primary structure is solely responsible for the interplay of forces that decides the tertiary structure. The alternative could be that the tertiary structure is formed on some template structure during or just after synthesis of the primary structure. The former proposition, however, receives support from the observation that denaturation is often reversible. Thus, for example, if disulphide bridges are reduced and then re-oxidized the tertiary structure and biological activities that were lost on reduction may often be recovered completely. Since this happens in the test tube far from any possible template, one must conclude that the native tertiary configuration is the most stable and the one to which the structure will naturally tend.

Even if the first view is correct (that primary structure is solely responsible), it is not yet possible to calculate the tertiary structure from knowledge of the primary structure, since there are so many interacting elements involved that the calculation is impossibly complex. We have to rely on experimental methods for determining the tertiary structure that a molecule has actually achieved. The most important of these is X-ray crystallography (see the end of this chapter).

## Co-factors and prosthetic groups

A discussion of protein structure and function would not be complete without an account of the class of small (non amino acid) molecules and ions that are occasionally found associated with the protein in its functioning state. They are usually called *co-factors* if weakly bound and *prosthetic groups* if strongly (usually covalently) bound. Certain substances may be co-factors or prosthetic groups depending on the particular protein. There is in any case no rigid line of distinction between the two terms.

There are tasks undertaken by proteins that are beyond even the quite respectable range of abilities of amino-acid side chains, and to tackle these it is essential for other molecules to be used. Co-factors and prosthetic groups, when bound in sites specifically intended to enhance their usefulness, may behave in a way which would not be expected from their properties in free solution. This is

**Table 4.2** Co-factors and prosthetic groups and their derivatives

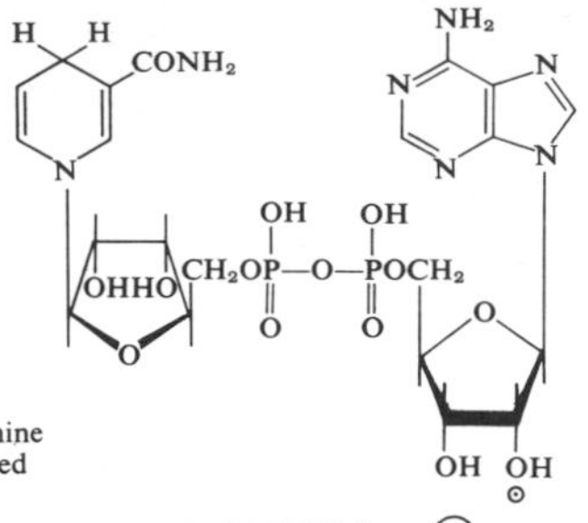

Nicotinamide Adenine
Dinucleotide-reduced
(NADH₂)
(see p. 234)

NADP, NADPH₂ have $P$ at ⊙

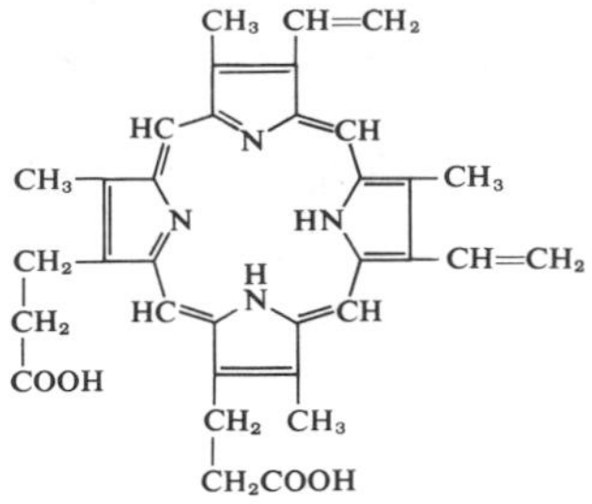

Porphyrin derivatives
(see pp. 104–108 and
235–236)

With $Fe^{2+}$ : haemoglobin

$Fe^{2+} \rightleftharpoons Fe^{3+}$ cytochromes

With $Mg^{2+}$ chlorophyll
(see p. 250 f.)
(Various side chains and
substitutions)

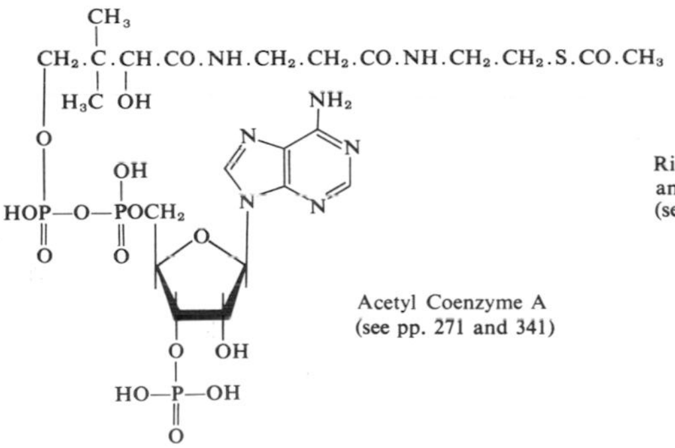

Acetyl Coenzyme A
(see pp. 271 and 341)

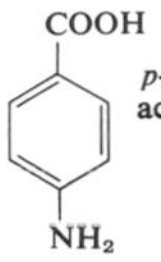

*p*-amino benzoic
acid (see p. 386)

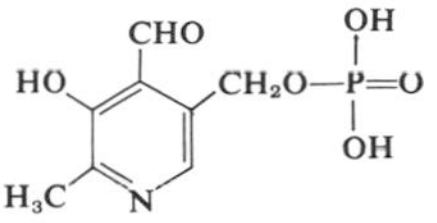

Pyridoxal phosphate
(see p. 358)

Riboflavin
and derivatives
(see pp. 234 and 278)

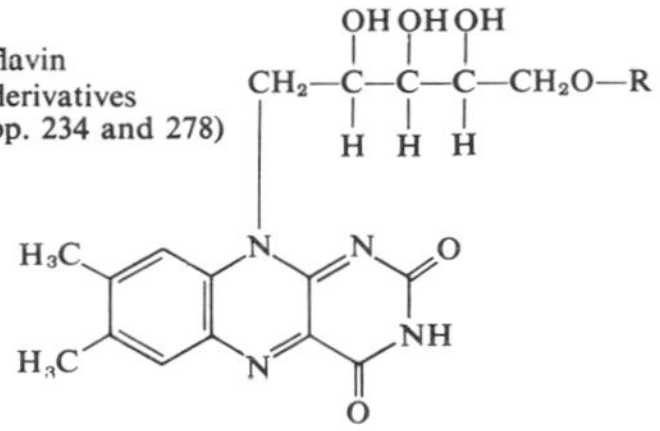

R=H: Riboflavin

R=$P$: Flavin
mononucleotide (FMN)

R = —$P$—$P$—ribose-adenine:
Flavin adenine
dinucleotide (FAD)

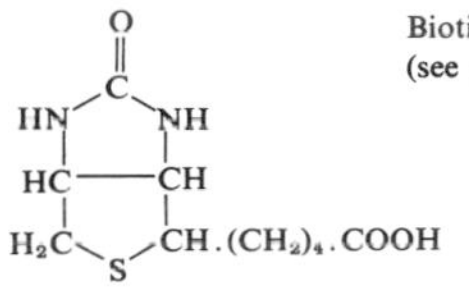

Biotin (vitamin H)
(see p. 342)

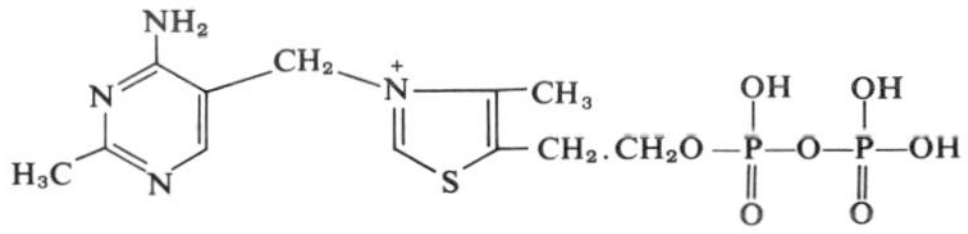

Thiamine pyrophosphate
(TPP)
(see pp. 272 and 297)

analogous to the situation existing for the amino-acid side chains themselves.
The protein therefore accommodates the co-factors in two ways: it provides a
specific site and it controls the chemical environment of that site.

Table 4.2 lists the more common co-factors and prosthetic groups. The specific
way in which they work is explained for each case on the page indicated. It will be
seen that some of the structures given are, for small* molecules, quite complex.
Many organisms, man included, are not able to synthesize all of these structures
for themselves, and they must then be obtained in the diet, that is, by consuming
organisms that have not lost the necessary synthetic ability. Apart from external

* When you look at the table, you may feel that 'small' is a rather odd term to apply to molecules such as the haem ring, which has over seventy atoms in it and a molecular weight of
over 600. Certainly, molecules like this are large compared to most of those that you will
have met in non-living chemistry, since molecules encountered there usually have no more
than a dozen or so atoms and molecular weights of the order of 100. We are using the term
'small' here as biologists, and comparing the molecules in the table with those of a protein
(even small proteins have 2000–3000 atoms and molecular weights of 12 000 or more) and a
nucleic acid, the largest of which may have millions of atoms and molecular weights in the
hundreds of millions.

sources, we must not forget the internal ones. The bacteria of the gut provide more of these requirements than is often realized. It is for this reason that we must supplement the vitamin intake of patients on prolonged antibiotic therapy, since the beneficial bacteria will often suffer along with those at which the therapy is aimed.

Many vitamins and so-called trace elements fall into the category of co-factors and prosthetic groups. That only very small quantities of many of these substances are required can be explained by the fact that the proteins with the need for the co-factor or prosthetic groups are very likely to be present themselves in only very small, catalytic amounts. It follows from this that the co-factor and prosthetic group with a much lower equivalent weight will be required in even lower quantities than this.

We can contrast this situation with the dietary requirement for the amino acids themselves. In the rat for example, ten of the protein amino acids cannot be synthesized and have to be obtained in the diet (p. 360). There is no question here of a need for trace amounts at a specific site; instead there is a general requirement for substances that are constituents of all proteins, and considerably larger amounts must be ingested.

You should by now be acquiring the habit of looking at the structure of molecules to see what can be predicted of their properties. For example, a glance at Table 4.2 will show that many of the molecules have charged or hydrogen-bonding groups and will be soluble in water. Others, such as vitamin K (related to ubiquinone, Fig. 12.4) are predominantly hydrophobic and will be found only in solid or liquid solution in fats or oils. The latter type must be brought into the body by eating the substances in which they are dissolved. Similarly, we may note from the table that the porphyrin derivatives (which are usually not vitamins since most of them can be synthesized by most organisms) are fairly hydrophobic and will sit firmly on hydrophobic patches or in hydrophobic clefts in proteins, which is in fact what their function requires (pp. 104 and 236).

## Mimicry at the molecular level

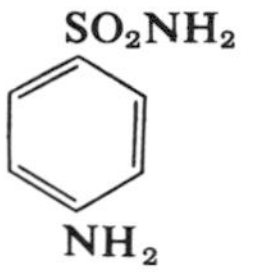

$SO_2NH_2$

$NH_2$

**Fig. 4.12**  Sulphonamide

Having discussed the co-factors and prosthetic groups, it is quite appropriate to look at the phenomenon of *mimicry*, which occurs when a compound resembles another sufficiently to undertake some of its functions. Co-factors and prosthetic groups are frequently mimicked in this way. Thus, the sulphonamide compounds (Fig. 4.12) are sufficiently similar to *p*-amino benzoic acid (Table 4.2) to be bound in the site designed to receive the *p*-amino benzoic acid. However, sulphonamide is insufficiently similar to this compound to mediate in the actual chemical process for which *p*-amino benzoic acid is required as a co-factor (for details of this process see p. 384). As a consequence the sulphonamide occupies binding sites to no purpose that would have otherwise been usefully engaged in binding the true co-factor. Some bacteria are found to be exceptionally sensitive to this property of the sulphonamides, which can prevent their growth as a consequence. These compounds, the 'sulpha' drugs, were among the first successful chemotherapeutic agents.

Mimicry also occurs in the substrates of enzymes. We shall see on p. 130 how malonate ion successfully occupies a site on an important enzyme that is intended for succinate ion, and blocks the access of the true compound. The antibiotic properties of penicillin may be due to the fact that it acts as just such a blocking agent in the synthesis of the bacterial cell wall (p. 180).

An example of mimicry drawn from a different branch of biology is the ability of certain compounds to cause genetic mutation because of their close resemblance to the nucleotide bases that carry the genetic information between the generations (chs. 7 and 24). Here again the resemblance is close enough to enable the compound to be incorporated into a biological system, but not close enough to enable it to function properly when there.

If the structural resemblance is so very great that the mimic compound is able to undertake the function of the compound that it supplants, it will not be toxic. Methionine and nor-leucine, which were mentioned earlier in this chapter, afford an interesting example of this. Nor-leucine is close enough in structure to be incorporated in place of methionine *and* close enough to be able to undertake the role of methionine in stabilizing protein structure by hydrophobic contacts. Micro-organisms that incorporate nor-leucine in place of methionine produce fully functioning proteins and are unharmed as a result.

## THE EXPERIMENTAL DETERMINATION OF PROTEIN STRUCTURE

We have already made a great deal of use of the knowledge that has recently been acquired of the structures of proteins, and will be making even more use of it later on. We therefore conclude this chapter with a brief account of the principal experimental methods by which the structural information has been gained. We consider first the determination of primary structure and then the determination of tertiary structure.

### Methods for the determination of primary structure

These methods fall into two main classes. These may be called *overlapping* and *stepwise cleavage*, respectively. Both, as usual in the chemical analysis of structures, no matter what the size, involve degradation of the original molecule.

In the *overlap* method this degradation is not extensive enough to give complete cleavage of all peptide bonds. For example a brief treatment with 6N-HCl at 100°C brings about a complete cleavage at a few particularly susceptible points in the sequence and incomplete or no cleavage at other points. Thus a part of insulin, the protein hormone, might give the mixture of peptides shown in Fig. 4.13. (Insulin was the first protein to have its amino-acid sequence determined. The work was carried out by Sanger and his colleagues over a period of approximately ten years ending in 1955.)

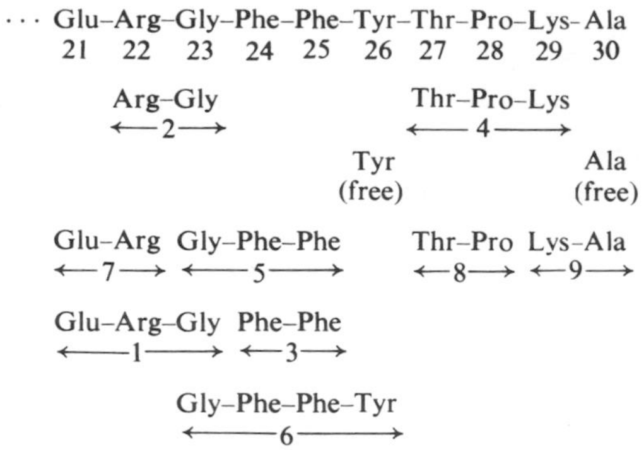

··· Glu–Arg–Gly–Phe–Phe–Tyr–Thr–Pro–Lys–Ala
21 22 23 24 25 26 27 28 29 30
Arg–Gly
2
Thr–Pro–Lys
4
Tyr (free)
Ala (free)
Glu–Arg
7
Gly–Phe–Phe
5
Thr–Pro
8
Lys–Ala
9
Glu–Arg–Gly
1
Phe–Phe
3
Gly–Phe–Phe–Tyr
6

**Fig. 4.13** The peptides obtained by partial hydrolysis of a section of insulin. The section shown consists of residues 21–30 of the longer of the two chains of the molecule (see Fig. 5.10). The numbering system for the peptides is arbitrary.

The amino-acid compositions of these can be determined by total hydrolysis (achieved by a longer acid treatment), followed by separation and estimation of the resulting amino acids (Appendix II). Note that no fragment is found in which the peptide bond 26–27 survives, but 24–25 is untouched. Note also that it is possible to deduce the sequence simply by the consideration of the amino-acid compositions of the fragments. This is done by a repetition of arguments of the type 'arginine is next to glycine (peptide 2) and arginine is next to glutamic (peptide 7), therefore arginine is between glutamic and glycine'. The term 'overlap' should now be clear. We do not yet know whether the sequence has to be read from left to right or from right to left. This can be decided by identifying the N- or C-terminal residue of at least one of the fragments. The methods by which it is done will be described in a moment.

In this example a sequence of ten residues has given rise to a mixture of eleven smaller fragments. Such mixtures can be resolved (see Appendix 2), but the task becomes unmanageable unless the parent sequence is small. Thus this method is suitable for very small proteins like insulin, which has two chains of 21 and 30 residues which can be sequenced separately. Larger proteins, which may easily contain 300 residues, are not amenable to this technique. In determining the sequence of these, fragments must be produced by complete cleavage at a relatively limited number of sites. We can achieve this kind of cleavage by the use of proteolytic enzymes, some of which are relatively specific for peptide bonds involving just one or two types of residue. Thus the sequence in Fig. 4.14 will be

(trypsin)
(trypsin)
Val.Leu. Asp.Phe Val.Glu.Lys Ala.Ser.Tyr Asp.Arg Ser
(chymotrypsin)
(chymotrypsin)
(chymotrypsin)

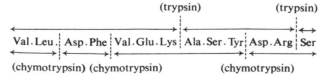

**Fig. 4.14** This figure shows the points at which trypsin and chymotrypsin cleave a typical peptide. The double-headed arrows indicate the fragments produced

cleaved as shown by the enzymes trypsin (specific for the two long-chain, basic amino acids lysine and arginine) and chymotrypsin (fairly specific for aromatic and a few other types of side chain). The fragments obtained may then be dealt with by the partial-acid type of treatment or by stepwise methods (see below). It is clear that, by comparing the amino-acid compositions of the tryptic fragments with those of the chymotryptic ones, and using the overlap principle, the order of these fragments may be determined relative to one another. Rather fewer peptides are produced when enzymes are used than is the case with the acid cleavage, and in higher yield. This is because we are not obliged to settle for partial cleavage at most bonds, but can select conditions that will usually split a given bond completely or not at all.

For really large proteins, and intractable small ones, an even more specific cleavage is employed to give even fewer (but now very large) fragments. Like partial acid cleavage this is a chemical method, but it cleaves 100 per cent at only one residue, methionine. The reagent is cyanogen bromide and its effect on methionyl peptides is shown in Fig. 4.15.

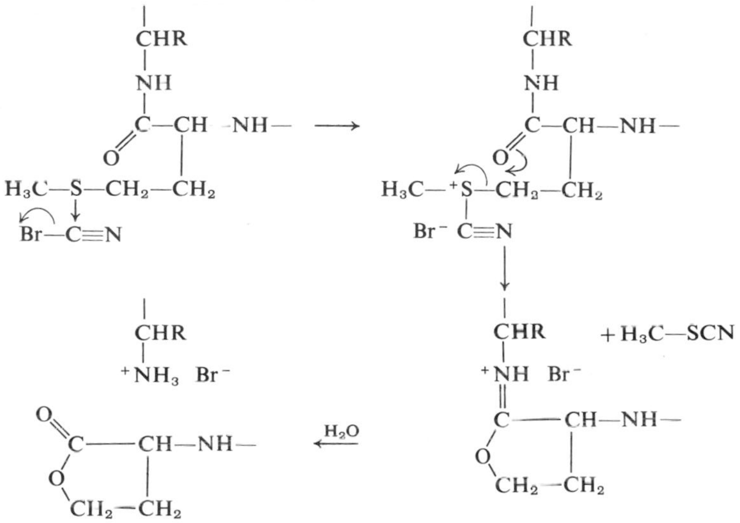

Fig. 4.15   The cleavage (by cyanogen bromide) of a peptide chain at methionine

Since methionine occurs relatively infrequently along the sequence of most proteins, a few large fragments are formed. These can then be separated and further degraded, either by the other overlap methods or by stepwise methods.

All these variations are dependent for their success on the separation of the fragments. Most fragments are separable (see Appendix 2) but a few are not. The difficulty is usually due to the presence of highly insoluble fragments in the mixtures to be separated. Sequence methods are now developed to the point that any protein of 300 residues or less may be successfully attempted provided that no extensive insoluble fraction (the so-called 'core') exists in the digest. A 300-residue protein without a 'core' problem could still take two or three people two to three years to complete.

*Stepwise* methods can be either enzymic or chemical. Proteolytic enzymes exist that cleave residues from either the N terminus or C terminus (exopeptidases), in contrast to those already mentioned that cleave at points along the chain (endopeptidases). These enzymes may therefore be used to identify sequences near the ends. The best stepwise method, however, is the Edman reaction (Fig. 4.16) because it can be operated with near complete efficiency one step at a time, with pauses if necessary to identify the residue removed.

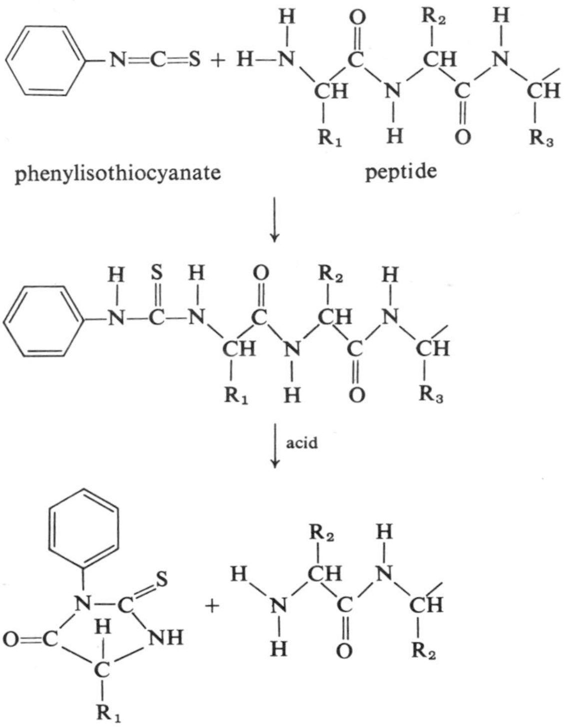

**Fig. 4.16** The removal of an amino-acid residue from a peptide chain by means of the Edman reaction. Note that a new amino acid is ready for removal by another cycle at the end

The phenylthiocarbamyl peptide first formed is cleaved by cyclizing the first residue to a phenylthiohydantoin, which may be identified by paper chromatography (Appendix 2). The advantage is that the rest of the peptide is left intact with the $\alpha$-amino group of the second residue ready for a repeat of the same process, so that the second residue can now be identified and the third residue unmasked. Provided that the yields in all steps are very high, the cycle can be repeated many times, sometimes by automation. Diminished yields usually bring about a halt after about thirty residues, and so this is not, as it might be thought, the universal answer to all sequencing problems.

A useful adjunct to either overlap or stepwise procedures is a simple method for determining the nature of the N-terminal residue. In this, a compound R—X (where —X is a halogen of enhanced reactivity) is reacted with the peptide.

$$R—X + NH_2—peptide \rightleftharpoons R—NH—peptide + HX \qquad (4.6)$$

R is chosen both to activate the halogen and to provide an R—NH bond more resistant to total acid hydrolysis than the remaining peptide bonds. Thus, on hydrolysis, the peptide is left as a mixture of the non-terminal amino acids and the R-derivative of the terminal one. If R imparts colour, or, for greater sensitivity, fluorescence in ultra-violet light, this residue can be identified by visual inspection of chromatograms. Two such R—X reagents are shown in Fig. 4.17.

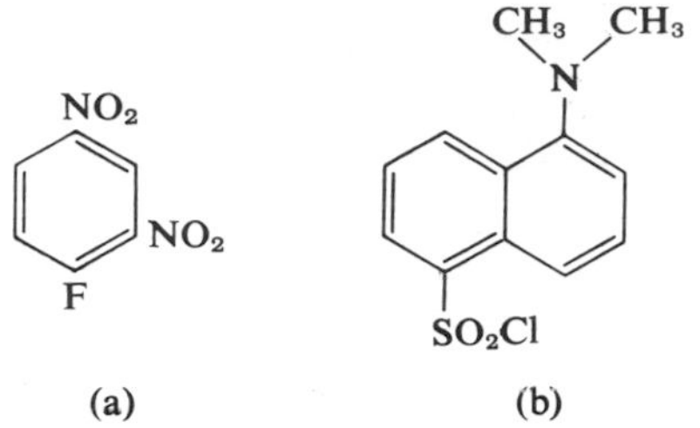

(a)  (b)

**Fig. 4.17**  (a) 1-fluoro-2,4-dinitrobenzene, (b) dansyl chloride

The aryl fluoride imparts a yellow colour to its derivatives – the naphthalene-sulphonyl derivative imparts fluorescence. Such is the sensitivity of fluorescence methods (unlike other visual methods, all the light that one sees is 'signal' and very little is 'noise') that $10^{-11}$ mole of amino-acid derivative can be identified.

A last point on the technique of sequence determination arises from the fact that there is a tendency for cysteine residues to form disulphide bridges (p. 69), or, if they have already formed such bridges, to exchange partners. This can be a nuisance. This problem can be resolved by reducing all the bridges and masking the cysteine residues by reaction with iodoacetate

$$-SH + ICH_2COOH \rightleftharpoons - S—CH_2COOH + HI. \tag{4.7}$$

This reagent may also react with pre-existing —SH groups, which is the basis of its use as an inhibitor of metabolic pathways (p. 263).

## Methods for the determination of tertiary structure

Of the experimental methods available for the determination of tertiary structure, X-ray crystallography is unrivalled in power. The original development of X-ray diffraction methods is due to L. R. and W. L. Bragg. The application of the method to proteins was begun in the 1930s by Bernal, and has been continued from that period until the present by a number of workers, notably Perutz and D. Crowfoot-Hodgkin. Several proteins, including myoglobin, haemoglobin, several enzymes and insulin, have been the subject of successful analysis to a resolving power high enough (2–3 Å) to enable the position of individual atoms to be deduced (Figs. 4.10, 6.8).

We consider X-ray diffraction by analogy with the way in which we view objects by normal light. We are able to perceive the structure of macroscopic objects by the way that light (electromagnetic radiation of wavelength of 4000–7000 Å) is *scattered* from them. Similarly we can, though with considerably more reliance on machinery, do the same for much smaller objects from the way that X-rays (electromagnetic radiation of wavelength about 1 Å) are scattered.

The reliance on machinery stems partly from the difference of scale and from the inability of the retina to detect X-rays in any useful way. However, the real difficulty is that X-rays cannot be focused by any known optical equipment. Thus it is necessary to intercept the X-radiation scattered by the object, record it, and do by computation that which a lens does automatically, namely, produce an image.

Radiation, considered as a wave-form, is completely specified by noting its amplitude and phase and the co-ordinates of the point at which these quantities are measured. The amplitude is easy to deduce from the degree of exposure of a photographic emulsion, or from the rate at which incoming quanta are detected in a carefully aligned Geiger counter. The positional information can be obtained by measurement on the film, or noting the position of the counter. Phase is a more difficult quantity, both to define and to measure. It appears as the quantity $\phi$ in the equation that describes wave motion:

$$A_0 = A \sin(\nu T - \phi) \tag{4.8}$$

where $A_0$ is the amplitude at a particular point, $A$ the maximum amplitude, $\nu$ the frequency and $T$ is time. The value of $\phi$ decides whether one experiences the

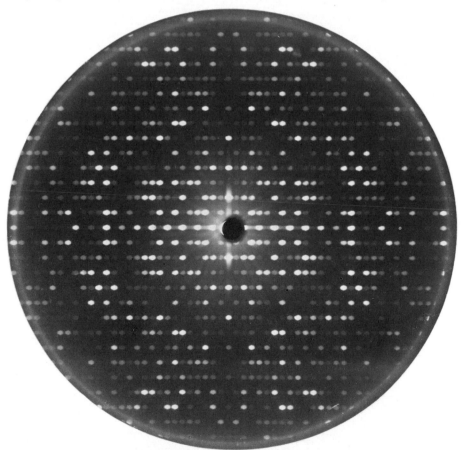

**Fig. 4.18**  An X-ray diffraction photograph of a single crystal of a protein

peak, trough or some intermediate portion of the wave. Knowledge of the phase is lost when the intensity is measured. (Intensity $= A^2$; it can be shown that $\phi$ is lost in squaring the trigonometrical part of equation (4.8).) Thus in order to reconstruct the complete description of the scattered radiation the phases must be restored by, in effect, solving simultaneous equations using additional sets of information. The additional information is obtained by repeating the X-ray experiments on closely similar material. The best source of such material is molecules in which only one atom has changed; provided that it is an atom, such as a metal, that is sufficiently heavy (i.e. electron dense) to make an appreciable difference to the pattern of the scattered radiation.

Crystals are used in X-ray diffraction more often than a powder or solution because a crystal lattice holds the molecules in one or just a few orientations with respect to the observer. This means that the pattern of scattered radiation from each molecule reinforces the patterns from the others to give a simpler, more intense diffraction pattern (Fig. 4.18). Thus X-ray crystallography, while capable of giving immensely detailed and important results (e.g. see Figs 5.13 and 6.8) relies completely, when globular proteins are involved, on the successful production of crystals of the natural form of the proteins and crystals of the proteins with one or a few heavy atoms. The heavy-atom crystals must be of the same crystal form as the native ones to be useful, when they are known as 'isomorphous'. (It is rare that fibrous proteins can be crystallized, but they are often aligned in the artificial or natural fibres that contain them and some useful results can then be obtained.)

Even if no great difficulties are encountered in any of this work, several man-years will almost certainly be needed, as is true for the primary structure, before a detailed structure is obtainable.

X-ray diffraction at high resolution ought in principle to supplant chemical methods of determination of primary structure because it should be possible to identify the amino acids by the characteristic shapes of their side chains. In fact the pictures that are obtained are rarely clear enough to enable one to see more than the outline of the side chain and its orientation to the rest of the molecule

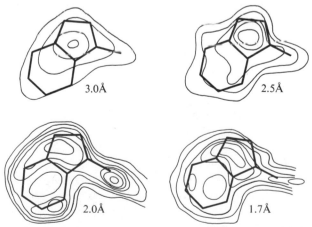

**Fig. 4.19** This figure shows the difficulty of fitting an amino-acid side chain to the electron–density contours calculated from X-ray diffraction data of various degrees of resolution.

(e.g. Fig. 4.19). This is usually insufficient to allow the nature of the amino acid to be determined without the help of chemical data. The structures that we shall use in later chapters are, for this reason, all composite products of studies on primary and tertiary structure.

## The interpretation of X-ray diffraction patterns – diffraction by helices

The computations that replace the lens to produce the images of the type shown in Fig. 4.19 are extremely complex, particularly for large molecules. They call for advanced computers if they are to be completed in a reasonable time. Nonetheless the novice can tell a certain amount from direct inspection of the X-ray patterns shown in this book.

First of all, the scale of the pictures is inverse or *reciprocal*. A large distance in space gives rise to a small interval on the diffraction pattern. Conversely, small distances in space give large distances on the pattern.

Secondly, *helical structures* are often found in biological macromolecules, and these regular structures are well suited to the production of particularly clear, simple diffraction patterns.

Consider a helical structure, such as that shown in Fig. 4.20g, in which the chain is made up of a series of repeating units. This structure may represent an α-helical polypeptide chain (p. 72), in which case the repeating units are the amino acids; it may be the double helix of DNA (p. 146) in which the repeating units are the nucleotides; or it may be a helical virus (p. 162) in which case the repeating units are based on the individual protein molecules that make up the helix. We shall see in succeeding chapters that the diffraction patterns from all these structures have much in common.

It is possible to use this simplified picture to gain a semi-quantitative explanation of the nature of the diffraction patterns obtained from a helix. To do so, we must remember that diffraction methods are best suited to repeating structures because these cause an intensification of the pattern by superposition of the separate diffraction patterns from each element.

The most striking repetitive feature in Fig. 4.20 is the run of parallel sections of the chain (accentuated in Fig. 4.20b). The diffraction pattern of such a series of lines is a series of spots (Fig. 4.20c). These lie on a line inclined to the axis of the helix (and thus the axis or *meridian* of the X-ray pattern) at an angle equal to that of the angle of pitch ($\theta$, Fig. 4.20b) of the helix. Since the X-rays will interact with the back of the helix as well as the front, there will be a run of parallel lines (accentuated in Fig. 4.20d) which will give rise to a second line of spots, inclined to the meridian by an angle of $-\theta$ (Fig. 4.20e). These two lines will appear together in the final pattern, to give rise to the characteristic *helix cross* (Fig. 4.20f), which we will see in the diffraction patterns of structures in later chapters. The spacing between the dots ($a$, Fig. 4.20f) is proportional to $1/p$ where $p$ is the pitch of the helix (Fig. 4.20b). We call $a$ the *layer line spacing*.

The second most striking repetitive feature is that of the subunits along the chain (accentuated in Fig. 4.20g). The repeating distance here ($t$, Fig. 4.20g) is analogous to the height of rise of individual steps in a spiral staircase – in contrast

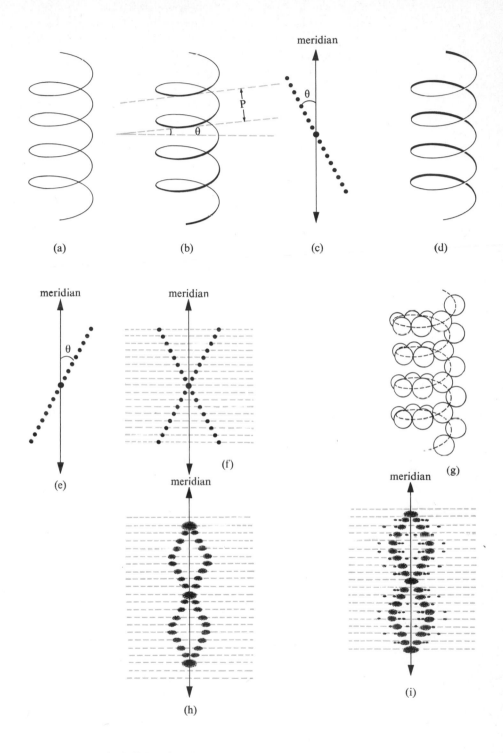

**Fig. 4.20** Helical diffraction (see text)

to $p$ which is the height that one would rise in a complete turn of the staircase. $t$ is therefore smaller than $p$ and so, in the reciprocal fashion of diffraction patterns, the spacing that represents it ($b$, see below) will be greater. It can be shown that the most important effect of the repeat of $t$ will be to move the spots in the helical cross progressively back toward the meridian of the picture (Fig. 4.20h). (Another effect is to increase the number of spots on each layer line (Fig. 4.20i).) The bent-back cross passes through the meridian again after $n$ layer lines where $n$ is the number of subunits per turn of the helix. It is fairly easy to see that spacing $b$ that we defined above is the distance of this first strong meridional spot from the origin of the pattern.

This is a very simple description of a complex computation, but we shall see in later chapters that it will be sufficient to help us to deduce the main structural features of several important biological macromolecules from their diffraction patterns.

Now that we are equipped with a knowledge of the features that give proteins their properties, and of how structures are obtained, we will turn, in the next two chapters, to a detailed examination of individual types of protein.

# Chapter 5   Protein function I

We have studied the way in which the various forces are combined to stabilize protein *structure*. We shall now see how it is that the same forces acting between the same set of elements manage to produce such a wide range of *behaviour*.

For convenience we may divide the proteins into three classes, depending on the way in which they contribute to the activities of living matter. The first class consists of the food-storage proteins, the second of proteins that have a structural or mechanical role and the third of those that act by binding to other molecules.

## Food-storage proteins

In order to synthesize a protein an organism must have an adequate supply of amino acids. Some organisms may be able to meet all their requirements by their own efforts, making use of the biosynthetic pathways described in Ch. 22. Should demand exceed this internal supply, or should an organism be totally incapable of synthesizing a particular amino acid (e.g. see p. 404), an external source is necessary. In such cases food proteins may be consumed, and then digested to amino acids by means of hydrolytic enzymes (p. 111).

Almost any protein may serve as a source of amino acids in this way, but specialized molecules have been developed for particular purposes. For example, mammalian young are dependent on their mothers' milk for, among other things, a supply of amino acids to meet the severe nutritional problems posed by their early growth. This demand is met by a number of milk proteins, in particular by the caseins. Caseins, which are soluble when secreted, are found to be exceptionally easily denatured in the gut. In newborn ruminants this process is greatly aided by a proteolytic enzyme called rennin, which is produced in the second stomach. This enzyme denatures the proteins by a few judicious cleavages, producing insoluble, disorganized structures that are more easily attacked by the other digestive enzymes. The intestinal mucosa of calves, rich in rennin, are dried and sold for use in the making of cheese and junket, under the name of rennet.

We shall see below (p. 106) that it is difficult to be sure that one has arrived at a full understanding of the function of a protein. However, as far as is known, this ease of denaturation and digestion is a major function of requirement for the casein molecule. A second function of casein may be the transport of phosphorus into the body, since it contains a significant proportion of the phosphorus content of milk in the form of phosphate groups esterifying serine residues in the protein.

With these rather simple functional requirements, the structure of casein is fairly loosely controlled by natural selection. Thus there is considerable variation among the caseins from different species of mammals. This variability of structure contrasts with some other protein molecules, where the function is sensitive to small changes in structure and there is tight genetic control of structure.

## Structural proteins

Some of the more important structural proteins are summarized in Table 5.1. In the table we see a number of protein structures adapted to the different requirements of the organism. Hair must be elastic (the alternative would be that it would be brittle*) while horn and hoof must be hard-wearing. Bone (which has, of course, lost much of its protein by the time it is fully formed) must be mechanically strong to maintain the integrity of vertebrates and enable them to bear loads.

**Table 5.1**   Structural proteins

| Type | ←Less elastic | | More elastic→ |
|---|---|---|---|
| Keratin derivatives | Horn <br> Hoof | | Hair |
| Collagen | Bone (extensively mineralized) | Tendon | |
| β-Pleated sheet structures | | Silk. | Stretched hair |
| Sclero-proteins | Insect exoskeleton (tanned) | | |
| Elastin-type structures | | | Elastin (vertebrates) <br> Resilin (insects) |

Individual bones are themselves beautifully designed macroscopic structures, and transmit enormous loads, up to a ton in some cases, without fracturing.

The insects do not employ bones, but solve rather similar problems by the possession of a hard, protein-based exoskeleton.

Tendons are the cables by which stresses are transmitted from one part of the animal body to another; they must therefore be very strong. A collagen fibre is as strong as a steel cable of the same weight and has some elasticity.

The great mechanical strength of silk is exploited commercially, but of course the need that first evoked this strength was for an insect larva to be able to build itself a protective cocoon for the period of metamorphosis in unfavourable conditions.

* The hair of the Bushman *is* brittle. As a consequence it breaks off readily and makes for a sparse, 'pepper-pot' coverage. This sparseness has been said to be an advantage in that it allows much more rapid heat loss under desert conditions.

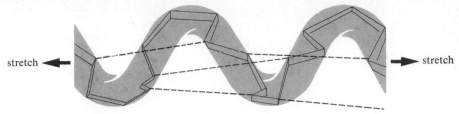

**Fig. 5.1**  Stretching an α-helix

Hinge-like structures are needed to anchor moving parts of animal bodies together (the attachment of wings of insects to the main body, for example). These structures need to possess the highest possible degree of elasticity. This is provided by proteins of the elastin and resilin type. These are superior in elastic properties to many of the most advanced synthetic rubbers.

These structural proteins are frequently of the high axial ratio (fibrous) type. A closer look at the molecular structures involved helps us to understand how all these requirements are met.

*Keratin* (found in hair, horn and hoof) relies for both its strength and its elasticity on the α-helical structure, one of the regular arrangements of peptide–backbone hydrogen bonds (Fig. 4.5). It will be seen that the hydrogen bonds, although they are weak individually, maintain a spring-like structure in which they are parallel. They may therefore be expected both to have some collective strength and, since they are individually somewhat elastic (they may be stretched a little more than covalent bonds may before they break), confer elasticity (Fig. 5.1).

The X-ray diffraction pattern of keratin (Fig. 5.2) shows some departures from that of a typical helix. If we look at the structure that is proposed for the α-helix we would expect to see a helix-cross pattern like that shown in Fig. 4.20 (refer to the treatment of helical diffraction at the end of the last chapter). We

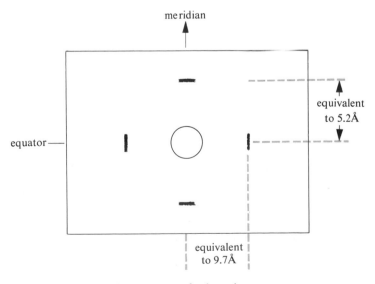

**Fig. 5.2**  The X-ray diffraction pattern of α-keratin

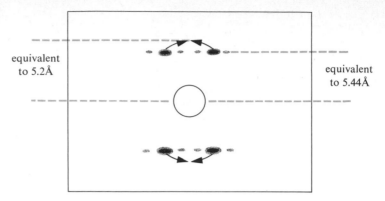

equivalent to 5.2Å

equivalent to 5.44Å

**Fig. 5.3**  The theoretically predicted X-ray diffraction pattern of the α-helix. The arrows show the position to which the remaining spots of the helix cross move (see text)

can explain the differences as follows. The first is that when there are rather few repeating units per turn of the helix, there is less superposition of radiation and the spots are less intense. This can be shown to have the effect of, in particular, diminishing the spots in the helix-cross that are well removed from the meridian. The result of this, if keratin had the strict α-helical structure, would be that the only clear spots would be those shown in Fig. 5.3. This is close to the actual picture, but not perfectly so. We have to take one further fact into account: the helix itself is not a perfectly straight rod, but slightly coiled itself (Fig. 5.4). It can be shown that this has the effect of moving the remaining helix-cross spots on to

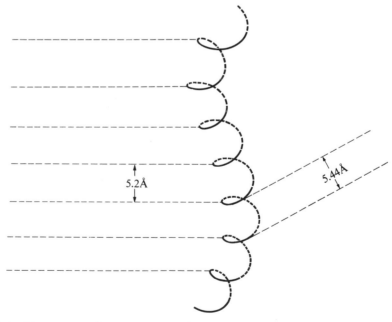

5.2Å

5.44Å

**Fig. 5.4**  The coiled-coil arrangement of the α-helices in keratin, showing the effect on the apparent pitch of the helix

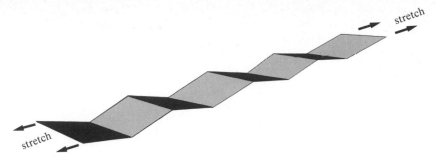

**Fig. 5.5** Stretching a β-pleated sheet

the meridian (Fig. 5.3). Finally, the 9.7 Å spacing on the equator corresponds to the distance between adjacent helices in the fibre.

The inelastic protein *α-fibroin* occurs in certain silks including that of the common silk moth. It typifies the β-pleated sheet structure (Fig. 4.6), another of the hydrogen-bonded conformations of the peptide backbone. Here the hydrogen bonds are at *right angles* to the direction of stretching. They can therefore play no part in the response of the silk to stretching save that they simply hold the bundle of adjacent polypeptide chains together. As a consequence, extension is resisted by the full strength of the covalent bonds which unlike the hydrogen bonds lie not perpendicular to but along the direction of stretching (Fig. 5.5). If

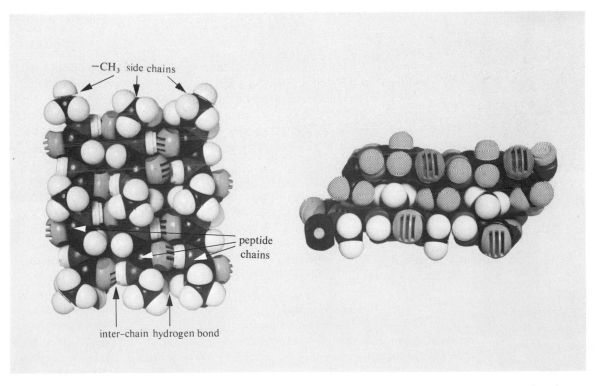

**Fig. 5.6** The interlocking that is possible between the side chains of adjacent β-pleated sheets

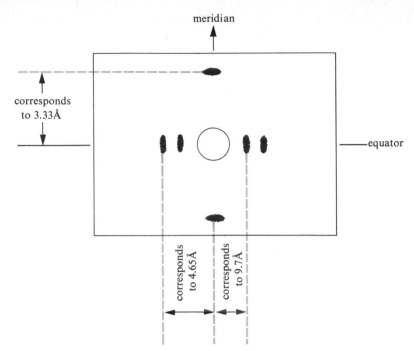

Fig. 5.7    The X-ray diffraction pattern of a fibre made up of β-pleated sheets

one could hold a silk fibre along its edges, rather than at its ends, then it would presumably show much more stretch.

Apart from the characteristic 2–space–2–space pattern of hydrogen bonds along the sheet there is another conspicuous feature – the fact that the side chains are perpendicular to the plane of the sheet. Therefore if the sheets are stacked on top of one another the side chains interlock and add greatly to the strength of the whole assembly (Fig. 5.6).

(Hair too can, if it is stretched sufficiently, pass from the α-helical structure to the β-pleated sheet form. We can regard this far more rigid form as the last defence against breaking.)

When examined by X-ray diffraction the β-pleated sheet, even though it is not helical, has repeated features and gives some spots that are easily interpretable. The 3.33 Å spacing on the meridian and the 4.65 Å spacing on the equator are at once recognizable (Fig. 5.7). The 9.7 Å spacing is that between adjacent planes.

*Collagen* has a more complex structure based on three helices twining about each other in a so-called superhelix (Fig. 5.8). The amino-acid sequence has to be rather odd to make this arrangement possible. Every third residue is usually glycine and one of the other two is usually proline. The proline residues sometimes have an —OH group added to the ring after the proline is incorporated into the chain. This conversion of proline to 3-hydroxy proline is presumably designed to influence the interaction of the chains with each other and their common environment by imparting a hydrophilic character to an otherwise hydrophobic region.

Fig. 5.8    The supercoiled structure of collagen

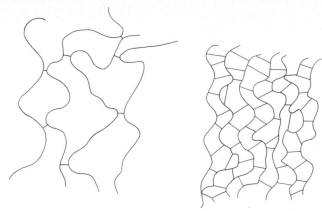

**Fig. 5.9**  Cross-linking of peptide chains

*Cross-linking*

Elasticity or rigidity may be enhanced by covalent cross-linking of the protein chain. A relatively open network of cross-links confers elasticity, while a tighter, more numerous, system of cross-links confers rigidity (Fig. 5.9). The covalent bonds used may be S—S bridges (p. 68) or may arise in other ways, such as by means of free-radical reactions (often aromatic in nature) initiated by enzymes.

For example tyrosyl residues may join two chains by joining themselves to produce

OH              OH

(to chain)   (to chain)
  I                II

In hair, the natural elasticity of α-keratin is enhanced by embedding it in a matrix of protein chains loosely cross-linked by disulphide bridges. In horn, another keratin-containing structure, the matrix has more —S—S— bridges and is rigid. Elastin and resilin do not have a matrix but employ direct cross-linking between chains. Here again the chains themselves are probably α-helical and therefore elastic and the cross-links are sufficiently few to enhance rather than suppress (see below) this elasticity.

Free-radical reactions are also used in the formation of rigid structures. Insect exoskeletons, for example, consist of protein that is extensively cross-linked in this way. It was only by developing this method of cross-linking that insects were able to develop a light exoskeleton which was nonetheless sufficiently rigid to sustain the stresses of flight. Spiders, which did not do so and remain soft-bodied, have never achieved flight. Insects are complex, segmented animals, and the segments themselves must have an intricate system of pores for the passage of air and other substances. The construction system that they use is the extrusion of a pasty mass of protein into the right shapes (reminiscent of the thermoplastic resin industry), followed by hardening by the cross-linking process (*tanning*). This is clearly an ideal way of making up such structures.

Rigid structures may also be obtained by embedding the protein in a matrix of some material other than protein. The best known example here is (as we have already mentioned) bone, in which the matrix is mineral.

In considering the strength of all these materials, attention is now turning, as in so many fields of biochemistry, to even higher orders of organization. It is beginning to emerge that the quite extraordinary advantages that the basic molecular structures have over non-living materials are matched by various ingeniously contrived features of the way in which they are arranged to form the finished fibre, bone or tendon. For example, it has long been known that materials finally fracture when they can no longer store the strain energy by, for example, stretching. A small crack then appears, and unless the consequent local weakening can be compensated in some way, it rapidly increases in size and propagates explosively through the structure. We are coming to realize that all the biological materials that have been mentioned, in addition to being designed for strength in the ways indicated, have quaternary and higher-order structures that can prevent the first small crack from spreading. These structures store the energy by unwinding (breaking non-covalent bonds). They thus make the required local compensation and stave off failure by preventing this initial crack from propagating.

Some of these common-sense principles of construction are applied, as one might expect, to the other class of macromolecule that undertakes structural duties, the carbohydrates. We shall see in Ch. 8 how they are adapted to fit the different circumstances obtaining when sugar residues are used instead of amino acids. We shall also treat there the cell walls of bacteria since, although they have polypeptide chains, they also have a considerable carbohydrate component.

### Muscle proteins

The principal structural components of muscle cells are bundles of the fibrous proteins actin and myosin. Muscular contraction occurs when the bundles slide between one another, diminishing the size of each muscle cell and thus of the muscle as a whole. This sliding is thought to be the result of a conformational change on the surface of one of the two proteins. The energy required for this change is almost certainly brought about by an interaction between the protein surface and the energy-storage compound ATP (see p. 24). Although myosin is in most of its characteristics a typical fibrous structural protein, it is unusual in that it also possesses enzymic activity, a property more usually associated with globular proteins. The enzymic activity in question here is the catalysis of the breakdown of the ATP and is clearly connected with the contraction mechanism. The molecular basis of muscular contraction is discussed more fully in Ch. 18.

### Fibrin and blood clotting

Fibrin is an interesting structural protein. In its native form (fibrinogen) it is a soluble protein of considerable axial ratio (see p. 71). When it is necessary to form a blood clot, two breaks are made in the protein by an enzyme, thrombin, and the balance of forces that stabilizes the fibrous structure is disrupted. The peptides that are lost are strongly negatively charged and their presence in fibrinogen may prevent an interaction that then becomes possible in the product of the cleavage (fibrin). Certainly, a shift occurs to a more globular structure and residues that were previously buried now come to the surface. Many residues are now able to join in hydrophobic and other interactions between molecules and stabilize a great number of *inter*-molecular links where before they were involved in *intra*-molecular links. (The analogies with the denaturation process (p. 76) should be obvious.) As a result of these changes the solubility of the protein is drastically lowered. The consequent precipitation causes a *clot* to form, that is the precipitate immobilizes hundreds of times its own weight of water. The final event is the cross-linking of the molecules by covalent bonds – in particular disulphide bridges and also (most unusually) peptide-type bonds between *side-chain* amino and carboxyl groups.

The phenomenon of activation by removal of a part of the chain will also be met with elsewhere. It occurs when it would be an embarrassment to have a protein expressing its full function before it was time to do so. Other examples are some hormones, and enzymes that degrade cellular constituents.

*Structural proteins of low axial ratio*

Globular proteins are sometimes used to solve structural problems, although not so frequently as are the fibrous proteins. An example of the use of globular proteins is the encapsulation of virus nucleic acid (ch. 7) into a rigid, protective quaternary structure (see p. 162).

## Proteins that bind other molecules

We come now to the third, most intensively studied class. These proteins exploit to the full the possibilities of combination of the different amino-acid side chains, producing a site on their surface which has both a specific shape and a specific array of forces. We indicate in Ch. 2 that this site will bind a particular molecule or part of a molecule with great tenacity. This tenacity derives from the fact that it is a perfect fit both in the geometrical sense and in terms of the chosen forces meeting just those parts of the incoming molecule upon which they can best act. Thus the binding is both powerful and reasonably specific, since even a small change in the structure that is to be bound will be likely to spoil the fit and upset the interplay of forces (see also enzyme specificity, p. 109).

If the sole function of the protein is to bind, as is to some extent the case with the immunoproteins (see below), the matter ends there. Most often, however, it is necessary not only to bind a molecule but also to provide a special environment in which its properties can be modified. Haemoglobin (see below) is an example; so too are the enzymes (Ch. 6), which enhance the reactivity of the molecules that they bind.

Table 5.2 shows a convenient classification of the better known members of this class.

**Table 5.2**  Proteins that bind to other molecules

| | |
|---|---|
| Hormones* | Transmit instructions for the control (chs 28 & 29) of the rate of metabolic activity. |
| Immunoproteins | Bind to and inactivate foreign materials invading the body |
| Enzymes | Catalyse biochemical reactions |
| Carrier proteins | Transport molecules, ions or electrons from one place to another either within the cell or over greater distances within the organism |

\* Not all hormones are proteins.

Although we propose to discuss a few enzymes and just one of the other types of proteins (haemoglobin) in detail, as examples of all the others, a few notes on Table 5.2 may be helpful.

*The hormones*, whether or not they are proteins, are thought to influence metabolic rates by their ability to bind at a specific site. This site may be on a membrane, which has its properties modified as the result of the binding. The rate of passage of metabolites through the membrane might be significantly altered; this effect would obviously influence the rate of metabolic activity within the cell or other structure bounded by the membrane. Alternatively, a hormone may bind to a specific enzyme and induce a change in the tertiary structure in such a way as to alter the catalytic activity. (We shall see (Ch. 29) that a change of activity of just one enzyme can frequently influence the rates of a large number of biochemical reactions.) Finally it is not impossible that hormones may act on the genetic apparatus and influence the rate of synthesis (p. 488) of an enzyme. To change the amount of an enzyme in a cell is equivalent in many ways to changing its activity and will have similar consequences.

One of the best known protein hormones is insulin, which is secreted by certain specialized cells in the pancreas. It is practically the smallest protein known, having only fifty-one amino acids. It has profound effects on, in particular, carbohydrate metabolism, and these are thought to result, at least in part, from interactions between the hormone and the external membrane of the cell.

Insulin has recently been shown to be one of those proteins that are synthesized in an inactive form and are subsequently activated by a limited cleavage by proteolytic enzymes. The inactive form is called proinsulin (Fig. 5.10a) and it is activated by cleavage to give the two-chain molecule (Fig. 5.10b) which has been known for some time.

*The immunoglobulins* deal with invading substances in the body and are responsible for such phenomena as immunity and transplant rejection. The system works in the following fashion. In some way, the body is able to recognize the presence of certain classes of alien molecules known as antigens. Antigens are usually macromolecules: protein, nucleic acid, polysaccharide, lipid or a combination of any of these. This definition effectively includes the external surfaces of most viruses and bacteria and many biologically produced toxins. The system can hardly be expected to discriminate between these and those artificial cases in which the 'invading' material is intended to benefit the host. Blood transfusions

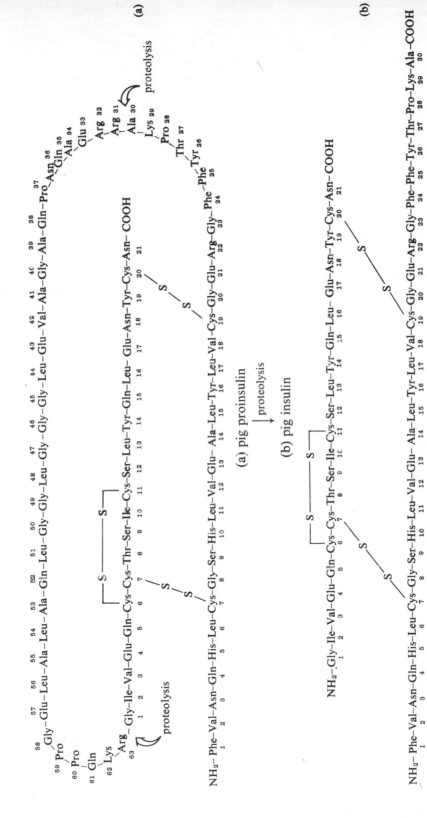

**Fig. 5.10** The activation of pig proinsulin (a) to insulin (b). Note the disulphide bridges and the points of action of the activating enzyme or enzymes.

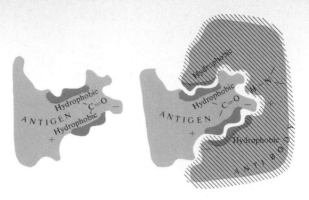

**Fig. 5.11** (a) A protein antigen (see text), (b) the antigen–antibody combination

and tissue transplants come most easily to mind, but other therapeutic substances such as some antibiotics and hormones may be antigenic towards some hosts. (We often speak of a patient's having an *allergy* to the substance. This term is accurate, since allergy is usually an immunological phenomenon. One of the most common forms, hay fever, is due to the antigenic properties of protein-polysaccharide complexes in plant pollen.)

Once the presence of the antigen is recognized, the large-scale production of the immunoprotein (antibody) begins. A large number of different antibodies will be produced in response to any one antigen. All will contain a binding site (see above) which will fit and bind to one of the structural features of the antigen. (Clearly only those features of the antigen that are not found in any of the molecules belonging to the host organism will evoke antibodies, or the organism would be continually at war with itself.) Suppose that there is an area of the invading antigen with the shape and combination of forces shown in Fig. 5.11a. Somehow the most suitable antibodies are selected from the many designs available to the organism. 'Most suitable' means an antibody with a binding site as close as possible to that which would provide a perfect fit (Fig. 5.11b) both sterically and in terms of the non-covalent bonding. Since it appears that an organism has millions of possible structures at its disposal, the best of these are, for most conceivable antigenic features, likely to fit very well indeed. The chosen antibodies will then bind with great strength to the antigen.

The antigen thus complexed is likely to lose its capacity to act. A bacterium, a virus or a tissue transplant covered in antibodies will probably cease to function. *Immunization* relies on the fact that once the body has learnt to synthesize an antibody it is able to produce large quantities very much more quickly should it ever be called upon to do so again. Rapidly multiplying viruses or bacteria will then have less chance of gaining the upper hand before the antibody concentration reaches an adequate level to deal with them. This *antigenic memory* may last for a lifetime or for a shorter period. In the latter case immunization has to be repeated at intervals.

The study of antibody structure is now one of the most promising areas of protein chemistry. It appears that one antibody differs from another only in a limited part of the molecule (which presumably includes the binding site), considerable regions of the structure being markedly less variable.

This fact may give a clue to the means by which the body acquires the ability to produce so many different antibody structures (the number is perhaps of the order of 1 million). One possibility is that each has its own sequence dictated by its own gene (p. 414). An alternative idea would involve *permutation*. If, for example, the first part of the structure were controlled by one gene and this could be combined with a gene for the second half, a million structures could be obtained by permuting 10 100 genes, say 100 for one half and 10 000 for the other.

Attention is now turning, too, to the elucidation of the primary and tertiary structure of the binding site itself and to the means by which the organism contrives to produce only those particular antibodies with binding sites that will fit the antigen invading at that particular moment. Although the mechanism that achieves this limitation is not understood, it is clearly necessary that only a selection of the possible antibodies should be circulating in the effective concentration at any one time. If this were not so, the circulatory systems would be supersaturated with the great mass of protein required.

*The enzymes*, catalysts of biochemical reactions, are among the most important constituents of living matter. They are considered in Ch. 6.

Some *carrier proteins* exist that transfer molecules, ions or electrons over quite small distances only. The membranes of the cell have some of the properties of the semi-permeable membranes found in ordinary chemistry but add to them a number of features for which it is believed that protein-lipid (p. 199) complexes are responsible. Figures 5.12a and 5.12b show the difference between *passive* transport of all non-living and some living semi-permeable membranes and *mediated* transport. (We shall discuss these systems in detail on pp. 200 ff. and 316 ff.). In this second case, the curve indicates that the transport mechanism is at first under-used and is later saturated so that no further increase in exterior concentration can increase the rate of transport. Perhaps the most remarkable feature of biological membranes is the ability of some mediated systems to promote *active transport* in which solute molecules are forced uphill into areas of higher concentration (see Ch. 18).

Biological membranes show great specificity in terms of which molecules they will transport and the direction in which they will transport them. Membranes exist, for example, which will accept D-sugars and reject L-sugars, and certain nerve membranes will concentrate $Na^+$ ions on one side and $K^+$ ions on the other. We shall discuss the structure of membranes, and the specificity of transport across them, in much greater detail in Chs 8 and 18.

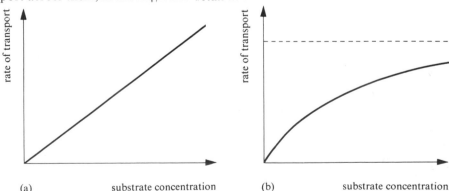

**Fig. 5.12** The relation between rate of transport and substrate concentration in (a) passive transport and (b) mediated transport. The dotted line indicates the maximum possible rate.

The cytochromes (Ch. 12) transport electrons over short distances in the cell, and many enzymes may be said to have short-range transport functions as well as catalytic ones. This is because in several metabolic pathways (p. 207) the enzymes concerned are arranged in close and defined proximity to each other in assemblies built into the membrane. The speed of operation of some of the assemblies is such that there is apparently not enough time for the reactants to diffuse in free solution from one site to another; they must be passed directly from one enzyme surface to another.

Other protein carriers exist to transport a substance from one part of an organism to another distant part. These may frequently act as stores, holding the substance carried until it is required. The leading example of such proteins is haemoglobin, an oxygen carrier (see below). There are also several proteins that carry particular metal ions about the body as well as many that carry specific molecules. All rely on the usual covalent and non-covalent forces to provide an area of high affinity for the substances to be carried.

Apart from any power that they may have of delivering the substance carried to a specific site, transport by these latter carrier proteins is superior in another way to the free diffusion of the substances alone. Many of the substances carried have their molecular weights low enough for there to be a danger of their being confused with normal waste products such as urea $((NH_2)_2CO)$. If this were to happen they would be lost by passive transport through the kidney into the urine, unless their effective molecular weight were raised by complexing to a protein.

## Haemoglobin: biological phenomena explained at the molecular level

In order to show in the greatest possible detail how the principles of Chs 3 and 4 may be applied to explain the activity of proteins we have chosen two examples. One is a small group of enzymes which will be dealt with in the next chapter; the other is haemoglobin. Haemoglobin is chosen to represent the non-enzymic proteins partly because of its medical and biological significance and partly because it presents one of the most rewarding instances of the study of biology in molecular terms.

The site of interaction between oxygen and haemoglobin is an atom of ferrous iron. It is prepared for its role in oxygen binding by being held in an environment containing five nitrogen atoms at precisely the correct orientations and distances for co-ordination to occur.

We have seen (p. 79) that the porphyrin ring is designed to provide four of these nitrogen atoms in the necessary planar configuration. The ring is capable of binding to a hydrophobic patch on the molecule; hydrogen bonding is also noticeable in fixing the ferro-porphyrin ring (haem) to the protein (globin). The nitrogen of a histidine side chain of globin takes up the fifth co-ordinating position (Fig. 5.13). The sixth position, which is called for by the geometry of co-ordinated iron, is found to be particularly suitable for binding oxygen. This binding is strong enough for the carrier function to be fulfilled, but not too strong to prevent the oxygen from being given up when required. Carbon monoxide, by contrast, will bind in the same place, but much more strongly and will thus not make room for oxygen when required. This phenomenon accounts for the toxi-

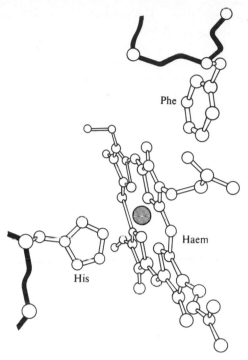

**Fig. 5.13**  The environment of the iron atom in haemoglobin

city of carbon monoxide. It is analogous to competitive inhibition in enzymes (p. 129) and the other cases of mimicry already referred to (p. 80).

Haemoglobin has been further developed, by the vertebrates at least, into an extremely subtle instrument. We can now appreciate how this has been done since the structure of the molecule has been determined in atomic detail by methods involving amino-acid sequencing and X-rays. It is very much in the interests of an animal to have an oxygen carrier that does not have a straight-

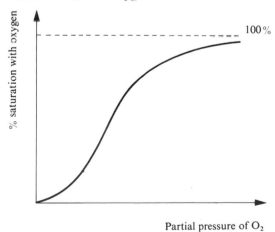

**Fig. 5.14**  The relation between partial pressure of oxygen and the degree of saturation of the haemoglobin molecule with oxygen

line relation between the amount of oxygen available (or required) and the percentage oxygenation of the carrier, but a so-called sigmoid relationship (Fig. 5.14). The advantage of the sigmoid curve is that almost the full oxygen capacity of the carrier can be called on with only a small drop in the oxygen concentration. If a straight-line relationship existed, the carrier would be hanging on to some of its oxygen at very low oxygen levels, or would still be inclined to give up some of its oxygen at very high oxygen levels. The fact that vertebrate haemoglobin shows this sigmoid relationship allows the vertebrate cell to exist in an even environment without the great fluctuations in oxygen level that would be necessary to utilize the capacity of a less sophisticated carrier to the full. These advantages must be a contributing factor to the success of the vertebrates as a group.

We can now consider how this greater efficiency of vertebrate haemoglobin is achieved. The active part of vetebrate haemoglobin is still the porphyrin ring (Fig. 5.13). The improvement in performance is managed by means of a new refinement in the way that the protein side chains influence its environment. What has happened is that four protein molecules (two of one kind ($\alpha$) and two of another, closely similar kind ($\beta$)) have aggregated into a specific quaternary

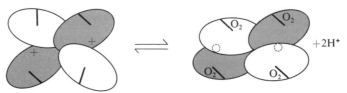

**Fig. 5.15**  The tetrameric structure of haemoglobin. An idealized view of the exposure of positively charged side-chains following the change in shape of the tetramer on **deoxygenation** of the haem rings

structure (Fig. 5.15). This structure is so arranged that when one ring is oxygenated, the disturbance to the protein structure at that point is transmitted by re-arrangement of the amino-acid side chains to the site of another haem ring. The consequent change in the environment of this second ring is such as to increase its affinity for oxygen. Once one oxygen molecule has gone on to the tetramer another is much more likely to follow. This gives the up-turn in the oxygenation curve which produces the sigmoid shape. (The phenomenon of activation (or de-activation) of a site on a quaternary structure by binding of a molecule at a distant site is also exploited in the regulation of the catalytic activity of enzymes (see p. 128).)

Haemoglobin has another important biological activity which could easily be overlooked; it is a major source of the buffering power of blood. The dissociation equilibria of its ionizable side chains (many of which project into the solution) control pH very effectively in the same way as the buffering substances used in non-biological chemistry. Even more subtly, the molecule is arranged so that, when the tetrameric structure is disturbed by de-oxygenation, different residues are exposed to compensate for the pH changes that would result from the production of carbonic acid when the oxygen is used up (Fig. 5.15).

The structure has such remarkable properties that it is clear that even small variations would be likely to have unfavourable results. A number of mutations

(p. 403) are known which have given rise in certain individuals to haemoglobin in which one amino acid has been substituted for another. If the change, as is often the case, does result in impairment of function, the individual will suffer from an anaemia. The precise clinical symptoms of this anaemia will depend on the nature and severity of the impairment of function of the molecule. Many anaemias which were first known solely as clinical patterns, sometimes of great complexity, have now been explained completely in molecular terms – a hopeful augury for other diseases. By knowing the location of particular amino-acid substitution in the haemoglobin of the patients concerned, it is becoming possible to predict the alteration in the properties of the protein from the nature of the substitution.

For example one of the residues at the haem-binding, hydrophobic patch is valine, the side chain of which is

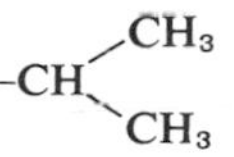

$$-\text{CH} \underset{\text{CH}_3}{\overset{\text{CH}_3}{<}}$$

(see p. 62). In some individuals a mutation has occurred in which a change of nucleotide base (involving only a few atoms) has occurred in the DNA (uracil has been replaced by cytosine: see Tables 7.1 and 26.1). The effect of this (p. 416) is to change just one of the residues in the $\beta$-chain from valine to alanine: the side chain is shortened to $-\text{CH}_3$. Thus the two methyl groups that were in contact with the ring are no longer there and the two good hydrophobic contacts made between them and the ring are lost. In Ch. 4 we valued hydrophobic contacts at about 1000 cal/mole each and so, very approximately, 2000 cal/mole is lost in the binding energy between two of the protein subunits and the haem rings. What will happen? 

We saw in Ch. 2 that

$$\Delta G^\circ = -1364 \log_{10} K$$

where $\Delta G^\circ$ is the standard free energy change of a reaction with an equilibrium constant $K$.

We can now calculate the effect on the equilibrium constant of the binding of the haem to the chain of the change from valine to alanine:

$$\text{change in } \Delta G^\circ = 1364 \text{ (change in } \log_{10} K)$$

therefore

$$\text{change in } \log_{10} K = \frac{2000}{1364} = 1.4.$$

Therefore change in $K \simeq 35$, that is, the equilibrium constant is changed by a factor of 35, enough to make all the difference between effective binding and no appreciable binding at all. Such a patient will therefore have only two usable oxygen-carrying sites per tetramer (those on the $\alpha$-chain, which is not affected by the mutation), and will lose the beneficial properties of the sigmoid curve. This fact can be related to the physical symptoms of this particular type of anaemia.

We should reflect that we have traced, by recognizable steps, the effect of a molecular rearrangement through to the symptoms of a disease that changes the whole life of an individual. It is the wish for such insights and the hope of the next stage, therapeutic methods based on them, that brings many people to work in biochemistry.

In concluding this chapter we must stress that haemoglobin has been selected solely as an example of the way in which it is becoming possible to explain quite complex biological phenomena in terms of a few relatively simple ideas. Information of this sort, though possibly a little less complete, exists for a number of other proteins, and much more is likely to become available in the near future.

# Chapter 6   Protein function II – the enzymes

The enzymes are the essential catalysts of biochemical reactions. Table 6.1 gives some idea of the range of reactions in which they are involved. No substance with the properties of an enzyme has been found which is not a protein, although many enzymes employ co-factors or prosthetic groups of a non-protein nature. It is now agreed that should such a substance ever be found, it will not be called an enzyme.

## The unique features of enzymes

The enzymes exercise their crucial role in living processes by means of three properties, in all three of which they excel the abilities of non-biological catalysts to a spectacular degree. These three are their great catalytic power, by which they increase the rates of reactions by very many orders of magnitude; their power to discriminate between closely related substances; and their amenability to control of the extent to which they will accelerate reactions.

The *catalytic power* of the enzymes is prodigious – frequently several orders of magnitude better than the corresponding non-biological catalysts. For example, one molecule of catalase, an enzyme which destroys hydrogen peroxide, is able to deal with approximately 5 million molecules of $H_2O_2$ per minute. This represents an acceleration of at least $10^{14}$ over the rate of the reaction without a catalyst. It is also many times better than the non-living catalysts of the reaction, such as finely divided platinum.* This is the typical pattern: enzymes bring about an extremely large increase in the rate of reaction, much larger even than that possible with the best non-living materials. Two other examples may be given: the enzymes hexokinase and phosphorylase which catalyse important steps in the breakdown of carbohydrate molecules (Ch. 14). It has been calculated that hexokinase accelerates its reaction (the phosphorylation of glucose (p. 259)) by at least $1.3 \times 10^{10}$ times and phosphorylase accelerates its reaction (the liberation of glucose units from the polymeric form in which they are stored (p. 268)) by at least $4 \times 10^{12}$ times.

The ability of enzymes to discriminate is equally striking. We have already mentioned (p. 49) a case in which living processes distinguish between different

* Catalase depends for its catalytic activity on an atom of iron, which is activated by the amino-acid side chains around it. The degree of activation is shown by the fact that 1 mg of iron in catalase is as effective a catalyst of the decomposition of $H_2O_2$ as $10^4$ kg of inorganic iron.

**Table 6.1**  A systematic classification of the enzymes. Taken from the recommendation of the International Union of Biochemical Societies

1. *Oxidoreductases*
   1.1   Acting on the $>$CH—OH group of donors
   1.2   Acting on the aldehyde or keto group of donors
   1.3   Acting on the $>$CH—CH$<$ group of donors
   1.4   Acting on the $>$CH—NH$_2$ group of donors
   1.5   Acting on the $\geqslant$C—NH—group of donors
   1.6   Acting on reduced NAD or NADP as donor
   1.7   Acting on other nitrogenous compounds as donors
   1.8   Acting on sulphur groups of donors
   1.9   Acting on haem groups of donors
   1.10  Acting on diphenols and related substances as donors
   1.11  Acting on H$_2$O$_2$ as acceptor
   1.12  Acting on hydrogen as donor
   1.13  Acting on single donors with incorporation of oxygen (oxygenases)
   1.14  Acting on paired donors with incorporation of oxygen into one donor (hydroxylases)

2. *Transferases*
   2.1   Transferring one-carbon groups such as methyl or carboxyl groups
   2.2   Transferring aldehydic or ketonic residues
   2.3   Acyltransferases
   2.4   Glycosyltransferases
   2.5   Transferring alkyl or related groups
   2.6   Transferring nitrogenous groups
   2.7   Transferring phosphorus-containing groups
   2.8   Transferring sulphur-containing groups

3. *Hydrolases*
   3.1   Acting on ester bonds
   3.2   Acting on glycosyl compounds
   3.3   Acting on thioether bonds
   3.4   Acting on peptide bonds (peptide hydrolases)
   3.5   Acting on C—N bonds other than peptide bonds
   3.6   Acting on acid-anhydride bonds
   3.7   Acting on C—C bonds
   3.8   Acting on halide bonds
   3.9   Acting on P—N bonds

4. *Lyases* (bond-breaking reactions)
   4.1   Carbon–carbon lyases
   4.2   Carbon–oxygen lyases
   4.3   Carbon–nitrogen lyases
   4.4   Carbon–sulphur lyases
   4.5   Carbon–halide lyases
   4.99  Other lyases

5. *Isomerases*
   5.1   Racemases and epimerases
   5.2   *Cis–trans* isomerases
   5.3   Intramolecular oxidoreductases
   5.4   Intramolecular transferases
   5.5   Intramolecular lyases
   5.99  Other isomerases

6. *Ligases* (bond-forming reactions)
   6.1   Forming C—O bonds
   6.2   Forming C—S bonds
   6.3   Forming C—N bonds
   6.4   Forming C—C bonds

optical isomers, such a distinction being almost the ultimate problem of ordinary preparative chemistry. This ability is an example of the discriminating power, or *specificity*, of enzymes. There are many other examples of the ability to distinguish between possible substrates, however closely related they may be. The enzyme urease, for instance, possesses exceptionally high specificity. It is very active in catalysing the breakdown of urea according to the equation

$$CO(NH_2)_2 + H_2O \rightleftharpoons CO_2 + 2NH_3.$$

Many compounds exist in which substituents (such as methyl and other alkyl groups) are placed on the —$NH_2$ groups of urea. In others the oxygen atom is replaced by sulphur. None of these modified substrates is touched by the enzyme.

Other enzymes have a somewhat broader specificity while still showing a preference for a particular group of substrates. For example, consider the digestive enzymes that catalyse the breakdown of proteins by hydrolysis of the peptide bond.

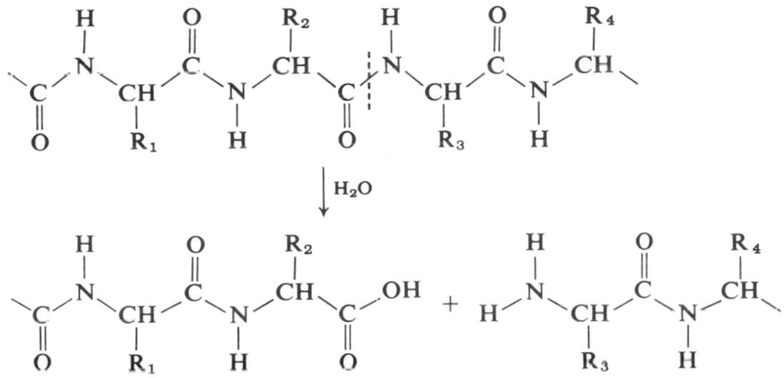

Fig. 6.1   Cleavage of the peptide bond by proteases (see text)

Such enzymes tend to show specificity in that they will cleave only peptide bonds in which certain classes of amino acids are involved. We saw on p. 83 that trypsin will hydrolyse only those peptide bonds formed by the carboxyl groups of the long-chain basic amino acids lysine and arginine (Fig. 6.2). Chymotrypsin, on the other hand, rejects these peptide bonds and attacks those formed by the carboxyl groups of a number of other amino acids. Its preference is largely that $R_2$ (Fig. 6.1) should be the side chain of certain of the hydrophobic amino acids, notably the aromatic ones. (Other proteolytic enzymes, for instance pepsin, show specificity for $R_3$ rather than $R_2$.)

Finally, there may be a relative lack of specificity. Some enzymes exist which will cleave nearly all peptide bonds, others will hydrolyse nearly all esters, and so on.

It should be noted that the specificity extends to the *products*. That is, in almost all cases a given enzyme will catalyse only one type of reaction with a particular substrate: it will take it through to one product only. (This does not mean that an enzyme can never catalyse more than one type of reaction when given more than one type of *substrate*; e.g. proteolytic enzymes usually become esterases when confronted with peptide or amino acid esters.)

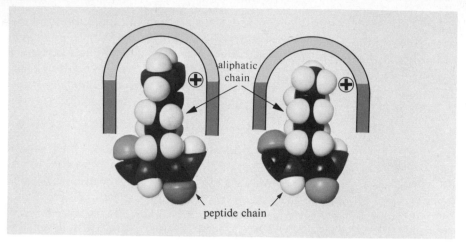

**Fig. 6.2**  The specificity of trypsin (see text)

The potential of this great catalytic power coupled to equally striking specificity has extremely important consequences. There is an almost infinite number of reactions which it would be energetically possible for a chemical compound (glucose, say) to undergo. The products of many of these reactions in their turn may each react in a large number of different ways, and so on for their products (Fig. 6.4). As we saw in Ch. 2, enzymes cannot make energetically unfavourable reactions happen; they can only accelerate reactions that are possible. We can speak of the kinetic constants of one of the possible reactions being greatly increased over those of the reactions that might compete for the same substrate. This acceleration is so great that it amounts to a selection process, and in the presence of suitable enzymes the starting material is steered through the maze of reactions as though the ever-multiplying possibilities did not exist.

This ability of enzymes to organize the chemical processes of living organisms, by virtue of their catalytic power and specificity, is increased still further by their amenability to *control* of the efficiency with which they accelerate reactions. We

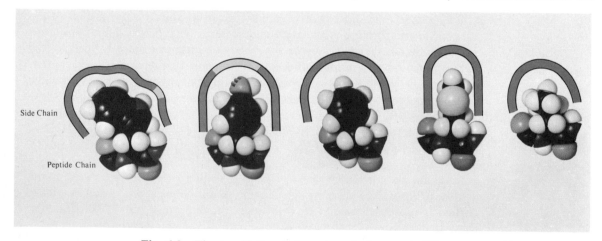

**Fig. 6.3**  The specificity of chymotrypsin (see text). (The tones used in these figures have their usual meanings (see Fig. 3.7).)

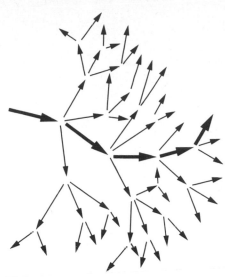

**Fig. 6.4** The ability of an enzyme system to steer a substrate along a particular series of reactions

shall refer to this characteristic in more detail later (p. 125 ff. and Ch. 29). For the moment we may note that if mechanisms exist that will turn the catalytic power of enzymes on and off, they will greatly enhance the capability of living systems to make the most of the networks of the type shown in Fig. 6.4. This will be particularly true if, as is often the case, the signals that turn them on and off are changes in concentration of later products in the chain of reactions; such an arrangement can lead to the efficient, self-regulating system that we know living matter to be. We might add here that the control of the activity of enzymes can also reside in their spatial arrangement within the cell. Many of the series of enzymes that catalyse a chain of reactions are bound together in multi-enzyme assemblies for greater efficiency in handing the intermediate compounds down the chain (p. 351).

## How enzymes work

We shall now look in turn at these three features of the enzymes in terms of the ideas of Ch. 4, much as we did in Ch. 5 for the properties of haemoglobin. Most enzymes are globular proteins and, at a brief glance, their molecular structures (where these are known) are not strikingly different from those of the non-enzymic proteins that we have considered.

However, we shall find, in making use of the dramatic improvement in the quality and quantity of our knowledge of enzyme structure that has been achieved in recent years, that some quite reasonable explanations can now be put forward for the phenomena that we have described.

We shall take as our principal examples the proteolytic enzymes which we have already mentioned, and another enzyme, lysozyme, which attacks the cell walls of certain bacteria (Fig. 6.5).

All of these happen to be extra-cellular enzymes of hydrolysis. This choice may

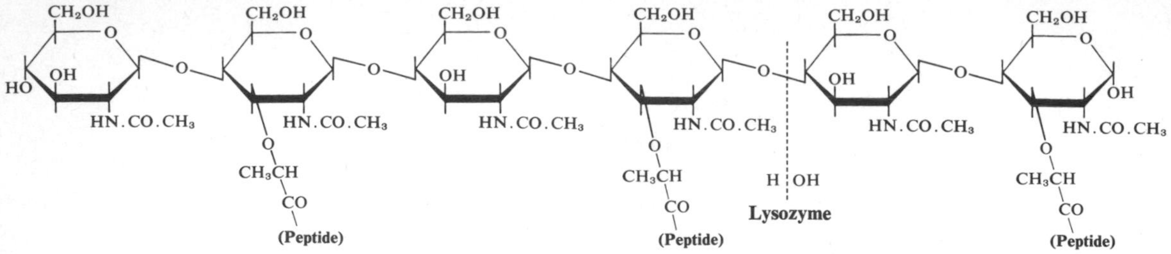

**Fig. 6.5** The cleavage of the mucopeptide chain by lysozyme (see text and compare Fig. 8.12c)

appear to involve an arbitrary concentration on a rather restricted class of enzymes at the expense of, say, the enzymes catalysing the reactions of intermediary metabolism which will be described in later chapters. Unfortunately it is unavoidable for two reasons. The first is that since the extra-cellular ones must exist in a more harsh environment, they are tougher and are therefore easier to handle. Secondly they are usually also produced by the organism in larger quantities and are therefore more easily available. However, we shall try, wherever possible, to bring in some of the information that is becoming available about the intracellular enzymes of metabolism. We shall refer, for example, to triose phosphate isomerase which we met in Ch. 2 (pp. 19 and 37) and which is seen in its proper metabolic context in Chs 14 and 19. It is important to include some examples of these intracellular enzymes, since many, in distinction to most of the extracellular ones, have a quaternary structure (p. 75) and receive from this fact the same sort of enhancement of competence that we saw for haemoglobin.

We shall observe that the substrate interacts with the enzyme at a specially tailored surface, the *active site*. Any specificity for the substrate will be conferred by the selective binding properties of the *specificity site*, which will usually be adjacent to or partly included in the active site. Control of enzyme efficiency is associated with the binding of other substances at one or other of these sites or, in the more highly developed enzymes, at other special *allosteric sites* designed to receive them.

## The basis of catalytic power

We have already laid the foundation for this section in Ch. 2. You will recall that the rate of a chemical reaction is governed by the equation:

$$\text{Rate} = k_{+1} \times \text{product of concentrations of reactants} \qquad (6.1)$$

that is

$$\text{Rate} = \frac{RT}{Nh} \times e^{-\Delta G \ddagger /RT} \times \begin{array}{c}\text{product of concentration}\\\text{of reactants}\end{array} \qquad (6.2)$$

which it is sometimes convenient to expand to

$$\text{Rate} = \frac{RT}{Nh} \times e^{\Delta S \ddagger /R} \times e^{-\Delta H \ddagger /RT} \times \begin{array}{c}\text{product of concentration}\\\text{of reactants}.\end{array} \qquad (6.3)$$

This equation contains all the factors that an enzyme can modify in order to increase the rate of a reaction. We must, therefore, seek an explanation of the action of enzymes in terms of these factors. Leaving aside the true constants, and noting that temperature remains more or less constant in biological systems, we are left with $\Delta G^{\ddagger} (= \Delta H^{\ddagger} - T\Delta S^{\ddagger})$ and the concentration as the only factors that can be changed. We shall pick out a number of ways in which the factors are indeed found to be variable. Although we list them separately, it will become obvious that some are to a certain extent just different aspects of others. Some of the detail of this very new branch of knowledge may have to be revised. Nonetheless we feel that the treatment that we give here does not misrepresent the situation too much.

### *Enhancement of rate (i) – lowering the barrier and raising the ground state*

Figure 6.6 recalls the position relating to $\Delta G^{\ddagger}$. Clearly, as equation 6.2 shows, for the substrate to pass more easily from A to B, $\Delta G^{\ddagger}$, the net height of the barrier must be lowered. This could be done in several ways. First, the peak could be lowered – this is, the available free energy of the activated state could be lowered. Since $\Delta G^{\circ}$ is a measure of the tendency of a process (in this case the decomposition of the intermediate) to occur, 'lowering the available free energy' means 'stabilize' as it does in everyday experience. To 'stabilize' means to bind, that is to tie down (think of a tent on a mountain top).

We would therefore expect to see enzymes *binding* the activated complex, since this represents the peak. They should be bound *more* strongly than the substrate itself or the substrate position on the energy diagram will be lowered even further than the peak and a *net increase* in the height of the peak will result. We used equation (2.39) on p. 39 to show that we would have an approximately $10^{10}$-fold increase in rate if we lowered the height of the peak by 14 000 cal/mole. This means, in the terms that we have just discussed the matter, that if the transition state is bound to the enzyme with a binding energy of 14 000 cal/mole, we will achieve the increase of rate of $10^{10}$. Our survey of the strength of interactions in Ch. 4 suggests that a combination of a few hydrophobic interactions, hydrogen bonds and van de Waals' forces acting at a very close range could achieve binding energies of this order. However, the forces will have to be

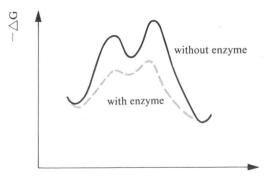

**Fig. 6.6** A typical example of the effect of an enzyme on the free-energy changes that occur during a chemical reaction.

deployed with great precision in order to engage properly with the intermediate, which is small compared with the enzyme.

In general, when we consider the structure of the enzymes we will look for structural features that appear to bind the suggested reaction intermediate – we shall see in a moment that just such features are recognizable. Unfortunately the activated complex cannot be synthesized and used for experiments to check these ideas because of its intrinsically unstable position on the peak of the energy diagram. However, stable compounds can sometimes be made that mimic both the arrangement of atoms in the complex and, to some extent, the displaced charges.

An example of the use of such compounds occurs in studies on the enzyme triose phosphate isomerase. Here, the substrate (*a*, Fig. 6.7) binds to the active

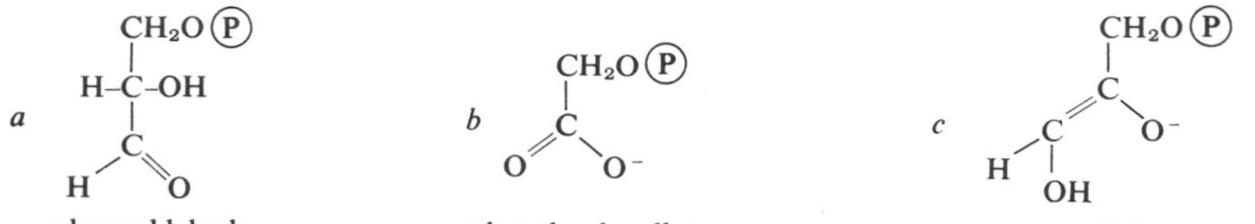

**Fig. 6.7** Triosophosphate isomerase: stable analogues of the reaction intermediate (see text). Compare this figure with the proposed reaction mechanism for this enzyme (p. 37)

site with a dissociation constant (p. 14) of approximately $10^{-3}$ M. It is therefore very interesting to note that the substance *b* (Fig. 6.7), which although a stable compound resembles in many respects the proposed intermediate state *c*, binds with even greater tenacity than the substrate, the constant being approximately $10^{-6}$ M.

Since we can reduce the overall height of the energy barrier by lowering the peak, it is reasonable to ask if we can reduce the barrier still further by raising the energy level of the starting state.

Converting this to the language of stability and binding, as before, this can be translated to mean that the starting state would have to be made *less* stable, that is it must be bound in a strained, or deformed, state. (This too can be done when tying down a tent, but we do not recommend it.)

Note that the conformation of the enzyme will change when it locks on to the substrate as well as *vice versa*: the change in the conformation of the one will bring about a change in that of the other. This has the effect of transferring some of the conformational energy of the enzyme to the substrate and provides one means of raising the energy of the starting state, as required.

We are now in a position to appreciate some of the features of the interaction between the active and binding sites of lysozyme on the one hand, and its substrate on the other, that lead to catalysis.

When the substrate binds to the enzyme (Fig. 6.8) the convoluted surface and the many forces that await specific parts of the incoming substrate tie it down

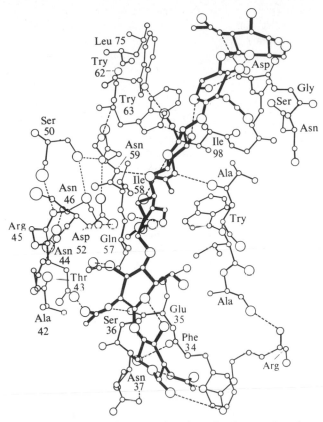

**Fig. 6.8** A computer-drawn diagram showing the interaction between a hexa-saccharide substrate (bold lines) and the amino acid side chains of the binding site of lysozyme. Oxygen atoms are shown as large circles and nitrogen atoms by slightly smaller ones. Hydrogen atoms are omitted. Hydrogen bonds are represented by dotted lines. Note the complexity and specificity of the interaction. Some of the individual bonds are easily traced, e.g. the hydrogen bond between the lengthy substituent on the second sugar ring from the bottom and the side chain of threonine 43. Others are less easy to follow without a stereoscopic drawing

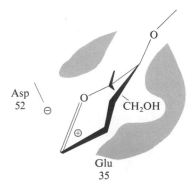

**Fig. 6.9** The stabilization of the carbonium-ion intermediate on the surface of the active site of lysozyme

firmly and strain the fourth of the six sugar rings. It is bound halfway between the chair and the boat configuration (Fig. 8.7). This is very similar to the abnormal configuration of the transition state Fig. 6.9. We see here an example of the raising of the ground state on binding of the substrate to the enzyme.

In Fig. 6.9 the chain has been cleaved at the site of the strained ring by a proton donated by the carboxylic acid side chain on the right. (This side chain belongs to a glutamic acid residue. It is the thirty-fifth residue from the amino terminus, there being 129 residues in all. We shall deal with its special properties below.) As a result of the cleavage, we have the activated complex in Fig. 6.9. It is stabilized by the good van der Waals' contacts that it makes with the residues behind the active site (upon which the ring was bent to its deformed shape in the first place) and, probably, by the proximity of the negative charge of the left-hand carboxylic acid group, which allows an ionic bond to form. (This group belongs to an aspartic acid residue. It is number 52 of the 129, but is brought near number 35 by the folding of the chain. We shall deal with it below, also.) Here then we have a good example of that binding of the transition state which we expected to find. As might have been expected the synthetic

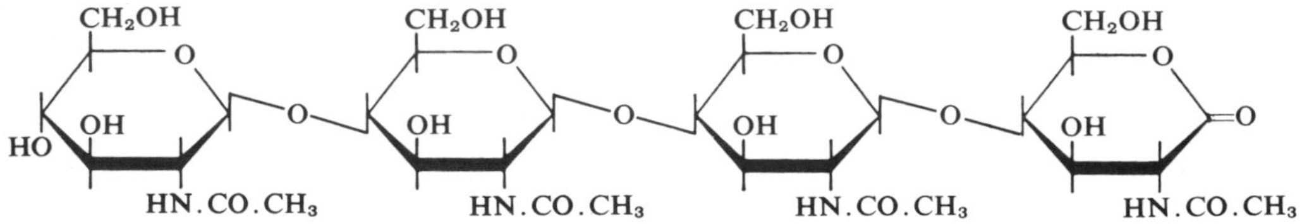

**Fig. 6.10** Lysozyme: a stable analogue of the reaction intermediate (see text)

compound in Fig. 6.10, which resembles the activated complex as far as a stable compound may, is strongly bound to the active site.

Both these effects on $\Delta G^{\ddagger}$ rely on the properties of enzymes as binding *surfaces*, serving to emphasize the fact that the enzymes, like some non-biological materials (such as platinum), belong to the class of catalysts that depend on surface properties. The superiority of the enzymes derives from the much better fit that their surfaces make with their substrates. The improved fit results from the fact that the enzyme surface, unlike that of the non-biological catalysts, has been continuously modified and improved for its specific task over thousands of millions of years of evolution (Ch. 27).

*Enhancement of rate (ii) – regions of abnormal reactivity*

The interplay of bonding forces from residues on the enzyme must not be thought of as acting solely on the part of the complex that derives from the substrate. The groups on the enzyme that become most closely involved with the complex are themselves frequently in unusual environments and are subjected to intense bonding from other groups in the molecule. Such bonding is able to modify their properties drastically.

We may cite several examples of this. The carboxylic acid group of residue 35 of lysozyme is in an exceptionally hydrophobic environment for a polar group. In such an environment the dielectric constant (p. 52) is lower than in water and therefore from equation (3.1), p. 52 the force restraining the $H^+$ from leaving is much greater than usual. That is, it is more difficult for the —COOH group to ionize. It is believed that the p$K$ has been raised from about 4.5 to about 6.5. (If you will compare for a moment the sections in Ch. 2 on p$K$ and on binding constant, you will see that this p$K$ change is equivalent to a change in the dissociation constant for protons from $3 \times 10^{-5}$ (antilog 4.5) to $3 \times 10^{-7}$ (antilog 6.5).) Thus the group will be able to hold on to its proton at pH values that will cause all others to have lost theirs. This proton is therefore available for use in the hydrolysis, even at neutral pH. Conversely, aspartic acid 52 makes an unhindered contact with water, ionizes freely (p$K$ less than 3.5) and is thus very ready to form the ionic bond shown in Fig. 6.9.

Another example of enhanced reactivity is provided by certain of the proteolytic enzymes. Trypsin and chymotrypsin (and also, among other enzymes, thrombin, p. 98) catalyse the breakdown of the protein chain in the three stages shown in Fig. 6.11. You will see that what is involved is essentially the acylation of the active-site serine by the left hand of the substrate chain, to form an ester, and then the hydrolysis of this ester by water. This serine is one of thirty-four in trypsin and there are in addition of course other, closely similar, threonine residues. Nonetheless as a result of the influences to which it is subjected by other parts of the enzyme molecule it is the only one to show the abnormal chemical reactivity required.

Triose phosphate dehydrogenase catalyses an important reaction in carbohydrate metabolism (pp. 262 and 322). It has at its active site the sulphur analogue of serine, namely cysteine. The part that the acylation and deacylation of this residue play in the reaction is shown on p. 262.

(The abnormal reactivities can be exploited by the experimentalist both to locate the active-site residues within the amino-acid sequence of the enzyme and as a means of inhibiting a particular enzyme while leaving others able to operate. Both of these applications, which will be described in full in later pages, depend on the fact that as a result of the enhanced reactivity of these residues, they will combine with reagents that leave all the other similar, but non-activated, side chains alone.)

*Enhancement of rate (iii) – orientation effects*

The effects on rate that we have been considering so far have been on $\Delta G^\ddagger$ as a whole, with no easily discernible emphasis on either $\Delta H^\ddagger$ or $\Delta S^\ddagger$ (equation (6.3), above). There are, however, effects that are more purely entropic in character.

We touched on one such effect briefly in Ch. 2 (p. 33) in which water molecules were displaced from the binding site by the incoming substrate and acquired greater freedom to move. This is a *decrease* in order on forming the enzyme-substrate complex, in other words an *increase* in $\Delta S^\ddagger$. This will lead to a decrease in $\Delta G^\ddagger$ and so an increase in $k_{+1}$.

There is an additional example of an entropic contribution in which surface properties are well situated to play a role. In order to understand it, we should realize that when two molecules collide in free solution, they need not react even if there is both a possible reaction for them to undergo and sufficient energy in the collision to send them over the barrier. This is because there is a further factor that has to be satisfied, namely that the groups on the reacting molecules are in the correct orientation to one another. There is much argument about how severe this restriction is. In some cases the colliding molecules have to approach each other along a very precise straight line indeed. In the most extreme cases a deviation from this line of perhaps 1/1000 of a degree effectively prevents the overlap and hybridization of the molecular orbitals that is essential to the formation of a chemical bond. As a result only one in approximately $10^4$ collisions

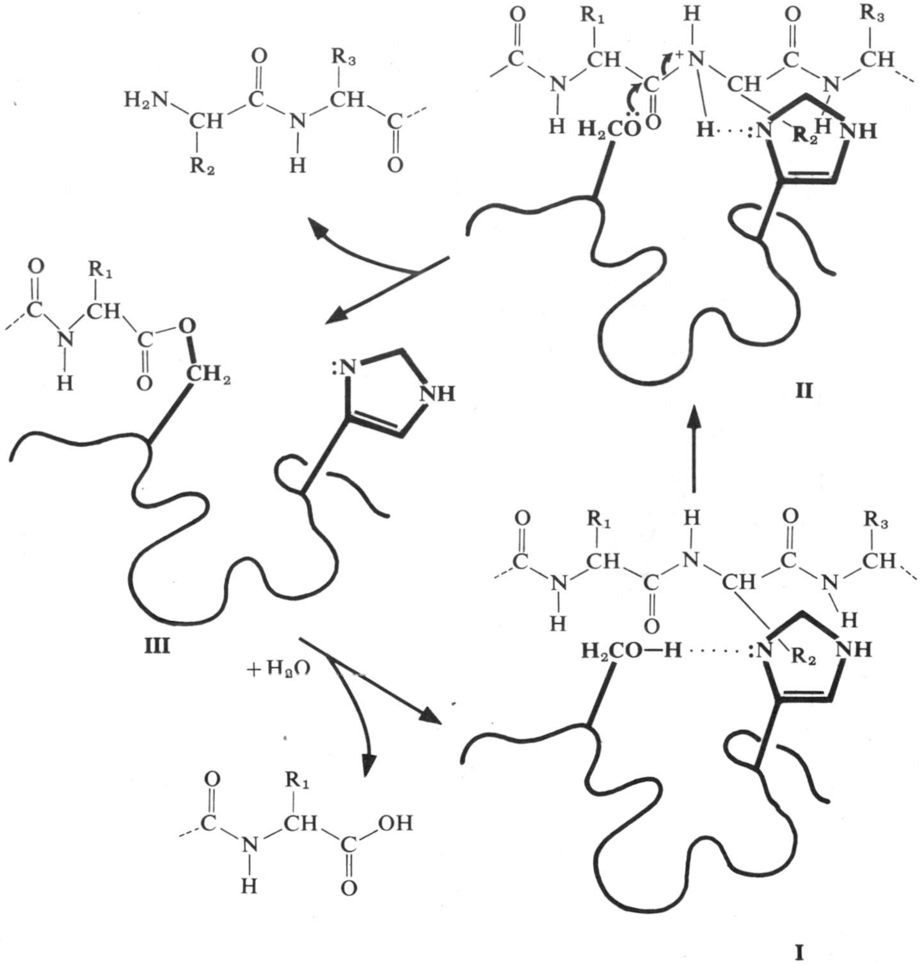

**Fig. 6.11** A proposed mechanism of action of serine proteases. In I the substrate approaches the enzyme. There is a hydrogen bond between the serine —OH and the imidazole nitrogen atom of histidine 57

In II the hydrogen bond transfers to the amino group of the substrate. This places the $>$C$=$O in a position to be attacked by the —OH group of the serine. The right-hand fragment of the substrate is released as a free polypeptide. The left-hand fragment is attached as an acyl group on the serine (III). This acyl group is then removed by hydrolysis and liberated as a free polypeptide. The enzyme is then ready to act on another molecule of substrate.

(Single arrows have been used to indicate the order of events; this does not mean that the reactions are irreversible)

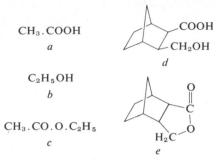

$$CH_3.COOH$$
*a*

$$C_2H_5OH$$
*b*

$$CH_3.CO.O.C_2H_5$$
*c*

**Fig. 6.12**  An intramolecular reaction compared with the analogous intermolecular reaction

are useful. (It must be said that in many other cases much greater freedom is permissible.)

Experimental evidence for the intensity of the restriction in some cases is found by studying the behaviour of small organic compounds. Compounds *a* and *b* (Fig. 6.12) which only react when colliding along the proper line produce an ester (compound *c*). Compound *d* undergoes the same reaction *intramolecularly* to form a cyclic ester or *lactone, e*. The constraints introduced by the bridged ring system hold the two reacting groups (the equivalent of *a* and *b*) tightly in exactly the correct configuration so that reaction occurs virtually every time the groups collide. The $k_{+1}$ is now about $10^4$ greater. Precise interpretations of this type of result differ, but it is an enhancement of rate of the approximate order of size that we might have predicted if all, rather than 1 in $10^4$ collisions, are along the straight line.

Whether the restriction is severe or not, whenever the enzyme surface can steer the molecules toward each other in the correct orientation, this will increase the number of useful collisions and thus increase the reaction rate. The extent of the increase will depend on the degree of restriction. In lysozyme the directional factors are principally concerned with holding the chain in the right position to receive the proton donated from residue 35 and in providing a route for the attack on the activated complex by $OH^-$. In the serine proteases, the directional factors are presumably concerned with steering the incoming chain in to acylate the active site, and providing the means by which the acylated intermediate is hydrolysed.

Thus we have a true entropic contribution to the reduction of $\Delta G^{\ddagger}$ (eqn. 6.3). We saw on p. 32 that to restrict the number of ways that molecules can arrange themselves is to reduce entropy. We saw further that such a restriction, or decrease in entropy, gives rise to an increase in $\Delta G$. In other words the steering together of reactants in a fixed orientation lowers the entropy of the starting state and $\Delta S^{\ddagger}$ is raised. As a result the effective $\Delta G^{\ddagger}$ decreases and $k_{+1}$ increases.

### Enhancement of rate (iv) – increasing the local concentration

Having listed a number of ways in which $\Delta G^{\ddagger}$ can be varied we are left with the manipulation of concentration. In one sense the orientation effect that we have dealt with above is equivalent to an increase in the effective local concentration of the reactants. However, there are even more direct instances that we may mention. On p. 262 we shall describe a case in which an enzyme, triose phosphate dehydrogenase, is able to bind one of two participants in a reaction and store it until the other reactant arrives. This has the effect of raising the effective local concentration of the reactant (which is scarce) and this will contribute to the enhancement of the reaction rate.

Many similar examples are known. In particular, reactions that involve protons as the active agent may be assisted by the existence of some specific site on the protein that binds the proton. In lysozyme we saw that as a consequence of an unusual environment the carboxylic residue of glutamic acid 35 has its pK shifted from about 4.5 to about 6.5. We can look at this a little more closely now and see that it is relevant also in the context of effecting an increase in concentration. Thus at pH 6.5 (at which the enzyme is active) the group remains 50 per cent protonated. This 50 per cent protonation would hold for the group in a normal environment at pH 4.5, whereas at pH 6.5 only one of the residues in 100 would be protonated (p. 12). The effect on the behaviour of the protein is thus equivalent to a drop in pH of 2 units, but without the disastrous consequences that a drop of this magnitude would have if it were general throughout the cell.

It is worth pausing to remember that the concentration of protons at physiological pH is quite small: $10^{-7}$ mole/litre at pH 7 by definition. The volume of the cell is so small that this will be equivalent to a hundred or so protons per cell. If, as is the case, there are at any instant thousands of enzymes in the cell all needing a proton, some enhancement of this sort will be extremely useful, even though fresh protons are available from the dissociation of $H_2O$.

All these factors may be summarized in Table 6.2, together with a very rough estimate of the magnitude of the contribution that they might make to the speeding up of chemical reactions. These are general estimates and are not intended to refer specifically to the enzymes that we have used as examples.

**Table 6.2**   Some of the factors that enable an enzyme to accelerate reactions

| | |
|---|---|
| Binding of activated complex (lowering the peak) | |
| Raising the ground state: enhanced reactivity of active site groups | easily $10^5$ |
| Maintaining the correct orientation of reactants | $10$–$10^3$ |
| Storing reactants | $1$–$10^3$ |

Multiplying these together we obtain an approximate value for the sort of enhancement of the rate of an enzyme-catalysed reaction over the rate in the absence of an enzyme. This is of the order of $10^9$–$10^{11}$. This is admittedly lower than the actual enhancements (as we have seen, rate enhancements of $10^{12}$–$10^{14}$ have been quoted), but close enough to show that the factors that we have considered may be among the most important ones.

Before leaving the question of the enhancement of rate let us stress again that the enzymes do not alter the position of equilibrium, simply the rate, since

$$\frac{k_{+1}}{k_{-1}} = K \text{ (eqn 2.37)}$$

the enzyme must speed up the reverse reaction to an equal extent. This effect is sometimes important as the organism often needs to operate processes in either direction and thus would need to make use of the accelerating power in either direction.

## Specificity

This, too, is a matter of binding between enzyme and substrate. We saw in Ch. 3 that a great variety of forces exist, some highly directional, some less so, some short range, some long range. We saw also that different types of force were effective in bringing together different types of groups. The enzymes, with their complex structure, are superbly adapted to deploy these forces so as to bind very strongly to a substrate of the desired shape and constitution and not to any other. The analogy has often been made between this binding and a key (the substrate) fitting into its particular lock (the enzyme), much as we saw for the antigen–antibody combination. As in that case, the analogy must be extended to allow the lock to wrap itself around the key, since the conformation of both will be flexible to some extent, and each will change somewhat in response to the other.

We have seen (Fig. 6.8) a little of the intricate pattern of bonding between enzyme and substrate in lysozyme. It is difficult to imagine how any substrate that differed at all greatly from the natural one could be accepted in its place.

We might expect trypsin and chymotrypsin to be interesting examples since their specificities differ from one another in ways that are well understood (p. 112). This expectation is justified, and recently obtained crystallographic results enable us to see how the structures of the two enzymes (which are, owing to their common genetic origin (Ch. 27), very similar overall) vary in just the way necessary to bring about the difference in specificity (Fig. 6.13).

The compound indole (Fig. 6.14a) resembles the side chain of tryptophan, one of the amino acids preferred by chymotrypsin. An elegant confirmation of the ideas that we have just put forward is provided by the effect of indole on the action of chymotrypsin. It is found that it stimulates the enzyme to cleave a much wider range of substrates, including simple aliphatic esters. We must conclude that the indole binds to the specificity site and that the 'lock folding around the key', of which we spoke, then occurs to produce the active configuration at the active site (Fig. 6.14b).

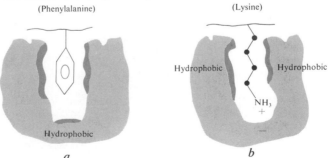

**Fig. 6.13**  The specificity sites of (a) chymotrypsin and (b) trypsin

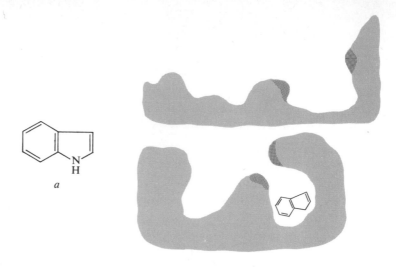

*a*

*b*

**Fig. 6.14** (a) Indole, (b) idealized view of the folding of the binding site of chymotrypsin around its substrate and the resulting change in the configuration of the active site. The residues forming the active site are symbolized by the darker regions

We have thus returned to the concept of a *specificity site* (which determines which molecules will be allowed to react), and an *active*, or *catalytic site*, which determines how and at what speed the reaction takes place. These two sites, of course, may largely overlap. In the lysozyme molecule, as in many other enzymes that have long chain substrates, the specificity site is the deep groove that is built into the face of the molecule, and the active site, the region near residues 35 and 52, lies within it (Fig. 6.15).

We should remember that the residues that make up these two types of site are taken from various parts of the amino-acid sequence, brought together by the convolution of the peptide backbone. We describe on p. 129 how these sites are located in the molecule.

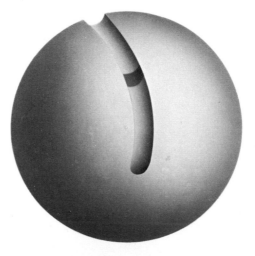

**Fig. 6.15** The active and specificity sites of lysozyme

## Factors that control the efficiency of enzyme action

We mentioned at the beginning of the chapter that the enzymes were amenable to control of various types, and that this amenability greatly increased their usefulness to the living organism. We will now discuss and seek to explain in molecular terms a number of the means by which control is managed. In some cases, it is once again a matter of considering the interaction between enzyme and substrate. However, we will more frequently find that some external factor is operating: we shall discuss, under this heading, the influence of activators, inhibitors, temperature and pH. Some of the more mathematical aspects of enzyme kinetics are left to an appendix. We are concerned here to describe and explain the phenomena that are encountered in simple terms and to place them in their proper biological context.

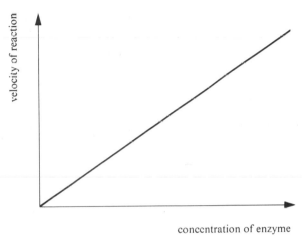

**Fig. 6.16** The relationship between the velocity of an enzyme-catalysed reaction and the concentration of enzyme

### *The quantity of enzyme*

This effect of varying enzyme concentration is simply described (Fig. 6.16). We shall see in Ch. 28 that control of metabolic processes is often achieved by varying the effective concentration of enzyme.

### *The quantity of substrate*

It will be seen in Appendix I that the effect of substrate concentration (Fig. 6.17) is a matter that helps us more in the experimental determination of some of the important parameters of enzyme action than in understanding how reactions are organized in the cell. For the moment we should simply note that while reactions increase in rate in proportion to the amount of enzyme present, there is, for a fixed amount of enzyme, a concentration of substrate beyond which a further increase has no effect. This is because the enzyme sites are then saturated with substrate. This behaviour obviously has consequences in biological systems, because it sets an upper limit to the rate of throughput of material. We shall

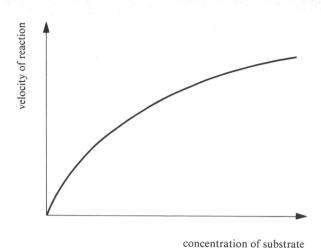

**Fig. 6.17**   The relationship between the velocity of an enzyme-catalysed reaction and the concentration of substrate

meet this saturation phenomenon again when we come to look at certain types of transport across biological membranes (Ch. 18). The failure to respond to a further increase in concentration of substrate is due to the fact that all available enzyme molecules, or, in the case of the membranes, transporting systems, have been brought into continuous operation by the high levels of substrate already existing.

*Activators*

Frequently, the interaction between enzyme and substrate is mediated by some small ion or molecule. We saw a number of molecules of this type in Table 4.2. We may add to this list the particular ions needed by the reaction shown on pp. 259, 264 and 315. There is often a degree of specificity shown for these ions, e.g. if $Cl^-$ will do, $Br^-$ often will not. This specificity is due to the carefully tailored size of the binding site, or the need for an electric field of a particular intensity. An enzyme that requires one of these ions or co-factors will not function if it is not present and the cell has to ensure a continuing supply, replenishing any losses by ingestion or, in appropriate cases, by synthesis.

Thus, operations that tend to remove these activators, such as dialysis (p. 544) or the addition of a chelating compound (Fig. 6.18, below), will inhibit enzyme action. In discussing the role of ions in activation, we must distinguish cases of *stabilization*. $Ca^{2+}$, in particular, binds to special negatively-charged sites on a number of enzymes (trypsin, an enzyme that operates in the denaturing

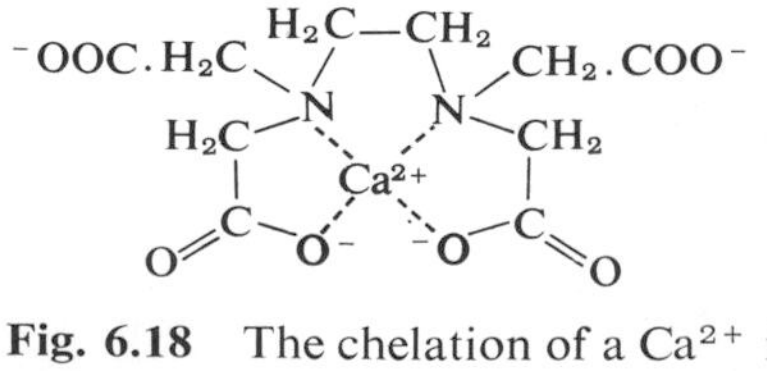

$$^-OOC.H_2C \quad \overset{H_2C-CH_2}{\underset{N}{\diagdown}} \quad CH_2.COO^-$$

**Fig. 6.18**   The chelation of a $Ca^{2+}$ ion by ethylenediamine tetraacetate

and digesting environment of the gut, is an example). Binding at these sites does not affect the activity directly, but helps to lock up the native configuration against denaturing agents.

Another type of activator has been recognized in recent years. This results from the action of the so-called *stimulatory effector*. You should by now have very well in mind the fact that the three-dimensional structure of a protein is flexible and responds to the binding of any other molecule. In many (but probably by no means all) cases, the resting structure of the molecule is not the active one. In lysozyme, for example, the whole of one side of the cleft swings over about 0.6 Å when the substrate is bound, and only then is the configuration correct and catalysis able to occur. This is a case of the substrate doing the job of inducing the correct configuration for itself. The movement is small, but as we have said the fit may involve short-range forces and must be a very good one indeed. Stimulatory effectors are thought to induce changes in structure that lead to enhanced activity. As a consequence of the new pattern of forces resulting from the binding of the effector, the enzyme assumes its correct shape for catalysis and becomes more active. These effectors may be the substrate molecule itself (in which case the phenomenon is called *homotropic regulation*), or some other molecule (in which case the phenomenon is called *heterotropic regulation*). Some enzymes show both types of regulation.

We have already seen a case of a stimulatory homotropic effect in the oxygen binding of haemoglobin (here the site of action of the effector is the 'active site' as well). We stressed, in that example, that it was changes in the quaternary

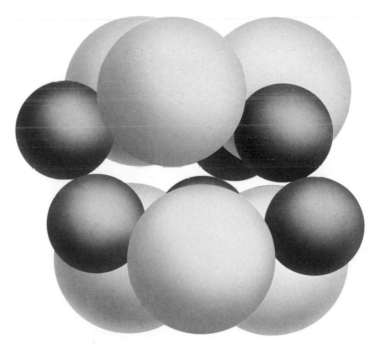

**Fig. 6.19** The possible subunit structure of aspartate transcarbamylase, an allosterically controlled enzyme (p. 131, Ch. 29). The regulatory subunits are dark in tone, the catalytic subunits are light

structure that were used to enable the regulatory signal to go from one site to another. So far, it has been found that all enzymes that experience such regulation have quaternary structure (for example, look at the fascinating array of subunits in Fig. 6.19) and it appears that structure at this level plays an essential part in the process. Not all enzymes with quaternary structure have these properties, of course.

The kinetic curves of allosteric enzymes are often quite complex. Figure A1.8, p. 520, shows a typical example in which a sigmoid curve is obtained. It is strikingly reminiscent of the oxygen-binding curve of haemoglobin and quite unlike the simple rectangular hyperbola of Fig. 6.17.

Stimulatory effectors, like the converse class of *allosteric inhibitors* (see below) are of the greatest importance in achieving a balance between the many interconnected reactions in the cell. They are often used when a set of enzymes controls a chain of reactions, like those in Fig. 29.5, p. 501. In such chains they relay 'stop' and 'start' messages back from one of the later members of the chain to one of the earlier. The impact of this on the economy of the living state is discussed more fully in Ch. 29.

## Inhibitors

The class of inhibitors with the crudest mode of action are those that cause a gross denaturation of the tertiary structure of the enzyme. Extremes of temperature and pH are in this category, which is why boiling or treatment with acid or alkali are used to kill unwanted organisms. We have already discussed the toxicity of heavy metals (p. 31) such as lead, and saw that they can act in a similar fashion. Heavy metals that act in this way belong to the class of effectively irreversible inhibitors. (All chemical reactions are reversible, but as we have seen before some, for kinetic or thermodynamic reasons, are not reversed to a detectable extent.)

Many effectively irreversible inhibitors bind covalently at the active site. These may simply exploit the enhanced chemical reactivity of the active site residue (see p. 118). Alternatively they may both exploit the reactivity and, by virtue of a resemblance to the substrate, be steered toward the reactive residues by the other residues of the binding site. Di-isopropyl fluorophosphate (DFP) (Fig. 6.20a) is an example of an inhibitor relying mainly on reactivity. It will acylate the active-site serine of trypsin and chymotrypsin (Fig. 6.20b), but cannot be readily de-acylated as in the normal reaction. The same reaction occurs with the active-site serine of the enzyme acetylcholine esterase which mediates the transmission of signals across the interface between nerves and muscles. DFP is a potent poison therefore, because there are few such interfaces in the body and therefore few molecules of inhibitor are required to send every voluntary muscle into rigor. DFP was an early 'nerve gas'. Iodoacetate (pp. 85 and 263) is another example of an inhibitor relying mainly on reactivity.

Bromohydroxy-acetone phosphate (Fig. 6.20c) is a good example of an inhibitor that uses the normal substrate-binding forces to supplement the simple chemical reactivity. It resembles dihydroxyacetone phosphate (Fig. 6.20d) sufficiently well to be bound by the active site of triosephosphate isomerase, for which dihydroxyacetone phosphate is a substrate. It attacks a glutamate

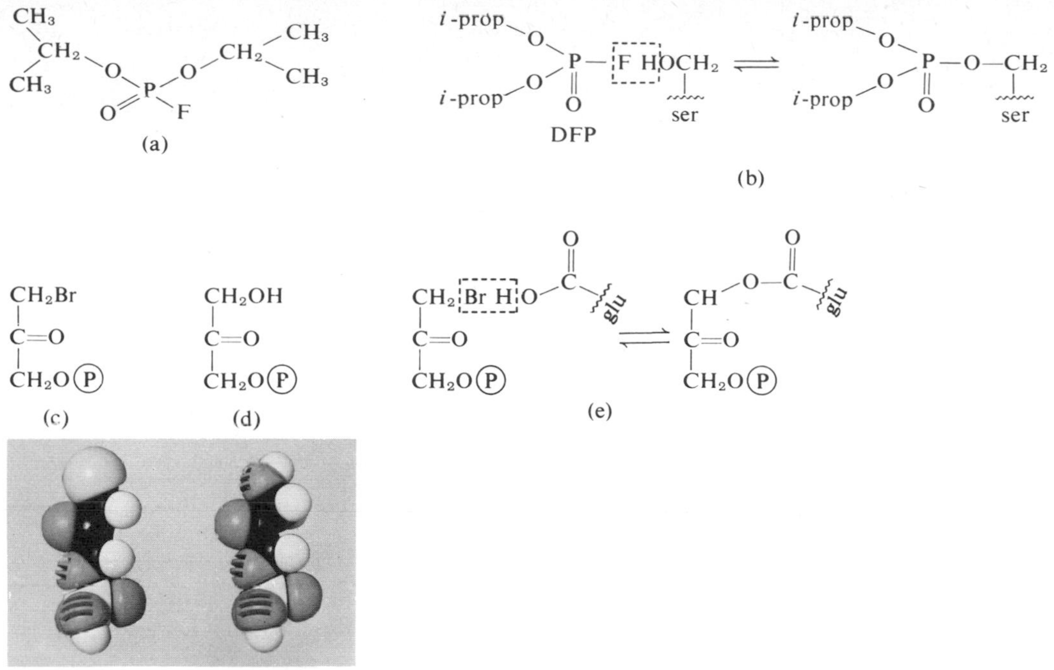

**Fig. 6.20** Two active-site-directed inhibitors (see text)

—COO⁻ group (Fig. 6.20e) which is thought to be the base, —B⁻, that appears in the proposed mechanism of action of this enzyme (p. 37).

These inhibitors, and many like them, are strongly and covalently bound to the active site residue and they will remain in position even when the enzyme is fragmented. This reaction then provides the means (mentioned on p. 124) of identifying the active-site residue by determining the amino-acid sequence of the fragment that still has the inhibitor bound to it. The task of identifying the fragment can be made easier by incorporating a radioactive atom into the molecule. $^{32}P$ would be suitable for DFP, $^{14}C$ for iodoacetate.

Some inhibitors, in contrast to those we have considered so far, do act in a manner which is more easily seen to be reversible. These are divided into *competitive* and *non-competitive* reversible inhibitors.

The effect of *competitive inhibitors* is diminished if we raise the substrate concentration, that is the inhibitor and the substrate compete for the enzyme. They are found to be compounds that resemble the substrate molecules sufficiently well to form some of the proper interactions with the binding site and yet are not sufficiently similar to take part in the reaction and to be released. The wrong key will not turn in the lock, but if it is close enough to the right shape, it may be inserted. Since we are dealing with a competitive process the probability of this occurring rather than the correct interaction with the substrate is increased the greater the concentration of inhibitor. An example is the enzymic conversion of succinate ion to fumarate ion, an important step in the breakdown of many biochemical compounds (Fig. 6.21a and see p. 278).

Malonate ion (Fig. 6.21b) will bind to the appropriate site, since the spacing of

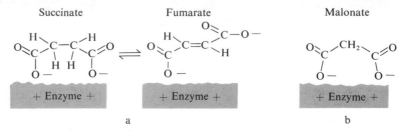

Succinate        Fumarate        Malonate

+ Enzyme +     + Enzyme +     + Enzyme +

a                  b

**Fig. 6.21** The effect of malonate ion on the catalysis of the dehydrogenation of succinate (see text)

its carboxyl groups is not greatly different from those of succinate. When succinate is transformed to fumarate it is released from the enzyme surface. Malonate clearly cannot undergo such a change and so it continues to occupy the binding site. While it remains, the enzyme is unable to accept any further molecules. Malonate is therefore a powerful inhibitor of the enzyme.

Compounds of this type are often extremely poisonous. Very small quantities are sufficient for the same reason that held good for the co-factors and prosthetic groups (p. 80). The very high toxicity of certain fluorinated compounds, for example, is due to the fact that the fluorine atom is small enough not to change the steric properties of the molecule a great deal but, because of its great electronegativity, confers a completely different chemical reactivity on the molecule.

Many antibiotics are competitive inhibitors (e.g. Fig. 6.22 a and b). They owe

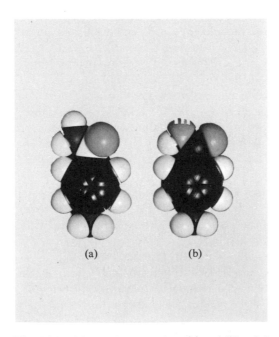

(a)            (b)

**Fig. 6.22** (a) p aminobenzoic acid and (b) sulphonamide, an antibiotic which acts as a competitive inhibitor. (See also pp. 80 and 386.)

their usefulness to the chance that they inhibit a bacterial enzyme more effectively, or with more serious metabolic consequences, than they inhibit the enzyme of the host. Alternatively, they may inhibit an enzyme in the bacteria that has no counterpart in the host – penicillin is an example.

We have already mentioned in this chapter another type of competitive inhibitor that binds non-covalently. These compounds, rather than mimicking the substrate directly, derive their inhibitory power from their resemblance to the transition state. We have seen that the thermodynamics of the catalytic process demand that the transition state is more tightly bound than the substrate – thus a compound that resembles this state at all faithfully will be a very good (i.e. strongly bound) inhibitor.

*Non-competitive inhibitors* are so named because they are not necessarily directed to the active site, and so do not have to compete with substrate for access to the enzyme surface. The kinetics of non-competitive inhibition are dealt with in Appendix I.

Some inhibitors appear to compete with the substrate and yet do not resemble it in molecular structure. These inhibitors have been found to bind at a site away from the specificity and binding sites. On binding to their site, these *allosteric inhibitors* induce a conformational change in the enzyme that depresses the efficiency of catalysis. Once again there are homotropic and heterotropic types and all the allosterically regulated enzymes so far studied have been found to have several subunits. Allosteric inhibitors are at least as common as stimulatory effectors and are used for analogous purposes in the control of enzyme activity.

We can now see that inhibitors and activators have an important place in biology. We have said something of their metabolic significance and will return to this in much greater detail in Ch. 29. We may also mention that many hormones are, or produce, enzyme activators. Again, many substances, natural and synthetic, that are poisonous to us or to other organisms (in the latter case we include them in the classes of specific pesticides, herbicides or antibiotics, as appropriate) are enzyme inhibitors, relying for their efficiency either on the principle of mimicry or on the heightened reactivity of some residue of the active site.

Finally we may point out that inhibitors have contributed greatly to our knowledge of the nature of enzyme action. This is true both of our knowledge of their mechanism of action, as you will have realized from the preceding pages, and also (as will be discussed in Ch. 11) of our knowledge of their interrelations in intermediary metabolism.

*Other factors*

The pH will affect the efficiency of an enzyme and, usually, there will be a pH at which activity is at a maximum. The activity will fall off on either side of this value (Fig. 6.23). Gross departures of pH from the optimum condition will denature the enzyme (p. 77) and thus inactivate it. This denaturation may be reversible or effectively irreversible, but the result in either case is to impose a range of pH outside which the enzymes will not be active. However, less brutal factors usually determine the nature of the main part of the curve. If the *substrate*

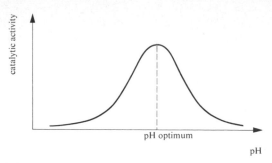

**Fig. 6.23** A typical curve showing the relation between the velocity of an enzyme-catalysed reaction and pH

is capable of undergoing a pH-controlled equilibrium, the chances are that only one of the two forms, charged or uncharged, will be suitable for the binding to the active site. Thus, for trypsin, the p$K$ of the $\epsilon$–$NH_2$ group of lysine is about 10; at pH's of 11 or above, the side-chain amino group will be almost all in the —$NH_2$ form (Table 2.1, p. 12) and not —$NH_3^+$. It is of course necessary that it be in the —$NH_3^+$ for binding in the specificity site (Fig. 6.13).

The *specificity site* itself may include groups that are subject to a pH-controlled equilibrium. In trypsin again, the carboxyl group in the specificity site must obviously be in the —$COO^-$ form rather than the —$COOH$ form. The p$K$ of this transformation is approximately 3 and so the binding will start to weaken below pH 2.5. In such a case activity will tend to fall off as we leave the range pH 2.5–11. This range is narrowed even further by the fact that interactions in the *active site* also involve titratable groups (Fig. 6.11) and these are, similarly, liable to perturbation by changes in pH. The p$K$ values here are even nearer to 7, and this is why the pH-activity curve is narrower.

There is little variation in pH from one part of a cell to another (although we have seen already how an enzyme can contrive a local area of what might be thought of as 'abnormal pH' for itself). pH variations are more marked outside cells and, for example, different proteolytic enzymes are required to work in the different regions of pH in the gut. In the peptic region, where the pH is about 2, pepsin (p. 111) which has the appropriate pH optimum is the principal agent. In the duodenum, where the pH is mildly alkaline, we find, as we would expect, trypsin, chymotrypsin and other similar enzymes.

Since the pH within cells, and also in many extra-cellular fluids, is closely controlled to near neutrality (p. 122), the effects of changes in pH are less often of importance in considering the properties of enzymes. However, we have seen that cases do exist in which pH is an important factor. We can readily realize also that artificially introduced changes in pH could enable us to learn something of the p$K$ values of groups involved in the catalytic mechanism. Such knowledge is sometimes of assistance in exploring the nature of the active site.

*The effect of temperature* is again of less direct biological interest since many animals control their body temperatures just as rigorously as they do their pH. Even those creatures that do not control their temperatures are unlikely to experience a temperature range greater than a few degrees about a mean value which is of the order of 300°K. Nonetheless there are a few biological lessons to be learned, and, as with pH, much light is thrown on the mode of action of enzymes by artificially achieved changes.

We saw on p. 40, Ch. 2, that the rate of a chemical reaction is accelerated as a result of a 10° increase in temperature, by a factor of approximately $e^{-\Delta H^{\ddagger}/19\,000}$. We gave on p. 40 a table of $\Delta H^{\ddagger}$ for some reactions that take place at measurable rates in the absence of enzymes. You might find it interesting to compare these values with those obtained for the same reactions when catalysed by enzymes (Table 6.3).

**Table 6.3** $\Delta H^{\ddagger}$ and $Q_{10}$ values for a number of enzyme-catalysed reactions (compare Table 2.3)

|  | $Q_{10}$ | $\Delta H^{\ddagger}$ |
|---|---|---|
| Hydrolysis of sucrose catalysed by invertase | 1.5 | 7 400 |
| Hydrolysis of casein catalysed by pepsin (p. 111) | 2.3 | 16 000 |
| Decomposition of $H_2O_2$ catalysed by catalase (p. 109) | 1.2 | 2 000 |
| Hydrolysis of urea catalysed by urease (p. 111) | 1.7 | 10 000 |

We can now obtain an idea of the size of the enthalpic contribution to the lowering of $\Delta G^{\ddagger}$ and thus to the enhancement of reaction rate. The contribution in the case of catalase is particularly marked, as one might have anticipated from the huge catalytic power of this enzyme (p. 109).

The first effect of temperature, then, is to accelerate the rate of enzyme-catalysed processes, although to a smaller extent than for processes occurring in the absence of enzymes. (This has interesting consequences for hibernation. When animals hibernate, the body temperature falls, the rates of metabolic reactions are lowered, and the food stores of the body are conserved. But we have shown that the rates of enzyme-catalysed reactions vary about half as sharply with temperature as do non-enzyme-catalysed ones. Therefore, a considerable drop in temperature (from the normal body temperature to the winter air temperature) is needed to produce a real economy. However, the rate of depletion of food reserves thereafter is likely to be relatively insensitive to such fluctuations in temperature that might be experienced during the hibernation period (the difference between a warm day in winter and a normal one) as a result of climatic changes. This must be a significant aid to survival, since there are very great dangers in these fluctuations. This is shown by the fact that, even with this diminished response to temperature, hibernating animals are often killed by an unusual *rise* in temperature of the environment, which causes them to exhaust their food reserves prematurely, rather than being frozen to death by an unusual *fall* in temperature.)

The enzymes of some marine invertebrates show an even more striking effect. These animals are those that have to survive large changes in sea temperature and their enzymes are so constructed as to impart a $Q_{10}$ of very near 1 to the reactions that they catalyse.

The second effect of temperature is to accelerate the rate of denaturation of the enzymes themselves. We saw in Ch. 4 that the $Q_{10}$ of the process, because of its extremely high $\Delta H^{\ddagger}$, is itself very high, of the order of 20.

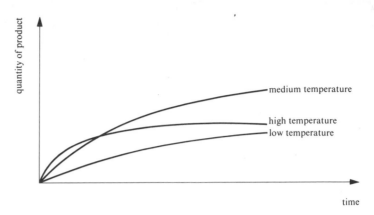

**Fig. 6.24** The effect of temperature on enzyme-catalysed reactions

We can now explain the graphical presentation of the effect of temperature on enzyme activity in Fig. 6.24. You will see that the initial velocity of the reaction (given by the shape of the curves at $t = 0$) steadily increases with temperature. However, after a certain temperature is passed, the cessation of activity comes earlier and earlier, with the net effect that less product is formed. There is thus a somewhat ill-defined optimum region of temperature, which is that at which the two factors of increased initial rate and decreased active life of the enzyme are balanced to produce the most product in a reasonable time. It is not easy to determine an exact value for the optimum temperature because it is a somewhat vague concept, and will depend on the length of time over which the measurements are made. However the approximate values obtained often show a clear correlation with the body temperatures of the organisms from which the enzyme came. Thus mammalian enzymes often have optimum temperatures in the range 35–45° while the enzymes from the bacteria that live in volcanic hot springs may have optima of 80°.

We have now completed this account of the behaviour of enzymes, and of the biological role of the features that we have described. We believe that this chapter, like the previous one, shows how modern knowledge of protein structure, interpreted in the light of organic and physical chemistry, can produce quite reasonable explanations of what might at first sight appear to be utterly mysterious aspects of the behaviour of living organisms.

# Chapter 7    Structure and function of the nucleic acids

You will have formed the impression in the last chapter that the proteins have an important part to play in every distinctive activity of living processes save one. The apparent omission was reproduction. We shall see that, even here, proteins do have their part to play, but the molecules most directly involved are the nucleic acids. This is indeed the only major function of the nucleic acids, and this fact explains why they possess many fewer types of monomer than there are types of amino acid in the proteins. We shall also see that the restricted range of functions has necessitated the production of far fewer types of nucleic acid than there are types of protein.

The fundamental unit of nucleic-acid structure is the *nucleotide* (p. 46). Like the amino acids, nucleotides make up the chain polymer by condensation (p. 45), and also like the amino acids, they were discovered as products of hydrolysis of the parent macromolecule. (The term for a short length of chain is *oligonucleotide*. It is equivalent to the term peptide in protein chemistry.) As we stated in Ch. 3, a mononucleotide consists of a heterocyclic compound (known as a 'base') linked to a sugar which in its turn is linked to a phosphate group. The five bases that are most frequently met with are adenine, cytosine, guanine, uracil and thymine. The nucleotide base–sugar combination *without* the phosphate group is a *nucleoside*. The names of the nucleosides are formed from the names of the bases: adenosine, cytidine, guanosine, uridine, thymidine. The nucleotides are named as nucleoside monophosphates (e.g. adenosine monophosphate) or as acids: adenylic acid, cytidylic acid, guanylic acid, uridylic acid and thymidylic acid.

The nucleotide bases are related either to the purine ring system or to the pyrimidine ring (Fig. 7.1). The ones that are principally used are shown in

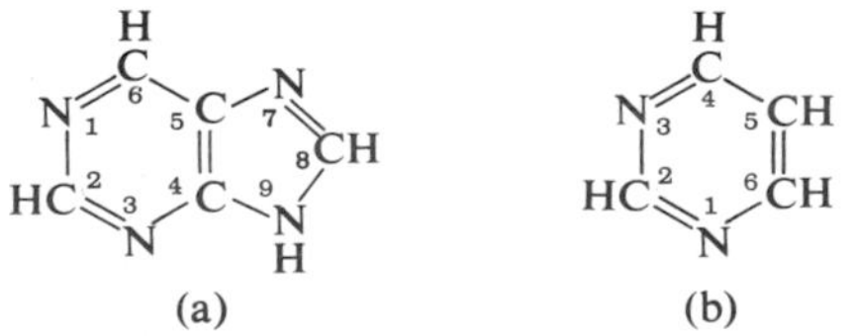

**Fig. 7.1**    (a) Purine, (b) pyrimidine

Table 7.1. In addition to these, chemical modification produces a whole range of so-called minor bases (of which a number are shown in Table 7.2). They are

**Table 7.1**  The major bases used in nucleic acids

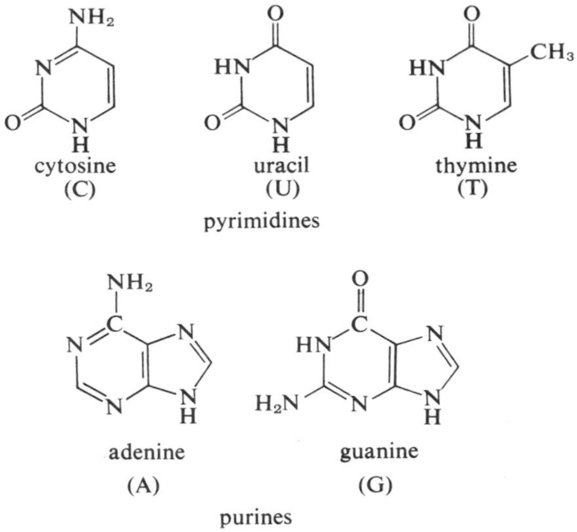

cytosine (C)     uracil (U)     thymine (T)

pyrimidines

adenine (A)     guanine (G)

purines

called 'minor' because they occur far more seldom than the others (see below). The work of Levine and his colleagues in the earlier part of this century showed that only two types of sugar are involved in nucleic acid (Fig. 7.2). Ribose

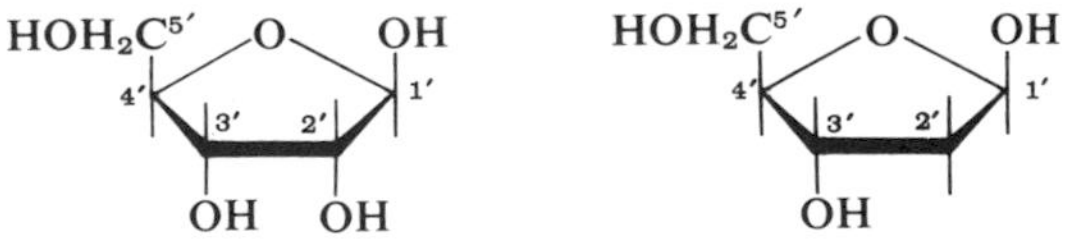

(a)                    (b)

**Fig. 7.2**  (a) $\beta$-D-ribose, (b) $\beta$-D-deoxyribose

gives its name to *ribonucleic acid (RNA)*, the class of nucleic acids in which it occurs. Deoxyribose, similarly, gives its name to the *deoxyribonucleic acids (DNA)*.

In most molecules of both DNA and RNA, only the major bases are used to any great extent. RNA contains adenine, cytosine, guanine and uracil, and DNA the same with the important exception that uracil is replaced with the closely related base thymine. In DNA from most sources there are also small but significant amounts of 5-methyl cytosine (see Table 7.4, p. 144). This Table also shows that in certain of the viruses that attack bacteria, cytosine is replaced completely by 5-hydroxymethyl cytosine.

## The phosphodiester bond

This bond performs for the nucleic acids the functions that are performed for the proteins by the peptide bond. As we saw in Ch. 3, it owes its name to the fact that it is formed by the esterification of two of the three —OH groups of phosphoric acid by the —OH groups on the pentose sugar. It is to the proton-donating ionization of the remaining —OH that the nucleic acids owe their acidic character and name.

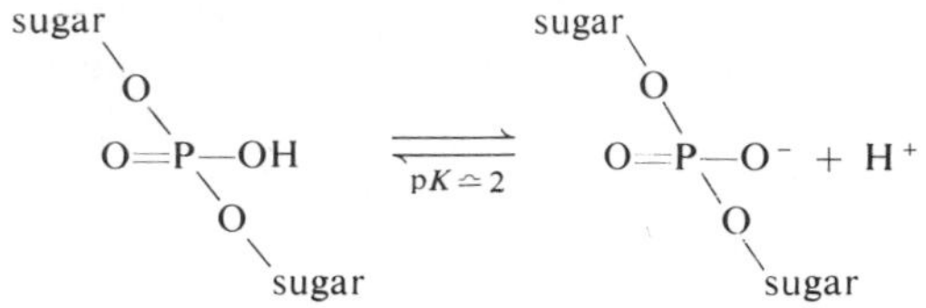

$$\text{sugar} \quad O=\overset{\overset{\displaystyle O}{|}}{\underset{\underset{\displaystyle \text{sugar}}{|}}{P}}-OH \quad \xrightleftharpoons[pK \simeq 2]{} \quad \text{sugar} \quad O=\overset{\overset{\displaystyle O}{|}}{\underset{\underset{\displaystyle \text{sugar}}{|}}{P}}-O^- + H^+$$

The individual groups in this linkage are able to rotate relatively freely about their bonds; this fact contributes to the ability of the nucleic acid chain to assume complicated convolutions when needed (see below).

There are only two —OH groups in deoxyribose, those on the C'-3 and the C'-5 carbon atoms, that are not involved in the formation of the sugar ring, and these must therefore be the ones involved in the phosphodiester linkage in DNA. The simplest way in which this could occur is shown in Fig. 3.2, p. 46. That

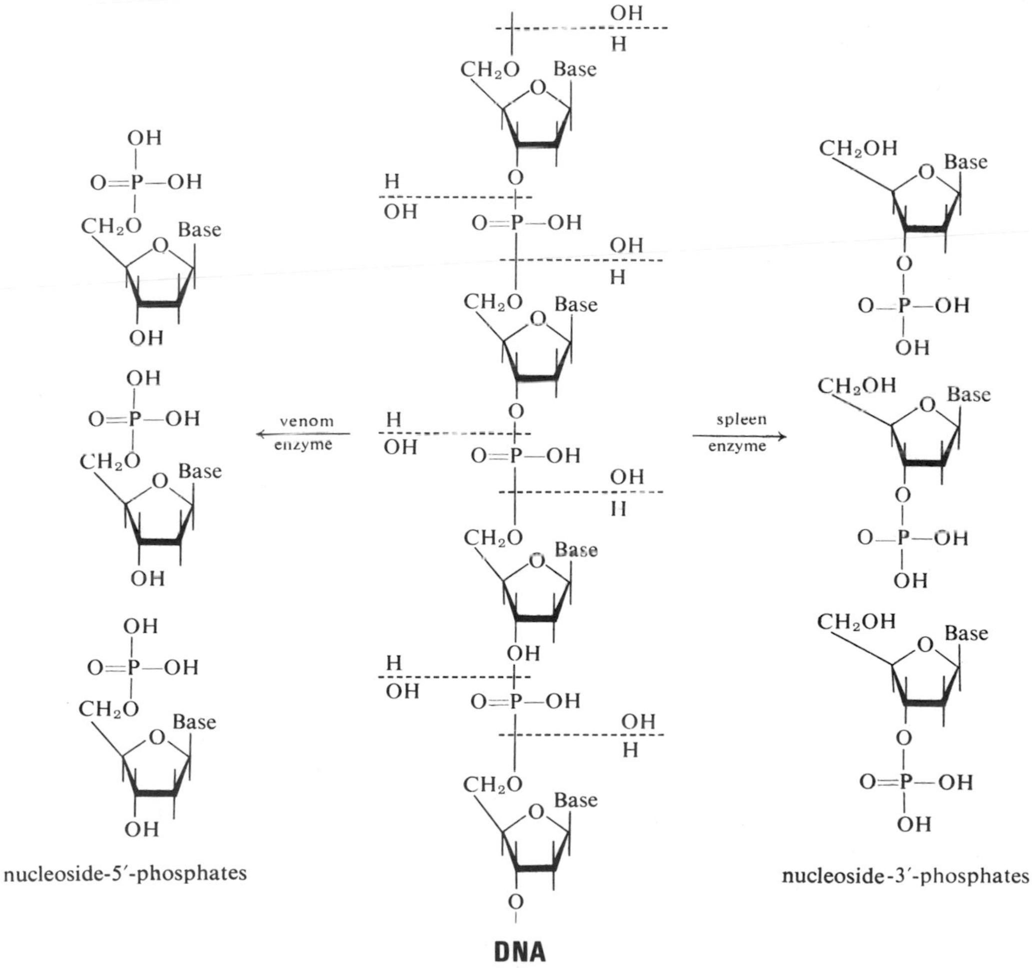

nucleoside-5'-phosphates     venom enzyme ←    DNA    → spleen enzyme     nucleoside-3'-phosphates

**DNA**

Fig. 7.3    The hydrolysis of DNA by enzymes (see text)

this is indeed the mode of linkage of a DNA chain is confirmed by a number of lines of evidence: for example Markham and his colleagues showed that treatment of oligodeoxyribonucleotides with an enzyme from snake venom (*venom diesterase*) yields deoxyribonucleoside-5′-phosphates, while treatment with a similar diesterase from animal spleen liberates deoxyribonucleoside-3′-phosphates (Fig. 7.3). You will recall from Ch. 3 that the end that carries the 3′-hydroxyl groups that is not involved in the linkage is called the 3′ end. The other end, for similar reasons, is called the 5′ end. These terms, which are useful in discussing nucleic acid structure, are analogous to the terms amino and carboxyl terminus in proteins.

In spite of the fact that ribose possesses an additional —OH group on the 2′ carbon atom, RNA also is bridged only by linkage between the 3′ and the 5′ carbon. (Branching or cross linking, which are theoretically possible in RNA, do not, as far as we know, occur.) The evidence for this conclusion is much the same for RNA as that for DNA; for example the spleen and venom diesterases give analogous ribonucleotide products. There was some initial confusion when both nucleoside-2′-phosphates and nucleoside-3′-phosphates were isolated from an alkaline hydrolysate of RNA. However, Todd and his collaborators showed that the 2′-phosphate, which would not have been expected if the structure shown in Fig. 3.2 is correct, is formed nonetheless, together with the 3′-phosphate, from the 2′,3′-cyclic phosphate (Fig. 7.4) which is produced initially in the hydrolysis.

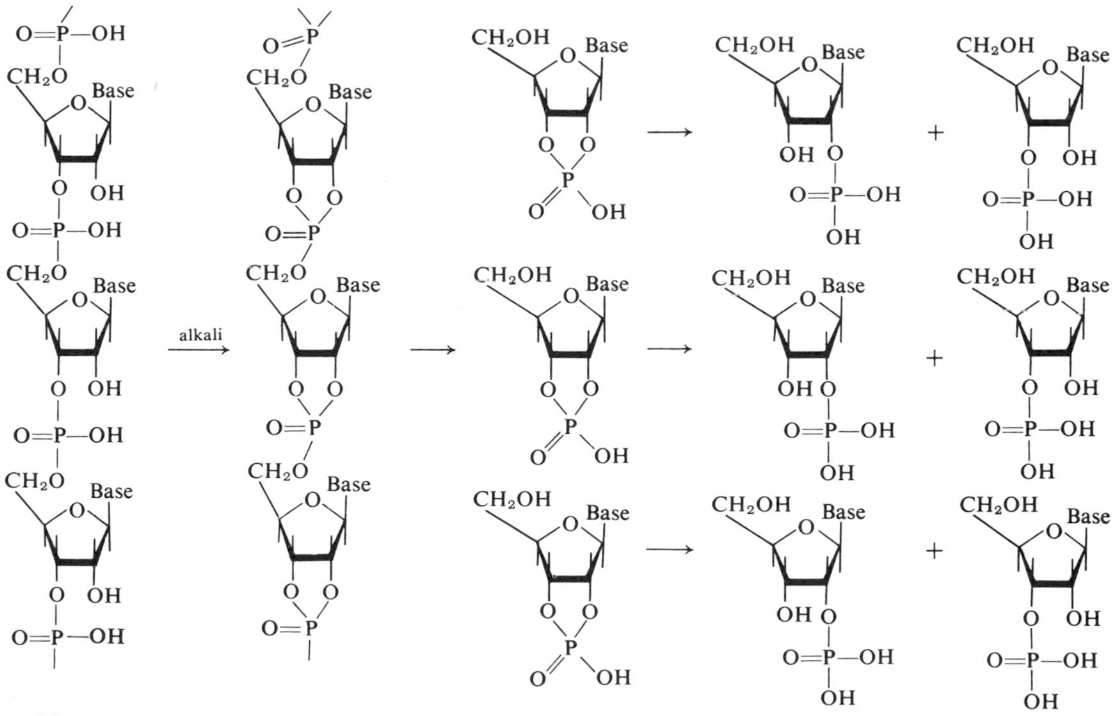


**RNA** Cyclic intermediate Cyclic nucleoside-2′,3′-phosphates Mixture of 2′- and 3′-nucleotides

**Fig. 7.4** The stages in the alkaline hydrolysis of RNA

In spite of the fact that both DNA and RNA contain the $3' \rightarrow 5'$ linkage we shall see (p. 149 ff.) that the bulky oxygen atom on the $2'$ position has an important effect in restricting the conformations that the RNA chain can take up by comparison with the DNA chain. Certain other differences are discernible. The extra —OH group of the ribose ring confers some additional degree of solubility in water on RNA. Moreover, as a consequence of the ability to participate in the mechanism shown in Fig. 7.4, RNA is attacked by alkali while DNA, which cannot form the cyclic phosphate, is not hydrolysed. Furthermore, enzymes are easily able to discriminate between nucleic acids of the two sorts, and organisms have developed quite distinct *ribonucleases* and *deoxyribonucleases* to catalyse the hydrolysis of the two forms. (These enzymes are *endo*nucleases, that is, they cleave the chain at various points along its length; it happens that the spleen and venom enzymes mentioned above, which do not discriminate between DNA and RNA, are *exo*nucleases, that is they are enzymes cleaving by stepwise removal of nucleotides from the end of the chain.)

The formation of the linkage between the sugar and the base, like the formation of the phosphodiester link, involves condensation. The bond that is formed was shown by Levine to be a glycosidic (p. 171) link between the C-1 of the sugar and the N-9 of the purines or N-1 of the pyrimidines. Todd and his group showed that it had the $\beta$-configuration (p. 169).

We can now see the repeating sugar–phosphate–sugar–phosphate–sugar phosphate linkage as the backbone of the nucleic acid chain, analogous to the repeating peptide backbone of the proteins. Amino-acid side chains are here replaced by the bases; once again the sequence of incorporation of the residues during synthesis of the molecule (the base sequence in the present case) is absolutely controlled. As with the proteins, it is the combination of properties of these side chains that must determine the specific properties of each different nucleic acid. We shall now examine these bases to see how far we can use their structures to explain their properties and predict the contribution that they might make to the structure and function of the completed polymer.

## The nucleotide bases – their general properties and their contribution to the stabilization of structure

When we compare Table 7.1 with Table 4.1 we can see that the restricted range of functions of the nucleic acids is matched not only by a restricted number of types of base, but also by a restricted range of properties of those bases. It is, nonetheless, useful to go through their properties in the same order as we did for the amino acids. We shall therefore first consider stereoisomerism and ionization.

As a result of aromatic unsaturation the rings themselves and the bonds that lead to their substituent positions lie in a single plane. This gives rise to a plane of symmetry and as a result there is no stereoisomerism in the principal bases. We shall see below that asymmetry does occur in the sugar portion of the structure.

Most bases have ionizable groups. The —NH$_2$ groups can undergo the ionization:

$$-NH_2 + H^+ \rightleftharpoons -NH_3^+$$

with pK values (between 4 and 5 for groups in the 6 position, between 2 and 3 for groups in the 2 position) much lower than for the corresponding ionizations in the amino acids. Because of these values there is no significant proportion of the bases in the charged form at pH 7. (This is a proton-accepting, and therefore basic, ionization. However, the purine and pyrimidines owe their classification as bases not to these —NH$_2$ groups (uracil and thymine do not, in any case, have such groups) but to the nitrogen atoms of the ring. Even when these are not masked by the attachment of the sugars, they are only very feebly basic.)

The *keto* groups undergo rapid *keto–enol* tautomerism (Fig. 7.5). (The alternative, and more proper, name is lactim-lactam tautomerism.)

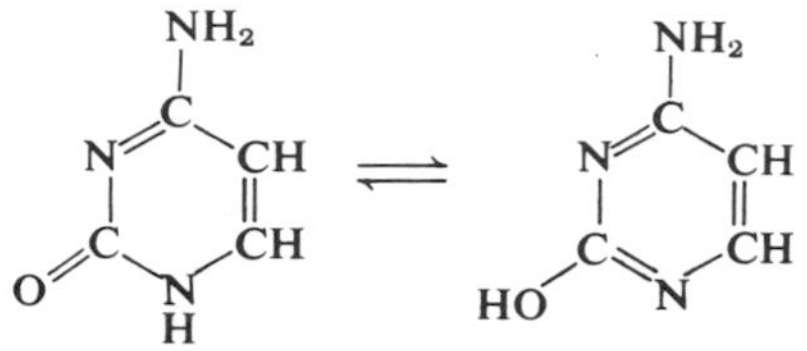

**Fig. 7.5** *keto–enol* tautomerism in a nucleotide base.

The groups can ionize

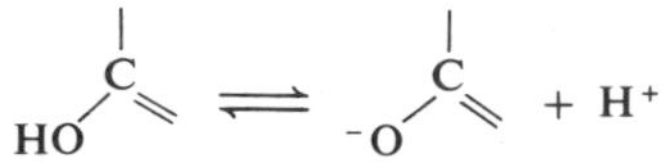

The pK values (9 or above) ensure that this ionization, also, is not significant at physiological pH. Thus neither the ionization of this group nor that of the —NH$_2$ group plays a part in the biological activity of the nucleic acids. (However, they are, together with the slightly differing degree of hydrophobicity between the bases, among the few properties that may be used to separate the otherwise very similar oligonucleotides during the operations required (which are analogous to those used with proteins) to determine the primary structure* of the nucleic acids (p. 163 and Appendix II).)

The *hydrophobic*, or *apolar*, properties of the bases have a considerable role to play in nucleic acid structure and function. A glance at the three-dimensional model of the bases shows that while other, polar, interactions are possible – fanning out from the edges of the base in the plane of the ring – the *faces* of the rings are unable to form such bonds. The faces will therefore not interact strongly with water but will be able to form strong, hydrophobic interactions with other bases when all are stacked together. The adhesion between the members of such stacks is further intensified by interchange between the electrons that circulate in the $\pi$-orbitals above and below the plane of each ring. We shall see in a moment that the combined apolar interactions between stacked bases play a much more important role in the stabilization of secondary structure than is often realized.

The *hydrogen-bonding* properties of the bases, in contrast, have received a

---

* The terms primary, secondary and tertiary structure have much the same meaning as they did for proteins. The primary structure is the order of the nucleotides in the chain; the term secondary structure relates to regions of regular conformation of the chain, stabilized by regular, repeating interactions (e.g. the double helix of DNA (p. 146)) and the tertiary structure is the overall conformation of the chain.

great deal of attention when the stabilization of nucleic acid structure is discussed.

The $>C{=}O$ and $-NH_2$ groups of the bases are able to form hydrogen bonds both with water and with corresponding groups on other bases. In particular the structures of adenine and thymine have been found to allow two good inter-base bonds and those of guanine and cytosine have been found to allow three (Fig. 7.6). You will probably know that the potential for specific base

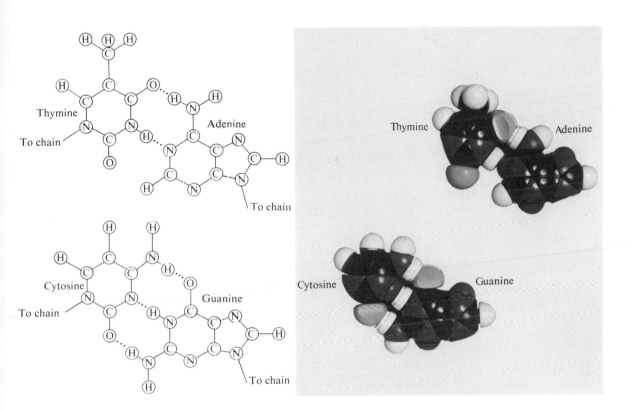

**Fig. 7.6** The hydrogen-bonded base pairs between (a) thymine and adenine and (b) cytosine and guanine

pairing is of crucial importance to the most vital property of nucleic acids. This is that a given nucleotide sequence should be able to recognize (i.e. bind to) one, but only one, other sequence of bases. In other words, it should have a binding (association) constant for one sequence that is high enough to overcome random thermal motion and a binding (association) constant for all others that is less than sufficient to overcome random thermal motion. We shall return to this matter in detail in a moment.

**Table 7.2**  Some minor bases used in nucleic acids. (The abbreviations are those that are used in Fig. 7.13)

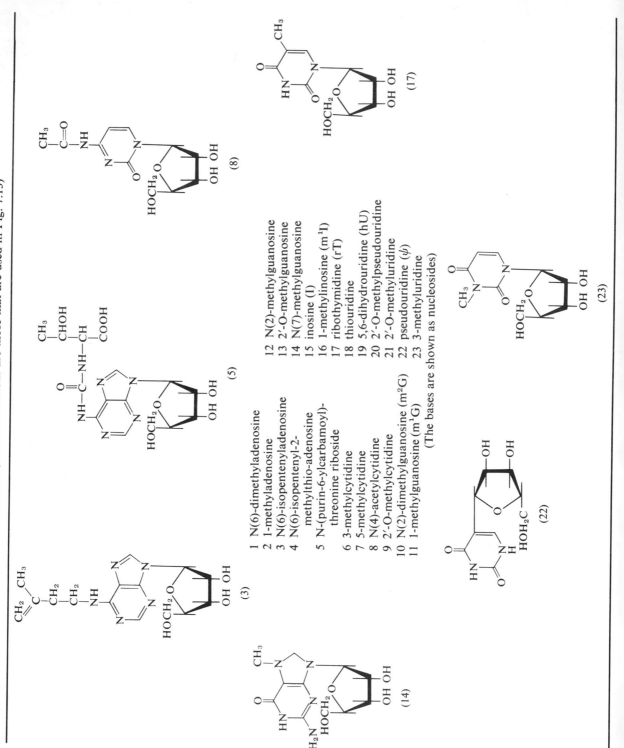

1  N(6)-dimethyladenosine
2  1-methyladenosine
3  N(6)-isopentenyladenosine
4  N(6)-isopentenyl-2-
   methylthio-adenosine
5  N-(purin-6-ylcarbamoyl)-
   threonine riboside
6  3-methylcytidine
7  5-methylcytidine
8  N(4)-acetylcytidine
9  2'-O-methylcytidine
10 N(2)-dimethylguanosine (m²G)
11 1-methylguanosine (m¹G)

12 N(2)-methylguanosine
13 2'-O-methylguanosine
14 N(7)-methylguanosine
15 inosine (I)
16 1-methylinosine (m¹I)
17 ribothymidine (rT)
18 thiouridine
19 5,6-dihydrouridine (hU)
20 2'-O-methylpseudouridine
21 2'-O-methyluridine
22 pseudouridine ($\psi$)
23 3-methyluridine

(The bases are shown as nucleosides)

These bases are almost exclusively restricted to a single class of nucleic acid – the *amino-acyl transfer RNAs* (p. 158). If we look at their structures (Table 7.2) we are at once reminded of the situation that obtains with the proteins. We now have a range of shapes and properties as wide as, or wider than, those that we encountered with the amino acids. It is significant therefore that the amino-acyl transfer RNAs have to behave most like a protein in terms of the versatility and specificity of the binding interactions that they are called on to undertake (see below, p. 159).

As we can see from the table, the minor bases could be formed by chemical modification of the major bases. This is thought actually to be the way in which they are made; enzyme systems exist that carry out the modifications to the chains once they are completed (see below).

Some of the modifications confer quite unusual steric properties, e.g., in the table, structure 5. Some decrease or modify polar properties (e.g. structures 5 and 8). Some interfere with the base-pairing (e.g. structure 23); some do not (e.g. structure 22).

# Nucleotides that do not occur in nucleic acids

Table 7.3   Some nucleotide co-factors

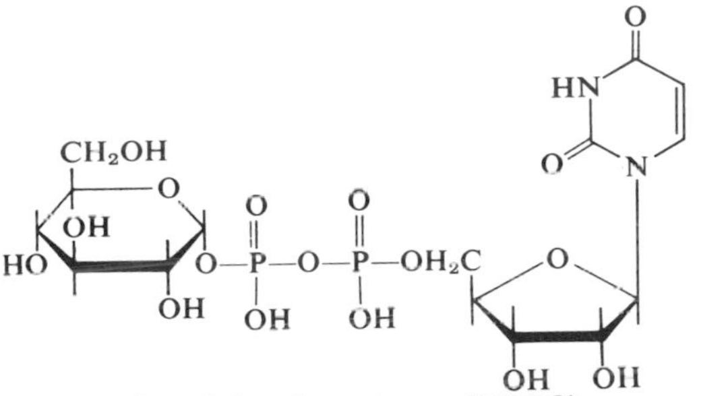

Uridine diphosphate glucose (UDPG)
(used as a donor of glucose in the
synthesis of polysaccharides (p. 334 ff.))

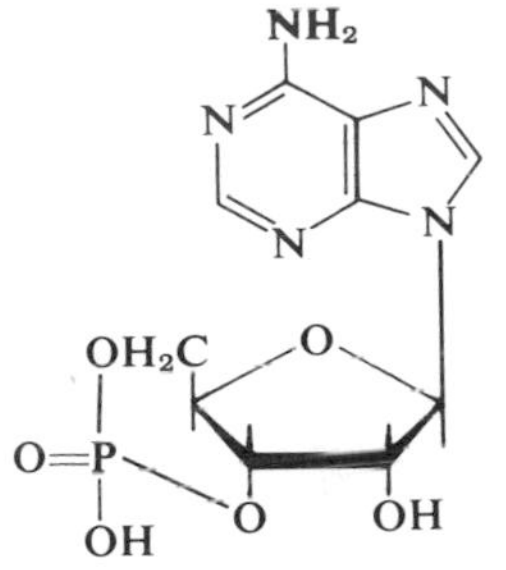

Adenosine-3′, 5′-monophosphate
(used for the control of metabolism
(pp. 480 and 505))

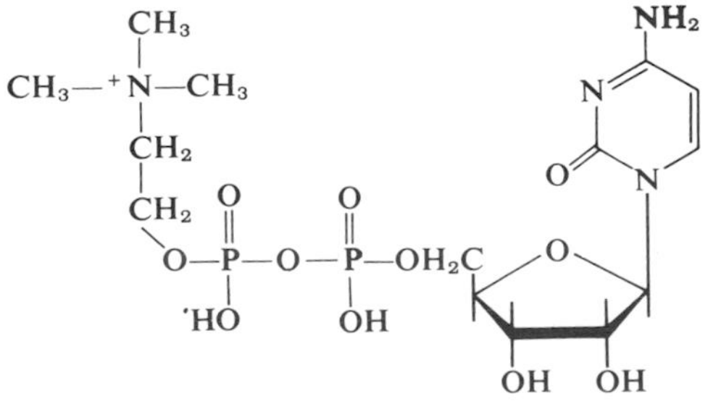

Cytidine diphosphate choline
(used as a donor of choline in the
biosynthesis of lipids (p. 348))

It is convenient to mention here the large number of co-factors and prosthetic groups that are in part nucleotides. A number have already been shown in Table 4.2, p. 79, and some others are shown in Table 7.3 – note particularly the curious 3′,5′-cyclic nucleotide of adenosine. We shall meet this compound again in later chapters, where we shall see that it plays a central role in the control of metabolism. Like all the other nucleotide structures in the two tables it acts by binding to specially prepared sites in a number of proteins.

## Secondary structure in DNA

### Clues from the base ratios

Chargaff pointed out in 1950 that there are certain interesting regularities in the base compositions of DNA from most sources. (Before reading on, you might like to see if you can deduce for yourself what they were, using Table 7.4.†)

**Table 7.4**  The base compositions of some samples of DNA

| Source of DNA | Adenine | Guanine | Cytosine | Thymine | 5-Methyl-cytosine | 5-Hydroxy-methyl cytosine* |
|---|---|---|---|---|---|---|
| Bovine thymus | 28.2 | 21.5 | 21.2 | 27.8 | 1.3 | — |
| Bovine spleen | 27.9 | 22.7 | 20.8 | 27.3 | 1.3 | — |
| Bovine sperm | 28.7 | 22.2 | 20.7 | 27.2 | 1.3 | — |
| Rat bone marrow | 28.6 | 21.4 | 20.4 | 28.4 | 1.1 | — |
| Herring testes | 27.9 | 19.5 | 21.5 | 28.2 | 2.8 | — |
| Paracentrotus lividus | 32.8 | 17.7 | 17.3 | 32.1 | 1.1 | — |
| Wheat germ | 27.3 | 22.7 | 16.8 | 27.1 | 6.0 | — |
| Yeast | 31.3 | 18.7 | 17.1 | 32.9 | — | — |
| E.coli 22.7 | 26.0 | 24.9 | 25.2 | 23.9 | — | — |
| Mb. tuberculosis | 15.1 | 34.9 | 35.4 | 14.6 | — | — |
| phage φX174 | 24.3 | 24.5 | 18.2 | 32.3 | — | — |
| phage T2r | 32.4 | 18.3 | 32.4 | — | — | 17.1 |

* The —OH of the hydroxymethyl group (—CH$_2$OH) can form a glycosidic linkage (p. 46) with glucose –in the example shown here 75% of the residues are glucosylated.

The regularities were, first

Number of purine bases = number of pyrimidine bases

and

Number of bases with —NH$_2$ in position 6 or 4 = number with keto group in position 6 or 4.

The latter relationship means

adenines + cytosines = thymines + guanines

while the former means

adenines + guanines = cytosines + thymines.

† Do not attempt to use the data for φX174. This virus represents an anomolous class, which was unknown in 1950 (see later).

Also:

$$G = C \quad \text{and} \quad A = T$$

There was no reason *a priori* why these relationships should have held – they do not do so for most RNA samples. The explanation had to await some further work which we shall now describe.

## Clues from X-ray diffraction

Figure 7.7 shows an X-ray diffraction photograph of an artificially produced fibre of a gel of DNA in water. It is clear from the sharpness of the spots that in such a fibre the long chains of DNA have fallen into parallel lines and that the pattern of X-rays scattered from each chain is reinforcing that from all the

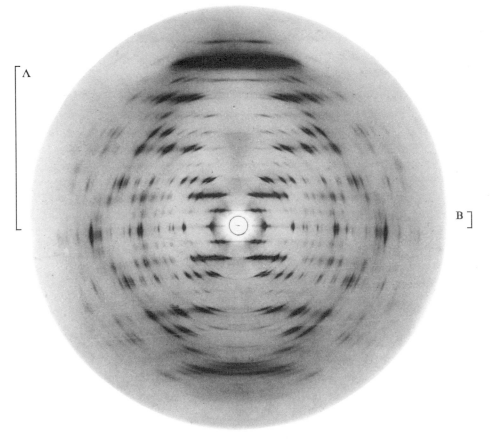

**Fig. 7.7** An X-ray diffraction photograph of DNA. The fibre was slightly tilted with respect to the X-ray beam and as a result only the top half of the picture is suitable for interpretation. Scale A is equivalent to 3.4 Å, Scale B is equivalent to 34 Å

others. These photographs, which were first obtained in a useful form by Franklin and Wilkins in 1953, show the helix cross (p. 88) as a dominant factor, and if you have read the section on helical diffraction you should be able to deduce for yourself that the pitch of the helix is 34 Å and that there are ten subunit structures per turn. The repeat distance of the subunit is therefore 3.4 Å, which is of the same order of size as a single nucleotide might occupy in the chain. (The maximum distance that there can be between the nucleotides in a fully extended chain is 7 Å. However, the chain need not be fully extended and we must recall that the 3.4 Å repeat is the *rise* on moving from one subunit to another and this is bound to be less than the distance *between* subunits along the circumference of the chain.)

It was Watson and Crick in 1953 who, when they came upon the X-ray pictures, combined the inferences that could be drawn from them with the observations of Chargaff and proposed the *double helical* structure of DNA (Fig. 7.8).

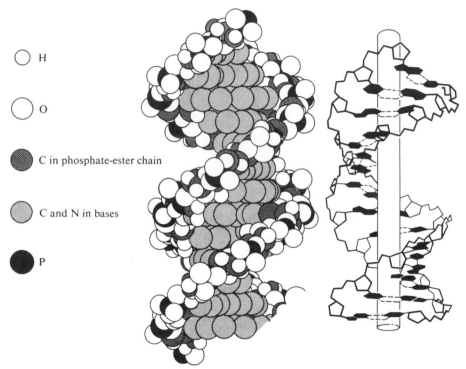

○ H

○ O

● C in phosphate-ester chain

● C and N in bases

● P

**Fig. 7.8**  The double-helical structure of DNA

We shall now describe a few of the main features of the structure, so that we can appreciate how the individual parts of the nucleotides help to stabilize it and also draw attention in a preliminary way to those characteristics that fit the structure for its role in the genetic process. (We should not, incidentally, be surprised to see a helical structure making an appearance again, as it did in the proteins. If a molecule is to be linear in broad outline, then the helix is one of the simplest ways in which the effects of intramolecular forces acting at regular intervals can be accommodated.)

The first feature is that pairs of bases face each other edge-on across the axis of the helix, to which they are perpendicular. It was as a result of formulating the double helical structure that Watson and Crick were led to realize that the hydrogen-bonded pairs adenine–thymine and cytosine–guanine (which we have already mentioned briefly (p. 141)) are the only stable ones and that, for example, adenine–cytosine or adenine–guanine are not stable. This conclusion is consonant both with Chargaff's observations and with the results of the examination of models of all the possible base pairs.

The sugar rings lie in planes parallel to the axis of the helix and hold the phosphate groups to the outside. Looked at overall, the structure is rod-like, scored with two parallel helical grooves, one deep and one shallow.

The division between the hydrophobic core of the stacked bases and the highly hydrophilic exterior with its charged sugar phosphate groups is very striking (Fig. 7.9). The mutual repulsion between the phosphate groups forces

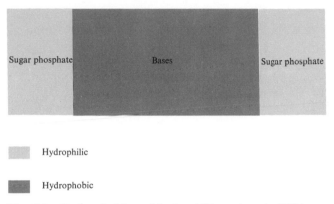

Hydrophilic

Hydrophobic

**Fig. 7.9**  Hydrophobic and hydrophilic regions in DNA

the chain into a straight-line configuration. However, if the charges are screened by salt-bridge formation between the phosphate groups and some other basic group, the chain becomes more flexible. The basic groups can be the basic side chains of proteins (see below). Alternatively, there are certain low molecular weight substances (Fig. 7.10) that are found in association with some nucleic acids (where a highly convoluted structure is required). As with the proteins the structure shows that the hydrophobic interactions are likely to make a significant contribution to the stabilization of the whole molecule, although they are less directional than hydrogen bonds and so will not assist in discriminating between correct and incorrect base pairs.

The hydrophobic interaction accounts perhaps for the bulk of the (approximately) 1000 cal/mole per base pair that holds the double helix together. There is no conflict between saying this and saying that the hydrogen-bonded base pairing is crucial in deciding whether or not the structure is stable. The hydrogen bonds will not form until *after* the hydrophobic interaction has squeezed out the water between the two helices. Moreover their free energy is required to give the final margin of stability to the structure: they are therefore decisive, irrespective of the relative size of the contribution of the two types of force.

We shall see later that the biological function of the nucleic acids calls for the unwinding of the double-helical structure. However, we have just seen that the separation of the components of a single pair of bases requires 1000 cal/mole. If you look for a moment at Table 7.5 you will see that a nucleic acid may have thousands of base pairs per chain. The free energy required to unwind such a molecule (and therefore the equilibrium constant against the unwinding) is huge.

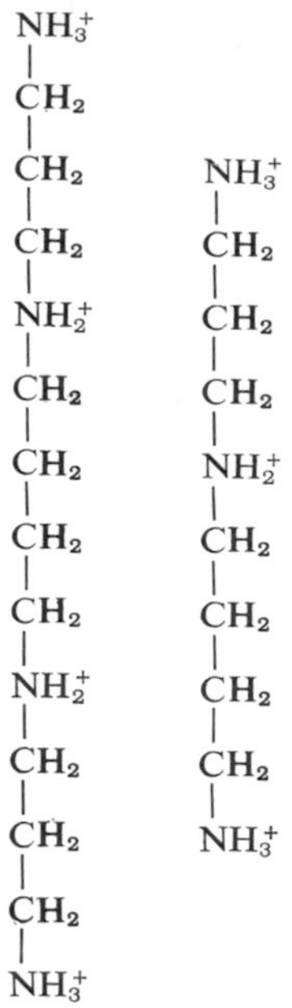

$$NH_3^+ - CH_2 - CH_2 - CH_2 - NH_2^+ - CH_2 - CH_2 - CH_2 - CH_2 - NH_2^+ - CH_2 - CH_2 - CH_2 - NH_3^+$$

spermine

$$NH_3^+ - CH_2 - CH_2 - CH_2 - NH_2^+ - CH_2 - CH_2 - CH_2 - CH_2 - NH_3^+$$

spermidine

**Fig. 7.10**   Polyamines that associate with some types of nucleic acid (see text)

The paradox – the need for unwinding and its apparent impossibility – is resolved when we learn that only a small section of the chain unwinds at a time. The relatively few base pairs that are separated soon close up again, as other, adjacent base pairs unwind. In this way a disturbance, involving only a few bases at a time, is propagated up the whole length of the double helix. All the bases will thus pass through the single-stranded condition, but not at once. The energy requirement is therefore not so large since, once the moving disturbance is established, much of the energy required to separate bases is supplied by that made available by bases reuniting. (Where the biological function calls for it, the bases can reunite, not with those with which they were originally paired but with freshly synthesized chains (p. 436). The principle is the same, since only a few bases are unpaired at a time and the energy made available by bases pairing up balances that required by those that break apart.)

We now come to an important feature of the double-helical structure: one that is crucial in fitting the molecule for its task of acting in the reproductive process. This is that once the sequence of one strand is specified, so, as a result of the base-pairing rules, is the sequence of the other strand. (These are then known as *complementary* sequences.) Turn for a moment to the examples on p. 428 and satisfy yourself that this is so – if you wish, you can construct similar examples of your own.*

* Note also an odd consequence of the use of the phosphodiester bond in DNA. This is that if we follow one chain from the 5′ end to the 3′ (these terms were defined on p. 46) we are simultaneously following the complementary strand from its 3′ end to its 5′ end. These chains are said to be *antiparallel*, and this is of importance in biological function in distinguishing one chain from the other (p. 431). The antiparallel nature of the chains does *not* of course mean that the base sequence of one strand is the base sequence of the other read backwards.

This property is the basis of the way in which DNA is fitted for its task as the repository of genetic information. We shall describe how this is so in outline in Ch. 24. We shall see in particular that the genetic information is encoded in the *sequence* of the bases, in much the same way that the information on the printed page is encoded in a sequence of letters, and that the expression of the information is almost always to determine the order in which amino acids are incorporated into proteins. It is the combination of the various capabilities of the proteins that then largely determines the gross inherited features of an organism – size, shape, colour, metabolic competence – that we can recognize by direct inspection. We shall limit ourselves here to saying that the way in which the structure permits DNA both to provide for its own replication and to specify the amino-acid sequence of proteins is among the most elegant examples of a structure-function relationship that we shall find in biology – as befits the molecule that confers upon living matter what is perhaps its major characteristic, the passing on of the features of one generation to the next. Indeed you may well feel that although the insight that led to our present understanding was a brilliant one, now that we know the details of how the structure is used it is very difficult to conceive of any other that would do.

## The secondary structure of RNA

In contrast to DNA, analysis of the base composition of RNA seldom suggests that it is in a double-stranded structure. Nonetheless, there are a few examples of completely double-stranded RNA molecules. Even where the structure is fundamentally single-stranded, there is a possibility that short tracts of bases will encounter tracts of complementary sequence by folding back of the chain. (Uracil replaces thymine perfectly well in the base pairs (Fig. 7.11).) In these

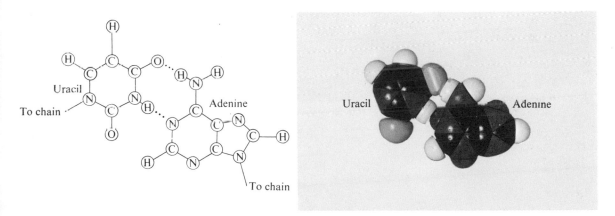

**Fig. 7.11** The hydrogen-bonded base pair between uracil and adenine

cases, will a double helix of the type shown in Fig. 7.8 be possible? We mentioned on p. 140 that the bulk of the additional oxygen atom at the 2′ position of the ribose could restrict the conformation of the chain. This restriction is in fact found to exclude the double helical structure as we have so far illustrated it.

The effect of the extra oxygen atom is to push the sugar ring out of the plane that it might have occupied had the structure closely followed that of Fig. 7.8. This in turn pulls the planes of the bases out of the perpendicular to the axis. The result is a somewhat deformed structure (Fig. 7.12), which nonetheless retains the stabilization by hydrophobic interactions and hydrogen-bonding, and, most important of all (see below), the ability of one sequence of bases to specify the other. We shall see later that these features are needed for certain of the functions of RNA, as they were for all the functions of DNA. The close relationship between the two types of double helix is underlined by the fact that, although there is some evidence that the structure in Fig. 7.8 is the natural one for DNA (it is called the 'B' form) as well as the one adopted under the artificial conditions of the gel used in the X-ray work, Franklin only had to add less water in making up the gel to obtain the so-called A form of DNA. This, it has been shown, is closely similar to the RNA structure of Fig. 7.12a.

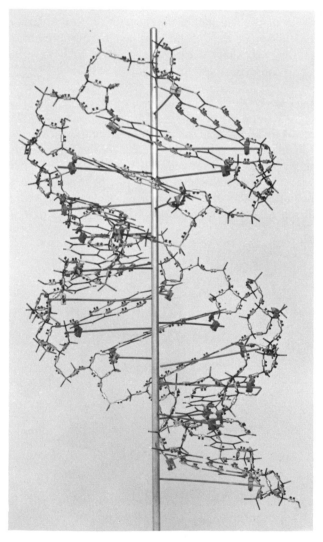

**Fig. 7.12a**  Double-stranded RNA

When RNA is single-stranded it will have (by virtue of the folding back mentioned above) many heavy, rod-like regions of double stranding interspersed with lighter single-stranded regions which will be randomly coiled. (Natural design often tends to maximize the number of complementary regions in a single chain (see below), but even if it did not chance alone would ensure that there would be many. The following calculation shows that there is a very high probability of quite large tracts of complementary sequence being able to find one another and enter into base-pairings. If there were equal numbers of all four bases, the chances of finding a given base in a specific position is 1 in 4. Therefore the chances of finding, say, a tetranucleotide sequence that was complementary to any given tetranucleotide sequence elsewhere in the chain is $\frac{1}{4} \times \frac{1}{4} \times \frac{1}{4} \times \frac{1}{4}$, namely 1 in 256. That is, one tetranucleotide sequence in every 256 bases would, on average, be the correct one. Many RNA molecules are much longer than this

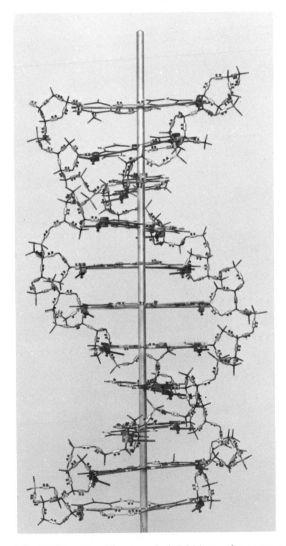

**Fig. 7.12b**   Double-stranded DNA on the same scale

(a)

(b)

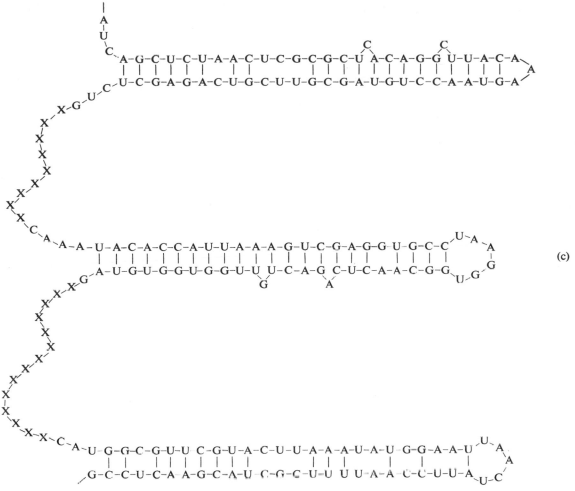

**Fig. 7.13** The experimentally obtained nucleotide sequences of (a) an amino-acyl transfer RNA, (b) 5S ribosomal RNA, (c) a virus RNA. The sequences have been arranged arbitrarily so as to maximize base pairing. X indicates bases that are not yet identified. The abbreviations for minor bases are defined in Table 7.2

and the probability of finding the complementary sequence becomes, correspondingly, very large.)

The result of this mixture of the two types of structure in one molecule is shown in Fig. 7.13. We should not think of it solely as a static structure, because under the random bombardment of solvent and other molecules it will be threshing about vigorously.

## The different types of DNA and RNA

Table 7.5 summarizes some of the principal facts concerning the different types of nucleic acid met with in living matter. For convenience in arrangement we have listed them by the places in which they are found. This has meant that we have had to look ahead somewhat to Ch. 10 in which we describe the cell as

**Table 7.5** The various types of nucleic acid

| SITE | Inside the nucleus (or nuclear zone in those cells, e.g. bacteria, that do not have a clearly defined nucleus, see 'procaryotes' and 'eucaryotes', Ch. 10) | | The remainder of the cell | | Outside the cell (for part of the time) |
|---|---|---|---|---|---|
| | | | Within organized structures, such as chloroplasts and mitochondria | Not within organized structures | |
| **DNA** | | | | | |
| Type | Chromosomal DNA (bacteria) | Chromosomal DNA (higher cells) | | | DNA of viruses (those attacking some animals and bacteria, but not plants) — Small viruses $\sim 5 \times 10^3$; Large viruses up to $10^5$ |
| Approx. number of bases | $\sim 4 \times 10^6$ base pairs | $\sim 5 \times 10^9$ base pairs in total chromosomal complement of one cell | $\sim 10^4$ base pairs | | |
| Topology |  (circle) |  (not known if linear or circular) |  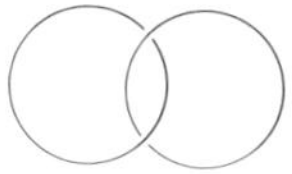 (circle) or (two circles) |   |   $\phi$X174 or $\lambda$ phage; $\phi$X174; T5 phage; T4 phage |
| | | | (traces) | | |
| Associated with protein? | Small quantities of protein only | Large quantities of protein (mostly basic but some neutral or acidic) | Yes | | Yes |
| Function and remarks | Storage of genetic information in the cell 99% or more of the DNA in the cell | | Minor component of DNA of cell | | Carries all essential genetic features of the virus |

**Table 7.5**  The various types of nucleic acid

| SITE | Inside the nucleus (or nuclear zone in those cells, e.g. bacteria, that do not have a clearly defined nucleus, see 'procaryotes' and 'eucaryotes', Ch. 10) | | The remainder of the cell | | | Outside the cell (for part of the time) |
|---|---|---|---|---|---|---|
| | | | Within organized structures, such as chloroplasts and mitochondria | Not within organized structures | | |
| **RNA** | | | | | | |
| Type | Messenger RNA (m-RNA) | Nucleolar RNA (n-RNA; eucaryotes only) | Ribosome-like objects (see next column) are found in these structures | Ribosomal RNA (r-RNA) 3 types:* 5S 16S 23S | Amino-acyl transfer RNA (t-RNA) | RNA of viruses (plant viruses, some animal and bacterial viruses) |
| Approx number of bases | $10^3 - 10^4$ | | | 120  1600  3200 | 70–80 | $10^3$–$10^4$ |
| Topology | 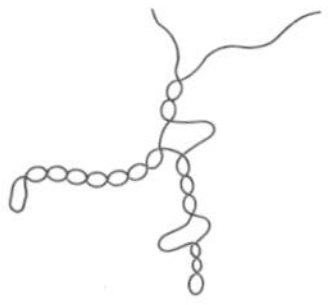 | | 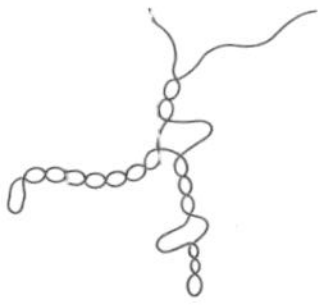 |  |  | Usually   except when replicating but occasionally   Wound Tumour Virus |
| Associated with protein? | See pp. 158, 438ff | | | Yes (see p. 437) | See p. 440 | Yes (exceptions exist, see p. 159) |
| Function and remarks | m-RNA is synthesized here, then migrates to the ribosome (q.v.), carrying the genetic information required for protein synthesis | Non-messenger RNA is synthesized in the nucleus and is assembled in a zone (visible only during some stages the life cycle of the cell) called the nucleolus (p. 218). Ribosomal RNA (q.v.) predominates | | Site of synthesis of protein | Transfers activated amino-acid to ribosomal-messenger RNA complex | As for DNA Secondary and tertiary structure controlled by the virus protein (p. 159) |

* These are sedimentation coefficients for RNA from bacterial ribosomes. Ribosomes from higher cells contain RNA with slightly different sedimentation coefficients (see p. 438).

containing a nucleus and, in the extra-nuclear volume, a number of organized bodies such as the mitochondrion and the chloroplast together with other less organized membrane structures and cellular fluid. As a measure of the relatively restricted range of functions, of which we spoke before, compare the table with Tables 5.1 and 5.2. In contrast to those tables, which mentioned only a few of the many types of proteins, Table 7.5 lists *all* the major forms of the nucleic acid of which we are aware. We intend only to survey the functions in this chapter since they are best dealt with when we come to discuss molecular genetics as a whole in Ch. 24 and protein synthesis in Ch. 26*.

The *number of bases* that are shown are only approximate and vary from one source of a given type of nucleic acid to another. The molecular weights of some of the types are extremely large, since each base pair contributes a molecular weight of about 650. As a consequence, for example, the single molecule that makes up the DNA of the bacterium *Escherichia coli* (see Ch. 10) has a molecular weight of about $3 \times 10^9$. (If it were stretched out in a straight line it would be 1.4 mm long, about 1000 times the largest dimension of the cell itself.) The DNA of an animal cell is vastly larger, the sum of the molecular weight of the DNA from all the chromosomes of a single cell being of the order of $10^{11}$–$10^{12}$. (The length of the DNA in a human cell, when unwound, would be 30 metres!) It is extremely difficult to observe very high values because of the extreme fragility of the molecule toward the shearing forces that it experiences on being handled in such operations as pipetting and stirring. RNA molecules are noticeably shorter, and are as a result rarely subject to mechanical shearing (they are, however, highly susceptible to cleavage by the small traces of nucleolytic enzymes that abound in animal and plant fluids).

The *topology* of DNA is often very strange. DNA from *E. coli* appears, when intact, to have neither a 3′ nor a 5′ end. The explanation, which is confirmed both by electron microscopy and genetic mapping (p. 409 ff.), is that the structure is an unbroken circle of double helix. As you will see from the table, several other types of DNA have this sort of structure – the topology of the DNA of animal chromosomes is much less clear. Virus DNA is usually of the circular form (the bacterial viruses λ and T5 are among the exceptions). In some of the very small viruses, the DNA can be single-stranded.† The DNA in bacteriophage T5 is distinguished from the normal double helix by breaks in each strand. Since the breaks on the two strands do not occur adjacent to one another, the molecule does not fall apart.

The symbol used to indicate the topology of the RNA molecules is intended to represent a single-strand structure alternating between randomly coiled and double-helical regions as described above. The sequences of messenger RNA, transfer RNA and at least one of the ribosomal RNAs show, in theory at least, a great potential for double stranding (Fig. 7.13). Some types of transfer-RNA molecules have been crystallized and the tertiary structure of one has been determined. All the X-ray diffraction studies show a large content of helix. The X-ray photograph looks rather like that of a globular protein, but there is so much helical structure that, at certain orientations of the crystal to the beam of X-rays, there is a helix cross clearly visible, superimposed on the rest of the pattern.

* We suggest that you re-read the next few pages after you have finished Chs. 24 and 26.
† e.g. that of φX174 (this fact explains why the base ratios in Table 7.4 do not follow Chargaff's rules).

The question of *association with protein* is raised in the table in order to emphasize how frequently the nucleic acids are found in complexes with substantial quantities of protein. We shall now consider it along with commenting on the *functions* of the various types of nucleic acid.

*Bacterial DNA* does not complex with substantial quantities of protein, but we shall see in Chs 25 and 28 that association with a certain number of specifically designed protein molecules is required for the DNA to function in the genetic process. This relative austerity in the use of proteins is in contrast with the chromosomal *DNA of higher cells*. Here, the DNA is associated with perhaps five times its own weight of protein. Much of this consists of small, highly basic proteins called histones. These molecules (the amino-acid sequence of one is shown in Fig. 7.14) abound in basic amino acids; it is very likely that the histones

Acetyl–Ser–Gly–**Arg**–Gly–**Lys**–Gly–Gly–**Lys**–Gly–Leu–Gly–**Lys**–Gly–Gly–Ala–
10

–**Lys**–**Arg**–**His**–**Arg**–**Lys**–Val–Leu–**Arg**–Asp  Asn–Ile–Gln–Gly–Ile–Thr–
20                                                                30

–**Lys**–Pro–Ala–Ile–**Arg**–**Arg**–Leu–Ala–**Arg**–Gly–Gly–Val–**Lys**–**Arg**–**Arg**–
40

–Ile–Ser–Gly–Leu–Ile–Tyr–Glu–Glu–Thr–**Arg**–Gly–Val–Leu  **Lys**–Val–
50                                                                60

–Phe–Leu–Glu–Asn–Val–Ile–**Arg**–Asp  Ala–Val–Thr–Tyr–Thr–Glu–**His**–
70

–Ala–**Lys**–**Arg**–**Lys**–Thr–Val–Thr–Ala–Met  Asp–Val–Val–Tyr–Ala–Leu–
80                                                                90

–**Lys**–**Arg**–Gln–Gly–**Arg**–Thr–Leu–Tyr–Gly–Phe  Gly–Gly
100

**Fig. 7.14** The amino-acid sequence of a histone. Basic residues are shown in bold type

are held in the grooves that we mentioned on p. 147 by a great number of ionic bonds between the positively-charged arginine and lysine residues on the one hand and the negatively-charged phosphate groups on the other.

It is very tempting to assign to these proteins a role that would explain one of the puzzling features of multicellular organisms. The problem arises if we take for granted for the moment a conclusion that we stress in Ch. 28, that the DNA of *each* cell encodes in its base sequence the total genetic complement of the whole organism. Any cell, one in the liver, for example, has within it the genetic competence to form the structures proper to a retinal, a pancreatic, a hair-secreting cell and so on, and yet it restricts itself to making use of only that part of its genetic complement that is proper to its assigned function. The protein molecules associated with DNA could conceivably be implicated in all this by binding with those parts of the DNA that must not be used and preventing access by the machinery that enables genes to be expressed. Associated with this question of how cells regulate their expression of their genes are such matters as the molecular basis of the growth and differentiation of the embryo and the molecular basis of malignant disease.

The three molecules of *ribosomal RNA* are also in intimate contact with protein molecules to form the functioning unit at which synthesis of protein occurs, the *ribosome* (for a more exhaustive discussion of ribosomal function, see p. 446 ff.). Each ribosome, in addition to its three strands of RNA, contains about fifty-five molecules of protein, all different in sequence and most of them basic to a greater or lesser extent. The proteins are held in a definite spatial relationship to one another and to the RNA, which winds in and out of the assembly, both binding it together and also presenting the proper portions of nucleotide sequence to bind the other types of nucleic acid which as we shall shortly see are involved in the synthesis of protein. The ribosome also has to catalyse the synthesis of each peptide bond. Taking all these functions into account, it is hardly surprising that the structure is an extremely intricate one, probably more intricate than any other structure that is dealt with in this book. The elucidation of the detailed structure will be a staggering task (each of the protein molecules is of the same order of size as those pictured in Fig. 4.10), but, as a glimpse of an order of structural complexity higher than anything so far achieved, it is to be expected that it will be well worth the effort.

We have said that the amino-acid sequence of proteins is specified by the sequence of nucleotides in DNA. The sequence of the DNA is *transcribed* by base-pairing into RNA. This *messenger RNA* then migrates to the ribosome, the site of protein synthesis, and is used there to direct the order of incorporation of amino acids into the growing chain. There is, however, no easy way in which a specific nucleotide sequence can recognize and bind to a specific amino acid. A nucleotide sequence can recognize and bind to another nucleotide sequence by base pairing, but the shapes and available pattern of forces of nucleotides and amino acids are very dissimilar. They are too dissimilar in fact for anything like base-pairing to occur. Nonetheless it is necessary for molecules of the one class to recognize those of the other if nucleotide sequence is to specify amino-acid sequence. This difficult gap is bridged by the *amino-acyl transfer-RNA* molecules which, we shall see in a moment, while being true nucleic acids possess to some extent properties that are more usually associated with proteins. As a result of this dual nature, they retain the ability to recognize other nucleotide sequences by base-pairing while being able to build recognition sites much more like those that we discussed in Chs 5 and 6, capable of absolute discrimination between amino acids.

We shall see in Ch. 26 that transfer-RNA molecules function in protein synthesis as follows: each amino acid is activated ready for incorporation into the polypeptide chain by means of the formation of a high-energy acyl phosphate bond in its carboxyl group (see p. 441). It is then loaded on to the amino-acyl transfer RNA by the same enzyme that catalysed the activation. (This enzyme is called the amino-acyl transfer RNA synthetase – specific ones exist for each amino acid.) The carboxyl group of the amino acid is transferred from the acyl-phosphate linkage to a simple ester linkage with the 3' end of the transfer RNA chain. Amino-acyl esters have a free energy of hydrolysis of about 8000 cal/mole and so the activation is retained.

We have seen in Chs 4 and 5 that great care must be taken to see that amino acids are not incorporated into the wrong places in proteins. For this reason the process that we have just described has to be very specific. A given amino acid will be bound and activated only by its own specific synthetase, and will then be condensed only to a transfer-RNA molecule that is also specific to it. Once the amino acid is on the correct transfer RNA it can then be recognized by the messenger RNA (see above) by means of base-paired interactions with an identifying sequence (the anticodon) that each transfer RNA carries and which is different for each type.

We have now described the main functional constraints on the transfer-RNA molecules and can begin to trace the features of the structure that have been developed in response. An amino acid recognizes its specific transfer RNA, not by the nucleotide *sequence* to which the amino acid is to be attached (—C—C—A irrespective of the type of transfer RNA) but by an interaction between the *tertiary structure* of the nucleic acid *as a whole* on the one hand and that of the temporary complex between the activated amino acid and its specific transfer RNA synthetase on the other. The transfer RNA is no doubt assisted in forming an appropriate recognition site for the synthetase by the possession of the wide range of modified side chains (see Table 7.2). This interaction between protein and nucleic acid is very much more like that between protein and protein that we described in Ch. 5 than that between two nucleic acids. This unusual role of the transfer RNA molecules possibly accounts for the fact that they are strikingly close to proteins in size (and strikingly dissimilar in this respect from most RNA molecules) and have almost as many types of base as there are amino acids, rather than the four which are the general rule in other types of nucleic acids.

The lobed structure that has been proposed can also be imagined as being of service in the interactions that are necessary with other nucleic acids, since the lobes are single stranded and thus able to enter into base pairing interactions. The anticodon – the point of recognition between transfer RNA and messenger RNA – is always prominent on the second unpaired stretch, counting clockwise from the 5' end, and the other lobes may well be of use in attaching the transfer RNA to the ribosomal RNA.

*Virus nucleic acids* carry all the genetic information that the virus possesses (see p. 223). We saw that bacterial DNA was smaller than that of higher animals – this is partly because bacteria have fewer hereditable characteristics. Viral nucleic acids are smaller yet again, because viruses entrust most of their functions to the machinery of the cell that they invade and require only the minimum genetic information to specify their next generation. Viral *RNA* is noteworthy as the only use of RNA for *storage* of genetic information (see p. 486).

The viral nucleic acids are almost always complexed with protein (a notable exception is potato ring necrosis virus, which Cadman showed to occur occasionally in the wild state as nucleic acid only). The function of the proteins (which, surprisingly after what we have learned of the histones, are seldom rich in basic amino acids) is to protect the nucleic acid from the mechanical and enzymic damage that we referred to on p. 156. The quaternary structure of the protein – of which, as you can see from Fig. 7.15, there are a great variety of types – determines the tertiary structure of the nucleic acid. The nucleic acid is threaded through gaps left when the protein molecules aggregate together. This can be shown, for example, by removing the nucleic acid from a virus of icosahedral symmetry (Fig. 7.15a) and combining it in the test tube with the protein separated from a virus of helical symmetry (Figs 7.15c and 7.17). The morphology of the resulting nucleoprotein particle and therefore the secondary structure of the nucleic acid is that of the donor of the protein.

Note that a helical form of structure is represented once again. Figure 7.16 shows the diffraction photograph of an orientated gel of tobacco mosaic virus (TMV). The helical nature of the particle should be clear to you at once. The

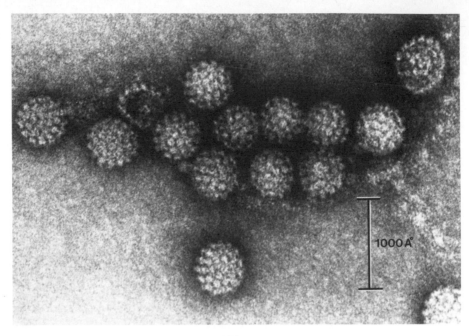

(a)

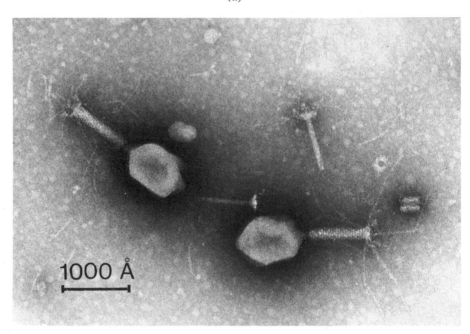

(b)

**Fig. 7.15** (a) Shope papilloma virus. This virus has icosahedral symmetry. LP indicates an incomplete particle. (b) Bacteriophage T4. This virus has a very complex structure. M and R indicate various incomplete particles. (c) Tobacco rattle virus. This virus has helical symmetry. The dark lines indicate that the particles have an axial hole

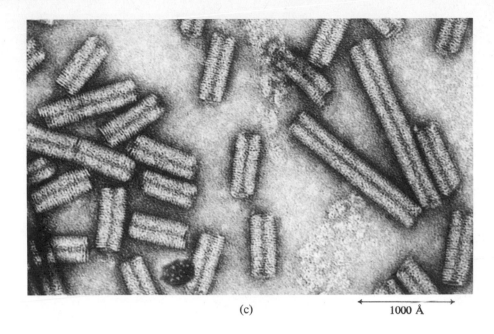

(c)                                    ←———— 1000 Å ————→

near-meridional repeat occurs after three layer lines and, from p. 88, it would appear that there are three subunits per turn of the helix. In fact a more sophisticated analysis of the pattern shows that in this case the 'repeating subunit' is the pitch itself (Fig. 7.17). What then is the repeat corresponding to three turns? The answer lies in the fact that there is not a whole number of protein molecules per turn, but there is a whole number per three turns. The three-pitch repeat is the distance along the axis that one has to go before one comes to a subunit in the same orientation as that from which one started.

It is not known if it is mere coincidence that in the two viruses for which the handedness has been determined the helix is a left-handed screw (Figs 7.17a and 17b) and not a right-handed one (Fig. 7.17c).

## EXPERIMENTAL METHODS OF DETERMINING NUCLEIC ACID STRUCTURE

We shall return in later chapters to much of what we have described in the last few pages. We will conclude, as we did in Ch. 5, with some account of the experimental methods used in determining the structure of nucleic acids. Since the determination of tertiary structure by X-ray methods does not involve any new principles we shall restrict this account to a comparison of the method of determination of the *primary sequences* of proteins and nucleic acids.

Once again the methods fall into the *overlap* or the *stepwise degradation* category. The factors that make the overlap method quite different for the nucleic acids to that for the proteins are, first, that there are fewer types of nucleotide than there are of amino acid, and secondly that the nucleotides are much more similar to each other.

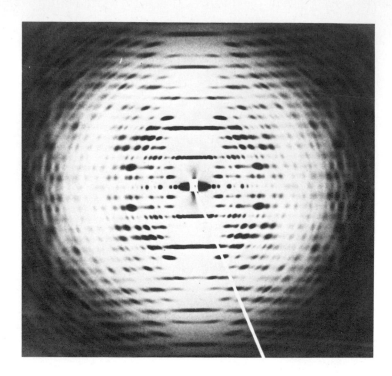

**Fig. 7.16** An X-ray diffraction photograph of tobacco mosaic virus, a virus of helical symmetry. (The white streak is the shadow of a wire which supports a screen (the shadow of which can also be seen, in the centre of the photograph) which absorbs the main, un-diffracted beam of X-rays)

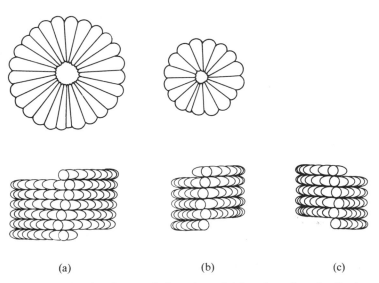

(a)  (b)  (c)

**Fig. 7.17** A drawing in plane and elevation of (a) a short length of tobacco rattle virus (see Fig. 7.15c) and (b) tobacco mosaic virus (see Fig. 7.16). Both structures are left-handed helices. (c) shows what the right-handed structure of tobacco mosaic virus would look like

The consequence of there being *fewer types* is that there are fewer possible fragments. There are 20 different amino acids, $20 \times 20$ different dipeptides, $20 \times 20 \times 20$ different tripeptides, and so on. There are, in contrast, only 4 nucleotides (in most cases), $4 \times 4$ dinucleotides, $4 \times 4 \times 4$ trinucleotides and so on. This factor, particularly when the greater length of nucleic acid chain is taken into account, makes the overlap method more difficult. If we obtained the tripeptide phe–phe–tyr from insulin then we could be fairly sure that it came from a unique site in the molecule: the peptide phe–tyr–pro– would almost certainly be a true overlap and we would deduce phe–phe–tyr–pro– and so on. However, a calculation similar to that on p. 151 will show that U–U–C, say, will on average occur once every sixty-four residues (thus perhaps 20–30 times in 16*S* RNA). U–C–A would be quite useless as an overlap since the sequence U–U–C–A would be expected to occur five or six times anyway. U–C–A could in any case have come equally well from A–U–C–A, C–U–C–A, and G–U–C–A, all of which should occur many times.

It might be thought that the occasional minor base would be very helpful – if there was only one pseudo uridine (see table 7.2: symbol $\psi$) in the molecule, then the sequences U–$\psi$–C and $\psi$–C–A would be bound to overlap and give U–$\psi$–C–A. This is a real advantage, but workers who have determined sequences of transfer-RNA molecules have found that a counteracting disadvantage is the need, if the sequence is to be a complete description of the covalent structure, to elucidate the structure of each new minor base as it is discovered. Such elucidation can often be exceedingly difficult (imagine trying to determine the structure of N(6)-isopentenyl adenosine (Table 7.2) with, at most, a few micrograms of material) and the consequent labour outweighs the advantage of an unambiguous overlap.

The similarity of bases poses a number of additional problems. First, specific nucleases are more difficult to find, since the ability to discriminate would tend not to arise during the evolution of a protein unless there were some biological necessity for it. Only one enzyme (ribonuclease $T_1$ from fungi, which cleaves at G) is as useful as trypsin has been for the proteins and others are sorely needed, especially for work on DNA. Secondly, the similarity of bases means that the fragments are extremely similar in properties and correspondingly difficult to separate.

Opportunities for sequence determination of nucleic acids on the same scale as for proteins had to await the sophisticated separation procedures developed by Sanger and his co-workers (see Appendix 2) and the development of methods for restricting the number of fragments and thus increasing their size in such a way that there was some chance that they represented unique sequences. The two methods of achieving this last aim have been a very quick exposure at low temperature (say 30 seconds at $0°$) to the most specific endonuclease available, and the chemical modification of some bases (e.g. methylation) which rendered them no longer recognizable by the specificity site of the enzyme.

Progress in the use of overlap methods has been greatly helped by a fortunate fact that was discovered when the stepwise degradation approach to sequence was tried. The spleen diesterase, which we have already mentioned (p. 139), breaks down a nucleotide chain by progressive removal of nucleoside-3′-phosphate from the 5′ end. It happens that, under the right conditions, it leaves untouched a small quantity of the oligonucleotide formed after each step.

If these intermediate oligonucleotides are separated (Appendix II) and analysed, their base compositions give the sequence at once. It is by such methods that the sequences shown in Fig. 7.13 were obtained. Many more are in the process of being determined, and we may look forward soon to being able to draw detailed conclusions about structure–function relationships, as we have done for the proteins in Chs 5 and 6, and about the evolution of nucleic-acid structure as we shall do for the proteins in Ch. 27.

# Chapter 8  Structure and function of polysaccharides and lipids

We have now moved from the proteins and nucleic acids, the two classes of macromolecule for which a great quantity of structural data is (or is becoming) available, to classes of which rather less is known. This relative ignorance is partly, as we shall see, because the structures are in some cases less closely defined than those that we have dealt with before, but partly also because of the greater difficulties that lie in the way of the experimenter. Little X-ray information is available, for example, for the precise tertiary structure of any polysaccharide and only scanty X-ray data exist for the lipid assemblies.

These difficulties should not cause us to forget that members of both of the types of macromolecule that we shall describe in this chapter have essential roles to play in the maintenance of life, and their mode of action must depend on the same principles that we have proposed and exemplified in earlier chapters. The rapidly growing body of knowledge about their structure and behaviour (which has reached a stage similar to that which the proteins and nucleic acids occupied perhaps only twenty years ago) makes it possible for us to begin to trace, in some of the simpler cases, the way in which these principles are employed. We believe that this chapter presents the type of data with which we must be armed if we are to be able to follow the relationship between structure and function. We hope that the day is not too distant when sufficient new data become available for you to have the opportunity, at least in part, of verifying for yourselves whether or not this belief is justified.

## THE POLYSACCHARIDES

The principal functions of the polysaccharides mirror those of those proteins that we discussed early in Ch. 5 – they can act as *food stores* or as *structural* components. (It is also becoming clear that some are able to behave like the other class of proteins discussed in that chapter, by undertaking *specific binding* interactions.)

As usual we will start our discussion of the polymer with an examination of the types of monomer that are available, and the characteristics that each can contribute to the final structure.

### The monosaccharides

The monosaccharides have the empirical formula $C_nH_{2n}O_n$, in which $n$ ranges from 3 to (in biological materials) 7, and in which the carbon atoms are adjacent to one another and all but one of the oxygen atoms form part of an —OH group. This remaining atom forms part of either a keto or an aldehyde group. This structure gives great scope for *isomerism* and we must look briefly at the various types that occur.

First, the monosaccharides in which the non —OH oxygen is in a keto group are called *ketoses*, and those with an aldehyde group are called *aldoses*. The parent structures (trioses, $n=3$) for the two types are dihydroxyacetone and glyceraldehyde respectively. All the other monosaccharides in which we are interested can (formally) be derived from these by adding —CHOH units next to the $>C=O$ of the keto- or aldo-group to give *tetroses, pentoses, hexoses* and so on.

This is an important type of isomerism in sugars. Here, as in the proteins, the biosynthetic machinery is very discriminating and only the correct stereoisomers are used in the macromolecules.

Glyceraldehyde has an asymmetric carbon atom and thus those structures that are derived from it will have optical activity. It is D-glyceraldehyde (Fig. 8.1)

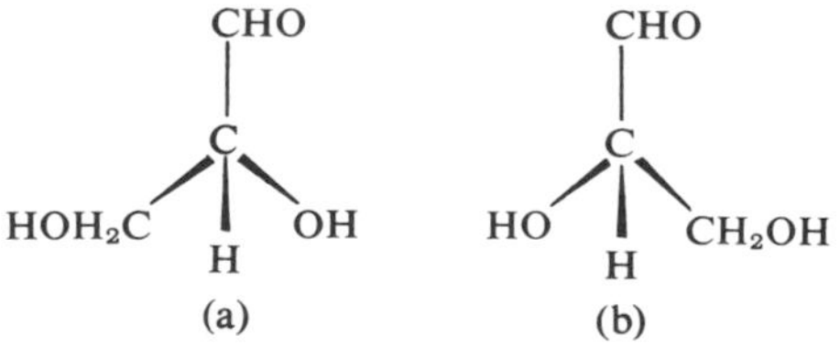

**Fig. 8.1**   (a) D-Glyceraldehyde, (b) L-glyceraldehyde

that is, formally, the parent structure for nearly all natural aldoses. Once again (cf. p. 50) we have a choice that appears to have been arbitrary, but which had to be kept to ever after. Dihydroxyacetone is not optically active, but as soon as one adds a —CHOH next to the carbonyl group a centre of asymmetry is created. The configuration here can then be related to that of glyceraldehyde and it is found that once again all ketoses are D (Fig. 8.2). (The actual direction of rotation (right or left) of polarized light for any molecule is no more predictable from the structure of a monosaccharide than it was for the amino acids.)

Each subsequent addition of —CHOH groups to extend the carbon skeleton introduces another asymmetric centre. The consequent stereoisomerism at these centres is termed *epimerism* and the two possible molecules are called *epimers* of each other.

In distinction to the strict choice of stereoisomerism of the centre that is formally related to glyceraldehyde, nature tolerates the existence of most epimers of the tetroses, pentoses and hexoses. In fact the multiplying possibilities that are created by the existence of epimers supplies much of the variety between monosaccharides.

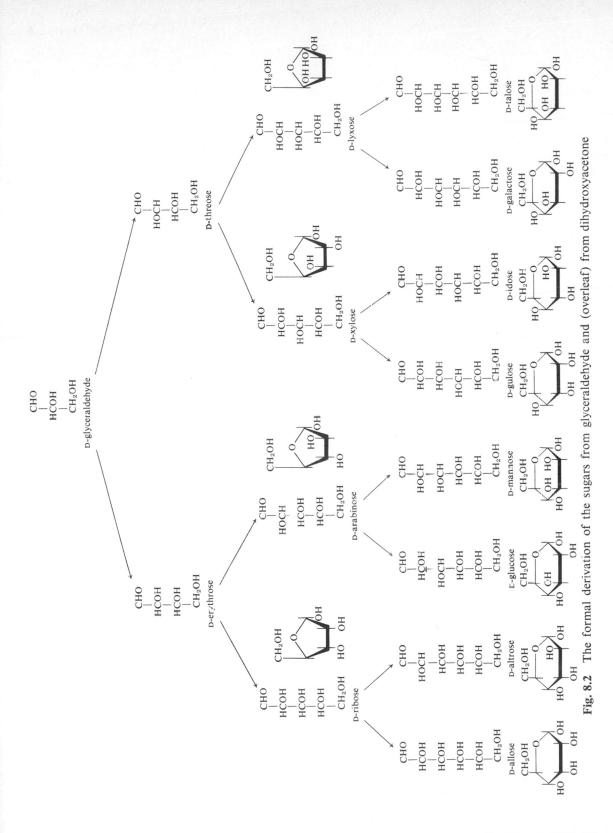

**Fig. 8.2** The formal derivation of the sugars from glyceraldehyde and (overleaf) from dihydroxyacetone

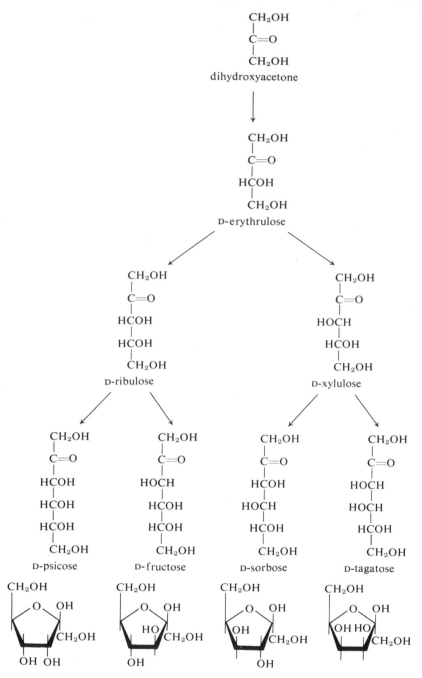

Fig. 8.2  Continued

# Isomerism that depends on ring structure

*Anomerism*

Note that we have shown the aldo-pentoses and all the hexoses as ring structures. The reason is that the reactivity of the $\rangle C{=}O$ in both aldose and ketose is such that it will readily react with an —OH group provided that (as with many of the ring closures met with in organic chemistry) a five- or six-membered ring can be formed. This restriction excludes the trioses and the tetroses and also, as a moment's inspection will show, the keto-pentoses.

The formation of these rings introduces a new asymmetric centre and a new type of stereoisomerism. Stereoisomerism on this new centre is called *anomerism* and the two possibilities are called *anomers* of each other. The two anomeric forms of any sugar are called α and β, the α-form being that in which the anomeric —OH group points furthest from the —CH$_2$OH group (Fig. 8.3). These terms are used in distinction to 'epimerism' and 'epimers' (though the type of isomerism is similar in principle) because of the readiness, to which we referred on p. 36, with which anomers interchange.

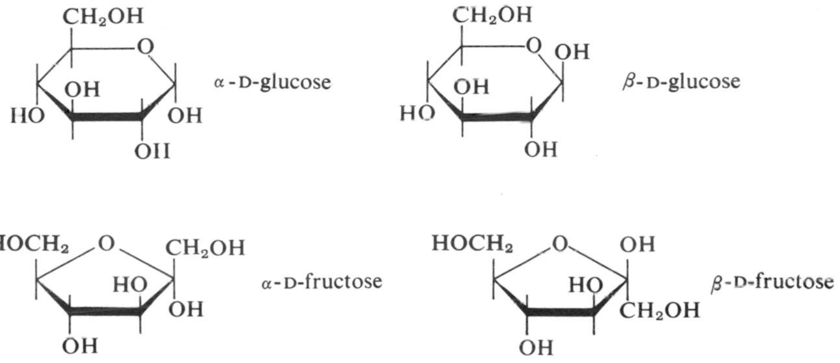

**Fig. 8.3**  The α- and β-configuration of glucose and fructose

These substances interconvert due to a dynamic equilibrium (Fig. 8.4). Although the ring form predominates, each molecule spends some time in the open form, when the asymmetry is lost. When the open-chain form passes back to the ring form it is possible for it to give rise to either anomer. Thus, a pure sample of an anomer only remains so when in the solid state. In solution, particularly at

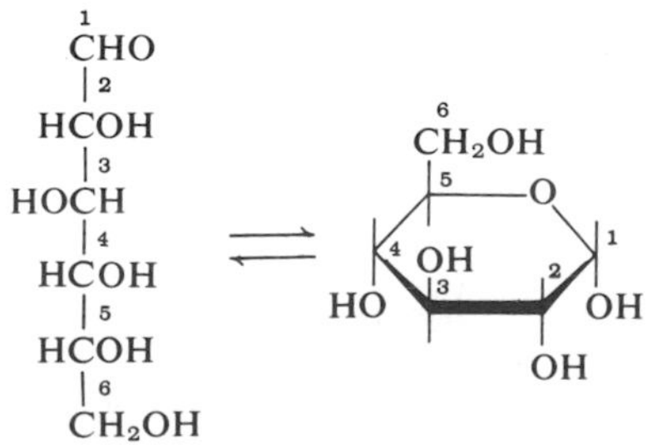

**Fig. 8.4**  The equilibrium between the straight-chain and the ring structure in the sugars. Note the conventional system of numbering

alkaline pH (in which the interchange between ring and open structure is facilitated) a pure sample of one anomer will change to an equilibrium mixture of the two forms. There need not be equal quantities of the two forms at equilibrium because there are other centres of asymmetry and so anomers, like *allo* and *threo* pairs of amino acids, do not have identical physical properties (see p. 50) including $\Delta G°$ of formation. Equilibration between the anomers is called *mutarotation* and it can be followed by observing optical rotation (Fig. 8.5).

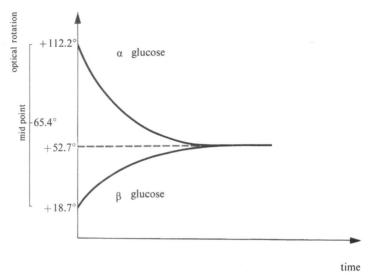

**Fig. 8.5**   The experimental detection of mutarotation (see text)

We can now see that the epimers do not interchange so easily because there is, in contrast to the anomeric centre, no easy mechanism for opening of the ring next to the epimeric centres.

*Pyranose–furanose isomerism*

The aldo-hexoses, unlike all the other types of monosaccharide shown so far, can form both five-membered and six-membered rings (Fig. 8.6). The five-membered

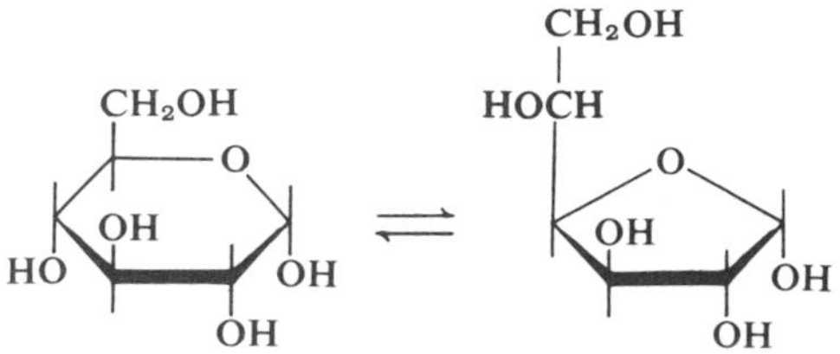

**Fig. 8.6**   The equilibrium between the pyranose and furanose form of an aldohexose

form, like the keto-hexose ring that it resembles, is called the *furanose* form. The six-membered form is called the *pyranose* form. The pyranose form has a lower $\Delta G°$ of formation than the furanose form and since the aldo-hexoses can exist in either form the pyranose is that usually met with.

The pyranose ring can exist in two configurations: the 'chair' and 'boat' (Fig. 8.7). One may be deformed into the other without any covalent changes.

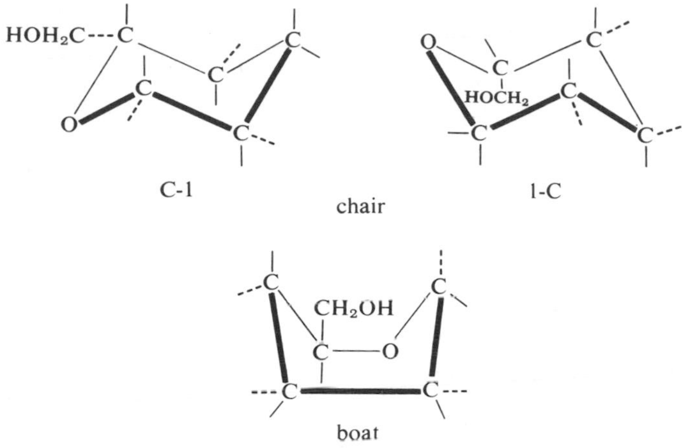

**Fig. 8.7**  The chair and boat forms of the pyranose ring

The 'chair' form is more stable and is the one usually found. There are two ways in which a chair can be made from a pyranose ring. In glucose, form C-1 is the more stable. Less stable than either is the form intermediate between the chair and boat, which we met on p. 118. Note that the —OH groups may be *axial* or *equatorial* and that, when the molecule passes from one structure to another, an —OH group may change from equatorial to axial and *vice versa*. This can affect the environment of the group and the interplay of forces between a mono-saccharide and other molecules.

## The glycosidic linkage

The same reactivity of the aldo- or keto-group that allows the formation of the ring structures in Fig. 8.2 is also the basis of the linkage between the mono-saccharides. We saw in Ch. 3 that the sugars are joined by the *glycosidic* linkage that forms as a result of the elimination of a molecule of water between the —OH group on what was (before the ring closure) the keto- or aldo-carbon atom, and the —OH group on any of the carbon atoms in the next monosaccharide. A great deal of the functional variations that we shall discuss derive from permuting two types of possibilities. On the one hand there are the many possibilities provided by the existence of so many —OH groups with which the reactive —OH might condense, and on the other hand there are the two anomeric positions that this reactive —OH group can adopt. Figure 8.8 shows a few of the possible structures (the nomenclature is self explanatory). We shall see in a moment that these apparently trivial variations in structure have a drastic effect on the properties of the polysaccharides – for example a straight chain of glucose molecules provides great mechanical strength if the linkage is 1 → 4-$\beta$ (cellulose, p. 176), an insoluble but hydrated granule if it is 1 → 4-$\alpha$ (starch, p. 174) or an extremely sticky jelly if it is 1 → 6-$\alpha$ (dextrans, p. 174).

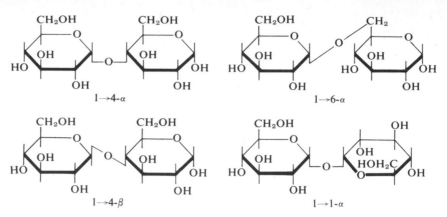

**Fig. 8.8**   Various disaccharides of glucose

## The general characteristics of the monosaccharides

About ten of the monosaccharides that we have so far discussed are met with in the polysaccharides (although, as we shall see, it very frequently happens that no more than one or two of these will occur in any given polysaccharide).

There seems to be little difference between the monosaccharides – there is nothing like the range of classes that we met with in the amino acids or even the more limited range of functional groups found among the nucleotides. At first sight we might expect the only forces by which they could interact to be the hydrogen bond and the short-range dispersion forces (Ch. 3). These forces are indeed important when we seek to explain the properties of polysaccharides, but we should not neglect the possibility of hydrophobic contacts. This might seem strange in molecules in which —OH groups are so abundant, but a moment's examination of, say glucose (which is the monosaccharide most widely used in polysaccharides), will show how it can occur. Glucose, when in the C-1 chair form, has all its  OH groups equatorial (Fig. 8.7) and there are therefore parts of the surface of the molecule that resemble in some degree a cyclic hydrocarbon. Some degree of hydrophobic interaction might be possible between these areas and another hydrophobic structure.

Nonetheless, the possibilities for permutation are much fewer than we have met with before in the other macromolecules, and they are further reduced by the way in which many polysaccharides limit themselves to only one type of mono-saccharide per polymer. This limitation is clearly connected with the much narrower range of functions that the simple polysaccharides undertake, and it incidentally saves the cell from having to elaborate a system for the control of sequence such as that we touched on in the last chapter. We shall see in a moment that the deployment of most of the types of forces that we have become familiar with in previous chapters is made possible by one or other of the natural chemical modifications of the monosaccharides. Modifications are known that introduce charged, hydrogen-bonding or hydrophobic groups. We shall, how-ever, be better able to appreciate the need for these modifications if we consider the simple types first. We shall follow the same classification that we used in the proteins – storage, structural and binding. All the storage polysaccharides with which we shall deal and many of the structural ones manage without modified monomeric units.

## Food-storage polysaccharides

We have already mentioned that the oxidative degradation of carbohydrate is one of the principal tertiary sources (p. 23) of metabolic energy. We shall explore the reaction schemes that are employed for this in Part 2 of this book. We shall see that the degradative pathways usually begin with glucose, or some sugar that is readily inter-convertible with it. Now in times of surplus of carbohydrate foodstuffs it is desirable to be able to store the excess material. We shall see that it is possible to convert excess carbohydrate to fat (pp. 273 and 340), but it is found that (see p. 291) it is undesirable to store *all* the excess carbohydrate in this form. It would, in theory, be possible to keep the glucose in the monosaccharide form, but it would then be readily soluble. A high concentration of mono-saccharide thus presents problems of osmotic pressure and the compounds would be perpetually diffusing from their sites of storage. Such diffusion would either bring about the loss of monosaccharide or allow it to interfere with other metabolic reactions.

The storage polysaccharides provide the answer to this problem. When mono-saccharides are polymerized the molecular weight is raised, a number of —OH groups become engaged in linking the monomers, and the chances of strong, non-covalent, inter-molecular interactions are increased. As a result of these factors solubility decreases and a quantity of monosaccharide that would have been excessively troublesome if it had been in solution can be stored in a convenient granular form.

This, then, is the first functional constraint on the storage of polysaccharides – that they should be sufficiently insoluble to form granules. This requirement is met by having between a few hundred and a few thousand monosaccharide residues in the polymer. The second is that the granule should have a sufficiently loose tertiary structure to allow access to the enzymes that add sugar residues to the chains or subtract them from it. These needs are met by having the polymer as a single chain (e.g. amylose, see below) or as a reasonably open, branched structure (e.g. glycogen, see below). (Recent work has shown that some at least of the enzymes that we have just mentioned are themselves bound on to the surface of storage granules in a mosaic structure. It seems that these granules, long thought to be essentially random structures, do in fact possess at least a little of the organized nature of most biological material.)

Even when all these needs are taken into account the requirements that the structures have to fulfill are not so precise or demanding as those that we noted for proteins or nucleic acids. We find, as a consequence, that considerably less control is exerted over the structures of these macromolecules. The number of monomers in the chain, the number of branching points and the positions at which they occur all seem to be allowed a considerable variation. The macro-molecules are thus what is termed a *population*, with structural characteristics varying about an average: when we describe any given structural property we shall be describing the average value.

*Starch* is the principal store for carbohydrate in plants. Starch is intended to nourish the plant during periods when photosynthesis is impossible and therefore the plants have the means to break it down when necessary. Animals too have hydrolytic enzymes that enable them, on ingestion of plant matter, to use these stores for themselves. (The plant kingdom as a whole is paid back, as the animals eventually release the carbon atoms into the atmosphere as $CO_2$. As things stood before industrialization began to release large additional quantities of $CO_2$ into the atmosphere, each molecule of $CO_2$ went through the cycle of photosynthesis–plant–animal–release on average about once every 2000 years. The plants depend on the $CO_2$ released by animals as much as the animals depend on the carbohydrate fixed by plants, although we might think that animals are in a far more dependent position. It is a moot point which kingdom would last the longer if the other were suddenly to disappear.)

Starch consists of a mixture of two types of glucose polymer. The first is *amylose*, a straight chain of between a hundred and a thousand residues joined by $1 \rightarrow 4$-α-linkages (Fig. 8.9a); the other is *amylopectin*. The $1 \rightarrow 4$-α-link is still prominent in amylopectin, but $1 \rightarrow 6$-α-links are employed in addition to produce the branches of which we spoke (Fig. 8.9b and 8.9d). Of the 5000 or so glucose residues in the average molecule of amylopectin, as many as 400 may be involved in the $1 \rightarrow 6$-α-branching linkage as well as in the $1 \rightarrow 4$-α-linkage. (These figures vary greatly depending on the source of the material.)

Neither amylopectin nor amylose is truly soluble in water, but both are sufficiently open structures to allow many of the —OH groups to interact in water. As a result the chains are not all clumped together and the enzymes are able to gain access to them as required. Amylose, being the simpler, more regular structure, must have the simpler conformation; it is in fact one of the few polysaccharides of which anything at all is known of the tertiary structure. It is an open, helical molecule (Fig. 8.9c). The inside diameter of the helix is just large enough to accommodate an iodine molecule and it is the consequent change in the light-absorbing properties of the halogen that is responsible for the blue colour given in the starch–iodine test. Most other polysaccharides, including amylopectin, give only dull, reddish brown colours.

*Glycogen* is the storage material from animals that has the role corresponding to that of starch in plants. It is like amylopectin, rather than amylose (for which there is no animal equivalent) but has somewhat more glucose residues per molecule and about one and a half times as many branching points. These differences do not alter the functional behaviour of the molecule to any significant extent.

*Other storage polysaccharides* are known. *Inulin*, for example is $2 \rightarrow 1$-β-poly-fructose.* The *dextrans* are $1 \rightarrow 6$-α-polyglucoses, with, in some cases, varying degrees of cross linking by means of $1 \rightarrow 4$-α and other linkages. They are usually considered to be storage polysaccharides for the organisms that produce them (yeasts and bacteria). However, it is probable that they have a structural role as well. For example, the bacterium *Leuconostoc mesenteroides* will discharge large quantities of dextran into its environment, so long as it is kept supplied with sugar. The $1 \rightarrow 6$-α-chain appears to allow a greater degree of hydration of the structure than was true of the $1 \rightarrow 4$-α-chain, and thus the dextran swells in water to a much greater extent and produces an extremely viscous (sometimes

* This is the storage polysaccharide of a number of plants. It is found, for example, in the Dahlia tuber.

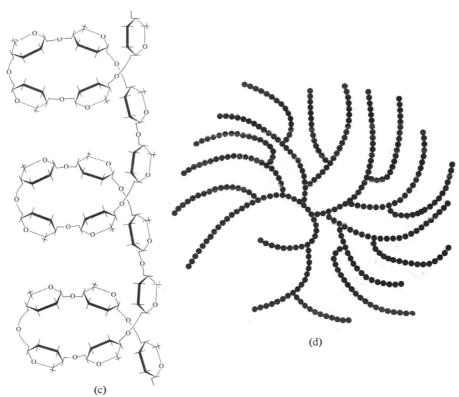

**Fig. 8.9** (a) Amylose, showing the mode of linkage of the chain; (b) amylopectin, showing a branch point; (c) the conformation of the chain in amylose; (d) the branched structure of the amylopectin and glycogen type of molecule

even solid) solution. *Leuconostoc* may use this property to immobilize its competitors: it certainly uses it to immobilize itself on the surface of the teeth of those unwise enough to give it the chance to do so by eating foods with added sugar in them. Other bacteria anchored by the dextran are then able to initiate dental decay, of which they are in fact thought to be the prime causes.

## Simple structural polysaccharides

If mechanical strength is required rather than the simple ability to store mono-saccharides, the polymer chains will usually be longer. In addition, configurations will be chosen that do not allow such extensive interaction between —OH groups and water as those that we have seen so far. These too are fairly simple requirements and so, once again, we see rather simple structures being used to meet them, in which some variation is permitted from one molecule to the next. We shall also see that, as with the structural proteins, strength or elasticity can be enhanced by cross linking, and by embedding in a matrix of some other substance.

*Cellulose* is the most important structural polysaccharide both in terms of abundance (over half the carbon atoms that form part of living material at any one time are in the form of cellulose) and in breadth of distribution.

The chemical structure is simply one of very long chains of glucose residues linked by the $1 \rightarrow 4$-$\beta$-glycosidic bond (Fig. 8.10.)

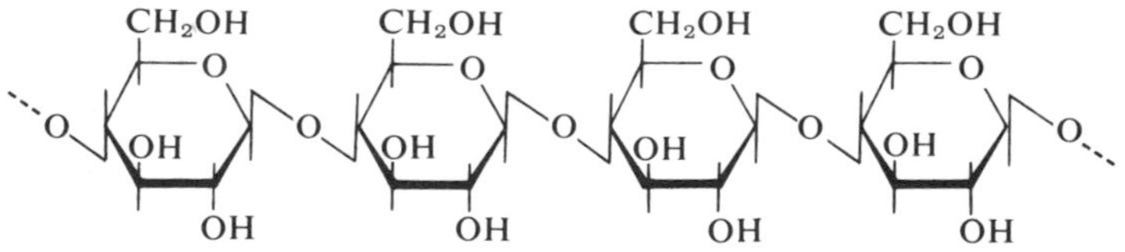

**Fig. 8.10**   Cellulose

Extraordinarily enough, in spite of the simplicity of the basic structure and the importance of the material, little is known about its three-dimensional structure, except that cellulose fibres consist of many chains laid parallel to one another. The $1 \rightarrow 4$-$\beta$-structure must allow a great deal of interchain bonding, because the access of water to the —OH group is totally denied, and even pure, uncomplexed cellulose (cotton for example) is totally insoluble in water and of considerable mechanical strength. Simple light microscopy shows that the chains must be very regularly disposed, since a single cellulose fibre has a near-transparent, crystalline appearance.

Higher animals have never developed enzymes that can hydrolyse cellulose. Those that can utilize cellulose do so by virtue of the cellulases secreted by micro-organisms in their intestines.

The properties of cellulose are usually enhanced by embedding in other materials. In plant cell walls, a collagen-like protein called *extensin* can be covalently bound to the chain at intervals, and this, where it occurs, may have the function of storing strain energy to prevent the propagation of cracks that was explained on p. 98. Other materials called *hemi-celluloses* appear to act in wood as a cement for the whole assembly, presumably because they are able to hydrate to some extent and form sticky, hydrogen-bonded complexes with water. *Xylans* (poly-1 → 4-β-xylose with some added arabinose molecules) and *pectins* (formed from hexose molecules modified to enhance their hydrophilic properties, see below) are the main components of the hemicellulose fraction of wood. Finally, when wood passes from the green, pliable stage, it is hardened by the interpenetration of a complex polymeric substance called *lignin*. This substance is formed by the cross-linking of a number of non-sugar, aromatic compounds (Fig. 8.11). The process of lignification has some chemical resemblance to that which strengthens the proteins of the insect cuticle (p. 97).

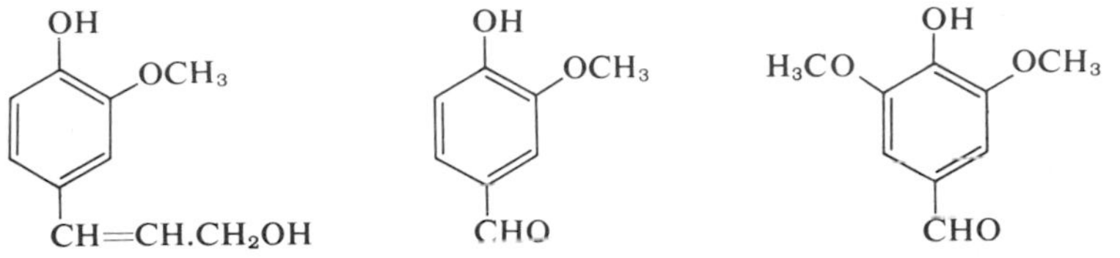

$$OH \qquad OCH_3 \qquad CH=CH.CH_2OH$$

coniferyl alcohol

$$OH \qquad OCH_3 \qquad CHO$$

vanillin

$$H_3CO \qquad OH \qquad OCH_3 \qquad CHO$$

syringaldehyde

**Fig. 8.11** Some of the aromatic compounds found in lignin

Other polymers of a single sugar (*homopolysaccharides*) are known besides those mentioned. Those of mannose (*mannans*) are found in the cell walls of both micro-organisms and plants, unlike those of xylose and arabinose which we have just mentioned and which are found principally in plants.

# Complex structural polysaccharides

These polysaccharides have their structural competence enhanced by chemical modification of the monosaccharides. Table 8.1 shows a number of the most important modified forms. Some modifications can introduce negative, or, very occasionally, positive charges. In neutral modifications the effect can be to enhance hydrophobicity or hydrogen-bonding ability. All types of modification change the steric properties of the molecule. Note that some of these modified forms are sufficiently useful to have been incorporated into other types of macromolecule. (Indeed we have, in the pair ribose–deoxyribose, one of the few examples (see the nucleic acids, p. 149 ff.) for which precise structural information is available that can be related to the properties of a sugar and the effects of chemical modification to its structure.) We shall mention a number of products, first from plants, then from bacteria and then from animal sources.

*Plant gums*

Plant gums help maintain the integrity of many vegetable structures. *Pectic acid* (which we have already met in hemicellulose (above), in the form of pectin, its methyl ester) is responsible for the solidification of jam. It contains galacturonic acid. *Gum arabic* contains galactose, galacturonic acid, arabinose and rhamnose. A gum needs to be easily hydrated, but not so much so that the ability to form non-covalent linkages between, on the one hand, the molecules that are

**Table 8.1** Modifications to the monosaccharides

| | Structure | related to | Remarks |
|---|---|---|---|

*Basic Modifications*
Amino sugars

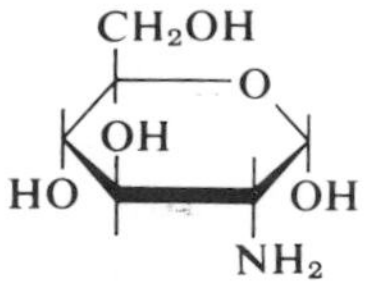

CH$_2$OH ... OH ... HO ... OH ... NH$_2$

D-glucosamine — glucose

Basic character usually suppressed by acylation (see below)

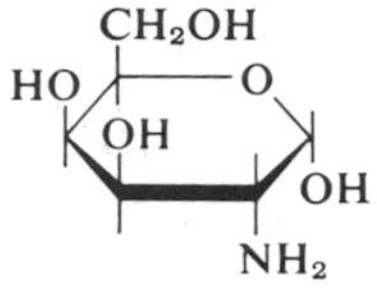

CH$_2$OH ... HO ... OH ... OH ... NH$_2$

D-galactosamine — galactose

*Neutral Modifications*
N-acyl amino sugars

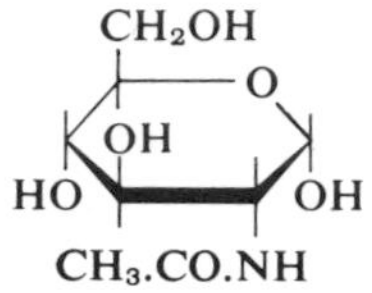

CH$_2$OH ... OH ... HO ... OH ... CH$_3$.CO.NH

N-acetyl-D-glucosamine

CH$_2$OH ... HO ... OH ... OH ... CH$_3$.CO.NH

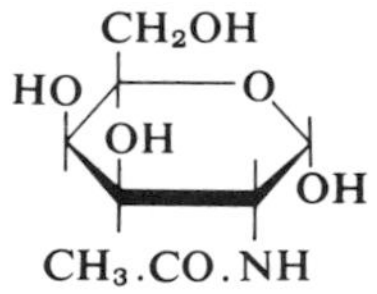

N-acetyl-D-galactosamine

The most usual form of the amino sugars. Very widely distributed. Included in some complex lipids. (The glycolyl group,

$$CH_2OH-\overset{\overset{O}{\|}}{C}-O-$$

which is slightly more hydrophilic, can replace the acetyl group in some cases.)

Reduction*

CH$_2$OH
|
CHOH
|
CH$_2$OH

glycerol — glyceraldehyde

Found in teichoic acids and lipids (q.v.)

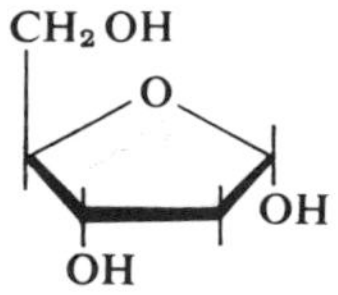

CH$_2$OH ... O ... OH ... OH

D-deoxyribose† — ribose

Found in DNA.

---

* See also the cyclic polyhydric alcohol *inositol* (p. 191).
† This sugar could have been called deoxyarabinose. However the name deoxyribose is preferable because the molecule is derived biosynthetically from ribose.

| | Structure | related to | Remarks |
|---|---|---|---|

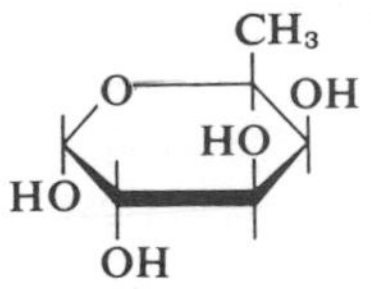

L-fucose (6-deoxy-L-galactose)

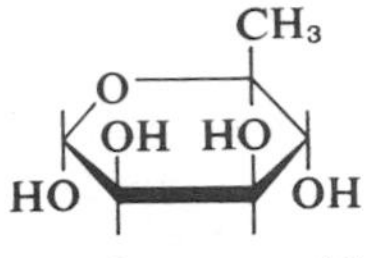

L-rhamnose (6-deoxy-L-mannose)

These compounds are among the very few which have the L-configuration

*Acidic Modifications*
Oxidation to sugar acids

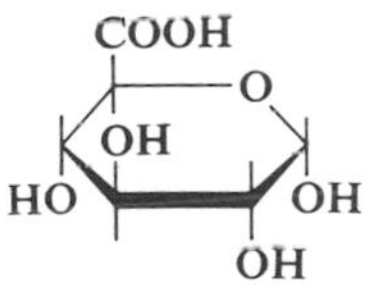

D-glucuronic acid

glucose

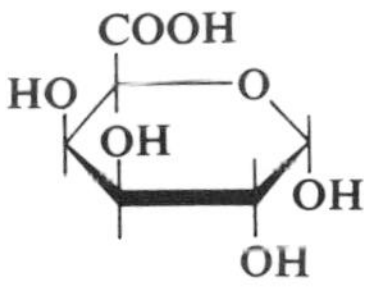

D-galacturonic acid

galactose

Widely distributed. Obtained by enzyme-catalysed oxidation of $CH_2OH$ group. The generic name is *uronic acids*.

Analogous compounds formed from mannose, idose and gulose

Sulphation

$CH_2O$—S=O

$CH_3$.CO.NH

$CH_2OH$

HO—S—O

$CH_3$.CO.NH

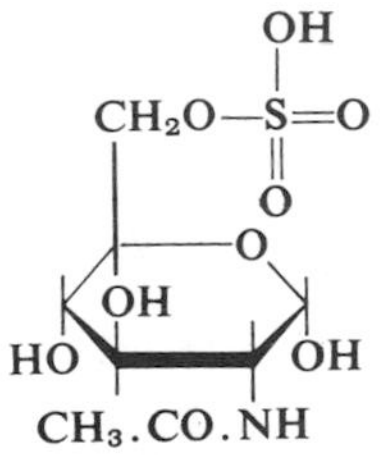

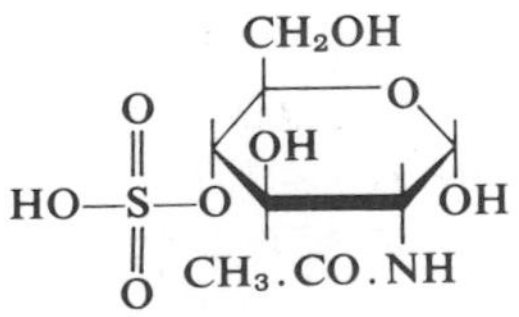

N-acetyl glucosamine sulphates

glucose

Sulphate usually on carbon 6 or 4 of an amino sugar. More rarely on the —$NH_2$ group itself (to give —$NHSO_2OH$, the *sulphamic acid* group)

**Table 8.1** (continued)

| | Structure | related to | Remarks |
|---|---|---|---|
| Muramic and Neuraminic acid | 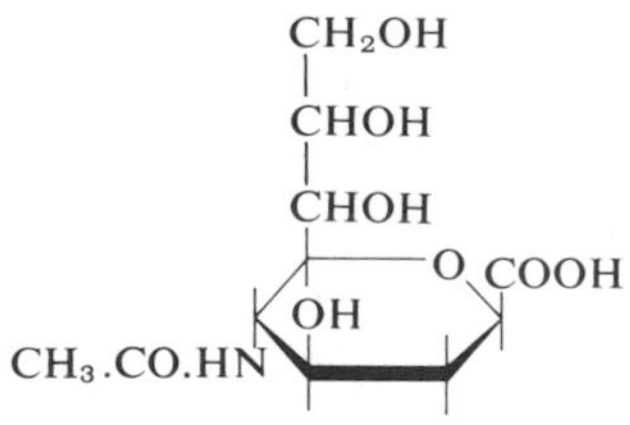 $CH_2OH$ <br> \| <br> $CHOH$ <br> \| <br> $CHOH$ <br><br> N-acetyl neuraminic acid | pyruvic acid and N-acetyl mannosamine | Also found in association with lipids |
| | 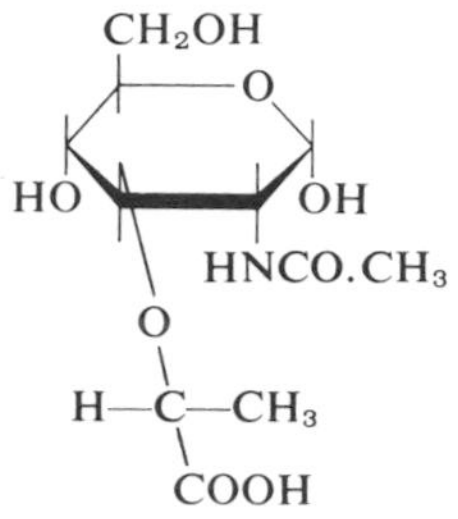 N-acetyl muramic acid | Lactic acid and N-acetyl glucosamine | |

to be cemented and, on the other, individual molecules of the gum itself is impaired. The combination of acidic, —OH and methyl groups in these molecules must be so controlled as to keep this balance.

## The bacterial cell wall

The bacterial cell contains a highly concentrated solution of small and large molecules. The combined osmotic pressure that these molecules exert can be very high. Unless the bacterium is in a medium that is as highly concentrated as the cell's contents water will tend to rush into the cell, and cause it to explode. This tendency could be overcome in three ways – by preventing the ingress of water, by elaborating means of pumping the water out again, and by building an extremely strong cell wall. Bacteria often live in very dilute solutions and tend to rely more on the first and last of these possibilities to overcome the problem of osmotic pressure. The structure that provides them with a strong cell wall is an ingeniously contrived polysaccharide assembly which we will now describe.

We may imagine the basic unit of this assembly as a very long chain of structure (N-acetyl glucosamine $(1 \rightarrow 4$-$\beta)$ N-acetyl muramic acid)$_n$. This chain is wound round and round the cell, much as one might wrap a potentially fragile object with steel wire (Fig. 8.12a). We find that the adjacent turns of the 'wire' are cross linked by *covalent bonds* (Fig. 8.12b) with all the strength which that

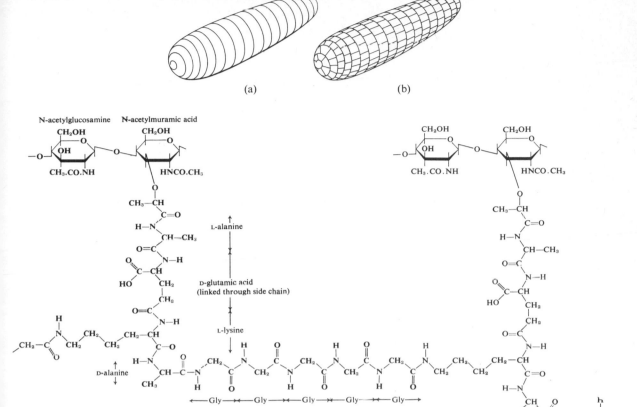

**Fig. 8.12** The reinforcement of the bacterial cell wall (see text), (a) the first stage, (b) the crosslinking stage, (c) a detailed view of a cross-link

implies. Thus the entire circumference of the cell is one single, seamless molecule. The mode of cross-linking varies slightly from cell to cell but the essential feature is that a short peptide joined to the N-acetyl muramic acid (it has the structure L-alanine-D-glutamic acid–L-lysine*-D-alanine) is cross-linked to a similar peptide on the adjacent turn of the basic chain. You will recall from Table 8.1 that N-acetyl muramic acid has a free —COOH group. This group forms a peptide link with the $\alpha$-NH$_2$ group of the L-alanine, and this link anchors the peptide to the polysaccharide chain. Cross-linking can then occur by the insertion of a chain running between the $\alpha$-COOH group of the D-alanine of one peptide and the $\epsilon$-amino group of the lysine of another. In *Staphylococcus aureus* this cross-chain is simply Gly–Gly–Gly–Gly–Gly. The resulting structure, which is called a mucopeptide (Fig. 8.12c), resists the tendency to burst with the full strength of thousands of covalent bonds. Because of the seamless nature of the construction, it has no weak points.

* In some species diaminopimelic acid (see p. 366) replaces lysine.

The mesh is sufficiently wide to allow pores for the passage of materials in and out of the cell, although the precise nature of the pores depends on the filling that is laid down on the basic framework that we have described. Walls of the *Gram-positive* bacteria (those that stain dark purple in Gram's procedure) are largely comprised of mucopeptides. In addition they contain small quantities of other polymers – such as *teichoic acid* (Fig. 8.13), which consists of hexose,

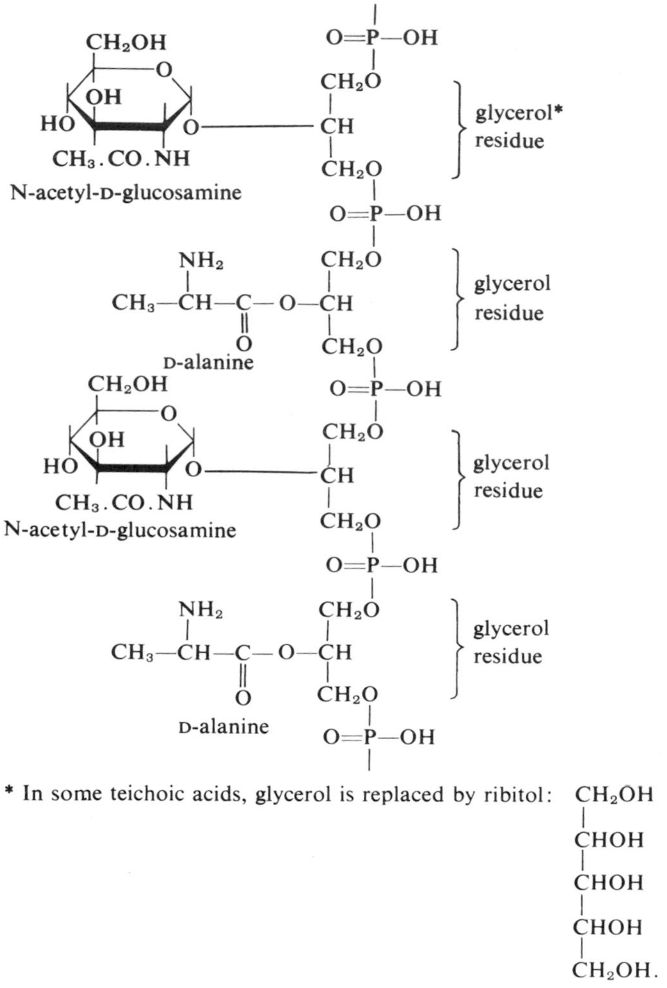

Fig. 8.13   Teichoic acid

sugar alcohols and amino acids (note the phosphodiester linkage), various polymers of sugar and sugar alcohols, and true polypeptide materials. *Gram-negative* bacteria have a much lower proportion of mucopeptide and contain larger quantities of polysaccharides, and, in particular, of complex lipids.

We can now explain the bacteriolytic action of lysozyme (p. 114). This enzyme cleaves the N-acetyl glucosamine–N-acetyl muramic acid polymer and can thus disrupt the whole structure of the cell wall, especially in the Gram-positive bacteria, which have little other supporting material in their walls.

*Chitin* is a material much used for structural purposes by invertebrate animals. It is an analogue of cellulose, in that it is a $1 \rightarrow 4$-$\beta$-linked polymer of N-acetyl glucosamine, and it appears to share many of its structural advantages. The insect wing, for example, is made of practically pure chitin and it has a strength and lightness that should be the envy of mechanical engineers.

*The cell surfaces of animals* do not have to be built to withstand great internal pressures, since animal cells are either capable of pumping out water or bathed in solutions of equal osmotic pressure (these are termed *isotonic* solutions). Animal cells are therefore free of the necessity to produce rigid structures, and they have been able to concentrate instead on the cultivation of various sophisticated means both of responding to the presence of adjacent cells to form a co-ordinated organism and of passing materials in and out. Many of these tasks appear to be undertaken by lipids (q.v.) and proteins, but polysaccharides also appear to play their part, either as free polymers or complexed with proteins.

The ease of hydration of some of the polysaccharides made from modified monomers place them in particular demand both as lubricants and to promote adhesion (the polysaccharide-based lubricant of the joints of vertebrates, for example, has a more favourable relationship between viscosity and rate of working than most industrially produced lubricants).

*Hyaluronic acid* (Fig. 8.14) is one such compound. It hydrates readily to form

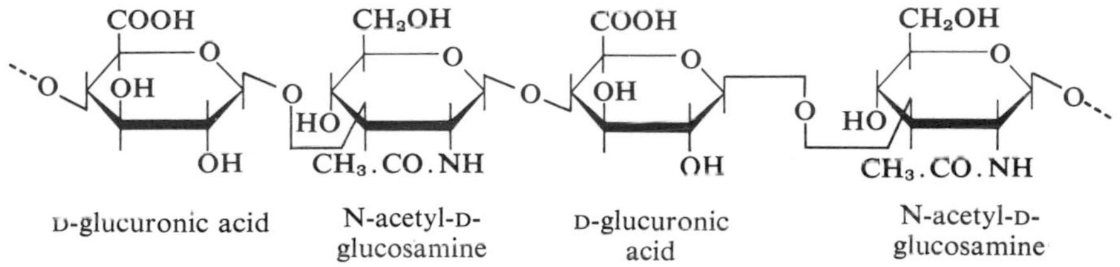

**Fig. 8.14**  Hyaluronic acid

a sticky substance which both lubricates and protects the integrity of the exterior of most animal cells. *Chondroitin*, which is usually found as *chondroitin sulphate* (Fig. 8.15), is a similar substance, with rather less lubricant properties than hyaluronic acid. Both substances are termed *mucopolysaccharides*.

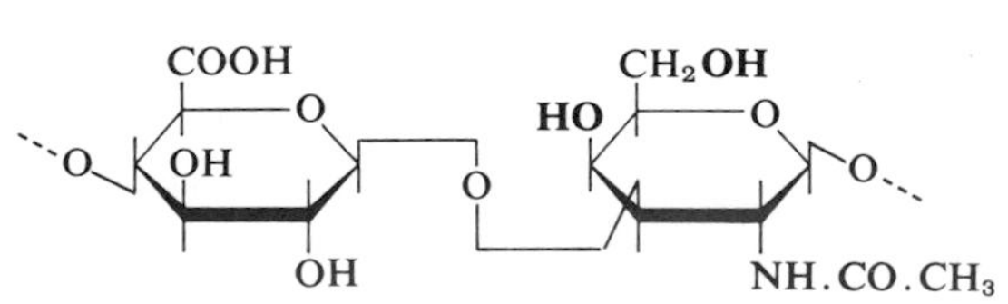

**Fig. 8.15**  Chondroitin. In chondroitin sulphate one or other of the —OH groups shown in bold type is esterified with a sulphate group

*Glycoproteins* consist of proteins to which polysaccharide (frequently muco-polysaccharide) chains are covalently linked. The site of attachment is usually a serine —OH or an aspartic acid —COOH group. These compounds therefore combine the sophistication and specificity of structure of the protein with the favourable mechanical properties of the polysaccharides. Little is so far understood of their structure–function relationships, but knowledge gained here would be worthwhile, particularly in helping us to understand the way in which cells interact with one another.

The carbohydrate portion of glycoproteins may contain hexoses, hexosamine, the sugar alcohols, L-fucose and N-acetyl neuraminic acid.

Apart from the purely structural roles that these compounds have (such as acting as the lubricant in saliva) and their possible role in cell-to-cell recognition, we can point to one phenomenon of this type which is probably a model for the more complex interactions – the role of the mucopolysaccharides in the *antigenic blood groups*. It should not surprise us that polysaccharides and muco-polysaccharides are antigenic. Since they figure so prominently in the coats of bacteria – one of the principal types of invading body – it would be surprising if it were not so. We can in fact understand the body as being hypersensitive to the intrusion of these materials, even when they are not parts of pathogens.

Now that we have completed our review of the polysaccharides, we will reiterate what we said at the start. The principles that link the structure of the poly-saccharides to their function are certainly those that apply in the case of the other macromolecules. The more rudimentary state of our knowledge of these molecules (although it is not much more rudimentary than was that of proteins and nucleic acids twenty years ago) makes it more difficult to give detailed explanations of individual pieces of behaviour. Nonetheless our knowledge is improving; the polysaccharides are so important, particularly to the developing science of the study of the interrelationship between cells, that we may expect our understanding of their structure and function to grow rapidly in the next few years.

## THE LIPIDS

The lipids are distinguished as a class by their exceptionally high solubility in non-polar solvents. They form structures of very high molecular weight, but they do so not by covalent linkages but by non-covalent, mainly hydrophobic interaction. As our studies of non-covalent interactions would lead us to suspect, the resulting structures are more pliable and less precisely defined than the ones that we have considered so far, in which there is always a covalent backbone to give some overall direction to the conformation.

Although this plasticity is often of the greatest value in enabling the macro-molecular assemblies of lipids to carry out their functions, it has made them very difficult to study. There is less firm knowledge of the chemistry of the individual components, let alone of the assemblies themselves, than there is even for the carbohydrates. However, as with the carbohydrates, more information is gradually becoming available. Also, as with the carbohydrates, the biological importance of these assemblies is immense, particularly in the higher levels of organization of the cell and of its organelles (p. 214), a level now being seriously explored in molecular terms for the first time.

Although it may not be worth while to dwell on the lipids for too long, because they are still rather poorly understood, there are some points that can usefully be made. We shall show that as with the monomers of the other classes, there are excellent possibilities for *permutation* of structures, by which the various inter-molecular forces can be deployed. We shall review briefly the food-storage and structural lipids, and then conclude with a brief account of the cell membranes, on the properties of which much of the organization of living processes depends. (The specific binding properties of lipids would probably make a fascinating story, but unfortunately the relevant research has yet to be done.)

## Lipid molecules

While provision is made for the deployment of all types of forces, hydrophobic ones predominate. These derive principally from long-chain aliphatic molecules (Tables 8.2 and 8.3). Each lipid monomer has usually at least two such molecules

**Table 8.2** Fatty acids

| Carbon atoms | Common name | Systematic name | Structure |
|---|---|---|---|
| Saturated fatty acids | | | |
| 12 | lauric | *n*-dodecanoic | $CH_3(CH_2)_{10}COOH$ |
| 14 | myristic | *n*-tetradecanoic | $CH_3(CH_2)_{12}COOH$ |
| 16 | palmitic | *n*-hexadecanoic | $CH_3(CH_2)_{14}COOH$ |
| 18 | stearic | *n*-octadecanoic | $CH_3(CH_2)_{16}COOH$ |
| 20 | arachidic | *n*-eicosanoic | $CH_3(CH_2)_{18}COOH$ |
| Unsaturated fatty acids | | | |
| 16 | palmitoleic | | $CH_3(CH_2)_5CH{=}CH(CH_2)_7COOH$ |
| 18 | oleic | | $CH_3(CH_2)_7CH{=}CH(CH_2)_7COOH$ |
| 18 | linoleic | | $CH_3(CH_2)_4CH{=}CHCH_2CH{=}CH(CH_2)_7COOH$ |
| 18 | linolenic | | $CH_3CH_2CH{=}CHCH_2CH{=}CHCH_2CH{=}CH(CH_2)_7COOH$ |
| 18 | *trans*-vaccenic | | $CH_3(CH_2)_5CH{=}CH(CH_2)_9COOH$ (*trans*) |
| 20 | arachidonic | | $CH_3(CH_2)_4CH{=}CHCH_2CH{=}CHCH_2CH{=}CHCH_2CH{=}CH(CH_2)_3COOH$ |

linked together at one end by covalent bonds, although the free compounds are sometimes known in nature. Covalent linking, if it occurs, may involve the sugar alcohol glycerol as a bridging compound: alternatively, the long-chain molecules may link together by means of their own functional groups (Fig. 8.16). In either case, the combined molecule has another —OH group to which is bound either the grouping that donates whatever additional, non-hydrophobic property is required (Table 8.4) or a further long-chain, hydrophobic molecule.

With these principles to guide us,* we can very rapidly put in perspective what

* Other minor lipids, such as the fat-soluble vitamins, do not fit exactly into this scheme (although they are often close enough for it to help in understanding the principles of their construction).

**Table 8.3**  Other lipid molecules

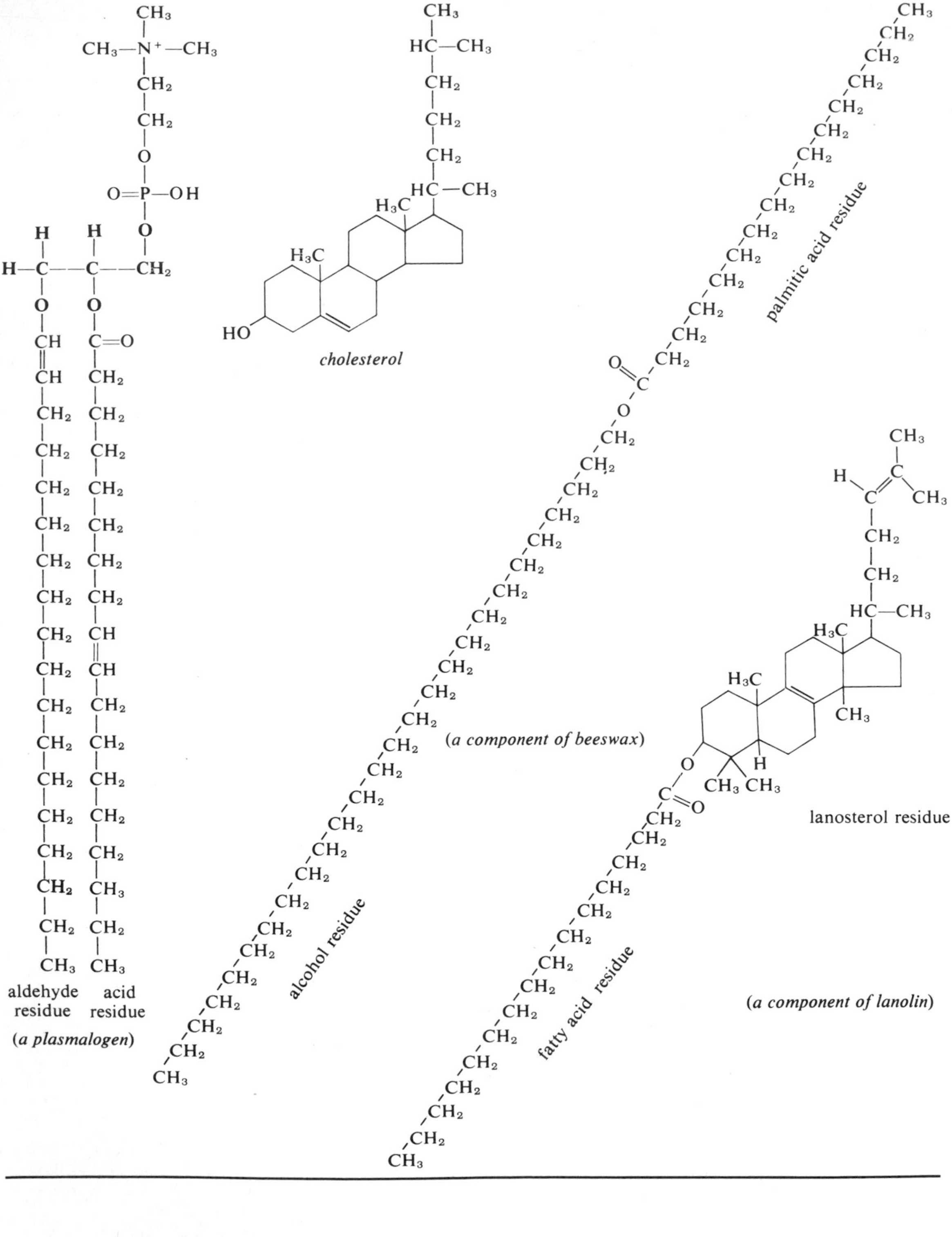

cholesterol

palmitic acid residue

(a component of beeswax)

alcohol residue

lanosterol residue

(a component of lanolin)

fatty acid residue

aldehyde   acid
residue   residue

(a plasmalogen)

might at first seem a bewildering array of compounds and an equally bewildering nomenclature. We shall now survey components of lipids, defining the major classes of lipid monomer on the way and paying particular attention to the possibilities of permutation of the various characteristics.

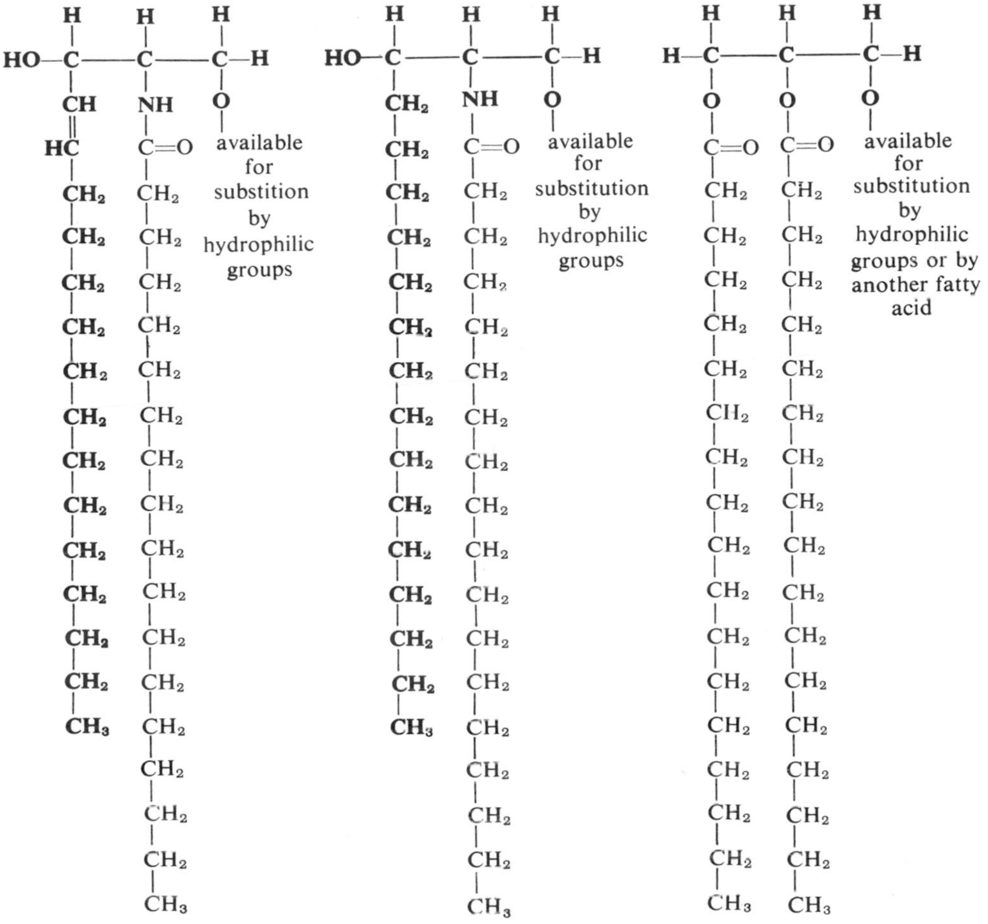

```
   H     H     H              H     H     H              H     H     H
   |     |     |              |     |     |              |     |     |
HO─C─────C─────C─H         HO─C─────C─────C─H          H─C─────C─────C─H
   |     |     |              |     |     |              |     |     |
   CH    NH    O              CH₂   NH    O              O     O     O
   ‖     |     |              |     |     |              |     |     |
   HC    C=O   available      CH₂   C=O   available      C=O   C=O   available
   |     |     for            |     |     for            |     |     for
   CH₂   CH₂   substition     CH₂   CH₂   substitution   CH₂   CH₂   substitution
   |     |     by             |     |     by             |     |     by
   CH₂   CH₂   hydrophilic    CH₂   CH₂   hydrophilic    CH₂   CH₂   hydrophilic
   |     |     groups         |     |     groups         |     |     groups or by
   CH₂   CH₂                  CH₂   CH₂                   CH₂   CH₂   another fatty
   |     |                    |     |                     |     |     acid
   CH₂   CH₂                  CH₂   CH₂                   CH₂   CH₂
   |     |                    |     |                     |     |
   CH₂   CH₂                  CH₂   CH₂                   CH₂   CH₂
   |     |                    |     |                     |     |
   CH₂   CH₂                  CH₂   CH₂                   CH₂   CH₂
   |     |                    |     |                     |     |
   CH₂   CH₂                  CH₂   CH₂                   CH₂   CH₂
   |     |                    |     |                     |     |
   CH₂   CH₂                  CH₂   CH₂                   CH₂   CH₂
   |     |                    |     |                     |     |
   CH₂   CH₂                  CH₂   CH₂                   CH₂   CH₂
   |     |                    |     |                     |     |
   CH₂   CH₂                  CH₂   CH₂                   CH₂   CH₂
   |     |                    |     |                     |     |
   CH₃   CH₂                  CH₃   CH₂                   CH₂   CH₂
         |                          |                     |     |
         CH₂                        CH₂                   CH₂   CH₂
         |                          |                     |     |
         CH₂                        CH₂                   CH₂   CH₂
         |                          |                     |     |
         CH₃                        CH₃                   CH₃   CH₃

bridging by sphingosine    bridging by sphinganine    bridging by glycerol
```

Fig. 8.16 Bridging between the components of lipid molecules

The *straight-chain* molecules can be of several types. Many are *fatty acids*, either saturated or unsaturated (Table 8.2). They usually have even numbers of carbon atoms (for the reason, see Ch. 20), most often from 14 to 24 carbon atoms. Those of 16 and 18 carbon atoms predominate. It is convenient, and the usual practice, to write the structural formula, say of the saturated $C_{18}$ acid, as $CH_3.(CH_2)_{16}.COOH$. However we shall not do so here in case it causes us to underrate the possibilities for hydrophobic bonding presented by a structure that is really

$$CH_3.CH_2.CH_2.CH_2.CH_2.CH_2.CH_2.CH_2.CH_2.CH_2.CH_2.CH_2.CH_2.CH_2.CH_2.CH_2.CH_2.COOH$$

The unsaturated fatty acids (there can be more than one unsaturated linkage in the chain) are almost always in the *cis* form rather than in the *trans* form. This seems sensible since the *trans* form is sterically very similar to the saturated acid, while the pronounced bend in the *cis* form (Fig. 8.17) will diminish the ease with

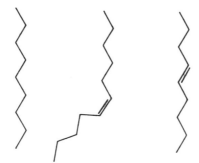

**Fig. 8.17**   *cis* and *trans* fatty acids

which inter-chain hydrophobic bonds might form and thus add a new variable that can be permuted in addition to chain length, degree of unsaturation, and the other characteristics that we shall mention.

Some rough indication of the impairment of chain–chain interaction caused by this bending of the chain is given by the melting point of the purified fatty acids, since melting points are clearly connected with the strength of inter-molecular bonds. The unsaturated fatty acids melt, on average, at a temperature 50° or more below that at which the corresponding saturated acid melts (many are liquid and are responsible for the fluidity of plant and animal oils). The sole exception among natural products is the one natural *trans* unsaturated acid, *trans* vaccenic acid, which is much closer in melting point to the saturated acids. We shall return to this question of the fluidity* of lipids in a biological context later in the chapter.

Several other types of long-chain molecule are found in lipids, all superficially resembling the fatty acids. These are shown in Table 8.3 and Fig. 8.16. The first type comprises the corresponding aldehydes. The second are the alcohols, which form esters with long chain acids. The third kind, *sphingosine* and *dihydrosphing-anine* (in which the *trans* unsaturated residue of sphingosine is reduced) are much like the saturated fatty acids for most of their length. The grouping at the end replaces the functions of the bridging compound glycerol (see below). The fourth kind, though more complex, can be seen from the Table to have some long-chain aliphatic character by virtue of their substituents. They are the *sterols*, which are represented in Table 8.3 by cholesterol, the principal form found in animals. Sterols are also known in plants, but, together with the sphingosine derivatives, they form one of the few major classes of biochemical compound

* Lipids are unusual in that they begin to melt at temperatures well below that at which there is a visible transition between solid and liquid. (The evidence for this comes from proton magnetic-resonance spectroscopy and the observation that lipids begin to take up their latent heat of fusion about 50° below their visible melting point.) This means that lipid molecules will have some freedom of migration even in an apparently solid structure (cf. p. 198).

that are not represented at all in the bacteria. The —OH group in sterols is either left free or engaged in bridging to a fatty acid by esterification.

## Bridging and permutation in the lipids

So far we have seen a great number of variations, some involving fine distinctions, some involving major ones, which can be permuted. The next stage is to see how these possibilities are multipled by bridging. The bridging function is most frequently undertaken by glycerol. If all three of its —OH groups are esterified by fatty acids, the resulting compound is known as a *neutral fat* (see Table 8.5). The acyl groups can all be of the same length and degree of saturation, in which case they are called *simple triglycerides*, or they may contain two or three different acyl groups, in which case they are called *mixed triglycerides*.

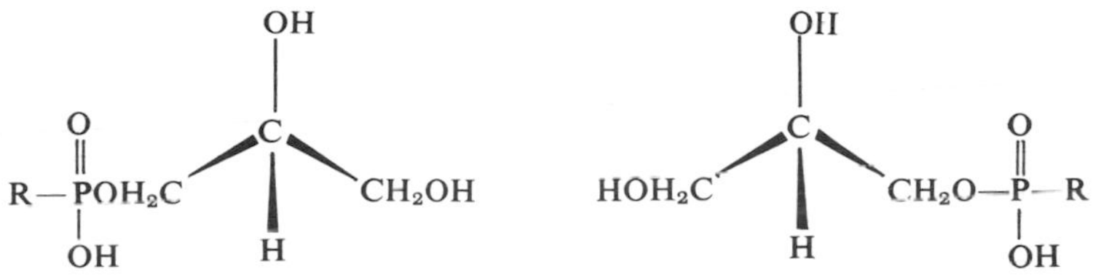

natural form        un-natural form

**Fig. 8.18** The stereochemistry of phosphoglycerides. For the nature of R see Table 8.4)

If, in structural materials, only two of the —OH groups are acylated, the third —OH group (always that of the —$CH_2OH$ group that allows the stereochemical configuration shown on the left of Fig. 8.18) is free. Members of this group can be phosphorylated to give *phosphatidic acids* (Fig. 8.19), the parent compounds of a large and important group of lipids which have dual (polar and hydrophobic) characteristics. They are called the *phosphoglycerides* (see Table 8.5), and they take their characteristics from various substituents that are placed on one of the phosphate —OH groups (Table 8.4).

If the —$CH_2OH$ group is not phosphorylated, it can enter into a glycosidic linkage with a sugar, normally galactose. We then have, again, a dual property but one in which the polar character is more mild. These compounds are known as *glycolipids* (see Table 8.5).

If only one of the —OH groups of glycerol is acylated, the other usually combines with the aldehyde-type of chain to give an $\alpha,\beta$-unsaturated ether linkage. The third —OH group is again phosphorylated and available for substitution as before. The resulting compounds, the *plasmalogens* (see Table 8.5) are clearly very closely related to the phosphoglycerides and similarly possess both polar and hydrophobic character.

Compounds that contain one of the sphingosines have no need of glycerol to act as a bridge. A fatty-acid molecule forms an amide with the —$NH_2$ group (rather than an ester with an —OH group as has been usual in the compounds that we have discussed up to now). The —$CH_2OH$ group of such compounds (*ceramides*) (Fig. 8.20) is available for phosphorylation and for further sub-

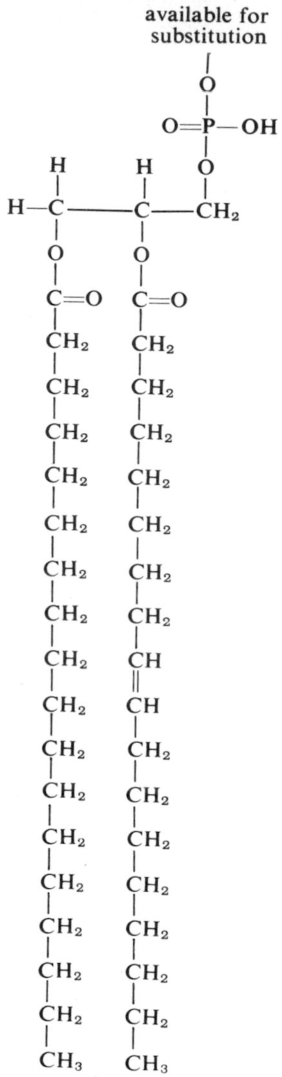

available for
substitution
|
O
‖
O=P—OH
|
O

H        H     O
|        |     |
H—C———C——CH$_2$
|        |
O        O
|        |
C=O   C=O
|        |
CH$_2$     CH$_2$
|        |
CH$_2$     CH$_2$
|        |
CH$_2$     CH$_2$
|        |
CH$_2$     CH$_2$
|        |
CH$_2$     CH$_2$
|        |
CH$_2$     CH$_2$
|        |
CH$_2$     CH$_2$
|        |
CH$_2$     CH
|        ‖
CH$_2$     CH
|        |
CH$_2$     CH$_2$
|        |
CH$_2$     CH$_2$
|        |
CH$_2$     CH$_2$
|        |
CH$_2$     CH$_2$
|        |
CH$_2$     CH$_2$
|        |
CH$_2$     CH$_2$
|        |
CH$_2$     CH$_2$
|        |
CH$_3$     CH$_3$

**Fig. 8.19**  Phosphatidic acids

stitution of the phosphate, as in many of the glycerol-bridged compounds. The class of compounds resulting from such substitutions are called *sphingolipids* (see Table 8.5). If the —CH$_2$OH is not phosphorylated, it can be substituted with galactose, as with the glycerol-bridged compound. These compounds are called *cerebrosides* and the polar characteristics may be enhanced by sulphation of carbon atom 2 of the galactose (cf. chondroitin sulphate, p. 183).

Alternatively if galactose is not used a more complex polysaccharide head can be fitted to the —CH$_2$OH by means of the glycosidic link, and a *ganglioside*, one of the most intricate of the types of lipid monomer, is thus produced (Fig. 8.21, Table 8.5).

All other bridged compounds, like the sphingosine derivatives, use their own functional groups to attach the long-chain acids and do not require glycerol. We

**Table 8.4** Some of the common substituents on the phosphate group of phosphatidic acid. The bold —**OH** group is the one that joins the substituent to the phosphate group

| *Phosphoglyceride* | *Alcohol component* |
|---|---|

Cardiolipin

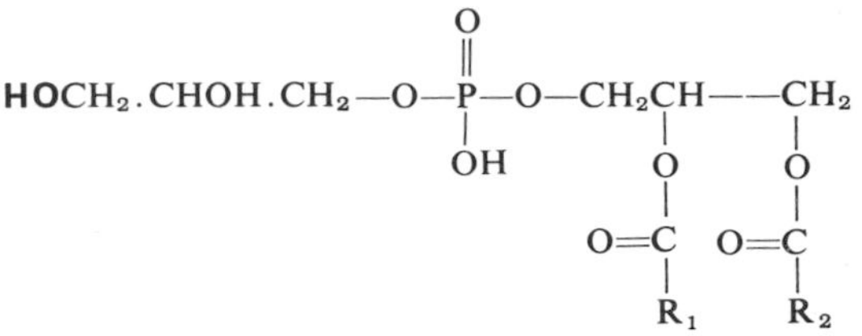

$$\mathbf{HO}CH_2.CHOH.CH_2\!-\!O\!-\!\overset{\displaystyle O}{\overset{\|}{\underset{\underset{OH}{|}}{P}}}\!-\!O\!-\!CH_2CH\!-\!\!-\!CH_2$$

with the substituents $O=\overset{|}{\underset{R_1}{C}}$ and $O=\overset{|}{\underset{R_2}{C}}$ attached below.

Phosphatidyl choline     $\mathbf{HO}CH_2.CH_2\overset{+}{N}(CH_3)_3$

Phosphatidyl ethanolamine     $\mathbf{HO}CH_2.CH_2NH_2$

Phosphatidyl glycerol     $\mathbf{HO}CH_2.CHOH.CH_2OH$

Phosphatidyl inositol     (inositol ring: OH, OH, OH, **HO**, OH, OH)

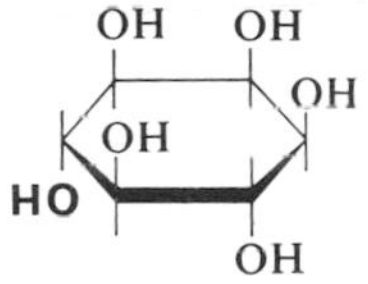

Phosphatidyl serine     $\mathbf{HO}CH_2.CHNH_2.COOH$

Phosphatidyl 3′-O-aminoacyl glycerol     $\mathbf{HO}CH_2.CHOH.CH_2O\!-\!\overset{\displaystyle O}{\overset{\|}{C}}$ with $R\!-\!CH$ and $NH_2$ attached below.

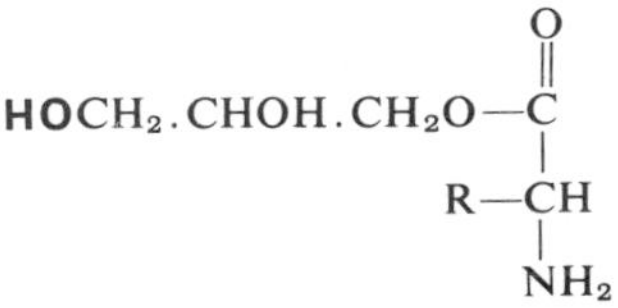

have already mentioned the long-chain acylation of the C-3 —OH group of the sterols and the acylation, by similar groups, of the long-chain monohydroxy alcohols. The products in both cases are the *waxes*, which waterproof a number of the external structures of plants* and animals.

Since we began to consider bridging we have brought to light a large number of additional variables that can be permuted. If we multiplied all of them together we would by now have a very large number of possible structures (you might like, before reading the next sentence, to try and work out the approximate number). Assuming that there are twelve common long-chain fatty acids (a conservative estimate) the number of possible different structures that could be formed from the molecules so far discussed lies between 1000 and 2000. As important as the number is the fact that these structures have a very finely graduated set of hydrophobic properties and also a reasonable range of polar properties, any of which can be called on to meet any particular need. We will now see by considering Table 8.4 both how the number of possible types is further multiplied and how the range of properties is widened. The table shows

* It is the natural wax on fruit that makes the use of some organic pesticides so dangerous. Simple pesticides can be washed off fruit, but complex non-polar ones will form hydrophobic interactions with the wax and be retained. The only safe thing nowadays is to discard the peel.

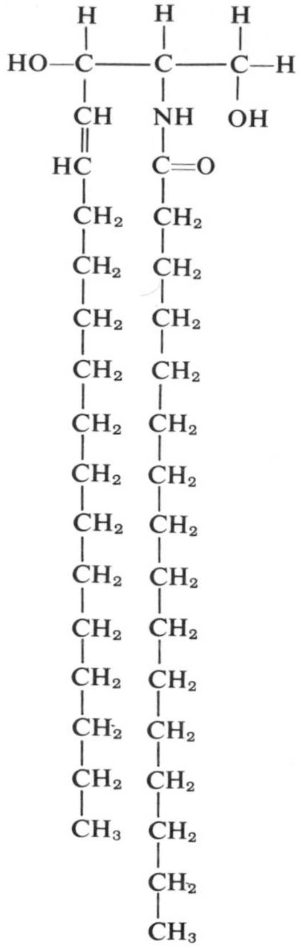

H   H   H
|   |   |
HO—C——C——C—H
|   |   |
CH  NH  OH
‖   |
HC  C=O
|   |
CH₂ CH₂
|   |
CH₂ CH₂
|   |
CH₂ CH₂
|   |
CH₂ CH₂
|   |
CH₂ CH₂
|   |
CH₂ CH₂
|   |
CH₂ CH₂
|   |
CH₂ CH₂
|   |
CH₂ CH₂
|   |
CH₂ CH₂
|   |
CH₂ CH₂
|   |
CH₂ CH₂
|   |
CH₃ CH₂
    |
    CH₂
    |
    CH₃

**Fig. 8.20**  A ceramide

a number of the alcoholic molecules that are able to esterify the phosphate group in the phosphoglycerides and sphingolipids.

The number of possible lipid structures is probably increased by this additional permutation to something of the order of $10^4$. Remember that these are only *monomers* that are used to make up the higher order, non-covalently associated, structures. Note that the range of properties has also been widened greatly by this final multiplication to include almost every type of non-covalent bond conceivable. Certain of the structures (notably phosphatidyl choline, phosphatidyl ethanolamine, cardiolipin, sphingomyelin and cholesterol) predominate in many natural lipid structures, and we do not suggest that all the permutations exist. However, a good number of the other structures either are present in smaller amounts or are available for special requirements as these arise. Table 8.5 summarizes the major classes of lipid monomer that we have covered.

## Food-storage lipids

As we mentioned when dealing with the carbohydrates, living organisms are often able to lay down reserves of food. *Fats*, being totally insoluble, are excellent for this purpose, as they can be deposited as droplets in a convenient part of the cell. They will remain there inert until the energy stored within them is required, when they can be degraded. The biosynthesis of fats is described in Ch. 20, their degradation in Ch. 17.

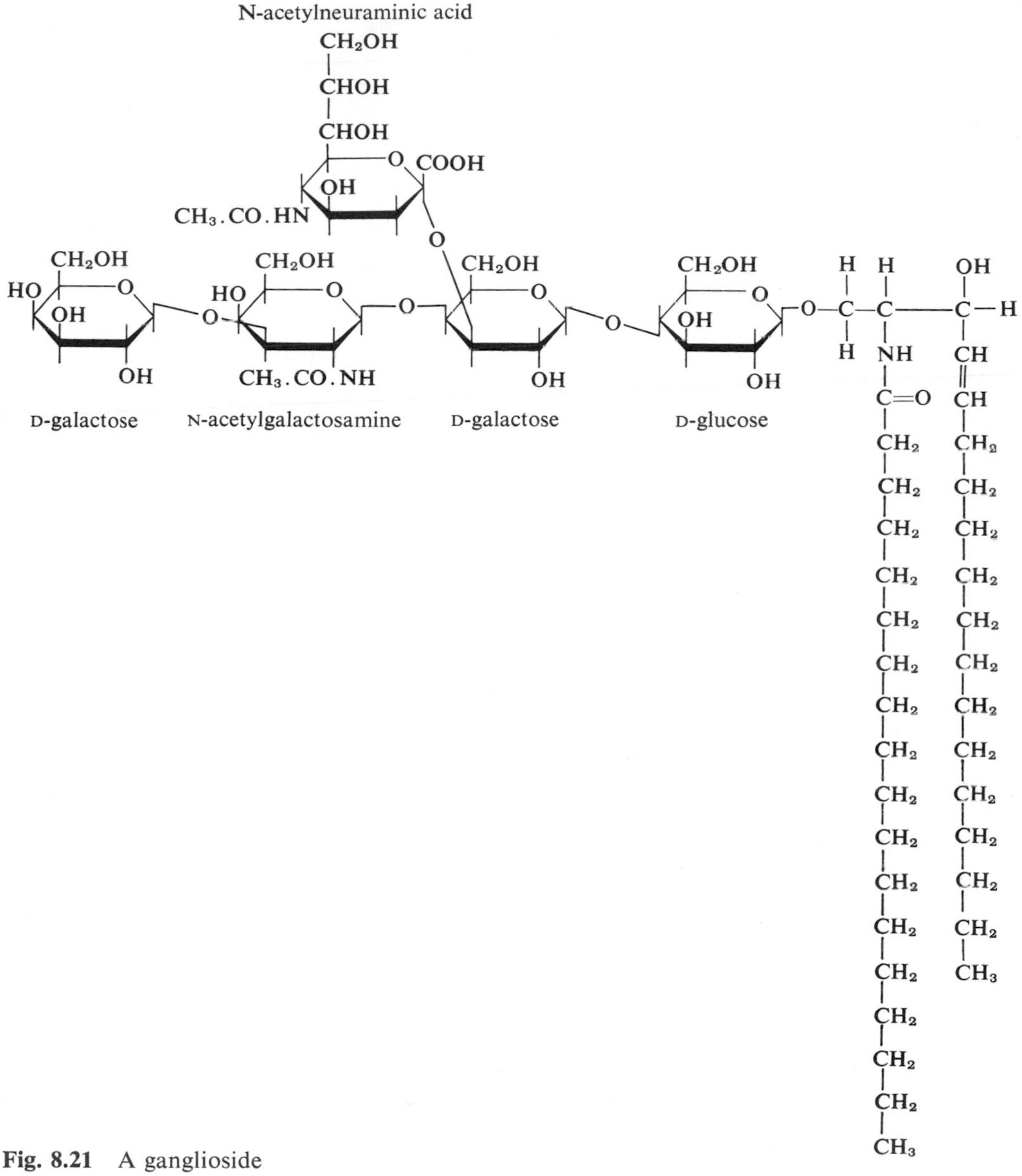

**Fig. 8.21** A ganglioside

The free fatty acids are almost totally insoluble unless the —COOH group is stabilized in the COO⁻ form by forming a salt with (usually) Na⁺ and K⁺. The resulting *soaps* are prepared by alkaline hydrolysis of triglycerides and their role in the removal of dirt has already been described on p. 58. Because of their biological origin, the conventional soaps have excellent bio-degradable properties.

**Table 8.5** A summary of the major types of lipid monomer

| Type | Principal function or site of occurrence | Most common, or typical members | |
|---|---|---|---|
| Fats* | Food stores | | 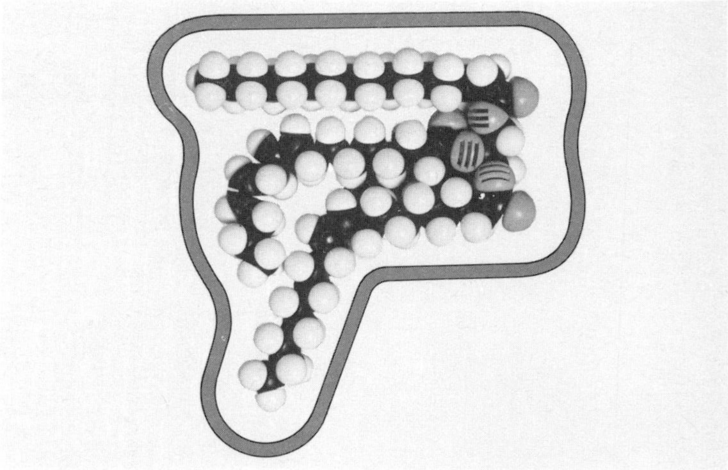 |
| Waxes | Waterproofing and general protection of external components of plants and animals (leaves, fruit, skin or exoskeleton, fur or feathers). | One of the components of lanolin (sheep wool) | 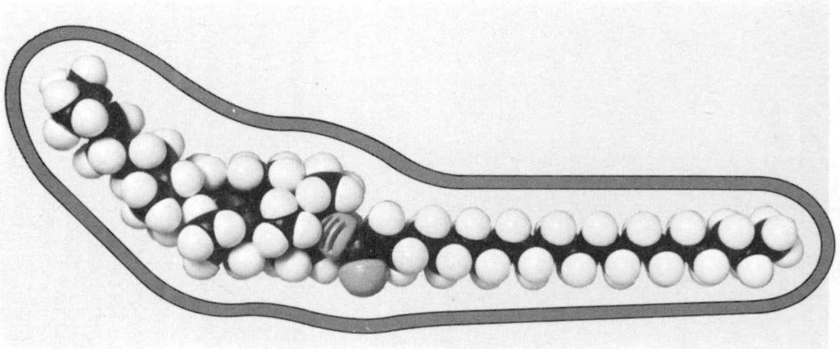 |
| Sterols | Found in external membranes of cells: enhances fluidity of these structures. | Cholesterol | 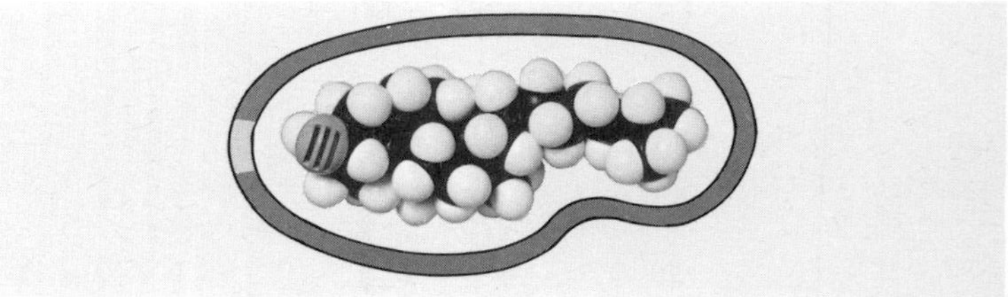 |

* In animals, most typically, two or three of the acyl groups are unsaturated.

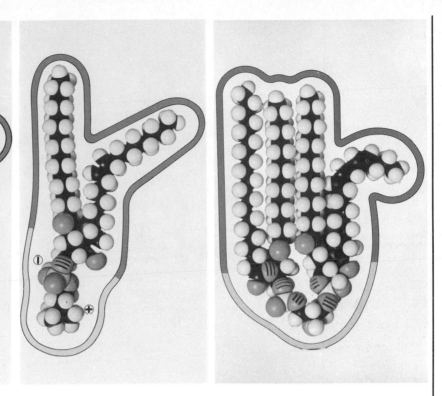

Phospho-
glycerides

Major lipid component
of cell membranes

Phosphatidyl
ethanolamine (cephalin)

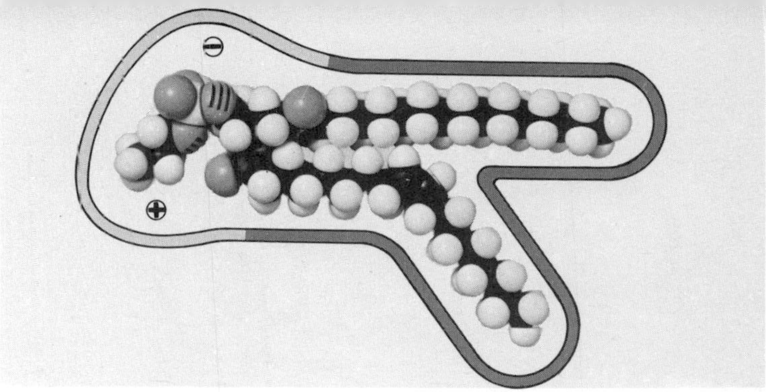

Phosphatidyl choline
(lecithin)

Cardiolipin

**Table 8.5** (continued)

| Type | Principal function or site of occurrence | Most common, or typical members |
|------|------------------------------------------|--------------------------------|
| Plasmalogens | Analogous to phospho-glycerides. Found es-pecially in nerves and muscle. | 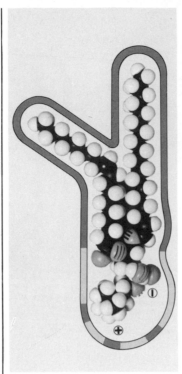 |
| Glycolipids | Component of non-animal membranes | 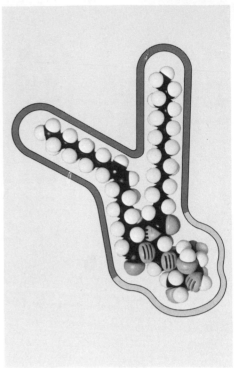 |

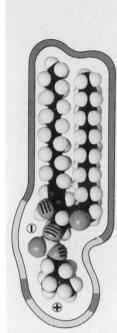 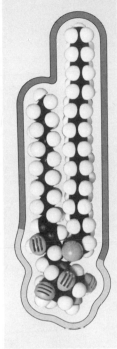 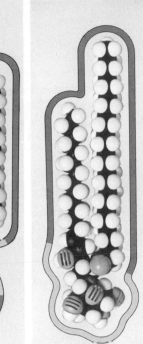

Sphingolipids    Membranes of brain and nervous tissues    Sphingomyelin

Cerebrosides    Membranes of myelin sheath of nerve tissue

Gangliosides    Similar in function to the sterols, found especially in brain and nervous tissue    See Fig. 8.21

## Structural lipids – the membranes

Biological membranes consist of, on average, 60 per cent protein (by weight) and 40 per cent lipid, a good deal of which is phosphorylated. Some of the protein molecules are enzymes, and in membranes from the appropriate parts of the cell (p. 217) they include many of the integrated multi-enzyme systems that catalyse the metabolic pathways that we shall describe in Part 2 of this book. Others of the protein molecules appear to have a role in promoting the transport of substances across the membranes, as we shall see below. Still others are concerned with the transport of reducing power and electrons in oxidative metabolism (see Ch. 12).

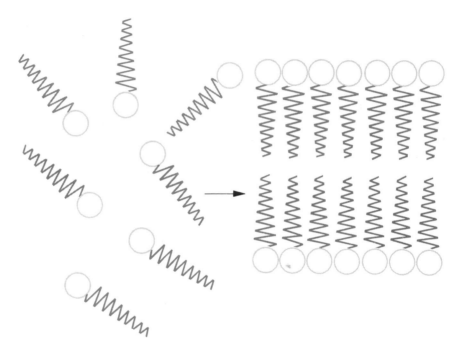

**Fig. 8.22**   The lipid double layer

Most models of the gross anatomy of the membranes postulate a *bi-layer* structure which arises as follows: The lipid fraction of the membrane material will be driven to associate by means of the *hydrophobic interaction*. All the monomers will try to align their non-polar chains for the maximum degree of side-to-side contact and will present their polar parts to the surrounding water. Contact between apolar parts and water will be diminished and stability will be enhanced if two of these structures form a sandwich (Fig. 8.22). Once again, therefore, we have a micelle, in this case a very thin, flat one (see p. 58). As there is no covalent backbone the micelle can be very large. It can also be of quite complex structure (p. 214 ff and Fig. 10.3). It is particularly important to notice when considering membranes that any one component may be *mobile* and travel quite widely within the half of the bi-layer structure in which it finds itself.

If the associations between the lipid monomers are all of one type, what are the various types of polar property for? Some will be needed to stabilize the interface between the micelle and water; some will interact specifically with each other to give rise to regions with particular geometrical or chemical properties, and some will, singly or in combination, form binding sites for the large numbers of different proteins that give the membranes their unique properties.

Correspondingly, the proteins are likely to have stretches of hydrophobic amino acids which will penetrate the non-polar part of the layer. It will then be possible for the polar parts of the protein to add their own hydrophilic contribution to stabilizing the interface with the surrounding water (Fig. 8.23).

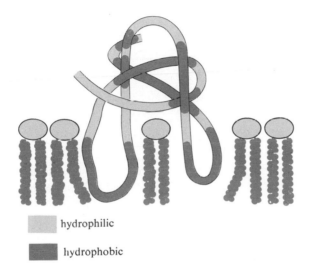

☐ hydrophilic

■ hydrophobic

**Fig. 8.23** The interaction between protein and lipid (idealized)

(Non-covalently bound lipoprotein complexes of this type are also known in other situations than the membrane, for example in the transport of fats between different tissues of the body.)

Although little is known as yet about the fine structure of the membrane surface there is every reason to suppose that it will be a *mosaic* of protein assemblies bound to the lipid matrix, some retained in fixed positions, others free to some extent to wander along the surface of the membrane. A need for such mobility of components on the membrane surface may be the reason for the following observation: when organisms experience lower body temperatures than usual, they produce a lipid fraction that contains acids with a higher than usual proportion of unsaturated chains. As we saw on p. 188, this would tend to increase the fluidity of the membrane.

Recent studies with the electron microscope have led to some controversy over the detailed morphology of the membranes. However, the picture in Fig. 8.22 is probably quite close to the truth and sufficient for a discussion of the principal function of the membranes – to allow, or, when necessary, to prevent, the passage of substances from one side of them to another.

You will be familiar, from the chemistry of non-living materials, with the phenomenon of *semi-permeability*. This property is shown by certain substances which form sheets with microscopic pores. Molecules that are small enough can pass through the pores while larger molecules are prevented from doing so. We shall describe in Appendix 2 the technique of *dialysis* in which sheets that contain pores large enough to allow free passage to water and other small molecules are used to allow the small molecules to escape while retaining macromolecules. The driving force for this tendency to *free diffusion* is entropic (Ch. 2): if molecules remained in one small corner when they could fill the whole volume, the system would be more organized than if they did not.

Biological membranes also allow free diffusion based on size or solubility in lipid. In addition they contain *mediated transport* systems, often much more discriminating than the diffusion process. Mediated transport may be *passive*, that is *with* the thermodynamic gradient or (a striking feature of living matter), *active*, in which substances are constrained to move uphill *against* the thermodynamic gradient.

*Free diffusion*

It can be shown that the passive movement of inorganic ions across the outer membranes of certain cells can be explained by considering the membrane to be penetrated by pores approximately 8 Å in diameter. This is large enough to permit the free passage of, for example, $Na^+$, $K^+$ and $Cl^-$, even in the hydrated form. It is not, however, sufficient for the passage of, for example, bicarbonate or sulphate, which are much larger. Certain cyclic antibiotics (e.g. *nonactin*, Fig. 8.24) confer permeability on membranes that do not have such pores and which are impermeable to the ions that we have mentioned. You can see that nonactin has a hydrophobic rim, which presumably embeds itself in the lipid of the membrane. $K^+$ can be held in the hole in the centre of the molecule, which can then sink into the hydrophobic surface and carry the ion through to the other side. Other similar molecules exist that permit a greatly enhanced rate of diffusion of various ions. These molecules effectively convert the transport process into one of passive mediated transport (see below).

A second type of free diffusion depends not on the existence of pores, but on the degree of hydrophobicity of the substance that has to pass. A molecule that is sufficiently apolar will merge with the membrane and can leave it on the other side if the concentration there is lower. For instance, diacetyl glycerol passes across cell membranes nearly ten times faster than monoacetyl glycerol, which in turn passes more than ten times faster than glycerol itself. The decrease in rate of passage here corresponds quite well to the progressive decrease in hydrophobicity. If we decrease hydrophobicity still further (e.g. by phosphorylating an —OH group) the resulting compound will hardly diffuse through at all.

Certain substances pass through the membranes much more rapidly than either their size or lipid solubility would lead one to expect. We shall now examine the systems that permit this to occur.

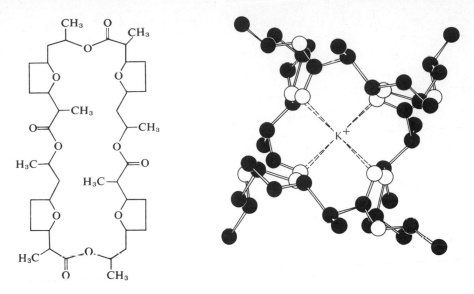

**Fig. 8.24** Nonactin

## Mediated transport systems

If we were to plot the rate of inward passage of substance against concentration on the 'input' side of the membrane, we would obtain, for passive diffusion, a graph like that in Fig. 8.25a. Many substances, however, show concentration–velocity curves like that in Fig. 8.25b. This will remind us at once of the curve

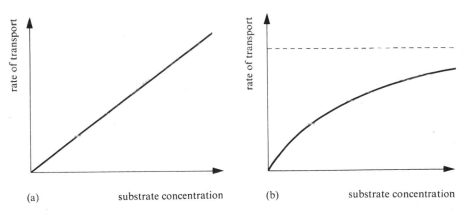

**Fig. 8.25** The relationship between rate of transport across membranes and the concentration of substrate (a) non-mediated transport, (b) mediated transport

linking the initial velocity of an enzyme-catalysed reaction with concentration of substrate. The conclusion is irresistible that, in the present case as with the enzymes, some association is occurring between the added substance and a limited number of specific receptor structures. In the higher regions of concentration the receptors are *saturated*–that is, all of them are fully employed all the time so that the addition of further substrate material can make no difference (cf. p. 125).

We may conclude that we have discovered the function of at least some of the protein molecules in the membrane. This conclusion is strengthened by the fact that, as we shall see in a moment, the selectivity of the transport system requires just the high degree of specificity of bonding at which we know, from Chs 5 and 6, the proteins excel. It has, in fact, become possible to isolate and characterize some of the transport proteins (see Ch. 18).

Consider, for example, the mediated transport of sugars. This can be a *passive* process – that is, it occurs only so long as the concentration on the input side is greater than the concentration on the output side. It is this concentration difference that provides the free-energy gradient that permits the transport to occur. (We touched on the relationship between $\Delta G$ and concentration on p. 17.) In such processes, when the concentrations on the two sides are equal, the rates of passage are equal and opposite and hence no *net* transport is observed (cf. the approach to chemical equilibrium p. 17). The sugars that are allowed by the structure of the membrane to use this means of passing across it include those in Table 8.6; some sugars that are either not allowed to use it or do so with difficulty

**Table 8.6**   The transport of sugars

| increasing tendency to assume the C—I configuration | | decreasing ease of transport |
|---|---|---|
| | 2-deoxy-D-glucose | |
| | D-glucose | |
| | D-mannose | |
| | D-galactose | |
| | D-xylose | |
| | L-arabinose | |
| | D-ribose | |

L-Glucose, L-Galactose, L-xylose, L-rhamnose, L-Fucose and D-arabinose are largely in the I—C form and are hardly transported at all.

are also shown. It appears that the essential feature is the possession of the C-1 chair form of the pyranose ring (p. 171) with equatorial —OH groups. This kind of preference is precisely that that one would expect of a protein specificity site. We can imagine hydrogen bonds running from amino-acid side chains to the equatorial —OH groups and possibly some hydrophobic contact with the under-surface of the sugar, which, as we saw on p. 172 is free of —OH. This specificity has a use beyond simply deciding which sugars shall enter the cell; it can also prevent them leaving again. Thus glucose, the major form in which carbohydrate enters the cell to be metabolized, is phosphorylated upon arrival and is therefore trapped in a form in which it can no longer pass back across the membrane.

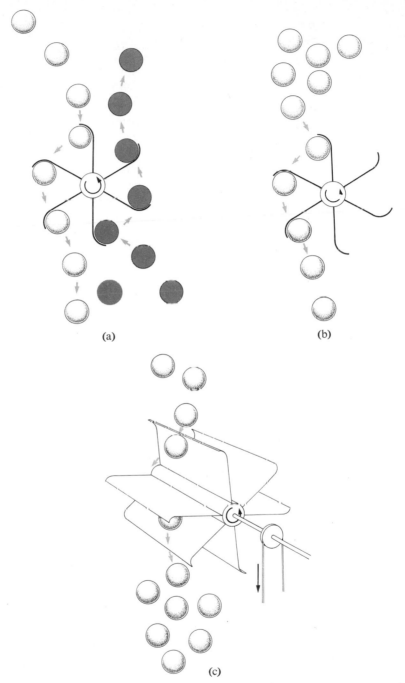

(a)

(b)

(c)

**Fig. 8.26** Idealized picture of the membrane-bound transport systems acting as rotating shuttles (a) exchange diffusion, (b) passive, non-coupled mediated transport, (c) active transport

We have emphasized that the thermodynamic driving force for the passage of a substance has so far been a gradient of concentration, and thus a gradient of free energy down which it can fall. If, on the other hand, we start with equal concentrations on both sides of the membrane, or even with concentrations producing a tendency to move against the direction in which net transport is required (no $\Delta G$ or an unfavourable $\Delta G$), how are we to proceed? The answer is, as it was with the thermodynamically unfavourable chemical reactions in Ch. 2, by *coupling*. Coupling can be to the free energy change associated with the changes in concentration of another substance (p. 319). (In this case the process is called *exchange diffusion* and transport in cannot occur without transport out.) Alternatively, as is so often done in biology to overcome energetic problems, coupling can be to the free energy of hydrolysis of high energy compounds (notably ATP), in which case the process is called *active transport*.

We can imagine the various mediated transport processes in terms of a rotating shuttle (Fig. 8.26). In passive, non-coupled transport, the shuttle will rotate irrespective of whether both sides are loaded or not, so long as the net transport is downhill. In exchange diffusion both sides must be loaded and the shuttle rotates because one side is allowing flow in the downhill direction. In active transport the shuttle is driven as if by a motor.

Many exchange-diffusion and active-transport systems exist. They all have the high specificity that is characteristic of transport processes, each system being limited to a single compound or to a narrow range of related compounds. It is hardly possible to over-estimate the importance to living matter both of the ability to be selective and the ability to *concentrate* materials from a dilute solution.

We shall meet examples of exchange diffusion on pp. 319 and 353, and we shall deal with active transport more fully when we consider the various uses of ATP, of which it is an example (Ch. 18).

# Part 2  Intermediary metabolism

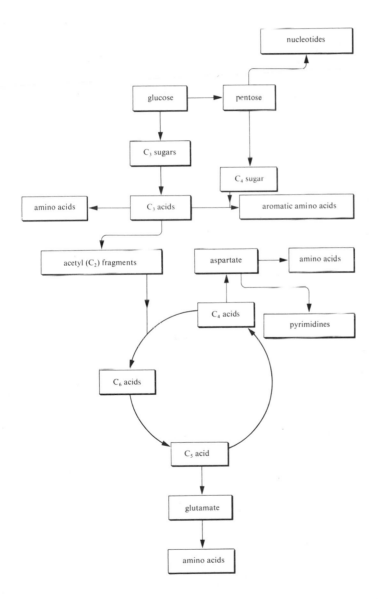

# Chapter 9    Introduction to intermediary metabolism

We have now set the scene for a discussion of some aspects of intermediary metabolism. A striking characteristic of the chemical transformations involved in metabolism is that they occur in very small, discrete steps, each of which is catalysed by a separate enzyme. Thus intermediary metabolism consists of a large number of reactions, by means of which molecules are gradually modified and shaped. The various reactions are generally thought of as comprising *pathways*, along which the compounds flow while undergoing these gradual changes. However, the pathways constantly converge and diverge, so that an intermediate formed as a result of one reaction may often have a choice of two or more subsequent paths to follow.

One reason why metabolic transformations involve so many steps will become clear if we consider the mode of action of enzymes. If metabolism is to be at all vigorous, it is necessary for enzymes to be as active as possible. In order to increase the rate of biochemical reaction the enzyme must bind to the substrate and to the reaction intermediate (p. 116). We have seen that for the increase in rate to be appreciable, the substrate and subsequently the reaction intermediate must be very strongly bound through a number of firm, specific contacts with the enzyme. It is remarkable that a linear polymer of amino acids can do so well in providing a binding surface even for normal enzyme-catalysed reactions, in which only a few atoms are involved: to provide a surface that could promote (for example) the simultaneous oxidation of all six carbon atoms of glucose would be considerably more difficult.

Thus it is at least in part as a result of the immense catalytic power that enzymes possess that the chemical change that any one enzyme reaction brings about is correspondingly slight. It seems to be for this reason that the pathways of intermediary metabolism involve such gradual modification of the reacting molecules.

The fact that any one enzyme does such a specific, small-scale job necessitates the presence in organisms of a large number of different enzymes. One might think that, even despite the points that we made above, it would be better for organisms not to have to synthesize so many enzymes. For example, we shall see in Ch. 14 that the breakdown of glucose to lactate requires eleven separate enzymes, each of which must be made by the specific protein-synthesizing machinery of the cell (Ch. 26). Would it not be more economical to have a single enzyme that achieves the production of lactate from glucose, even if that enzyme had a much more complex structure and, perhaps, less catalytic power than the enzymes that actually exist?

The answer is that this division of metabolic reactions into small steps is admirably suited to fulfilling two separate functions undertaken by intermediary metabolism. These are to provide energy in the form of ATP and to produce the building blocks from which macromolecules are synthesized. In the degradation of glucose to lactate several intermediates are formed which are used in other pathways – for example glucose-6-phosphate which can be converted to glycogen (Ch. 19), glyceraldehyde-3-phosphate which can be converted to glycerol and thus used in the synthesis of fats (Ch. 20), and pyruvate which (among many other uses) can be converted to alanine for the synthesis of proteins (Ch. 22). A similar multiple use is made of every pathway in intermediary metabolism. If glucose were degraded to lactate in one step the essential intermediates would have to be formed by other routes, each of which would require its own separate enzyme or enzymes. So the advantage that would seem to result if enzymes were designed to achieve much more radical transformations would in fact be no advantage at all.

On the contrary, there is yet another advantage in splitting up metabolic pathways into small steps. The economy of the cell demands that the processes of intermediary metabolism be carefully controlled, so that the ATP and intermediates that they provide should be available in the right quantities at the right time. Now, by dividing metabolic pathways into a large number of steps, the cell can effect a very fine control of pathways that deal with several different substrates by altering the rate of just a single reaction. For example, we shall see that the degradation of glucose to lactate involves a reaction catalysed by the enzyme phosphofructokinase (p. 260). This enzyme, however, is needed equally for the degradation of lactose via galactose, and of glycogen, in each case to lactate. So by blocking the activity of phosphofructokinase the cell can block the degradation of *all* carbohydrate (p. 504) without at the same time affecting any other metabolic process.

With these considerations in mind, we can now take a general look at the pathways of intermediary metabolism themselves. We shall first discuss the way in which these pathways can be classified, and then give a brief account of intermediary metabolism which will serve as an introduction to the remainder of Part 2.

For many years it was customary to classify pathways of intermediary metabolism solely as *catabolic* and *anabolic*. Catabolic pathways are those in which carbon compounds are broken down to products that have less free energy; some of the energy released is used to synthesize ATP. Anabolic pathways are those that are used to synthesize molecules from smaller precursors at the expense of ATP. More recently an additional classification has been introduced, that of *amphibolic* pathways. These are the pathways that are used to provide energy in the form of ATP, but which have intermediates (as we explained above) that are used as starting points for biosynthesis.

In one sense all catabolic pathways can be regarded as being amphibolic, but the term is generally reserved for those pathways in which intermediates are *extensively* used for synthetic reactions. The most obvious example is a pathway known as the *tricarboxylic acid cycle* or *Krebs cycle*. This, as we shall see, is used for the final oxidation of all carbon sources, whether carbohydrate, fat or protein. In addition many of the compounds that are intermediates in this process of oxidation are precursors of essential cell constituents – many amino acids, all pyrimidines and all pyrrole compounds are derived from intermediates of the Krebs cycle.

*Catabolic pathways*

Figure 9.1 gives an outline of the chief catabolic pathways that are used for the degradation of *carbohydrate* and of *fat*. Polysaccharides (Ch. 8) are degraded to their constituent sugars by means that we shall discuss in Ch. 14; the principal food-storage polysaccharides are polymers of glucose, which are degraded to a derivative of glucose itself. The chief catabolic pathway for glucose passes through derivatives of three-carbon sugars to three-carbon acids. Glycerol, a

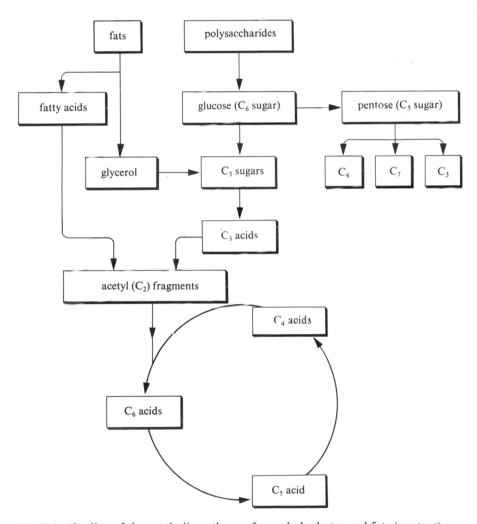

**Fig. 9.1** Outline of the catabolic pathways for carbohydrates and fats (see text)

product of hydrolysis of fats (Ch. 8) is also oxidized to three-carbon acids. These three-carbon acids, whether derived from carbohydrate or from glycerol, are now oxidized to acetyl (two-carbon) fragments. Fatty acids, which are produced, with glycerol, from hydrolysis of fats, are also oxidized to acetyl fragments (Ch. 16).

Thus acetyl fragments are the common products of oxidation of carbohydrate and of fats. They are now oxidized via the Krebs cycle (Ch. 15), to which we have already referred. The acetyl group ($C_2$) joins with a four-carbon acid to give a six-carbon acid; this is then successively oxidized via a five-carbon acid to give a four-carbon acid again which can once more accept an acetyl group. This cyclical series of reactions thus completely oxidizes the acetyl fragments produced either from carbohydrate or from fat.

During the course of these degradations, pairs of hydrogen atoms are removed from the intermediates in several reactions of the pathway. The pairs of hydrogen atoms are passed to a carrier called NAD (see Ch. 2). We shall discuss the details of these reactions, and the subsequent oxidation of the hydrogen atoms, in Ch. 12.

An alternative pathway for the degradation of glucose involves the oxidation of one carbon atom to give carbon dioxide and to leave a derivative of a pentose. From pentose, other sugars containing three, four or seven carbon atoms can be formed. These reactions are discussed in Ch. 17.

### Anabolic pathways

Each of the anabolic pathways by which *carbohydrates* and *fats* are synthesized employs some reactions that are different from those used in the catabolic pathways (Chs 19 and 20). In animals, glucose is synthesized chiefly from three-carbon acids – either three-carbon acids that have themselves accumulated during metabolic activity or three-carbon acids that are derived, sometimes via four-carbon acids, from amino acids (see Fig. 9.2). In plants and other photosynthetic organisms, glucose is synthesized from three-carbon sugars which are formed early in photosynthesis. Fatty acids are synthesized from acetyl fragments by a pathway that is not simply the reverse of that by which fatty acids are broken down; the formation of fats from fatty-acid derivatives is, similarly, not the reverse of the hydrolysis of fats.

The anabolic pathways by which *amino acids* (Ch. 22) are synthesized take as their starting points intermediates from the catabolic and amphibolic pathways that we have already mentioned (see Fig. 9.3). The three-carbon acids formed from carbohydrates give rise to several amino acids, including the aromatic amino acids which are formed from a condensation product of a three-carbon acid and a four-carbon sugar. The five-carbon acid that is an intermediate in the Krebs cycle is converted to glutamate, which is the precursor of some other amino acids. Similarly a four-carbon acid that is another intermediate in the Krebs cycle is converted to aspartate, which is the precursor of several other amino acids. Aspartate also gives rise to *pyrimidines* (Ch. 23). *Purines* are formed piecemeal from a large number of precursors (Ch. 23). The ribose phosphate and deoxyribose phosphate that are constituents of nucleotides are formed from the pentose produced by oxidation of glucose.

The remainder of Part 2 of the book describes in far more detail the pathways that we have mentioned here. You will find that these chapters too are extensively cross-referenced. It is our intention to stress that the metabolic pathways that we discuss are closely inter-related, and thus to show how intermediary metabolism combines the two functions that we have mentioned – providing energy and supplying intermediates for biosynthesis.

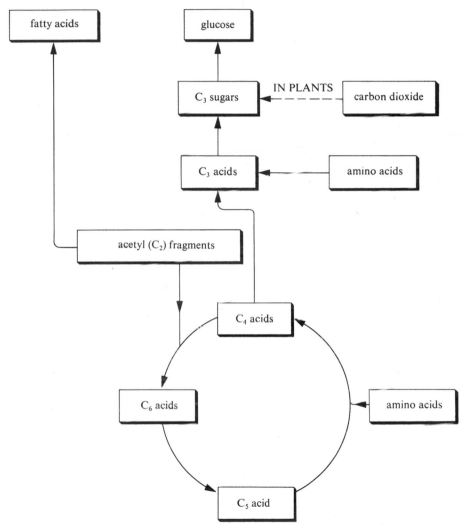

**Fig. 9.2**   Outline of the anabolic pathways for glucose and fatty acids (see text)

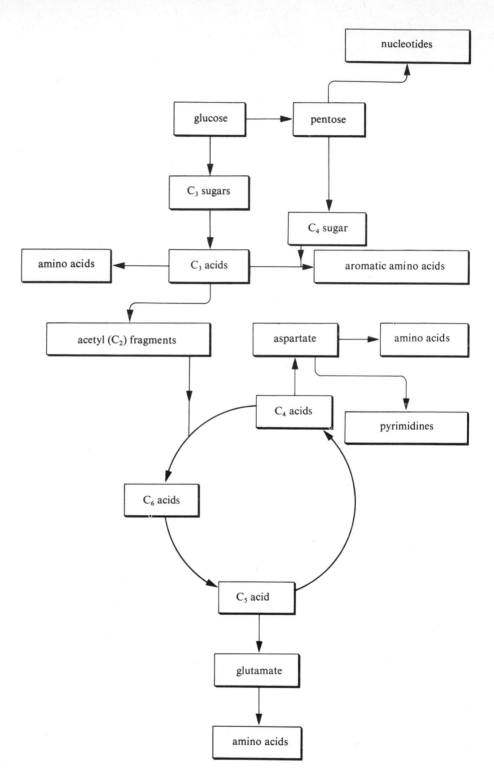

**Fig. 9.3**  Outline of the anabolic pathways for amino acids and nucleotides (see text)

# Chapter 10  The cell

All the reactions that we discuss in this book take place through the chemical activities of living *cells*. In this chapter we shall describe some of the features of cells that are evident either on inspection through microscopes or by chemical analysis. We shall also introduce some of the more important classes of biochemical activities that are associated with the various parts of the cell, although we shall reserve a more detailed account of biochemical specialization within cells to Ch. 21.

Cells can be divided into two types, *procaryotic* and *eucaryotic*, which can be distinguished in the following way. While all cells (with the exception of some curious micro-organisms about which it can be disputed whether the term 'cell' is applicable) are contained within a cell membrane, procaryotic cells have little or no membranous material *within* the cell, whereas eucaryotic cells have clear intracellular membranes. In particular eucaryotic cells have a well-defined nucleus enclosed in a *nuclear membrane*. Eucaryotic cells are characteristic of higher organisms (plants and animals) and of protozoa, fungi and some algae. Procaryotic cells are found in blue-green algae, bacteria and some other micro-organisms such as the rickettsiae. We shall discuss the structure first of the eucaryotic and then of the procaryotic cell.

## Animal cells

Although all eucaryotic cells are distinguished, by the criteria we gave above, from procaryotic cells, there are many different classes of cell within the eucaryotes. In particular, animal cells are different in many respects from plant cells, and we shall describe these two classes of eucaryotic cell separately. Even when we have narrowed the class that we are discussing down as far as animal cells, we still have a large number of cell types that we might consider. Vertebrates, for example, contain dozens of specialized sorts of cells – reticulocytes (precursors of red blood cells), whose chief function is to make haemoglobin; various endocrine cells, which produce substances (hormones) that are secreted into the blood; exocrine cells, which synthesize compounds (such as digestive enzymes) that are secreted to spaces outside the blood; muscle cells, which are designed to be able to contract, and so on. The task of finding a 'typical' animal cell is therefore not an easy one. We might, however, regard as most 'typical' for our purpose a cell that has a very large range of biochemical activities, as opposed to the others that we have mentioned which, through specialization, have lost many of the capabilities that their ancestor cells possessed. On this criterion we shall choose the liver cell as the 'typical' animal cell, since the liver can carry out most of the activities that we shall be describing in later chapters.

Figure 10.1 shows a thin section through a liver cell, as seen in the electron microscope. You can see that the interior of the cell contains a wealth of visible structures (contrast the bacterial cell, which we shall discuss below). We shall describe the cell from the outside towards the centre, giving an outline of the appearance, composition and activities of each of the characteristic 'organelles' – by this term we mean the subcellular structures that are visible with the electron microscope and that carry out specialized functions.

The liver cell is surrounded by the *cell membrane*, sometimes called the plasma membrane, which constitutes the boundary of the cell and serves to regulate the movement of substances in and out. The membrane consists largely of lipid and protein, but its structure, despite many years of detailed and painstaking work, is rather poorly understood (see Ch. 8). When seen in the electron microscope after fixation with osmium tetroxide (which is one of the commonest fixatives used by electron microscopists) the cell membrane appears as a three-layered structure of two dark lines separated by a light line. This three-layered structure is believed to correspond with a particular organization of lipid and protein molecules within the membrane (see p. 199).

As we saw in Ch. 8 the cell membrane is the permeability barrier of the cell. Water, and a few uncharged substances of very low molecular weight, can cross the membrane quite freely, but the majority of the substances found in the body are unable to diffuse across. However, cells need to take up many of these substances, and liver cells in particular must be able to take in the digestion products of foodstuffs, conveyed from the intestine via the portal circulation. Consequently the cell membrane possesses a complex system of specific carriers, which enable particular compounds to be transported into the cell. The nature and the functioning of these carriers was discussed in Ch. 8.

The cell membrane encloses the non-particulate or *soluble fraction* of the cytoplasm, which is that part of the cell in which the other organelles are suspended. The non-particulate fraction of the cytoplasm contains much protein in solution, and has characteristic enzymic activities, particularly those concerned with the first part of the pathway of breakdown of glucose (see Ch. 14), and with the synthesis of fatty acids (Ch. 20), of amino acids (Ch. 22) and of nucleotides (Ch. 23).

Ramifying throughout the soluble fraction are numerous tubules which connect with each other and probably form a single system of channels penetrating the whole cytoplasm. These tubules are bounded by membranes, and make up a network called the *endoplasmic reticulum*. The tubules have various shapes: some are more or less cylindrical, while others are very much flattened to form plates ('cisternae'). The membranes enclosing the cylindrical tubules are smooth, but those enclosing the flattened compartments are often studded on the outside with small particles called *ribosomes* (see Chs 7, 11 and 26). These systems are known as the 'smooth' and the 'rough' endoplasmic reticulum; their inner spaces are continuous with one another. The membranes that surround the tubules have the same type of structure that we mentioned above.

The endoplasmic reticulum does not contain very many enzyme-mediated

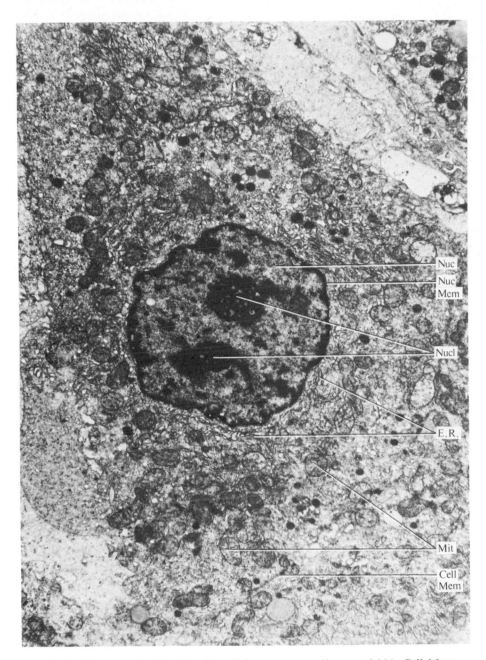

**Fig. 10.1** Electron micrograph of a cell from a mouse liver. × 6 000. Cell Mem = cell membrane; E.R. = rough endoplasmic reticulum; Mit = mitochondria; Nuc = nucleus; Nucl = nucleoli; Nuc Mem = nuclear membrane. (The photograph was kindly supplied by Dr J. C. F. Poole, Sir William Dunn School of Pathology, University of Oxford)

activities, but the esterification of fatty acids (see Ch. 20) does appear to occur within it. The chief function of the endoplasmic reticulum is to act as an intra-cellular transport system. In those cells of the pancreas which secrete proteins, the rough endoplasmic reticulum is highly developed. It has been shown that proteins are synthesized at the ribosomes of the rough endoplasmic reticulum, that they enter the cisternae, and that they appear outside the cell as granules. We may infer that there is a movement of proteins along the channels of the endoplasmic reticulum from their site of synthesis towards the exterior of the cell.

Quite similar to the smooth endoplasmic reticulum is an organelle called the *Golgi complex*. This too consists of cisternae enclosed by membranes, but the membranes lie closer together than do those of the endoplasmic reticulum (see Fig. 10.2). A feature of the Golgi complex is the presence of cylindrical or

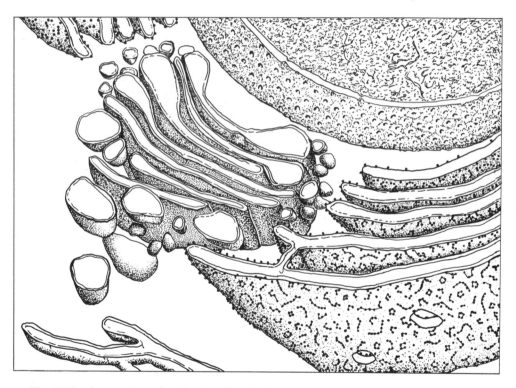

**Fig. 10.2** Impression of endoplasmic reticulum and Golgi complex. 'Rough' endo-plasmic reticulum (bottom right) is studded with ribosomes, but 'smooth' endo-plasmic reticulum (bottom left) is not. The cisternae of the Golgi complex (above the smooth endoplasmic reticulum) are close together, and part of the Golgi complex consists of vacuoles. All these organelles are surrounded by membranes (after Du Praw, *Cell and Molecular Biology*, Academic Press)

spherical vacuoles, formed from the cisternae by 'pinching off'. The function of the Golgi complex is not very clear. There is some evidence that the vacuoles at any rate are active, like the endoplasmic reticulum, in secreting proteins and other compounds from the cell, and this view is supported by the fact that the compartments within the Golgi complex are continuous with those of the endoplasmic reticulum. It has been suggested that substances synthesized in the cell and designed to be secreted are transported first into the vacuoles of the Golgi complex, and that the membranes of these vacuoles then fuse with the cell membrane so that the contents of the vacuoles are liberated outside the cell. Another possible function of the Golgi complex is to form the lysosomes, which we shall now consider.

*Lysosomes* are roughly spherical or egg-shaped bags bounded by a membrane. They seem to arise from the cisternae of the Golgi complex or the endoplasmic reticulum. The characteristic feature of lysosomes is that they contain a high concentration of hydrolytic enzymes – ribonuclease, deoxyribonuclease, proteases, etc. These act to break down large molecules that enter the cell by phagocytosis: the vacuole formed when phagocytosis occurs fuses with a lysosome to form a digestive vacuole in which breakdown of the contents can occur, and the small molecules thus produced are transported across the lysosomal membrane into the soluble fraction of the cytoplasm. Another function of the lysosome is to act in auto-digestion of injured cells: in response to various stimuli the contents of the lysosomes, which are normally kept segregated from the rest of the cell, may be released and break down some of the cellular components.

*Mitochondria* are among the best studied of the cytoplasmic organelles. They too are bounded by membranes that are composed of lipid and protein, but unlike the membranous organelles that we have hitherto considered, mitochondria do not form part of the continuous membrane system that extends throughout the cytoplasm. In fact mitochondria have *two* membranes. The outer one is smooth and regular, but the inner one is highly convoluted inwards into folds called cristae, which have the effect of increasing the surface area of the membrane (see Fig. 10.3). The cristae project into the mitochondrial matrix, which is a gel containing a high concentration of protein.

Both of the mitochondrial membranes are composed of lipid and protein. The inner membrane contains many enzymic activities, and it can be assumed that the protein of this membrane is composed in part of such enzymes. The activities associated with the membrane are those involved in the electron-transport chain and oxidative phosphorylation (see Ch. 12). The components appear to be arranged in assemblies (compare the assemblies of chlorophyll molecules and associated enzymes in the chloroplast, p. 251) each of which contains all the constituents of the respiratory chain. In this way each assembly is independently capable of electron transport and of oxidative phosphorylation.

Other important enzymic activities are also located in the mitochondria, particularly those of the tricarboxylic acid cycle (Ch. 15) and fatty-acid oxidation (Ch. 16). These two pathways are quantitatively much the most significant sources of electrons that can be passed along the respiratory chain. Since the bulk of the relevant enzymes are located in the matrix of the mitochondrion, which (owing to the cristae) is in intimate contact with the inner membrane, the mitochondrion is admirably adapted both to abstracting electrons from substrates and to using them to synthesize ATP. Consequently most of the ATP that the cell obtains is manufactured in the mitochondrion; we shall see in Ch. 21 how it is transferred to the soluble fraction of the cytoplasm.

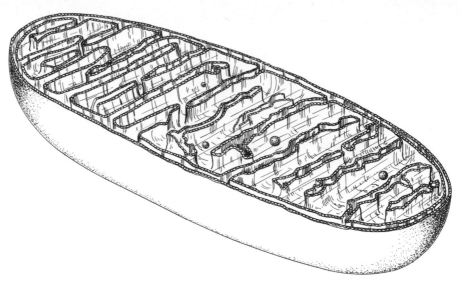

**Fig. 10.3** Impression of a mitochondrion. The outer membrane is unfolded, but the inner membrane is extensively invaginated to form cristae which project into the mitochondrial matrix (After Du Praw, *Cell and Molecular Biology*, Academic Press)

The largest of the organelles in a liver cell is the *nucleus*. The nucleus is enclosed by a double membrane system; both the inner and the outer membrane seem to be three-layered membranes as we described above. The membrane is penetrated by small pores through which material can pass; it is not known what controls the passage of molecules into and out of the nucleus, but it is apparently not simply their size. The interior organization of the nucleus is very complicated; the nucleus contains many different kinds of proteins, as well as some lipid, but the most characteristic of its constituents is DNA. We shall see in Ch. 24 that DNA is the genetic material of the cell, and since the nucleus contains almost all of the DNA it is evident that the nucleus is the guardian of the genetic information that is needed to control all cellular activity.

The DNA of the nucleus is associated with a number of proteins. One well-known class are the basic proteins known as histones (p. 157); it has been suggested that histones are involved in the control of genetic expression (p. 488), but this idea has not yet been proved. There may also be some RNA associated with the DNA–protein complexes. When the nucleus is not dividing, the DNA is organized into a diffuse mass of thin filaments, in which little structure can be seen. At division, the nuclear membrane disappears, and the DNA condenses into much more tightly folded strands – the chromosomes – which can be clearly seen even with the light microscope. The chromosomes divide and may undergo recombination (see Ch. 24) before the daughter chromosomes are segregated in the formation of two daughter nuclei.

Within the nucleus a dense body (or sometimes more than one) can often be seen. This is known as the *nucleolus*; it is usually conspicuous when the cell is not dividing, so long as protein synthesis is actively occurring, but it disappears during division, and bcomes small and inconspicuous at any time when the cell is not active in synthesis. The nucleolus is very rich in RNA, and it appears to represent the site in the nucleus at which RNA is synthesized (or, if RNA is synthesized all over the nucleus, as is possible, then the site at which newly synthesized RNA is collected and processed). In particular, the RNA that is destined to be included in the ribosomes (p. 157) is accumulated in the nucleolus and there cut into molecules of the appropriate size.

## Plant cells

To find a typical plant cell is as difficult as to find a typical animal cell, since plant tissues are no less specialized than animal tissues. For our purpose, however, it is convenient for illustrating the characteristics of plant cells to consider a photosynthesizing cell, since this shows the activity that is characteristic of green plants. Accordingly, we shall describe some of the features of a cell from the leaf parenchyma of a flowering plant (see the electron micrograph shown in Fig. 10.4).

The *cell membrane, endoplasmic reticulum, Golgi complex, mitochondria* and *nucleus* of the plant cell are similar in most respects to the corresponding organelles of the animal cell, and we shall not consider them in detail. We might mention, however, that the mitochondria are fewer in number and contain fewer cristae in the plant than in the animal cell; the bulk of the ATP synthesized in a photosynthetic cell is made in the chloroplast, and oxidative phosphorylation is used only in the dark. The characteristic organelles of the parenchymal cell are the cell wall, the vacuole and the chloroplast. We shall defer a detailed discussion of the chloroplast till Ch. 13, since it is more convenient to describe its structure at the same time as we discuss the reactions of photosynthesis. For the present we shall note only that the chloroplast is a highly membranous organelle and contains photosynthetic pigments.

The *cell wall* of the plant cell has no analogy in animal cells. Whereas in animals that are so large as to require mechanical support, specialized organs (such as the skeletons of vertebrates) have been evolved to meet the need, in plants support is provided for each cell by the rigid, thick wall. This lies outside the cell membrane which, in plants as in animals, is the permeability barrier of the cell; the cell wall protects the membrane of the plant cell, especially from the possibility of osmotic shock, which is an ever-present danger in some habitats. The cell wall is composed chiefly of polysaccharides; the most characteristic is cellulose (see p. 176), but other polymers such as xylans, pectins and lignin (see p. 177) also occur.

The interior *vacuole* of the parenchymal cell is a space bounded by a membrane; it contains sugars, salts and other substances in solution. In young cells the central vacuole is small, or there may be several very small vacuoles; these grow and coalesce as the cell ages. The fact that the solutes inside the vacuole are highly concentrated ensures that the cell retains a high osmotic pressure and readily takes up water. The vacuole serves, too, as a kind of excretory organ: materials that the cell does not need may be deposited within the vacuole and sometimes even crystallize there. Since the vacuole occupies the central area of the cell, it forces the cytoplasm to the edge where it is in intimate contact with the cell membrane and can readily absorb nutrients and oxygen.

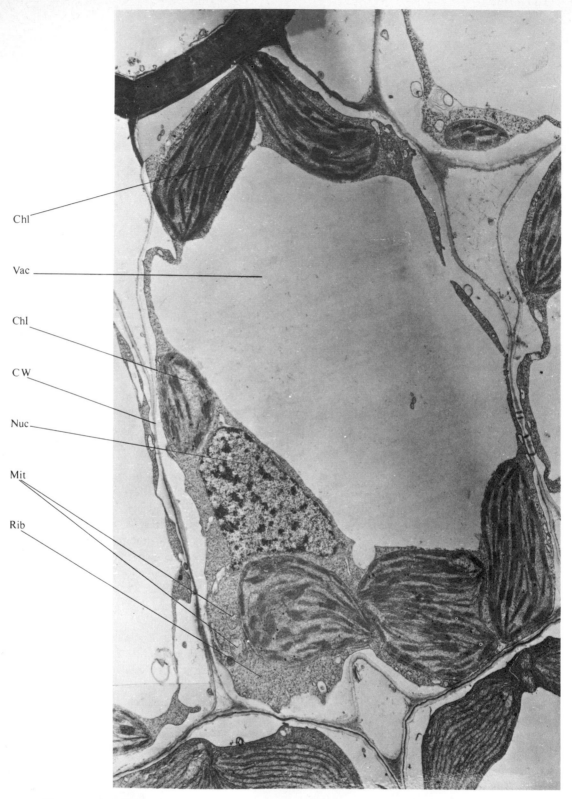

Chl

Vac

Chl

CW

Nuc

Mit

Rib

**Fig. 10.4**  Electron micrograph of a cell from the leaf parenchyma of maize.  × 8 700. Chl = chloroplasts; CW = cell wall; Mit = mitochondria; Nuc = nucleus; Rib = ribosomes; Vac = vacuole. (The photograph was kindly supplied by Dr J. Brangeon and Dr B. E. Juniper, Department of Botany, University of Oxford)

Bacteria are the best-known members of the great group of procaryotic organisms. As an example of a procaryotic cell we shall briefly describe the bacterium *Escherichia coli*, the structure, biochemistry and genetics of which have been studied in far more detail than those of any other procaryote.

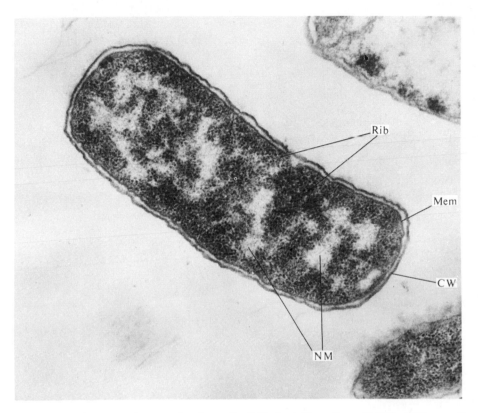

**Fig. 10.5** Electron micrograph of a cell of *Escherichia coli*. ×51 000. CW = cell wall; Mem = bacterial membrane; NM = nuclear material; Rib = ribosomes (The photograph was kindly supplied by Dr D. Kay, Sir William Dunn School of Pathology, University of Oxford)

As we mentioned earlier (p. 213) the procaryotic cell has no membranes internal to the cell membrane (but see below for a qualification) and its structure is correspondingly far simpler than that of eucaryotic cells. In fact, cells of *E. coli* have little visible internal structure (see Fig. 10.5). There is no endoplasmic reticulum, Golgi complex or lysosomes, there are no mitochondria and there is no discrete nucleus. On the other hand there is a thick cell wall, and the cell membrane has more functions than that of animal or plant cells.

The *cell wall* of *E. coli* is similar in principle to that of a plant cell wall. The rigid backbone is composed of a polymer of amino-sugars (pp. 180 ff.) (compare the polymers of sugars, such as cellulose, in plant cell walls), which provides mechanical strength to prevent osmotic rupture. Additional strength is conferred on the polymer by cross-linking peptide chains. However, the cell wall of *E. coli* is more flexible than that of plants, since it contains, in addition to the amino-sugar polymer, complex lipopolysaccharides (whose structure is not well understood) and lipoproteins. Like that of plants, the cell wall of bacteria is highly permeable.

The *bacterial membrane* contains lipids and proteins, and resembles to some extent the cell membrane of animal and plant cells. It is the permeability barrier of the *E. coli* cell, and it contains numerous specific carrier systems that transport substances into and out of the cell. But the *E. coli* membrane has other functions as well: in fact it takes over many of the functions that, in eucaryotic cells, are carried out by the mitochondria. The enzymes of the electron transport chain

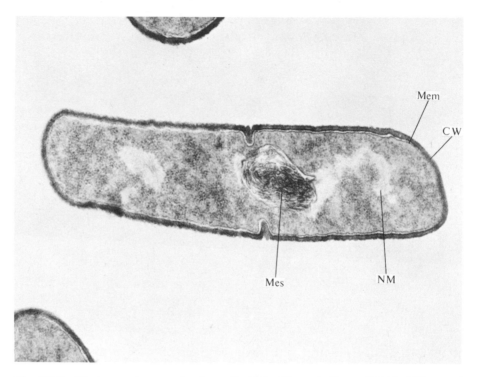

**Fig. 10.6** Electron micrograph of a cell of *Bacillus subtilis.* × 45 000. CW = cell wall; Mem = bacterial membrane; Mes = mesosome; NM = nuclear material. Notice that newly synthesized membrane and wall material are beginning to divide the cell into two (The photograph was kindly supplied by Dr D. Kay)

and of oxidative phosphorylation are located in the bacterial membrane, and so too are many of the enzymes involved in the oxidation of reduced substrates. Furthermore, the enzymes necessary for synthesizing the cell-wall molecules are also found in the membrane.

The membrane of many bacterial cells, especially those of the Gram-positive species (see p. 182), shows complex invaginations to form membranous whorls called *mesosomes* (see Fig. 10.6). The function of the mesosomes is not known for certain. It has been suggested that they are sites of synthesis of new membrane or of cell-wall material, but this idea has not been proved.

The *soluble cytoplasm* of *E. coli* is comparable with that of eucaryotic cells and contains many of the cell's enzyme systems. Suspended in the cytoplasm are the *ribosomes*, which are smaller than those of eucaryotic cells (see p 438); some ribosomes, however, are attached to the membrane (see above) just as in eucaryotic cells some ribosomes are attached to the membranes of the endoplasmic reticulum.

The *nuclear material* of *E. coli* is not confined within a discrete nucleus as it is in higher cells. Instead there is a rather diffuse arrangement of DNA stretching through much of the cell. The DNA is not generally associated with basic proteins like histones, although a few highly specific protein molecules, acting as enzymes or as repressor molecules (see Chs 25 and 28), bind to the DNA at particular times or in particular sites. The DNA consists of a single, very long molecule of double-helical structure (see Ch. 7), which is elaborately coiled. It functions as the genetic material of the cell, in a way that we shall discuss in detail in Ch. 24.

## Viruses

We shall conclude this chapter with a mention of *viruses*, which, although they are not cells, may conveniently find a place here (see also Ch. 7).

All known viruses can exist in two forms. They can be found inside cells, as active parasites, or outside cells, in a kind of dormant state. The most striking characteristic of viruses, and the one that distinguishes them from all cellular organisms, is that they are incapable of independent metabolism. In order to take part in any metabolic activity, a virus must be inside a cell (whether an animal, a plant or a procaryotic cell); when present inside a cell it can grow and reproduce, but only by subverting the metabolism of its host.

The extracellular form of a virus, therefore, is adapted merely to the purpose of moving from one host cell to another. To this end, the virus particle has a coat, which always contains protein and is most often composed of protein alone (although some animal viruses have lipid present in their coats). This coat surrounds the essential component of the virus, which is a single molecule of *nucleic acid*, either DNA or RNA. Sometimes there are other structures present, such as fibres that are used to help the virus to adsorb to the host cell. But we can regard all of these envelope structures as having just two functions: to protect the viral nucleic acid and to ensure its penetration into a new host cell. (An experiment demonstrating that it is the nucleic acid rather than the protein that does penetrate into the host is described on p. 403.)

Intracellularly, the virus exists, for the most part, as free nucleic acid. After the nucleic acid has entered the host cell it is repeatedly replicated (Ch. 25), so that finally a hundred or more copies may be formed from the single copy that penetrated the cell. After several rounds of replication of nucleic acid, viral envelopes too start to be made inside the cell. The host cell becomes full of molecules of viral nucleic acid, and of coats, and complete virus particles are then assembled. Finally, the cell lyses and the particles are released.

The chief interest in viruses for the biochemist is this. It is plain that the extracellular form of the virus has (if any enzymic activity at all) quite insufficient enzymic activity to sustain an independent metabolism – at most only a few different kinds of protein are present in the virus particle. Consequently the virus must make use of the metabolic pathways of its host for its supply of energy and of intermediates. However, the cell that is infected with a virus soon turns to making viral components (nucleic acid and viral envelopes) in place of its own macromolecules. These processes are directed by the viral nucleic acid that has entered the cell. A virus-infected cell is therefore an unusually convenient form of biological material for studying both the relation between nucleic acid and the proteins formed at its direction, and also the replication of the nucleic acid itself. To put this another way (a way that we shall justify in Ch. 24) a virus-infected cell is an unusually convenient form of biological material for studying the expression and the replication of the genetic substance.

# Chapter 11 **Methods in studying metabolism**

In this chapter we shall give an outline of some of the experimental methods that have been generally used in metabolic (as opposed to structural) studies in biochemistry. We shall have occasion to refer to these methods when discussing, in subsequent chapters, the pathways by which molecules are degraded and synthesized in living organisms. For the most part we shall restrict our attention to those methods that are characteristic of biochemical studies (rather than, for example, of organic chemistry). We shall first describe the preparation of biological material that is the source of the enzymic activities needed for metabolic studies, and then the methods used for working out the pathways themselves.

## BIOLOGICAL PREPARATIONS

The history of biochemistry has shown a gradual refinement in the level of complexity of the biological material that forms its object of study. The earliest biochemical studies were generally done with intact organisms, and as time went on these were taken apart – first to the level of the organ, then to slices of tissue, then to the broken cell, then to the organelle and finally to the purified enzyme preparation. Nonetheless, all these levels of organization are still used in experiments with particular aims in mind.

*Intact organisms.* Much valuable information has been gained from studies of whole animals, especially those designed to establish nutritional requirements (see pp. 360 f.). One might at first sight think that the complexity of organization of the whole animal would preclude any possibility of determining metabolic pathways by these means. In fact, however, although the finer details of a pathway cannot be established, the general outlines often can. For instance one can show that the chief excretion product of nitrogen metabolism is ammonia in some animals, urea in others and uric acid in still others (see p. 388), and these differences can often be related to the habitats of the animals concerned. Again, a surprisingly accurate account of fatty-acid catabolism was obtained by studies of the excretion products of particular fatty acids (see pp. 283 f.). Whole animals are still used in investigations of the effect of certain diets on the enzymic composition of particular organs (see p. 488).

Work with micro-organisms often employs the intact organism. Whole bacteria, for example, often metabolize substrates that are dissolved in the medium in which they are suspended, and sometimes excrete the metabolic products. If the bacteria are growing, they may incorporate an added substrate (or part of its molecule) into their own macromolecules – for example many species of bacteria will incorporate amino acids into protein, or nucleotide bases into nucleic acid; if they are suspended in a medium (such as a carbon-free buffer) in which they cannot grow, they may exhibit other metabolic activities – for example they will decarboxylate some amino acids to form amines.

Returning now to higher organisms, once it has been established that a whole animal is capable of a certain metabolic activity, one next wants to know what organ is responsible. For instance, mammals excrete the nitrogen from amino acids as urea: in which organ is the urea formed? Questions of this kind can be answered by experiments with *organ perfusion*. If an organ such as the liver or kidney is isolated with its own circulation system intact, one can perfuse the blood vessels either with blood (which has been treated to prevent clotting) or with artificial solutions of chemicals, and analyse both the input and the output fluid (Fig. 11.1). In this way it can be established, for instance, that the liver is the chief source of urea. If amino acids are added to fluid that is used to perfuse the liver they are deaminated, and their nitrogen atoms are incorporated into urea, which appears in the output fluid.

Although the biochemical era in which perfusion was used to ascribe reactions to particular organs is now ended, the technique is still employed a great deal for other purposes. Modern studies with the method are often concerned with the control of metabolic pathways. Thus a good part of the information that we shall present on the control of degradation and synthesis of carbohydrates (pp. 502 ff.) is derived from work with perfused organs.

The technique which, probably more than any other, has been responsible for most of our knowledge of intermediary metabolism, is that of the *tissue-slice*. Animal or plant tissues can be sliced with sharp razors to yield slices that are thin enough to allow oxygen and substrates from the medium to penetrate to the cells. One can then incubate the slices with metabolites, and study the removal of substrates and the accumulation of products by analysing the medium at various times during incubation. This technique was used, for example, by Krebs to establish the details of the reactions by which urea is synthesized in liver.

An important drawback to all methods in which the cells of the material remain intact is that there is often a barrier to the entry of the substance under study. This difficulty is as often encountered with micro-organisms as it is with cells of higher organisms. In all types of cell, the cell membrane (see pp. 198 f. and 214) may prevent certain compounds from penetrating the cytoplasm; this fact has sometimes been overlooked, with the result that it has been erroneously concluded that particular substances are not metabolized. For instance, citric acid does not penetrate cells of many species of bacteria, and the fact that the oxidation of citric acid could not be demonstrated was taken to mean that these bacteria have no tricarboxylic acid cycle – a conclusion that is quite erroneous.

With animal cells, the permeability barrier can easily be broken by homogenizing the tissue, and *homogenates* of liver, kidney, brain, etc. are now probably the most commonly used preparations of these tissues in biochemical work. Homogenates are made by placing the tissue, suspended in sucrose solution to maintain the appropriate osmotic pressure, in a strong glass tube which is provided with a rotating, close-fitting pestle; the shearing action breaks the cell

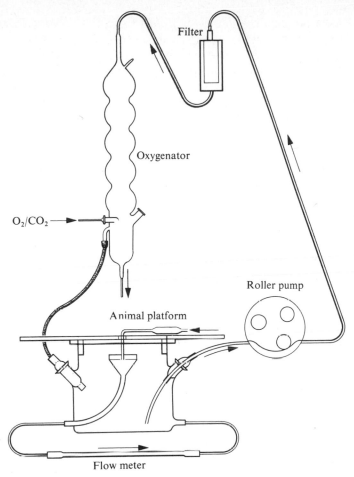

Filter

Oxygenator

$O_2/CO_2$ →

Roller pump

Animal platform

Flow meter

**Fig. 11.1** Apparatus used for perfusion of the liver. Perfusing medium is oxygenated and is made to enter through the portal vein, the flow being driven by a pump. The medium leaves the liver by the vena cava and is collected in a reservoir from which samples may be taken for the estimation of biochemical intermediates and other compounds of interest. The whole apparatus is kept in a cabinet at constant temperature

membrane and liberates the contents of the cells. The homogenate is usually then strained through muslin to remove connective tissue.

With plant and bacterial cells, which have thick, rigid cell walls (see pp. 219 and 221), breaking the cells is much more difficult. Some of the methods that have been used are: grinding the cells with alumina or quartz, shaking them with glass beads at high frequency of oscillation, forcing them under high pressure through a narrow orifice, or exposing them to ultrasonic vibration. These methods have the common disadvantage of rupturing not only the cell wall but also much of the intracellular structure. To overcome this difficulty, more gentle methods based on enzymic digestion of the cell wall are sometimes used.

A homogenate of animal cells can be regarded as a complex mixture of all the organelles of the cell, together with a few unbroken cells. For studies of the metabolic activities of individual organelles, it is necessary to separate them. The most commonly used method is *differential centrifugation* (see also Appendix 2). If the tissue homogenate is centrifuged, the organelles will sediment in order of decreasing size. For example, a liver homogenate may first be centrifuged at 600*g* for 10 minutes; this treatment will sediment nuclei and any unbroken cells. The supernatant liquid is then centrifuged at 10 000*g* for 10 minutes; mitochondria and lysosomes will now sediment. (Unfortunately it is almost impossible to separate mitochondria and lysosomes by differential centrifugation.) The remaining supernatant liquid is finally centrifuged at 100 000*g* for 60 minutes; the pellet is conventionally called the *microsome fraction*, and it consists of pieces of endoplasmic reticulum with ribosomes attached to some of them (see p. 437). The final supernatant liquid represents the soluble fraction of the cytoplasm.

A shortened procedure suffices to obtain ribosomes from bacterial cells. After a low-speed centrifugation (500*g* for 10 minutes) to remove unbroken cells and the debris of cell walls and membranes, ribosomes are prepared by centrifugation at 100 000*g* for 120 minutes.

We saw in Ch. 10 that nuclei and mitochondria of eucaryotic cells are surrounded by membranes, which have permeability properties similar to those of the cell membrane. It may therefore happen that in a preparation of (for example) mitochondria from broken cells, penetration of an added substrate still presents difficulties, and this fact can complicate the interpretation of experiments with isolated organelles. Consequently, subcellular organelles may sometimes have to be themselves disrupted (for example by treatment with detergent) for studies of the metabolism of some substrates.

The final stage in fractionating the metabolic activities of the cell is to isolate pure enzymes either from crude preparations of broken cells or from purified organelle fractions. The techniques used are those of protein fractionation which we described in Ch. 9.

An important part of the biochemical understanding of metabolic activities is to put them in their context. We may be tempted to feel that we have completed our study of a particular reaction when we have isolated in pure form the enzyme responsible for it. But a complete understanding can only come if we show how the enzymic activity forms part of the metabolic capability of the whole cell, and, indeed, of the whole organism. The enzyme's activity may be affected not only by its own substrates and products but also by compounds that are metabolically far removed from these; such effects may be important in the control and integration of metabolism (see Ch. 29). Thus a full understanding of metabolism is not reached until we have not only purified enzymes from whole cells but also reconstructed the intermediate pathways as completely as possible.

## METHODS OF ELUCIDATING METABOLIC REACTIONS

We have already seen (pp. 112 and 207) that intermediary metabolism depends on a complex network of interlocking reactions. The complexity of the system makes the prospects for unravelling any of the reaction pathways seem at first sight rather dim. How can one hope to trace the path of a molecule, within a cell that contains thousands of other compounds, as it is continually modified and manipulated on its way to some distant product? In particular, how can one hope to identify intermediates of reaction pathways when these can be presumed, on grounds of the economy of the cell, to be present in very low concentrations?

There are in essence two main methods that can be used to overcome these difficulties. First, one can *perturb* the system by blocking one pathway or one reaction, and then see what intermediates accumulate behind the block. Secondly, one can *label* a compound in such a way that it can be recognized among the complex mixture of molecules in an actively metabolizing cell, and then see what labelled products result from its metabolism.

Several means are available for blocking particular metabolic reactions. Many enzymes are poisoned by *inhibitors*, which, as we saw in Ch. 6, affect one or other of the kinetic characteristics of an enzyme reaction. In cells, or cell extracts, that are treated with an inhibitor of a specific enzyme, the flow of material along a pathway is dammed, and earlier intermediates accumulate and may be spilled out. The use of inhibitors to elucidate metabolic pathways is illustrated on p. 264, where we discuss the effect of fluoride on glycolysis, and on pp. 274 f., where we discuss the effect of malonate on the tricarboxylic acid cycle.

Analogous to the use of inhibitors is the *removal of an essential cofactor* for a reaction. Since cofactors are generally small molecules, they can often be removed by dialysis (p. 544), and the dialysed cell extract will then lose its ability to carry out the reaction for which the soluble cofactor is necessary. Addition of the diffusible cofactor back to the extract restores the activity. As with extracts that have been treated with inhibitors, dialysed extracts can often be made to accumulate intermediates which can be identified. The use of dialysis has the further advantage that it achieves a fractionation of the components of an enzyme reaction, so that a cofactor can often be easily isolated from the diffusate and thus identified. We shall mention the use of dialysis in studying glycolysis on p. 258.

A third means of blocking a reaction is the use of microbial *mutants*. As we shall see on pp. 404 ff., it is possible to isolate strains of bacteria (and of fungi, and to a lesser extent of other organisms) that have lost the genetic capacity to carry out a particular reaction. Provided that such a mutation does not wreck the whole functioning of the cell (as a complete block in the synthesis of, say, DNA would do), the strain will be viable in certain circumstances. Its cells will function as if they were inhibited at the point where the missing enzyme normally acts: if they are supplied with the end-product of the blocked pathway they will grow normally, but they will tend to accumulate intermediates of the pathway prior to the metabolic block and excrete these into the medium.

Consider, for example, the biosynthetic sequence

$$A \rightarrow B \rightarrow C \rightarrow D \rightarrow E$$

where E can be regarded as the end-product of the pathway (for instance an amino acid). In a bacterial mutant that is missing the enzyme that converts C to D, growth can be supported by the addition of the amino acid E to the culture, and the cells will tend to excrete A, B and C, which can then be identified as precursors of E.

Bacterial mutants can be used to give even more information than this. Often it may be possible to isolate several different kinds of mutants with a single requirement. To use the same pathway as an example, one may be able to find a mutant blocked in the conversion of D to E, another blocked in the conversion of C to D, and another blocked in the conversion of B to C. All of these will grow if supplied with the amino acid E. The first will grow only if supplied with E, since even if it is given A, B, C or D it will be unable to convert these to E. The second, however, will grow if supplied with D or E, since from D it can make E, and similarly the third will grow if supplied with C or D or E. Thus a set of mutants can provide information about the *sequences* of intermediates in a pathway.

There are snags in the interpretation of experiments with blocked pathways – whether the block is due to inhibitors, to removal of a cofactor or to mutation – that we should mention. Because the whole of metabolism forms an interlocking network, the product of one reaction is often the substrate of two or more other reactions. Thus, if an intermediate accumulates behind a block, it may not remain unchanged but may be acted on by an enzyme other than that which is blocked. Hence what is isolated as an intermediate of a blocked pathway may, in fact, be a *secondary product* of the true intermediate. Again, in our hypothetical biosynthetic pathway, what may actually happen is:

$$A \rightarrow B \rightarrow C \rightarrow D \rightarrow E$$
$$\uparrow$$
$$X$$

so that in a mutant blocked in the enzyme that converts B to C, the requirement for the amino acid E will be satisfied not only by C and D (which are true intermediates) but also by X which is not a true intermediate but which can be converted to C by a reaction away from the main pathway.

The second chief method of studying metabolic reactions depends on *labelling* precursors in such a way that their products can be easily identified in the cell. An ingenious early example of labelling was that by which the principles of fatty-acid oxidation were worked out (see p. 283). It depends on making phenyl derivatives of fatty acids and examining the products that are excreted when these derivatives are administered in the diet; since the phenyl group is not itself metabolized the products can be readily recognized. However, such a drastic chemical change as the introduction of a phenyl group may often block the activity of the enzymes that are being studied – the pathway of fatty-acid oxidation happens to be particularly insensitive to this kind of inhibition.

Chemical labelling has now been almost entirely superseded by *isotopic labelling*. The uncommon isotopes of the various elements present in living material – C, H, O, N, S and P in particular – can be used to prepare molecules that are chemically almost identical to normal precursors or intermediates in metabolic reactions, but distinguishable by physical means. Especially useful are the radioactive isotopes: $^{14}C$, $^{3}H$, $^{35}S$ and $^{32}P$. These are very often used in biochemical work, and the elucidation of a large number of reactions and reaction mechanisms has depended upon them. It is a comparatively easy matter, for example, to prepare sugars or amino acids labelled with $^{14}C$, to add these to metabolically active cells or cell extracts, and to identify the compounds in which radioactivity appears.

There are so many examples of the use of radioactive compounds in bio-chemistry that it is hard to know which to cite. One especially common use of radioactive amino acids in recent years has been in the study of protein synthesis: the early systems capable of protein synthesis were very feebly active, and no information could have been gained from them had it not been possible to detect the incorporation of very small amounts of radioactivity from labelled amino acids. Earlier examples of the use of radioactive isotopes in intermediary metabolism are the elucidation of the pathway of carbon-dioxide fixation in photosynthesis (pp. 321 ff.) and the discovery that aspartic acid is a precursor of pyrimidines (pp. 390 ff.). The use of radioactive isotopes has also contributed greatly to our understanding of molecular genetics: an important example is the experiment in which Hershey and Chase proved that DNA (labelled with $^{32}P$) and not protein (labelled with $^{35}S$) is the genetic material of bacteriophage (p. 403).

Non-radioactive isotopes are more difficult to detect and assay than radio-active isotopes, and it is possible that this is the reason why the biochemistry of nitrogen and oxygen, which have no convenient radioactive isotopes, is less well understood than that of carbon. There are, however, some important examples of the use of the stable isotopes $^{15}N$ and $^{18}O$ that have yielded valuable information. Schoenheimer fed $^{15}N$-labelled amino acids to adult rats and found that they were incorporated into protein at a rate that could be explained only if it were assumed that the animal's protein is constantly being degraded and resynthesized. He concluded that body constituents are in a *dynamic state* of metabolic turnover, rather than static. One of the most famous examples of the use of $^{18}O$ is the demonstration that in photosynthesis the oxygen produced is derived from water rather than from carbon dioxide (p. 248), a discovery that had a profound effect in elucidating the reactions of photosynthesis.

A valuable variation on the use of radioactive precursors by the methods we have mentioned is the technique of *isotope competition*. If we refer again to the scheme

$$A \rightarrow B \rightarrow C \rightarrow D \rightarrow E$$

(p. 229), we can readily see that if radioactively labelled A is supplied to growing organisms, they will incorporate the radioactivity, in the form of the amino acid E, into their protein. If now non-radioactive E is supplied as well, the specific activity (that is, the radioactivity per mole) of E will fall. Similarly the provision of non-radioactive B, C or D will (provided that they can penetrate the per-meability barrier of the cell) dilute the radioactivity in E that is derived from A. An example of the use of this method is the discovery of 'families' of amino acids that we refer to in Ch. 22: it is found, for instance, that the addition of non-radioactive aspartic acid to cultures of *Escherichia coli* growing on radio-active glucose depresses the specific activity (not, notice, the total amount synthesized) of lysine, methionine and threonine (pp. 365 ff.).

As with the interpretation of blocked pathways, there are snags in the inter-pretation of isotopic experiments which we must mention. First, the detection of radioactivity is so extremely sensitive that a very minor pathway can easily, if supplied with a radioactive precursor, give rise to what may appear to be sub-stantial quantities of product. A more serious difficulty is this: it is possible for radioactivity from one compound to be incorporated into another even though there is no *net* metabolic flow in that direction. For example, an important principle of animal biochemistry is that glucose cannot be synthesized from fatty acids (p. 339). However, if $^{14}$C-labelled fatty acids are fed to an animal, radioactivity does appear in the glucose residues of glycogen. The explanation is that the carbon atoms of the fatty acids enter the tricarboxylic acid cycle (Ch. 15), hence find their way into oxaloacetic acid and are therefore incorporated into glucose (Ch. 19), even though there is no pathway by which the total *amount* of carbohydrate synthesized can be increased by supplying fatty acids. Such difficulties have, in the past, seriously misled biochemists in their studies of intermediary metabolism, but they are now sufficiently well understood for major misinterpretations to be unlikely to recur.

# Chapter 12 The respiratory chain: oxidative phosphorylation

We have seen in Ch. 2 that oxidative processes are prominent among the energy sources available to the cell. Oxidations of organic compounds in which molecular oxygen is used as the final electron acceptor will yield as much as 50 000 cal/g atom of oxygen. In biological oxidations (unlike normal combustion in non-living systems) this yield of free energy is coupled to chemical synthesis and is not primarily expressed in the production of heat. The principal biochemical medium of energy exchange is ATP (p. 24); its synthesis from ADP requires considerably less than 50 000 cal/mole and thus one would expect that several molecules of ATP could result from one, efficiently coupled, oxidation. A process known as *oxidative phosphorylation*, which occurs in several stages, exists for this purpose. To carry out the process in stages has the advantage that the energy may be taken off in steps nearer in size to those required for the production of a single molecule of ATP. We shall see that in fact three molecules of ATP are produced by the full process, so that the energy yield is nearly 50 per cent. It is noteworthy that this process, so central to the living state, is so efficient. It compares favourably with the thermal efficiency of the reciprocating steam engine (about 7 per cent) and with the overall thermal efficiency of the steam turbine–dynamo couple (about 30 per cent).

Other modes of biological oxidation do exist and some involve energy coupling. However, oxidative phosphorylation is of unique importance. It is a common adjunct to the majority of degradative reaction pathways described in this book and is, in aerobic organisms, the principal source of the ATP required by the cell.

In the cells of higher organisms oxidative phosphorylation takes place, as was mentioned on p. 217, in an organized structure within the cell known as the *mitochondrion*. We shall see that there are many similarities between the oxidative production and the photosynthetic production of ATP. The latter takes place in an analogous organelle in photosynthetic cells, the chloroplast. In both cases the principal catalysts are proteins that contain prosthetic groups and are embedded in a matrix of lipid. We shall also see that it is possible that lipids have an active role in the phosphorylation process; they are certainly extremely important in establishing and controlling the critical three-dimensional interrelationships between the other components.

We shall now describe the commonly accepted features of the pathway that transports hydrogen and electrons and some of the theories that have been put forward to explain the mechanism of the coupling between the transport down this pathway and the synthesis of ATP from ADP. Almost certainly much more remains to be discovered, but it is unlikely that our understanding of the transport pathway will change in any fundamental way. The theories about the mechanism of oxidative phosphorylation are not nearly so settled, but it is instructive to examine the present stage that the argument has reached and to become familiar with the terms in which it is being conducted.

## The hydrogen and electron transport pathway

The oxidation of an organic compound may be written

$$AH_2 + \tfrac{1}{2}O_2 \rightleftharpoons A + H_2O. \tag{12.1}$$

For the majority of biochemical compounds the process is divided into several stages. The first stage is

$$AH_2 + NAD_{(oxidized)} \rightleftharpoons A + NAD_{(reduced)} \tag{12.2}$$

where NAD is the hydrogen carrier that was mentioned on p. 29 and which we shall discuss in detail in a moment. This stage will be an enzyme-catalysed reaction. The enzyme may either be absolutely specific for A or may catalyse the dehydrogenation of a range of similar compounds.

NAD (p. 79, Table 4.2) is usually bound non-covalently to the enzyme that catalyses reaction (12.2). (The dissociation constant of the binding (see p. 14) is usually of the order of $10^{-3}$ M, a value which is more characteristic of a cofactor than of a prosthetic group.) The part of the molecule that is concerned in reduction–oxidation is the nicotinamide ring (Fig. 12.1). (It should be possible to

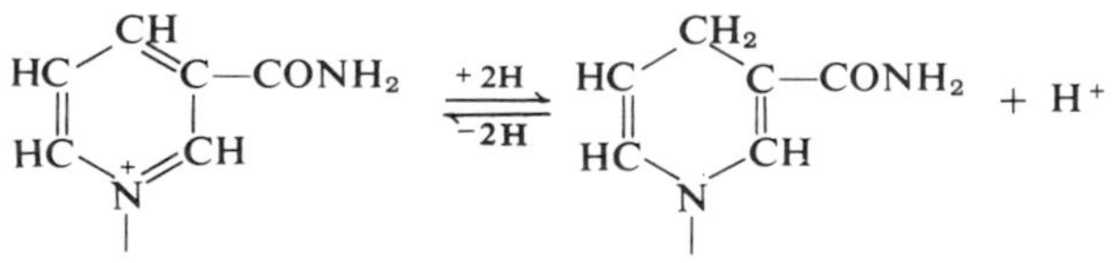

**Fig. 12.1** The reduction and oxidation of the nicotinamide ring

see from this figure why the oxidized form of NAD is sometimes written as $NAD^+$ and the reduced form NADH, with a similar notation for NADP. We prefer the notation $NADH_2$ for the reduced form and NAD for the oxidized form.)

Most animals are unable to synthesize nicotinamide. The vital place of NAD and NADP in metabolism explains why nicotinamide is a vitamin (see also p. 80).

If coupling to a reducing reaction (p. 29) is not to occur, the next stage in the transport of hydrogen towards oxygen is the reduction by $NADH_2$ of another ring system. The ring is found in nature as part of two molecules, flavin mononucleotide (FMN) and flavin adenine dinucleotide (FAD) (see Table 4.2), which are attached to proteins by non-covalent forces. (The attachment is strong enough for them to be termed prosthetic groups.) The FMN- and FAD-protein complexes are known as flavoproteins.

Like nicotinamide, riboflavin, the parent compound of FMN and FAD, cannot be synthesized by man and is a vitamin.

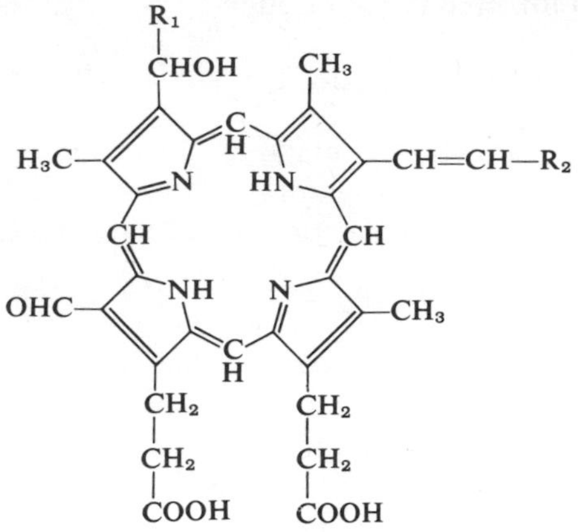

Cytochrome *a*

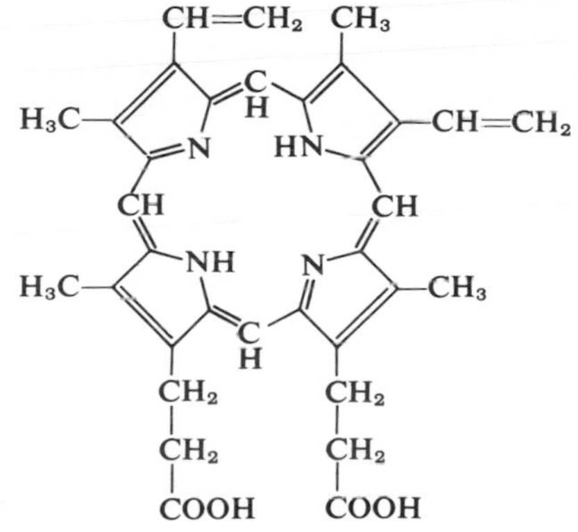

Cytochrome *b*

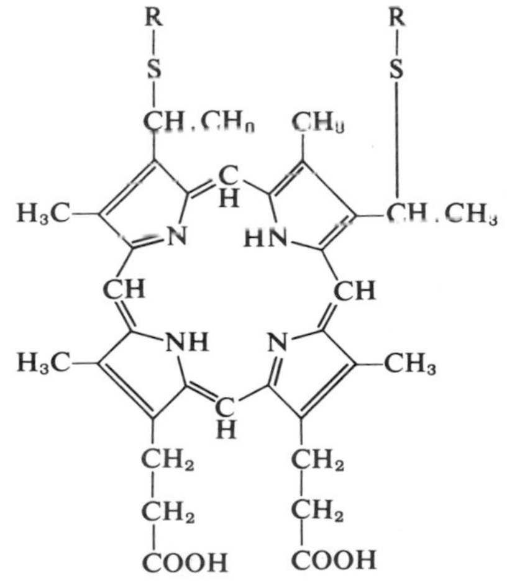

Cytochrome *c*

**Fig. 12.2**  The substituents on the haem ring in the cytochromes

The flavoprotein involved in the reaction

$$\text{NADH}_2 + \text{flavoprotein}_{\text{(oxidized)}} \rightleftharpoons \text{NAD} + \text{flavoprotein}_{\text{(reduced)}} \quad (12.3)$$

can be thought of as an enzyme catalysing the dehydrogenation of $\text{NADH}_2$. Some flavoproteins are, in fact, dehydrogenase enzymes in the more usual sense – for example, that catalysing the desaturation of aliphatic carbon chains

$$R_1.CH_2.CH_2.R_2 \rightleftharpoons R_1.CH{=}CH.R_2$$

(see pp. 278 and 285). Such dehydrogenations do not need to use NAD since the enzyme involved is able to introduce the hydrogen atoms into the pathway at the point that is appropriate to flavoproteins.

The early stages of the pathway thus involve the transfer of *hydrogen* atoms between complex ring molecules bound to proteins. If we now turn to the final stages, we find that these consist of *electron* transfer down a chain of complexes that are formed between protein and prosthetic groups (the cytochromes, see below), ultimately arriving at oxygen.

The cytochromes rely on ferro-porphyrin rings similar to those used by haemoglobin. The three main families of cytochromes (cytochromes *a*, *b* and *c*) differ from one another by virtue of their selection of substituents at the corners of the ring (Fig. 12.2). Individual members of a family differ from one another in their amino-acid sequence and not necessarily in the ring substitution.

Cytochrome *c* is the best studied of the cytochromes and is probably typical in its structure and behaviour. In distinction to the ring of haemoglobin (p. 105), the ring of cytochrome *c* is covalently as well as non-covalently bound to the protein. The sixth co-ordination position of the iron, which in haemoglobin was free to receive oxygen, is permanently occupied by an amino-acid side chain. Rather than being exposed on the surface of the protein, the ring is buried in a crevice and only one edge lies on the surface (Fig. 12.3). This arrangement

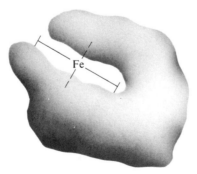

**Fig. 12.3**   The buried haem ring of cytochrome *c*

serves to emphasize that the cytochromes act as conductors of electrons and do not bind ions or molecules in the way that haemoglobin does.

The central event in the conduction process is the reaction

$$Fe^{3+} + e \rightleftharpoons Fe^{2+}$$

We have said that the pathway involves two systems, hydrogen transfer and electron transfer. How they are linked is not quite clear. There is almost certainly another ring compound, ubiquinone, between the two systems, which accepts hydrogens from the isoalloxazine ring of the flavoprotein (Fig. 12.4). (The long hydrocarbon chain of ubiquinone would favour hydrophobic bonding to the lipid of the mitochondrion.) This is the last easily recognized transfer of hydrogen atoms down the chain; thereafter a step occurs which may formally be written as:

$$\text{2H (from the hydrogen transfer pathway)} \rightleftharpoons 2H^+ + 2e^- \qquad (12.4)$$

and electron transfer commences.

We have already seen that the whole process is taking place in a highly unusual environment, a lipid–protein matrix. Therefore, while eqn (12.4) and eqn (12.5) below are useful for a general understanding of what takes place, we need not necessarily assume that the protons and electrons ever exist as identifiable separate entities. Again, the succession of reactions is so fast that we are led to assume that there is no free diffusion of intermediates from one protein to another. The active part of the carriers must be almost in contact, and the product of one reaction handed directly on to the catalyst of the next.

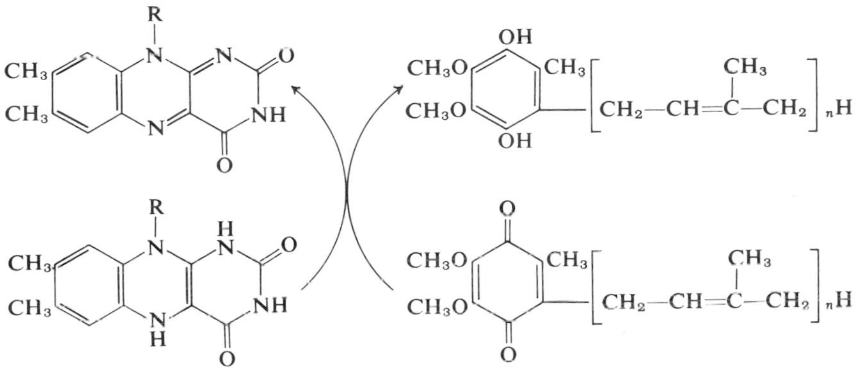

(For the nature of R see Table 4.2)
*iso*alloxazine part of FMN or FAD                    ubiquinone

**Fig. 12.4** The reaction between ubiquinone and the *iso*alloxazine ring of the flavo-proteins

How reaction (12.4) is brought about is far from clear except that other, metal-containing, proteins appear to be involved. Thereafter, matters appear to be straightforward with the transfer of electrons from one cytochrome to another. The order of steps in the process as a whole is given in Fig. 12.5.

Finally either cytochrome $a$ or $a_3$ (you should note that the separate existence of cytochrome $a$ and $a_3$ has been questioned; they may be simply the same molecule in different environments) catalyses, with the help of a $Cu^{2+}$ ion, the final reaction in which the electrons are donated to molecular oxygen

$$2e^- + 2H^+ + \tfrac{1}{2}O_2 \rightleftharpoons H_2O \qquad (12.5)$$

and the oxidative pathway is at an end.

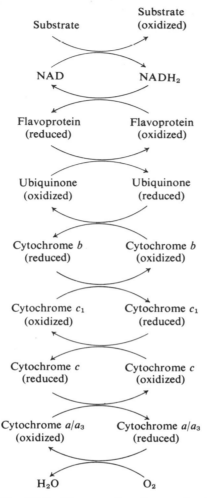

Substrate     Substrate (oxidized)

NAD     $NADH_2$

Flavoprotein (reduced)     Flavoprotein (oxidized)

Ubiquinone (oxidized)     Ubiquinone (reduced)

Cytochrome $b$ (reduced)     Cytochrome $b$ (oxidized)

Cytochrome $c_1$ (oxidized)     Cytochrome $c_1$ (reduced)

Cytochrome $c$ (reduced)     Cytochrome $c$ (oxidized)

Cytochrome $a/a_3$ (oxidized)     Cytochrome $a/a_3$ (reduced)

$H_2O$     $O_2$

**Fig. 12.5** The overall pathway of oxidative transport

*Methods for studying the pathway*

The processes that we have just described are summarized in Fig. 12.5. We shall now describe some of the evidence that has been obtained to support it. The scheme resembles a multienzyme pathway of the type that we described in Chs 9 and 11. In those pathways there were at each step intermediate products capable of a separate existence which could be isolated and studied. Such intermediates do not exist in the present case, but, nonetheless, some of the methods for study of the true multienzyme pathway are applicable here. Notable among these is the use of *inhibitors*. You will recall that if an agent can be found that will inhibit a given step while leaving all the others untouched, products *before* the inhibited step in the chain of reactions will accumulate, while products *after* the inhibited step will drain away (p. 229). If the intermediate stages were to involve definite chemical compounds, it is easy to see how chemical analysis of the inhibited system would enable us to place the intermediates as being either before or after the inhibited step in the chain. If, as in the present case, there are no true intermediate compounds other than the carriers themselves, how can inhibitors be of assistance? The answer is that, fortunately, all the carriers involve conjugated ring systems and as a result the transition between the oxidized and reduced states is accompanied by a characteristic change in the absorption of light. In an inhibited system, we can often observe only one of the two states of the carriers and we can therefore draw conclusions from our observations about the site of inhibition.

Thus, consider the effect of the inhibition of the chain between cytochrome $b$ and cytochrome $c_1$ (this can be accomplished by the addition of one molecule of the antibiotic *Antimycin A* per molecule of cytochrome $c_1$). NAD, the flavoprotein, ubiquinone and cytochrome $b$, having nowhere to discharge their reducing power, will function for as long as possible until they are all in the reduced form. No more oxidation of substrate can then take place. On the other hand cytochromes $c_1$, $c$ and $a/a_3$ will pass any available reducing power (electrons) to molecular oxygen and, having no source of fresh reducing power, they will themselves remain in the oxidized state. All this can readily be checked by spectroscopy – searching the visible and near-ultraviolet spectrum of the inhibited system for the characteristic absorption bands of the oxidized and reduced carriers.

Other inhibitors act at other sites. For example *rotenone* (a compound used to poison fish) acts between the flavoprotein and ubiquinone; and a number of toxic compounds ($CN^-$, $CO$, $NH_3$) act between cytochrome $c$ and oxygen. The spectroscopic changes that are observed with the various inhibitors confirm that the sequence of reactions is as given in Fig. 12.5.

This represents the first type of use of specific inhibitors in studying the oxidative pathway. We shall describe other uses of such compounds in a moment.

### The sites of synthesis of ATP

As we have said, the transport of the equivalent of 2H down the pathway to oxygen is accompanied by the phosphorylation of ADP to produce ATP (oxidative phosphorylation). We have already said that the number of molecules of ATP produced per atom of oxygen reduced (called the *P/O ratio*) is three. However, P/O ratios are difficult to determine. Oxidative phosphorylation will take place only in the intact mitochondrion, or a reasonably representative subfragment of it, and such preparations abound with systems that will seize on the ATP the moment that it is produced and use it for their own purposes. There are also a number of powerful ATPases. Therefore, although it is a reasonably straightforward matter to measure the oxygen uptake, it is quite impossible to assess the amount of ATP produced by direct analysis for ATP. Usually, values for the P/O ratio are obtained by adding a large quantity of some compound that inhibits most ATP-using enzymes, together with some enzyme (one less sensitive than most to the inhibitor) which, in the presence of an adequate substrate, takes the ATP through into a reasonably inert form which can be estimated. For example the inhibitor might be $F^-$ (p. 229), the enzyme hexokinase (p. 259), the substrate glucose and the product actually analysed glucose-6-phosphate. This procedure clearly makes use of conditions that are far from physiological, and furthermore the endogenous ATP-using processes are difficult to suppress completely. There is also the problem that $NADH_2$, which would be the ideal substrate for a model P/O determination, cannot pass the mitochondrial membrane. It is therefore only recently that whole-number values have been obtained for P/O ratios. Previously when values of, say, 2.6 were obtained it was assumed that the most likely value was 3. Now that values of 3.0 are obtained there remains the outside possibility that the true value is 4. However, for the moment, it is reasonable to accept a P/O ratio of 3 as a basis for further study.

We should note that not all substrates allow P/O ratios of 3 on oxidation. Those like succinate, which are oxidized directly by a flavoprotein enzyme (cf. p. 236) and therefore miss out NAD, appear to give P/O ratios of 2. This immediately suggests that one ATP molecule of the three is produced somewhere between the removal of 2H from the substrate and the donation to a flavoprotein ring, and that the remaining two are produced somewhere else. Having obtained some idea of the location of one site, it is natural to wish to confirm the position of this site and obtain similar information about the other two.

The *energetics* of the system are found to be a useful guide. It is easy to use electrolytic and spectroscopic means to obtain the oxidation–reduction electrode potential of each of the purified carriers. We saw on p. 23 that the electrode potential is directly convertible into $\Delta G^{\circ\prime}$, and so a graph of the form of Fig. 12.6 can be drawn up. The correction of the true concentrations to 1 M, made necessary by the definition of $\Delta G^{\circ\prime}$ and the difference between free dilute solution and the surface of the mitochondrial membrane, makes it desirable to approach such a graph with some caution. Correction to the values of $\Delta G^{\circ\prime}$ obtained is also required as a result of the system's being displaced from

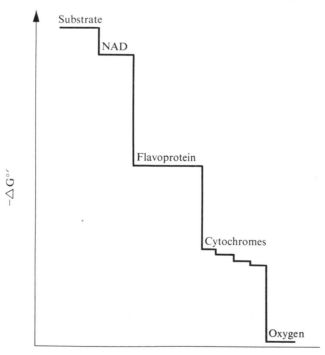

**Fig. 12.6**   The free energy changes associated with the oxidative pathway

equilibrium (p. 21). However, as it stands, the figure suggests that there are some stages that involve large steps in $\Delta G^{\circ\prime}$ and some that do not. There appear to be three points at which the $\Delta G^{\circ\prime}$ would be sufficiently large to be coupled to the synthesis of ATP. The first is between NAD and flavoprotein (which tends to confirm our tentative conclusion that this is indeed the site of one of the phosphorylations), the second somewhere between the ubiquinone–cytochrome $b$ region and cytochrome $c_1$ and the last on the oxygen side of cytochrome $c$.

It may enhance your faith in the usefulness of $\Delta G^{\circ\prime}$ determinations in general to learn that in spite of the drawbacks that were mentioned in the preceding paragraph, all other experimental evidence confirms that these are indeed the three sites of phosphorylation of ADP.

One line of approach towards confirming the location of the sites makes use of the inhibitors that we met before. They would obviously be of no use on their own, since they simply bring the whole system, including phosphorylation, to a standstill. If, however, they are supplemented with *artificial donors* or *acceptors of electrons*, they become extremely useful. We can demonstrate how this is so by supposing that we have blocked the transport pathway between cytochrome $b$ and $c_1$ with Antimycin A (see above). If we start the pathway working again by adding substrate and using an acceptor able to take, and retain, electrons from the reduced form of cytochrome $b$ (which will have piled up in this form above the site of inhibition), we find that the P/O ratio is 1. This result confirms the general location of the first site yet again. If on the other hand, we *add* electrons *below* the site of inhibition we can explore the other sites. For example, ascorbate (Fig. 12.7) can be oxidized by cytochrome $c$ (i.e. it can donate electrons to cytochrome $c$), and in this case the P/O ratio is again 1, suggesting once more that the thermodynamic means of assignment is correct. These and analogous experiments all produce the same result. They constitute our second example of the use of inhibitors.

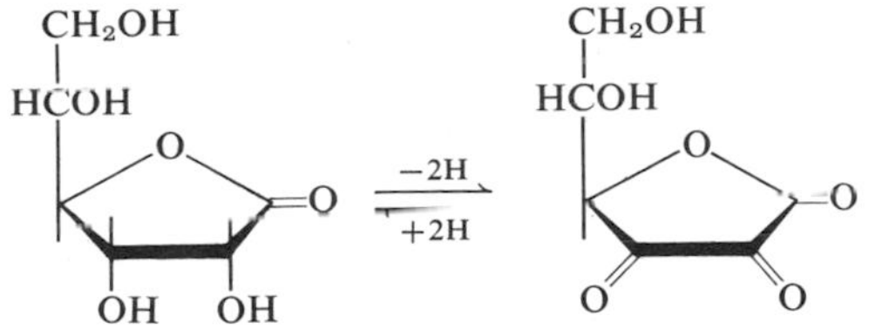

**Fig. 12.7**  The oxidation of ascorbic acid

Precisely similar conclusions can be reached by depriving the system of ADP. Such experiments depend on the finding that electron transport and phosphorylation are *coupled* (see also below), and that if phosphorylation is prevented – in this case through the lack of anything to phosphorylate – then so too is electron transport. Chance and Williams found that the spectroscopic changes that resulted from deprivation of ADP pointed to inhibition at three sites and that these sites were the ones already referred to.

# The mechanism of oxidative phosphorylation

We may have located the *sites* of phosphorylation, but we have not yet said anything about the *mechanism*. The transfer of hydrogen and electrons seems at first sight to have little to do with the reaction:

$$\text{ADP} + \text{phosphate} \rightleftharpoons \text{ATP}.$$

In fact all chemical reactions are electron processes and we should not find the connection too hard to imagine. Imagine is the word, unfortunately, because in spite of much experimental work no definite conclusions have emerged. However, it is instructive to examine the present state of the major hypotheses.

## The chemical-coupling hypothesis

You will recall that we mentioned on p. 27 that, in addition to the process of *oxidative* phosphorylation which we are now considering, some phosphorylations of ADP take place that are not at all connected with the electron-transport pathway, but are part of the mechanism of straightforward chemical reactions. We shall give an example of these *substrate-level* phosphorylations (glance for a moment at p. 262), in which the complete chemical process appears to be well understood. It is therefore tempting to look for similar chemical transformations of intermediate compounds which might be associated with the hydrogen and electron transport pathway.

In the example to which we have just referred you will see that oxidation of a molecule that is not a high-energy compound (a thiohemiacetal, $R.S.CHOH.R$) produces a compound (a thioester), which, as we saw in Ch. 2, is a high-energy compound. Could not similar transformations be taking place in the mitochondrion, with the original low-energy compound being regenerated after the high-energy bond is used to make ATP?

That is

$$C_{red} + I \rightleftharpoons C_{ox} + I^*$$
$$I^* + ADP + P_i \rightleftharpoons ATP + I$$

where C is the appropriate carrier in the transport chain and I is the intermediate compound (written as $I^*$ when in its activated state). I is called the *auxiliary coupling factor*.

This states the matter at its simplest. After much work it is clear that if the hypothesis is to be tenable at all the scheme must be at least as complex as

$$C_{red} + I + C'_{ox} \rightleftharpoons C_{ox} \sim I + C'_{red}$$
$$C_{ox} \sim I + \text{Enzyme} \rightleftharpoons C_{ox} + \text{Enzyme} \sim I$$
$$\text{Enzyme} \sim I + P_i \rightleftharpoons I + \text{Enzyme} \sim P$$
$$\text{Enzyme} \sim P + ADP \rightleftharpoons ATP + \text{Enzyme}$$

where $\sim$ indicates a high-energy bond between two substances.

That is, this hypothesis would state that as the reduced carrier, $C_{red}$, is oxidized (with the consequent reduction of the next carrier C′) it forms a high-energy chemical bond with I. I is then transferred to an enzyme, with retention of the high-energy bond. This is the enzyme that forms ATP and it does so by first exchanging the high-energy bond to I for a high-energy phosphate bond and then donating this phosphate group to ADP.

This model fits most of the known experimental facts, as indeed it was formulated to do. For example it explains the tight coupling between electron transport and phosphorylation – if ADP is not phosphorylated, the enzyme cannot acquire more ~I from $C_{ox}$~I and so $C_{ox}$ will not be available to be converted once more to $C_{red}$ for the next cycle.

The behaviour of those compounds that are known as *uncouplers* can also be accounted for on this model. Uncouplers permit transport of reducing power and electrons to occur even if ADP is not phosphorylated. They even increase its rate over that obtaining in the coupled state. Their action can be compared with that of depressing the clutch of an automobile engine. With the clutch engaged the engine cannot turn (electron and hydrogen transport cannot occur) unless the wheels go round (ADP is phosphorylated). With the clutch depressed the engine will be able to turn independently of the wheels and, freed of its load, will turn even faster. The best known uncoupler is 2,4-*dinitrophenol* (DNP).* Its action, it is suggested, would be to catalyse the hydrolysis of $C_{ox}$~I (or Enzyme~I).

Some of the inhibitors of oxidative transport appear to act at one of the later steps of the coupling process rather than on the oxidation–reduction itself. *Oligomycin*, for example, brings respiration to a halt, but electron transport is resumed if DNP is added (phosphorylation of ADP does not, however, recommence). We would assume that oligomycin acts at one of the enzymic transfer steps in our scheme.

Quite a large amount of other evidence may be reconciled with the model. However, the most intensive search over some years has failed to detect any of the high-energy intermediates (~I, ~P compounds) in a direct experiment. This failure may simply be due to the reliance of the system on a high degree of structural organisation, which tends to be lost in the kind of test-tube experiment that would be necessary to isolate ~I compounds. Experiments on fragments of mitochondria show that the system does indeed rely on a measure of organization, but to so great a degree (see below) as to give rise to the suspicion that there is more to the story than we have so far considered. The spatial organization of the system may, as we shall see in a moment, be rather more directly concerned in the coupling process than would be necessary merely to keep the components of our putative sub-reactions in the correct orientation. An emphasis on structural organisation leads, in fact, to an entirely different hypothesis, which we shall now examine.

* You will recall that one of the purposes of the oxidative production of ATP is to couple the degradation of foodstuffs to the biosynthetic storage of compounds, such as fats. One might therefore expect ingestion of DNP to enable an individual to eat as much as he likes without getting fat. Consequently, in spite of the closely integrated role of ATP in so many other biochemical processes, a biochemically alert manufacturer decided to sell DNP as a slimming pill. It was indeed effective for this purpose. Sales were discontinued after it was noticed that, among other side effects, people taking it went blind.

In order to appreciate this alternative hypothesis we must review our knowledge of the structure of membranes (Ch. 8) and apply it in the special circumstances that obtain in the mitochondrion. You will remember from Ch. 10 that the outer mitochondrial membrane is different in appearance from the inner one. The outer membrane has a higher lipid content (particularly of cholesterol and phosphatidyl inositol, p. 191) than the inner one and a simpler, smoother structure. It is also simpler in a functional sense. Unlike the inner membrane (which is impervious even to $H^+$), it allows the passive entry of a large number of solutes of small molecular weight and, mirroring its lower protein content, it has fewer specific types of enzymic activity associated with it. The inner membrane in contrast has a number of specific transport systems (cf. p. 217). Moreover, it is in the inner membrane that the carrier system for the oxidative pathway is located. (Many other integrated, multiprotein systems are also found here.) The regular disposition, all over the inner membrane, of assemblies of the protein carriers is striking. Now it is found that if the inner membrane is broken up, phosphorylation is no longer possible *unless the pieces are able to reform themselves into closed bodies entirely surrounded by membrane.*

It is this last observation that provides powerful support for the *chemi-osmotic* hypothesis of Mitchell. Briefly stated, the hypothesis is as follows:

As reducing power and electrons are transported down the chain, $H^+$ ions are expelled from the outer surface of the inner membrane. Six protons are expelled for the passage of the equivalent of two electrons and the protons accumulate outside the membrane. We saw on p. 17 that to concentrate a substance requires free energy; the free energy of the fall of the electrons down the chain is therefore, if this model is correct, stored in the form of the concentration gradient of $H^+$ that the expulsion creates.

One possible mechanism by which this gradient of $[H^+]$ could be set up would be that the steps of the transport pathway that involve $H^+$ (the first three and then the terminal step) are catalysed by carriers that are so oriented that they can *take up* protons only from the *inside* of the membrane and *discharge* them only to the *outside*. The lowered pH on the outside, i.e. an excess of $H^+$, would be balanced by a raised pH on the inside, i.e. an excess of $OH^-$. All that we have learned of the structures of proteins and of membranes makes it easy for us to believe that this sort of thing is quite possible. The appropriate determinations of pH are extremely difficult, but such results as have been obtained appear to confirm that a detectable concentration gradient of $H^+$ is indeed set up. (The concentration gradient may be replaced to a considerable extent by a gradient of electric *charge*. We can see from p. 17 and p. 23 that a difference in electrical potential is equivalent, in terms of free energy, to a difference in concentration.)

If we accept for the moment that the energy of the fall of electrons to oxygen is first stored as a concentration gradient, how is this then used for the phosphorylation of ADP?

Mitchell points out that the phosphorylation of ADP, which for convenience is usually written as

$$ADP + P_i \rightleftharpoons ATP$$

actually involves the condensation of the $P_i$ to form the high-energy phosphoric anhydride bond with the exclusion of water. The equation could be written more fully as

$$\text{Adenosine}-\overset{\overset{\displaystyle O}{\|}}{\underset{\underset{\displaystyle OH}{|}}{P}}-O-\overset{\overset{\displaystyle O}{\|}}{\underset{\underset{\displaystyle OH}{|}}{P}}-OH \;+\; OH-\overset{\overset{\displaystyle O}{\|}}{\underset{\underset{\displaystyle OH}{|}}{P}}-OH \;\rightleftharpoons\;$$

$$\text{Adenosine}-\overset{\overset{\displaystyle O}{\|}}{\underset{\underset{\displaystyle OH}{|}}{P}}-O-\overset{\overset{\displaystyle O}{\|}}{\underset{\underset{\displaystyle OH}{|}}{P}}-O-\overset{\overset{\displaystyle O}{\|}}{\underset{\underset{\displaystyle OH}{|}}{P}}-OH \;+\; H_2O \qquad (12.6)$$

but for the various reasons that we gave in our note on abbreviations and conventions (p. xi) this is not normally done.

The $\Delta G^{\circ\prime}$ of this reaction is such as to produce an equilibrium lying very far over to the left; under normal circumstances we could not expect to be able either to push or pull the reaction to the right simply by the manipulation of concentrations (pp. 19 ff.). But the highly organized hydrophobic environment of the mitochondrial membrane is not 'normal' in this sense, and there is some evidence to suggest that, just as Mitchell's proposition demands, the membrane system does pull the reaction over by, in effect, manipulating the concentration of water, and that it makes use of the $[H^+]$ gradient to do so.

It is proposed that the enzyme catalysing reaction (12.6) produces water not as $H_2O$, as we wrote it, but in the form of $H^+$ and $OH^-$. If, this time, the $H^+$ is so channelled that it tends to move *inward* and the $OH^-$ channelled so that it tends to move *outward*, then $H^+$ could be taken up by the excess of $OH^-$ that we have postulated to be on the inside of the membrane and the $OH^-$ could be taken up by the excess of $H^+$ on the outside. In both cases the reaction is

$$H^+ + OH^- \rightleftharpoons H_2O.$$

We can see from p.13 that the equilibrium constant for this process is $10^{14} \, M^{-2}$ in favour of the formation of $H_2O$. The $H^+$ and $OH^-$, once they meet the $OH^-$ and $H^+$, will therefore be most unlikely to return to the site of reaction as the ions, since they will be converted to $H_2O$ to an overwhelming extent. They might return as $H_2O$, but we could imagine that the specific permeability properties of the membrane could be such as to prevent the return in this form.

This then could be the manipulation of concentration of $H_2O$ that we sought. $H_2O$ is removed in the form of the ions, trapped by combination and not allowed to return to the site of synthesis of ATP. The energetics of such a process are not at all unreasonable: if the effective pH on the inside is 9.5, and that on the outside is 4.5, eqn (2.14) on p. 17 shows that there is sufficient energy to pull the reaction over to net synthesis of ATP.

All this might seem far-fetched in comparison with the chemical-coupling hypothesis but, if it does, it is probably because we bring prejudices to the subject that we have acquired from the study of non-living matter and even from the study of the simpler biochemical processes. It is easy to feel that a set of simple, definable chemical reactions is more plausible than an appeal to seemingly vague directional properties of the system. There is, however, abundant evidence that the living cell is fully capable of producing such directional components if necessary. The need for an intact membrane around any structure that is to carry out oxidative phosphorylation, and the failure to isolate the ~I or ~P intermediates do nothing to enhance our faith in the 'straightforward' nature of the process. Concentration (or charge) gradients of the type postulated are often observed, and the action of uncouplers and inhibitors will fit both hypotheses equally well. For example, DNP would now act, not as a catalyst of hydrolysis of ~P or ~I compounds, but as a carrier of $H^+$ through the membrane:

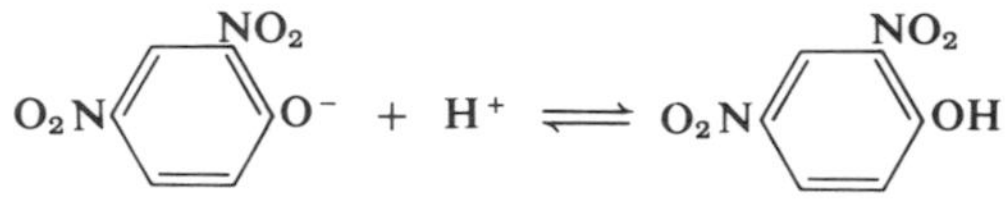

$$O_2N \cdots O^- \; + \; H^+ \; \rightleftharpoons \; O_2N \cdots OH$$

2,4-dinitrophenylate ion          2,4-dinitrophenol

The right-hand compound would make an excellent, lipid-soluble proton carrier and could easily disrupt the concentration gradient upon which the whole coupling process would depend.

A final adjudication between the two types of hypothesis has yet to be made.

Leaving now the details of the coupling, we shall close this chapter by repeating the main facts, none of which are seriously in dispute. These are that: (i) the vital task of trapping the free energy generated by the fall of electrons to molecular oxygen is undertaken by a tightly integrated chain of carriers; (ii) the first clearly recognizable chemical form in which the energy resides is ATP, the major energy carrier of the cell; (iii) the number of molecules of ATP produced from the oxidation of an amount of substrate possessing the equivalent of 2H is most probably 3, an overall efficiency of about 50 per cent.

# Chapter 13  The light reaction: photosynthetic phosphorylation

All life on earth depends on the ability of certain organisms to carry out photosynthesis, that is, the synthesis of organic molecules (particularly polysaccharides) from simple organic compounds by the use of radiant energy from the sun. These organisms all possess characteristic photosynthetic pigments. When photosynthetic organisms are mentioned we tend to think first of green plants; in fact, however, photosynthesis occurs in a wide range of organisms which includes not only the higher green plants, the eucaryotic red, green and brown algae and such unicellular organisms as *Euglena*, but also the procaryotic blue-green algae and several species of bacteria. In this chapter and Ch. 19 we shall describe how the photosynthetic fixation of carbon dioxide to form polysaccharide is carried out. Most biochemistry textbooks cover photosynthesis in a single chapter; the reason for our dividing the discussion into two parts will become clear later.

If we consider the conversion

$$CO_2 \longrightarrow [CH_2O]_n$$

where $[CH_2O]_n$ represents a carbohydrate, we can see that the fixation of carbon dioxide depends on the provision of two agents. In the first place, a carbohydrate such as starch is a much larger molecule than carbon dioxide and the synthesis must therefore require ATP. Secondly, a carbohydrate is a much more reduced molecule than carbon dioxide, and the synthesis must therefore require a reducing agent. Thus the problems faced by photosynthetic organisms are: How can the energy of sunlight be used for the generation of ATP and reducing power, and how can ATP and reducing power be used for the fixation of carbon dioxide?

Before we tackle these questions directly, let us look a little more closely at the conversion of carbon dioxide to carbohydrate as it is carried out both by green plants and by photosynthetic bacteria. It is well known that in plants water is essential for photosynthesis and oxygen is evolved as a by-product of the synthesis. These facts led, many years ago, to the formulation of photosynthesis in plants as:

$$H_2O + CO_2 \longrightarrow [CH_2O] + O_2. \tag{13.1}$$

In photosynthetic bacteria, by contrast, oxygen is not produced and the above equation is obviously not applicable. Equation (13.1) suggests that in plants water is used as a donor of hydrogen which reduces carbon dioxide; in bacteria carbon dioxide is reduced by other compounds which act as donors of hydrogen according to the general equation:

$$2H_2D + CO_2 \longrightarrow [CH_2O] + H_2O + 2D \qquad (13.2)$$

where $H_2D$ represents any of a number of reduced compounds, inorganic or organic.* Apart from the fact that in both equations $CO_2$ is reduced to carbohydrate, the two look at first sight quite different, and it was for a long time believed that photosynthesis in green plants and in bacteria was fundamentally dissimilar. In 1940, however, van Niel brilliantly unified the two by rewriting equation (13.1) in the form

$$2H_2O + CO_2 \longrightarrow [CH_2O] + H_2O + O_2. \qquad (13.3)$$

This formulation is obviously analogous to equation (13.2), and suggests that photosynthesis in plants and bacteria is similar. The water molecules that appear on the *left*-hand side of eqn (13.3) are the source of the reducing fragments necessary to synthesize carbohydrate from carbon dioxide; the corresponding oxidizing fragments are, in plants, liberated as molecular oxygen. (Hence eqn (13.3) predicts that the oxygen evolved by plants is derived from water rather than from carbon dioxide – a prediction that was soon confirmed by the use of isotopically labelled water (see p. 231).) The water molecule that appears on the *right*-hand side of equations (13.2) and (13.3) is one of the products formed by reducing carbon dioxide, the other being carbohydrate itself.

In the light of this discussion we can now focus more sharply on the questions: How can the energy of sunlight be used for the generation of ATP and reducing power, and how can ATP and reducing power be used for the fixation of carbon dioxide? Equation (13.2) implies that the hydrogen donor $H_2D$ is split to a reducing moiety and oxidizing moiety; the reducing moiety is used to reduce carbon dioxide to carbohydrate and water, and the oxidizing equivalents are liberated as the oxidized form D. This set of reactions is plainly complex, and it seems likely that we could make progress in analysing it most easily if we could consider it bit by bit. Fortunately our wishes in this respect correspond with experimental reality; for it was discovered in 1937 by Hill that the liberation of reducing equivalents and oxidizing equivalents in photosynthesis can be entirely separated from the fixation of carbon dioxide. Hill showed that cell-free preparations from photosynthetic plant tissue, in the presence of light, will reduce an electron carrier at the expense of water. This so-called Hill reaction does not require carbon dioxide, and we can represent it as:

$$2H_2O + 2NADP \longrightarrow 2NADPH_2 + O_2. \qquad (13.4)$$

(See Ch. 2 for a discussion of the role of NADP in biosynthesis.)

Reaction (13.4) thus accounts for the formation of one of the two prerequisites of carbon dioxide fixation. What of the formation of the other prerequisite, namely ATP? In 1954 Arnon and his collaborators found that cell-free preparations of leaves, when illuminated, will synthesize ATP from added ADP and inorganic phosphate. Like the Hill reaction, this formation of ATP does not require carbon dioxide.

---

* Examples of compounds used by photosynthetic bacteria as hydrogen donors are $H_2S$, which is oxidized to elemental sulphur, and lactate, which is oxidized to pyruvate.

Thus the *production* of the ATP and NADPH$_2$ needed to fix carbon dioxide is independent of their utilization. It also turns out that the fixation of carbon dioxide with the aid of ATP and NADPH$_2$ can take place in the absence of light. These facts enable us to divide photosynthesis into a *light reaction*, as a result of which ATP and NADPH$_2$ are formed, and a *dark reaction*, in which they are

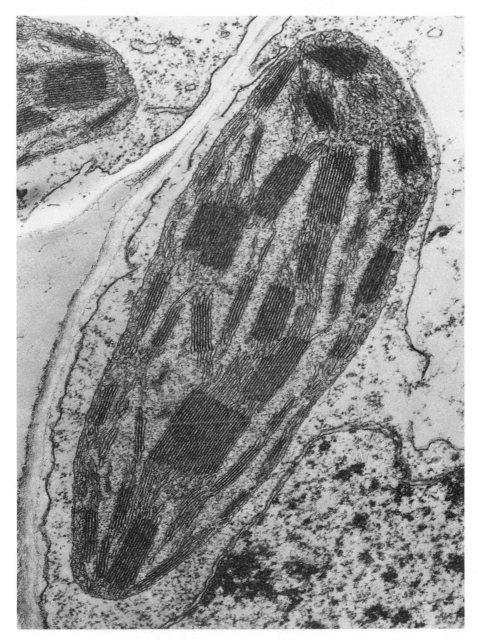

**Fig. 13.1** Electron micrograph of a chloroplast in a leaf cell of maize. × 31 300. (The photograph was kindly supplied by Dr J. Brangeon and Dr B. E. Juniper, Department of Botany, University of Oxford)

used to synthesize carbohydrate. The light reaction is peculiar to photosynthetic tissues, whereas the dark reaction has much in common with the synthesis of carbohydrate in non-photosynthetic tissue and, in fact, uses most of the same steps as those that are employed for the synthesis in animals. We shall devote the rest of this chapter to the light reaction; in Ch. 19 we shall discuss the synthesis of carbohydrate both in plants and in animals, treating the dark reaction of photosynthesis as a special case of the synthesis of carbohydrate.

### The photosynthetic apparatus

In eucaryotic photosynthetic organisms the site of photosynthesis is the chloroplast, a pigmented organelle that is bounded by a double membrane. As in mitochondria (see p. 217) the inner membrane is highly convoluted, being arranged to form stacks of discs. Each stack is called a *granum*, and the grana are embedded in a soluble matrix called the *stroma* (see Fig. 13.1).

Isolated chloroplasts can carry out both the light and the dark reaction of photosynthesis, but the apparatus peculiar to the light reaction is all located in

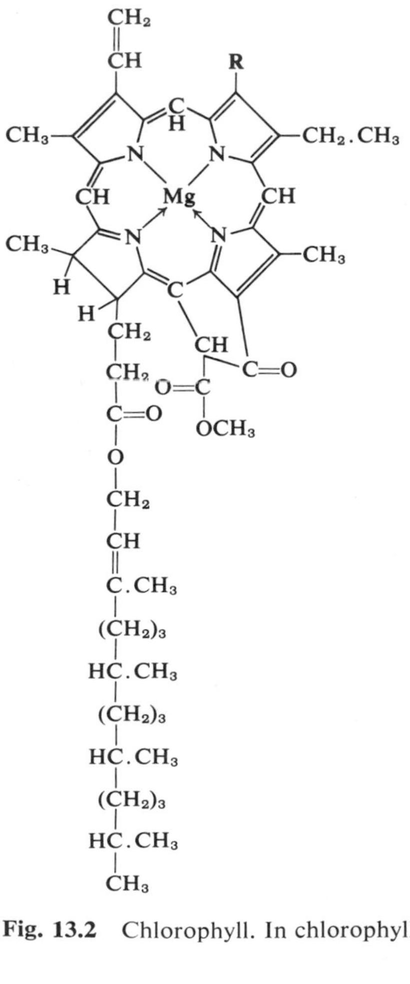

**Fig. 13.2** Chlorophyll. In chlorophyll *a* R is —CH$_3$. In chlorophyll *b* R is —CHO

the grana. The most conspicuous part of this apparatus is the set of photo-synthetic pigments which are responsible for the colour of photosynthetic tissues. The predominant pigment is chlorophyll, and each organism that is capable of photosynthesis contains at least one molecular species of chlorophyll. The commonest species, which is found (in association with other pigments) in green plants, is called chlorophyll *a*, the structure of which is given in Fig. 13.2. Green plants contain in addition another form called chlorophyll *b* (Fig. 13.2), whereas other photosynthetic eucaryotes have different characteristic chloro-phylls. Eucaryotes also have two other types of chlorophyll, called P700 and P670, which are present in very small quantities but are thought to be of great importance in the light reaction. Apart from chlorophylls, many photosynthetic organisms contain other pigments, notably yellow *carotenoids* and, in algae, blue or red *phycobilins*.

The pigments of the grana are organized into arrays which are called *pigment systems* or *photosystems*. The array known as 'pigment system I' contains several hundred molecules of chlorophyll *a*, several dozen molecules of caro-tenoid and one molecule of P700. 'Pigment system II' contains some hundreds of molecules of chlorophyll *a* and, in green plants, several hundred molecules of chlorophyll *b*, together with one molecule of P670 and some associated manganese.

Photosynthetic bacteria do not possess chloroplasts, but have instead cell membranes that carry out photosynthesis. These membranes contain a charac-teristic bacterial chlorophyll, and the molecules of this too are organized into an array corresponding to pigment system I of higher plants. However, bacteria lack pigment system II, a fact which has interesting consequences that we shall discuss later.

## Details of the light reaction

The light reaction of photosynthesis can be summarized quite simply. The primary event is the promotion of an electron, by the action of light on chloro-phyll, to a higher energy level (see p. 23). Electrons at this high energy level may be used either to reduce NADP to $NADPH_2$, or, by passage down an electron transport chain similar to that found in mitochondria, to synthesize ATP. We shall now consider the details of the processes.

*Pigment system I*

As we mentioned above, the chlorophyll molecules of the granum are organized into arrays, which may be thought of as 'funnels' for collecting light and passing it to a reactive centre. When light falls on any of the molecules of chlorophyll *a* in an array of pigment system I, it may be absorbed, and the energy from the absorbed light is then spread throughout the array of molecules. The result is that the specialized chlorophyll molecule P700, which is the reactive centre of the array, becomes excited. In the excited state P700 readily loses an electron to an acceptor which the unexcited P700 molecule would not normally be capable of reducing. In this way the energy of light is able to *reverse* the usual direction of flow of electrons by promoting them to an extremely energy-rich situation.

The identity of the compound to which the excited molecule of P700 loses an electron is not known, and we shall call it X. As a result of this primary photo-reaction, X has gained an electron and could be written as $X^-$, and pigment system I has lost one and could be written $(PSI)^+$. Now at this stage there are two possible paths of electron flow from $X^-$. Under some conditions isolated chloroplasts (or grana prepared from them) can catalyse the return of an electron from $X^-$ to $(PSI)^+$. The flow of electrons along this pathway is stimu-lated by compounds that are involved in electron transport, such as FMN (see p. 234), and it results in the restoration of both $X^-$ and $(PSI)^+$ to their normal electronic states. The most important point, however, is that it is accompanied by the *esterification of ADP to form ATP*. Thus a cyclic flow of electrons, first from PSI to X in the presence of light and then from $X^-$ back via an electron-transport chain to $(PSI)^+$, is responsible for a form of phosphorylation; it is called cyclic photophosphorylation (see Fig. 13.3).

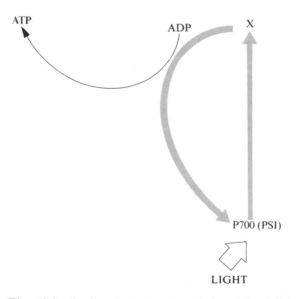

**Fig. 13.3** Cyclic photophosphorylation. The flow of electrons is represented by thick grey arrows

An alternative pathway for the flow of electrons from $X^-$ leads not to their return to $(PSI)^+$ but instead to NADP. $X^-$ does not, however, reduce NADP directly, but via an intermediate carrier called ferredoxin. Ferredoxin is a small protein which contains iron but no haem groups (contrast haemoglobin, p. 104), and in photosynthesis it becomes reduced by $X^-$ and in turn reduces NADP – the latter reaction is catalysed by an enzyme called ferredoxin-NADP reductase. (Figure 13.4 represents the position we have now reached.) But the result of all these events, although very satisfactory from the point of view of producing $NADPH_2$, is to leave pigment system I electron-deficient; we must therefore now ask how an electron can be obtained to restore it to its original state. The answer in bacteria is that the electron is obtained by oxidizing $H_2D$ (equation (13.2), p. 248) to **D**. What is the electron donor in green plants?

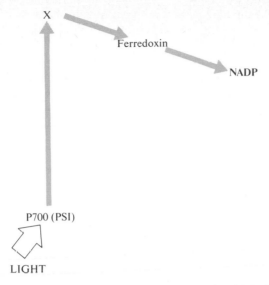

**Fig. 13.4** Photosynthetic reduction of NADP. The flow of electrons is represented by thick grey arrows. Notice that PSI is left deficient in electrons by the flow depicted here – see also Fig. 13.5

*Pigment System II*

We have mentioned previously that the pigment molecules in chloroplasts are organized into two types of array, pigment systems I and II. Now there are two lines of experimental evidence which suggest that these two systems have different functions. First, studies have been made of the way in which the efficiency of photosynthesis in plants varies with the wavelength of the incident light. It turns out that the overall efficiency drops greatly at wavelengths greater than about 680 nm; however, if this light of long wavelength is accompanied by light of shorter wavelengths, the efficiency of photosynthesis is increased in a greater than additive way – in other words light at the shorter wavelength enhances the use of light at the longer wavelength. This result suggests that the two photosystems, responsive to light of different wavelengths, are operating in concert. The second relevant fact is that photosynthetic bacteria lack (as we said before) pigment system II, and since bacterial photosynthesis does not result in the evolution of oxygen we may tentatively conclude that it is pigment system II that is involved in the use of water as a reductant (see equation (13.2) on p. 248).

In describing pigment system I we showed how the loss of an electron from an excited molecule of P700 could lead either to the formation of ATP or to the reduction of NADP. Pigment system II has different functions: it is responsible for supplying an electron to pigment system I (which was, you will remember, left electron-deficient by the original light-induced event), and for accepting in turn an electron from water. In the course of these reactions, a further supply of ATP is generated.

The array of chlorophyll molecules in pigment system II, like its counterpart in pigment system I, acts as a funnel for collecting light. (Its absorption maximum is at a shorter wavelength than that of pigment system I.) As a result of the absorption of light, P670, the reactive centre of the system (analogous to the P700 of pigment system I), becomes excited and loses an electron to an unidentified acceptor called Q. This electron appears to be at a higher energy level than (PSI)$^+$, and it can therefore flow downwards to restore (PSI)$^+$ to its original electronic configuration. As in cyclic photophosphorylation, the passage of the electron from Q$^-$ to (PSI)$^+$ occurs via an electron-transport chain and *is accompanied by the synthesis of ATP*. Details of the carriers in the electron-transport chain are known rather less well for chloroplasts than they are for mitochondria (see pp. 234–238). One of the carriers is believed to be a specialized plant quinone, analogous to ubiquinone (see p. 237), called plastoquinine, which is probably close to Q in the electron transport chain; another is the specialized plant cytochrome (see p. 236) called cytochrome *f*, which is close to pigment system I.

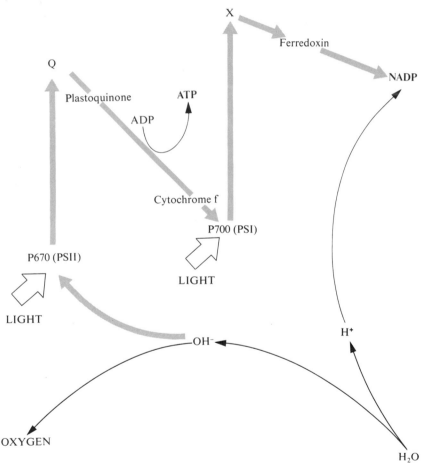

**Fig. 13.5** Coupling of photosystems I and II. The flow of electrons is represented by thick grey arrows. Notice the participation of water and the evolution of molecular oxygen

PSI has now regained an electron to replace the one that it lost in the original photoreaction, but its gain is PSII's loss, and we have to consider how (PSII)$^+$

can be restored to its proper complement of electrons. We have to consider also a further point, about which we were discreetly silent before: although NADP has received an electron from X via ferredoxin, it has not yet received a $H^+$ ion. In fact both the need of $(PSII)^+$ for an electron and the need of $NADP^-$ for a proton can be met from the products of ionization of water. One of these products, the proton from water, can, of course, be used directly to react with $NADP^-$. The other product of ionization of water, $OH^-$, transfers an electron via an unknown carrier system to reduce $(PSII)^+$, in the process becoming converted to water and molecular oxygen. The details of this reaction are obscure and probably very complex; but they can be written as:

$$4OH^- \longrightarrow 2H_2O + O_2 + 4e.$$

This equation helps to illustrate the source of the molecular oxygen evolved by green plants.

## Conclusion

Figure 13.5 depicts the events that we have described. It does not include the flow of electrons in cyclic photophosphorylation, and in fact the details of this process are not very clear. It may be that electrons from $X^-$ are shunted along the carriers that we have mentioned – e.g. from $X^-$ to ferredoxin and then (via intermediate carriers) to cytochrome $f$. Alternatively it may be that cyclic photophosphorylation involves a separate set of carriers and that electrons join the chain from pigment system II to pigment system I only near its end. Whatever its mechanism, cyclic photophosphorylation serves to increase the quantity of ATP produced above that that would be formed if the flow of electrons followed only those pathways that are shown in Fig. 13.5.

# Chapter 14 **Glycolysis**

We pointed out in Ch. 2 that the *tertiary source* of free energy that is available to living organisms is represented by the food reserves. These reserves are synthesized in the first instance by autotrophic organisms at the expense of the ATP that is produced in photosynthesis (see Ch. 13). Thus the reserves can be used by the autotrophic organisms themselves as sources of free energy during times when the primary source of free energy – the radiant energy of the sun – is not available. Heterotrophic organisms acquire food reserves by preying, directly or indirectly, on the autotrophic organisms. But almost all organisms, whether autotrophic or heterotrophic, employ the same biochemical pathways to catabolize their food and to gain the free energy that is stored in it.

The chief food-storage molecules are carbohydrates and lipids (see Ch. 8), and it is usual in describing intermediary metabolism to begin with a discussion of the carbohydrates. There are several reasons for according carbohydrates this priority. Carbohydrates are the primary products of photosynthesis (see Ch. 19), and are stored, in the form of starch, in quite large quantities in plants. The equivalent animal product, glycogen, is also stored in quite large quantities both in muscle and in the liver; in muscle it provides a reserve that can be immediately drawn on as a source of energy for muscular contraction, and in the liver it acts as a reservoir that is used to maintain the concentration of blood glucose. Lipids, by contrast, serve as stores against longer-term needs. Again, although many animal tissues can use either carbohydrate or lipid as a source of energy, there are some, notably erythrocytes and the brain, that do not normally catabolize lipid and must be constantly supplied with glucose (but see pp. 289 ff.).

Since starch and glycogen (p. 174) are polymers of glucose, which, as we shall see later in this chapter, can be easily broken down to glucose itself, and since glucose is the form in which carbohydrate is supplied to the tissues of the animal body by the blood, it is useful to discuss carbohydrate catabolism in terms of glucose catabolism. Later in this chapter (pp. 267–269) we shall see that other carbohydrates are quite readily converted to compounds that can enter the pathway of glucose catabolism.

Both aerobic and anaerobic organisms break down glucose, and one might imagine that the catabolic pathway would be quite different in the two kinds of organisms. In fact, however, the same pathway is used almost universally in the initial stages of glucose breakdown. This pathway involves a *fermentation* of glucose, that is to say a splitting of the molecule into smaller molecules without the net reduction of an external oxidizing agent; it is a process in which one part of the molecule is oxidized at the expense of another part which thereby becomes reduced. A process of this sort is obviously the appropriate means of breaking down substrates in organisms which, being anaerobic, have no supply of molecular oxygen to which electrons can be donated. The fact that aerobic organisms use the same pathway is at first sight more surprising. We need not find it too astonishing, however, if we bear two points in mind. First, we have reason to believe that aerobic organisms evolved from anaerobic organisms, and in doing so they may have found it convenient to retain some of the metabolic stock-in-trade of their anaerobic ancestors. Secondly, it would have been not merely convenient but positively advantageous to retain a pathway which has the function of providing both energy from the catabolism of glucose and also several useful intermediates. In Ch. 9 we stressed the fact that intermediary metabolism commonly serves these two purposes, and we shall see later in this chapter how the degradation of glucose leads to the formation of compounds that serve as starting points for other pathways. If aerobic organisms had evolved a wholly aerobic pathway to release the energy from glucose, they would then have been faced with the problem of finding alternative means of making these other compounds that are necessary for synthetic reactions.

In both aerobic and anaerobic organisms, then, the fermentation of glucose is used to supply energy and certain needed metabolites. But whereas in anaerobic organisms the end-products of fermentation cannot be further metabolized, are of no use to the cell and are simply discarded, in aerobic organisms the end-product of the fermentation reaction serves as the starting point for oxidative metabolism. In other words, by making use of molecular oxygen the aerobic organism can continue the catabolism of products that, to the anaerobic organism, represent simply waste. In this chapter we shall see that the fermentation of glucose leads to a very incomplete breakdown of the molecule and yields comparatively little energy; the next chapter will show that by using oxygen the aerobic organisms can combust the molecule completely and gain far more energy from it.

The range of fermentations of glucose that are available to anaerobes is large and seems at first bewilderingly diverse. But the diversity is more apparent than real. The pathway of glucose fermentation is common to all organisms through almost all of its steps; the diversity is introduced only at the end of the sequence. As we have said above, the end-product of fermentation is discarded by the organism, and its nature depends only on the final one or two enzymic steps of the pathway, which are simply different mechanisms that organisms have evolved for waste disposal.

We can therefore illustrate the pathway of glucose fermentation by describing a single fermentation in detail. Later we shall mention briefly some of the alternative products that are formed in other fermentations, and show how these fermentations diverge from the illustrative example only in the final stages.

Our example will be a fermentation that has a wide distribution throughout living organisms – the splitting of glucose to two molecules of lactate. This so-called *homo-lactic* fermentation occurs among many species of micro-organism. It is also characteristic of animal cells, especially animal muscle. Its occurrence in animals exemplifies the principle that we mentioned earlier, namely that even aerobic cells employ the reactions of an anaerobic pathway in the initial stages of glucose catabolism; they then continue the catabolism via the aerobic pathway that we shall describe in the next chapter. Muscle is a tissue in which fermentation assumes particular importance, since muscle cells can actually exist anaerobically for quite long periods of time. When they are actively working, their requirement for energy exceeds their capacity for oxidative breakdown of carbohydrate, since the rate of this oxidation is limited by the rate at which oxygen can be supplied in the blood. The result is that active muscle, unlike other tissues, produces large quantities of lactate, which spills into the blood and returns to the liver to be reconverted to carbohydrate (Ch. 19). Any lactate remaining in the muscle at the cessation of exercise is oxidized via the aerobic pathway (see next chapter) as oxygen becomes available from the circulation.

The overall equation for the formation of lactic acid from glucose is:

$$C_6H_{12}O_6 \longrightarrow 2C_3H_6O_3$$

and this equation at once makes clear that the process is a true fermentation, involving neither net oxidation nor net reduction. Although, as we shall see, many steps are involved in the pathway, we can understand more easily what is happening if we bear in mind that the total reaction, as far as the glucose molecule is concerned, is merely a splitting into equal halves. This splitting involves a substantial loss of free energy, and the result is that two molecules of ATP can be synthesized.

We have written lactic acid above as $C_3H_6O_3$, in order to make clear its relation to glucose, but its structural formula is actually $CH_3.CHOH.COOH$. Looking at this formula we can see that one end of the molecule is comparatively reduced and the other comparatively oxidized, as opposed to glucose in which the hydrogen and oxygen atoms are spread more or less evenly along the molecule. This rearrangement is characteristic of a fermentation, in which, as we mentioned above, one part of the molecule is oxidized at the expense of another part which thereby becomes reduced.

The fermentation of glucose is often known as *glycolysis*, and the pathway by which it occurs is called the Embden–Meyerhof pathway after the two biochemists who completed the elucidation of its reactions in the 1930s. Studies on the mechanism of glycolysis began with the discovery, by Büchner and Büchner at the end of the last century, that the fermentation of glucose could be carried out with cell-free extracts of yeast; it was shown some years later that the same was true of cell-free extracts of muscle. (The final products of fermentation are rather different in yeast and muscle, as we shall see.) It was then shown by Harden and Young that the extracts could be separated by dialysis into a heat-labile fraction, which we now know to consist of the enzymes of glycolysis, and a heat-stable fraction, which we now know to contain NAD (see below and refer to p. 234) and ADP and ATP. Harden and Young also found that phosphate was involved in glycolysis and that a hexose diphosphate was an intermediate in the sequence.

Subsequent studies made use largely of enzyme inhibitors, and the elucidation of the glycolytic pathway in detail is a classical example of the value of inhibitors in working out metabolic sequences (see Ch. 11). The addition of iodoacetate to a fermentation mixture causes an accumulation of both the hexose diphosphate described by Harden and Young (subsequently identified as fructose-1,6-

diphosphate) and also two triose phosphates. The addition of fluoride, another inhibitor, leads instead to the accumulation of glycerate-3-phosphate and glycerate-2-phosphate. All of these compounds thus appeared to be intermediates in the fermentation, and the enzymic steps leading to their formation, and leading from glycerate-2-phosphate to lactate, were finally elucidated by the work of Embden, Meyerhof, Warburg and others.

Studies with cell-free systems (see Ch. 11) have shown that the enzymes for the whole glycolytic sequence appear to be located in the soluble fraction of the cytoplasm.

## Glucose to lactate

1. *Phosphorylation of glucose*

The first step in the fermentation of glucose is phosphorylation by ATP, yielding glucose-6-phosphate and ADP. $Mg^{2+}$ is needed as co-factor, as it generally is for reactions involving ATP, presumably because the magnesium salt of ATP is the reacting species.

(14.1)

glucose  glucose-6-phosphate

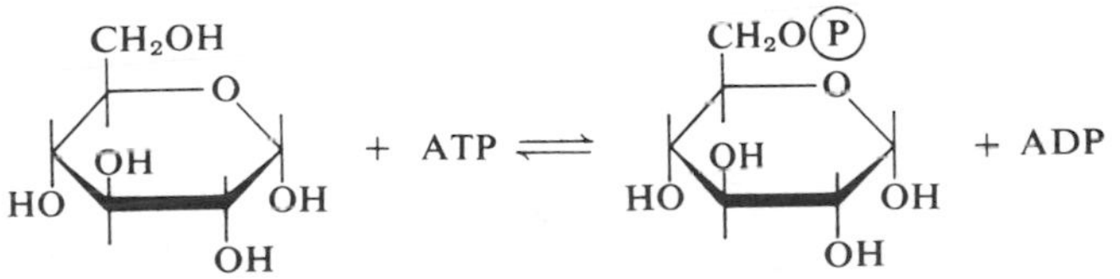

$\Delta G^{\circ\prime} = -4300$ cal/mole

The enzyme commonly involved in the reaction is called hexokinase, and this is very widely distributed in nature; another enzyme, glucokinase (which has different kinetic properties) is found in a few tissues (especially liver) and it catalyses the same reaction. The large negative $\Delta G^{\circ\prime}$ overwhelmingly favours the formation of glucose-6-phosphate. One might imagine that this fact would preclude the release of glucose from glucose-6-phosphate; but that would be disastrous, at any rate for animals. Glucose 6-phosphate, like most phosphate esters, does not pass readily across cell membranes, and if glucose could not be formed from glucose-6-phosphate it would be impossible for glucose to be released from the liver to maintain the concentration of glucose in the blood. In fact animals have solved this problem by developing another enzyme, glucose-6-phosphatase, which catalyses in the liver the hydrolysis of the phosphate ester.

glucose-6-phosphate  glucose

$\Delta G^{\circ\prime} = -3300$ cal/mole

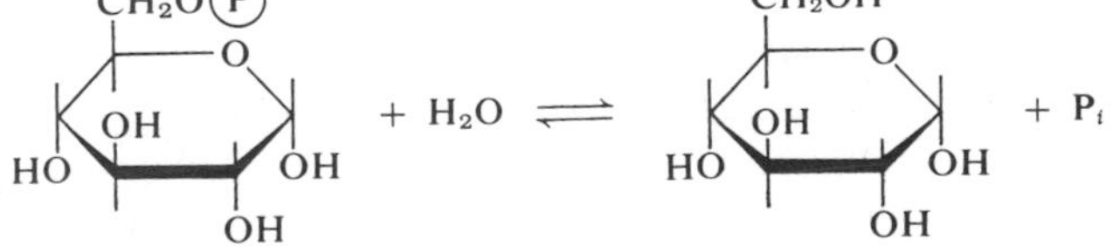

One might ask where the liver gets its supply of glucose-6-phosphate – surely not from glucose directly, or it would be involved in an absurd and wasteful cycle? The answer is, from *glycogen*, which can be converted to glucose-6-phosphate by reactions that we shall discuss below (p. 267) and hence can be used to supply glucose to the blood.

## 2. *Isomerization of glucose-6-phosphate*

Glucose-6-phosphate is converted to its isomer fructose-6-phosphate in a reaction catalysed by hexose phosphate isomerase.

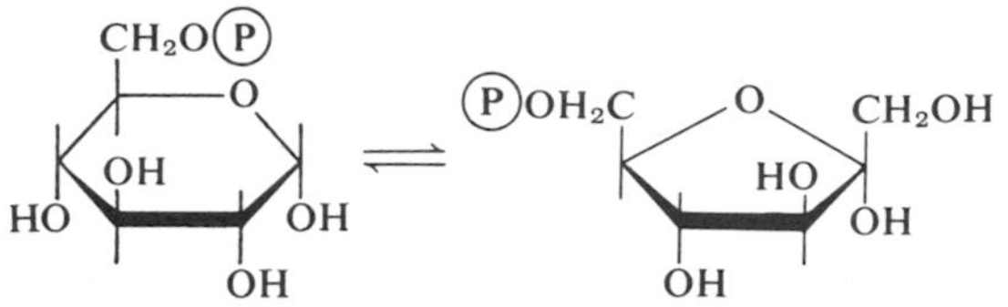

glucose-6-phosphate      fructose-6-phosphate

$$\Delta G^{\circ\prime} = +500 \text{ cal/mol}$$

(14.2)

As we discussed on p. 19, the equilibrium constant of this reaction is not far from 1, and a useful net synthesis of fructose-6-phosphate can easily be achieved even though the reaction is energetically slightly disfavoured.

## 3. *Phosphorylation of fructose-6-phosphate*

A second molecule of ATP is now used up in further phosphorylating fructose-6-phosphate to yield fructose-1,6-diphosphate (the hexose diphosphate of Harden and Young).

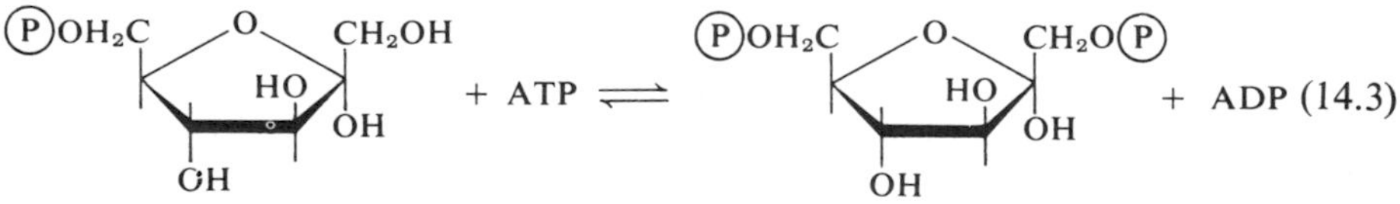

fructose-6-phosphate      fructose-1,6-diphosphate

$$\Delta G^{\circ\prime} = -4200 \text{ cal/mole}$$

(14.3)

This reaction is catalysed by the enzyme phosphofructokinase, and has features in common with the reaction catalysed by hexokinase (see (14.1) above). Owing to the cleavage of the terminal bond of ATP, a large quantity of free energy is lost and the reaction goes almost to completion; this in turn helps to pull reaction (14.2) above to completion (see p. 24). As with reaction (14.1), the breakdown of fructose-1,6-diphosphate to fructose-6-phosphate requires hydrolysis by a phosphatase, this time called fructose diphosphatase.

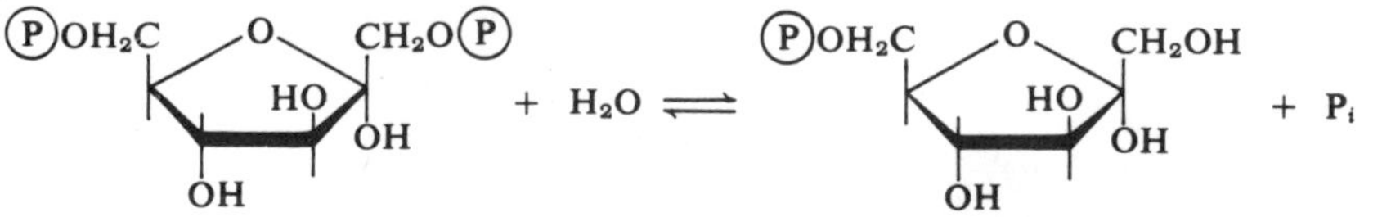

fructose-1,6-diphosphate      fructose-6-phosphate

$$\Delta G^{\circ\prime} = -3400 \text{ cal/mole}$$

The presence of this diphosphatase is essential in the synthesis of carbohydrate (see Ch. 19). Notice that the pathways of carbohydrate breakdown (glycolysis) and carbohydrate synthesis (gluconeogenesis) contain distinct reactions catalysed by separate enzymes – a point to which we shall return in Ch. 29.

## 4. Cleavage of fructose-1,6-diphosphate

The splitting of fructose-1,6-diphosphate yields two molecules of triose phosphate. This splitting is brought about by the reverse of an aldol condensation. It is catalysed by an enzyme called aldolase, and yields glyceraldehyde-3-phosphate and dihydroxyacetone phosphate.

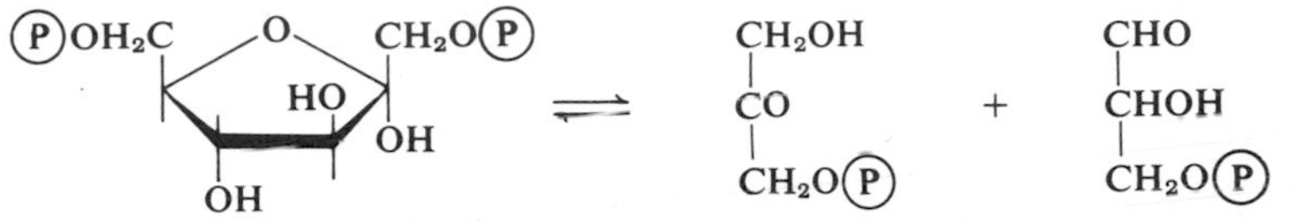

$$
\begin{array}{cc}
\text{fructose-1,6-diphosphate} & \rightleftharpoons \quad \text{dihydroxy-acetone phosphate} \quad + \quad \text{glyceraldehyde-3-phosphate}
\end{array}
\tag{14.4}
$$

$$\Delta G^{\circ\prime} = +5300 \text{ cal/mole}$$

The $\Delta G^{\circ\prime}$ of this reaction, taken alone, greatly favours its proceeding from right to left, and one might well wonder how a cleavage of fructose-1,6-diphosphate is actually achieved. The answer is that which we gave on p. 20. The reaction does not exist in isolation, but as part of a sequence of enzyme reactions that function together in the fermentation of glucose, the overall $\Delta G^{\circ\prime}$ of which is strongly negative. Thus the equilibrium of reaction (14.4) is constantly displaced as the triose phosphate is removed by subsequent reactions.

## 5. Interconversion of triose phosphates

The two products of reaction (14.4) are themselves interconverted by the enzyme triose phosphate isomerase, which we mentioned on pp. 37 and 116.

$$
\begin{array}{cc}
\text{dihydroxyacetone phosphate} & \rightleftharpoons \quad \text{glyceraldehyde-3-phosphate}
\end{array}
\tag{14.5}
$$

$$\Delta G^{\circ\prime} = +1800 \text{ cal/mole}$$

Dihydroxyacetone phosphate as such is not further metabolized by the glycolytic pathway (although it can be converted to $\alpha$-glycerophosphate, see p. 346), but glyceraldehyde-3-phosphate, as we shall see in the next reaction, is immediately oxidized. Again, then, the equilibrium of the reaction is displaced by the removal of the product, and the result is that the dihydroxyacetone phosphate produced in reaction (14.4) does actually proceed down the glycolytic pathway. (This point was discussed in some detail on p. 20). Thus, as a result of reactions (14.4) and (14.5), *all* of the carbon of fructose-1,6-diphosphate (and hence all of the carbon that was present originally in glucose) is now in the form of glyceraldehyde-3-phosphate. We shall see below (pp. 267–269) that the other compounds that are metabolized by glycolysis all funnel into glyceraldehyde-3-phosphate, and from this point the reactions of the pathway are common to all substrates.

## 6. *Oxidation of glyceraldehyde-3-phosphate*

We have now arrived at a most significant step, the reaction in which glyceraldehyde-3-phosphate is oxidized to glycerate-1,3-diphosphate.

$$
\begin{array}{c}
\text{CHO} \\
| \\
\text{CHOH} + \text{NAD} + \text{P}_i \;\rightleftharpoons\; \\
| \\
\text{CH}_2\text{O}\,\text{(P)}
\end{array}
\qquad
\begin{array}{c}
\text{COO}\,\text{(P)} \\
| \\
\text{CHOH} + \text{NADH}_2 \\
| \\
\text{CH}_2\text{O}\,\text{(P)}
\end{array}
\qquad (14.6)
$$

glyceraldehyde-          glyceric acid-
3-phosphate             1,3-diphosphate

$$\Delta G^{\circ\prime} = +300\ \text{cal/mole}$$

This reaction involves a substrate-level phosphorylation which, as we have seen (p. 27), is the formation of a high-energy compound during a process in which the passage of electrons down the respiratory chain (Ch. 12) does not participate. The mechanism of the reaction has been studied in great detail, in the hope that its elucidation would solve the problem of how the high-energy phosphate compound ATP is formed during oxidative phosphorylation. Unfortunately the hope has proved vain; nonetheless, the mechanism is of great interest.

The active centre of the enzyme, which is called triose phosphate dehydrogenase, contains a sulphydryl group, to which a molecule of glyceraldehyde-3-phosphate binds covalently. Nearby a molecule of NAD is also bound. We thus have the reaction:

$$
\text{E}\!\!\begin{array}{l}\nearrow \text{NAD} \\ \searrow \text{SH}\end{array}
\;+\;
\begin{array}{c}\text{CHO} \\ | \\ \text{CHOH} \\ | \\ \text{CH}_2\text{O}\,\text{(P)}\end{array}
\;\rightleftharpoons\;
\text{E}\!\!\begin{array}{l}\nearrow \text{NAD} \\ \searrow \text{S.CHOH.CHOH.CH}_2\text{O}\,\text{(P)}\end{array}
$$

A pair of hydrogen atoms is now transferred to the NAD, leaving a thioester. This type of structure is rather labile, and the thioester can be regarded as a high-energy compound (see Ch. 2 and compare acetyl coenzyme A, p. 273).

$$
\text{E}\!\!\begin{array}{l}\nearrow \text{NADH}_2 \\ \searrow \text{S}\sim\text{CO.CHOH.CH}_2\text{O}\,\text{(P)}\end{array}
$$

Next the bound $\text{NADH}_2$ is reoxidized by a molecule of NAD in solution, and the $-\text{S}\sim\text{C}\overset{\diagup}{\diagdown}$ bond is cleaved by inorganic phosphate. The lability of the bond is retained in the resulting glycerate-1,3-diphosphate, and this too is a high-energy compound (see p. 26).

$$
\text{E}\!\!\begin{array}{l}\nearrow \text{NADH}_2 \\ \searrow \text{S}\sim\text{CO.CHOH.CH}_2\text{O}\,\text{(P)} + \text{P}_i\end{array}
\;+\text{NAD}\;\rightleftharpoons\;
\text{E}\!\!\begin{array}{l}\nearrow \text{NAD} \\ \searrow \text{SH}\end{array}
\;+\;
\begin{array}{c}\overset{\text{O}}{\diagup\!\!\diagdown} \\ \text{C—CHOH.CH}_2\text{O}\,\text{(P)} \\ \text{O}\,\text{(P)}\end{array}
\;+\;\text{NADH}_2
$$

In this way, the oxido-reduction reaction generates a compound which can be used, as we shall see immediately below, to synthesize ATP. But during the reaction a molecule of NAD has become reduced, and this occurrence seems to conflict with what we said earlier about fermentation, namely that it is a break-down of the glucose molecule that is not dependent on the net reduction of an external oxidizing agent. In fact, we shall see later that the hydrogen is stored only temporarily as $NADH_2$; it will soon be returned to the main pathway to reduce an intermediate in the fermentation, and in fact its return is essential for the continuance of the fermentation.

The action of iodoacetate on the pathway (p. 258) can now be explained. Iodoacetate will alkylate the enzyme's —SH group (cf. p. 85):

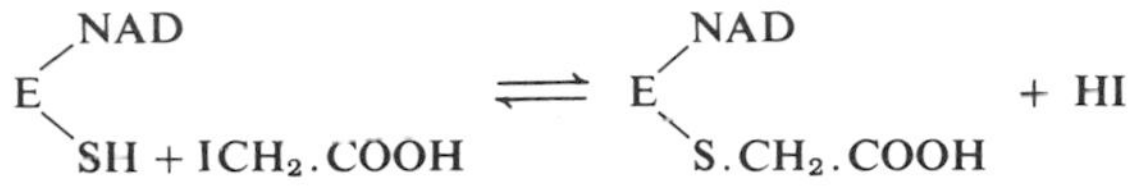

$$E\begin{smallmatrix}\diagup NAD \\ \diagdown SH\end{smallmatrix} + ICH_2.COOH \rightleftharpoons E\begin{smallmatrix}\diagup NAD \\ \diagdown S.CH_2.COOH\end{smallmatrix} + HI$$

and the resulting compound is not oxidizable. The consequent inhibition of the enzyme causes an accumulation of fructose-1,6-diphosphate and the triose phosphates (see p. 259).

### 7. Synthesis of ATP from glycerate-1,3-diphosphate

Glycerate-1,3-diphosphate has a greater free energy of hydrolysis than ATP (see p. 26), and in the presence of phosphoglycerate kinase it can transfer the phosphate group from the C-1 to ADP.

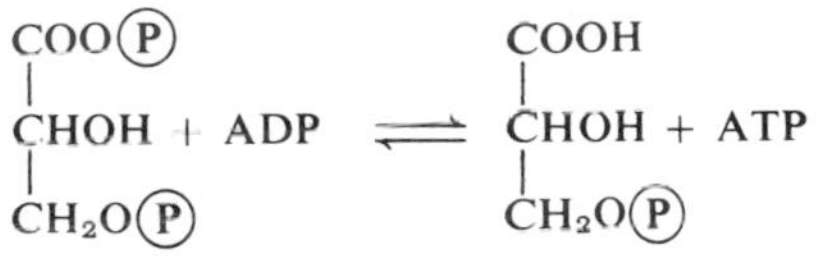

$$\begin{matrix} COO\textcircled{P} \\ | \\ CHOH \\ | \\ CH_2O\textcircled{P} \\ \text{glyceric acid-} \\ \text{1,3-diphosphate} \end{matrix} + ADP \rightleftharpoons \begin{matrix} COOH \\ | \\ CHOH \\ | \\ CH_2O\textcircled{P} \\ \text{glyceric acid-} \\ \text{3-phosphate} \end{matrix} + ATP \qquad (14.7)$$

$$\Delta G^{\circ\prime} = -4800 \text{ cal/mole}$$

The large energy change in this reaction helps to drive the preceding reactions to completion.

### 8. Isomerization of glycerate-3-phosphate

Glycerate-3-phosphate is converted to glycerate-2-phosphate.

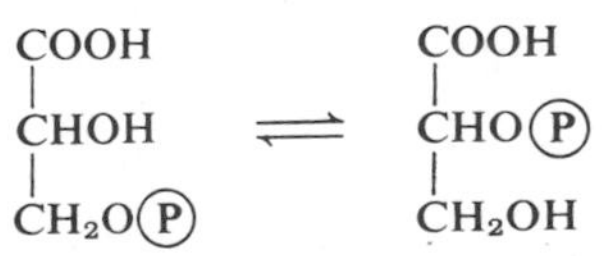

$$\begin{matrix} COOH \\ | \\ CHOH \\ | \\ CH_2O\textcircled{P} \\ \text{glyceric acid-} \\ \text{3-phosphate} \end{matrix} \rightleftharpoons \begin{matrix} COOH \\ | \\ CHO\textcircled{P} \\ | \\ CH_2OH \\ \text{glyceric acid-} \\ \text{2-phosphate} \end{matrix} \qquad (14.8)$$

$$\Delta G^{\circ\prime} = +1400 \text{ cal/mole}$$

The equilibrium is catalysed by phosphoglyceromutase, which requires glycerate-2,3-diphosphate as co-factor. The enzyme acts by transferring a phosphate group (marked with an asterisk below) from the C-3 of glycerate-diphosphate to the C-2 of glycerate-3-phosphate.

$$
\begin{array}{ccccccc}
\text{COOH} & & \text{COOH} & & \text{COOH} & & \text{COOH} \\
| & & | & & | & & | \\
\text{CHO}\,\textcircled{P} & + & \text{CHOH} & \rightleftharpoons & \text{CHO}\,\textcircled{P} & + & \text{CHO}\,\textcircled{P}* \\
| & & | & & | & & | \\
\text{CH}_2\text{O}\,\textcircled{P}* & & \text{CH}_2\text{O}\,\textcircled{P} & & \text{CH}_2\text{OH} & & \text{CH}_2\text{O}\,\textcircled{P}
\end{array}
$$

## 9. Dehydration of glycerate-2-phosphate

Glycerate-2-phosphate is dehydrated in the presence of an enzyme called enolase. $Mg^{2+}$ is required for the reaction.

$$
\begin{array}{ccc}
\text{COOH} & & \text{COOH} \\
| & & | \\
\text{CHO}\,\textcircled{P} & \rightleftharpoons & \text{CO}\,\textcircled{P} \;+\; \text{H}_2\text{O} \qquad (14.9) \\
| & & \| \\
\text{CH}_2\text{OH} & & \text{CH}_2
\end{array}
$$

glyceric acid-        phospho-*enol*-
2-phosphate         pyruvic acid

$$\Delta G^{\circ\prime} = -800 \text{ cal/mole}$$

The resulting compound, phospho-*enol*pyruvate, must be presumed to have suffered a profound rearrangement of its electrons, since, unlike glycerate-2-phosphate, it is a very high-energy compound indeed. As we pointed out on p. 27, the free energy of hydrolysis of phospho-*enol*pyruvate is far greater than that of ATP, and is in fact exceptionally high even for the high-energy compounds of biochemical systems. Thus this reaction, like reaction (14.6), has resulted in the formation of a compound that can be used to synthesize ATP, and is another example of a substrate-level phosphorylation. Fluoride inhibits the enolase reaction by removing the $Mg^{2+}$, and it is for this reason that fluoride causes an accumulation of glycerate-3-phosphate and glycerate-2-phosphate (see p. 259).

## 10. Synthesis of ATP from phospho-enolpyruvate

In the presence of pyruvate kinase, phospho-*enol*pyruvate readily phosphorylates ADP. The removal of the phosphate group from phospho-*enol*pyruvate leaves pyruvate, almost entirely in the *keto* form.

$$
\begin{array}{ccc}
\text{COOH} & & \text{COOH} \\
| & & | \\
\text{CO}\,\textcircled{P} \;+\; \text{ADP} & \rightleftharpoons & \text{CO} \;+\; \text{ATP} \qquad (14.10) \\
\| & & | \\
\text{CH}_2 & & \text{CH}_3
\end{array}
$$

phospho-*enol*-        pyruvic
pyruvic acid        acid

$$\Delta G^{\circ\prime} = -5600 \text{ cal/mole}$$

The free-energy change in reaction (14.10) is quite exceptionally large, and the $\Delta G^{\circ\prime}$ of pyruvate is much lower than that of any intermediate in glycolysis that we have hitherto encountered. This fact helps to explain why the aldolase reaction ((14.4) above) goes virtually to completion despite its positive $\Delta G^{\circ\prime}$ (see also p. 261).

As one might expect, pyruvate cannot be converted in useful quantities to phospho-*enol*pyruvate (compare the reaction on p. 20). Consequently the reversal of glycolysis (that is the synthesis of carbohydrate) requires a bypass for reaction (14.10); we shall describe this bypass in Ch. 19.

## 11. *Reduction of pyruvate*

The glycolytic sequence is almost complete, but one extremely important reaction remains. We saw in reaction (14.6) that the oxidation of glyceraldehyde-3-phosphate was accompanied by the reduction of NAD to $NADH_2$. Now NAD is a co-factor for many oxido-reduction reactions (see Ch. 12), but it is,

$$
\begin{array}{ccc}
\text{COOH} & & \text{COOH} \\
| & & | \\
\text{CO} + NADH_2 & \rightleftharpoons & \text{CHOH} + NAD \\
| & & | \\
\text{CH}_3 & & \text{CH}_3 \\
\text{pyruvic acid} & & \text{lactic acid}
\end{array}
\qquad (14.11)
$$

$$\Delta G^{\circ\prime} = -6000 \text{ cal/mole}$$

like most co-factors, present in only low concentration in the cell. If all of the comparatively few molecules of NAD in the cell were reduced to $NADH_2$, reaction (14.6) would be blocked and glycolysis would immediately come to a stop. Thus, in order to permit the fermentation of large quantities of glucose,

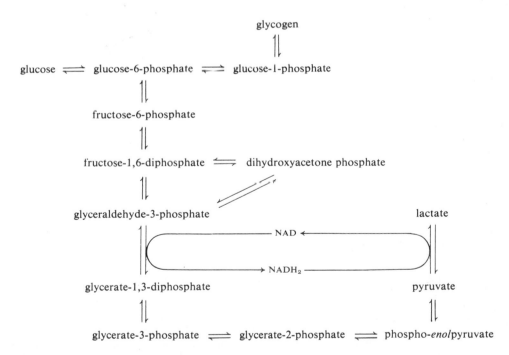

**Fig. 14.1**  Scheme of homolactic fermentation. Notice that $NADH_2$, produced by the oxidation of glyceraldehyde-3-phosphate to glycerate-1,3-diphosphate, is used to reduce pyruvate to lactate

the cell must find some way of reoxidizing $NADH_2$ to NAD so that it can take part again in reaction (14.6). In the fermentation that we are considering, $NADH_2$ is reoxidized by reaction with the pyruvate that was formed in reaction (14.10).

The splitting of glucose to two molecules of lactate (two because all the reactions from (14.6) onwards involve *both* molecules of triose phosphate) is now complete. The reaction scheme is summarized in Fig. 14.1. Because NAD has been restored to its oxidized form, this sequence of reactions can continue to function and can bring about the fermentation of large quantities of glucose. Under anaerobic conditions lactate will, as we mentioned earlier (p. 258), accumulate in the medium surrounding the cell, and thus the pH of the medium will gradually fall. It is this acidification that generally causes fermentation finally to come to a stop; were it not for this factor, anaerobic cells could continue to ferment glucose indefinitely since no external oxidizing agent need be supplied during glycolysis to accept electrons. (Contrast the need for oxygen for the functioning of the tricarboxylic acid cycle (Ch. 15).)

## Other products of fermentation

We have just seen that the reduction of pyruvate in reaction (14.11) serves the function of reoxidizing $NADH_2$ to NAD and thus permitting its continued use in reaction (14.6). Some organisms employ different means of reoxidizing $NADH_2$. Yeast, for example, first decarboxylates the pyruvate and then uses the resulting acetaldehyde to reoxidize $NADH_2$.

$$
\begin{array}{ccccc}
\text{COOH} \\
| \\
\text{CO} & \longrightarrow & CO_2 + CH_3CHO & \xrightarrow{\;\;NADH_2\;\;NAD\;\;} & CH_3CH_2OH \\
| \\
\text{CH}_3 \\
\text{pyruvic acid} & & \text{acetaldehyde} & & \text{ethanol}
\end{array}
$$

The coenzyme thiamine pyrophosphate (p. 271 )is needed for the decarboxylation of pyruvate. (Its mechanism of action will be described in Ch. 15.) In other micro-organisms, pyruvate may be split in other reactions and the products may then be recombined in various ways. As a result, a large number of organic compounds are produced in fermentations by different organisms – for example, the four-carbon compounds 2,3-butylene glycol and butanol are formed after condensation of two two-carbon fragments that are split out of pyruvate, and the three-carbon compounds acetone and isopropanol are formed by the loss of carbon dioxide from a four-carbon compound. All these and many other of the products of fermentation are of industrial importance to man. To the organisms that make them, on the other hand, they represent merely waste products. They are ingenious solutions to the problem of how to dispose of the reducing power carried in the form of $NADH_2$, and they thus enable the reoxidized NAD to continue to function in reaction (14.6) of glycolysis.

## Degradation of other carbohydrates and of glycerol

The Embden–Meyerhof pathway can be used for the breakdown of monosaccharides other than glucose, of polysaccharides such as glycogen and starch, and of glycerol. In each case, a few special reactions are necessary to bring the compound into the main pathway of glycolysis.

*Fructose* can be phosphorylated by hexokinase (see p. 259) to fructose-6-phosphate, which then directly enters the pathway at reaction (14.3). In liver a more important pathway for the utilization of fructose begins with its phosphorylation, in the presence of the specific enzyme fructokinase, at the C-1 position. The resulting fructose-1-phosphate is split by a specific aldolase into dihydroxyacetone phosphate and glyceraldehyde, and the glyceraldehyde is then phosphorylated with ATP.

fructose　　　　fructose-1-phosphate

dihydroxy-acetone phosphate　　　glyceraldehyde　　　glyceraldehyde-3-phosphate

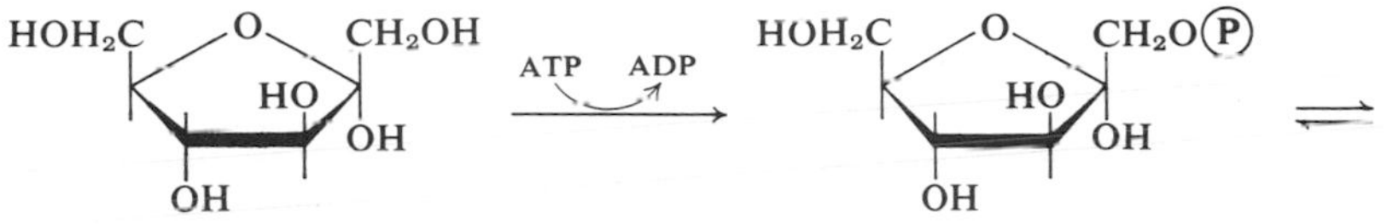

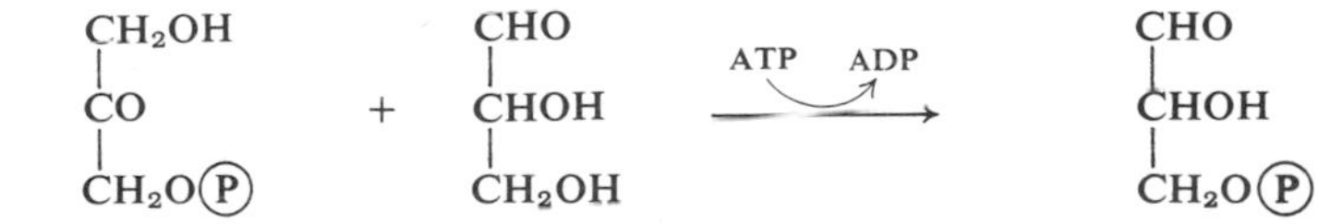

The dihydroxyacetone phosphate and glyceraldehyde-3-phosphate enter the Embden–Meyerhof pathway at reactions (14.5) and (14.6).

*Galactose* is also phosphorylated at the C-1 position with ATP, the enzyme being galactokinase. Galactose-1-phosphate is converted to glucose-1-phosphate in a reaction that involves the formation of a derivative of uridine diphosphate (UDP). (This reaction and the other uses of UDP-sugar derivatives are described in Ch. 19.) Glucose-1-phosphate is converted to glucose-6-phosphate, owing to the presence of a mutase (discussed immediately below), so that the carbon from galactose finally enters the Embden Meyerhof pathway at reaction (14.2).

*Glycogen* and *starch* have to be depolymerized before they can enter the glycolytic pathway, but they are not converted to free glucose. Instead they are broken down by reaction with inorganic phosphate, a process which is called *phosphorolysis* (by analogy with hydrolysis) and is catalysed by an enzyme called phosphorylase. The attack is on the non-reducing end of the chain (see p. 47), and it yields glucose-1-phosphate and a chain polymer one unit shorter than before.

Although the standard free-energy change in this reaction is small, the equilibrium is greatly displaced (see p. 21) and the concentration of glucose-1-phosphate is kept extremely low. The reaction is an extremely important point of control, as we shall see in Ch. 29.

Repeated phosphorolysis of the non-reducing end of the polymer will give rise to several molecules of glucose-1-phosphate and will, if amylopectin (p. 174) or

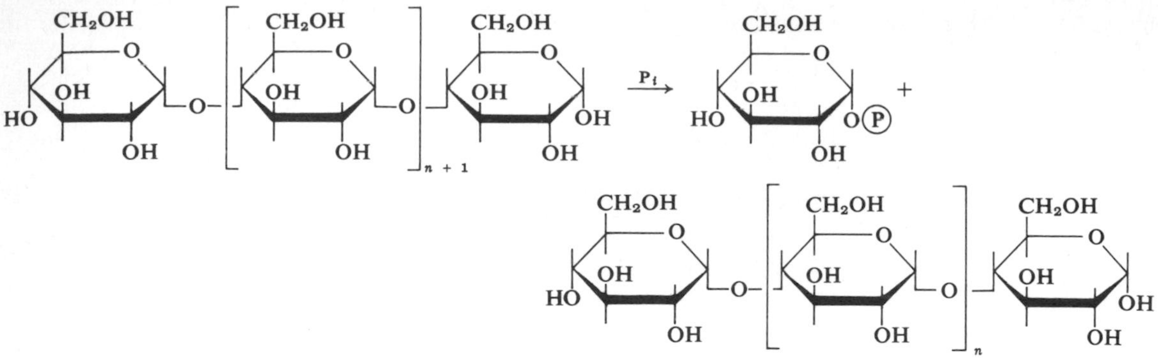

$$\Delta G^{\circ\prime} = +700 \text{ cal/mole}$$

glycogen (p. 175) is the substrate, finally leave a chain terminated with a $1 \rightarrow 6,\alpha$-link (see p. 171), which the enzyme cannot act on. This bond can be broken by a hydrolysis catalysed by a 'debranching' enzyme, to yield another length of $1 \rightarrow 4,\alpha$-linked polymer which can in turn be degraded by phosphorolysis. In this way glycogen and starch give rise to large quantities of glucose-1-phosphate, and this is now converted to glucose-6-phosphate in a reaction catalysed by phosphoglucomutase.

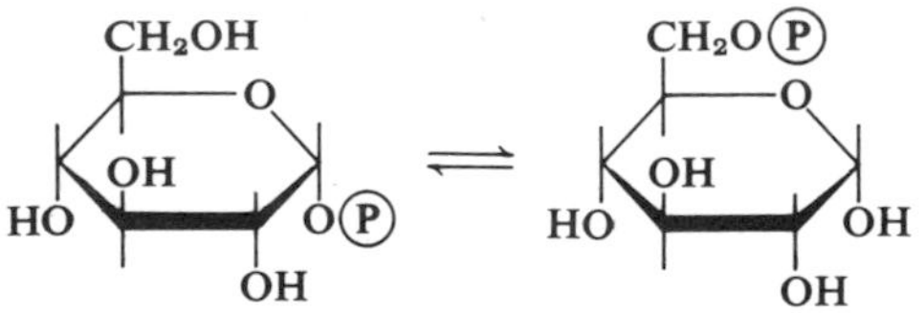

glucose-1-phosphate          glucose-6-phosphate

This reaction is analogous to the conversion of glycerate-3-phosphate to glycerate-2-phosphate (reaction (14.8)), and it requires an analogous co-factor, glucose-1,6-diphosphate. Glucose-6-phosphate joins the Embden–Meyerhof pathway at reaction (14.2).

The fermentation of glycogen is of particular importance in the actively working *muscle*, since it provides energy for contraction in such a way that the muscle does not have to depend on the arrival of glucose from the blood. During contraction the store of glycogen in muscle is gradually used up, and during rest it is replenished (see Ch. 19). The breakdown of glycogen in *liver* serves the function not so much of providing energy via glycolysis as of maintaining the concentration of glucose in the blood. Glucose-6-phosphate, formed from glycogen by the two reactions we have just described, is hydrolysed in the liver to glucose (see p. 259), and the glucose is secreted into the blood.

*Glycerol* is also catabolized via the Embden–Meyerhof pathway, after conversion to dihydroxyacetone phosphate. This conversion involves reaction with ATP (catalysed by glycerokinase) to form $\alpha$-glycerophosphate, and oxidation with glycerophosphate dehydrogenase.

The dihydroxyacetone phosphate enters the pathway at reaction (14.5). Note that in the breakdown of glycerol another molecule of NAD is reduced, in

$$\begin{array}{ccc}
\underset{\text{glycerol}}{\begin{array}{l}\text{CH}_2\text{OH} \\ | \\ \text{CHOH} \\ | \\ \text{CH}_2\text{OH}\end{array}}
\xrightarrow[\text{ATP} \quad \text{ADP}]{}
\underset{\alpha\text{-glycerophosphate}}{\begin{array}{l}\text{CH}_2\text{OH} \\ | \\ \text{CHOH} \\ | \\ \text{CH}_2\text{O}\,\textcircled{P}\end{array}}
\xrightarrow[\text{NAD} \quad \text{NADH}_2]{}
\underset{\begin{array}{c}\text{dihydroxyacetone}\\\text{phosphate}\end{array}}{\begin{array}{l}\text{CH}_2\text{OH} \\ | \\ \text{CO} \\ | \\ \text{CH}_2\text{O}\,\textcircled{P}\end{array}}
\end{array}$$

addition to that reduced in reaction (14.6). As a result there is a surplus of reducing power beyond that that can be disposed of in reaction (14.11), and thus glycerol cannot be metabolized in the complete absence of an electron acceptor.

## Energetics of glycolysis

In the fermentation of glucose, a molecule of ATP is used up in each of the reactions (14.1) and (14.3). Reaction (14.4) splits the molecule in halves, and one molecule of ATP is gained for each half-molecule of glucose in each of the reactions (14.7) and (14.10). Thus the net gain of ATP from the fermentation of glucose to two molecules of lactate (or other fermentation products) is two molecules per molecule of glucose. (The gain from the fermentation of the other sugars that we have mentioned is the same.) Some anaerobic organisms derive all of their ATP from fermentations of this kind. We shall see in Ch. 15 that the introduction of oxygen, which permits the complete combustion of carbohydrate by taking the three-carbon products on to carbon dioxide, vastly increases the yield of ATP.

In the degradation of glycogen and starch, the 'priming' reaction with ATP (reaction (14.1)) is not needed as the formation of a phosphate ester of glucose uses inorganic phosphate (see p. 267). Consequently, the net gain of ATP from the fermentation of these polysaccharides is three molecules per residue of glucose.

## Glycolysis and biosynthetic pathways

The Embden–Meyerhof pathway is not so rich a source of intermediates that are starting points of biosynthetic sequences as is the tricarboxylic acid cycle (see p. 280). Nonetheless, it bears out the principle that we proposed earlier (p. 208), namely that catabolic sequences serve the twin functions of yielding energy and providing intermediates for biosynthesis. For example, α-glycerophosphate, which is needed for the synthesis of lipids (see Ch. 20), is derived from dihydroxyacetone phosphate by the reverse of the reaction we described above. Again several amino acids needed for protein synthesis (see Ch. 26) are formed from intermediates of the Embden–Meyerhof pathway. Glycerate-3-phosphate can be converted to serine and hence to glycine or to cysteine, pyruvate is transformed into alanine, and phospho-*enol*pyruvate participates in the formation of the aromatic amino acids (see p. 373). We shall not describe the details of these syntheses here (see pp. 377 and 369), but we have said enough to illustrate the principle.

# Chapter 15  Oxidation of pyruvate: the tricarboxylic acid cycle

The lengthy sequence of reactions that we described in the last chapter yields only two molecules of ATP per molecule of glucose that is fermented. At the same time, this reaction sequence leaves a product (lactate) that still contains most of the energy present in the original glucose molecule. Under anaerobic conditions the remainder of the energy can obviously not be extracted; but by comparing the $\Delta G^{\circ\prime}$ of the following reactions one can see that it should be possible to obtain a far larger yield of energy by admitting oxygen and allowing complete combustion.

$$C_6H_{12}O_6 \longrightarrow 2CH_3.CHOH.COOH \qquad\qquad \text{(A)}$$
$$\underset{\text{glucose}}{} \qquad\qquad\qquad \underset{\text{lactic acid}}{}$$

$$\Delta G^{\circ\prime} = -47\,000 \text{ cal/mole}$$

$$C_6H_{12}O_6 + 6O_2 \longrightarrow 6CO_2 + 6H_2O \qquad\qquad \text{(B)}$$
$$\underset{\text{glucose}}{}$$

$$\Delta G^{\circ\prime} = -686\,000 \text{ cal/mole}$$

The difference that derives from making oxygen an acceptor of the hydrogen atoms in the glucose molecule is striking. But how in biochemical terms is this effect of oxygen exploited? Equation (B) above might suggest that the glucose is oxidized directly by molecular oxygen, but this is a shorthand for what actually happens. We must bear in mind that aerobic organisms have to solve precisely the same problem as that facing anaerobic organisms, namely how to regenerate NAD from the $NADH_2$ that is formed in reaction (14.6) of the glycolytic pathway. But whereas anaerobic organisms make pyruvate the final acceptor of the reducing power from $NADH_2$, aerobic organisms can pass the reducing power along the *respiratory chain* (see Ch. 12) to molecular oxygen, thus regenerating NAD and at the same time forming ATP by oxidative phosphorylation (p. 242). The effect of making molecular oxygen the final hydrogen acceptor is, as we shall see (p. 280), that the yield of ATP from oxidation of glucose is almost twenty-fold greater than from its fermentation.

In discussing the metabolism of carbohydrates under aerobic conditions, we may therefore take pyruvate as our starting point, since when the respiratory chain is operating the $NADH_2$ will not be available in high enough concentration to reduce much pyruvate to lactate (reaction (14.11) of glycolysis). We shall first discuss the complex reaction by which pyruvate is oxidized to a derivative of acetate and then the cyclic series of reactions by which the acetyl derivative is completely degraded. All these reactions occur in the mitochondrion, a fact which has some important consequences (see Ch. 21).

## Pyruvate to acetyl coenzyme A

We saw on p. 266 that yeast is capable of degrading pyruvate by decarboxylation to yield acetaldehyde; this was the penultimate step of an anaerobic fermentation of glucose. Organisms that metabolize pyruvate aerobically, by contrast (in other words organisms that metabolize *glucose* aerobically), degrade pyruvate by *oxidative* decarboxylation to yield a compound that is at the oxidation state not of acetaldehyde but of acetate. The overall reaction could be written:

$$CH_3.CO.COOH + R.H + NAD \longrightarrow CH_3.CO.R + CO_2 + NADH_2.$$

We must now consider both the nature of the acetyl carrier R and the mechanism of the reaction.

It has been known for many years that the mammalian body is capable of acetylating many foreign compounds, and that this is one of the means that is commonly employed for making toxic compounds harmless. The nature of the acetylating agent remained obscure until about 1950, when it was shown by Lipmann and his collaborators to be the complicated molecule depicted on p. 71, to which the name of acetyl coenzyme A was given. Coenzyme A contains, among other moieties, pantothenic acid, which is a vitamin (see p. 80), and the end of the molecule, to which an acetyl group can be attached, has a —SH group. (We shall abbreviate the structure of coenzyme A as CoA-SH, and the structure of acetyl coenzyme A as $CH_3.CO.S\text{-}CoA$.) Now in fact acetyl coenzyme A is not only an acetylating agent for detoxications, but a compound of widespread importance in metabolism. Its acetyl group is the oxidation product of pyruvate (see below); it is formed in the degradation of fats (Ch. 16) and of some amino acids (Ch. 22); it is the starting point for the synthesis of most fatty acids (Ch. 20), of several amino acids (Ch. 22), and of steroids and carotenoids.

The formation of acetyl coenzyme A from pyruvate is a multi-stage reaction which involves $Mg^{2+}$ and several co-factors and is catalysed by an enzyme complex called the pyruvate dehydrogenase system. In the first step pyruvate reacts with the coenzyme thiamine pyrophosphate. The structure of this coenzyme is given on p. 79; it contains the vitamin thiamine, deficiency of which leads to a condition called beri-beri – a serious disease characterized by inability to metabolize pyruvate and by the appearance of pyruvate in the blood. In the reaction with pyruvate it is the thiazole ring of thiamine pyrophosphate that is involved. We shall write the coenzyme as

$$R\overset{+}{-}N\underset{\underset{HC}{\|}}{\overset{}{\qquad}}C-CH_3$$

to make this point clear.

The nucleophilic carbon atom adjacent to the positive nitrogen atom in the ring attacks the α-carbon atom of pyruvate, and carbon dioxide is then lost from

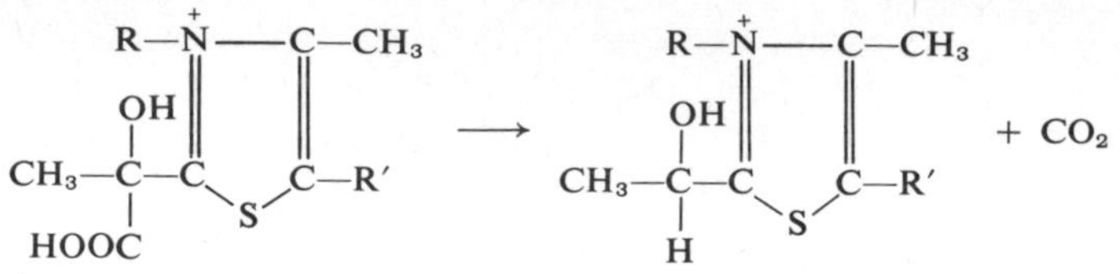

$$R-\overset{+}{N}-C-CH_3 \qquad\qquad R-\overset{+}{N}-C-CH_3$$

(chemical structure with OH, CH$_3$–C–C–C–R′, S, HOOC)  $\longrightarrow$  (chemical structure with OH, CH$_3$–C–C–C–R′, S, H)  $+ CO_2$

the complex. The pyruvate molecule has now become a hydroxyethyl group. In the oxidative metabolism of pyruvate that we are now considering, the hydroxyethyl group is next transferred to the carrier lipoic acid (see below); parenthetically we may point out that in the non-oxidative decarboxylation of pyruvate by yeast (see p. 266) acetaldehyde (which is isomeric with a hydroxyethyl group) is released directly from the complex depicted above and thiamine pyrophosphate is thus regenerated.

In oxidative metabolism the hydroxyethyl group is now removed from the complex with thiamine pyrophosphate and oxidized to an acetyl group by reaction with another coenzyme, this time the disulphide-containing compound lipoic acid, whose structure is

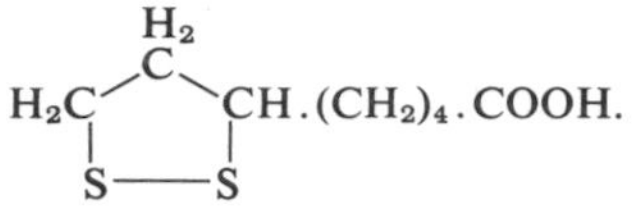

$$\begin{array}{c} H_2 \\ C \\ H_2C \quad CH.(CH_2)_4.COOH. \\ S\!-\!S \end{array}$$

(In the cell lipoic acid is covalently bound via its carboxyl group to the enzyme that catalyses the next reaction.)

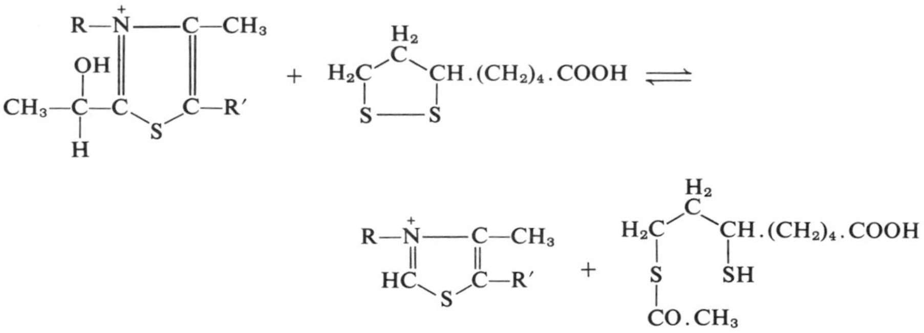

(chemical structure: R–$\overset{+}{N}$–C–CH$_3$, OH, CH$_3$–C–C–C–R′, S, H)  $+$  $\begin{array}{c} H_2 \\ C \\ H_2C \quad CH.(CH_2)_4.COOH \\ S\!-\!S \end{array}$  $\rightleftharpoons$

(chemical structure: R–$\overset{+}{N}$–C–CH$_3$, HC–C–R′, S)  $+$  $\begin{array}{c} H_2 \\ C \\ H_2C \quad CH.(CH_2)_4.COOH \\ S \quad SH \\ CO.CH_3 \end{array}$

This reaction regenerates the thiamine pyrophosphate, which can be used in further reactions, and leaves lipoic acid in its reduced form, carrying an acetyl group.

The oxidation of pyruvate to an acetyl group is thus complete, but two further reactions remain. First the acetyl group has to be transferred to coenzyme A:

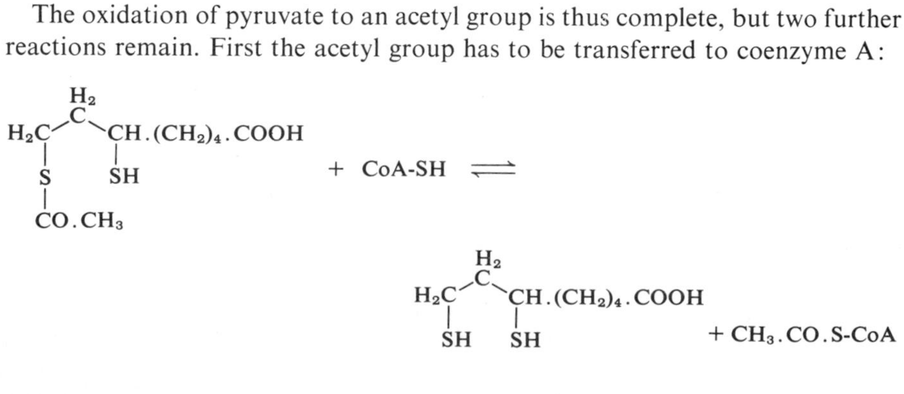

$$\begin{array}{c} H_2 \\ C \\ H_2C \quad CH.(CH_2)_4.COOH \\ S \quad SH \\ CO.CH_3 \end{array}$$  $+$  CoA-SH  $\rightleftharpoons$

$$\begin{array}{c} H_2 \\ C \\ H_2C \quad CH.(CH_2)_4.COOH \\ SH \quad SH \end{array}$$  $+ CH_3.CO.S\text{-}CoA$

and then the lipoic acid, which has been left in its reduced form, has to be re-oxidized so that it too can be used in further reactions. This oxidation of reduced lipoic acid is accomplished by NAD in a reaction catalysed by the enzyme dihydrolipoyl dehydrogenase.

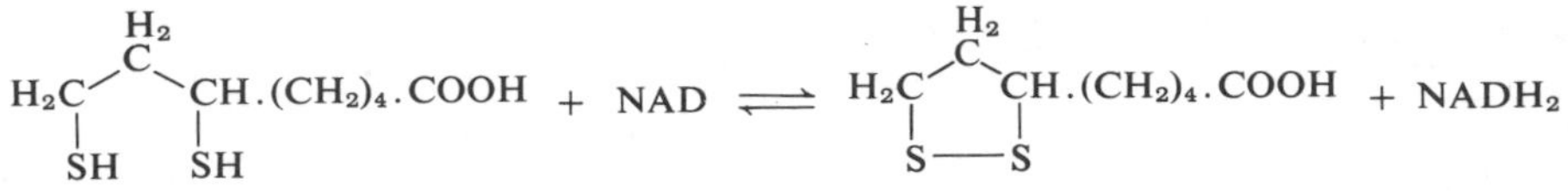

$$H_2C\underset{SH}{\overset{\overset{\displaystyle H_2}{C}}{}}CH.(CH_2)_4.COOH + NAD \rightleftharpoons H_2C\underset{S}{\overset{\overset{\displaystyle H_2}{C}}{}}CH.(CH_2)_4.COOH + NADH_2$$

The whole series of reactions we have just described takes place only in aerobic conditions, and $NADH_2$ can therefore be reoxidized via the respiratory chain.

The summary reaction

$$CH_3.CO.COOH + CoA\text{-}SH + NAD \rightleftharpoons CH_3.CO.S\text{-}CoA$$
$$+ CO_2 + NADH_2 \quad (15.1)$$

which is equivalent to the reaction we gave on p. 271, has a $\Delta G^{o\prime}$ of $-8000$ cal/mole. The $NADH_2$ is (as we have said) immediately reoxidized, so that pyruvate cannot be regenerated in the cell from acetyl coenzyme A. This fact has extremely important consequences, as we shall see when we discuss the synthesis of carbo-hydrate (p. 339).

Acetyl coenzyme A has a high-energy thioester bond (compare the thioester bond formed during the synthesis of glycerate-1,3-diphosphate, p. 262), the free energy of hydrolysis of which is much the same as that of ATP (see p. 26). It is for this reason that the compound is so highly reactive and readily donates acetyl groups. We shall now discuss how the acetyl group is oxidized.


## The tricarboxylic acid cycle

The tricarboxylic acid (sometimes called 'citric acid') cycle is always associated with the name of Krebs, who in 1937 proposed it as the final pathway of oxida-tion of carbohydrate. The elucidation of this final pathway had been for several decades one of the highest ambitions of biochemists, and we shall here outline some of the experimental work by which the existence of the cycle was proved.

The work of Thunberg in the early years of this century had demonstrated that animal tissues contain a number of dehydrogenase enzymes capable, if provided with an appropriate hydrogen acceptor, of oxidizing certain organic acids. (A convenient hydrogen acceptor is the dye methylene blue which be-comes converted to a colourless form on reduction.) One of the most active of the dehydrogenases acts on succinate, oxidizing it to fumarate, and this enzyme is competitively inhibited by malonate which closely resembles the substrate succinate (see p. 130).

During the 1930s the oxidation of carbohydrate by muscle (especially the very active pigeon breast muscle) was being studied by a large number of bio-chemists, and it was established that when minced muscle is incubated in air, oxygen is consumed, carbon dioxide is liberated and the cellular store of glycogen is gradually used up. The number of moles of oxygen consumed was equal to the number of moles of carbon dioxide liberated, a feature that is characteristic of the oxidation of carbohydrate:

$$C_6H_{12}O_6 + 6O_2 \longrightarrow 6CO_2 + 6H_2O$$

but not of other substrates such as fatty acids.

$$C_5H_{11}COOH + 8O_2 \longrightarrow 6CO_2 + 6H_2O$$

In such minced muscle preparations, the rate of respiration usually falls after a short time; it was, however, found that the rate could be restored by the addition of small quantities of succinate, and that the additional consumption of oxygen provoked by succinate was far greater than necessary just to oxidize the succinate itself. The addition of malonate (p. 229), on the other hand, blocked the normal respiration of the muscle. The obvious conclusion is that succinate dehydrogenase acts catalytically in the oxidation of carbohydrate. Next it was found that succinate was not alone in this effect, but that other organic acids such as fumaric and malic acids also stimulated the respiration of muscle. It is easy to arrange these compounds in a theoretical metabolic sequence:

$$
\begin{array}{ccccc}
\text{COOH} & & \text{COOH} & & \text{COOH} \\
| & & | & & | \\
\text{CH}_2 & & \text{CH} & & \text{HOCH} \\
| & \rightleftharpoons & \| & \rightleftharpoons & | \\
\text{CH}_2 & & \text{HC} & & \text{CH}_2 \\
| & & | & & | \\
\text{COOH} & & \text{COOH} & & \text{COOH} \\
\text{succinic} & & \text{fumaric} & & \text{malic} \\
\text{acid} & & \text{acid} & & \text{acid}
\end{array}
$$

and investigators were intrigued to see whether other organic acids, possibly interconvertible with these, would also act catalytically in respiration.

Krebs now found that several other acids did in fact have this property, namely the tricarboxylic acids citric, *cis*-aconitic and isocitric acids, and the dicarboxylic acids α-oxoglutaric, succinic, fumaric, malic and oxaloacetic acids. Once again these compounds can be arranged in sequences.

$$
\begin{array}{ccccccc}
\text{CH}_2.\text{COOH} & & \text{CH}.\text{COOH} & & \text{CHOH}.\text{COOH} & & \text{CO}.\text{COOH} \\
| & & \| & & | & & | \\
\text{HOC}.\text{COOH} & \rightleftharpoons & \text{C}.\text{COOH} & \rightleftharpoons & \text{CH}.\text{COOH} & \rightleftharpoons & \text{CH}_2 & \rightleftharpoons \\
| & & | & & | & & | \\
\text{CH}_2.\text{COOH} & & \text{CH}_2.\text{COOH} & & \text{CH}_2.\text{COOH} & & \text{CH}_2.\text{COOH} \\
\text{citric acid} & & \textit{cis}\text{-aconitic} & & \text{isocitric} & & \alpha\text{-oxoglutaric} \\
& & \text{acid} & & \text{acid} & & \text{acid}
\end{array}
$$

$$
\begin{array}{ccc}
\text{COOH} & & \text{COOH} \\
| & & | \\
\text{CH}_2 & & \text{CH} \\
| & \rightleftharpoons & \| \\
\text{CH}_2 & & \text{HC} \\
| & & | \\
\text{COOH} & & \text{COOH} \\
\text{succinic} & & \text{fumaric} \\
\text{acid} & & \text{acid}
\end{array}
$$

and

$$
\begin{array}{ccccccc}
\text{COOH} & & \text{COOH} & & \text{COOH} & & \text{COOH} \\
| & & | & & | & & | \\
\text{CO} & \rightleftharpoons & \text{HOCH} & \rightleftharpoons & \text{CH} & \rightleftharpoons & \text{CH}_2 \\
| & & | & & \| & & | \\
\text{CH}_2 & & \text{CH}_2 & & \text{HC} & & \text{CH}_2 \\
| & & | & & | & & | \\
\text{COOH} & & \text{COOH} & & \text{COOH} & & \text{COOH} \\
\text{oxaloacetic} & & \text{malic} & & \text{fumaric} & & \text{succinic} \\
\text{acid} & & \text{acid} & & \text{acid} & & \text{acid}
\end{array}
$$

In every case, the effect on the respiration of the muscle was blocked by malonate, which is a specific inhibitor of succinate dehydrogenase. A similar stimulation of respiration, and a similar inhibition by malonate, was found by Krebs when he studied the oxidation of pyruvate by muscle in place of the oxidation of endogenous glycogen.

A plausible conclusion would have been that the central reaction in carbohydrate metabolism was the oxidation of succinate, that this reaction was catalytic in carrying hydrogen from carbohydrate to oxygen, and that the other organic acids stimulated respiration only because they could be converted to succinate.

$$
\begin{array}{ccccc}
\text{C}_6\text{H}_{12}\text{O}_6 & & \text{fumaric acid} & & \text{H}_2\text{O} \\
& \bowtie & & \bowtie & \\
\text{CO}_2 & & \text{succinic acid} & & \text{O}_2
\end{array}
$$

Krebs showed that this conclusion was wrong. It was true that when malonate was added to muscle to block succinate dehydrogenase, the addition of the compounds from citrate to $\alpha$-oxoglutarate caused an accumulation of succinate. But Krebs found that in the presence of malonate the 4-carbon dicarboxylic acids (i.e. fumaric, malic and oxaloacetic acids) were also converted to succinate, which then accumulated. There must therefore be a link between the two sequences, and Krebs was able to show that in some conditions pyruvate and oxaloacetate could react to form citrate. In other words, the reaction sequence given above is in fact *cyclical*. All the organic acids mentioned act *catalytically* in the oxidation of pyruvate, and oxaloacetate is *regenerated* each time the cycle turns.

The only important modification that has to be made to these conclusions is due to the later discovery of acetyl coenzyme A. It is this compound, rather than pyruvate, that reacts with oxaloacetate to form citrate. We shall now describe the cycle in more detail, stressing that it serves for the complete oxidation of acetyl groups from acetyl coenzyme A and thus is responsible not only for the terminal oxidation of almost all carbohydrate but also for the oxidation of almost all fat (see next chapter) and many amino acids (see Ch. 22).

*Formation of citrate*

Acetyl coenzyme A reacts directly with oxaloacetate to form citrate.

$$CH_3.CO.S\text{-}CoA + \begin{array}{c} CO.COOH \\ | \\ CH_2.COOH \end{array} + H_2O \rightleftharpoons \begin{array}{c} CH_2.COOH \\ | \\ HOC.COOH \\ | \\ CH_2.COOH \end{array} + CoA\text{-}SH \quad (15.2)$$

| acetyl coenzyme A | oxaloacetic acid | citric acid |

$$\Delta G^{\circ\prime} = -7700 \text{ cal/mole}$$

This reaction, which is catalysed by citrate synthetase, goes essentially to completion and can not be used to form a useful quantity of oxaloacetate. As we have mentioned above, oxaloacetate is regenerated by the cycle, and, after the acetyl group has been completely oxidized, returns to react with another molecule of acetyl coenzyme A. In a sense, then, we can regard the oxaloacetate as carrying the acetyl group into the reactions of the tricarboxylic acid cycle; for the cycle to function, a supply of oxaloacetate is essential, and we shall see that there is an important reaction by which the supply can, if necessary, be replenished (p. 281). We shall also see that there is a separate reaction by which citrate can be cleaved to provide acetyl groups for fatty-acid synthesis (p. 354). The concentration of citrate appears to be of importance in regulating the activities of some enzymes (see Ch. 29).

*Isomerization of citrate*

Citrate is in equilibrium with its dehydration product, *cis*-aconitate, and with its isomer, isocitrate. The equilibrium is catalysed by an enzyme called aconitase, and it was previously thought that *cis*-aconitate was an *obligatory* intermediate between citrate and isocitrate. More recent evidence suggests that this is not so.

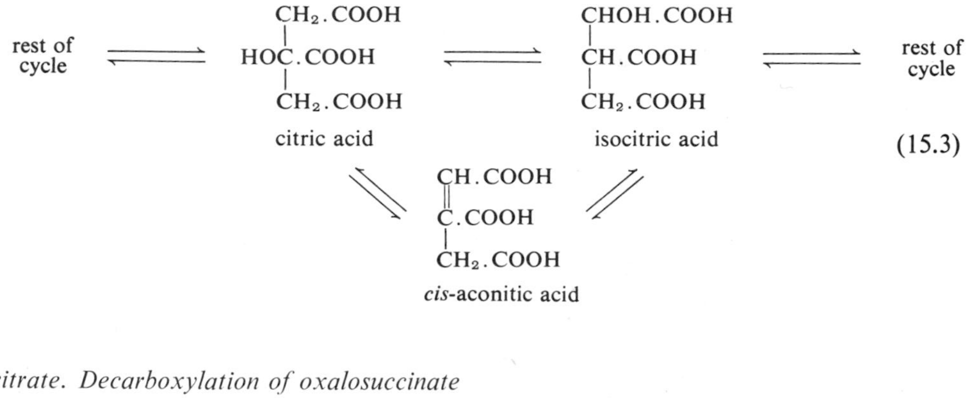

$$\begin{array}{ccccccc} \text{rest of} \\ \text{cycle} \end{array} \rightleftharpoons \begin{array}{c} CH_2.COOH \\ | \\ HOC.COOH \\ | \\ CH_2.COOH \end{array} \rightleftharpoons \begin{array}{c} CHOH.COOH \\ | \\ CH.COOH \\ | \\ CH_2.COOH \end{array} \rightleftharpoons \begin{array}{c} \text{rest of} \\ \text{cycle} \end{array}$$

citric acid        isocitric acid      (15.3)

$$\begin{array}{c} CH.COOH \\ || \\ C.COOH \\ | \\ CH_2.COOH \end{array}$$

*cis*-aconitic acid

*Oxidation of isocitrate. Decarboxylation of oxalosuccinate*

The enzyme isocitrate dehydrogenase catalyses two successive reactions by which isocitrate is converted to $\alpha$-oxoglutarate. The first involves the oxidation of isocitrate by NAD; the second is a decarboxylation of the highly unstable oxalosuccinate that results from the oxidation.

$$\begin{array}{l} \text{CHOH.COOH} \\ | \\ \text{CH.COOH} \\ | \\ \text{CH}_2\text{.COOH} \end{array} \quad \xrightarrow[\text{NAD} \quad \text{NADH}_2]{} \quad \begin{array}{l} \text{CO.COOH} \\ | \\ \text{CH.COOH} \\ | \\ \text{CH}_2\text{.COOH} \end{array} \longrightarrow \begin{array}{l} \text{CO.COOH} \\ | \\ \text{CH}_2 \\ | \\ \text{CH}_2\text{.COOH} \end{array} + \text{CO}_2 \qquad (15.4)$$

$$\qquad\quad \text{isocitric} \qquad\qquad\qquad\qquad \text{oxalosuccinic} \qquad\quad \alpha\text{-oxoglutaric}$$
$$\qquad\quad \text{acid} \qquad\qquad\qquad\qquad\qquad \text{acid} \qquad\qquad\qquad \text{acid}$$

We recall that these reactions are occurring in aerobic conditions, so that the $\text{NADH}_2$ is reoxidized via the respiratory chain. It used to be believed that the oxidizing agent for isocitrate was NADP. There have always been good theoretical grounds for thinking this suggestion unlikely (see pp. 29 ff.), and there is now clear experimental evidence that NAD is the normal physiological acceptor of hydrogen from isocitrate.

### Oxidative decarboxylation of α-oxoglutarate

$\alpha$-Oxoglutaric acid is an $\alpha$-keto acid, homologous with pyruvic acid, and it undergoes oxidative decarboxylation in the same way as we described for pyruvic acid. A succinyl group is split out of $\alpha$-oxoglutarate and passed successively to thiamine pyrophosphate, lipoic acid and coenzyme A. (The equations for these steps can be obtained by substituting $\text{HOOC.CH}_2\text{.CH}_2\text{.CO}-$ for $\text{CH}_3\text{.CO}-$ on p. 272.) The final product is called succinyl coenzyme A; it is exactly analogous to acetyl coenzyme A and contains a similar high-energy thioester bond. The sequence of transfer reactions of the succinyl group leaves lipoic acid in the reduced form, and the oxidized form is restored by reaction with NAD. The summary reaction

$$\text{HOOC.CH}_2\text{.CH}_2\text{.CO.COOH} + \text{CoA-SH} + \text{NAD} \rightleftharpoons$$
$$\qquad\quad \alpha\text{-oxoglutaric acid}$$
$$\qquad\qquad\qquad \text{HOOC.CH}_2\text{.CH}_2\text{.CO.S-CoA} + \text{CO}_2 + \text{NADH}_2 \quad (15.5)$$
$$\qquad\qquad\qquad\qquad\qquad \text{succinyl coenzyme A}$$

is analogous to reaction (15.1).

### Formation of succinate

The free energy of hydrolysis of succinyl coenzyme A is, like that of acetyl coenzyme A, much the same as that of ATP (see p. 26). In theory, then, it should be possible to form a molecule of ATP by coupling the phosphorylation of ADP with the cleavage of succinyl coenzyme A to succinate. The theory is made practice in a reaction catalysed by succinyl thiokinase – curiously it is GTP that is formed first, but this can immediately phosphorylate ADP so that the end result is no different.

$$\begin{array}{l} \text{CO.S-CoA} \\ | \\ \text{CH}_2 \\ | \\ \text{CH}_2 \\ | \\ \text{COOH} \end{array} \quad + \text{GDP} + \text{P}_i + \text{H}_2\text{O} \rightleftharpoons \begin{array}{l} \text{COOH} \\ | \\ \text{CH}_2 \\ | \\ \text{CH}_2 \\ | \\ \text{COOH} \end{array} + \text{CoA.SH} + \text{GTP} \qquad (15.6)$$

$$\quad\;\; \text{succinyl} \qquad\qquad\qquad\qquad\qquad\qquad\quad \text{succinic}$$
$$\quad\; \text{coenzyme A} \qquad\qquad\qquad\qquad\qquad\qquad \text{acid}$$

$$\text{GTP} + \text{ADP} \rightleftharpoons \text{GDP} + \text{ATP} \qquad\qquad (15.6a)$$

This reaction is an example of substrate-level phosphorylation (see p. 27 and compare the triose phosphate dehydrogenase reaction, pp. 262 f.).

Since we have stressed the analogies between acetyl coenzyme A and succinyl coenzyme A, you may well ask what happened to the free energy that was liberated when acetyl coenzyme A was split. The answer is that a high-energy phosphate bond was not formed on that occasion because the energy was used instead to promote the synthesis of citrate (reaction (15.2)).

## Oxidation of succinate

We have now arrived at the reaction in which succinate is oxidized to fumarate – that which is subject to inhibition by malonate (see pp. 130 and 274). The enzyme succinate dehydrogenase, contains FAD (p. 234), to which two hydrogen atoms are passed from succinate without the intervention of NAD.

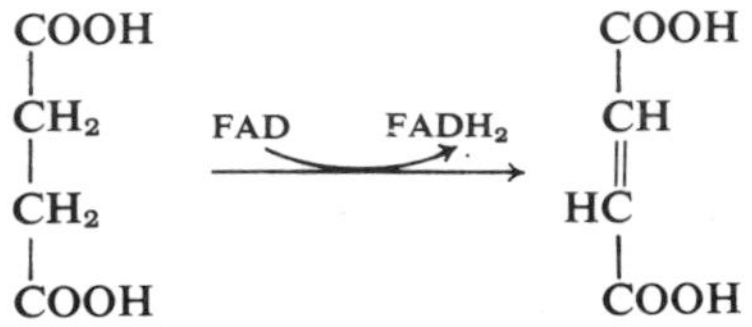

$$
\begin{array}{ccc}
\text{COOH} & & \text{COOH} \\
| & & | \\
\text{CH}_2 & \xrightarrow{\text{FAD} \quad \text{FADH}_2} & \text{CH} \\
| & & \| \\
\text{CH}_2 & & \text{HC} \\
| & & | \\
\text{COOH} & & \text{COOH} \\
\text{succinic} & & \text{fumaric} \\
\text{acid} & & \text{acid}
\end{array}
\qquad (15.7)
$$

Succinate dehydrogenase is built into a complex of electron carriers in the mitochondrial membrane (Ch. 10), and the reduced flavoprotein is immediately reoxidized by this complex so that it can function further in oxidation of succinate.

## Hydration of fumarate

In the presence of a hydratase called fumarase, fumarate accepts water and forms malate. The reaction is stereo-specific; only the L-isomer of malate is produced. (The asymmetric C atom is shown with an asterisk).

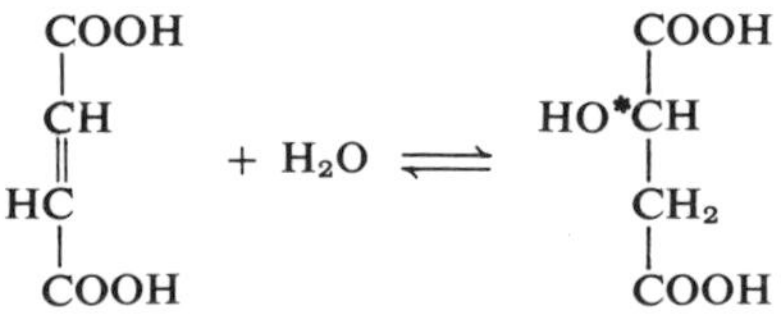

$$
\begin{array}{ccc}
\text{COOH} & & \text{COOH} \\
| & & | \\
\text{CH} & + \text{H}_2\text{O} \rightleftharpoons & \text{HO}^\bullet\text{CH} \\
\| & & | \\
\text{HC} & & \text{CH}_2 \\
| & & | \\
\text{COOH} & & \text{COOH} \\
\text{fumaric} & & \text{malic acid} \\
\text{acid}
\end{array}
\qquad (15.8)
$$

Malate is oxidized by NAD in a reaction catalysed by malate dehydrogenase.

$$
\begin{array}{c}
\text{COOH} \\
| \\
\text{HOCH} \\
| \\
\text{CH}_2 \\
| \\
\text{COOH}
\end{array}
\quad + \text{ NAD} \rightleftharpoons
\begin{array}{c}
\text{COOH} \\
| \\
\text{CO} \\
| \\
\text{CH}_2 \\
| \\
\text{COOH}
\end{array}
\quad + \text{ NADH}_2
\qquad (15.9)
$$

malic acid        oxaloacetic acid

This reaction completes the tricarboxylic acid cycle, regenerating oxaloacetate which can react once again with acetyl coenzyme A (reaction (15.2)).

The net effect of reactions (15.7), (15.8) and (15.9) is to achieve the oxidation of a carbon atom from —$CH_2$— in succinate to —CO— in oxaloacetate. This is an important conversion, which also occurs in the oxidation of fatty acids (see next chapter).

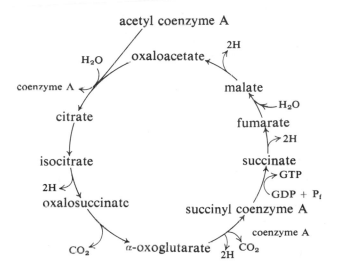

**Fig. 15.1** The tricarboxylic acid cycle

## Summary of reactions and energetics

During the course of the reactions of the tricarboxylic acid cycle an acetyl group from acetyl coenzyme A is incorporated into citrate, and subsequently two carbon atoms appear as carbon dioxide and four pairs of hydrogen atoms are released. We may summarize the oxidation of acetyl coenzyme A as follows:

$$CH_3.CO.S\text{-CoA} + 3H_2O \longrightarrow 2CO_2 + 4(H_2) + CoA\text{-SH}.$$

The oxidation of pyruvate involves not only the reactions of the tricarboxylic acid cycle but also the conversion of pyruvate to acetyl coenzyme A, and can be summarized:

$$CH_3.CO.COOH + 3H_2O \longrightarrow 3CO_2 + 5(H_2).$$

The three molecules of carbon dioxide are released in reactions (15.1), (15.4) and (15.5). Of the five pairs of hydrogen atoms, four are passed to NAD in reactions (15.1), (15.4), (15.5) and (15.9), and one directly to a flavoprotein in reaction (15.7); these all pass down the respiratory chain (Ch. 12).

We can use these facts to calculate the yield of ATP from the oxidation of acetyl coenzyme A or of pyruvate. Each of the pairs of hydrogen atoms from NAD that is oxidized via the respiratory chain gives rise to three molecules of ATP, and each of the pairs that is oxidized from FAD gives rise to two molecules of ATP (see chapter 12). In addition one molecule of ATP is gained in reaction (15.6). Consequently the oxidation of the acetyl group of acetyl coenzyme A should yield twelve molecules of ATP, and the oxidation of pyruvate fifteen molecules of ATP.

We are now in a position to calculate the yield of ATP from the complete oxidation of glucose. We saw in the previous chapter that the Embden–Meyerhof pathway gives rise to two molecules of pyruvate and yields two molecules of ATP and two of $NADH_2$. If each of these two molecules of $NADH_2$ yielded three molecules of ATP during the passage of hydrogen atoms down the respiratory chain (since under aerobic conditions $NADH_2$ is not used to reduce pyruvate), then the total yield of ATP per molecule of glucose oxidized would be $2 \times 15 + 2 + 6 = 38$ molecules. In practice the matter is slightly more complicated, because the $NADH_2$ produced during glycolysis will be in the soluble portion of the cell whereas the respiratory chain operates only in the mitochondrion; reducing power must therefore be passed from the soluble fraction into the mitochondrion, and the effect of this transfer is to reduce from six to four the number of molecules of ATP produced from the $NADH_2$ (see Chapter 21). Even so 36 molecules of ATP are produced from the oxidation of glucose – a striking contrast with the yield of two molecules of ATP per molecule of glucose under anaerobic conditions.

## The tricarboxylic acid cycle and biosynthetic pathways: replenishment of intermediates

The tricarboxylic acid cycle is the classical example of an 'amphibolic' pathway (see Ch. 9). Intermediates of the cycle are essential starting compounds for the synthesis of a host of products. We shall mention some of these syntheses now in order to illustrate the amphibolic nature of the cycle, although we shall defer a detailed description of the synthetic routes until later chapters.

In the first place, intermediates of the tricarboxylic acid cycle are essential for the synthesis of many amino acids. $\alpha$-Oxoglutarate can be converted to glutamate, and glutamate in turn can give rise to glutamine, proline and (though not in all animals) arginine. Oxaloacetate is converted to aspartate, from which asparagine can be formed and, in many organisms, lysine, methionine, threonine and isoleucine. The details of these conversions are described in Ch. 22.

A further use for the aspartate that is formed from oxaloacetate is to serve as a precursor of pyrimidines. This pathway is described in Ch. 23.

Yet another biosynthetic pathway that employs an intermediate of the tricarboxylic acid cycle is the synthesis of pyrrole derivatives, which are assembled into the porphyrin ring and are thus needed for haemoglobin, the chlorophylls and the cytochromes. The starting point for this synthesis is succinyl coenzyme A.

If the tricarboxylic acid cycle were used only as a pathway of oxidation, catalytic amounts of its intermediates would be sufficient for the oxidation of unlimited quantities of acetyl groups from acetyl coenzyme A. For each acetyl group that combines with oxaloacetate in reaction (15.2), one oxaloacetate molecule reappears after a turn of the cycle. But since the syntheses that we have mentioned actually remove intermediates from the cycle, it is plain that there must be a mechanism for replenishing them. Without such a mechanism, the concentration of all the intermediates including oxaloacetate would gradually fall and the mechanism for oxidizing acetyl groups from carbohydrate and fat would come to a halt.

In part, the tricarboxylic acid cycle can be replenished by the reverse of some of the reactions that we have just mentioned – for example α-oxoglutarate not only gives rise to glutamate but also can be formed from glutamate. But this means of replenishing the cycle would suffice only in conditions where biosynthetic reactions were exactly balanced by degradative reactions, and where there were no side reactions causing wastage of intermediates, and these conditions are never met in practice. In particular, during active growth (for example in young animals) biosynthetic reactions predominate. There must therefore be a means of forming tricarboxylic acid cycle intermediates from some other source. This indispensable need is met by a carboxylation of pyruvate.

$$CH_3.CO.COOH + CO_2 + ATP \rightleftharpoons \overset{\displaystyle CO.COOH}{\underset{\displaystyle CH_2.COOH}{|}} + ADP + P_i$$

pyruvic acid  oxaloacetic acid

$$\Delta G^{0\prime} = -500 \text{ cal/mole}$$

The enzyme that catalyses this reaction is called pyruvate carboxylase; it contains biotin, which is a vitamin (pp. 79–80), and we shall discuss its mechanism on p. 341. (Another carboxylation, forming malate from pyruvate, was previously thought to be involved in replenishing the tricarboxylic acid cycle, but this latter reaction is now believed to be quantitatively of much less significance than that catalysed by pyruvate carboxylase.)

One consequence of the necessity for replenishing the cycle by means of this carboxylation is of great importance. In order for the reaction to take place pyruvate must be present, and pyruvate is provided in substantial quantities only by the glycolytic pathway. Thus if carbohydrate metabolism is impaired replenishment of the tricarboxylic acid cycle will be difficult or impossible. We shall return to this point in discussing the oxidation of fatty acids in the next chapter.

# Chapter 16 **Oxidation of fats**

In many higher organisms, a large fraction of the ATP produced comes from the oxidation of fats. Tissues vary in the quantity of fat that they degrade – the liver, for example, degrades a good deal whereas the brain hardly degrades fat at all (see p. 256). But in the organism as a whole, as much as one-half of the energy produced may be derived from oxidation of fats.

We have seen (p. 192) that fats can associate into large structures from which water is excluded. This property enables fats to be stored in a highly concentrated form in the cell to serve as food reserves. Fats have an advantage, from this point of view of acting as food reserves, over carbohydrates: since they contain long-chain fatty acids of the form $CH_3.CH_2.CH_2.CH_2....COOH$ ($= C_nH_{(2n)}O_2$) their combustion requires more oxygen, and produces more ATP, per C atom than does the oxidation of carbohydrates which, as the formula $C_nH_{2n}O_n$ shows, already have a fair amount of oxygen (see also p. 274). The result is that the complete oxidation of fats yields some 9000 cal/gram and that of carbohydrates only about 4000 cal/gram.

Before they can be oxidized, fats must first be hydrolysed. There are several intracellular lipases that catalyse the hydrolysis of fats to yield glycerol and fatty acids.

$$
\begin{array}{ll}
CH_2O.COR & \qquad CH_2OH \qquad R.COOH \\
| & \qquad \quad | \qquad\qquad\qquad + \\
CHO.COR' \quad + 3H_2O \rightleftharpoons & CHOH \; + \; R'.COOH \\
| & \qquad \quad | \qquad\qquad\qquad + \\
CH_2O.COR'' & \qquad CH_2OH \qquad R''.COOH
\end{array}
$$

The glycerol can enter the Embden–Meyerhof pathway after phosphorylation by the specific enzyme glycerokinase (see p. 268), and we may now consider the oxidation of the fatty acids.

Experiments described by Knoop in 1904 provided an essential clue to the mode of oxidation of fatty acids, and led to the conclusion that the key feature of this oxidation is that the acids are degraded *two carbon atoms at a time*. Knoop's experiments depended on labelling fatty acids, feeding them to animals, and isolating the degradation products from the urine. Thus in principle Knoop made use of the technique that we described on p. 230, but whereas there we discussed labelling in terms of atomic isotopes, Knoop labelled his compounds by introducing a foreign chemical group which remained intact during metabolism.

Knoop prepared a series of fatty acids with various lengths of chain, each labelled with a phenyl group at the methyl end, thus

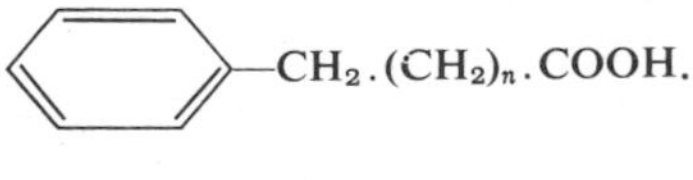

$$\text{—CH}_2.(\text{CH}_2)_n.\text{COOH}.$$

When these were fed to rabbits, one of two degradation products could always be isolated from the urine. If the number of carbon atoms in the chain was even, the degradation product was phenylacetic acid,

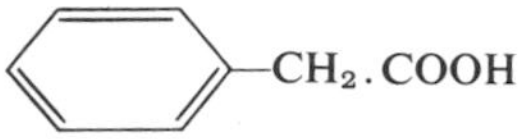

$$\text{—CH}_2.\text{COOH}$$

regardless of how long the fatty-acid chain was. When the number of carbon atoms in the chain was odd, the degradation product was benzoic acid,

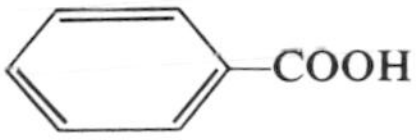

$$\text{—COOH}$$

regardless of how long the fatty-acid chain was. Knoop concluded that the fatty-acid chain was oxidized by the successive removal of two-carbon fragments from the —COOH end, leaving either phenylacetic acid or benzoic acid which could not be further metabolized.

Many years passed before a cell-free system capable of oxidizing fats was developed, but it was eventually discovered by Lehninger that the oxidation of fatty acids does not proceed in broken-cell systems unless ATP is present. Although this fact suggested that fatty acids must be activated before they can be oxidized (compare the activation of glucose by ATP, p. 259), the nature of the activated form of the fatty acid remained obscure until Lynen showed that coenzyme A was essential for the oxidation to proceed. Subsequently it became clear that not only must fatty acids be activated in the first instance, by esterification to coenzyme A, but in fact the whole oxidation sequence actually involves coenzyme A derivatives and yields several molecules of acetyl coenzyme A as a final product. Lehninger also showed that the oxidation of fatty acids occurs exclusively in the mitochondrion.

Most fatty acids occurring in nature have even-numbered carbon atoms in their chains (see Ch. 8). One of the commonest is stearic acid ($CH_3.(CH_2)_{16}.COOH$), which we shall use to illustrate the oxidation sequence. From the principles that we have just established – that oxidation of fatty acids proceeds by the successive removal of two-carbon fragments from the —COOH end, and that it gives rise to the formation of several molecules of acetyl coenzyme A – we could conclude that the oxidation of stearic acid proceeds via eight successive oxidations of —$CH_2.CH_2$— to —$CH_2.CO$—, to give nine molecules of acetyl coenzyme A. A shortened way of writing this would be:

$$CH_3.(CH_2)_{16}.COOH + 9CoA.SH + 8O \longrightarrow 9CH_3.CO.S\text{-}CoA + H_2O + 8(H_2)$$

and this equation does in fact summarize the oxidation of stearic acid that occurs in the cell. We shall now consider the details of the process.

## Stearic acid to acetyl coenzyme A

*Activation of the fatty acid*

The formation of the coenzyme A derivative is an essential first step in the metabolism of any fatty acid. The reaction is catalysed by an enzyme called thiokinase. (There are several thiokinases, each specific for a group of fatty acids with a certain range of chain lengths.) The reaction requires ATP, which is split to give AMP and pyrophosphate instead of the more usual products ADP and phosphate.

$$CH_3.(CH_2)_{16}.COOH + ATP + CoA.SH \rightleftharpoons CH_3.(CH_2)_{16}.CO.S\text{-}CoA + AMP + PP_i$$

$$\Delta G^{\circ\prime} \sim 0$$

This reaction is actually the sum of two half-reactions. First the fatty acid reacts with ATP, splitting out pyrophosphate and yielding an intermediate called fatty-acyl AMP (Fig. 16.1).

$$CH_3.(CH_2)_{16}.COOH + ATP \rightleftharpoons CH_3.(CH_2)_{16}.CO\text{-}AMP + PP_i$$
$$\text{stearoyl AMP}$$

This then reacts with coenzyme A to liberate AMP.

$$CH_3.(CH_2)_{16}.CO\text{-}AMP + CoA.SH \rightleftharpoons CH_3.(CH_2)_{16}.CO.S\text{-}CoA + AMP$$
$$\text{stearoyl coenzyme A}$$

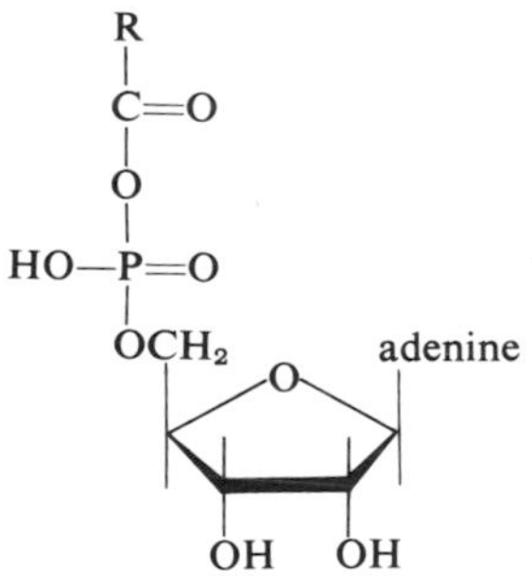

**Fig. 16.1**   Fatty-acyl AMP

When we discuss the activation of amino acids for protein synthesis (p. 440) we shall see that they too are activated via an acyl AMP derivative.

The activation of fatty acids, as illustrated above with stearic acid, is readily reversible. However, one of the products of the reaction is inorganic pyrophosphate, and this is split owing to the presence of the ubiquitous enzyme pyrophosphatase.

$$PP_i + H_2O \rightleftharpoons 2P_i$$
$$\Delta G^{\circ\prime} = -7000 \text{ cal/mole}$$

In practice, then, the concentration of inorganic pyrophosphate in the cell is extremely low, and there is no chance of AMP being rephosphorylated directly to ATP. This is an important example of that destruction of the product of a reaction that we mentioned on p. 20. Consequently we may regard the activation of the fatty acid as proceeding effectively to completion. Provided that we bear in mind that the activation reaction and the hydrolysis of pyrophosphate are actually separate, we might permit ourselves to emphasize this close approach to completion by adding the two reactions together and writing them like this.

$$CH_3.(CH_2)_{16}.COOH + ATP + CoA.SH + H_2O \rightleftharpoons$$
$$CH_3.(CH_2)_{16}.CO.S\text{-}CoA + AMP + 2P_i$$

We shall see that there are several reactions that, like the activation of fatty acids, produce pyrophosphate from ATP (see pp. 334, 435 and 440). The pyrophosphate produced is in each case instantly broken down by pyrophosphatase, and this invariably has the effect of making the total reaction go effectively to completion.

We may compare the activation of fatty acids with the activation of glucose (p. 259), of fructose and galactose (p. 267) and of glycerol (p. 269). In each case the reaction that brings the substance into the metabolic sequence cannot be reversed under physiological conditions and is thus a potential point of control (see Ch. 29).

## Dehydrogenation of acyl coenzyme A

The previous reaction yielded stearoyl coenzyme A, which, for reasons that will soon become clear, we shall now write as

$$CH_3.(CH_2)_{12}.CH_2.CH_2.CH_2.CH_2.CO.S\text{-}CoA.$$

This now undergoes a sequence of three reactions which are together called $\beta$-oxidation. The effect of these reactions is to oxidize the $\beta$-carbon atom of the fatty acyl coenzyme A from —$CH_2$— to —$CO$—, thus yielding a $\beta$-ketoacyl coenzyme A. It is this $\beta$-oxidation that accounts for the results obtained by Knoop (see p. 283). In the first reaction of the $\beta$-oxidation sequence, stearoyl coenzyme A is oxidized to the corresponding $\alpha,\beta$-unsaturated compound.

$$CH_3.(CH_2)_{12}.CH_2.CH_2.CH_2.CH_2.CO.S\text{-}CoA \xrightarrow{\text{FAD} \quad \text{FADH}_2}$$
$$CH_3.(CH_2)_{12}.CH_2.CH_2.CH{=}CH.CO.S\text{-}CoA$$

This reaction is catalysed by a fatty acyl coenzyme A dehydrogenase – an enzyme that contains FAD, to which the pair of hydrogen atoms is passed. The compound produced by the oxidation, an enoyl coenzyme A, has the *trans* configuration and should properly be written:

$$CH_3.(CH_2)_{12}.CH_2.CH_2.\overset{H}{\underset{H}{C}{=}C}.CO.S\text{-}CoA.$$

## Hydration of enoyl coenzyme A

The previous product accepts water in a reaction catalysed by enoyl hydratase to form the corresponding $\beta$-hydroxyacyl coenzyme A. The reaction is stereospecific; only the L-isomer of the $\beta$-hydroxyacyl coenzyme A is produced.

$$CH_3 . (CH_2)_{12} . CH_2 . CH_2 . \overset{\text{H}}{\underset{\text{H}}{C}}{=}C . CO . S\text{-}CoA + H_2O \rightleftharpoons$$

$$CH_3 . (CH_2)_{12} . CH_2 . CH_2 . \overset{\text{OH}}{\underset{\text{H}}{C}} . CH_2 . CO . S\text{-}CoA$$

*Dehydrogenation of β-hydroxyacyl coenzyme A*

A second oxidation step produces β-ketoacyl coenzyme A from the β-hydroxyacyl coenzyme A. The hydrogen acceptor in this reaction is NAD, and the enzyme is called β-hydroxyacyl coenzyme A dehydrogenase.

$$CH_3 . (CH_2)_{12} . CH_2 . CH_2 . \overset{\text{OH}}{\underset{\text{H}}{C}} . CH_2 . CO . S\text{-}CoA + NAD \rightleftharpoons$$

$$CH_3 . (CH_2)_{12} . CH_2 . CH_2 . CO . CH_2 . CO . S\text{-}CoA + NADH_2$$

We must pause at this point to remark on a striking similarity between these three reactions and the last three reactions of the tricarboxylic acid cycle (p. 279). The first reaction in each case is an oxidation of —$CH_2$—$CH_2$— to form the unsaturated compound —CH=CH—; in both the cycle and the β-oxidation of fatty acids the enzyme is a flavoprotein, and the product (fumaric acid or enoyl coenzyme A) is a *trans* compound. In the second reaction water is added across the double bond to form —CHOH.$CH_2$—; in both cases the product is the L-isomer of the hydroxyacid derivative. The third reaction oxidizes —CHOH.$CH_2$— to —CO.$CH_2$—; in both the cycle and the β-oxidation sequence NAD is used as the oxidizing agent at this stage, by contrast with the first reaction.

*Cleavage of β-ketoacyl coenzyme A*

Now that the β-carbon atom of the original fatty acid has been completely oxidized, the compound can be split to yield acetyl coenzyme A and leave a fatty-acid derivative two carbon atoms shorter than the starting acid. This splitting is brought about by reaction with the sulphydryl-containing compound coenzyme A and is therefore called thiolysis. The enzyme is a β-keto thiolase.

$$CH_3 . (CH_2)_{12} . CH_2 . CH_2 . CO . CH_2 . CO . S\text{-}CoA + CoA . SH \rightleftharpoons$$

$$CH_3 . (CH_2)_{12} . CH_2 . CH_2 . CO . S\text{-}CoA + CH_3 . CO . S\text{-}CoA$$

$$\Delta G^{\circ\prime} = -6100 \text{ cal/mole}$$

The free-energy change of this reaction greatly favours cleavage. We may note that the degradation sequence of fatty acids thus contains reactions (the activation step, see p. 284 above, and the β-keto thiolysis) that proceed virtually to completion, in just the same way as glycolysis does (see p. 261).

Palmitoyl coenzyme A, the product of the last reaction, now re-enters the β-oxidation sequence that we have just described, and its β-carbon atom is oxidized by the same reactions.

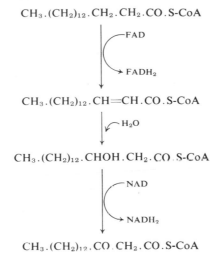

$$CH_3.(CH_2)_{12}.CH_2.CH_2.CO.S\text{-}CoA$$

FAD

FADH$_2$

$$CH_3.(CH_2)_{12}.CH{=}CH.CO.S\text{-}CoA$$

H$_2$O

$$CH_3.(CH_2)_{12}.CHOH.CH_2.CO.S\text{-}CoA$$

NAD

NADH$_2$

$$CH_3.(CH_2)_{12}.CO.CH_2.CO.S\text{-}CoA$$

The resulting β-ketoacyl coenzyme A is in its turn split by coenzyme A in the presence of β-keto thiolase:

$$CH_3.(CH_2)_{12}.CO.CH_2.CO.S\text{-}CoA + CoA.SH \rightleftharpoons$$
$$CH_3.(CH_2)_{12}.CO.S\text{-}CoA + CH_3.CO.S\text{-}CoA$$

myristoyl coenzyme A

and the myristoyl coenzyme A in turn goes back to the beginning of the β-oxidation sequence. In this way fatty acids, in the form of their coenzyme A derivatives, are broken down to form several acetyl coenzyme A units. Stearic acid, for example, undergoes eight rounds of β-oxidation and yields nine molecules of acetyl coenzyme A (see Fig. 16.2).

The oxidation of fatty acids clearly produces large quantities of acetyl coenzyme A. In normal circumstances this acetyl coenzyme A will enter the tricarboxylic acid cycle and the acetyl group will thus be completely oxidized to carbon dioxide and water. (We shall consider later the circumstances in which acetyl coenzyme A is not able to enter the cycle.)

## Summary of reactions and energetics

In each round of β-oxidation, fatty acyl coenzyme A releases one pair of hydrogen atoms to FAD and one to NAD. The summary reaction for the eight rounds of β-oxidation and cleavage that stearoyl coenzyme A undergoes is:

$$CH_3.(CH_2)_{16}.CO.S\text{-}CoA + 8CoA.SH + 8H_2O + 8FAD + 8NAD \longrightarrow$$
$$9CH_3.CO.S\text{-}CoA + 8FADH_2 + 8NADH_2$$

which is a more detailed and accurate way of writing the equation that we gave in rather unsophisticated form on p. 283.

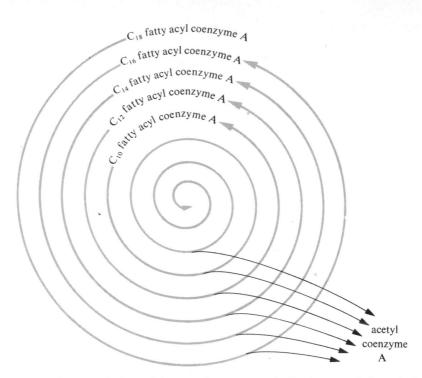

$C_{18}$ fatty acyl coenzyme A

$C_{16}$ fatty acyl coenzyme A

$C_{14}$ fatty acyl coenzyme A

$C_{12}$ fatty acyl coenzyme A

$C_{10}$ fatty acyl coenzyme A

acetyl coenzyme A

**Fig. 16.2** The degradation of fatty-acyl coenzyme A. Each turn of the spiral releases acetyl coenzyme A and leaves a fatty-acyl coenzyme A shortened by two carbon atoms

We can now calculate the yield of ATP from the complete oxidation of stearic acid. Since each of the pairs of hydrogen atoms passed to FAD yields two molecules of ATP, and each of the pairs passed to NAD yields three molecules of ATP (see p. 239), the eight rounds of β-oxidation should produce forty molecules of ATP. The nine molecules of acetyl coenzyme A liberated by β-oxidation will each give rise to twelve molecules of ATP during oxidation of the acetyl group by the tricarboxylic acid cycle (see p. 280). Thus the complete oxidation of the stearoyl group would be expected to yield $(40 + 9 \times 12)$ molecules $= 148$ molecules of ATP.

However, this calculation starts from stearoyl coenzyme A, and the formation of this compound from stearic acid required the use of two high-energy bonds (see p. 284), so that the net yield of ATP from a molecule of stearic acid is diminished to 146 molecules. Note that the β-keto thiolase reaction produces not a fatty acid shorter by two carbon atoms than the original acid, but a fatty acyl coenzyme A which can immediately undergo further β-oxidation; consequently only one activation with ATP is needed for the entire fatty acid.

We may therefore say that the complete oxidation of a molecule of stearic acid via β-oxidation and the tricarboxylic acid cycle yields about 146 molecules of ATP. As with the calculation that we made for the yield of ATP from oxidation of glucose (p. 280), we must stress that this is not intended as a precise figure.

Nevertheless it is instructive to compare the yield of ATP from oxidation of a molecule of stearic acid (containing eighteen carbon atoms) with that from the oxidation of three molecules of glucose (also containing eighteen carbon atoms). Three molecules of glucose would yield $3 \times 38$ (see p. 280) $= 114$ molecules of ATP. We mentioned at the beginning of this chapter (p. 282) that fatty acids, being more reduced compounds than carbohydrates, require more oxygen for their oxidation (see p. 274) and yield more ATP per C atom; we now see what the difference in yield of ATP actually amounts to.

## Odd-numbered fatty acids

Although even-numbered fatty acids are far commoner than odd-numbered, the latter do occur in some fats. Moreover the coenzyme A derivatives of odd-numbered fatty acids are formed in the degradation of some amino acids, as we shall see in Ch. 22. We must therefore consider how the oxidation of odd-numbered fatty acids is achieved.

$\beta$-Oxidation in odd-numbered fatty acids can proceed normally until the final thiolysis reaction (see p. 286) produces propionyl coenzyme A, which cannot be split further. This compound is carboxylated in the presence of a biotin-containing enzyme (see p. 281) to methylmalonyl coenzyme A.

$$\underset{\text{propionyl coenzyme A}}{CH_3.CH_2.CO.S\text{-}CoA} + CO_2 + ATP \rightleftharpoons \underset{\substack{\text{methylmalonyl} \\ \text{coenzyme A}}}{\overset{\displaystyle COOH}{\overset{\displaystyle |}{CH_3.CH.CO.S\text{-}CoA}}} + ADP + P_i$$

Methylmalonyl coenzyme A is isomerized in a curious reaction catalysed by methylmalonyl mutase, the coenzyme of which is deoxyadenosylcobalamin, a derivative of vitamin $B_{12}$. (This vitamin, in the form of its deoxyadenosyl derivative, is involved in very few biochemical reactions, and its mode of action is comparatively little understood; the vitamin is needed in exceptionally small amounts in the diet, and the vitamin $B_{12}$-deficiency disease pernicious anaemia is actually due to a failure to absorb vitamin $B_{12}$ from the gastro-intestinal tract.) The isomerization produces succinyl coenzyme A, an intermediate in the tricarboxylic acid cycle (p. 277).

$$\underset{\substack{\text{methylmalonyl} \\ \text{coenzyme A}}}{\overset{\displaystyle COOH}{\overset{\displaystyle |}{CH_3.CH.CO.S\text{-}CoA}}} \rightleftharpoons \underset{\text{succinyl coenzyme A}}{HOOC.CH_2.CH_2.CO.S\text{-}CoA}$$

## Formation and utilization of ketone bodies

As we mentioned at the beginning of this chapter, tissues differ in their rate of oxidation of fatty acids. In the liver, in particular, fatty acids can sometimes be degraded so quickly that the rate of formation of acetyl coenzyme A exceeds the capacity of the tricarboxylic acid cycle to oxidize it. In these circumstances acetyl coenzyme A starts to accumulate, and so too does acetoacetyl coenzyme A, which is the product of the final round of $\beta$-oxidation in which the four-carbon unit butyryl coenzyme A is oxidized.

$$CH_3.CH_2.CH_2.CO.S\text{-}CoA \longrightarrow CH_3.CH{=}CH.CO.S\text{-}CoA \longrightarrow$$
butyryl coenzyme A

$$CH_3.CHOH.CH_2.CO.S\text{-}CoA \longrightarrow CH_3.CO.CH_2.CO.S\text{-}CoA$$
acetoacetyl coenzyme A

Acetoacetyl coenzyme A is deacylated to acetoacetic acid in a rather curious two-stage reaction.

$$CH_3.CO.CH_2.CO.S\text{-}CoA + CH_3.CO.S\text{-}CoA \xrightarrow{\quad H_2O \quad CoA.SH \quad}$$
acetoacetyl        acetyl
coenzyme A        coenzyme A

$$\underset{\underset{CH_3}{|}}{\overset{\overset{OH}{|}}{HOOC.CH_2.C.CH_2.CO.S\text{-}CoA}} \longrightarrow CH_3.CO.CH_2.COOH + CH_3.CO.S\text{-}CoA$$

$\beta$-hydroxy-$\beta$-methylglutaryl      acetoacetic acid     acetyl coenzyme A
coenzyme A

We might note, incidentally, that $\beta$-hydroxy-$\beta$-methylglutaryl coenzyme A is an important intermediate, being a precursor of steroids. Our main concern here, however, is with the acetoacetate. This compound can undergo enzymic reduction at the expense of $NADH_2$ to yield $\beta$-hydroxybutyrate. Alternatively it may be (non-enzymically) decarboxylated to yield acetone. The three compounds, acetoacetate, $\beta$-hydroxybutyrate and acetone, are collectively known as *ketone bodies*. In conditions leading to rapid degradation of fats, these ketone bodies diffuse from the liver into the blood.

Acetone is not metabolized to any large extent and is simply excreted, but acetoacetate and $\beta$-hydroxybutyrate (which are interconvertible by use of the $NAD/NADH_2$ couple) are an important source of metabolic fuel for heart muscle, for the kidney and, especially, for the brain. In these tissues the tricarboxylic acid cycle can often accept a further supply of acetyl coenzyme A, even at times when the liver's tricarboxylic acid cycle is overloaded by rapid degradation of fat. The conversion of acetoacetate to acetyl coenzyme A in those tissues that metabolize it proceeds by a sequence of two reactions. First acetoacetate accepts coenzyme A from succinyl coenzyme A (see p. 277):

$$CH_3.CO.CH_2.COOH + \underset{\underset{\underset{COOH}{|}}{\overset{\overset{CH_2}{|}}{CH_2}}}{CO.S\text{-}CoA} \rightleftharpoons CH_3.CO.CH_2.CO.S\text{-}CoA + \underset{\underset{\underset{COOH}{|}}{\overset{\overset{CH_2}{|}}{CH_2}}}{COOH}$$

acetoacetic acid     succinyl       acetoacetyl coenzyme A    succinic
               coenzyme A                                acid

and then the resulting acetoacetyl coenzyme A is split by the usual thiolase reaction.

$$CH_3.CO.CH_2.CO.S\text{-}CoA + CoA.SH \rightleftharpoons 2CH_3.CO.S\text{-}CoA$$
acetoacetyl coenzyme A           acetyl coenzyme A

Acetyl coenzyme A then reacts with oxaloacetate in the normal way to yield citrate.

In what circumstances might the rate of degradation of fatty acids overwhelm the capacity of the liver's tricarboxylic acid cycle to oxidize the resulting acetyl coenzyme A? In order to answer this question, we shall have to anticipate that part of Ch. 19 that is concerned with the synthesis of carbohydrate in animals. The route to carbohydrate passes through oxaloacetate, and oxaloacetate is, you will remember (p. 281), produced by carboxylation of pyruvate. Let us consider what will happen if the liver's stores of glycogen are depleted, for instance after exercise or as a result of fasting. The rate of glycolysis will be diminished (so that little pyruvate is produced), and the liver will turn instead to oxidizing fatty acids, thus producing large quantities of acetyl coenzyme A. Normally this acetyl co-enzyme A would react with oxaloacetate, but in the present circumstances oxalo-acetate is scarce for two reasons. In the first place there is little pyruvate to be carboxylated; in the second place what oxaloacetate is available is being used by the liver for the synthesis of carbohydrate, to supply those tissues (brain and erythrocytes) to which glucose is indispensable.

The result is to set in train the formation of ketone bodies that we have just outlined. It is worth noting that the brain's ability to use ketone bodies as a fuel is a valuable one since it reduces the requirement for glucose at a time when car-bohydrate is scarce. Even so, the brain cannot function without *some* glucose.

The events that we have described occur whenever the liver's stores of glyco-gen are depleted, and they are in no way pathological. In some conditions, how-ever, such as prolonged starvation or diabetes, the concentration of ketone bodies in the blood rises greatly, giving rise to a condition of severe *ketosis*. Conventionally, this condition is explained as arising when the rate of produc-tion of ketone bodies exceeds the capacity of the brain, etc. to oxidize them. In reality the situation appears to be much more complicated than this, and we do not fully understand the mechanism of pathological ketosis or the biochemical basis of the serious symptoms to which the condition gives rise.

# Chapter 17 **The pentose phosphate pathway**

In Chs 14 and 15 we described how glucose and other sugars are metabolized via the glycolytic pathway and the tricarboxylic acid cycle. This route accounts for the bulk of carbohydrate metabolism in most organisms. There is, however, an alternative pathway for the oxidation of sugars, which we shall consider in this chapter. This pathway has several names: it is sometimes called the pentose phosphate pathway, sometimes the hexose monophosphate shunt, sometimes the phosphogluconate pathway and sometimes the Warburg–Dickens pathway. Although in most organisms this means of metabolizing carbohydrate is quantitatively of less importance than glycolysis and the Krebs cycle, its existence is absolutely essential to the proper functioning of metabolism; we shall shortly see why the pentose phosphate pathway, even though it appears at first sight to duplicate the Embden–Meyerhof and Krebs pathway, is in fact indispensable. The pentose phosphate pathway takes as its starting point glucose-6-phosphate, which is produced (you will recall) either by the phosphorylation of glucose (p. 259) or by phosphorolysis of glycogen or starch followed by isomerization of glucose-l-phosphate (p. 268). In this respect, as well as the fact that its enzymes are located in the soluble fraction of the cytoplasm, the pathway resembles the glycolytic sequence; but there the resemblance ends. The key feature of glycolysis is that it is a fermentation, in which the reducing power removed from the molecule early in the pathway is returned at the end (see p. 265). In the pentose phosphate pathway, on the other hand, the very first reaction is an oxidation of glucose-6-phosphate, and the hydrogen atoms removed are not returned to the pathway at a later stage.

The pentose phosphate pathway can be considered most easily in two parts. The first part, which is characteristic of the pathway in all circumstances, involves the oxidation of glucose-6-phosphate to pentose phosphate; the second, in which there are many possibilities for metabolic interconversions (see p. 207), involves manipulations of the pentose phosphate to form a variety of other sugars.

## Oxidation of glucose-6-phosphate to pentose phosphate

This section of the pentose phosphate pathway was elucidated by Warburg, Dickens, Lipmann, Horecker and their co-workers. The first reaction, which was discovered by Warburg, is a dehydrogenation of glucose-6-phosphate with

NADP as a co-factor. The enzyme, glucose-6-phosphate dehydrogenase, is distributed very widely among living organisms.

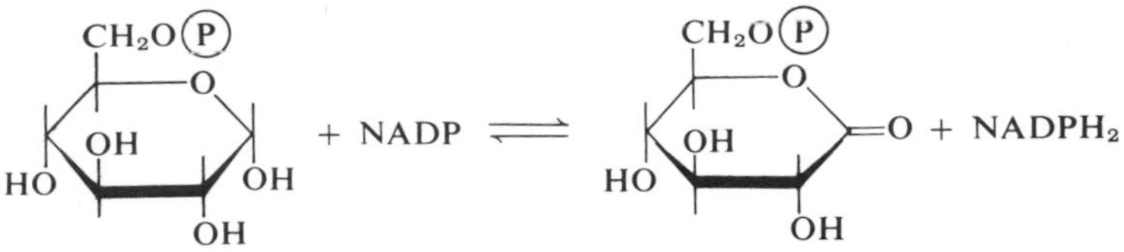

glucose-6-phosphate          gluconolactone-
                             6-phosphate

The product is a lactone, a highly unstable compound which would hydrolyse at an appreciable rate in the absence of an enzyme. The rate of hydrolysis is, however, greatly increased by the presence of an enzyme called lactonase.

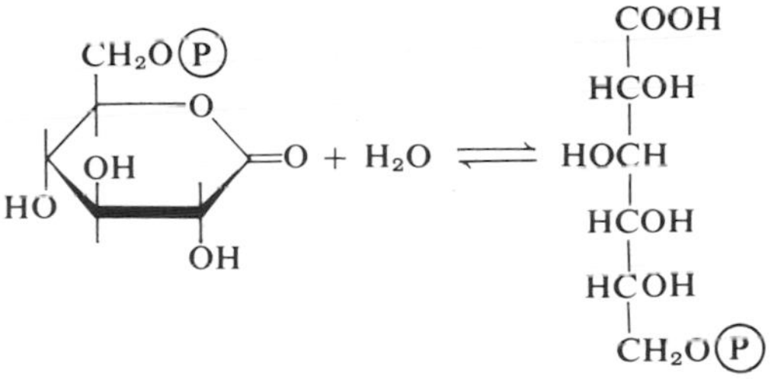

gluconolactone-              gluconic acid-6-
6-phosphate                  phosphate

$$\Delta G^{\circ\prime} = -5000 \text{ cal/mole}$$

The resulting gluconate-6-phosphate now undergoes a second dehydrogenation. The hydrogen acceptor is again NADP, and the enzyme is gluconate-6-phosphate dehydrogenase. The product, however, instead of being a keto acid as one might expect, is a phosphorylated *pentose*. In other words the dehydrogenation of gluconate-6-phosphate is accompanied by decarboxylation.

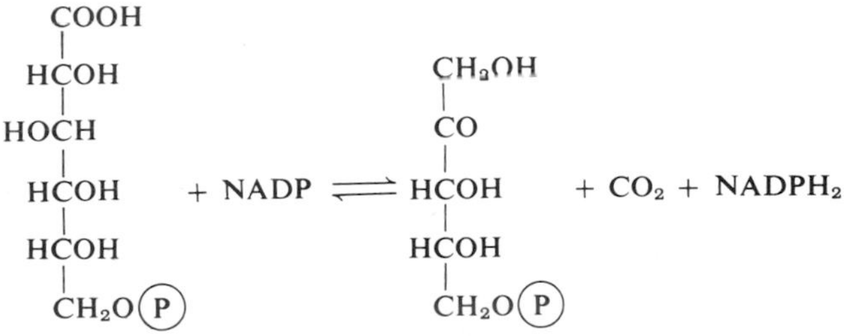

gluconic acid-6-            ribulose-5-
phosphate                  phosphate

You may have noticed that earlier we spoke rather evasively of the oxidation of glucose-6-phosphate to pentose phosphate without specifying which pentose phosphate was formed. In fact, although ribulose-5-phosphate is the first product, comparatively little of this compound is found in cells in which the pentose phosphate pathway is operating. Ribulose-5-phosphate is converted to two other sugars. The conversion to xylulose-5-phosphate is catalysed by an epimerase:

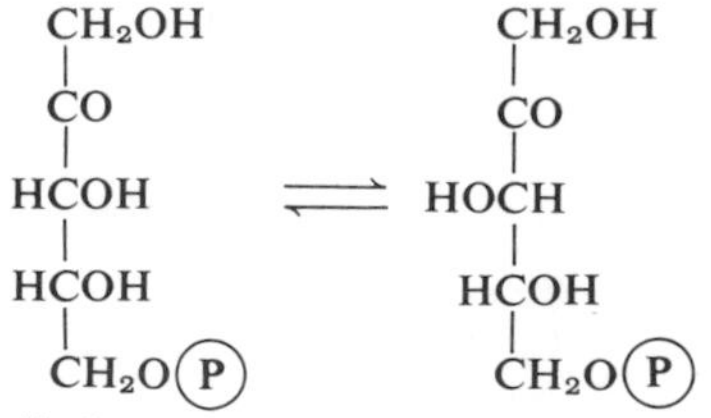

$$
\begin{array}{ccc}
\text{CH}_2\text{OH} & & \text{CH}_2\text{OH} \\
| & & | \\
\text{CO} & & \text{CO} \\
| & & | \\
\text{HCOH} & \rightleftharpoons & \text{HOCH} \\
| & & | \\
\text{HCOH} & & \text{HCOH} \\
| & & | \\
\text{CH}_2\text{O}\,\textcircled{P} & & \text{CH}_2\text{O}\,\textcircled{P} \\
\text{ribulose-5-} & & \text{xylulose-5-} \\
\text{phosphate} & & \text{phosphate}
\end{array}
$$

and the conversion to ribose-5-phosphate by an isomerase (pentose phosphate isomerase; compare the reaction catalysed by hexose phosphate isomerase (p. 260), which is exactly analogous):

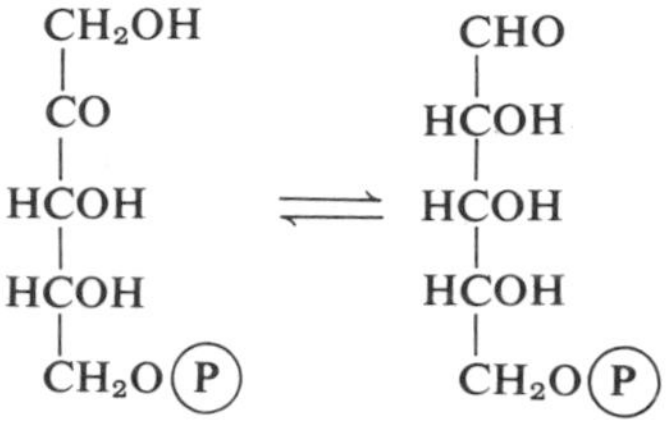

$$
\begin{array}{ccc}
\text{CH}_2\text{OH} & & \text{CHO} \\
| & & | \\
\text{CO} & & \text{HCOH} \\
| & & | \\
\text{HCOH} & \rightleftharpoons & \text{HCOH} \\
| & & | \\
\text{HCOH} & & \text{HCOH} \\
| & & | \\
\text{CH}_2\text{O}\,\textcircled{P} & & \text{CH}_2\text{O}\,\textcircled{P} \\
\text{ribuiose-5-} & & \text{ribose-5-} \\
\text{phosphate} & & \text{phosphate}
\end{array}
$$

which can be written (see p. 167)

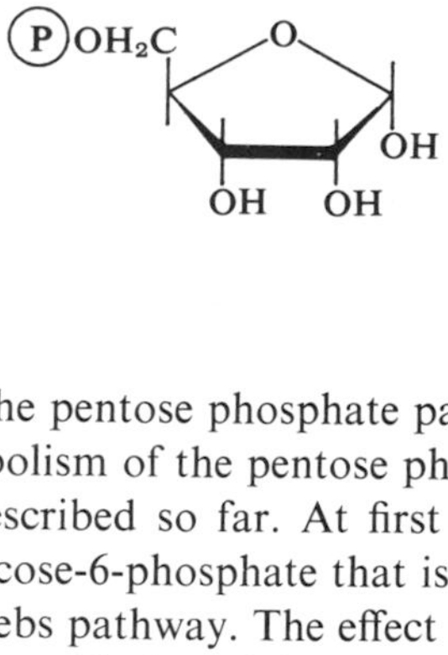

## Functions of the pathway

Now that we have outlined the first part of the pentose phosphate pathway, we may pause before pursuing the further metabolism of the pentose phosphate to consider the significance of the reactions described so far. At first sight they appear to represent a means of oxidizing glucose-6-phosphate that is simply an alternative to the Embden–Meyerhof and Krebs pathway. The effect of the first three reactions we have described is a complete oxidation of the C-1 carbon atom of glucose-6-phosphate to $CO_2$, with the concomitant reduction of two molecules of NADP; if by some means the remaining carbon atoms of glucose-6-phosphate were similarly oxidized, the net result of the process would be the total oxidation of the sugar with the formation of twelve molecules of $NADPH_2$, and if these were reoxidized via the respiratory chain thirty-six molecules of ATP would notionally be formed. For several years it was believed that the function of the pentose phosphate pathway could indeed be described in these terms. But there are good reasons for believing that such an account is oversimplified. In the first place, it is hard to see why an alternative to glycolysis and the Krebs pathway should be needed; the enzymes for glycolysis and the tricarboxylic acid cycle occur in practically all aerobic cells in sufficient concentration to account for glucose oxidation at a rate great enough to supply the ATP needed by the cell. A block in the pathway would be catastrophic for the cell and could certainly not be overcome by diverting glucose to an alternative metabolic pathway. Secondly, $NADPH_2$ is oxidized to only a slight extent by the respiratory chain, and the regeneration of NADP from $NADPH_2$ depends on reactions that are quite different from those that regenerate NAD from $NADH_2$ (see pp. 29–30).

What, then, are the specific functions of the pentose phosphate pathway? We can answer this question by considering the significance of the compounds that are produced in the reactions we have discussed so far.

In the first place, these reactions lead to the formation of pentoses and in particular to ribose-5-phosphate. Ribose-5-phosphate is a constituent of nucleotides – not only the nucleotides in RNA but also those that form part of many coenzymes, such as NAD and coenzyme A (see p. 79). Deoxyribose nucleotides, which are constituents of DNA, are formed from ribonucleotides (see p. 392), so that that represents a further use for ribose-5-phosphate. Thus the generation of pentose phosphate is indispensable to the organism, and we may regard this as one of the essential functions that the pentose phosphate pathway serves.

Secondly, as we hinted above, the provision of $NADPH_2$ is of great importance to the organism, and it is striking that the oxidation of glucose-6-phosphate to ribulose-5-phosphate yields not just one but two molecules of $NADPH_2$. All of the dehydrogenase reactions that we described when discussing glycolysis, the tricarboxylic acid cycle and the oxidation of fatty acids involve the reduction of either NAD or FAD, and we saw in Ch. 12 that the reducing equivalents from these co-factors are (in aerobic conditions) normally passed directly down the respiratory chain. By contrast we shall see that $NADPH_2$ is used as a reducing agent in *synthetic* reactions. Since the pentose phosphate pathway is by far the most important source of $NADPH_2$, a cell can control its rate of production of this reducing coenzyme by altering the proportion of glucose that is metabolized via the Embden–Meyerhof pathway and Krebs cycle as against the pentose phosphate pathway.

We can illustrate the use of $NADPH_2$ as a reducing agent by referring in outline to the synthesis of fatty acids. Just as the breakdown of fatty acids involves their oxidation (see Ch. 16), so their synthesis from simple precursors requires reduction. We shall see in Ch. 20 that fatty acids are synthesized by condensing two-carbon units derived from acetyl coenzyme A; since the acetyl group is $CH_3.CO—$, whereas fatty acids have long reduced chains of the type $CH_3.CH_2.$ $CH_2.CH_2.CH_2....$, it is evident that reduction as well as condensation is required in their synthesis. (We shall describe the details of the synthesis in Ch. 20.) $NADPH_2$ is also used in the synthesis of steroids and in other important reductions.

In any given tissue, the quantity of $NADPH_2$ and the quantity of pentose phosphate required for synthetic purposes will often not be in balance. In particular a tissue that is actively synthesizing fats or steroids will require a great deal of $NADPH_2$, and in order to produce this $NADPH_2$ will be forced to oxidize a large quantity of glucose-6-phosphate. To synthesize one molecule of stearic acid (for example) requires sixteen molecules of $NADPH_2$ (see p. 344), and this $NADPH_2$ must be supplied by the oxidation of eight molecules of glucose-6-phosphate to pentose phosphate. It is improbable that tissues that are synthesizing fats will find a use for the large amount of pentose phosphate produced as a by-product of reducing NADP.

Thus if pentose phosphate were a dead-end product of metabolism, the quantity produced in the pentose phosphate pathway might well be an embarrassment to a tissue that is active in reductive synthesis. In fact, however, as we have stressed before (pp. 112 and 208), there are very few dead-ends in intermediary metabolism, and pentose phosphate can actually be converted to many other sugars. These conversions form the second part of the pentose phosphate pathway, which we shall now consider.

## Conversion of pentose phosphate to hexose phosphate

These conversions of sugar phosphates involve two special enzymes, each of which catalyses the transfer of some portion of one sugar to another. *Transketolase* is responsible for the transfer of a two-carbon unit, $CH_2OH.CO—$, and *transaldolase* for the transfer of a three-carbon unit, $CH_2OH.CO.CHOH—$. In both types of reaction the donor is always a ketose phosphate, and the acceptor an aldose phosphate. Between them the two enzymes can bring about the interconversion of tetrose, pentose, hexose and heptose sugars. It is conventional to write the second part of the pentose phosphate pathway as accomplishing the conversion of three molecules of pentose phosphate into two molecules of hexose phosphate and one of triose phosphate:

$$\mathbf{C_5} + \mathbf{C_5} \xrightleftharpoons{\text{transketolase}} C_3 + C_7$$

$$C_7 + C_3 \xrightleftharpoons{\text{transaldolase}} C_4 + C_6$$

$$\mathbf{C_5} + C_4 \xrightleftharpoons{\text{transketolase}} C_3 + \mathbf{C_6}$$

(where **bold type** signifies the original pentose phosphate molecules, and the final hexose phosphate and triose phosphate molecules). We must, however, emphasize that the precise fate of the carbon atoms in the original sugar phosphates varies with circumstances, and there is by no means any rigid adherence to the scheme suggested above. On the other hand, this scheme is a useful means of illustrating the kinds of conversion that are in fact possible, so it is worth our examining it in more detail.

In the first transketolase reaction, the two-carbon fragment is transferred from xylulose-5-phosphate to ribose-5-phosphate.

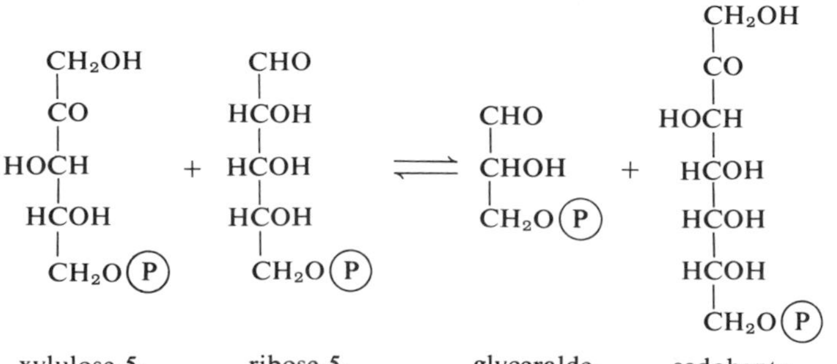

|  |  |  |  |
|---|---|---|---|
| xylulose-5-phosphate | ribose-5-phosphate | glyceraldehyde-3-phosphate | sedoheptulose-7-phosphate |

Transketolase needs as co-factor the vitamin derivative thiamine pyrophosphate, which, as we saw earlier (p. 272), is also involved in the transfer of a two-carbon fragment from pyruvate to lipoic acid. The mechanism of the two reactions is very similar: in the transketolase reaction the dihydroxyethyl derivative of thiamine pyrophosphate is formed as an intermediate in the transfer, and this is

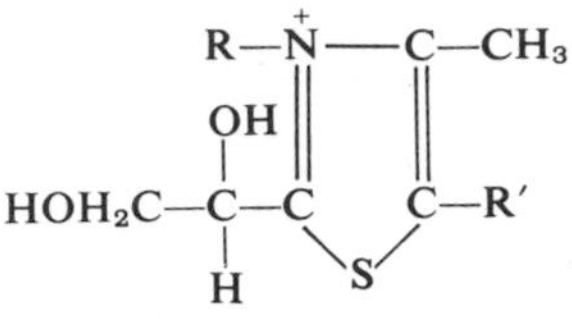

analogous to the hydroxyethyl derivative involved in the decarboxylation of pyruvate.

Glyceraldehyde-3-phosphate, formed in the above reaction, is an intermediate in the Embden–Meyerhof pathway (see p. 261), and it may be further metabolized via that pathway or may possibly give rise to $\alpha$-glycerophosphate (see p. 346). Alternatively it may react with the sedoheptulose-7-phosphate formed in the last reaction.*

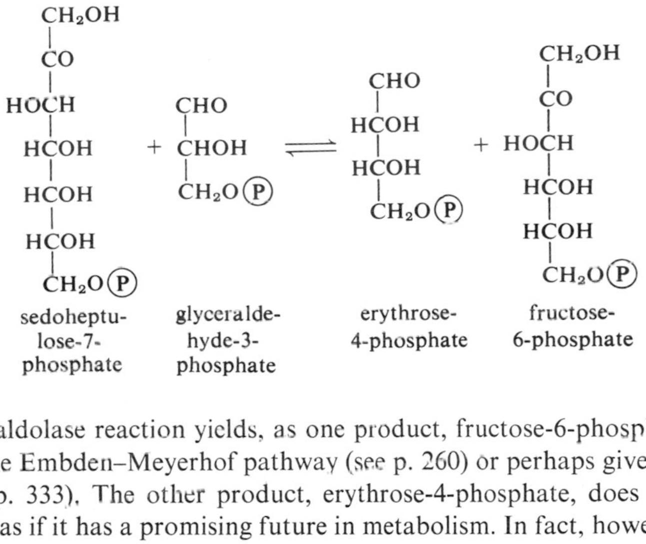

This transaldolase reaction yields, as one product, fructose-6-phosphate, which can join the Embden–Meyerhof pathway (see p. 260) or perhaps give rise to glucose (see p. 333). The other product, erythrose-4-phosphate, does not at first sight look as if it has a promising future in metabolism. In fact, however, it is an important compound, as it is the starting point for the synthesis of the aromatic amino acids (see p. 373) and other aromatic compounds.

In a second reaction catalysed by transketolase, erythrose-4-phosphate can accept a $CH_2OH.CO-$ fragment from another molecule of xylulose-5-phosphate; thiamine pyrophosphate is, of course, again involved as co-factor.

---

* The net effect of these two reactions is to convert two pentose phosphates into one hexose phosphate and one tetrose phosphate. You might think that this conversion could be brought about more simply in a single reaction by transfer of a one-carbon fragment. Curiously enough the transfer of single-carbon units occurs in only a few reactions which involve a special set of co-factors (see p. 383), and is never used in the interconversion of sugars.

$$
\begin{array}{c}
\underset{\substack{\text{xylulose-5-}\\\text{phosphate}}}{
\begin{array}{l}
\text{CH}_2\text{OH}\\
|\\
\text{CO}\\
|\\
\text{HOCH}\\
|\\
\text{HCOH}\\
|\\
\text{CH}_2\text{O}\,\textcircled{P}
\end{array}}
\quad + \quad
\underset{\substack{\text{erythrose-}\\\text{4-phosphate}}}{
\begin{array}{l}
\text{CHO}\\
|\\
\text{HCOH}\\
|\\
\text{HCOH}\\
|\\
\text{CH}_2\text{O}\,\textcircled{P}
\end{array}}
\quad \rightleftharpoons \quad
\underset{\substack{\text{glyceralde-}\\\text{hyde-3-}\\\text{phosphate}}}{
\begin{array}{l}
\text{CHO}\\
|\\
\text{CHOH}\\
|\\
\text{CH}_2\text{O}\,\textcircled{P}
\end{array}}
\quad + \quad
\underset{\substack{\text{fructose-6-}\\\text{phosphate}}}{
\begin{array}{l}
\text{CH}_2\text{OH}\\
|\\
\text{CO}\\
|\\
\text{HOCH}\\
|\\
\text{HCOH}\\
|\\
\text{HCOH}\\
|\\
\text{CH}_2\text{O}\,\textcircled{P}
\end{array}}
\end{array}
$$

The products are fructose-6-phosphate and glyceraldehyde-3-phosphate, both of which can enter the Embden–Meyerhof pathway.

In theory it is possible, by doubling all these equations, to account for the formation of four molecules of fructose-6-phosphate and two molecules of glyceraldehyde-3-phosphate from six molecules of pentose phosphate. The two molecules of glyceraldehyde-3-phosphate can give rise to a molecule of fructose-1,6-diphosphate by the reverse of reactions (14.5) and (14.4) of the Embden–Meyerhof pathway (p. 261), and the fructose-1,6-diphosphate can then be cleaved (p. 260) to fructose-6-phosphate. We thus have a total of five molecules of fructose-6-phosphate formed from six molecules of pentose phosphate, and we could even imagine that these might be converted back to five molecules of glucose-6-phosphate.

A flow-sheet summarizing these conversions is given in Fig. 17.1, and we should again emphasize that, while it represents one formal possibility for the further metabolism of the pentose phosphate produced by oxidation of glucose-6-phosphate, the metabolic routes actually followed in cells are much more fluid and involve many more points of intersection with other pathways than the scheme suggests. The chief importance of the pentose phosphate pathway is that it provides pentoses and NADPH$_2$; the subsequent interconversion of the sugar phosphates is ancillary to that function.

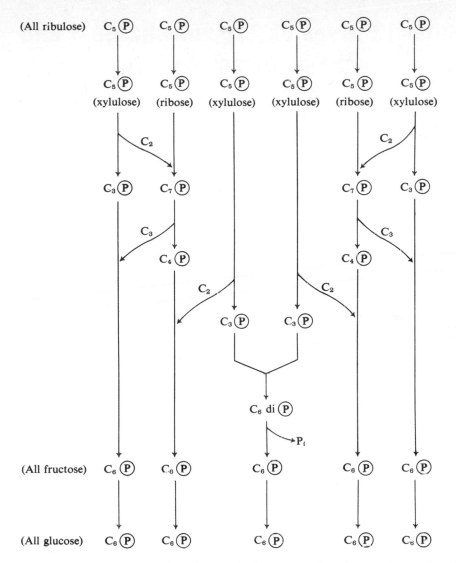

**Fig. 17.1** The interconversion of sugars in the pentose phosphate pathway. Transfer of two-carbon fragments is catalysed by transketolase and transfer of three-carbon fragments by transaldolase. The flow-sheet shows the conversion of six five-carbon sugars to five six-carbon sugars, but this is only a formal illustration of the conversions that are possible

# Chapter 18 Uses of ATP: mechanical and chemi-osmotic work

We have followed, in Chs 12–17, various means that living organisms employ to produce ATP and reducing power. These are two essential things that they require to carry out those functions – growth, movement, self-organization, reproduction – by which we most easily recognize living as distinct from non-living matter. We now begin a series of chapters that survey the ways in which ATP and reducing power are used. The present chapter deals with the role of ATP in other than purely chemical reactions: in the promotion of movement, and in the transport of solutes against an osmotic gradient.

## THE TRANSFORMATION OF CHEMICAL ENERGY INTO MOVEMENT

### The source of chemical energy

There are many anatomical structures in animals, micro-organisms, and even plants that induce movement of one kind or another. Nearly all these structures are devices for transforming chemical energy into mechanical energy, and in virtually all cases the last strictly chemical form in which the energy resides before it is transformed into mechanical work is ATP.

In the muscle used by vertebrates to execute voluntary movements (which is the best understood of these structures, and the one upon which we intend to concentrate), the ultimate sources of ATP are the catabolic reactions that we have just finished describing, especially those of glycolysis, which is a very active process in muscle. Although mechanisms exist (see Ch. 29) that enable the organism to switch from a low, resting rate of catabolism to a very much higher, active rate (up to a hundredfold that of the resting rate), this increase in metabolic rate takes a few seconds to become effective. The resting level of ATP in the muscle cell (about $5 \times 10^{-6}$ mole/gram of muscle) would supply only sufficient energy for contraction for less than one second. An awkward gap could, therefore, arise between the exhaustion of the resting level of ATP and the generation of much larger quantities by the accelerated catabolic processes of the active state. This gap is bridged by an intermediate high-energy compound creatine phosphate* (see Table 2.2, p. 26). This compound is able to re-phosphorylate ADP:

$$\text{creatine phosphate} + \text{ADP} \rightleftharpoons \text{creatine} + \text{ATP}$$

* This is the compound used by vertebrates. There are several other molecules that can act in this way (the generic name for them is *phosphagens*), notably arginine phosphate. Any one species of animal generally uses only one phosphagen.

and replenish the ATP. (The enzyme that catalyses this reaction is called *creatine phosphokinase*.) There is, in the resting muscle, sufficient creatine phosphate (about $2 \times 10^{-5}$ mole/gram of muscle) to supply an amount of ATP adequate for 2–3 seconds' contraction, which is all that is needed before other sources build up their supply to the necessary level.

## The structures that convert chemical energy into mechanical work

Before continuing our discussion of the skeletal muscle of vertebrates we shall first look at a number of simpler structures.

One of the simplest of such structures is the *bacterial flagellum* (plural: *flagella*) (Fig. 18.1). The flagella move with a spiral, whip-like motion with a frequency of

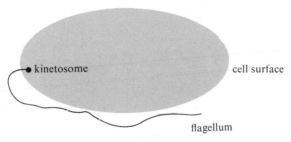

**Fig. 18.1** The bacterial flagellum

up to 60 times per second, and this movement imparts movement to the cell as a whole. Each flagellum consists of a triple-stranded helix of the protein *flagellin* (Fig. 18.2a). There is no external membrane.

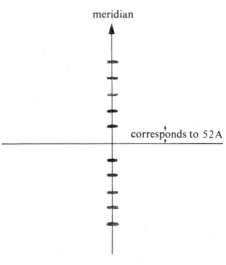

**Fig. 18.2** (a) The X-ray diffraction pattern of the flagellin fibre
**Fig. 18.2** (b) The flagellin helix

The gyratory movement of the flagellum is generated in a *basal granule*, one of which lies at the terminus of each flagellum, just below the surface of the cell. ATP is consumed during movement, and it is believed that (as is true also for the more complex structures that we shall consider) the chemical energy is first used to alter the conformation of some protein assembly in the granule. It appears that, in distinction to the main structural proteins of most of the other types of assembly, flagellin itself has no direct role in the energy transformation but simply brings the forces generated in the basal granule to bear against the medium surrounding the cell.

The *cilia and flagella of eucaryotes*, on the other hand, have made a considerable advance in complexity, and here the structural proteins also take some part in the energy transformation. These structures are widely distributed in the animal kingdom. The characteristic '9 + 2' pattern of fibres (Fig. 18.3) is seen in

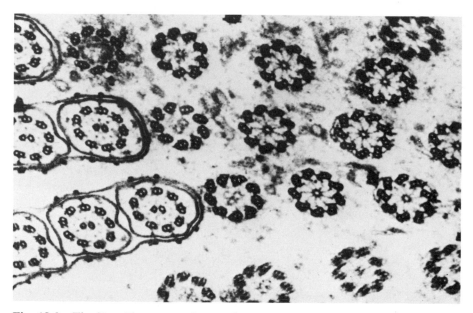

**Fig. 18.3** The '9 + 2' structure (see text)

the transverse section of cilia and flagella from cells ranging from the protozoa to mammalian sperm. They are also much larger than bacterial flagella and they have an external membrane that fuses with that of the cell.

The longitudinal section also shows great differences in comparison to bacterial flagella (Fig. 18.4). The basal granule is replaced by a cylindrical structure called the *kinetosome* which is partly made up of the extension of the nine outer fibres into the cytoplasm. The kinetosome appears to collect ATP (and possibly, in some cells, receives primitive versions of nervous impulses) and transmit it up into the body of the flagellum. The structure moves when the outer fibres on one side contract relative to those on the other side. This will obviously have the effect of bending the flagellum toward the side on which contraction has taken place. This process, like the one that promotes the movement of bacterial flagella, is mediated by ATP.

*Vertebrate muscle* is of two main types: *smooth*, or *involuntary* muscle of such anatomical features as the uterus, the contractions of which are not under conscious control, and the *striated* or *voluntary* muscle that moves the skeleton in obedience to the wishes of the animal. Heart muscle is morphologically intermediate between the two types, and it is interesting that some individuals can control their pulse rate by a conscious act of will.

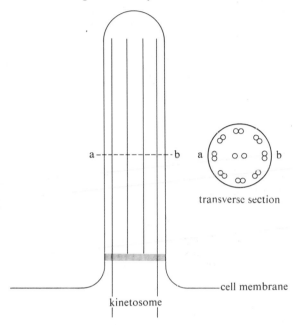

**Fig. 18.4**  A cilium seen in longitudinal and transverse section

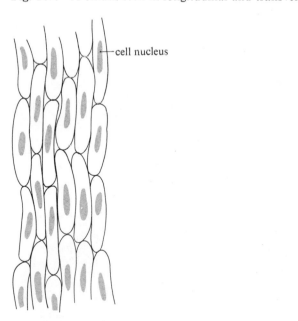

**Fig. 18.5**  Smooth muscle

The structure and behaviour of smooth muscle is less well understood than that of the striated form, although in its chemistry smooth muscle is believed to be very similar to striated muscle. Smooth muscles consist of an aggregate of roughly ovoid cells (Fig. 18.5). Contractions occur in waves and are the result of the shortening of the length of each cell.

Striated muscle, in contrast, has been intensively studied because from the moment that one first examines it in the light microscope its structure offers great promise of a full explanation of the contractile mechanism.

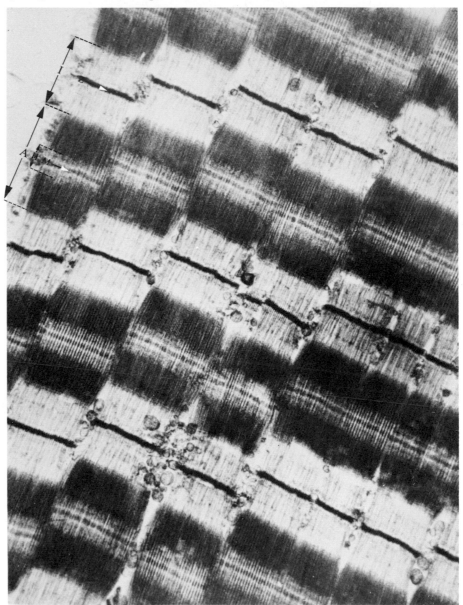

**Fig. 18.6** (a)   Electron micrograph of a longitudinal section of skeletal muscle. I = I band; A = A band; Z = Z line; M = M line; H = H zone (see text). Photograph by H. E. Huxley

Having already mentioned the source of the chemical energy, we shall first discuss the results of investigations of striated muscle from the point of view of morphology, using the results of light microscopy, electron microscopy and X-ray diffraction. We shall then describe the structural chemistry of the proteins that make up so much of the muscle fibre. Finally we shall attempt to reconcile the morphological and chemical approaches, and say something of the initiation of contraction by the nerve impulse.

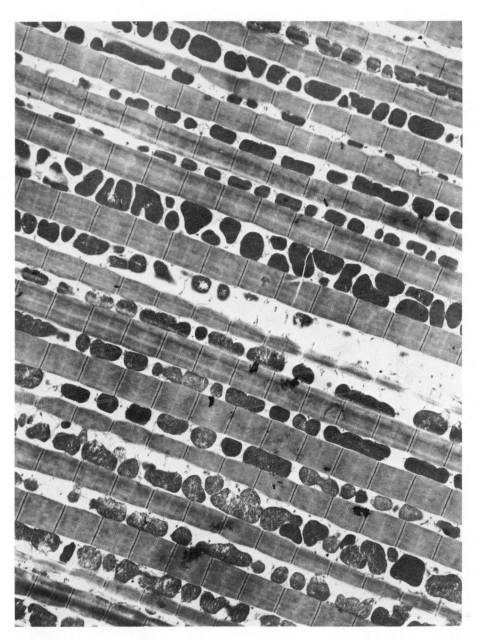

**Fig. 18.6** (b) Longitudinal section of insect flight muscle. Photograph by H. E. Huxley

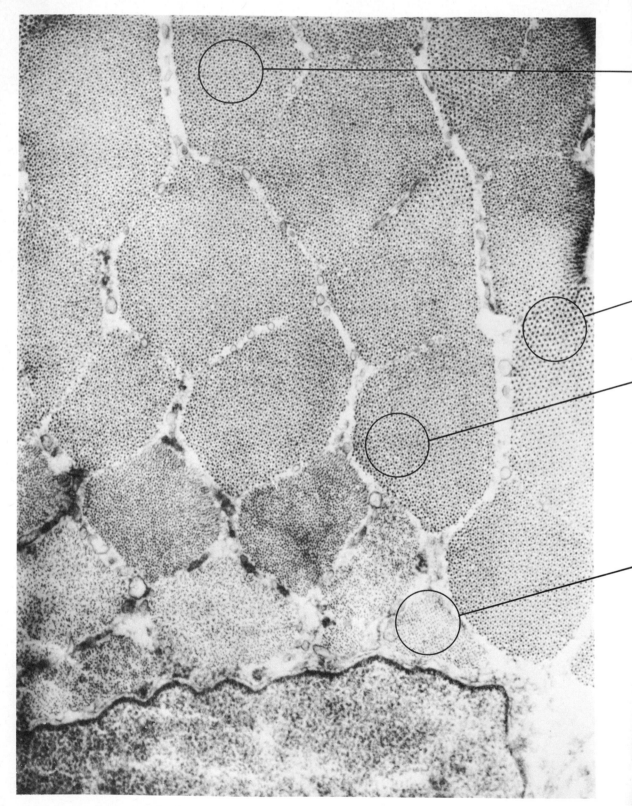

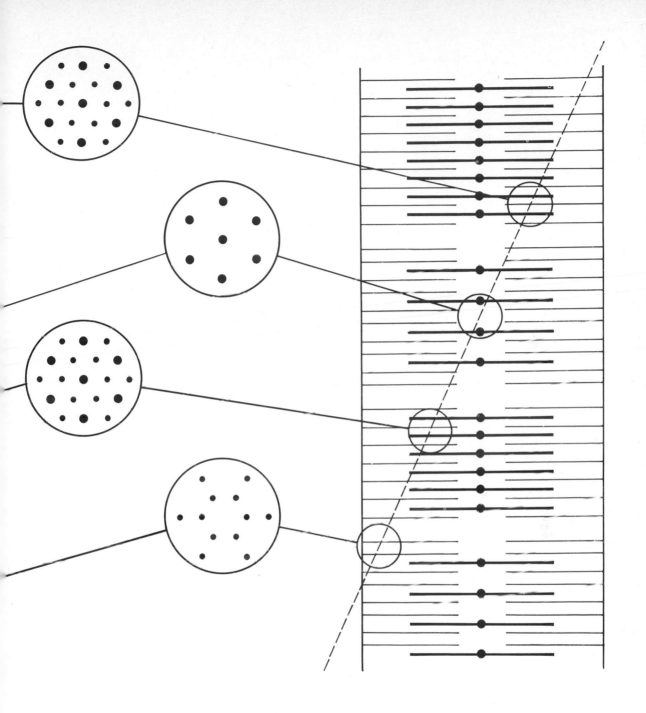

**Fig. 18.7** (a)   A transverse section of vertebrate muscle. The angle of section was in fact slightly less than 90° to the long axis of the myofibril. The consequences of this are illustrated by the drawings above. Photograph by H. E. Huxley

Vertebrate striated muscle is composed of bundles of fibres – each fibre being bounded by a plasma membrane (called, in muscle, the *sarcolemma*) and constituting a single giant cell with many nuclei and mitochondria. The structure that gives striated muscle its characteristic appearance and properties is a series of parallel bundles of contractile fibres (*myofibrils*) (Fig. 18.6a). These have, seen from the side, a striped appearance even in the light microscope (hence the term 'striated'). This appearance is found, on increased magnification in the electron microscope, to be due to a complex banding of light and dark areas (Fig. 18.6a, note that closely similar structures exist in insect flight muscle, Fig. 18.6b). The limits of the repeating unit (called the *sarcomere*) are defined by two of the thin, dark zones (called Z lines). Notable features within the sarcomere are the *A band* (so called because the physical properties of this band are *anisotropic* – that is they vary with the direction of measurement); the *M band*, a dark line in the middle of the lighter *H zone* comprising the boundary zones of adjacent sarcomeres, and the *I band* (its properties are *isotropic*, that is they do not vary with the direction of measurement). (While we are defining the special terminology of the muscle cell, we may as well mention that the terms 'endoplasmic reticulum' and 'cytoplasm' (see Ch. 10) are replaced by *sarcoplasmic reticulum* and *sarcoplasm* respectively.)

Some clue about the nature of the dark and light zones is obtained from electron micrographs of *transverse* sections of muscle fibres (Fig. 18.7a). (These are particularly useful if the section is taken, not at 90° to the axis of the myofibril, but at an angle of one or two degrees less than 90° – so that we obtain a section that gradually progresses through the different levels of the sarcomere and permits us in particular to examine the A bands at different levels in the same field of view.) We find three principal types of pattern. Most commonly there is a hexagonal array of thick objects each of which is surrounded by six thinner ones (compare the insect structure in Fig. 18.8). These objects are called *myofilaments*. It is possible to pick out regions in which there are only the thin myofilaments and others in which there are only the thick myofilaments.

The diagram in Fig. 18.9 presents the generally accepted interpretation of these findings. The A bands are now seen as consisting of interpenetrating myo-

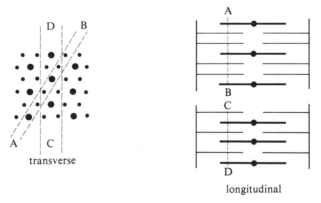

transverse

longitudinal

**Fig. 18.7** (b)　The reason for the two types of longitudinal view of the sarcomere (see text)

filaments of both the thick and thin type. The areas containing only one of the types of myofilament appear lighter in the microscope. (The crossbridges shown between the thick and thin filaments are discussed below.) The H zone has only thick myofilaments, and the I band has only the thin myofilaments, which grow in both directions out of the Z line. That this is a good model for the structure is indicated by the fact that one would predict two types of transverse view, depending on whether the cut was made parallel to the lines marked A in Fig. 18.7b or parallel to those marked B, and both of these are found. (Compare the sarcomere shown in the central portion of the lower part of Fig. 18.11 with the ones partly visible at the left and right of the picture.)

The function of these structures became apparent when electron micrographs were taken of muscle that was either contracted or stretched and these were compared with those of resting muscle. As the muscle expanded and contracted, so

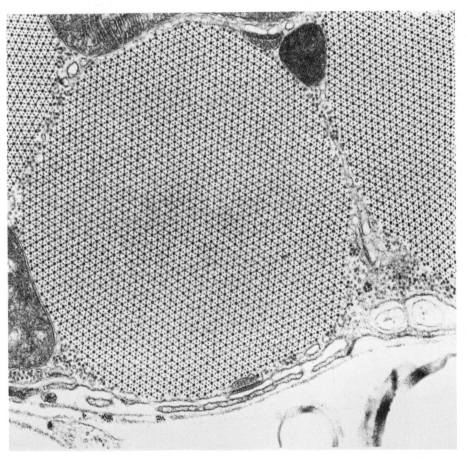

**Fig. 18.8** Transverse section of insect flight muscle. Photograph by Dr M. Cullen, Department of Zoology, Oxford

did the length of the sarcomere. The A band remained of the same size, but the I band varied in length (Fig. 18.10). The conclusion is irresistible that the thin myofilaments shorten the sarcomere by sliding between the thick myofilaments. This effect is clearly visible if we compare Fig. 18.10 with Fig. 18.11; in addition these pictures show the *crossbridges* formed by outgrowths from the thick myofilaments. These crossbridges are thought to provide the impetus for the sliding process by being made and broken progressively with a series of points along the thin filaments, the progression of points being in the desired direction for sliding.

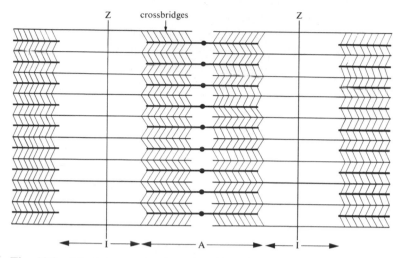

**Fig. 18.9**   The interpenetration of actin and myosin filaments

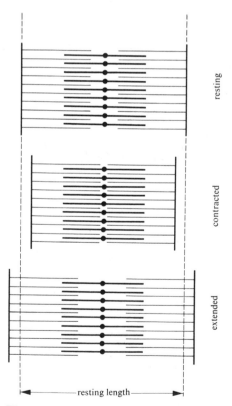

**Fig. 18.10**   The changes in the length of the sarcomere during contraction

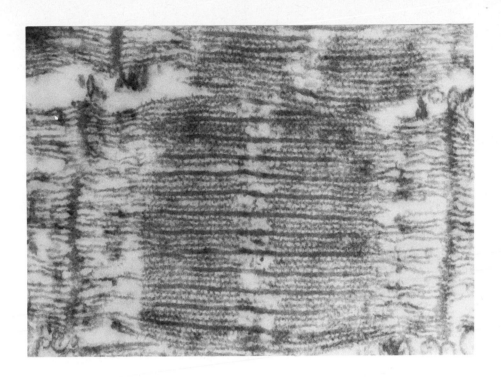

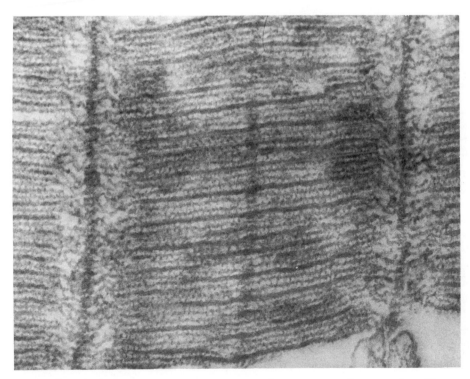

**Fig. 18.11** An enlarged view of the sarcomere in two stages of contraction. Photographs by H. E. Huxley

These crossbridges are also visible on X-ray diffraction patterns of muscle. (Striated muscle, as a periodic structure, is well suited to X-ray analysis.) A diffraction diagram is shown in Fig. 18.12 in which the X-ray beam was perpendicular to the fibre axis. This diagram is interpreted in terms of a pattern of

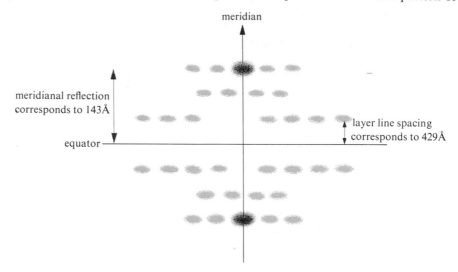

**Fig. 18.12**  The X-ray diffraction pattern of muscle (idealized)

disposition of the crossbridges known as the '6/2' helix (Fig. 18.13). This helix has just the characteristics necessary to enable crossbridges to be formed from each of the thick filaments to the six adjacent thin filaments.

### The proteins of the myofibril

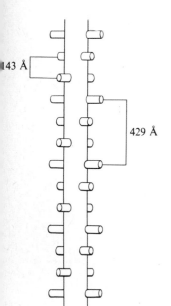

This is, for the moment, as far as the purely visual examination of muscle can take us. In order to approach a little more closely to the central event of contraction – the first appearance of the $\Delta G$ of hydrolysis of ATP in a non-chemical form – we must look at the chemistry of the proteins of which the myofilaments are composed. It turns out that the thick myofilaments are made up of a protein known as *myosin* and the thin myofilaments of a protein known as *actin*. It is possible to extract these proteins separately from muscle with salt solutions of appropriate concentrations, and to characterize them. (At first they were extracted as a complex, which was called *actomyosin* until the existence of the two separate components was recognized.)

The fundamental subunit of actin is a protein molecule of about 220 amino acids. These globular subunits (*G-actin*) aggregate into a fibrous quaternary structure (*F-actin*) under the conditions normally found in the cell. F-actin probably consists of two chains of end-to-end aggregates of G-actin, wound about one another in a double helix, each such double helix (in association with other proteins, see below) constituting a single thin myofilament (Fig. 18.14).

◄ **Fig. 18.13**  The '6/2' helix (see text)

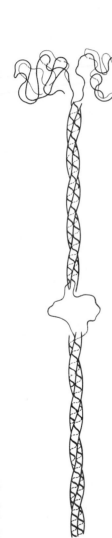

Myosin, the protein of the thick myofilament, is much more complex. Its fundamental structural unit is again a protein double-helix, but this time not one made up of end-to-end aggregates of a small globular molecule. Each of the two intertwined chains of the main stem of the myosin molecule is an unbroken length of polypeptide chain of about 2000 amino acids (the longest polypeptide sequence known), largely in the α-helical configuration. The stem is thus a double-coiled coil (Fig. 18.15). One region of the stem near the middle is highly susceptible to proteolytic agents, and this region presumably has a more open structure. The two chains continue beyond the main stem and, after passing through another loose, open region, assume a globular configuration. There are probably other, shorter protein chains associated with the globular regions. Myosin is unusual among structural proteins in that it also possesses an enzymic activity (which, as can be shown by testing the various fragments that are produced by proteolytic agents, is associated with the globular regions of the head). It is of the greatest significance that the enzymic activity is the catalysis of the hydrolysis of ATP – the very compound that we know to be central to the contractile process. (This ATPase activity is very low in the absence of actin, a fact which, as we shall see, is of importance in understanding how muscular contraction is controlled by the nervous system.)

Electron micrographs of dried solutions of purified myosin show clusters of molecules the form of which has been found to be characteristic also of the arrangement in living matter.

There are several other types of protein in the myofibril. Notable among these are two which associate with the thin myofilament. These are the globular protein *troponin* and the fibrous protein *tropomyosin*. The two types of protein alternate in the groove between the chains of the double-helical G-actin structure (Fig. 18.16). From this position, these two proteins are able to control whether or not the enzymically active part of the myosin molecule is allowed to come into contact with, and thus be stimulated by, actin. However, the inter-relationships between actin, myosin and ATP dominate the behaviour of muscle, and since all three can be purified, the way is obviously clear for test-tube studies on simplified systems aimed at collecting information that can be used to interpret the behaviour of the more complex, living tissue.

The simplest artificial system, which has the very minimum of the ordered nature of the original, is a dilute solution of the actin and myosin. When ATP is added to this solution the viscosity falls at once – a fall in viscosity normally means a fall in molecular weight of the solute, and in the present case it is interpreted to mean that the actomyosin complex, of which we spoke before, has dissociated into actin and myosin. This conclusion is confirmed by other physico-chemical methods of estimating molecular weight. The viscosity slowly returns to the original value and this increase is accompanied by the hydrolysis of the ATP, catalysed by the myosin ATPase.

Remember the conclusions that we drew from the ultra-structure of the myofibril, namely that contraction depends on the sliding of the thin myofilaments

◄ **Fig. 18.15**   The myosin molecule

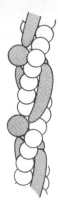

◄**Fig. 18.16** The association between F-actin, troponin and tropomyosin in the absence of $Ca^{2+}$

(actin) over the thick myofilaments (myosin) mediated by crossbridges between the two. The finding that ATP affects the association between actin and myosin is therefore very significant. Matters are made even more clear if the soluble acto-myosin preparation is formed into artificial, solid fibres by squirting it into a solution in which actomyosin is insoluble: these fibres contract when dipped into a solution of ATP. Once again there is physico-chemical evidence that ATP momentarily dissociates the actomyosin complex which then re-forms during the contraction stage while the ATP is hydrolysed.

The ATPase activity that we saw in free solution thus appears to be only a part of the interaction between ATP and the protein. It is simply the only remaining aspect of it that can still be observed when we have disrupted the structure of the myofibril in extracting the protein. The mechanical changes that the interaction exists to bring about are, of course, no longer possible.

### Actin–myosin relationships in the myofibril

The impression that we obtain from these and other results is as follows. The lower parts of the main stems of the myosin molecules aggregate side-to-side to build up the body of the thick myofilament. The open, flexible parts of the stem provide flexible joints in the remainder of the straight part of the molecule, which appears to act as a stalk carrying the globular (and enzymically active) head of the molecule. This head, and its stalk, form the crossbridges that we were able to see in the electron micrographs and X-ray patterns. The stalk is long enough to allow the globular head to make contact with the actin of the thin myofilament.

Before we seek to give a description of the molecular interactions that occur between ATP and the crossbridges and which promote contraction, we need to consider briefly the way in which the nervous system controls the process as a whole.

### The initiation of contraction by the nervous impulse

We said above that the troponin–tropomyosin complex was able to prevent the ATPase-containing head of the myosin crossbridge from binding to the actin. We also mentioned that the ATPase activity was very low if actin and myosin were not present together. If the catalysed breakdown of ATP is an integral part of the contractile mechanism, any agent that temporarily prevents troponin and tropomyosin from interposing between actin and myosin will, in permitting expression of the ATPase activity, allow contraction to commence.

Such an agent exists, and it is well suited to act as a chemical message from the nervous system. The agent is $Ca^{2+}$. It has been shown that the incoming nervous impulse releases $Ca^{2+}$, which was previously trapped in the sarcoplasmic reticulum. Measurement shows that the effect of this is that the concentration of $Ca^{2+}$ quite suddenly increases about tenfold. Now troponin has the property of binding $Ca^{2+}$. At the high concentration of $Ca^{2+}$ that follows the arrival of the nervous impulse, a sufficient quantity of $Ca^{2+}$ binds to troponin to initiate a major conformational change in the molecule. It has been deduced from X-ray diffraction studies and electron micrography that the effect of this change is to swing the adjacent tropomyosin molecule out of the grove in the F-actin (Fig. 18.17). Since tropomyosin is a fibrous protein (that is, it is much longer than it is broad) one molecule of it is able to overlay several of the globular G-actin sub-units that constitute the F-actin chain (Fig. 18.16). The number of G-actin units that are exposed when the tropomyosin swings out of position is seven, so that several myosin heads can bind to the actin for every tropomyosin molecule that was affected. Thus $Ca^{2+}$ enhances the ATPase activity of the myosin by permitting the interaction with actin that is necessary to stimulate it. In addition to this, the stimulation is further amplified by the fact that each molecule of troponin that is influenced by $Ca^{2+}$ allows several actin–myosin contacts.

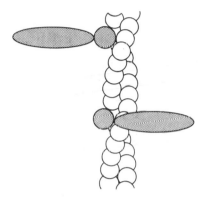

**Fig. 18.17** The association between F-actin, troponin and tropomyosin in the presence of $Ca^{2+}$. For clarity, the drawing exaggerates the outward movement of the tropomyosin molecules

When the nervous stimulation ceases, $Ca^{2+}$ is rapidly sequestered once again in the sarcoplasmic reticulum and the concentration of the free ion rapidly drops. The $Ca^{2+}$ falls away from the troponin, the G-actin units are once more masked and the actin–myosin interaction ceases to be possible.

*A model of the events that accompany contraction*

We can now picture the main events in muscular contraction, which are as follows.

1. In the resting state, the crossbridges between actin and myosin are not formed, but the myosin heads are already poised (a legacy of the last ATP molecule that was broken down before the previous contraction phase ceased – see below) in a strained configuration.
2. The nervous impulse permits the crossbridge to form and the ATPase to be activated, as described above.
3. A molecule of ATP is then able to bind to the actin–myosin complex. (Note that ATP does not bind to the myosin before the complex with actin is formed.) In the act of being hydrolysed, it transfers its available free energy to the myosin. This induces an otherwise impossible conformational change in the myosin which breaks the crossbridge and places its head in a position in

which it can (if permitted by the tropomyosin) attach itself to an actin monomer that it could not previously have reached.

4. This induces a mechanical strain, which is relieved by the thin filament's being pulled past the thick one until the new site of attachment to the thin filament reaches the same position relative to the myosin molecule that was occupied by the previous site.

5. A fresh ATP molecule breaks the bridges and 'energizes' the myosin, and the rest of the cycle – re-attachment and relief of the consequent mechanical strain by sliding – is repeated.

6. Many repetitions of this process would be required for a complete muscular movement.

7. When the nervous impulses cease, the myosin heads, although 'energized', can no longer attach themselves to the actin because the tropomyosin is once again in the way. The muscle fibre therefore *relaxes* and can be pulled back to its original, resting length. This is done either by some external load on the body of the animal or by the action of opposed muscle fibres. The state of *rigor* can be understood as resulting from there being sufficient $Ca^{2+}$ to allow crossbridges but insufficient ATP to break them again.

We thus have a reasonably complete picture, in molecular terms, of the events of muscular contraction. We can see that the picture is one of a beautifully controlled set of interactions of the type that we saw in Chs 3, 4 and 5. There are of course some rather ill-defined features – for example, it would be highly desirable to know more of the precise way in which the ATP is able to strain the conformation of the proteins at the point of crossbridging. This and other problems associated with the contractile process are under intensive study in many laboratories.

## ACTIVE TRANSPORT OF SOLUTES

We saw in Ch. 2 (eqn 2.14, p. 17) that if we remove a solute from a region in which it is at a concentration $C_1$ and bring it into a region in which it is at a concentration $C_2$, the free energy change per mole of solute is given by the relationship:

$$\Delta G = RT \log_e \frac{C_2}{C_1}. \qquad (18.1)$$

This shows that if $C_2$ is less than $C_1$, $\Delta G$ is negative and the solute will therefore tend to migrate spontaneously – as common experience teaches that it would.

If on the other hand $C_2$ is greater than $C_1$, $\Delta G$ is positive and the solute will not migrate 'uphill' unless driven – again a commonsense conclusion.

The passage of many substances across biological membranes is, as we saw in Ch. 8, often of the passive, downhill type. Sometimes, however, cells and cell organelles concentrate solutes within themselves by abstracting them from a more dilute external medium, and energy must then be supplied by coupling. Occasionally, also, they deplete their internal stock of a certain solute by expelling it into a *more* concentrated external medium – this operation, which is used, among other things, for removing unwanted solutes that diffuse in, is equally energy-requiring.

We saw on p. 204 that the energy requirements of such processes can sometimes be met by coupling to the downhill transport of a solute in the opposite direction. The positive and negative $\Delta G$ terms which derive from the equation on the previous page would then be made to balance. We give examples of such *exchange–diffusion* processes on p. 353.

However, there is a large class of energy-requiring transport processes that are coupled, not to the downhill flow of some other solute, but to the hydrolysis of the terminal phosphate group of ATP. These *active-transport* processes are involved in many of the most important activities of living matter, in particular those that depend on the organization of different operations into separate compartments in the cell. In addition active-transport processes are responsible for the control of the balance between the various inorganic ions across membranes.

An imbalance between ions on either side of a membrane may be expressed in terms of molar concentrations, or, in the special case of $H^+$, in terms of the equivalent function pH (see p. 11). In such cases free energy calculations follow also eqn (18.1). (It is sometimes more convenient to express the imbalance as an electrical potential, since that is often what is actually measured. The establishment of such potentials requires coupling to a free-energy source and free-energy calculations can be made from the relationship $\Delta G° = z\mathscr{F}V$ ($z$ is the number of charges of the particular ion; $\mathscr{F}$ is the faraday, 96 500 coulombs per gram equivalent, and $V$ is the potential difference). The potential difference met with across biological membranes is most often of the order of 100 mV. This, for a monovalent ion, is equivalent to a $\Delta G°'$ of 2300 cal/mole. Since this equivalence between $V$ and $\Delta G$ exists, we shall continue to use $\Delta G$ in all cases.)

## The characteristics of active-transport systems

Active-transport systems resemble in many respects the other mediated transport processes that we studied in Ch. 8. In particular they show the phenomenon of *saturation*, a hyperbolic relationship between rate of transport and concentration of material available for transport (refer to Fig. 8.25 and pp. 201 ff.), and *specificity* with regard to the substances that can or cannot be transported (p. 202). Their distinguishing characteristic is the obligatory role of ATP, which may be demonstrated by their sensitivity to inhibitors that suppress the formation of ATP. (When ATP is absent, some but not all of the active transport systems can act as passive transport systems for their specific substrates – that is they mediate their transfer *down* a concentration gradient.)

The activities of living cells require that many classes of compound be transferred against osmotic gradients. These include inorganic ions and neutral molecules of low molecular weight.

## The transport of ions – sodium pumps

An example of the active transport of an inorganic ion that is so familiar as to be easily overlooked is the secretion of $H^+$ into the peptic region of the gut. This is a true active-transport process because the protons have to be abstracted from the tissue fluid at about pH 7 ([$H^+$] approximately equal to $10^{-7}$M) and discharged to a final pH of about 2 ([$H^+$] approximately $10^{-2}$ M). You can use equation (18.1) to show for yourself that the energy required for transfer of a single proton is of the order of that supplied by the hydrolysis of a molecule of ATP.

Many other ions are actively transported (including $Ca^{2+}$, when it is returned to the sarcoplasmic reticulum after muscular contraction ceases), but probably the most significant processes, in animal cells at least, are those that involve the transport of $Na^+$ and $K^+$. On the whole, cells keep a concentration of $K^+$ which is about ten times higher than that of the surrounding liquid ($K^+$ is a co-factor in several important biochemical reactions). This high concentration is maintained by active transport inward in the face of passive leakage out again. It is desirable that this inward transport be balanced by the outward transport of another positively charged species if the osmotic pressure difference across the membrane is not to be intolerable. (Note the distinction between this *preference* for a counter flow of some substance and the *requirement* for counter flow that is seen in exchange diffusion: in this case the counter flow is not to supply energy, since that is supplied by ATP, but for reasons less closely connected with $\Delta G$.)

The outgoing cation is $Na^+$. The transport of this is also an active process, because it takes place in the face of a concentration of $Na^+$ outside the membrane that is approximately ten times that inside the membrane.

It is significant for the study of this *$Na^+$–$K^+$ coupled pump* that a ATPase has been found in the membrane protein fraction that is stimulated only when $Na^+$ and $K^+$ are present together. Experimental study of this enzyme is proceeding, and has led to the interim conclusion that it is first phosphorylated by the ATP and that in this activated configuration it then binds $Na^+$. (No other cation can replace $Na^+$ in this binding.) The approach of $K^+$ causes the enzyme–phosphate–$Na^+$ complex to break up. It is presumed that the lipid–protein complex has some of those directional properties that we inferred that the electron-transport assembly might have. We presume further that the change in conformation that occurs when the $Na^+$ is discharged from the complex is able to force the $Na^+$ *outside* the membrane even though it arrived from inside and force the $K^+$

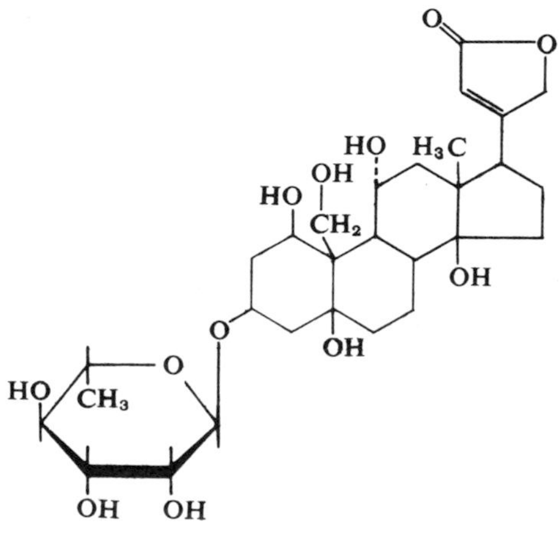

Fig. 18.18   Ouabain

inside the membrane. The glycosidic compound *ouabain* (Fig. 18.18) which is a specific inhibitor of $Na^+$ transport, is believed to act by competing with $K^+$ for access to the enzyme–phosphate–$Na^+$ complex.

In distinction to the coupled $Na^+$–$K^+$ pump, a pump exists that handles $Na^+$ alone and extrudes it from the cell. It is able to act without inward passage of another cation provided that it does not transport too much $Na^+$. This particular activity is one source of the electrical charge differences, of the order of 100 mV, that are essential for the function of nerve and muscle cells.

## The transport of neutral substances

Many neutral substances are actively transported. Water itself is one of these. The contractile vacuoles of, for example, the fresh-water protozoa are continually working to expel water from the cell in order to save themselves from osmotic rupture. There is evidence that the contractile vacuoles are filled from the cytoplasm of the cell by an ATP-requiring, active-transport process.

Sugars, amino acids and some vitamins are also concentrated as a result of active-transport processes. Once again, there is a specific carrier system for each group of compounds, and this binds its substrate by the usual complex of forces. This ability to concentrate substances is vital to many organisms that have to obtain needed compounds from a very dilute solution. Some bacteria, for example, rely on such systems to concentrate sugars from their environment (cf. p. 203). Sometimes there are distinct carrier systems for different sub-groups of compounds (e.g. acidic, basic and various types of neutral amino acids). However it appears that the inward transport is, in many such cases, not driven directly by an ATP-induced conformational change of the carrier. What is believed to happen is that the carrier systems for many of these classes of compound are also *passive* carriers for $Na^+$. The carriers shuttle back and forth between the outer and inner surfaces, being driven to do so by the fall of $Na^+$ down the *concentration gradient* which, as we saw a moment ago, is produced by an ATP-requiring, active-transport extrusion. However the carriers also have binding sites for the other substances we mentioned, and these are able to take a ride, so to speak, into the region in which they are at a higher concentration on the back of the downward fall of $Na^+$. The dependence on ATP of this type of transport of neutral substances is thus indirect, but nonetheless real.

Some neutral substances are trapped inside the membrane by being transformed into a substance that is unable to diffuse out. For example glucose is phosphorylated to glucose-6-phosphate which cannot pass the membrane (p. 259) Such processes often involve the use of ATP to bring about a phosphorylation.* However they are not true active-transport processes, because they do not involve the transport of a substance into a region in which it is *itself* at a higher concentration.

* A system that appears to be of widespread occurrence in bacteria uses, as a phosphorylating agent for trapping sugars, a phosphorylated *protein* called HPr. HPr itself acquires its phosphate group not from ATP but, remarkably enough, from another high-energy compound, phospho-*enol*pyruvate (pp. 26, 264); it then donates its phosphate group in a reaction catalysed by an enzyme that is specific to each sugar that is being phosphorylated.

The *behaviour* of a number of the main active-transport systems is now quite well described, particularly in terms of the two characteristics that we have already mentioned, namely, their specificity and the relationship between the rate of transport and the concentration of substrate available for transport. *Explanations*, particularly of the all-important directionality of the behaviour of the carriers, must await more detailed knowledge of the structures of the individual components of the system. Some of these components are now becoming available for study in purified (and sometimes even crystallized) form, and we can await with some confidence a picture of their operations of the more detailed type found in the earlier chapters of this book. The work involved should prove worthwhile because the transport systems appear to hold the key to many of the important aspects of the organization of living matter at the level of complexity that lies between the individual macromolecule on the one hand and the complete cell on the other. We may expect that it is in this region that many of the major advances in our understanding of biology will occur over the next few years (cf. also Ch. 21).

# Chapter 19  Synthesis of carbohydrate

As we mentioned in Ch. 13 (p. 250), the pathways of carbohydrate synthesis in photosynthetic and in heterotrophic organisms have most of their reactions in common. The differences between the two types of organism lie primarily in their starting materials. Whereas photosynthetic organisms can make carbohydrate from carbon dioxide, carbohydrate synthesis in heterotrophs is mainly at the expense of lactate, of Krebs cycle intermediates, or of those compounds (principally the amino acids) that can give rise to pyruvate or Krebs cycle intermediates. In this chapter we shall first discuss the pathway by which, in the dark reaction of photosynthesis, carbon dioxide is converted to triose phosphate. Next we shall discuss the conversion of lactate in animals and of Krebs cycle intermediates in almost all organisms to triose phosphate. Lastly, we shall show how the triose phosphate is used, by pathways that also are common to almost all organisms, to make monosaccharides, oligosaccharides and polysaccharides.

## SYNTHESIS OF TRIOSE PHOSPHATE BY THE CALVIN CYCLE

As a result of the processes that we described in Ch. 13, photosynthetic organisms are able to provide themselves with ATP and $NADPH_2$, and we must now see how these compounds are used to fix carbon dioxide. The pathway that is employed for the synthesis of organic compounds at the expense of carbon dioxide is often called the Calvin cycle: it was worked out by Calvin and his collaborators in a painstaking series of experiments around 1954. The method that Calvin principally used was based on isotopic labelling (see p. 231). Photosynthetic algae were exposed to radioactive carbon dioxide for short periods and quickly killed, and cell extracts were then examined by chromatographic techniques (see Appendix 2) to see which compounds had become labelled most rapidly.

We can summarize the important results in this way. Carbon dioxide reacts with a pentose phosphate to give two molecules of triose. If we consider this reaction ($C_1 + C_5 \rightarrow 2C_3$) as multiplied by three, the yield is six molecules of triose. Of these six molecules, one represents a net gain, corresponding to three molecules of carbon dioxide, and it is used to synthesize carbohydrate. The remaining five molecules of triose are rearranged to form three molecules of pentose, and these are used again in the reaction with carbon dioxide. These rearrangements are very similar to those of the pentose phosphate pathway (Ch. 17).

The key reaction is that in which carbon dioxide is fixed. The compound with which carbon dioxide reacts is the diphosphate derivative of ribulose (see pp. 20 and 293), and the reaction yields two molecules of glycerate-3-phosphate.

$$
\begin{array}{c}
\text{CH}_2\text{O}\,\text{\textcircled{P}} \\
| \\
\text{CO} \\
| \\
\text{HCOH} \quad + \text{CO}_2 + \text{H}_2\text{O} \rightleftharpoons \\
| \\
\text{HCOH} \\
| \\
\text{CH}_2\text{O}\,\text{\textcircled{P}}
\end{array}
\qquad
\begin{array}{c}
\text{CH}_2\text{O}\,\text{\textcircled{P}} \\
| \\
\text{CHOH} \\
| \\
\text{COOH} \\
\\
\text{COOH} \\
| \\
\text{CHOH} \\
| \\
\text{CH}_2\text{O}\,\text{\textcircled{P}}
\end{array}
\qquad (19.1)
$$

ribulose-1,5-
diphosphate

2 × glyceric
acid-3-phosphate

$$\Delta G^{\circ\prime} = -8400 \text{ cal/mole}$$

This reaction was discovered by Calvin as a result of his finding that one of the compounds that became labelled most rapidly when radioactive carbon dioxide was incorporated into photosynthesizing algae was glycerate-3-phosphate. The enzyme involved is called carboxydismutase, which is present in high concentration in the chloroplast.

Glycerate-3-phosphate is now converted to glyceraldehyde-3-phosphate by a series of two reactions. In the first, glycerate-3-phosphate is phosphorylated at the expense of ATP in the presence of a kinase:

$$
\begin{array}{c}
\text{COOH} \\
| \\
\text{CHOH} \quad + \text{ATP} \rightleftharpoons \\
| \\
\text{CH}_2\text{O}\,\text{\textcircled{P}}
\end{array}
\qquad
\begin{array}{c}
\text{COO}\,\text{\textcircled{P}} \\
| \\
\text{CHOH} \quad + \text{ADP} \\
| \\
\text{CH}_2\text{O}\,\text{\textcircled{P}}
\end{array}
\qquad (19.2)
$$

glyceric
acid-3-
phosphate

glyceric
acid-1,3-
diphosphate

and the resulting glycerate-1,3-diphosphate is then reduced by $NADPH_2$ in the presence of triose phosphate dehydrogenase.

$$
\begin{array}{c}
\text{COO}\,\text{\textcircled{P}} \\
| \\
\text{CHOH} \quad + \text{NADPH}_2 \rightleftharpoons \\
| \\
\text{CH}_2\text{O}\,\text{\textcircled{P}}
\end{array}
\qquad
\begin{array}{c}
\text{CHO} \\
| \\
\text{CHOH} \quad + \text{NADP} + \text{P}_i \\
| \\
\text{CH}_2\text{O}\,\text{\textcircled{P}}
\end{array}
\qquad (19.3)
$$

glyceric
acid-1,3-
diphosphate

glyceralde-
hyde-3-
phosphate

In essence, these last two reactions are the reverse of reactions (14.6) and (14.7) of the Embden–Meyerhof pathway (p. 263). That pathway is involved in the breakdown of carbohydrate; photosynthesis is devoted to the formation of carbohydrate. Reaction (19.2) of the Calvin cycle uses ATP; reaction (14.7) of the Embden–Meyerhof pathway synthesizes ATP. Reaction (19.3) of the Calvin cycle is a reduction; reaction (14.6) of the Embden–Meyerhof pathway is a corresponding oxidation.

We can therefore see how, in reactions (19.2) and (19.3) of the Calvin cycle, the ATP and $NADPH_2$ formed in the light reaction of photosynthesis are used in the dark reaction. (We shall discover soon (p. 328) that there is a further requirement for ATP in the rearrangement leading to ribulose-1,5-diphosphate.) And although the reactions leading from glycerate-3-phosphate to glyceraldehyde-3-phosphate are, at first sight, precisely the reverse of degradative reactions in glycolysis, there is one important difference. Reaction (14.6) of the Embden–Meyerhof pathway, like almost all degradative oxidations, uses NAD, whereas reaction (19.3) of the Calvin cycle, like almost all synthetic reductions, uses $NADPH_2$. Appropriately enough, the triose phosphate dehydrogenase of the chloroplast is specific for NADP.

These first three reactions of the Calvin cycle are in fact responsible for the *net synthesis* of triose phosphate from carbon dioxide. We can see this result most clearly by starting (as we did above, see p. 321) with three molecules of ribulose diphosphate. If we multiply reaction (19.1) by three:

$$3 \times \text{ribulose diphosphate} + 3CO_2 + 3H_2O \rightleftharpoons 6 \times \text{glyceric acid-3-phosphate}$$

and reactions (19.2) and (19.3) by six:

$$6 \times \text{glyceric acid-3-phosphate} + 6ATP \rightleftharpoons$$
$$6 \times \text{glyceric acid-1,3-diphosphate} + 6ADP$$

$$6 \times \text{glyceric acid-1,3-diphosphate} + 6NADPH_2 \rightleftharpoons$$
$$6 \times \text{glyceraldehyde-3-phosphate} + 6NADP + 6P_i$$

and add them up, we get:

$$3 \times \text{ribulose diphosphate} + 3CO_2 + 3H_2O + 6ATP + 6NADPH_2 \rightleftharpoons$$
$$6 \times \text{glyceraldehyde-3-phosphate} + 6ADP + 6NADP + 6P_i. \quad (19.4)$$

Of the six molecules of glyceraldehyde-3-phosphate (eighteen carbon atoms) that result, five molecules (fifteen carbon atoms) represent a rearrangement of the original three molecules of ribulose diphosphate. One molecule of glyceraldehyde-3-phosphate, however, has been synthesized from the three molecules of carbon dioxide entering in reaction (19.1), and it is this synthesis that is the characteristic achievement of photosynthesis.

We shall pursue later (pp. 332 ff.) the fate of the newly synthesized molecule of glyceraldehyde-3-phosphate that represents a net gain of organic carbon to the system. Meanwhile we must consider how the remaining five molecules of glyceraldehyde-3-phosphate are rearranged to give three molecules of ribulose-1,5-diphosphate capable of taking part again in reaction (19.1).

## Rearrangement of triose phosphate

The rearrangement has a good deal in common with that by which pentose phosphate can be converted into hexose phosphate (see pp. 296 ff.), and the transketolase involved in that pathway, together with its coenzyme thiamine pyrophosphate, is used here as well. In addition two enzymes involved in the

Embden–Meyerhof pathway are required. The first of these is triose phosphate isomerase, which catalyses the equilibrium between glyceraldehyde-3-phosphate and dihydroxyacetone phosphate (p. 261).

$$
\begin{array}{ccc}
\text{CHO} & & \text{CH}_2\text{OH} \\
| & & | \\
\text{CHOH} & \rightleftharpoons & \text{CO} \\
| & & | \\
\text{CH}_2\text{O}\,\textcircled{P} & & \text{CH}_2\text{O}\,\textcircled{P} \\
\text{glyceralde-} & & \text{dihydroxy-} \\
\text{hyde-3-} & & \text{acetone} \\
\text{phosphate} & & \text{phosphate}
\end{array}
\tag{19.5}
$$

The second is aldolase. We saw previously (p. 261) that aldolase can catalyse the breakdown of fructose-1,6-diphosphate to two molecules of triose phosphate, and we pointed out that in fact the change in free energy of this reaction greatly favours the condensation of the two triose phosphates to yield fructose-1,6-diphosphate. More generally, the enzyme catalyses the aldol condensation between dihydroxyacetone phosphate and aldehydes, especially aldol sugars.

We can summarize the rearrangement of five molecules of triose phosphate to yield three molecules of pentose phosphate in the following way (compare the scheme outlined on p. 296).

$$
\mathbf{C_3} + \mathbf{C_3} \xrightleftharpoons{\text{aldolase}} C_6
\tag{19.6}
$$

$$
C_6 + \mathbf{C_3} \xrightleftharpoons{\text{transketolase}} C_4 + C_5
\tag{19.7}
$$

$$
\mathbf{C_3} + C_4 \xrightleftharpoons{\text{aldolase}} C_7
\tag{19.8}
$$

$$
C_7 + \mathbf{C_3} \xrightleftharpoons{\text{transketolase}} C_5 + C_5
\tag{19.9}
$$

(where **bold type** signifies the original triose phosphate molecules).

In the first aldolase reaction, two molecules of triose phosphate (it is important to bear in mind that, as the enzyme triose phosphate isomerase is present, the two isomers are readily exchanged) condense together.

$$
\begin{array}{ccccc}
\text{CH}_2\text{OH} & & \text{CHO} & & \text{CH}_2\text{O}\,\textcircled{P} \\
| & & | & & | \\
\text{CO} & + & \text{CO} & \rightleftharpoons & \text{CO} \\
| & & | & & | \\
\text{CH}_2\text{O}\,\textcircled{P} & & \text{CH}_2\text{O}\,\textcircled{P} & & \text{HOCH} \\
& & & & | \\
& & & & \text{HCOH} \\
& & & & | \\
& & & & \text{HCOH} \\
& & & & | \\
& & & & \text{CH}_2\text{O}\,\textcircled{P} \\
\text{dihydroxy-} & & \text{glyceral-} & & \text{fructose-1,6-} \\
\text{acetone} & & \text{dehyde-} & & \text{diphosphate} \\
\text{phosphate} & & \text{3-phosphate} & &
\end{array}
\tag{19.6}
$$

The fructose-1,6-diphosphate thus formed is now hydrolysed by the enzyme fructose diphosphatase. (We saw on p. 260 that the glycolytic reaction by which fructose-1,6-diphosphate is *synthesized* from fructose-6-phosphate involves the breakdown of ATP and has a large negative $\Delta G^{\circ\prime}$ which favours synthesis of the diphosphate, and we remarked then that the breakdown of fructose-1,6-

diphosphate required a different enzyme. This fact is of importance in controlling the two pathways, as we shall see in Ch. 29.)

$$
\begin{array}{c}
\text{CH}_2\text{O}\,\text{P} \\
| \\
\text{CO} \\
| \\
\text{HOCH} \\
| \\
\text{HCOH} \\
| \\
\text{HCOH} \\
| \\
\text{CH}_2\text{O}\,\text{P}
\end{array}
\quad + \text{H}_2\text{O} \rightleftharpoons
\begin{array}{c}
\text{CH}_2\text{OH} \\
| \\
\text{CO} \\
| \\
\text{HOCH} \\
| \\
\text{HCOH} \\
| \\
\text{HCOH} \\
| \\
\text{CH}_2\text{O}\,\text{P}
\end{array}
\quad + \text{P}_i \qquad (19.6a)
$$

fructose-1,6-diphosphate     fructose-6-phosphate

Fructose-6-phosphate now donates a $\text{CH}_2\text{OH}.\text{CO}-$ group to another molecule of glyceraldehyde-3-phosphate, in a reaction catalysed by transketolase.

$$
\begin{array}{c}
\text{CH}_2\text{OH} \\
| \\
\text{CO} \\
| \\
\text{HOCH} \\
| \\
\text{HCOH} \\
| \\
\text{HCOH} \\
| \\
\text{CH}_2\text{O}\,\text{P}
\end{array}
\; + \;
\begin{array}{c}
\text{CHO} \\
| \\
\text{CHOH} \\
| \\
\text{CH}_2\text{O}\,\text{P}
\end{array}
\; \rightleftharpoons \;
\begin{array}{c}
\text{CHO} \\
| \\
\text{HCOH} \\
| \\
\text{HCOH} \\
| \\
\text{CH}_2\text{O}\,\text{P}
\end{array}
\; + \;
\begin{array}{c}
\text{CH}_2\text{OH} \\
| \\
\text{CO} \\
| \\
\text{HOCH} \\
| \\
\text{HCOH} \\
| \\
\text{CH}_2\text{O}\,\text{P}
\end{array}
\qquad (19.7)
$$

fructose-6-phosphate   glyceraldehyde-3-phosphate   erythrose-4-phosphate   xylulose-5-phosphate

One of the products of this reaction is a pentose phosphate which (apart from an epimerization to ribulose-5-phosphate) does not take part in any further rearrangements. The other product is erythrose-4-phosphate, and this can now undergo condensation with another molecule of dihydroxyacetone phosphate in the presence of aldolase. The reaction is analogous to the condensation that formed fructose-1,6-diphosphate (reaction (19.6) above).

$$
\begin{array}{c}
\text{CH}_2\text{OH} \\
| \\
\text{CO} \\
| \\
\text{CH}_2\text{O}\,\text{P}
\end{array}
\; + \;
\begin{array}{c}
\text{CHO} \\
| \\
\text{HCOH} \\
| \\
\text{HCOH} \\
| \\
\text{CH}_2\text{O}\,\text{P}
\end{array}
\; \rightleftharpoons \;
\begin{array}{c}
\text{CH}_2\text{O}\,\text{P} \\
| \\
\text{CO} \\
| \\
\text{HOCH} \\
| \\
\text{HCOH} \\
| \\
\text{HCOH} \\
| \\
\text{HCOH} \\
| \\
\text{CH}_2\text{O}\,\text{P}
\end{array}
\qquad (19.8)
$$

dihydroxyacetone phosphate   erythrose-4-phosphate   sedoheptulose-1,7-diphosphate

The sedoheptulose-1,7-diphosphate thus formed is hydrolysed in a reaction analogous to reaction (19.6a) above:

$$
\begin{array}{ccc}
\begin{array}{c}
\text{CH}_2\text{O}\,\textcircled{P} \\
| \\
\text{CO} \\
| \\
\text{HOCH} \\
| \\
\text{HCOH} \\
| \\
\text{HCOH} \\
| \\
\text{HCOH} \\
| \\
\text{CH}_2\text{O}\,\textcircled{P} \\
\text{sedoheptulose-1,7-} \\
\text{diphosphate}
\end{array}
& + \text{H}_2\text{O} \rightleftharpoons &
\begin{array}{c}
\text{CH}_2\text{OH} \\
| \\
\text{CO} \\
| \\
\text{HOCH} \\
| \\
\text{HCOH} \\
| \\
\text{HCOH} \\
| \\
\text{HCOH} \\
| \\
\text{CH}_2\text{O}\,\textcircled{P} \\
\text{sedoheptulose-7-} \\
\text{phosphate}
\end{array}
& + \;\; \text{P}_t
\end{array}
\qquad (19.8a)
$$

and the resulting sedoheptulose-7-phosphate donates a $\text{CH}_2\text{OH}\cdot\text{CO}$— group to yet another molecule of glyceraldehyde-3-phosphate.

$$
\begin{array}{c}
\text{CH}_2\text{OH} \\
| \\
\text{CO} \\
| \\
\text{HOCH} \\
| \\
\text{HCOH} \\
| \\
\text{HCOH} \\
| \\
\text{HCOH} \\
| \\
\text{CH}_2\text{O}\,\textcircled{P} \\[4pt]
\text{sedoheptulose-} \\
\text{7-phosphate}
\end{array}
+
\begin{array}{c}
\text{CHO} \\
| \\
\text{CHOH} \\
| \\
\text{CH}_2\text{O}\,\textcircled{P} \\[4pt]
\text{glyceral-} \\
\text{dehyde-3-} \\
\text{phosphate}
\end{array}
\rightleftharpoons
\begin{array}{c}
\text{CHO} \\
| \\
\text{HCOH} \\
| \\
\text{HCOH} \\
| \\
\text{HCOH} \\
| \\
\text{CH}_2\text{O}\,\textcircled{P} \\[4pt]
\text{ribose-5-} \\
\text{phosphate}
\end{array}
+
\begin{array}{c}
\text{CH}_2\text{OH} \\
| \\
\text{CO} \\
| \\
\text{HOCH} \\
| \\
\text{HCOH} \\
| \\
\text{CH}_2\text{O}\,\textcircled{P} \\[4pt]
\text{xylulose-5-} \\
\text{phosphate}
\end{array}
\qquad (19.9)
$$

Reactions (19.8), (19.8a) and (19.9) are analogous to reactions (19.6), (19.6a) and (19.7), and finally result in the formation of one molecule of ribose-5-phosphate and one molecule of xylulose-5-phosphate. The ribose-5-phosphate produced in reaction (19.9) can be converted to ribulose-5-phosphate by the isomerase we discussed on p. 294; similarly the xylulose-5-phosphate produced in reactions (19.7) and (19.9) can be converted to ribulose-5-phosphate by the epimerase (p. 294). We have thus accounted for the formation of three molecules of ribulose-5-phosphate from five molecules of triose phosphate. A flow-sheet summarizing these conversions is given in Fig. 19.1.

Two of the reactions in the Calvin cycle involve the hydrolysis of phosphate esters of sugars, namely reactions (19.6a) and (19.8a). The $\Delta G^{\circ\prime}$ of each of these is approximately $-4000$ cal/mole, so that they can be regarded as helping the sequence as a whole to flow in the direction of formation of pentose phosphate. A further result of these hydrolytic reactions is that the ribulose formed by the rearrangements we have described is in the form of its monophosphate. However, the fixation of carbon dioxide in reaction (19.1) of the Calvin cycle requires ribulose diphosphate. Consequently each molecule of ribulose-5-phosphate must be phosphorylated by ATP, as we discussed on p. 24. The reaction is catalysed by a kinase, and it too has a large negative $\Delta G^{\circ\prime}$ and helps to pull the

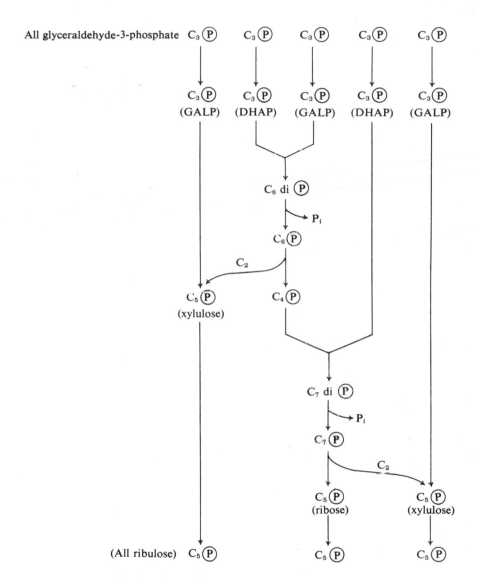

**Fig. 19.1** The interconversion of sugars in the Calvin cycle. GALP = glyceralde-hyde-3-phosphate. DHAP = dihydroxyacetone phosphate

sequence even more completely in the direction of formation of pentose from triose.

$$
\begin{array}{c}
\text{CH}_2\text{OH} \\
| \\
\text{CO} \\
| \\
\text{HCOH} \\
| \\
\text{HCOH} \\
| \\
\text{CH}_2\text{O}\,\text{P} \\
\text{ribulose-} \\
\text{5-phosphate}
\end{array}
\quad + \text{ ATP} \rightleftharpoons
\begin{array}{c}
\text{CH}_2\text{O}\,\text{P} \\
| \\
\text{CO} \\
| \\
\text{HCOH} \\
| \\
\text{HCOH} \\
| \\
\text{CH}_2\text{O}\,\text{P} \\
\text{ribulose-1,5-} \\
\text{diphosphate}
\end{array}
\quad + \text{ ADP} \qquad (19.10)
$$

$$\Delta G^{\circ\prime} = -5200 \text{ cal/mole}$$

Ribulose-1,5-diphosphate can re-enter the sequence at reaction (19.1).

## Summary of the Calvin cycle reactions

We can now look once more at the stoichiometry of the Calvin cycle, by summarizing each set of reactions to give three composite reactions. We have already summarized reactions (19.1), (19.2) and (19.3) as reaction (19.4).

$$3 \times \text{ribulose diphosphate} + 3\text{CO}_2 + 3\text{H}_2\text{O} + 6\text{ATP} + 6\text{NADPH}_2 \rightleftharpoons$$
$$6 \times \text{glyceraldehyde-3-phosphate} + 6\text{ADP} + 6\text{NADP} + 6\text{P}_i. \quad (19.4)$$

Reaction (19.5) accounts for the interconversion of the triose phosphates. The sum of reactions (19.6), (19.6a), (19.7), (19.8), (19.8a) and (19.9) gives:

$$5 \times \text{triose phosphate} \rightleftharpoons 3 \times \text{pentose phosphate} + 2\text{P}_i. \quad (19.11)$$

If we now multiply reaction (19.10) by three we get:

$$3 \times \text{pentose phosphate} + 3\text{ATP} \rightleftharpoons 3 \times \text{pentose diphosphate} + 3\text{ADP}.$$
$$(19.12)$$

The sum of reactions (19.4), (19.11) and (19.12) gives the complete reaction for the fixation of three molecules of carbon dioxide to form one molecule of triose phosphate.

$$3\text{CO}_2 + 3\text{H}_2\text{O} + 9\text{ATP} + 6\text{NADPH}_2 \rightleftharpoons$$
$$\text{glyceraldehyde-3-phosphate} + 9\text{ADP} + 8\text{P}_i + 6\text{NADP}$$

This summary reaction represents the first biosynthesis that we have considered in this book. Several points emerge clearly from it. In the first place, it uses ATP to build up a relatively large molecule from a very simple molecule. Secondly, it uses $\text{NADPH}_2$ (not $\text{NADH}_2$) to reduce a highly oxidized molecule to a more reduced molecule. (The ATP and $\text{NADPH}_2$ required were, of course, formed during the light reaction.) Thirdly, it includes several individual reactions ((19.1), (19.6a), (19.8a) and (19.10)) that have a large negative $\Delta G^{\circ\prime}$, so that the process as a whole is pulled strongly towards the biosynthetic product. We made these points in general terms when introducing the concepts of intermediary metabolism in Ch. 9; you can now see how they work out in detail in a single biosynthetic pathway.

## SYNTHESIS OF TRIOSE PHOSPHATE FROM GLUCONEOGENIC PRECURSORS

The ability to convert pyruvate and intermediates of the tricarboxylic acid cycle into carbohydrate – a process that is known as gluconeogenesis (see also p. 261) – is very widespread among living organisms. However, we have pointed out in Chs 14 and 15 that the concentrations of pyruvate and intermediates of the cycle that are actually present in cells are quite low; what purpose, then, is served by a pathway that uses these compounds as precursors? For part of the answer, we must anticipate a point that we shall develop in more detail in Ch. 22. Many amino acids are capable of giving rise, during the course of their degradation, to pyruvate or intermediates of the tricarboxylic acid cycle, with the result that carbohydrate can be formed from the carbon atoms of a large number of the amino acids. The second important precursor for gluconeogenesis is lactate – the use of this compound, unlike that of amino acids, is confined to animals, in which lactate is produced during muscular exercise (see Ch. 14).

In this section we shall consider the synthesis of triose phosphate from lactate. The pathway passes through pyruvate and also through oxaloacetate, so that it can account for the synthesis of triose phosphate from amino acids that give rise to pyruvate or to Krebs cycle intermediates. The uses of triose phosphate will be described in the next section.

Lactate accumulates in large quantities in the blood of higher animals during exercise, and its conversion to carbohydrate occurs chiefly in the liver. The first reaction is the oxidation of lactate by NAD in the presence of lactate dehydrogenase.

$$
\begin{array}{llll}
\text{COOH} & & \text{COOH} & \\
| & & | & \\
\text{CHOH} + \text{NAD} & \rightleftharpoons & \text{CO} & + \text{NADH}_2 \qquad (19.13) \\
| & & | & \\
\text{CH}_3 & & \text{CH}_3 & \\
\text{lactic} & & \text{pyruvic} & \\
\text{acid} & & \text{acid} &
\end{array}
$$

This reaction is simply the reverse of reaction (14.11) of the Embden–Meyerhof pathway (p. 265). We saw in that pathway that the $NADH_2$ used to reduce pyruvate was supplied by an oxidation earlier in the sequence; similarly we shall see shortly that the $NADH_2$ produced by the oxidation of lactate is used in a reduction later in the synthesis of triose phosphate (see reaction (19.19)).

When describing glycolysis we pointed out (on p. 264) that the formation of pyruvate from phospho-*enol*pyruvate was accompanied by a large loss of free energy ($\Delta G^{\circ\prime} = -7500$ cal/mole), and that consequently pyruvate could not be directly phosphorylated in the cell to form phospho-*enol*pyruvate. A bypass is therefore needed for the synthesis of phospho-*enol*pyruvate, and this is provided by a sequence of two reactions. The first is the carboxylation of pyruvate that we have discussed already (p. 281) in the context of the replenishment of Krebs cycle intermediates. The enzyme is pyruvate carboxylase.

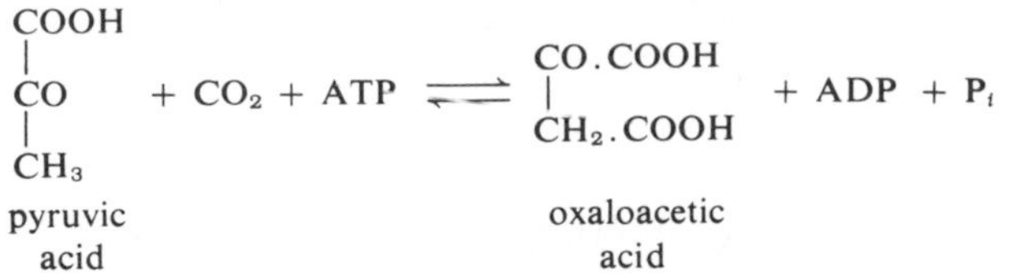

$$
\underset{\substack{\text{pyruvic} \\ \text{acid}}}{\overset{\displaystyle \begin{array}{c} \text{COOH} \\ | \\ \text{CO} \\ | \\ \text{CH}_3 \end{array}}{}} + CO_2 + ATP \rightleftharpoons \underset{\substack{\text{oxaloacetic} \\ \text{acid}}}{\overset{\displaystyle \begin{array}{c} \text{CO.COOH} \\ | \\ \text{CH}_2.\text{COOH} \end{array}}{}} + ADP + P_i \qquad (19.14)
$$

$$\Delta G^{\circ\prime} = -500 \text{ cal/mole}$$

The second reaction is a decarboxylation of oxaloacetate in the presence of a phosphate donor – most commonly GTP. The enzyme for this reaction is called phospho-*enol*pyruvate carboxykinase, and its product is phospho-*enol*pyruvate.

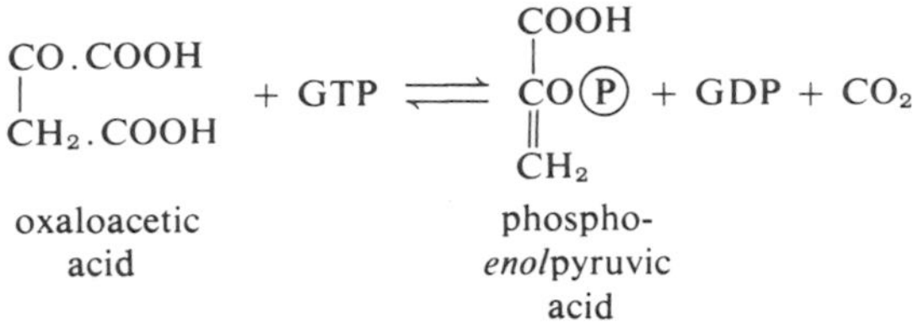

$$
\underset{\substack{\text{oxaloacetic} \\ \text{acid}}}{\overset{\displaystyle \begin{array}{c} \text{CO.COOH} \\ | \\ \text{CH}_2.\text{COOH} \end{array}}{}} + GTP \rightleftharpoons \underset{\substack{\text{phospho-} \\ \textit{enol}\text{pyruvic} \\ \text{acid}}}{\overset{\displaystyle \begin{array}{c} \text{COOH} \\ | \\ \text{CO}\,\textcircled{P} \\ \| \\ \text{CH}_2 \end{array}}{}} + GDP + CO_2 \qquad (19.15)
$$

$$\Delta G^{\circ\prime} = +700 \text{ cal/mole}$$

These two reactions thus synthesize phospho-*enol*pyruvate from pyruvate at the expense of two high-energy phosphate bonds.* The *overall* reaction, the sum of the two reactions just given, can be written:

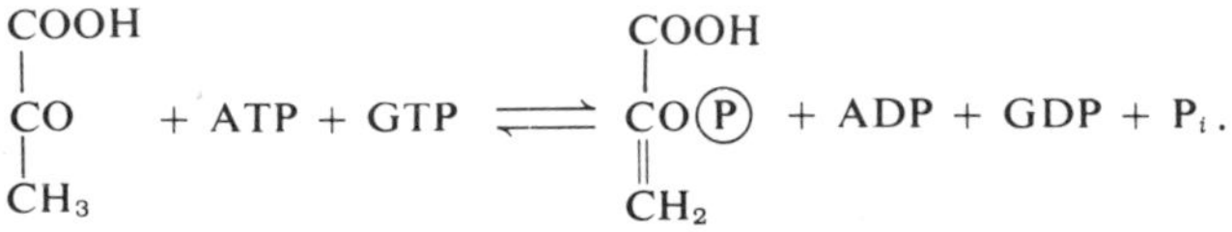

$$
\overset{\displaystyle \begin{array}{c} \text{COOH} \\ | \\ \text{CO} \\ | \\ \text{CH}_3 \end{array}}{} + ATP + GTP \rightleftharpoons \overset{\displaystyle \begin{array}{c} \text{COOH} \\ | \\ \text{CO}\,\textcircled{P} \\ \| \\ \text{CH}_2 \end{array}}{} + ADP + GDP + P_i.
$$

The $\Delta G^{\circ\prime}$ of this overall process is $+200$ cal/mole, and the reaction can easily proceed as written provided that the overall $\Delta G^{\circ\prime}$ of the subsequent reactions to carbohydrate is fairly large and negative (see Ch. 2). It will become clear in the remainder of this chapter that this condition is met.

Phospho-*enol*pyruvate is converted to glyceraldehyde-3-phosphate by a series of four reactions that are the reverse of the corresponding reactions, (19.6), (19.7), (19.8) and (19.9), of glycolysis and are catalysed by the same enzymes (see pp. 262–264). First phospho-*enol*pyruvate accepts water in a reaction catalysed by enolase, and the resulting glycerate-2-phosphate is converted to glycerate-3-phosphate in a reaction catalysed by phosphoglyceromutase (see p. 263 for details of the mechanism of action of this mutase).

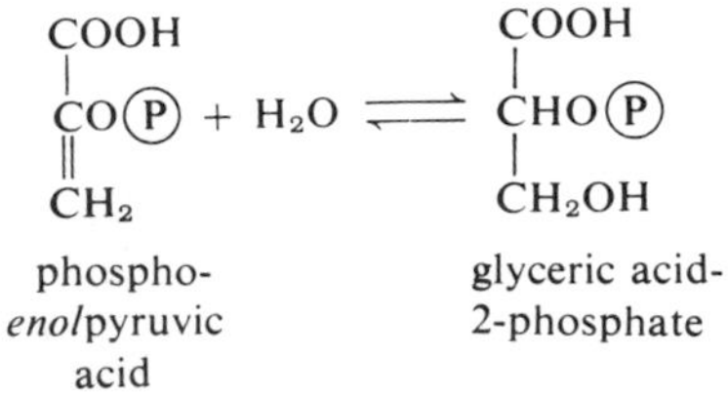

$$
\underset{\substack{\text{phospho-} \\ \textit{enol}\text{pyruvic} \\ \text{acid}}}{\overset{\displaystyle \begin{array}{c} \text{COOH} \\ | \\ \text{CO}\,\textcircled{P} \\ \| \\ \text{CH}_2 \end{array}}{}} + H_2O \rightleftharpoons \underset{\substack{\text{glyceric acid-} \\ \text{2-phosphate}}}{\overset{\displaystyle \begin{array}{c} \text{COOH} \\ | \\ \text{CHO}\,\textcircled{P} \\ | \\ \text{CH}_2\text{OH} \end{array}}{}} \qquad (19.16)
$$

$$\Delta G^{\circ\prime} = +800 \text{ cal/mole}$$

* Actually the situation is rather more complicated than it appears here, because oxaloacetate is synthesized in the mitochondrion but phospho-*enol*pyruvate is needed for carbohydrate synthesis in the soluble cytoplasm. This point is discussed in Ch. 21.

$$
\begin{array}{ccc}
\text{COOH} & & \text{COOH} \\
| & & | \\
\text{CHO}\,\textcircled{P} & \rightleftharpoons & \text{CHOH} \\
| & & | \\
\text{CH}_2\text{OH} & & \text{CH}_2\text{O}\,\textcircled{P}
\end{array}
\qquad (19.17)
$$

glyceric acid-    glyceric acid-<br>
2-phosphate    3-phosphate

$$\Delta G^{\circ\prime} = -1400 \text{ cal/mole}$$

Glycerate-3-phosphate must now be phosphorylated by ATP in the presence of phosphoglycerate kinase; the resulting glycerate-1,3-diphosphate is reduced by $NADH_2$ by the action of triose phosphate dehydrogenase (the mechanism of which we discussed on p. 262).

$$
\begin{array}{ccc}
\text{COOH} & & \text{COO}\,\textcircled{P} \\
| & & | \\
\text{CHOH} + \text{ATP} & \rightleftharpoons & \text{CHOH} + \text{ADP} \\
| & & | \\
\text{CH}_2\text{O}\,\textcircled{P} & & \text{CH}_2\text{O}\,\textcircled{P}
\end{array}
\qquad (19.18)
$$

glyceric acid-    glyceric acid-<br>
3-phosphate    1,3-diphosphate

$$\Delta G^{\circ\prime} = +4800 \text{ cal/mole}$$

$$
\begin{array}{ccc}
\text{COO}\,\textcircled{P} & & \text{CHO} \\
| & & | \\
\text{CHOH} + \text{NADH}_2 & \rightleftharpoons & \text{CHOH} + \text{NAD} + \text{P}_i \\
| & & | \\
\text{CH}_2\text{O}\,\textcircled{P} & & \text{CH}_2\text{O}\,\textcircled{P}
\end{array}
\qquad (19.19)
$$

glyceric acid-    glyceral-<br>
1,3-diphosphate    dehyde-3-phosphate

$$\Delta G^{\circ\prime} = -300 \text{ cal/mole}$$

Reactions (19.18) and (19.19) correspond to reactions (19.2) and (19.3) of the Calvin cycle (p. 322), but whereas reaction (19.3) uses $NADPH_2$ (produced in the light reaction of photosynthesis) to reduce glycerate-1,3-diphosphate, reaction (19.19) uses $NADH_2$. What is the source of this $NADH_2$?

We can discuss this question conveniently in conjunction with another. If we calculate the overall $\Delta G^{\circ\prime}$ for the four reactions (19.16), (19.17), (19.18) and (19.19), we find that it is $+3900$ cal/mole. How, then, can this set of reactions, which appears to be disfavoured energetically, be used to synthesize glyceraldehyde-3-phosphate?

A consideration of the circumstances in which triose phosphate is synthesized from lactate provides a complete answer to the first question and a partial answer to the second. Lactate accumulates during muscular exercise. At the end of exercise the ratio of lactate to pyruvate in gluconeogenic tissues (principally the liver) increases, and therefore, owing to the equilibrium shown in reaction (19.13), the concentration of $NADH_2$ will rise. It is under these reducing conditions that biosynthetic reactions are favoured (see Ch. 12); the $NADH_2$ formed in reaction (19.13) will help to drive reaction (19.19) to the right and hence all the reactions (19.16) to (19.19) to the right. In other words, even though the free-energy change of these reactions with all reactants at standard concentration disfavours synthesis of glyceraldehyde-3-phosphate, the flood of $NADH_2$

produced from lactate will displace the equilibrium sufficiently to favour synthesis.

This explanation, however, is only part of the answer. The other point is that the reactions that remove glyceraldehyde-3-phosphate to form carbohydrate (see next section) are accompanied by a large loss of free energy, so that once again the equilibrium is displaced in favour of synthesis.

The above discussion serves to point out that when the precursor for gluconeogenesis is not lactate but amino acids, carbohydrate will be synthesized only if quite a large amount of $NADH_2$ is available from other reactions for use in reaction (19.19). Once again we see that biosynthesis is favoured by reducing conditions (see Ch. 12); if the ratio of $NADH_2$ to NAD is low, carbohydrate will not be synthesized.

# SYNTHESIS OF CARBOHYDRATE FROM TRIOSE PHOSPHATE

## 1. Synthesis of glucose-6-phosphate and glucose

Now that we have described the formation of triose phosphate either from carbon dioxide or from gluconeogenic precursors, we can consider how it is used to form glucose-6-phosphate. We recall that the two isomers of triose phosphate, glyceraldehyde-3-phosphate and dihydroxyacetone phosphate, can be interconverted owing to the presence of triose phosphate isomerase (see pp. 261 and 324). Glyceraldehyde-3-phosphate and dihydroxyacetone phosphate can condense in the presence of aldolase to give fructose-1,6-diphosphate; this reaction occurs also in the Calvin cycle (p. 324), and is the reverse of reaction (19.4) of glycolysis.

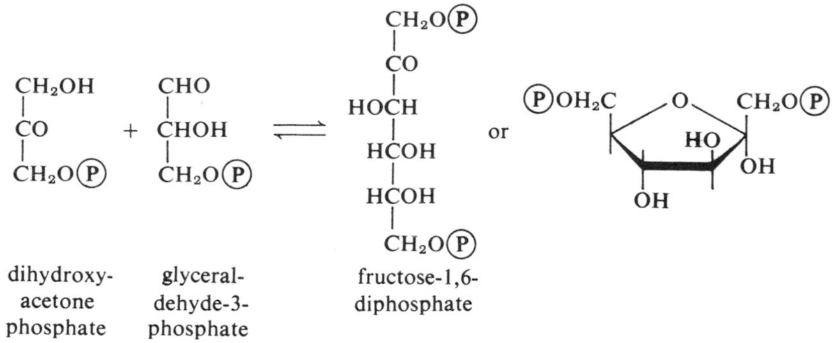

dihydroxy-
acetone
phosphate

glyceraldehyde-3-
phosphate

fructose-1,6-
diphosphate

$\Delta G^{\circ\prime} = -5300$ cal/mole

Fructose-1,6-diphosphate is hydrolysed in another reaction that also occurs in the Calvin cycle (p. 325) and is, as we pointed out there, not the reverse of a glycolytic reaction.

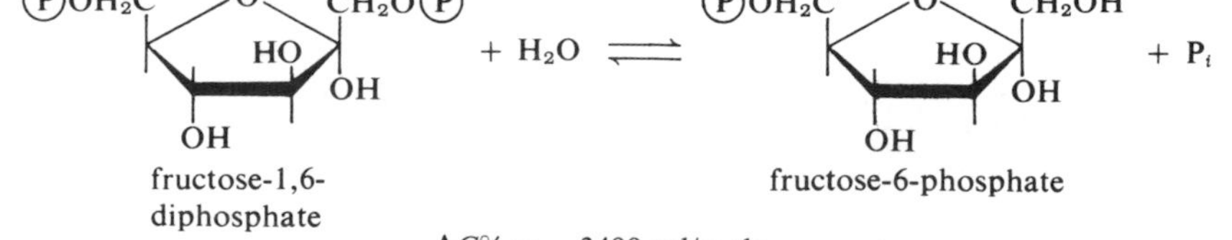

fructose-1,6-
diphosphate

fructose-6-phosphate

$\Delta G^{\circ\prime} = -3400$ cal/mole

The resulting fructose-6-phosphate is converted to glucose-6-phosphate owing to the presence of hexose phosphate isomerase, the enzyme that catalyses the reverse reaction in glycolysis (see p. 260).

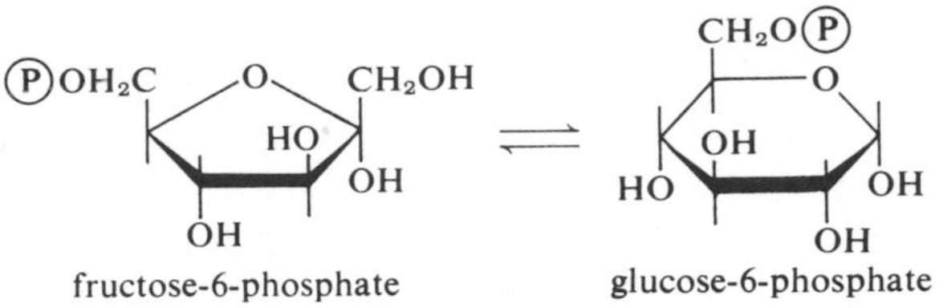

fructose-6-phosphate     glucose-6-phosphate

$$\Delta G^{\circ\prime} = -500 \text{ cal/mole}$$

The $\Delta G^{\circ\prime}$ of these three reactions taken together is $-9200$ cal/mole. This very high negative $\Delta G^{\circ\prime}$ easily pulls the sequence of reactions from lactate to glucose-6-phosphate virtually to completion, even though the $\Delta G^{\circ\prime}$ of the reactions from phospho-*enol*pyruvate to triose phosphate is positive (see p. 331). Part of this overall negative $\Delta G^{\circ\prime}$ is provided by the high negative $\Delta G^{\circ\prime}$ of the hydrolysis of fructose-1,6-diphosphate, and this is another of the advantages of the fact that this special reaction, different from that in glycolysis, is included in the pathway of carbohydrate synthesis (see p. 261).

In the liver, and to some extent in a few other tissues such as the kidney and the intestinal mucosa, glucose-6-phosphate can be split by its specific phosphatase (see pp. 259 and 268) so that free glucose is released.

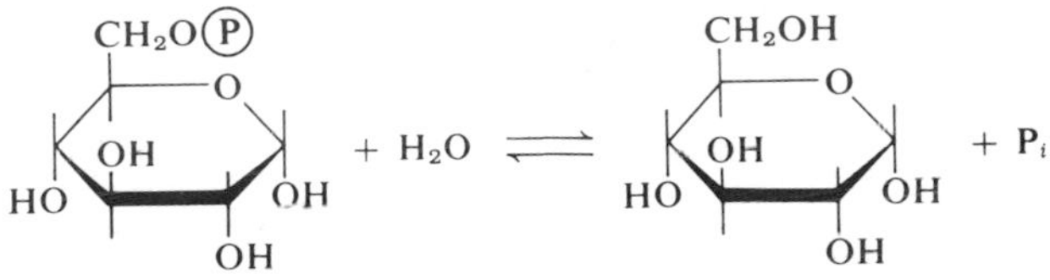

$$\Delta G^{\circ\prime} = -3300 \text{ cal/mole}$$

In most animal tissues, however, as well as in plants, glucose-6-phosphate is used almost exclusively for synthetic reactions – as, indeed, a substantial fraction of the liver glucose-6-phosphate is too.

## 2. Synthesis of uridine diphosphate glucose

Several important synthetic pathways take glucose-6-phosphate as their starting point, and these all begin by converting glucose-6-phosphate into a form of glucose that is especially reactive, having a high potential either for transfer or for conversion into another sugar. This compound is called uridine diphosphate glucose (UDP-glucose),* the formula of which is (see also Table 7.3):

* In some plants and micro-organisms the analogous compounds ADP-glucose, GDP-glucose and CDP-glucose are used in place of UDP-glucose.

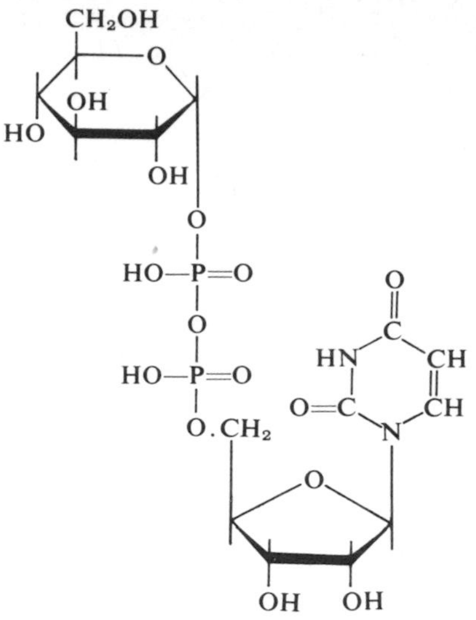

CH₂OH
OH
HO
OH
O
HO—P=O
O
HO—P=O
O.CH₂
O
HN
C
O=C
CH
CH
N
O
OH   OH

The first step in the synthesis of UDP-glucose is the conversion of glucose-6-phosphate into glucose-1-phosphate. The enzyme, phosphoglucomutase, was discussed on p. 268, where we showed how the two isomers of glucose phosphate were interconverted during the degradation of glycogen and starch.

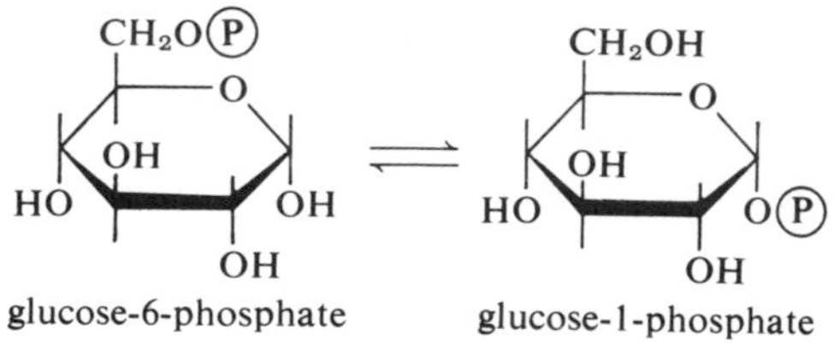

CH₂O(P)          CH₂OH
O              O
OH             OH
HO      OH     HO      O(P)
OH             OH
glucose-6-phosphate        glucose-1-phosphate

Glucose-1-phosphate is now made to react with uridine triphosphate (UTP). The enzyme is UDP-glucose pyrophosphorylase.

UTP + glucose-1-phosphate ⇌ UDP-glucose + PP$_i$

This activation of glucose is analogous in some ways to the activation of fatty acids that we discussed on p. 284. At first sight the analogy may seem far-fetched. However, we may point out that the effect of the reaction (like that for fatty acids) is to bring the glucose into a form in which it is exceptionally reactive; we shall shortly see that UDP-glucose is able to donate glucose to various acceptors in the synthesis of oligosaccharides and polysaccharides. The other point of similarity between the activation of glucose and the activation of fatty acids is that both reactions yield pyrophosphate. As we discussed in some detail on pp. 20 and 284, the liberation of pyrophosphate has the effect of pulling the reaction over towards completion.

UDP-glucose belongs to a class of compounds in which nucleoside diphosphates are attached to sugars. This type of nucleotide was discovered, and has been intensively studied, by Leloir. Nucleoside diphosphate sugars are used very widely in the synthesis of polysaccharides of all kinds – not only glycogen and starch, but also other sugar polymers such as xylans which occur in plants, polymers of amino-sugars which occur in bacterial cell walls, and even polymers such as chitin which occur in the exoskeleton of insects (see Ch. 8). Nucleoside diphosphate sugars are also used in the interconversion of monosaccharides and the synthesis of disaccharides. We shall now exemplify the synthesis of polysaccharide from UDP-glucose, and then we shall discuss a few of the other reactions in which UDP-sugars participate.

## 3. Synthesis of glycogen and starch

UDP-glucose contains a phosphoryl-glucoside bond, the free energy of hydrolysis of which is about the same as that in ATP, and it is therefore able readily to donate its glucose to a suitable acceptor. In the synthesis of glucose polymers, the acceptor is a 'primer', that is a chain consisting of a number of glucose residues. UDP-glucose transfers its glucose residue to the non-reducing end of such a chain, forming a new glycosidic link (see p. 46). We can exemplify this reaction by showing the synthesis of amylose, the straight-chain constituent of starch, which consists of glucose residues joined by $1 \rightarrow 4, \alpha$-links (see p. 174). The enzyme is called UDP-glucose glucosyl transferase.

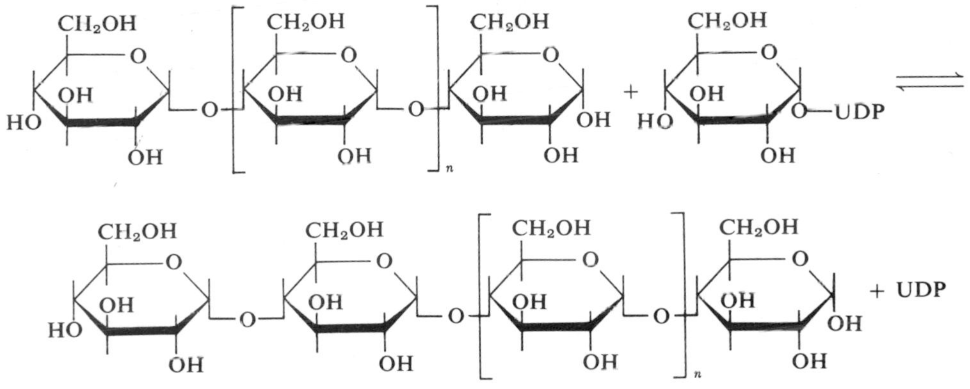

$$\Delta G^{\circ\prime} = -3200 \text{ cal/mole}$$

The UDP liberated in this reaction cannot take part in oxidative or substrate-level phosphorylation directly, but it can be phosphorylated by ATP.

$$\text{ATP} + \text{UDP} \rightleftharpoons \text{ADP} + \text{UTP}$$

$$\Delta G^{\circ\prime} \simeq 0$$

The resulting UTP can now react again with glucose-1-phosphate and thus be used once more in the synthetic reaction.

We are now in a position to calculate the requirement for ATP in the synthesis of glycogen from lactate. Reactions (19.14), (19.15) and (19.18) use up one molecule of ATP (or GTP) each; thus the synthesis of one molecule of triose phosphate from lactate requires three molecules of ATP, and the synthesis of one molecule of glucose-6-phosphate from two molecules of lactate requires six molecules of ATP. A further molecule of ATP is needed, as we have just seen, to rephosphorylate UDP: thus a total of seven molecules of ATP is needed to convert two molecules of lactate into a single residue of glucose polymerized in glycogen.

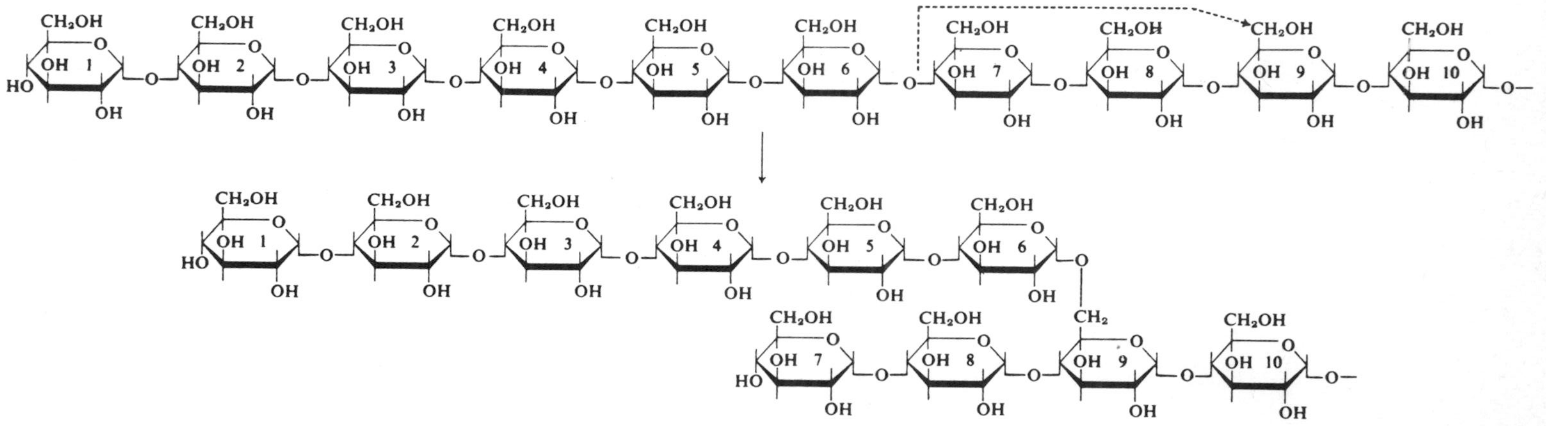

**Fig. 19.2** Branching of a polysaccharide. The substrate of the branching enzyme is a 1→4,α-linked polymer of glucose. The enzyme takes some six residues from the non-reducing end of the chain and attaches them via a glycosidic linkage to the C-6 position of a residue in either the same chain (as shown here) or a different chain

Strictly speaking, though, we have not formed glycogen but rather a straight chain $1 \rightarrow 4,\alpha$-linked glucose polymer. From this straight chain, amylopectin and glycogen are formed by the action of a branching enzyme. This enzyme removes a small portion of the chain (consisting of perhaps half-a-dozen glucose residues) and transfers it to the C-6 position of another glucose residue of the chain (see Fig. 19.2). The $\Delta G^{\circ\prime}$ of this reaction is small, and no further ATP is needed for it.

The synthesis of glycogen from UDP-glucose is regulated by a complicated control system in which hormones are involved. We shall discuss this, and the regulation of glycogen degradation, in Ch. 29.

The other polymers of sugars and of amino-sugars that we mentioned on pp. 173 ff. are synthesized in reactions closely analogous to those we have described for glycogen and starch.

## 4. Other uses of UDP-glucose

One rather unexpected use of UDP-glucose is its involvement in the metabolism of galactose. Galactose is an important sugar in the diet, especially of young mammals and some human adults, since it is formed by the intestinal hydrolysis of lactose from milk. As we mentioned on p. 267, galactose is phosphorylated at the C-1 position by ATP in the presence of galactokinase; the resulting galactose-1-phosphate is then converted to UDP-galactose by either of the following reactions.

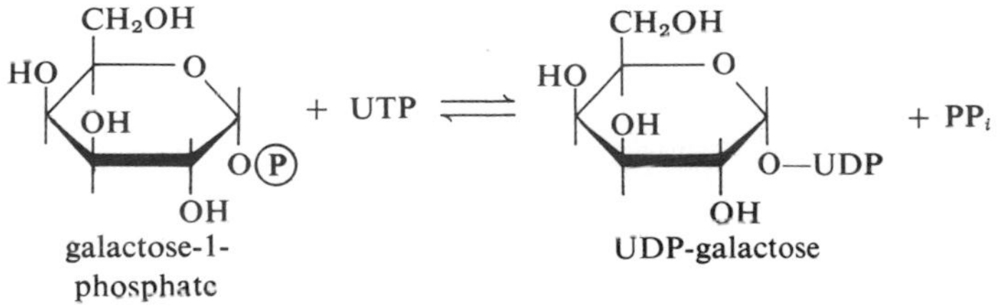

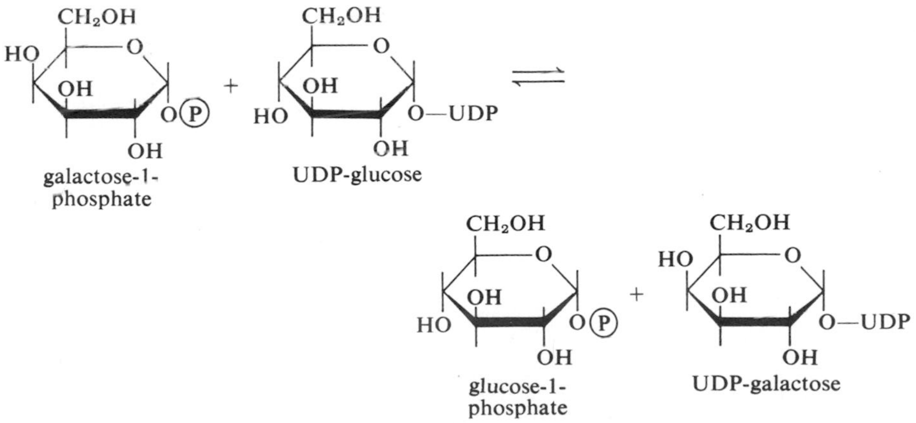

galactose-1-phosphate $+$ UTP $\rightleftharpoons$ UDP-galactose $+$ PP$_i$

galactose-1-phosphate $+$ UDP-glucose $\rightleftharpoons$ glucose-1-phosphate $+$ UDP-galactose

The UDP-galactose is then converted to UDP-glucose by an epimerase, and the resulting UDP-glucose reacts with another molecule of galactose-1-phosphate as shown above.

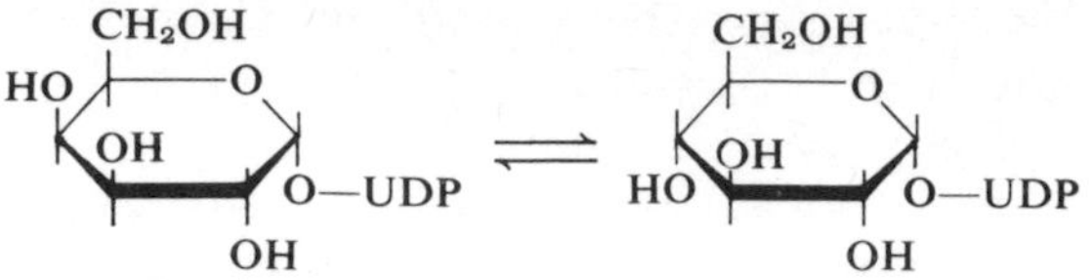

UDP-galactose ⇌ UDP-glucose

In this way the UDP-sugar acts as a shuttle for the conversion of galactose-1-phosphate into glucose-1-phosphate. The glucose-1-phosphate thus produced can be converted into glucose-6-phosphate by the mutase reaction (see p. 268).

The conversion of UDP-galactose to UDP-glucose is readily reversible, and it provides a means for the synthesis of a galactose residue from glucose. This conversion is important in the synthesis of lactose, a reaction which occurs in the mammary gland and which is catalysed by an enzyme complex called lactose synthetase.

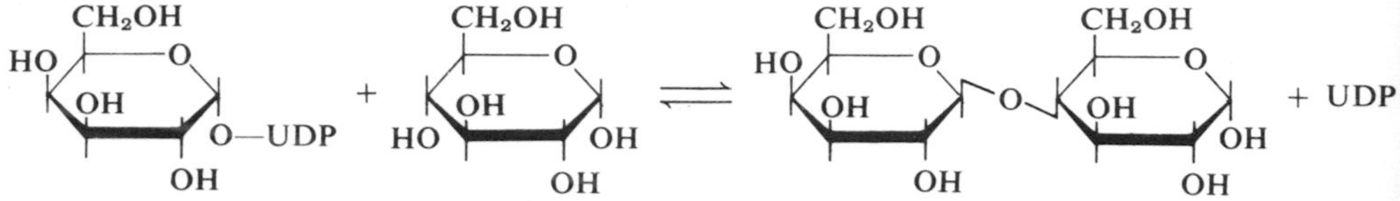

UDP-galactose + glucose ⇌ lactose + UDP

UDP-glucose is also used in the synthesis of sucrose in plants. First UDP-glucose reacts with fructose-6-phosphate to form sucrose-6-phosphate:

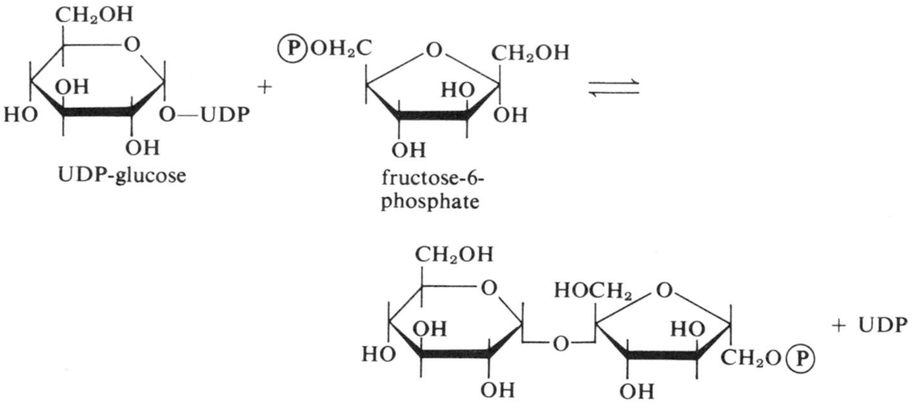

UDP-glucose + fructose-6-phosphate ⇌

sucrose-6-phosphate + UDP

and this is then hydrolysed by a phosphatase to yield free sucrose.

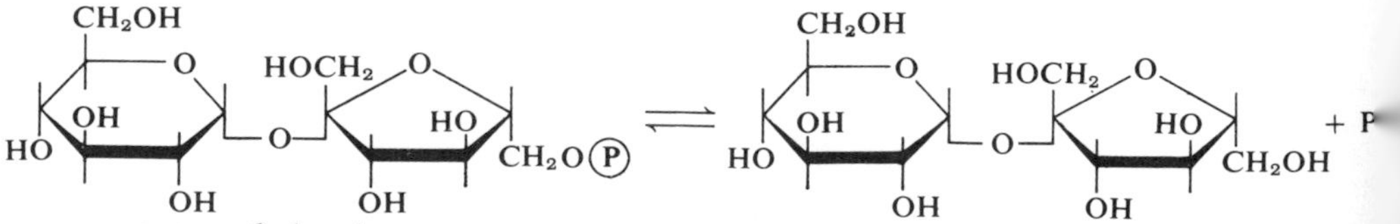

sucrose-6-phosphate ⇌ sucrose + P

# INTERRELATIONS BETWEEN FAT AND CARBOHYDRATE

On p. 321 we pointed out that compounds that can give rise to intermediates of the tricarboxylic acid cycle could act as precursors for gluconeogenesis. For example, those amino acids that are converted to α-oxoglutarate (see Ch. 22) can enter the synthetic sequence via oxaloacetate at reaction (19.15). There is, however, one important compound that enters the tricarboxylic acid cycle and yet cannot give rise to carbohydrate, namely acetyl coenzyme A. The reason why acetyl coenzyme A cannot be used to synthesize carbohydrates is this: when the acetyl group reacts with oxaloacetate citrate is formed (p. 276), but after one turn of the cycle only oxaloacetate is left and the two carbon atoms corresponding to the acetyl group have been completely oxidized. Thus there is no way in which the cycle can make use of the acetyl group for any reaction other than oxidation; acetyl coenzyme A does not add to the stock of intermediates in the cycle, and it cannot provide carbon to replenish these intermediates (see p. 291).

As a result, fatty acids cannot be used (unless they are odd-numbered fatty acids, see p. 289), for the net synthesis of carbohydrate, and this is a most important principle of animal physiology which we can scarcely emphasize too much. (Notice that this principle does not hold for plants and micro-organisms, which have a mechanism for converting acetyl coenzyme A to carbohydrate.) It is in part the fact that carbohydrate cannot be synthesized from fatty acids that makes the carboxylation reaction for the replenishment of Krebs cycle intermediates (p. 281) so exceptionally important.

# Chapter 20 Synthesis of lipid

The ability to synthesize lipids – which, for the present purpose, we shall take to mean triglycerides and phospholipids – seems to be almost universal among living organisms, and the pathways of biosynthesis appear to be more or less the same in micro-organisms, plants and animals. In this chapter we shall first describe the synthesis of fatty acids, and then show how fatty acids are used in the formation of triglycerides and of phospholipids. These processes illustrate extremely well the principles of biosynthetic reactions that we outlined in Chs 2 and 9 – the use of ATP and of $NADPH_2$, the inclusion in the pathways of reactions that are effectively operated in only one direction, and the fact that the sequence differs in several ways from that used for catabolism. This last point, in particular, is even more strikingly true of lipid biosynthesis as compared with lipid catabolism than it is of carbohydrate biosynthesis as compared with carbohydrate catabolism.

## Synthesis of fatty acids

One of the most important clues to the elucidation of the pathway of biosynthesis of fatty acids came from the discovery by Wakil in 1958 that carbon dioxide is catalytic in the synthesis – in other words that carbon dioxide is required for synthesis to proceed even though $^{14}CO_2$ added experimentally is not incorporated into fatty acids. Now it has been known for a long time that the major precursor for fatty-acid synthesis is acetyl coenzyme A – a fact which accounts for the well-known observation that animals fed with large quantities of carbohydrate rapidly become fat. But while it is easy in principle to imagine the condensation of acetyl coenzyme A units, accompanied by reduction, to give a long-chain fatty acyl coenzyme A:

$$(n + 1)CH_3.CO.S\text{-}CoA + n(H_2) \longrightarrow$$
$$CH_3.CH_2.(CH_2.CH_2)_{n-1}.CH_2.CO.S\text{-}CoA + nH_2O + nCoA.SH$$

it is hard to see why carbon dioxide should be involved in this process.

This difficulty was resolved by the discovery that acetyl coenzyme A can be carboxylated in the soluble cytoplasm of cells from tissues that synthesize fatty acids, to form malonyl coenzyme A ($HOOC.CH_2.CO.S\text{-}CoA$). (Notice that we say in the soluble cytoplasm: one of the most important differences between fatty-acid degradation and fatty-acid synthesis is that the former occurs in mitochondria and the latter in the soluble fraction.) It turned out that it is in fact this malonyl coenzyme A, rather than acetyl coenzyme A, that is the immediate precursor of fatty acids. During the course of the synthesis, the malonyl group loses the carbon dioxide that it previously acquired; it is for this reason that the carbon dioxide does not appear in the final product.

Thus the *synthesis* of fatty acids, in the soluble cytoplasm, uses malonyl coenzyme A as precursor; we have seen earlier (see Ch. 16) that the *degradation* of fatty acids, in the mitochondrion, yields acetyl coenzyme A as product. A further difference between the two processes is that whereas in degradation all of the intermediates react as their coenzyme A derivatives, in synthesis the intermediates occur as derivatives of a different carrier. This carrier is a small protein called 'acyl carrier protein' (ACP), which has been extensively studied by Vagelos; it has covalently linked to it a large prosthetic group called 4'-phosphopantetheine, identical to that that occurs in coenzyme A (see p. 79). (We shall here write it as ACP.SH.)

Yet another difference between the synthesis and the degradation of fatty acids is that the enzymes involved in synthesis are associated together in a tightly bound complex (at least in some organisms) so that the intermediates are never set free. As we shall see below, the reaction sequence resembles a spiral, in which the same set of enzymes participates again and again to lengthen the chain by two carbon atoms at a time; because of the tight association between the enzymes the intermediates are passed round and round this 'fatty acid synthetase' complex until the long-chain derivative is released at the end. Let us now look at the details of these processes.

*Formation of malonyl coenzyme A*

We have seen that acetyl coenzyme A arises not only from oxidation of fatty acids but also is the product of oxidation of pyruvate and hence of carbohydrate (see p. 273).* The carboxylation of acetyl coenzyme A is achieved in a reaction catalysed by acetyl coenzyme A carboxylase.

$$CH_3.CO.S\text{-}CoA + ATP + CO_2 \rightleftharpoons HOOC.CH_2.CO.S\text{-}CoA + ADP + P_t \quad (20.1)$$

acetyl coenzyme A  malonyl coenzyme A

Acetyl coenzyme A carboxylase, like pyruvate carboxylase (see p. 281), is dependent on the cofactor biotin, the formula of which is:

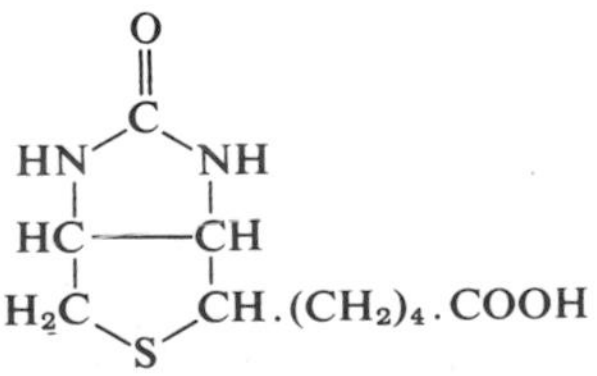

* However, this acetyl coenzyme A is in the mitochondrion, whereas for fatty-acid synthesis acetyl coenzyme A is needed in the soluble cytoplasm. The transfer from the mitochondrion to the cytoplasm is discussed in Ch. 21.

In the cell, biotin is covalently bound through its carboxyl group to the carboxylase enzyme (compare lipoic acid, p. 272). The carboxylation is a two-stage reaction. First biotin (bound to the enzyme) accepts carbon dioxide in a reaction driven by the hydrolysis of ATP.

$$
\begin{array}{c}
\text{O} \\
\parallel \\
\text{C} \\
\text{HN} \diagup \quad \diagdown \text{NH} \\
\text{HC} \!-\!\!-\! \text{CH} \\
\text{H}_2\text{C} \diagdown_{\text{S}} \diagup \text{CH}.(\text{CH}_2)_4.\text{COOH}
\end{array}
\quad + \text{CO}_2 + \text{ATP} \rightleftharpoons
$$

$$
\begin{array}{c}
\text{COOH} \quad \text{O} \\
\diagdown \quad \parallel \\
\text{C} \\
\text{N} \diagup \quad \diagdown \text{NH} \\
\text{HC} \!-\!\!-\! \text{CH} \\
\text{H}_2\text{C} \diagdown_{\text{S}} \diagup \text{CH}.(\text{CH}_2)_4.\text{COOH}
\end{array}
\quad + \text{ADP} + \text{P}_i
$$

The carboxybiotin then reacts with acetyl coenzyme A.

$$
\begin{array}{c}
\text{COOH} \quad \text{O} \\
\diagdown \quad \parallel \\
\text{C} \\
\text{N} \diagup \quad \diagdown \text{NH} \\
\text{HC} \!-\!\!-\! \text{CH} \\
\text{H}_2\text{C} \diagdown_{\text{S}} \diagup \text{CH}.(\text{CH}_2)_4.\text{COOH}
\end{array}
\quad + \text{CH}_3.\text{CO}.\text{S-CoA} \rightleftharpoons
$$

$$
\begin{array}{c}
\text{O} \\
\parallel \\
\text{C} \\
\text{HN} \diagup \quad \diagdown \text{NH} \\
\text{HC} \!-\!\!-\! \text{CH} \\
\text{H}_2\text{C} \diagdown_{\text{S}} \diagup \text{CH}.(\text{CH}_2)_4.\text{COOH}
\end{array}
\quad + \text{HOOC}.\text{CH}_2.\text{CO}.\text{S-CoA}
$$

The rate of synthesis of malonyl coenzyme A from acetyl coenzyme A by this reaction is under the allosteric control of intermediates of the tricarboxylic acid cycle (see Ch. 6). The details of this control system are discussed in Ch. 29. (See also p. 506 for mention of the control of pyruvate carboxylase.)

*Initiation of the synthesis*

Meanwhile a separate molecule of acetyl coenzyme A reacts with the acyl carrier protein, and the acetyl group is then transferred to covalent linkage with the —SH group of the *condensing enzyme,* one of the components of the fatty-acid synthetase complex.

$$\text{CH}_3.\text{CO}.\text{S-CoA} + \text{ACP}.\text{SH} \rightleftharpoons \text{CH}_3.\text{CO}.\text{S-ACP} + \text{CoA}.\text{SH} \tag{20.2a}$$

$$\text{CH}_3.\text{CO}.\text{S-ACP} + \text{E-SH} \rightleftharpoons \text{E-S}.\text{CO}.\text{CH}_3 + \text{ACP}.\text{SH} \tag{20.2b}$$

Next the acyl carrier protein accepts a malonyl group from malonyl coenzyme A in a reaction catalysed by malonyl transacylase.

$$\text{HOOC}.\text{CH}_2.\text{CO}.\text{S-COA} + \text{ACP}.\text{SH} \rightleftharpoons \text{HOOC}.\text{CH}_2.\text{CO}.\text{S-ACP} + \text{CoA}.\text{SH} \tag{20.3}$$
malonyl coenzyme A                             malonyl ACP

The condensing enzyme now catalyses a reaction between the malonyl and the acetyl units. This condensation reaction occurs by attack of the —$CH_2$-group of malonyl ACP (marked with a dagger below) on the carbonyl group of the acetyl residue (marked with an asterisk). Carbon dioxide is lost, and the product is acetoacetyl ACP.

$$HOOC.†CH_2.CO.S\text{-}ACP + E\text{-}S.*CO.CH_3 \rightleftharpoons$$
$$CO_2 + E.SH + CH_3.*CO.†CH_2.CO.S\text{-}ACP$$
$$\text{acetoacetyl ACP}$$

$$(20.4)$$

The $\Delta G^{\circ\prime}$ of this reaction tends greatly to favour the formation of acetoacetyl ACP (note that in the corresponding reaction in the *oxidation* of fatty acids (p. 286) the $\Delta G^{\circ\prime}$ was such as to favour cleavage to *form* acetyl coenzyme A). It is largely the elimination of carbon dioxide that produces this effect; this carbon dioxide derives from that that was incorporated into the molecule in reaction (20.1), so that we can regard reaction (20.4) as being in reality the condensation of two acetyl groups, with carbon dioxide serving a catalytic role in the condensation. The ATP used in reaction (20.1), therefore, has served to activate one molecule of acetyl coenzyme A and thus to promote reaction (20.4).

*Reduction of the acetoacetyl group*

The acetoacetyl group must now be reduced. The first step in this reduction is the formation of a $>CHOH$ group from the $>CO$ group.

$$CH_3.CO.CH_2.CO.S\text{-}ACP + NADPH_2 \rightleftharpoons CH_3.CHOH.CH_2.CO.S\text{-}ACP + NADP$$
$$\text{acetoacetyl ACP} \qquad\qquad\qquad \text{β-hydroxybutyryl ACP}$$
$$(20.5)$$

This reaction is catalysed by an enzyme of the synthetase complex called the β-ketoacyl ACP reductase. In principle it is the reverse of the oxidation reaction catalysed by β-hydroxyacyl coenzyme A dehydrogenase (p. 286). There are, however, two important differences, First, in accordance with the general rule that we have now established (see pp. 29, 294 and 322), $NADPH_2$ is used for the reduction whereas NAD is the acceptor of hydrogen in the oxidation. Secondly, the β-hydroxybutyryl group that is produced in reaction (20.5) is the D-isomer, whereas that that occurs in fatty-acid oxidation is the L-isomer (p. 285).

The β-hydroxybutyryl ACP is dehydrated in the presence of enoyl ACP dehydratase, and the resulting double bond is reduced (we mentioned this latter reaction on p. 29).

$$CH_3.CHOH.CH_2.CO.S\text{-}ACP \rightleftharpoons CH_3.CH{=}CH.CO.S\text{-}ACP + H_2O$$
$$\text{β-hydroxybutyryl ACP} \qquad\qquad \text{crotonyl ACP}$$
$$(20.6)$$

$$CH_3.CH{=}CH.CO.S\text{-}ACP + NADPH_2 \rightleftharpoons CH_3.CH_2.CH_2.CO.S\text{-}ACP + NADP$$
$$\text{crotonyl ACP} \qquad\qquad\qquad \text{butyryl ACP}$$
$$(20.7)$$

The unsaturated compound produced in reaction (20.6) has the *trans* configuration, like that produced during fatty-acid oxidation. On the other hand, the reducing agent in reaction (20.7) is, like that in reaction (20.5), $NADPH_2$; the oxidizing agent in the corresponding reaction in fatty-acid degradation is a flavoprotein (see p. 285), and this is yet another point of difference between the two processes. We may recall that the principal source of $NADPH_2$ is the direct oxidation of glucose-6-phosphate to pentose phosphate (p. 293); we can now see how $NADPH_2$ is used in large quantities in fatty-acid synthesis so that there is a consequent need for rapid oxidation of glucose-6-phosphate (see also p. 295).

### Repetition of the sequence

The reactions described so far have led to the production of a butyryl group from two acetyl groups. It is now possible for another condensation to occur, analogous to that in reaction (20.4). The butyryl group is transferred to the condensing anzyme, just as the acetyl group was in reaction (20.4), and it then condenses with another molecule of malonyl ACP; carbon dioxide is again eliminated.

$$CH_3.CH_2.CH_2.CO.S\text{-}ACP + E\text{-}SH \rightleftharpoons E.S.CO.CH_2.CH_2.CH_3 + ACP.SH$$

butyryl ACP

$$HOOC.CH_2.CO.S\text{-}ACP + E.S.*CO.CH_2.CH_2.CH_3 \rightleftharpoons$$

malonyl ACP

$$CO_2 + E.SH + CH_3.CH_2.CH_2.*CO.CH_2.CO.S\text{-}ACP$$

$\beta$-oxohexanoyl ACP

The $\beta$-oxohexanoyl ACP that results is now reduced by a repetition of reactions (20.5), (20.6) and (20.7) to hexanoyl ACP, which can once again react with malonyl ACP, and so on (see Fig. 20.1).

$$CH_3.CH_2.CH_2.CO.CH_2.CO.S\text{-}ACP$$

$\downarrow$ $NADPH_2$

$$CH_3.CH_2.CH_2.CHOH.CH_2.CO.S\text{-}ACP$$

$\downarrow$ $-H_2O$

$$CH_3.CH_2.CH_2.CH{=}CH.CO.S\text{-}ACP$$

$\downarrow$ $NADPH_2$

$$CH_3.CH_2.CH_2.CH_2.CH_2.CO.S\text{-}ACP$$

In this way long-chain fatty acyl groups are built up. We can summarize the synthesis of stearoyl ACP in this way:

$$CH_3.CO.S\text{-}ACP + 8HOOC.CH_2.CO.S\text{-}ACP + 16\,NADPH_2 \longrightarrow$$
$$CH_3.(CH_2)_{16}.CO.S\text{-}ACP + 8ACP.SH + 8CO_2 + 16NADP + 8H_2O.$$

This formulation emphasizes three facts. First, the *complex* of intermediates, enzymes and acyl carrier protein is a tight one, and the cycle of condensation and reduction continues in this complex, without liberation of intermediates, until the long-chain acyl group is completed with around eighteen carbon atoms – the length being determined, presumably, by the specificity of the synthetase complex. Secondly, each acyl group that lengthens the chain by two carbon atoms has to be *reduced* by two molecules of $NADPH_2$. Thirdly, all of the two-carbon

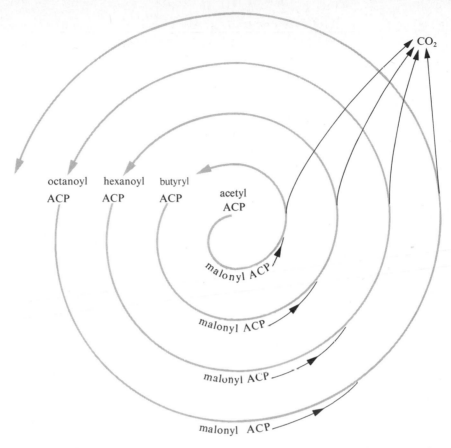

**Fig. 20.1** The synthesis of fatty-acyl ACP. Each turn of the spiral uses a malonyl residue to increase the length of fatty-acyl ACP by two carbon atoms and releases carbon dioxide. Compare Fig. 16.2, p. 287.

units except the first actually enter the sequence as *malonyl* groups; one consequence of this fact is that, in the final fatty acid produced, the two carbon atoms at the methyl end derive from acetyl ACP but all the rest are from malonyl ACP.

Another way of writing the synthesis of the stearoyl group is:

$$9CH_3.CO.S\text{-}CoA + 8CO_2 + 8ATP + 16NADPH_2 \longrightarrow$$
$$CH_3.(CH_2)_{16}.CO.S\text{-}CoA + 8CO_2 + 8ADP + 8P_i + 16NADP + 8CoA.SH + 8H_2O.$$

This formulation is intended to emphasize the fact that carbon dioxide is involved catalytically in the synthesis, and that ATP, in addition to $NADPH_2$, is needed for it. Reference to the summary reaction on p. 328 will remind you of how ATP and $NADPH_2$ are needed for the dark reaction of photosynthesis, another process in which a relatively large, reduced molecule is built up from small, highly oxidized precursors.

Stearoyl ACP is not the only possible product from the fatty-acid synthetase complex. Smaller acyl groups, such as myristoyl or palmitoyl ACP, are produced as well (see above). The acyl groups are finally set free from the acyl carrier protein, either by hydrolysis to the free fatty acid, or by reaction with coenzyme A by the reverse of a reaction similar to reactions (20.2) and (20.3).

$$CH_3.(CH_2)_n.CO.S\text{-}ACP + CoASH \rightleftharpoons CH_3.(CH_2)_n.CO.S\text{-}CoA + ACP.SH$$

## Synthesis of triglycerides and of phospholipids

*Triglycerides*

An intermediate in the synthesis of both triglycerides and phospholipids is *phosphatidic acid* (p. 190). You will remember that its formula is:

$$\begin{array}{l} CH_2O.CO.R \\ | \\ CHO.CO.R' \; . \\ | \\ CH_2O \, \textcircled{P} \end{array}$$

This is formed from $\alpha$-glycerophosphate and the coenzyme A derivatives of fatty acids.

$$\begin{array}{l} CH_2OH \\ | \\ CHOH \\ | \\ CH_2O\,\textcircled{P} \end{array} + \begin{array}{l} CH_3.(CH_2)_n.CO.S\text{-}CoA \\ + \\ CH_3.(CH_2)_m.CO.S\text{-}CoA \end{array} \rightleftharpoons \begin{array}{l} CH_2O.CO.(CH_2)_n.CH_3 \\ | \\ CHO.CO.(CH_2)_m.CH_3 \\ | \\ CH_2O\,\textcircled{P} \end{array} + 2CoA.SH$$

$\alpha$-glycero-
phosphate

phosphatidic acid

The $\alpha$-glycerophosphate required for this reaction may be supplied by either of two reactions. In some tissues, the liver for example, glycerol can be phosphorylated directly by ATP in the presence of glycerokinase (see p. 269).

$$\begin{array}{l} CH_2OH \\ | \\ CHOH \\ | \\ CH_2OH \end{array} + ATP \rightleftharpoons \begin{array}{l} CH_2OH \\ | \\ CHOH \\ | \\ CH_2O\,\textcircled{P} \end{array} + ADP$$

glycerol

$\alpha$-glycero-
phosphate

In other tissues, notably adipose tissue, glycerokinase is very feebly active and $\alpha$-glycerophosphate is formed by reducing some of the dihydroxyacetone phosphate that is formed during glycolysis (p. 261).

$$\begin{array}{l} CH_2OH \\ | \\ CO \\ | \\ CH_2O\,\textcircled{P} \end{array} + NADH_2 \rightleftharpoons \begin{array}{l} CH_2OH \\ | \\ CHOH \\ | \\ CH_2O\,\textcircled{P} \end{array} + NAD$$

dihydroxy-
acetone
phosphate

$\alpha$-glycero-
phosphate

The formation of triglycerides from phosphatidic acid occurs by a sequence of two reactions. First phosphatidic acid is hydrolysed in the presence of a phosphatase, and then the resulting diglyceride reacts with a third molecule of fatty-acyl coenzyme A.

$$\begin{array}{c}
\text{CH}_2\text{O.CO.(CH}_2)_n.\text{CH}_3 \\
|\\
\text{CHO.CO.(CH}_2)_m.\text{CH}_3 \ + \ \text{H}_2\text{O} \\
|\\
\text{CH}_2\text{O} \textcircled{P}
\end{array}
\rightleftharpoons
\begin{array}{c}
\text{CH}_2\text{O.CO.(CH}_2)_n.\text{CH}_3 \\
|\\
\text{CHO.CO.(CH}_2)_m.\text{CH}_3 \ + \ \text{P}_i \\
|\\
\text{CH}_2\text{OH}
\end{array}$$

phosphatidic acid                diglyceride

$$\begin{array}{c}
\text{CH}_2\text{O.CO(CH}_2)_n.\text{CH}_3 \\
|\\
\text{CHO.CO(CH}_2)_m.\text{CH}_3 \ + \ \text{CH}_3.(\text{CH}_2)_l.\text{CO.S-CoA} \\
|\\
\text{CH}_2\text{OH}
\end{array}
\rightleftharpoons
\begin{array}{c}
\text{CH}_2\text{O.CO.(CH}_2)_n.\text{CH}_3 \\
|\\
\text{CHO.CO.(CH}_2)_m.\text{CH}_3 \ + \ \text{CoA.SH} \\
|\\
\text{CH}_2\text{O.CO.(CH}_2)_l.\text{CH}_3
\end{array}$$

diglyceride                       triglyceride

Once again we may contrast this sequence of reactions with the corresponding degradative pathway. The degradation of triglycerides is by simple hydrolysis (p. 282) to free fatty acids and free glycerol.

*Phospholipids*

The formation of phospholipids is slightly more complicated. There are two pathways for the synthesis of phospholipids from phosphatidic acid, one of which occurs almost exclusively in higher animals and the other predominantly in plants and micro-organisms. Both of them involve activation by a nucleoside triphosphate to give a nucleoside diphosphate carrier analogous to UDP-glucose (see pp. 333 ff.). The nucleoside involved here is cytidine, rather than the uridine generally used in polysaccharide synthesis.

In the first pathway that we shall consider (that occurring chiefly in animals), the *bases* (ethanolamine and choline) that are to be inserted into the phospholipid are phosphorylated, and the phosphoryl derivatives then react with cytidine triphosphate (CTP) to give cytidine diphosphate ethanolamine (CDP-ethanolamine) and cytidine diphosphate choline (CDP-choline)(cf. p. 143).

$$\text{HOH}_2\text{C.CH}_2\text{NH}_2 + \text{ATP} \rightleftharpoons \textcircled{P}\text{OH}_2\text{C.CH}_2\text{NH}_2 + \text{ADP}$$

ethanolamine              phosphorylethanol-
                                amine

$$\text{HOH}_2\text{C.CH}_2\overset{+}{\text{N}}(\text{CH}_3)_3 + \text{ATP} \rightleftharpoons \textcircled{P}\text{OH}_2\text{C.CH}_2\overset{+}{\text{N}}(\text{CH}_3)_3 + \text{ADP}$$

choline                       phosphorylcholine

$$\textcircled{P}\text{OH}_2\text{C.CH}_2\text{NH}_2 + \text{CTP} \rightleftharpoons \text{CDP-ethanolamine} + \text{PP}_i$$

$$\textcircled{P}\text{OH}_2\text{C.CH}_2\overset{+}{\text{N}}(\text{CH}_3)_3 + \text{CTP} \rightleftharpoons \text{CDP-choline} + \text{PP}_i$$

Notice that this pathway is analogous to the formation of UDP-glucose via glucose-1-phosphate (p. 334).

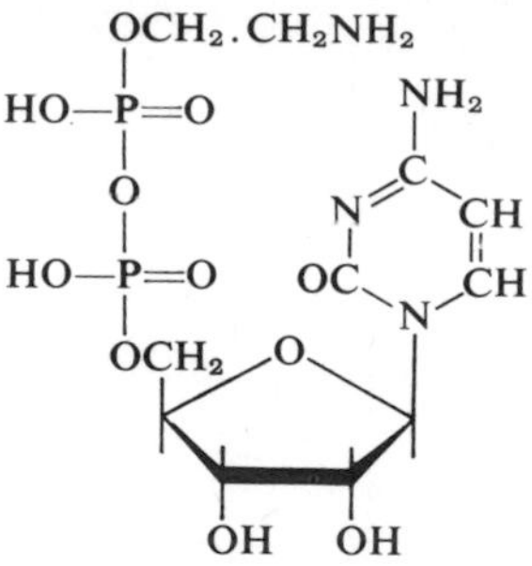

**CDP-ethanolamine** (structure, left)

$$OCH_2 . CH_2NH_2$$

$$HO-P=O$$

$$O$$

$$HO-P=O$$

$$OCH_2$$ — cytosine ribose ring with $NH_2$, $N=C$, $CH$, $OC$, $N$, $CH$

$$OH \quad OH$$

CDP-ethanolamine

**CDP-choline** (structure, right)

$$OCH_2 . CH_2\overset{+}{N}(CH_3)_3$$

$$HO-P=O$$

$$O$$

$$HO-P=O$$

$$OCH_2$$ — cytosine ribose ring

$$OH \quad OH$$

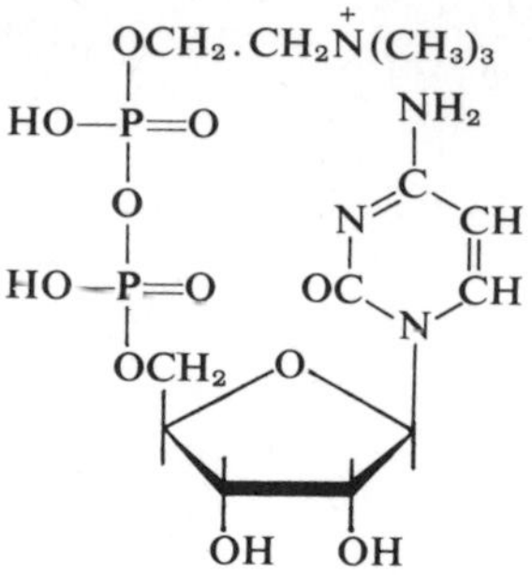

CDP-choline

CDP-ethanolamine and CDP-choline can now react with diglyceride to form phosphatidyl ethanolamine and phosphatidyl choline. (We give here the formation of phosphatidyl ethanolamine; the synthesis of phosphatidyl choline is precisely analogous.)

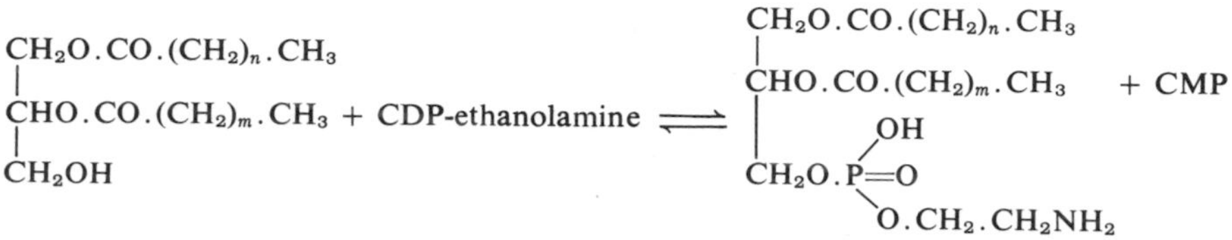

$$\begin{array}{l} CH_2O . CO . (CH_2)_n . CH_3 \\ | \\ CHO . CO . (CH_2)_m . CH_3 \ + \ CDP\text{-ethanolamine} \ \rightleftharpoons \\ | \\ CH_2OH \end{array}$$

diglyceride

$$\begin{array}{l} CH_2O . CO . (CH_2)_n . CH_3 \\ | \\ CHO . CO . (CH_2)_m . CH_3 \qquad + \ CMP \\ | \qquad\qquad OH \\ | \qquad\qquad / \\ CH_2O . P = O \\ \qquad\qquad\backslash \\ \qquad\qquad O . CH_2 . CH_2NH_2 \end{array}$$

phosphatidyl ethanolamine

In the other pathway of synthesis of phospholipid, CTP is used, interestingly enough, to activate the *other* reactant involved in the formation of phosphatidyl ethanolamine and phosphatidyl choline, that is the *phosphatidic acid*. The compound thus formed is CDP-diglyceride, the formula of which is:

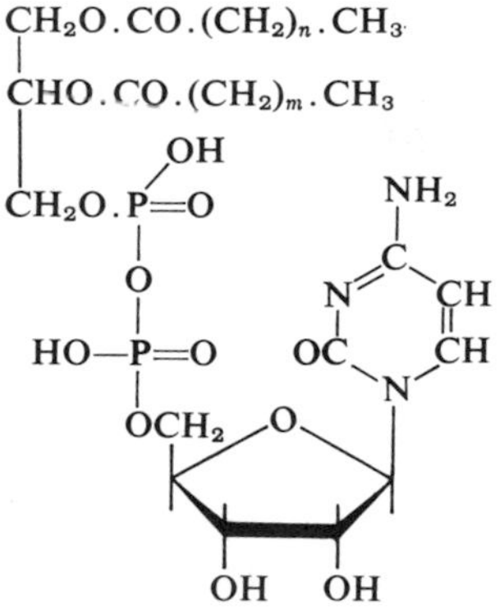

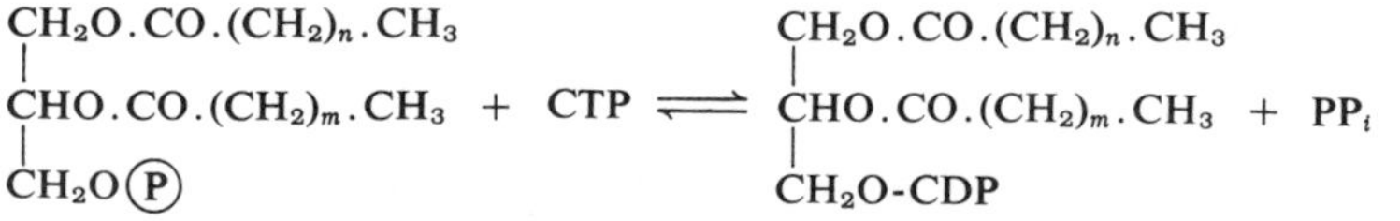

$$\begin{array}{l} CH_2O . CO . (CH_2)_n . CH_3 \\ | \\ CHO . CO . (CH_2)_m . CH_3 \\ | \qquad\qquad OH \\ | \qquad\qquad / \\ CH_2O . P = O \\ \qquad\qquad | \\ \qquad\qquad O \\ HO-P=O \qquad \text{(cytosine ribose ring with } NH_2\text{)} \\ | \\ OCH_2 \end{array}$$

$$OH \quad OH$$

$$\begin{array}{l} CH_2O . CO . (CH_2)_n . CH_3 \\ | \\ CHO . CO . (CH_2)_m . CH_3 \ + \ CTP \ \rightleftharpoons \\ | \\ CH_2O\,\textcircled{P} \end{array}$$

phosphatidic acid

$$\begin{array}{l} CH_2O . CO . (CH_2)_n . CH_3 \\ | \\ CHO . CO . (CH_2)_m . CH_3 \ + \ PP_i \\ | \\ CH_2O\text{-CDP} \end{array}$$

CDP-diglyceride

This can now react with serine to give phosphatidyl serine, which yields phosphatidyl ethanolamine by decarboxylation.

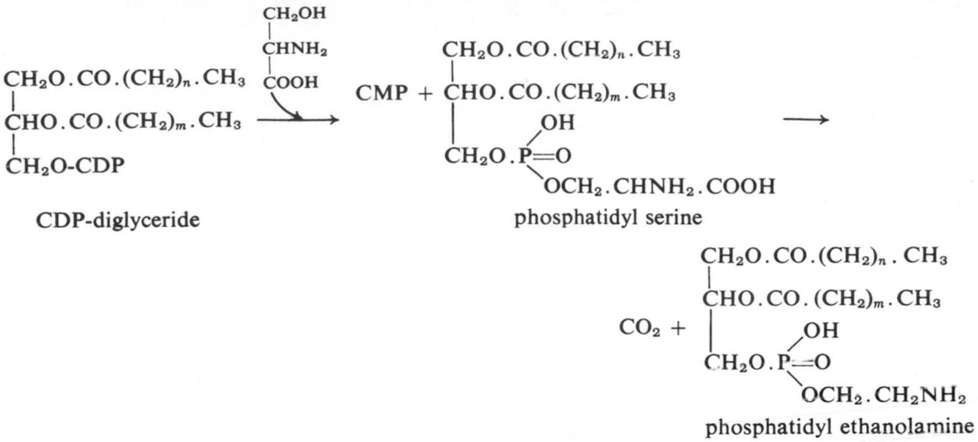

$$
\begin{array}{l}
\text{CH}_2\text{O.CO.(CH}_2)_n.\text{CH}_3 \\
| \\
\text{CHO.CO.(CH}_2)_m.\text{CH}_3 \\
| \\
\text{CH}_2\text{O-CDP}
\end{array}
\quad
\begin{array}{l}
\text{CH}_2\text{OH} \\
| \\
\text{CHNH}_2 \\
| \\
\text{COOH}
\end{array}
\longrightarrow
\text{CMP} +
\begin{array}{l}
\text{CH}_2\text{O.CO.(CH}_2)_n.\text{CH}_3 \\
| \\
\text{CHO.CO.(CH}_2)_m.\text{CH}_3 \\
| \qquad\quad \text{OH} \\
\text{CH}_2\text{O.P}{=}\text{O} \\
\qquad\quad \text{OCH}_2.\text{CHNH}_2.\text{COOH}
\end{array}
\longrightarrow
$$

CDP-diglyceride                     phosphatidyl serine

$$
\text{CO}_2 +
\begin{array}{l}
\text{CH}_2\text{O.CO.(CH}_2)_n.\text{CH}_3 \\
| \\
\text{CHO.CO.(CH}_2)_m.\text{CH}_3 \\
| \qquad\quad \text{OH} \\
\text{CH}_2\text{O.P}{=}\text{O} \\
\qquad\quad \text{OCH}_2.\text{CH}_2\text{NH}_2
\end{array}
$$

phosphatidyl ethanolamine

In the organisms in which this pathway operates, phosphatidyl ethanolamine can accept methyl groups from a methyl donor (such as S-adenosylmethionine, see p. 385), and the ethanolamine group is methylated in stages to choline.

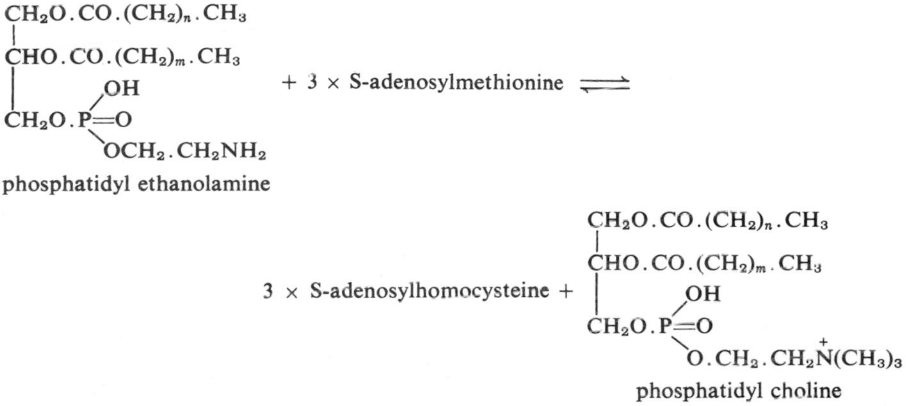

$$
\begin{array}{l}
\text{CH}_2\text{O.CO.(CH}_2)_n.\text{CH}_3 \\
| \\
\text{CHO.CO.(CH}_2)_m.\text{CH}_3 \\
| \qquad\quad \text{OH} \\
\text{CH}_2\text{O.P}{=}\text{O} \\
\qquad\quad \text{OCH}_2.\text{CH}_2\text{NH}_2
\end{array}
\quad + 3 \times \text{S-adenosylmethionine} \rightleftharpoons
$$

phosphatidyl ethanolamine

$$
3 \times \text{S-adenosylhomocysteine} +
\begin{array}{l}
\text{CH}_2\text{O.CO.(CH}_2)_n.\text{CH}_3 \\
| \\
\text{CHO.CO.(CH}_2)_m.\text{CH}_3 \\
| \qquad\quad \text{OH} \\
\text{CH}_2\text{O.P}{=}\text{O} \\
\qquad\quad \text{O.CH}_2.\text{CH}_2\overset{+}{\text{N}}(\text{CH}_3)_3
\end{array}
$$

phosphatidyl choline

Actually, in some organisms (particularly bacteria) only one or two methyl groups are used, and phosphatidyl monomethylethanolamine or phosphatidyl dimethylethanolamine occur in the phospholipid in place of phosphatidyl choline.

# Chapter 21 Intracellular compartmentation in carbohydrate and fat metabolism

In the previous chapters we have discussed the pathways of carbohydrate and fat metabolism rather as if the cell contained an homogeneous soup, in which reactions were equally likely to take place in any part of the cytoplasm. In fact, however, as we stressed in Ch. 10, the cell of a higher organism is a highly organized structure, and it contains many internal membranes which act as barriers to the free movement of material. In this chapter we shall examine the intracellular distribution of the enzymes that catalyse the pathways we have described, and show how the localization of these enzymes both benefits and poses problems to the cell as a whole.

The questions raised by intracellular organization have still been answered only very partially by biochemists, and a great deal remains to be discovered. We are far from understanding the distribution of enzymes needed for all metabolic processes, and far, too, from fully understanding the function of some of the specialized organelles of the cell. In those areas of metabolism that we are concerned with here – the degradation and synthesis of carbohydrates and fats – the problems are comparatively simplified: the distribution of the relevant enzymes is fairly well known, and we are concerned with only two intracellular compartments (since all of the reactions, with the exception of the synthesis of triglyceride in the endoplasmic reticulum, occur in the mitochondrion or the soluble fraction of the cytoplasm). The pathways of degradation and synthesis of carbohydrates and fats will therefore serve to illustrate some of the principles of intracellular organization; we shall make no attempt at a comprehensive treatment of all metabolism in terms of intracellular location.

For the present purpose, then, we shall regard the cell as containing only soluble cytoplasm in which mitochondria are suspended, and we shall ignore the other organelles which we described in Ch. 10. Mitochondria are bounded by a double membrane (p. 217), and the inner membrane, as we have seen (p. 244), prevents the free diffusion of essentially all metabolites, ions, co-factors and proteins. All substances of metabolic significance* that move either into or out of the mitochondrion do so because they are transported by specific *carriers* (see p. 201). The result is that the mitochondrion is a *compartment* in which the concentration of many metabolites is very different from that in the soluble cytoplasm.

This fact is of great benefit to the economy of the cell. By arranging for the

---

* With the exception of molecules of very low molecular weight such as water.

enzymes that catalyse particular pathways to be located within the mitochondrion, the cell can ensure that the relevant metabolic intermediates are present in high local concentration in the vicinity of the enzymes that deal with them. Again, the product of one reaction will need to diffuse only a short way before encountering the enzyme that catalyses the next reaction (cf. p. 237). In this way the cell can achieve a very rapid flux of material along a metabolic pathway by confining the enzymes of that pathway inside the mitochondrion.

There is a further advantage that derives from intracellular compartmentation. We have often referred to the fact that the pathways of degradation of most molecules are different from their pathways of synthesis. This arrangement facilitates the control of degradation and synthesis by mechanisms that we shall discuss in Ch. 29. Now often the enzymes of the degradative pathway are, in fact, located in a different cellular compartment from the enzymes of the synthetic pathway. (We have already seen that the enzymes involved in the degradation of fatty acids are confined to the mitochondrion and those involved in the synthesis of fatty acids are in the soluble cytoplasm.) Such spatial separation makes it still easier for the cell to control the pathways independently. If, for example, the concentration of ATP is important in regulating the activity of the metabolic pathways (see pp. 504 ff.), the concentration of ATP inside the mitochondrion might be maintained at a different level from that outside the mitochondrion, and these different concentrations may achieve an extremely fine control over the balance of degradation and synthesis.

At the same time the impermeability of the mitochondrial membrane presents the cell with the problem of how movement across it may be effected for those molecules that actually need to pass. For example, the mitochondrion is the organelle within which pyruvate (from glycolysis) and fatty acids (from the hydrolysis of lipids) are oxidized. But pyruvate and fatty acids are produced (by degradation of glucose and hydrolysis of fats) in the soluble cytoplasm – how can they pass across the mitochondrial membrane to the site of their oxidation?

There appear to be three mechanisms for ensuring transport of metabolites across the mitochondrial membrane. The first is transport of the metabolite itself by a more or less specific carrier of the type that we discussed on pp. 201 ff. (ATP, for example, is transported in this way.) The second involves the conversion of the metabolite to a form in which it can more freely cross the membrane. (Fatty-acyl coenzyme A is an example.) The third mechanism is more roundabout. Sometimes there is no direct means at all of transporting a particular molecule across the mitochondrial membrane. However, the molecule may donate its functional group to another metabolite which *can* be transported, and in this way the relevant function may pass across the membrane in a form different from that of the original molecule. (The transport of the reducing equivalent of $NADH_2$, for example, is effected in this way, as we shall see.)

With these points in mind, let us examine how the transport of substances involved in carbohydrate and fat metabolism occurs between the soluble fraction of the cytoplasm and the mitochondrion.

*Glycolysis* occurs in the soluble cytoplasm and produces two compounds which are oxidized in the mitochondrion, namely pyruvate and $NADH_2$. Not a great deal is known about the entry of pyruvate into the mitochondrion; we may assume that it occurs via a protein carrier located in the mitochondrial membrane – possibly one that is not highly specific for pyruvate. A good deal of attention, on the other hand, has been paid to the problem of how $NADH_2$ that is produced in the cytoplasm (whether from glycolysis or from other reactions) can be oxidized in the mitochondrion via the respiratory chain. The answer is an intriguing one.

It turns out that $NADH_2$, and for that matter $NADPH_2$, NAD and NADP, cannot pass across the mitochondrial membrane. But since glycolysis produces two molecules of $NADH_2$ for every molecule of glucose degraded (see p. 266) there must be some means, in aerobic conditions, of reoxidizing the $NADH_2$ via the respiratory chain. The re-oxidation is achieved by a system known as a *shuttle*, which works (at any rate in insect muscle where it has been carefully studied) in the following way (see Fig. 21.1). $NADH_2$ in the cytoplasm, in the

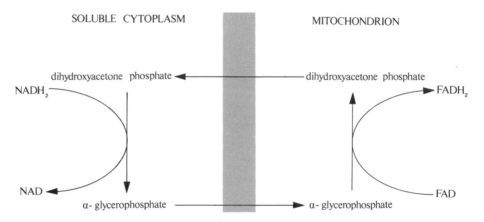

SOLUBLE CYTOPLASM    MITOCHONDRION

**Fig. 21.1** The transfer of reducing equivalents to the mitochondrion (see text)

presence of α-glycerophosphate dehydrogenase, reduces dihydroxyacetone phosphate to α-glycerophosphate (p. 346). The mitochondrial membrane contains a carrier for α-glycerophosphate, so that this passes into the mitochondrion and the hydrogen can then be oxidized via the respiratory chain.* In this way the reducing *equivalents* of $NADH_2$ can enter the mitochondrion even though the $NADH_2$ itself cannot. (The dihydroxyacetone phosphate produced returns to the soluble cytoplasm with the help of a carrier.) We shall see shortly that there is another shuttle of a similar sort that acts to transport the reducing equivalent of $NADH_2$ *out* of the mitochondrion when needed for gluconeogenesis.

The oxidation of *fatty-acyl coenzyme A* also occurs in the mitochondrion, but the hydrolysis of triglycerides that makes fatty acids available (p. 282) takes place outside the mitochondrion. We must therefore see in what form fatty acids can enter the mitochondrion. The membrane turns out to be completely impermeable

* The mitochondrial α-glycerophosphate dehydrogenase is a flavoprotein. In consequence, only two molecules of ATP per molecule of substrate are synthesized by the oxidation of α-glycerophosphate inside the mitochondrion (see p. 240). The result is to diminish by two molecules the yield of ATP that we calculated should be produced by the oxidation of one molecule of glucose (p. 280).

to fatty-acyl coenzyme A, but permeable to esters formed between fatty acids and a substance called *carnitine*, which has the formula

$$(CH_3)_3 . \overset{+}{N} . CH_2 . CHOH . CH_2 . COOH$$

A transferase enzyme in the mitochondrial membrane catalyses the reaction of fatty-acyl coenzyme A, formed in the soluble cytoplasm by the reaction we described on p. 284, with carnitine:

$$CH_3 . (CH_2)_n . CO . S\text{-}CoA + (CH_3)_3 \overset{+}{N} . CH_2 . CHOH . CH_2 . COOH \rightleftharpoons$$

$$(CH_3)_3 \overset{+}{N} . CH_2 . \underset{\underset{\displaystyle CO . (CH_2)_n . CH_3}{|}}{\overset{\overset{\displaystyle |}{\displaystyle O}}{CH}} . CH_2 . COOH \qquad + CoA . SH$$

and the resulting fatty-acyl carnitine now passes through the membrane into the mitochondrion, where the reverse of the above reaction takes place. Since fatty-acyl carnitine contains a high-energy bond, no expenditure of ATP is required for the two reactions.

As a result of the oxidation of pyruvate, fatty-acyl coenzyme A, and other compounds, large quantities of *ATP* are produced in the mitochondria. The bulk of this ATP, however, is needed outside the mitochondrion, for the synthesis of carbohydrates and fatty acids in the soluble cytoplasm, for the synthesis of proteins at the ribosomes (see p. 448) and for many of the other processes that we mentioned on pp. 24 ff. It turns out that the movement of ATP across the mitochondrial membrane is catalysed by a specific carrier – it was the study of this carrier, and its specific inhibition by atractylate, that contributed substantially to our understanding of the protein nature of membrane carriers (pp. 201 ff.). The function of the carrier is to promote an exchange of ATP with ADP: for every molecule of ATP that passes out of the mitochondrion into the soluble cytoplasm, a molecule of ADP passes from the cytoplasm into the mitochondrion. We can thus picture a cycle of ADP–ATP synthesis and degradation. ADP passes from the cytoplasm into the mitochondrion and is there phosphorylated; the resulting ATP moves out of the mitochondrion again and donates its terminal phosphate group in some energy-requiring process, and the cytoplasmic ADP is ready to start the cycle again.

The carrier that we have just described accounts for the presence in the cytoplasm of ATP, which is one of the prerequisites for *gluconeogenesis*. On p. 321 we remarked that the starting compound for gluconeogenesis is generally lactate or one of the amino acids that gives rise to an intermediate in the Krebs cycle; in either event one of the essential reactions for the synthesis of carbohydrate is the conversion of oxaloacetate to phospho-*enol*pyruvate (p. 330). A problem that we did not previously consider is that oxaloacetate, like the other Krebs cycle intermediates, occurs in the mitochondrion, whereas the enzymes needed for gluconeogenesis from phospho-*enol*pyruvate are in the cytoplasm.

The difficulty is circumvented in the following way (see Fig. 21.2). Oxaloacetate in the mitochondrion is reduced in the presence of malate dehydrogenase (p. 279) to malate. There is a specific carrier in the mitochondrial membrane which transports malate into the soluble cytoplasm, and there is a cytoplasmic form of malate dehydrogenase which produces oxaloacetate again. The phospho-*enol*pyruvate carboxykinase and other enzymes needed for gluconeogenesis are present in the cytoplasm, so that from this point the synthesis of carbohydrate poses no special problem in terms of localization of intermediates. However, gluconeogenesis needs not only cytoplasmic ATP and cytoplasmic oxaloacetate but also cytoplasmic $NADH_2$ (see p. 331). This need is met, with remarkable

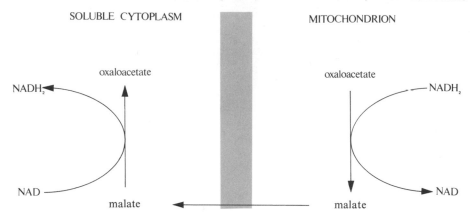

**Fig. 21.2** The transfer of oxaloacetate and reducing equivalents to the soluble cytoplasm (see text)

economy, by precisely the same system as we have just discussed. Inside the mitochondrion we have:

$$oxaloacetate + NADH_2 \rightleftharpoons malate + NAD.$$

The malate then passes out of the mitochondrion, and in the cytoplasm we have:

$$malate + NAD \rightleftharpoons oxaloacetate + NADH_2.$$

The net result is a transfer not only of oxaloacetate but also of $NADH_2$ from the mitochondrion into the cytoplasm. The system thus provides a 'shuttle' for transferring reducing power to the soluble cytoplasm.*

We shall conclude this brief sketch of the effect of intracellular compartmentation by referring to the *synthesis of fatty acids*. As we have seen (p. 340) the precursor for fatty acids is acetyl coenzyme A, which derives chiefly from the oxidation of pyruvate, derived in turn from glycolysis. Now the oxidation of pyruvate to acetyl coenzyme A occurs in the mitochondrion (p. 273), but the synthesis of fatty acids takes place in the soluble cytoplasm. Acetyl coenzyme A must therefore be transported across the mitochondrial membrane, but this process cannot occur directly as there is no carrier for acyl coenzyme A. Instead the mechanism is as follows. Acetyl coenzyme A reacts with oxaloacetate to form citrate (p. 276). The mitochondrial membrane contains a carrier which can transport citrate into the cytoplasm, and the citrate that appears there gives rise to acetyl coenzyme A in a reaction catalysed by an enzyme called the citrate cleavage enzyme.

$$citrate + coenzyme A + ATP \rightleftharpoons$$
$$oxaloacetate + acetyl coenzyme A + ADP + P_i$$

---

* The system that we have described is that known to operate in the rat. Other species sometimes have different shuttle systems.

The other prerequisites for fatty-acid synthesis are ATP, the transport of which we have already discussed, and $NADPH_2$. $NADPH_2$ is produced chiefly in the Warburg–Dickens pathway (pp. 292 ff.), the enzymes of which are themselves located in the soluble cytoplasm.

This sketch of some of the problems that result from the existence of impermeable barriers, and of the mechanisms that are used to overcome them, is obviously very far from exhaustive. It has been our intention merely to indicate a few of the reactions that the organization of the cell makes more complicated than they first appear. The metabolism of carbohydrate and fat provides useful illustrations of these complications, and although we could try to discuss all metabolic reactions in similar terms, such a discussion would probably prove more confusing than helpful. We have said enough in this chapter to illustrate the principles; there is no doubt that much biochemical work will be concentrated in this direction in the future.

# Chapter 22 **Metabolism of amino acids**

In Chs 5 and 6 we described some of the functions of proteins, and showed how important they are in almost all biological processes. In our discussion of inter-mediary metabolism, by contrast, we have so far confined our attention entirely to carbohydrates and fats and have made no mention of amino acids. The reason for this neglect of amino acids is that our focus hitherto has been on those processes in intermediary metabolism that are important in producing ATP – namely the degradation of carbohydrates and fats – and the means by which the reverse processes are accomplished. By comparison with carbohydrates and fats, amino acids are of rather little importance as fuel. They are, however, of over-whelming importance as constituents of protein, and our chief emphasis in this chapter will therefore be on the way in which organisms provide themselves with the twenty species of amino acid that are needed for the synthesis of proteins. At the same time we shall refer briefly to the degradation of amino acids, partly because they are used to *some* extent as sources of ATP and partly because some amino acids provide an exception to the general rule that pathways of synthesis are different from pathways of degradation; it is in fact convenient to discuss the synthesis and degradation of amino acids together.

The most important point that we wish to stress about amino-acid metabolism is that its reactions are linked with the metabolic sequences that we have described in previous chapters – that is, the metabolism of amino acids is closely connected with that of carbohydrates and fats. Now there are two aspects to this connection. The first is that there is a small number of central reactions which are of im-portance in the metabolism of amino acids in general. These are amphibolic reactions (see Ch. 9), involved both in the synthesis and in the degradation of amino acids, and one of them provides a direct link between amino-acid meta-bolism and the tricarboxylic acid cycle. The second aspect is that the synthesis of most individual amino acids takes as its starting point one or other of the intermediates in carbohydrate degradation that we have previously encountered, and the degradation of individual amino acids gives rise to other intermediates, either of carbohydrate metabolism or of fat metabolism.

We shall first describe the amphibolic reactions that are central to the metabolism of all amino acids. Next we shall mention some of the ways in which amino acids have been classified and show how the systems of classification help us to make generalizations about amino-acid metabolism. Finally we shall adopt one of the classificatory schemes, use it to describe the routes of synthesis of the amino acids and give some outline of their routes of degradation.

# CENTRAL REACTIONS

There is one reaction that is of crucial importance in amino-acid metabolism, namely the *synthesis of glutamate* by reductive amination of α-oxoglutarate.

$$
\begin{array}{l}
\text{COOH} \\
| \\
\text{CO} \\
| \\
\text{CH}_2 \\
| \\
\text{CH}_2 \\
| \\
\text{COOH}
\end{array}
+ \text{NH}_3 + \text{NADH}_2 \text{ or NADPH}_2
\rightleftharpoons
\begin{array}{l}
\text{COOH} \\
| \\
\text{CHNH}_2 \\
| \\
\text{CH}_2 \\
| \\
\text{CH}_2 \\
| \\
\text{COOH}
\end{array}
+ \text{H}_2\text{O} + \text{NAD or NADP} \quad (22.1)
$$

α-oxoglutaric acid     glutamic acid

This reaction is used in both directions in the cell, and the enzyme that catalyses it is named for its ability to catalyse the back reaction – glutamate dehydrogenase. The reaction is of special significance for two reasons. In the first place, it provides a link between the tricarboxylic acid cycle (and therefore the metabolism of carbohydrates and fats) and the metabolism of amino acids. Secondly it is, in many organisms including higher animals, the one reaction in which inorganic nitrogen can be fixed to form the α-amino group of an amino acid. We shall see (pp. 363 and 387) that there are two other reactions for the fixation of ammonia; neither of these, however, is used in the *de novo* synthesis of the α-amino group.

The amination of α-oxoglutarate to form glutamate is of even more widespread importance than may appear at first sight, since it serves to form the α-amino group not only of glutamate itself but also of most other amino acids. This effect comes about because the amino group, once introduced into glutamate, can be *transferred* to yield other amino acids. In this type of transfer reaction glutamate* reacts with an α-keto acid, and the two acids exchange the groups on the α-carbon atoms.

$$
\begin{array}{l}
\text{COOH} \\
| \\
\text{CHNH}_2 \\
| \\
\text{CH}_2 \\
| \\
\text{CH}_2 \\
| \\
\text{COOH}
\end{array}
+
\begin{array}{l}
\text{COOH} \\
| \\
\text{CO} \\
| \\
\text{R}
\end{array}
\rightleftharpoons
\begin{array}{l}
\text{COOH} \\
| \\
\text{CO} \\
| \\
\text{CH}_2 \\
| \\
\text{CH}_2 \\
| \\
\text{COOH}
\end{array}
+
\begin{array}{l}
\text{COOH} \\
| \\
\text{CHNH}_2 \\
| \\
\text{R}
\end{array}
\quad (22.2)
$$

glutamic acid     α-oxo-glutaric acid

This reaction is called a *transamination*. It has an equilibrium constant close to one, and can be used to form glutamate by donation of the α-amino group from an amino acid to α-oxoglutarate.

Transaminations are catalysed by a group of enzymes called transaminases,

---

* In some organisms the couple alanine-pyruvate or the couple aspartate-oxaloacetate replaces the couple glutamate-α-oxoglutarate in certain transaminations.

all of which employ pyridoxal phosphate as coenzyme. Pyridoxal phosphate has the formula:

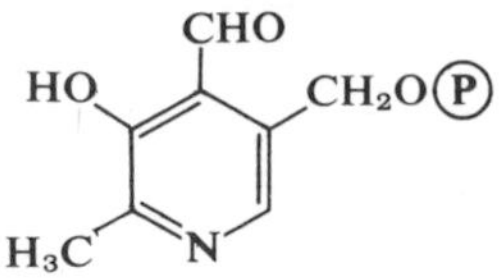

and is formed in the body from pyridoxine:

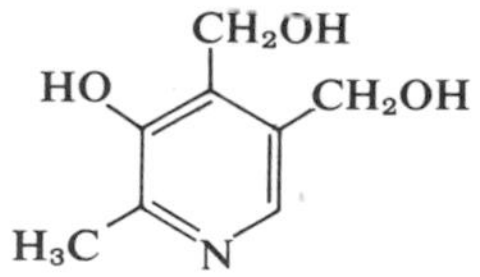

which is a vitamin (see p. 80). During the transamination reaction, pyridoxal phosphate is reversibly converted to its aminated form, pyridoxamine phosphate, via a Schiff's base.

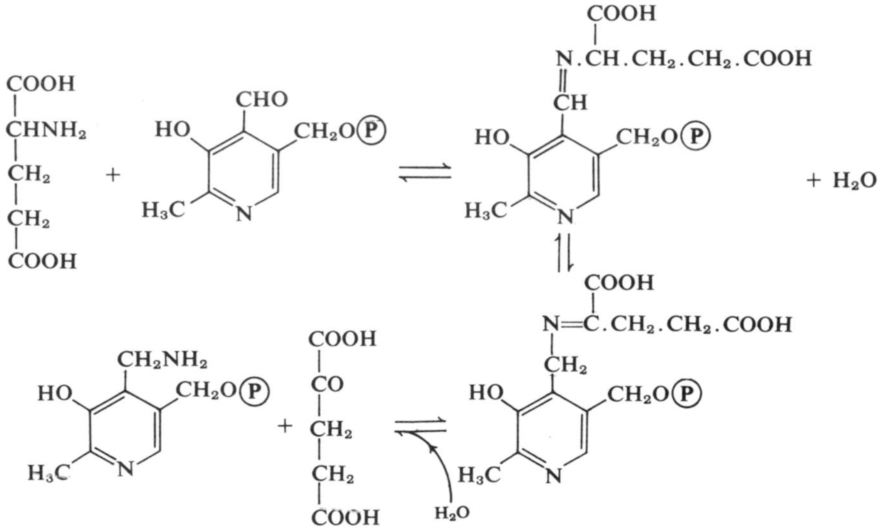

An $\alpha$-keto acid can now react with pyridoxamine phosphate by a reversal of this sequence, to restore pyridoxal phosphate and yield the amino acid that corresponds to the reacting $\alpha$-keto acid.

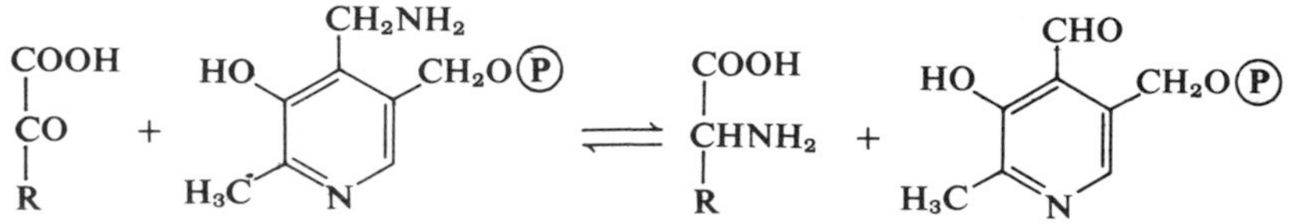

The $\alpha$-oxoglutarate that is liberated in the transamination can be converted again to glutamate by reaction (22.1), and this can then take part once more in transamination. In this way the couple glutamate-$\alpha$-oxoglutarate can function catalytically in the synthesis of an amino acid, indirectly fixing ammonia into the corresponding $\alpha$-keto acid.

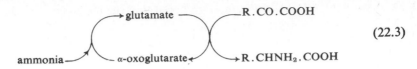

$$(22.3)$$

As we mentioned above, these reactions are amphibolic. We have so far concentrated on the way in which they are used for the synthesis of amino acids, but as reactions (22.1) and (22.2) are both freely reversible the couple glutamate-$\alpha$-oxoglutarate can equally be used in the degradation of amino acids.

$$
\begin{array}{c}
\text{glutamate} \leftarrow \qquad \rightarrow R.CO.COOH \\
\\
NH_3 \leftarrow \quad \rightarrow \alpha\text{-oxoglutarate} \qquad R.CHNH_2.COOH
\end{array}
\qquad (22.4)
$$

These reactions serve to convert an amino acid to its corresponding $\alpha$-keto acid by the liberation of ammonia. We shall follow the degradation of the $\alpha$-keto acids produced in this kind of reaction later in this chapter, and the fate of the ammonia in the next chapter.

The reason why glutamate features in all of the reactions we have mentioned so far is that the dehydrogenase specific for glutamate is a highly active enzyme of widespread metabolic importance. There is a non-specific L-amino acid oxidase present in some tissues which can in principle carry out reactions of the type:

$$
\begin{array}{ccc}
COOH & & COOH \\
| & & | \\
CHNH_2 + \tfrac{1}{2}O_2 & \rightleftharpoons & CO \quad + NH_3 \\
| & & | \\
R & & R
\end{array}
$$

but the enzyme is so weakly active that it is probably of negligible metabolic significance. In effect, therefore, all conversions between ammonia and $\alpha$-amino groups proceed *via* the glutamate-$\alpha$-oxoglutarate couple.*

Another use of that couple is in the transfer of an amino group from an amino acid to an $\alpha$-keto acid without the formation of free ammonia. This reaction is used when one amino acid is relatively abundant and another is in short supply.

$$
\begin{array}{c}
R'.CO.COOH \leftarrow \qquad \rightarrow \text{glutamate} \qquad R.CO.COOH \\
\\
R'.CHNH_2.COOH \qquad \alpha\text{-oxoglutarate} \leftarrow \rightarrow R.CHNH_2.COOH
\end{array}
\qquad (22.5)
$$

By the use of these various reactions, organisms can synthesize most of the amino acids that they need provided that they have the corresponding $\alpha$-keto acids, and similarly they can deaminate most amino acids to the corresponding $\alpha$-keto acids. For this reason the pathways of synthesis and degradation of amino acids can best be regarded as special pathways of *carbon* metabolism rather than of nitrogen metabolism. In the synthesis of an amino acid the amino

---

* The liver and kidney also contain a D-amino acid oxidase of great catalytic power. Obviously this is of no consequence in the metabolism of the normal (L) amino acids. Its function may be, as we suggested on p. 51, to destroy any D-amino acids that chance to occur in the body.

group is often introduced into the molecule fairly late in the process, and in the degradation of an amino acid the amino group is usually removed from the molecule fairly early in the process. It is for these reasons that the routes both of synthesis and of degradation of the amino acids are, as we remarked above, closely connected with the routes of metabolism of carbohydrate and fat. These connections will become evident as we follow the synthesis and degradation of the individual amino acids in the remainder of this chapter.

## CLASSIFICATION SCHEMES IN AMINO-ACID METABOLISM

A study of the metabolism of each of the twenty amino acids can at first sight present a rather bewildering complexity, and several schemes for classifying amino acids have been introduced in an attempt to make the subject less daunting. We have already classified amino acids (in Ch. 4) according to the contribution that each makes to the *structure* of proteins. In this section we shall classify them in three different ways by *metabolic* criteria. Each of the three classifications will enable us to discuss certain general features of amino-acid metabolism.

*Essential and non-essential amino acids*

Many organisms, including most plants and many micro-organisms, are capable of synthesizing for themselves all of the twenty amino acids that are normal constituents of proteins. In other organisms, particularly higher animals, the ability to synthesize some of the amino acids has been lost during evolution. (In Ch. 24 we shall discuss in outline how the loss occurs.) The result is that the amino acids that cannot be made by a given species of animal become essential constituents of the diet for that animal.

The precise list of essential amino acids varies from one animal species to another,* and is of no great theoretical significance although it is naturally of great practical importance. Some general points about essential amino acids can, however, be made. First, it is necessary to distinguish between essential amino acids and vitamins (p. 80 ff.). The difference is that essential amino acids are needed in substrate quantities as they are actually *constituents* in the synthesis of protein. By contrast vitamins are needed in catalytic quantities as they are used as cofactors of enzymic reactions and are therefore regenerated at the end of the reaction (see pp. 78 and 265). Secondly, the fact that some amino acids are essential has important implications for evaluating the quality of an animal's diet. If the protein constituents of the diet contain only small quantities of an essential amino acid, then the protein can be used by the animal only to the extent that it satisfies the requirement for that particular amino acid.† The rate of synthesis of the animal's own protein will be limited by the availability of the one essential amino acid; the other amino acids present will be of no use in this

* In the rat the essential amino acids are arginine, histidine, lysine, threonine, methionine, isoleucine, leucine, valine, phenylalanine and tryptophan.

† In general plant proteins are less valuable than animal proteins because they contain smaller quantities of essential amino acids.

respect, and since they cannot be used in protein synthesis they will be deaminated, generally by reaction (22.4) above. Since proteins in the diet never contain amino acids in precisely the proportions that the animal requires, there is always some deamination of amino acids and consequently a need to dispose of the α-keto acids that are released.

### Glucogenic and ketogenic amino acids

The nature of the final product of catabolism of such α-keto acids forms the basis of another classification of the amino acids. We can illustrate the general principles of catabolism by using the metabolic sequences we have described in Chs 15 and 16 to discuss the catabolic routes of α-keto acids, and shall then see what final products usually arise from them.

The deamination product of an amino acid is, as we saw from reaction (22.4), an α-keto acid. Now we have two precedents for the degradation of α-keto acids: pyruvate (p. 271), which undergoes oxidative decarboxylation to acetyl coenzyme A, and α-oxoglutarate (p. 277), which undergoes oxidative decarboxylation to succinyl coenzyme A. Similarly the α-keto acids derived by deamination of amino acids can be expected sometimes to undergo oxidative decarboxylation to yield the corresponding acyl coenzyme A derivatives. If this acyl coenzyme A has a relatively long carbon chain, it may be treated in the way that we have described for acyl coenzyme A derivatives of fatty acids (Ch. 16).

There are several possible end-products of these catabolic sequences. Some amino acids give rise to oxaloacetate or pyruvate. Others give rise to α-oxoglutarate. A third class gives rise to succinyl coenzyme A; a fourth to acetyl coenzyme A or acetoacetyl coenzyme A. Finally there is a class that yields products belonging to more than one of the other four classes. Table 22.1 summarizes the end-products of amino-acid catabolism.

**Table 22.1** End-products of amino-acid metabolism

| End-product | Oxaloacetate and pyruvate | α-Oxo-glutarate | Succinyl coenzyme A | Acetyl coenzyme A and acetoacetyl coenzyme A |
|---|---|---|---|---|
| Amino acid | Aspartate | Glutamate | Methionine | Lysine |
| | Asparagine | Glutamine | Threonine | Isoleucine |
| | Alanine | Proline | Isoleucine | Leucine |
| | Serine | Arginine | Valine | Tryptophan |
| | Glycine | Histidine | | Phenylalanine* |
| | Cysteine | | | Tyrosine* |

* Also yields fumarate.

Now as we have stressed earlier (pp. 281 and 321), there is an important difference between pyruvate and Krebs cycle intermediates on the one hand, and acetyl coenzyme A and acetoacetyl coenzyme A on the other hand. Pyruvate and Krebs cycle intermediates can be converted to carbohydrate by the pathway that we described in Ch. 19. By contrast acetyl coenzyme A and acetoacetyl coenzyme A cannot give rise to carbohydrate (see p. 339). This difference is the basis of the classification of amino acids as 'glucogenic' – those whose degradation products can be used as sources of glucose – and 'ketogenic' – those whose degradation leads to acetyl coenzyme A and/or to the production of ketone bodies (see pp. 289 ff.).

*'Families' of amino acids*

A different way of classifying amino acids is on the basis not of their degradation products but of their routes of synthesis. It turns out that most amino acids arise from one of five starting compounds, namely glutamate, aspartate, pyruvate, phospho-*enol*pyruvate or glycerate-3-phosphate – notice that when we say an amino acid 'arises from' one of these compounds we are referring not to the amino group (which generally comes from glutamate by transamination (see reaction 22.2)) but to the carbon skeleton. Thus if, for example, a culture of bacteria of the species *Escherichia coli* (see Ch. 24) is grown in the presence of $^{14}$C-labelled aspartate (see pp. 231 f.), radioactivity will be found in the carbon atoms of several amino acids but not in any of the others; those amino acids that arise from aspartate are said to belong to the aspartate 'family'.

As we are here more interested in the synthesis than in the degradation of amino acids, we shall adopt this classification as the basis of what follows. We shall also mention another point that can be used to distinguish amino acids, which will help in clarifying the discussion of the synthetic and degradative pathways. As we noted above, the routes of synthesis and of degradation are similar for some amino acids but not for others, and we shall refer to this fact when discussing the metabolism of the individual amino acids.

Table 22.2 gives the classification of the amino acids by families.

**Table 22.2**  'Families' of amino acids to show synthetic routes

| *Glutamate* | *Aspartate* | *Pyruvate* | *Phospho-*enol-*pyruvate* | *Glycerate-3-phosphate* |
|---|---|---|---|---|
| Glutamate | Aspartate | Alanine | Phenylalanine | Serine |
| Glutamine | Asparagine | Leucine | Tyrosine | Glycine |
| Proline | Lysine | Valine | Tryptophan | Cysteine |
| Arginine | Methionine | Isoleucine | | |
| | Threonine | (in part) | | |

# SYNTHESIS AND DEGRADATION OF THE INDIVIDUAL AMINO ACIDS

## 1. The glutamate family

As we have already explained in detail, the synthesis of *glutamate* from α-oxo-glutarate is readily reversible. The degradation of glutamate is therefore via the dehydrogenase reaction (p. 357) to α-oxoglutarate, which enters the tricarboxylic acid cycle.

*Proline* is synthesized from glutamate by a three-step pathway that involves two reductions by $NADH_2$, one before and one after the closure of the ring. The first step involves the synthesis of glutamic acid semialdehyde, an intermediate that we shall meet again when we consider the metabolism of arginine.

$$
\begin{array}{c}
\text{COOH} \\
|\\
\text{CHNH}_2 \\
|\\
\text{CH}_2 \\
|\\
\text{CH}_2 \\
|\\
\text{COOH} \\
\text{glutamic} \\
\text{acid}
\end{array}
\quad \xrightleftharpoons{\text{NADH}_2} \quad
\begin{array}{c}
\text{COOH} \\
|\\
\text{CHNH}_2 \\
|\\
\text{CH}_2 \\
|\\
\text{CH}_2 \\
|\\
\text{CHO} \\
\text{glutamic} \\
\text{acid} \\
\text{semialdehyde}
\end{array}
\quad \rightleftharpoons \quad
\begin{array}{c}
\text{H}_2\text{C}\text{---CH}_2 \\
|\qquad\quad| \\
\text{HC}_{\diagdown}\;\;_{\diagup}\text{CH.COOH} \\
\text{N} \\
\Delta\text{-pyrroline-} \\
\text{5-carboxylic acid}
\end{array}
\quad \xrightleftharpoons{\text{NADH}_2} \quad
\begin{array}{c}
\text{H}_2\text{C}\text{---CH}_2 \\
|\qquad\quad| \\
\text{H}_2\text{C}_{\diagdown}\;\;_{\diagup}\text{CH.COOH} \\
\underset{\text{H}}{\text{N}} \\
\text{proline}
\end{array}
$$

The intermediates formed during the degradation of proline are identical to those formed during the synthesis.

*Glutamine* is formed from glutamate by fixation of ammonia in a reaction driven by hydrolysis of ATP and catalysed by glutamine synthetase.

$$
\begin{array}{c}
\text{COOH} \\
|\\
\text{CHNH}_2 \\
|\\
\text{CH}_2 \\
|\\
\text{CH}_2 \\
|\\
\text{COOH} \\
\text{glutamic} \\
\text{acid}
\end{array}
\;+\; \text{NH}_3 + \text{ATP} \;\rightleftharpoons\;
\begin{array}{c}
\text{COOH} \\
|\\
\text{CHNH}_2 \\
|\\
\text{CH}_2 \\
|\\
\text{CH}_2 \\
|\\
\text{CONH}_2 \\
\text{glutamine}
\end{array}
\;+\; \text{H}_2\text{O} + \text{ADP} + \text{P}_i
$$

Apart from its use as a constituent of protein, glutamine is a compound of widespread metabolic importance. It is used as a store of amino groups, since (as we shall see immediately below) ammonia can be readily split out of it. For similar reasons it can be used to transport ammonia in a non-toxic form before its excretion (see Ch. 23). Glutamine also serves as a donor of amino groups in several synthetic reactions (see, for example, pp. 380 and 394).

Glutamine is broken down to glutamate by a simple hydrolysis catalysed by glutaminase.

$$
\begin{array}{c}
\text{COOH} \\
|\\
\text{CHNH}_2 \\
|\\
\text{CH}_2 \\
|\\
\text{CH}_2 \\
|\\
\text{CONH}_2 \\
\text{glutamine}
\end{array}
\;+\; \text{H}_2\text{O} \;\rightleftharpoons\;
\begin{array}{c}
\text{COOH} \\
|\\
\text{CHNH}_2 \\
|\\
\text{CH}_2 \\
|\\
\text{CH}_2 \\
|\\
\text{COOH} \\
\text{glutamic} \\
\text{acid}
\end{array}
\;+\; \text{NH}_3
$$

Strictly speaking, this reaction is not identical to that by which glutamine is synthesized, but it is convenient to regard glutamine as belonging to the group of amino acids that have common synthetic and degradative pathways.

The pathway of synthesis of *arginine* differs slightly from one organism to another; we shall give the route as it occurs in *Escherichia coli*, an organism in which the synthesis has been very extensively studied (see also p. 415).

Glutamate is first acetylated by acetyl coenzyme A, and the resulting N-acetyl glutamate is converted by reduction and transamination to N-acetyl ornithine.

$$
\begin{array}{c}
\text{COOH} \\
|\\
\text{CHNH}_2 \\
|\\
\text{CH}_2 \\
|\\
\text{CH}_2 \\
|\\
\text{COOH}
\end{array}
\xrightarrow[\text{CH}_3.\text{CO.S-CoA} \quad \text{CoA.SH}]{}
\begin{array}{c}
\text{COOH} \\
|\\
\text{CHNH.COCH}_3 \\
|\\
\text{CH}_2 \\
|\\
\text{CH}_2 \\
|\\
\text{COOH}
\end{array}
\xrightarrow[\text{NADPH}_2 \quad \text{NADP}]{}
\begin{array}{c}
\text{COOH} \\
|\\
\text{CHNH.COCH}_3 \\
|\\
\text{CH}_2 \\
|\\
\text{CH}_2 \\
|\\
\text{CHO}
\end{array}
\rightleftharpoons
\begin{array}{c}
\text{COOH} \\
|\\
\text{CHNH.COCH}_3 \\
|\\
\text{CH}_2 \\
|\\
\text{CH}_2 \\
|\\
\text{CH}_2\text{NH}_2
\end{array}
$$

glutamic acid     N-acetyl glutamic acid     N-acetylglutamic acid semialdehyde     N-acetyl ornithine

After hydrolysis, the terminal —NH$_2$ group of the resulting ornithine is converted to —NH.C(NH$_2$)=NH to yield arginine. This conversion is complicated but extremely important, and we shall discuss its mechanism in the next chapter (pp. 388 ff.).

$$
\begin{array}{c}
\text{COOH} \\
|\\
\text{CHNH.COCH}_3 \\
|\\
\text{CH}_2 \\
|\\
\text{CH}_2 \\
|\\
\text{CH}_2\text{NH}_2
\end{array}
\xrightarrow[\text{H}_2\text{O}]{}
\begin{array}{c}
\text{COOH} \\
|\\
\text{CHNH}_2 \\
|\\
\text{CH}_2 \\
|\\
\text{CH}_2 \\
|\\
\text{CH}_2\text{NH}_2
\end{array}
\longrightarrow
\begin{array}{c}
\text{COOH} \\
|\\
\text{CHNH}_2 \\
|\\
\text{CH}_2 \\
|\\
\text{CH}_2 \\
|\\
\text{CH}_2.\text{NH.CO.NH}_2
\end{array}
\longrightarrow \longrightarrow
\begin{array}{c}
\text{COOH} \\
|\\
\text{CHNH}_2 \\
|\\
\text{CH}_2 \\
|\\
\text{CH}_2 \qquad\quad \text{NH}\\
|\qquad\qquad \|\\
\text{CH}_2.\text{NH.C.NH}_2
\end{array}
$$

N-acetyl ornithine     ornithine           arginine

The degradation of arginine gives rise to glutamate and follows much the same route as the synthesis, except that the intermediates between ornithine and glutamate are not acetylated.

$$
\begin{array}{c}
\text{COOH} \\
|\\
\text{CHNH}_2 \\
|\\
\text{CH}_2 \\
|\\
\text{CH}_2 \qquad\quad \text{NH}\\
|\qquad\qquad \|\\
\text{CH}_2.\text{NH.C.NH}_2
\end{array}
\longrightarrow
\begin{array}{c}
\text{COOH} \\
|\\
\text{CHNH}_2 \\
|\\
\text{CH}_2 \\
|\\
\text{CH}_2 \\
|\\
\text{CH}_2\text{NH}_2
\end{array}
\rightleftharpoons
\begin{array}{c}
\text{COOH} \\
|\\
\text{CHNH}_2 \\
|\\
\text{CH}_2 \\
|\\
\text{CH}_2 \\
|\\
\text{CHO}
\end{array}
\rightleftharpoons
\begin{array}{c}
\text{COOH} \\
|\\
\text{CHNH}_2 \\
|\\
\text{CH}_2 \\
|\\
\text{CH}_2 \\
|\\
\text{COOH}
\end{array}
$$

arginine     ornithine     glutamic acid semialdehyde     glutamic acid

The first step in this sequence is an hydrolysis catalysed by arginase, and it yields (in addition to ornithine) urea; we shall discuss it further in the next chapter.

Glutamate, proline and glutamine can all be synthesized by almost all organisms. Arginine, however, is essential in most mammalian species because its rate of synthesis is too low to supply the requirement for arginine in protein synthesis and because such arginine as is synthesized is rapidly broken down (see p. 389). All four give rise to α-oxoglutarate on degradation and are therefore glucogenic.

## 2. The aspartate family

*Aspartate* is synthesized by the transamination of oxaloacetate, and is degraded by the reverse of the same reaction.

$$
\begin{array}{c}
\text{COOH} \\
| \\
\text{CHNH}_2 \\
| \\
\text{CH}_2 \\
| \\
\text{CH}_2 \\
| \\
\text{COOH} \\
\text{glutamic} \\
\text{acid}
\end{array}
+
\begin{array}{c}
\text{COOH} \\
| \\
\text{CO} \\
| \\
\text{CH}_2 \\
| \\
\text{COOH} \\
\text{oxaloacetic} \\
\text{acid}
\end{array}
\rightleftharpoons
\begin{array}{c}
\text{COOH} \\
| \\
\text{CO} \\
| \\
\text{CH}_2 \\
| \\
\text{CH}_2 \\
| \\
\text{COOH} \\
\text{α-oxoglutaric} \\
\text{acid}
\end{array}
+
\begin{array}{c}
\text{COOH} \\
| \\
\text{CHNH}_2 \\
| \\
\text{CH}_2 \\
| \\
\text{COOH} \\
\text{aspartic} \\
\text{acid}
\end{array}
$$

*Asparagine* is analogous to glutamine, and is formed in a reaction analogous to that catalysed by glutamine synthetase:

$$
\begin{array}{c}
\text{COOH} \\
| \\
\text{CHNH}_2 \\
| \\
\text{CH}_2 \\
| \\
\text{COOH} \\
\text{aspartic} \\
\text{acid}
\end{array}
+ \text{NH}_3 + \text{ATP} \rightleftharpoons
\begin{array}{c}
\text{COOH} \\
| \\
\text{CHNH}_2 \\
| \\
\text{CH}_2 \\
| \\
\text{CONH}_2 \\
\text{asparagine}
\end{array}
+ \text{H}_2\text{O} + \text{ADP} + \text{P}_i
$$

and broken down in a reaction analogous to that catalysed by glutaminase.

$$
\begin{array}{c}
\text{COOH} \\
| \\
\text{CHNH}_2 \\
| \\
\text{CH}_2 \\
| \\
\text{CONH}_2 \\
\text{asparagine}
\end{array}
+ \text{H}_2\text{O} \rightleftharpoons
\begin{array}{c}
\text{COOH} \\
| \\
\text{CHNH}_2 \\
| \\
\text{CH}_2 \\
| \\
\text{COOH} \\
\text{aspartic} \\
\text{acid}
\end{array}
+ \text{NH}_3
$$

(The enzymes responsible for these reactions are asparagine synthetase and asparaginase.) In some organisms, particularly plants, asparagine has the function of storing amino groups that, in other organisms, is fulfilled by glutamine.

Aspartate and asparagine are both non-essential, as might be expected since they can so easily be made from readily available precursors. Since they are degraded to oxaloacetate, both are glucogenic.

The aspartate family also includes lysine, methionine and threonine, all of which are synthesized in most plants and bacteria from aspartic acid semi-aldehyde (compare glutamic acid semialdehyde, p. 363). This compound is made from aspartate in two steps; the first is catalysed by aspartokinase.

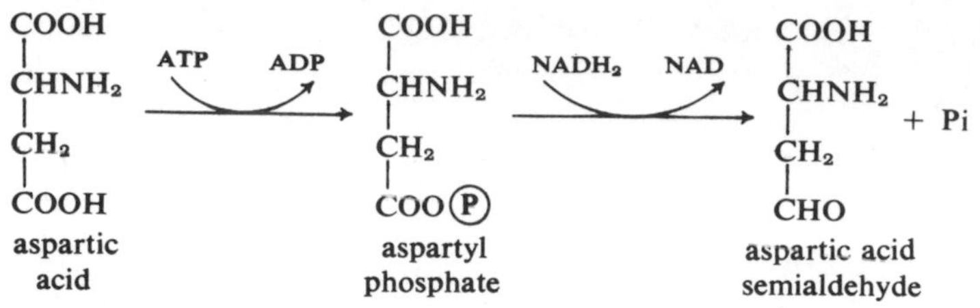

COOH | CHNH_2 | CH_2 | COOH
aspartic acid

$\xrightarrow{\text{ATP} \quad \text{ADP}}$

COOH | CHNH_2 | CH_2 | COO(P)
aspartyl phosphate

$\xrightarrow{\text{NADH}_2 \quad \text{NAD}}$

COOH | CHNH_2 | CH_2 | CHO  + Pi
aspartic acid semialdehyde

In the biosynthesis of *lysine*, aspartic acid semialdehyde reacts with pyruvate to yield a cyclic compound: this is then reduced, the ring is opened, and eventually one of the isomers of diaminopimelic acid is formed. (Diaminopimelic acid, in addition to its role as an intermediate in the synthesis of lysine, also occurs in large quantities in some bacterial cell walls, see p. 181.) Lysine is produced by decarboxylation of diaminopimelic acid.

COOH | CHNH_2 | CH_2 | CHO
aspartic acid semialdehyde

+ COOH | CO | CH_3
pyruvic acid

$\longrightarrow$

dihydropicolinic acid

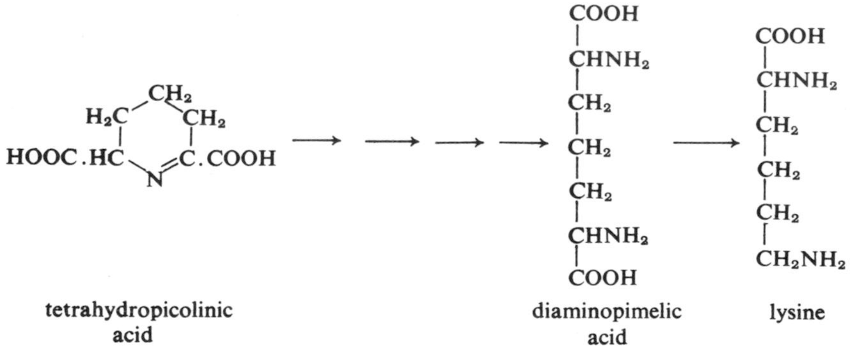

$\longrightarrow$

tetrahydropicolinic acid

$\longrightarrow \longrightarrow \longrightarrow \longrightarrow$

COOH | CHNH_2 | CH_2 | CH_2 | CH_2 | CHNH_2 | COOH
diaminopimelic acid

$\longrightarrow$

COOH | CHNH_2 | CH_2 | CH_2 | CH_2 | CH_2NH_2
lysine

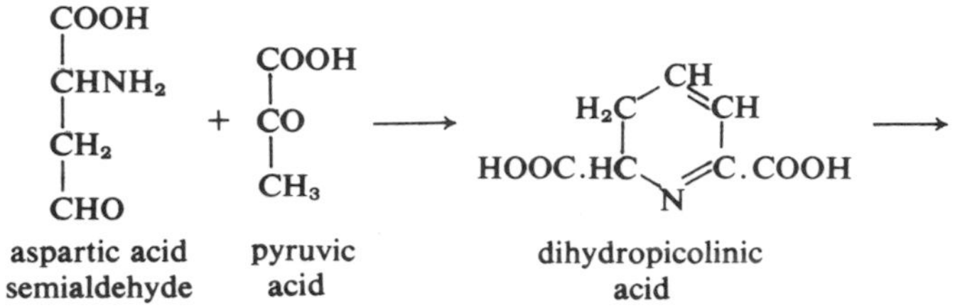

The pathway of degradation of lysine is extremely complicated and we shall not consider it in detail. Interestingly enough, the degradation proceeds, like the synthesis, via a series of cyclic compounds, this time including pipecolic acid. At a later stage α-oxoadipic acid is formed.

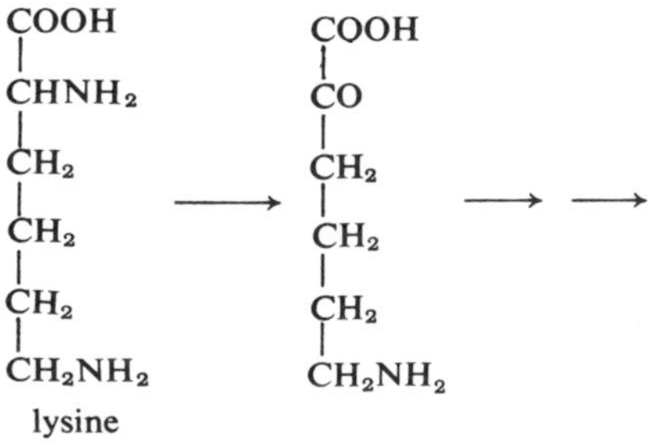

COOH | CHNH_2 | CH_2 | CH_2 | CH_2 | CH_2NH_2
lysine

$\longrightarrow$

COOH | CO | CH_2 | CH_2 | CH_2 | CH_2NH_2

$\longrightarrow \longrightarrow$

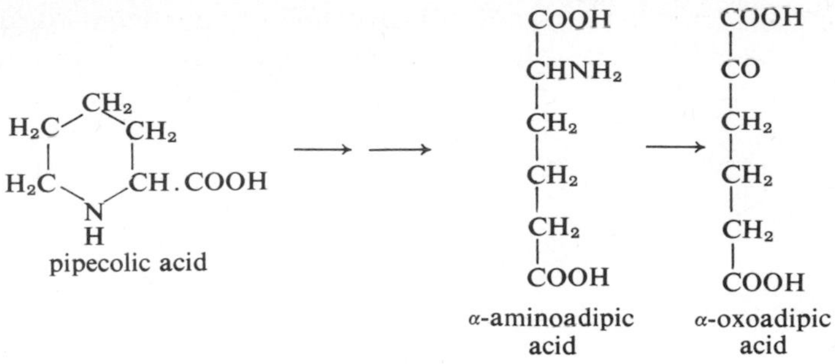

pipecolic acid → → α-aminoadipic acid → α-oxoadipic acid

The further metabolism of α-oxoadipic acid is much more readily comprehensible (see p. 361). It undergoes oxidative decarboxylation and then oxidation, and finally acetoacetyl coenzyme A is produced.

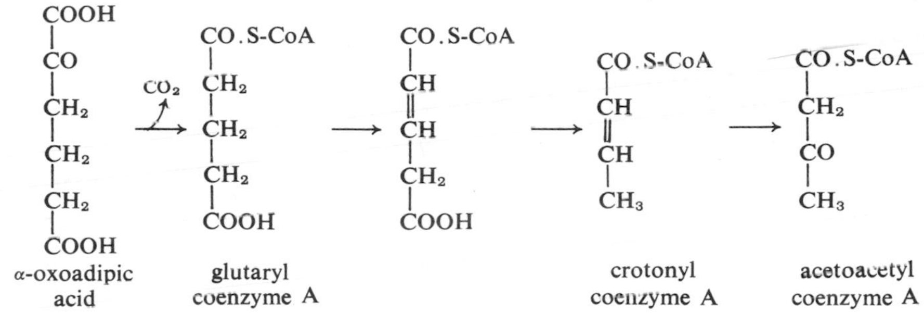

α-oxoadipic acid → ($CO_2$) glutaryl coenzyme A → → crotonyl coenzyme A → acetoacetyl coenzyme A

Lysine is therefore ketogenic.

In the synthesis of *methionine* from aspartic acid semialdehyde, the first step is a reduction to homoserine. This then reacts with succinyl coenzyme A (p. 277).

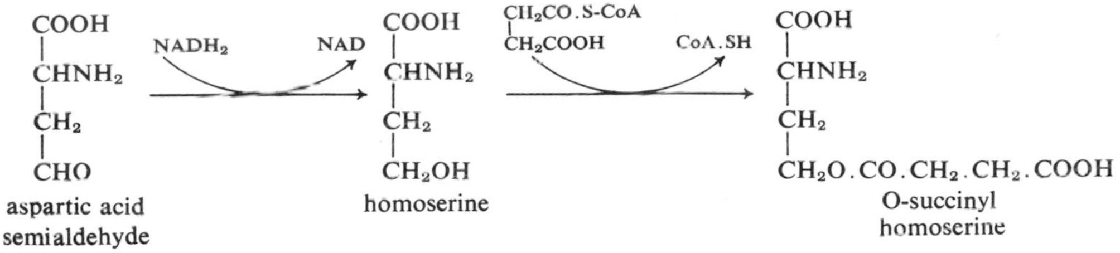

aspartic acid semialdehyde → ($NADH_2$ → $NAD$) homoserine → ($CH_2CO.S\text{-}CoA$, $CH_2COOH$; $CoA.SH$) O-succinyl homoserine

The succinyl group is replaced by reaction with cysteine to form a compound called cystathionine, which consists of two amino acids joined by a $>C-S-C<$ bridge.

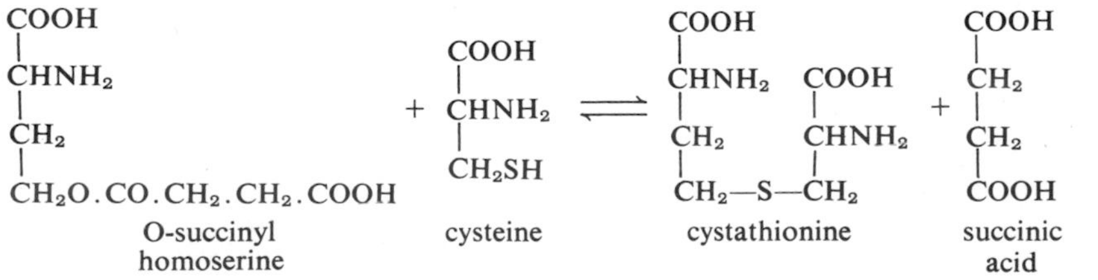

O-succinyl homoserine + cysteine ⇌ cystathionine + succinic acid

Cystathionine is an important intermediate in the metabolism of the sulphur-containing amino acids. It can be formed (as here) from a three-carbon sulphydryl-containing compound and a four-carbon hydroxyl-containing compound (the latter is actually esterified in this case), or from a four-carbon sulphydryl-containing compound and a three-carbon hydroxyl-containing compound as in the synthesis of cysteine (see below). Similarly it can undergo cleavage on either side of the sulphur atom and is involved, as we shall see directly, in the degradation of methionine.

In the synthesis of methionine, cystathionine loses pyruvate and ammonia, leaving homocysteine, and the latter is then methylated. The mechanism of methylation is of importance, and we shall defer discussion of it until the section on the transfer of one-carbon groups (pp. 383 ff.).

$$
\begin{array}{cccc}
\text{COOH} & & & \\
| & & & \\
\text{CHNH}_2 \quad \text{COOH} & \xrightarrow{\text{NH}_3 \quad \text{CH}_3} & \text{COOH} & \text{COOH} \\
| \qquad\quad | & & | & | \\
\text{CH}_2 \qquad \text{CHNH}_2 + \text{H}_2\text{O} & & \text{CHNH}_2 & \longrightarrow \quad \text{CHNH}_2 \\
| \qquad\quad | & & | & | \\
\text{CH}_2-\text{S}-\text{CH}_2 & & \text{CH}_2 & \text{CH}_2 \\
& & | & | \\
& & \text{CH}_2\text{SH} & \text{CH}_2\text{S}.\text{CH}_3 \\
\text{cystathionine} & & \text{homocysteine} & \text{methionine}
\end{array}
$$

(with COOH–CO–CH₃ as the leaving group above NH₃ CH₃)

Thus in those organisms (most plants and microorganisms) that can make methionine, the four-carbon skeleton arises from aspartate and the sulphur atom from cysteine. The origin of the sulphur atom of cysteine is discussed on p. 378.

In the degradation of methionine, the methyl group is first transferred to an acceptor (see p. 385); the resulting homocysteine is converted to cystathionine in a reaction that we shall discuss in detail when considering the synthesis of cysteine (see p. 378). Cystathionine undergoes the opposite kind of cleavage to that we described above, to yield homoserine, and this is deaminated and dehydrated to α-oxobutyrate (compare the dehydration deamination of threonine below). α-Oxobutyrate undergoes the expected oxidative decarboxylation (see p. 361) to give propionyl coenzyme A, which is converted to succinyl coenzyme A by the mechanism we described on p. 289.

$$
\begin{array}{cccc}
\text{COOH} & \text{COOH} & & \text{COOH} \\
| & | & \text{COOH} & | \\
\text{CHNH}_2 & \text{CHNH}_2 & | & \text{CHNH}_2 \\
| \longrightarrow & | \longrightarrow & \text{CHNH}_2 \quad \text{COOH} \longrightarrow & | \longrightarrow \\
\text{CH}_2 & \text{CH}_2 & | \qquad\quad | & \text{CH}_2 \\
| & | & \text{CH}_2 \qquad \text{CHNH}_2 & | \\
\text{CH}_2 & \text{CH}_2 & | \qquad\quad | & \text{CH}_2\text{OH} \\
| & | & \text{CH}_2-\text{S}-\text{CH}_2 & \\
\text{S}.\text{CH}_3 & \text{SH} & & \\
\text{methionine} & \text{homocysteine} & \text{cystathionine} & \text{homoserine}
\end{array}
$$

$$
\begin{array}{ccc}
\text{COOH} & \text{CO}_2 \quad \text{CO}.\text{S-CoA} & \text{CO}.\text{S-CoA} \\
| & | & | \\
\text{CO} & \text{CH}_2 & \text{CH}_2 \\
| \nearrow & | \longrightarrow & | \\
\text{CH}_2 & \text{CH}_3 & \text{CH}_2 \\
| & & | \\
\text{CH}_3 & & \text{COOH} \\
\text{α-oxobutyric} & \text{propionyl} & \text{succinyl} \\
\text{acid} & \text{coenzyme A} & \text{coenzyme A}
\end{array}
$$

Methionine is therefore glucogenic.

The synthesis of *threonine* also takes homoserine as a precursor (see under methionine). Homoserine is phosphorylated by ATP, and homoserine phosphate yields threonine in a curious reaction catalysed by an enzyme that uses pyridoxal phosphate as co-factor.

$$
\underset{\text{homoserine}}{
\begin{array}{c}
\text{COOH} \\
| \\
\text{CHNH}_2 \\
| \\
\text{CH}_2 \\
| \\
\text{CH}_2\text{OH}
\end{array}}
\quad \xrightarrow[\text{ATP} \quad \text{ADP}]{}\quad
\underset{\substack{\text{homoserine} \\ \text{phosphate}}}{
\begin{array}{c}
\text{COOH} \\
| \\
\text{CHNH}_2 \\
| \\
\text{CH}_2 \\
| \\
\text{CH}_2\text{O}\,\textcircled{P}
\end{array}}
\quad \xrightarrow[\text{H}_2\text{O} \quad \text{P}_i]{}\quad
\underset{\text{threonine}}{
\begin{array}{c}
\text{COOH} \\
| \\
\text{CHNH}_2 \\
| \\
\text{CHOH} \\
| \\
\text{CH}_3
\end{array}}
$$

The first step in the degradation of threonine is a reaction called dehydration deamination. The enzyme, threonine dehydratase, which uses pyridoxal phosphate as co-factor, removes water and forms an imino acid. This reacts again with water to give $\alpha$-oxobutyrate, which is converted to succinyl coenzyme A by the sequence we described under methionine.

$$
\underset{\text{threonine}}{
\begin{array}{c}
\text{COOH} \\
| \\
\text{CHNH}_2 \\
| \\
\text{CHOH} \\
| \\
\text{CH}_3
\end{array}}
\xrightarrow[\text{H}_2\text{O}]{}
\begin{array}{c}
\text{COOH} \\
| \\
\text{CNH}_2 \\
\| \\
\text{CH} \\
| \\
\text{CH}_3
\end{array}
\xrightarrow[\text{H}_2\text{O} \quad \text{NH}_3]{}
\underset{\substack{\alpha\text{-oxobutyric} \\ \text{acid}}}{
\begin{array}{c}
\text{COOH} \\
| \\
\text{CO} \\
| \\
\text{CH}_2 \\
| \\
\text{CH}_3
\end{array}}
\longrightarrow \longrightarrow
\underset{\substack{\text{succinyl} \\ \text{coenzyme A}}}{
\begin{array}{c}
\text{CO.S-CoA} \\
| \\
\text{CH}_2 \\
| \\
\text{CH}_2 \\
| \\
\text{COOH}
\end{array}}
$$

Threonine is therefore also glucogenic.

The synthesis of lysine, methionine and threonine depends on the formation of aspartic acid semialdehyde, which cannot be made in most higher animals. Consequently all three of these amino acids are essential in mammals. Moreover methionine is used, in mammals, to supply the sulphur atom to cysteine (see p. 378), whereas in those organisms that can make methionine for themselves it is cysteine (as we have just seen) that supplies sulphur to methionine.

# 3. The pyruvate family

*Alanine* is formed directly by transamination of pyruvate. The reaction is reversible (see p. 359) so that the degradation of alanine leads to pyruvate and hence into the main pathways of carbohydrate metabolism (pp. 271 and 330.)

$$
\underset{\substack{\text{glutamic} \\ \text{acid}}}{
\begin{array}{c}
\text{COOH} \\
| \\
\text{CHNH}_2 \\
| \\
\text{CH}_2 \\
| \\
\text{CH}_2 \\
| \\
\text{COOH}
\end{array}}
+
\underset{\substack{\text{pyruvic} \\ \text{acid}}}{
\begin{array}{c}
\text{COOH} \\
| \\
\text{CO} \\
| \\
\text{CH}_3
\end{array}}
\rightleftharpoons
\underset{\substack{\alpha\text{-oxoglutaric} \\ \text{acid}}}{
\begin{array}{c}
\text{COOH} \\
| \\
\text{CO} \\
| \\
\text{CH}_2 \\
| \\
\text{CH}_2 \\
| \\
\text{COOH}
\end{array}}
+
\underset{\text{alanine}}{
\begin{array}{c}
\text{COOH} \\
| \\
\text{CHNH}_2 \\
| \\
\text{CH}_3
\end{array}}
$$

Isoleucine, leucine and valine are all members of the pyruvate family. Isoleucine, however, has a more complicated genealogy than the other two, being a member of both the pyruvate and the aspartate families.

*Isoleucine* is derived in part from α-oxobutyrate, which, as we have just seen, is the product of dehydration deamination of threonine. In the first reaction α-oxobutyrate reacts with a hydroxyethyl group, carried by thiamine pyrophosphate and derived from pyruvate (see p. 272), to yield α-aceto-α-hydroxybutyrate.

$$
\begin{array}{c}
\text{COOH} \\
| \\
\text{CO} \\
| \\
\text{CH}_2 \\
| \\
\text{CH}_3
\end{array}
\quad + \quad \text{CH}_3.\text{CHOH}.\text{TPP} \quad \longrightarrow \quad \text{TPP} \quad + \quad
\begin{array}{c}
\text{COOH} \\
| \\
\text{CH}_3.\text{CO}.\text{COH} \\
| \\
\text{CH}_2 \\
| \\
\text{CH}_3
\end{array}
$$

α-oxobutyric acid            α-aceto-α-hydroxybutyric acid

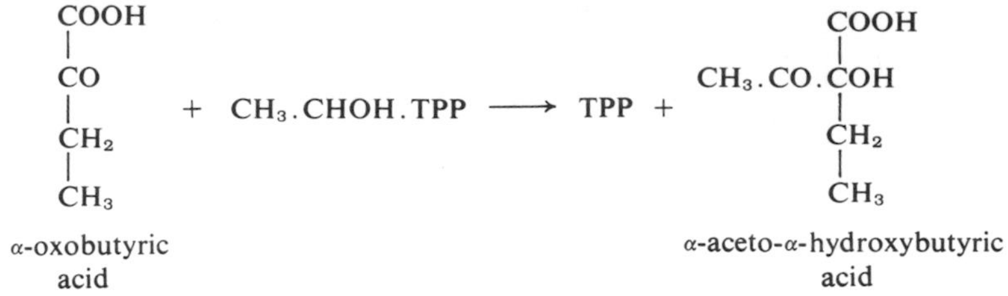

The first reaction in the synthesis of *valine* is exactly analogous: pyruvate (the next lower homologue of α-oxobutyrate) reacts with a hydroxyethyl group in just the same way, and α-acetolactate is formed.

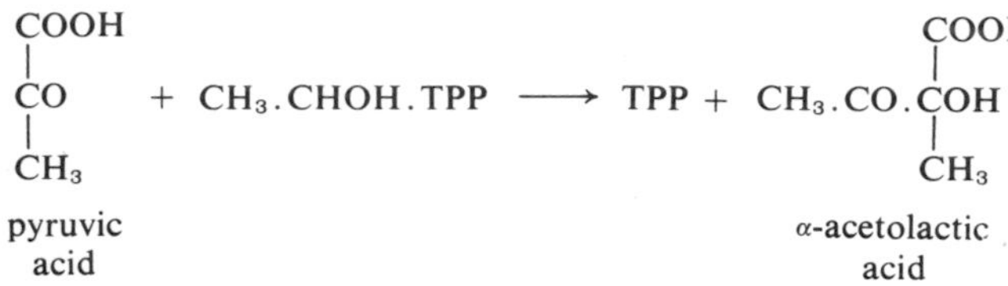

$$
\begin{array}{c}
\text{COOH} \\
| \\
\text{CO} \\
| \\
\text{CH}_3
\end{array}
\quad + \quad \text{CH}_3.\text{CHOH}.\text{TPP} \quad \longrightarrow \quad \text{TPP} \quad + \quad
\begin{array}{c}
\text{COOH} \\
| \\
\text{CH}_3.\text{CO}.\text{COH} \\
| \\
\text{CH}_3
\end{array}
$$

pyruvic acid            α-acetolactic acid

We can now consider the synthesis of isoleucine and valine in parallel. α-Aceto-α-hydroxybutyrate and α-acetolactate are reduced by $NADPH_2$, and the alkyl group migrates. A keto-acid is formed by dehydration, and finally isoleucine and valine are produced by transamination.

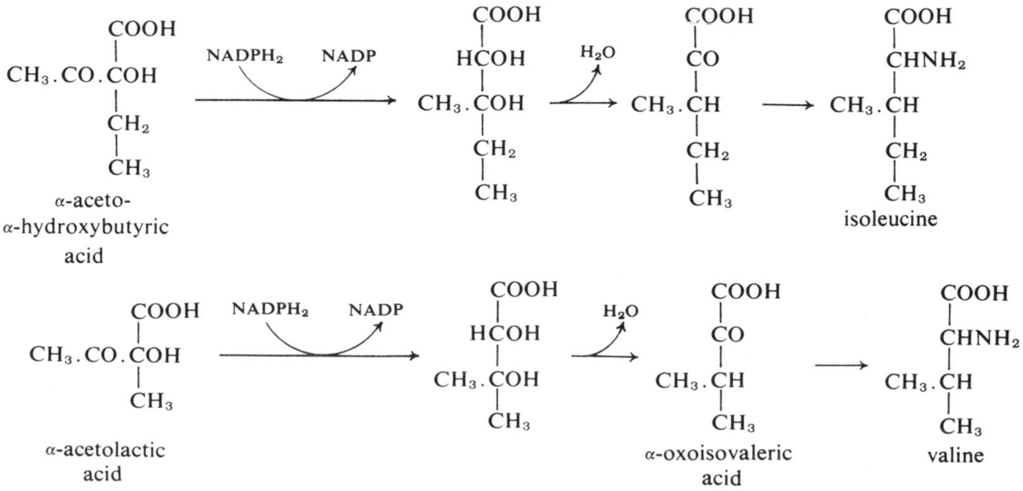

α-aceto-α-hydroxybutyric acid     $NADPH_2$   NADP     $H_2O$     isoleucine

α-acetolactic acid     $NADPH_2$   NADP     $H_2O$     α-oxoisovaleric acid     valine

We can also consider the degradation of isoleucine and valine in parallel. Both follow quite faithfully the general principles of catabolism that we outlined on p. 361. The products of transamination of isoleucine and valine undergo oxidative decarboxylation to yield acyl coenzyme A derivatives, and these then follow the pathway of fatty acyl coenzyme A degradation.

$$
\begin{array}{c}
\text{COOH} \\
\text{CHNH}_2 \\
\text{CH}_3.\text{CH} \\
\text{CH}_2 \\
\text{CH}_3 \\
\text{isoleucine}
\end{array}
\longrightarrow
\begin{array}{c}
\text{COOH} \\
\text{CO} \\
\text{CH}_3.\text{CH} \\
\text{CH}_2 \\
\text{CH}_3
\end{array}
\xrightarrow{\text{CO}_2}
\begin{array}{c}
\text{CO.S-CoA} \\
\text{CH}_3.\text{CH} \\
\text{CH}_2 \\
\text{CH}_3
\end{array}
\xrightarrow[\text{FAD} \quad \text{FADH}_2]{}
\begin{array}{c}
\text{CO.S-CoA} \\
\text{CH}_3.\text{C} \\
\| \\
\text{CH} \\
\text{CH}_3
\end{array}
\xrightarrow{\text{H}_2\text{O}}
\begin{array}{c}
\text{CO.S-CoA} \\
\text{CH}_3.\text{CH} \\
\text{CHOH} \\
\text{CH}_3 \\
\alpha\text{-methyl-}\beta\text{-} \\
\text{hydroxybutyryl} \\
\text{coenzyme A}
\end{array}
$$

$$
\begin{array}{c}
\text{COOH} \\
\text{CHNH}_2 \\
\text{CH}_3.\text{CH} \\
\text{CH}_3 \\
\text{valine}
\end{array}
\longrightarrow
\begin{array}{c}
\text{COOH} \\
\text{CO} \\
\text{CH}_3.\text{CH} \\
\text{CH}_3
\end{array}
\xrightarrow{\text{CO}_2}
\begin{array}{c}
\text{CO.S-CoA} \\
\text{CH}_3.\text{CH} \\
\text{CH}_3
\end{array}
\xrightarrow[\text{FAD} \quad \text{FADH}_2]{}
\begin{array}{c}
\text{CO.S-CoA} \\
\text{CH}_3.\text{C} \\
\| \\
\text{CH}_2
\end{array}
\xrightarrow{\text{H}_2\text{O}}
\begin{array}{c}
\text{CO.S-CoA} \\
\text{CH}_3.\text{CH} \\
\text{CH}_2\text{OH} \\
\beta\text{-hydroxyisobutyryl} \\
\text{coenzyme A}
\end{array}
$$

$\alpha$-Methyl-$\beta$-hydroxybutyryl coenzyme A (from isoleucine) continues to follow this pathway and is oxidized by NAD and then split by the $\beta$-ketothiolase (see p. 286). The final products are acetyl coenzyme A and propionyl coenzyme A (which yields succinyl coenzyme A – see p. 289).

$$
\begin{array}{c}
\text{CO.S-CoA} \\
\text{CH}_3.\text{CH} \\
\text{CHOH} \\
\text{CH}_3 \\
\alpha\text{-methyl-}\beta\text{-} \\
\text{hydroxybutyryl} \\
\text{coenzyme A}
\end{array}
\xrightarrow[\text{NAD} \quad \text{NADH}_2]{}
\begin{array}{c}
\text{CO.S-CoA} \\
\text{CH}_3.\text{CH} \\
\text{CO} \\
\text{CH}_3
\end{array}
\xrightarrow{\text{CoA.SH}}
\begin{array}{cc}
\text{CO.S-CoA} & + \; \text{CO.S-CoA} \\
\text{CH}_3.\text{CH}_2 & \quad \text{CH}_3 \\
\text{propionyl} & \text{acetyl} \\
\text{coenzyme A} & \text{coenzyme A}
\end{array}
$$

Isoleucine is therefore both glucogenic and ketogenic.

$\beta$-Hydroxyisobutyryl coenzyme A (from valine) yields methylmalonyl coenzyme A and hence succinyl coenzyme A (see p. 289).

$$
\begin{array}{c}
\text{CO.S-CoA} \\
\text{CH}_3.\text{CH} \\
\text{CH}_2\text{OH} \\
\beta\text{-hydroxyiso-} \\
\text{butyryl} \\
\text{coenzyme A}
\end{array}
\longrightarrow
\begin{array}{c}
\text{COOH} \\
\text{CH}_3.\text{CH} \\
\text{CH}_2\text{OH}
\end{array}
\xrightarrow[\text{NAD} \quad \text{NADH}_2]{}
\begin{array}{c}
\text{COOH} \\
\text{CH}_3.\text{CH} \\
\text{CHO} \\
\text{methylmalonyl} \\
\text{semialdehyde}
\end{array}
$$

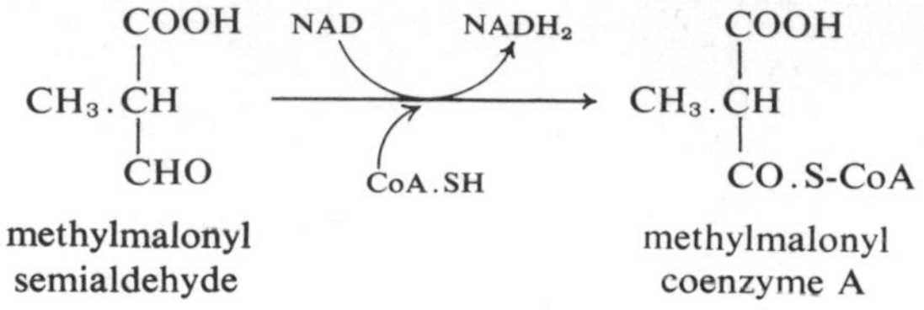

$$\underset{\text{methylmalonyl}\atop\text{semialdehyde}}{\underset{\displaystyle\overset{\displaystyle\text{COOH}}{\underset{\displaystyle\overset{\displaystyle|}{\underset{\displaystyle\text{CH}_3.\text{CH}}{\underset{\displaystyle|}{\text{CHO}}}}}{}}}\quad\xrightarrow[\text{CoA.SH}]{\text{NAD}\quad\quad\text{NADH}_2}\quad\underset{\text{methylmalonyl}\atop\text{coenzyme A}}{\underset{\displaystyle\overset{\displaystyle\text{COOH}}{\underset{\displaystyle|}{\underset{\displaystyle\text{CH}_3.\text{CH}}{\underset{\displaystyle|}{\text{CO.S-CoA}}}}}{}}$$

Consequently valine is glucogenic.

The synthesis of *leucine* begins with α-oxoisovalerate, which is the penultimate compound in the synthesis of valine. This condenses with acetyl coenzyme A in a reaction reminiscent of the formation of citrate (p. 276); the subsequent steps are reminiscent of the degradation of citrate to α-oxoglutarate, and the synthesis is completed by transamination.

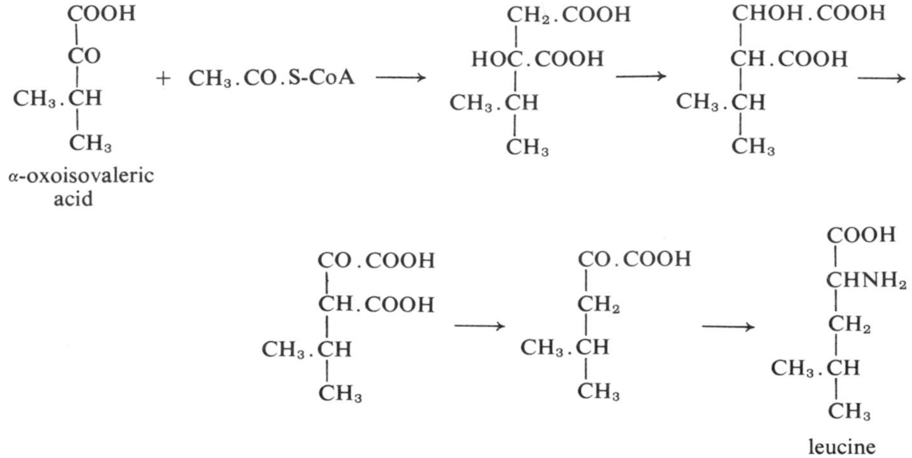

In the degradation of leucine, transamination is followed by the expected oxidative decarboxylation and oxidation (see p. 361).

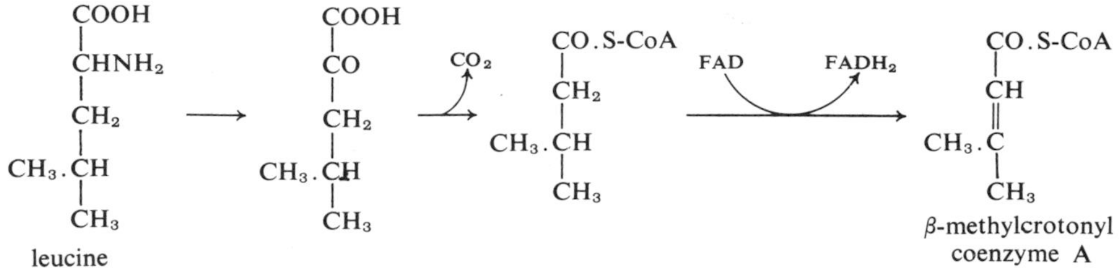

The resulting β-methylcrotonyl coenzyme A obviously cannot undergo β-oxidation since the β-carbon atom cannot carry both —H and —OH. Instead it is carboxylated and hydrated to yield β-hydroxy-β-methylglutaryl coenzyme A, which is an intermediate in the formation of ketone bodies (see p. 290) and is split to acetyl coenzyme A and acetoacetate.

$$
\begin{array}{c}
\text{CO.S-CoA} \\
| \\
\text{CH} \\
\| \\
\text{CH}_3.\text{C} \\
| \\
\text{CH}_3
\end{array}
\quad\xrightarrow{\;\text{CO}_2\;}\quad
\begin{array}{c}
\text{CO.S-CoA} \\
| \\
\text{CH} \\
\| \\
\text{CH}_3.\text{C} \\
| \\
\text{CH}_2 \\
| \\
\text{COOH}
\end{array}
\quad\xrightarrow{\;\text{H}_2\text{O}\;}\quad
\begin{array}{c}
\text{CO.S-CoA} \\
| \\
\text{CH}_2 \\
| \\
\text{CH}_3.\text{COH} \\
| \\
\text{CH}_2 \\
| \\
\text{COOH}
\end{array}
\quad\longrightarrow\quad
\begin{array}{c}
\text{CO.S-CoA} \\
| \\
\text{CH}_3 \\
\text{acetyl CoA} \\
+ \\
\text{CH}_3\text{—CO} \\
| \\
\text{CH}_2 \\
| \\
\text{COOH}
\end{array}
$$

β-methylcrotonyl       β-hydroxy-     acetoacetic
coenzyme A      β-methylglutaryl    acid
            coenzyme A

Leucine is therefore ketogenic.

The branched-chain amino acids cannot be made in most mammalian species and they are therefore all essential amino acids. In the organisms in which they are made – most bacteria, for example – the pathways of synthesis of all three (as you can see from the above discussion) are closely interconnected, and the same enzymes sometimes serve in the synthesis of two amino acids. The parallel reactions for the synthesis of valine and isoleucine (p. 370) are, in some species, catalysed by exactly the same set of enzymes, with a curious consequence that we shall discuss on p. 502.

## 4. The phospho-*enol*pyruvate family

The three aromatic amino acids phenylalanine, tyrosine and tryptophan are formed from phospho-*enol*pyruvate, which is an intermediate in the Embden–Meyerhof pathway. The first several steps in the sequence are common to the biosynthesis of all of the aromatic amino acids, and we shall describe these first; later we shall outline the paths that lead to the individual amino acids.

In the first step, phospho-*enol*pyruvate reacts with erythrose-4-phosphate, which is, you recall, an intermediate in the rearrangement of pentose phosphate to hexose phosphate (p. 297). After cyclization and dehydration, the product is reduced to shikimate, an important intermediate in the synthesis not only of these three amino acids but also of many other aromatic compounds (particularly in plants).

$$
\begin{array}{c}
\text{COOH} \\
| \\
\text{CO}\,\text{P} \\
\| \\
\text{CH}_2
\end{array}
\;+\;
\begin{array}{c}
\text{CHO} \\
| \\
\text{CHOH} \\
| \\
\text{CHOH} \\
| \\
\text{CH}_2\text{O}\,\text{P}
\end{array}
\quad\xrightarrow[\;\;\;\;\;\;\;\;\;\;\;\;]{\text{H}_2\text{O}\qquad \text{P}_i}\quad
\begin{array}{c}
\text{COOH} \\
| \\
\text{CO} \\
| \\
\text{CH}_2 \\
| \\
\text{HOCH} \\
| \\
\text{HCOH} \\
| \\
\text{HCOH} \\
| \\
\text{CH}_2\text{O}\,\text{P}
\end{array}
\quad\xrightarrow{\;\;\text{P}_i\;\;}\quad
$$

phospho-*enol*-  erythrose-
pyruvic acid  4-phosphate

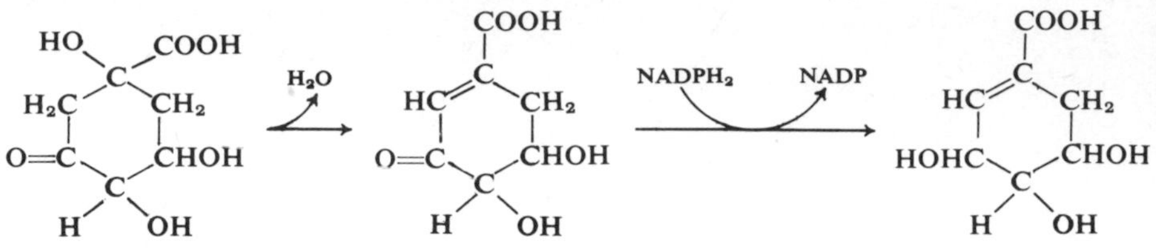

COOH ... (reaction scheme with structures)

shikimic acid

Shikimate is phosphorylated and then reacts with a further molecule of phospho-*enol*pyruvate to yield chorismate, which is the last compound of the common aromatic sequence.

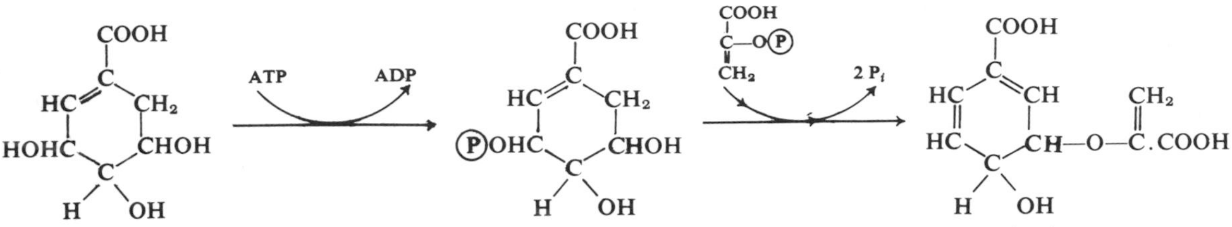

ATP    ADP    2 P$_i$

chorismic acid

In the synthesis of *phenylalanine* and *tyrosine*, chorismate is rearranged to form prephenate, and this is then decarboxylated and reduced. The resulting phenylpyruvate is transaminated to yield phenylalanine, which can be hydroxy-lated to tyrosine.

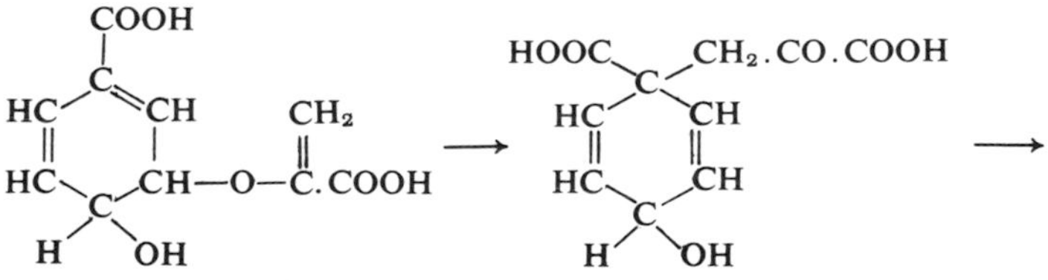

chorismic acid        prephenic acid

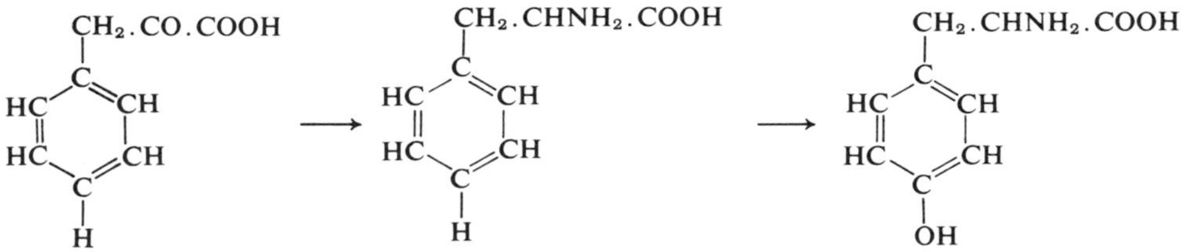

phenylpyruvic acid        phenylalanine        tyrosine

The synthesis of *tryptophan* from chorismate is more complicated. In the first step, glutamine (p. 363) is used to form anthranilate in a reaction catalysed by anthranilate synthetase (see p. 500):

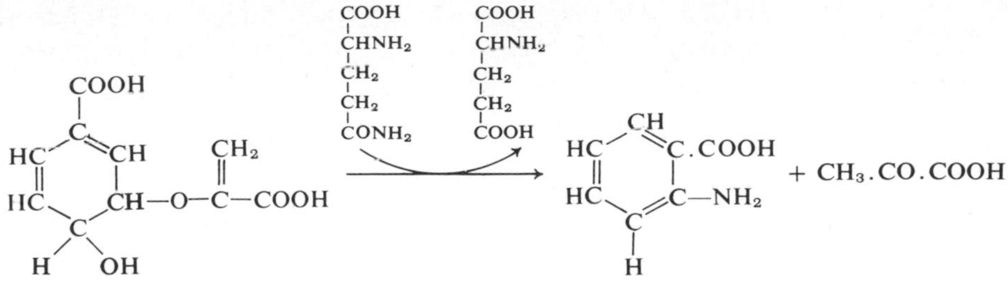

$$\text{COOH}$$
$$\text{CHNH}_2$$
$$\text{CH}_2$$
$$\text{CH}_2$$
$$\text{CONH}_2$$

$$\text{COOH}$$
$$\text{CHNH}_2$$
$$\text{CH}_2$$
$$\text{CH}_2$$
$$\text{COOH}$$

$$+ \text{CH}_3.\text{CO}.\text{COOH}$$

chorismic acid

anthranilic acid

Meanwhile, a compound called 5-phosphoribosyl pyrophosphate is formed by the transfer of a pyrophosphate group from ATP to ribose-5-phosphate (p. 294). We shall see that 5-phosphoribosyl pyrophosphate has several important uses in other synthetic reactions (pp. 380 and 390).

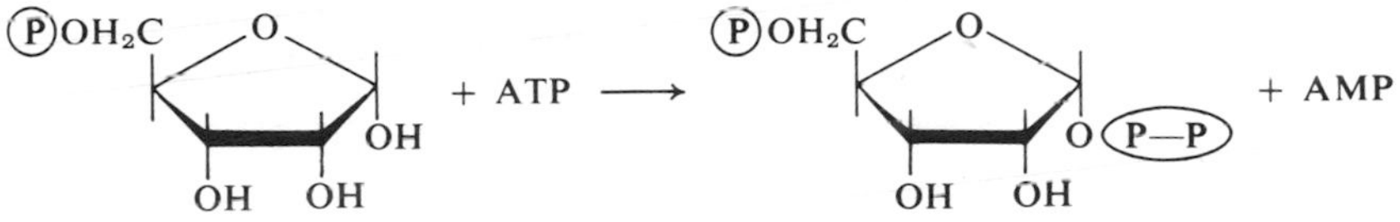

$$+ \text{ATP} \longrightarrow \qquad + \text{AMP}$$

ribose-5-phosphate

5-phosphoribosyl pyrophosphate

These two react to give 5-phosphoribosylanthranilate, and indole-glycerophosphate is formed from this compound.

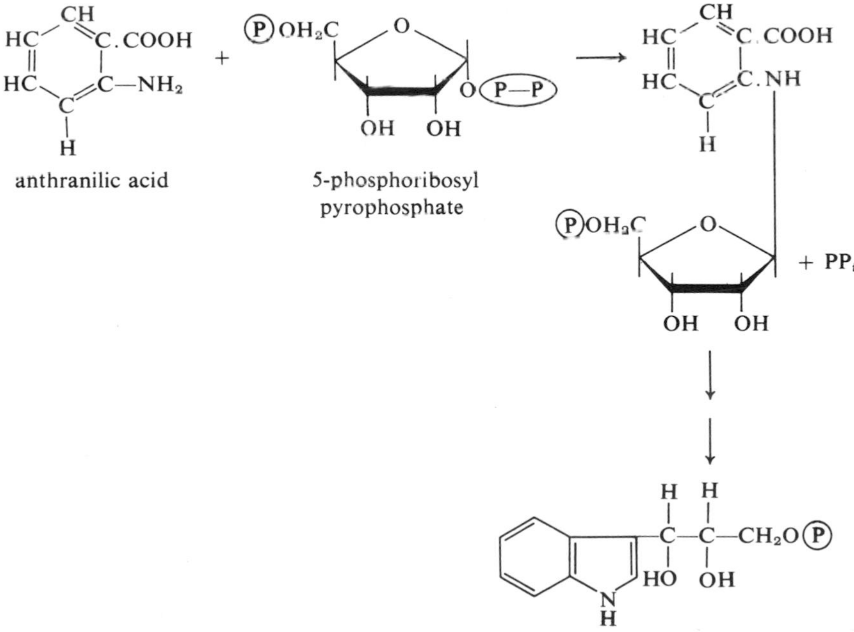

anthranilic acid

5-phosphoribosyl pyrophosphate

$$+ \text{PP}_i$$

indole-glycerophosphate

Indole-glycerophosphate finally reacts with serine in the presence of the enzyme tryptophan synthetase (see p. 414) to form tryptophan. Glyceraldehyde-3-phosphate is eliminated.

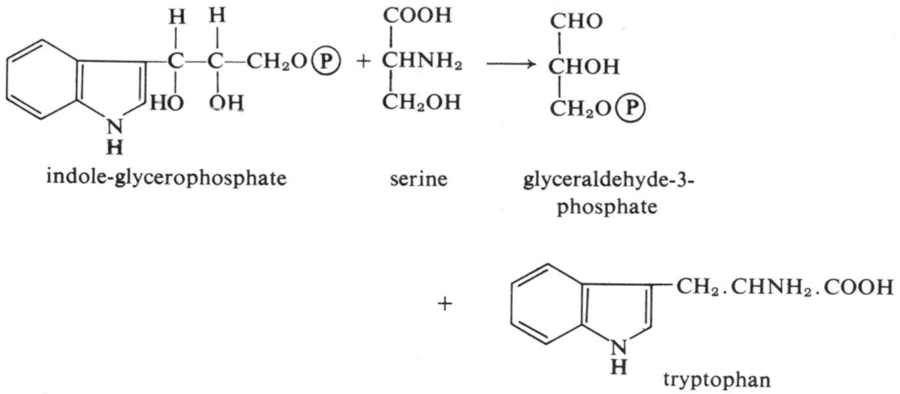

indole-glycerophosphate      serine      glyceraldehyde-3-phosphate

tryptophan

Many of the reactions involved in the synthesis of the aromatic amino acids do not occur in animals. All three of these amino acids are therefore essential in mammalian species. However, most species can accomplish the hydroxylation of phenylalanine to form tyrosine, so that the requirement for tyrosine can be met by supplying phenylalanine.

We shall not discuss the degradation of the aromatic amino acids in any great detail, but shall just outline the pathways. In degradation phenylalanine is first hydroxylated to tyrosine, and this is then transaminated to form *para*-hydroxyphenylpyruvate, which eventually yields fumarylacetoacetate. Cleavage to fumarate and acetoacetate follows.

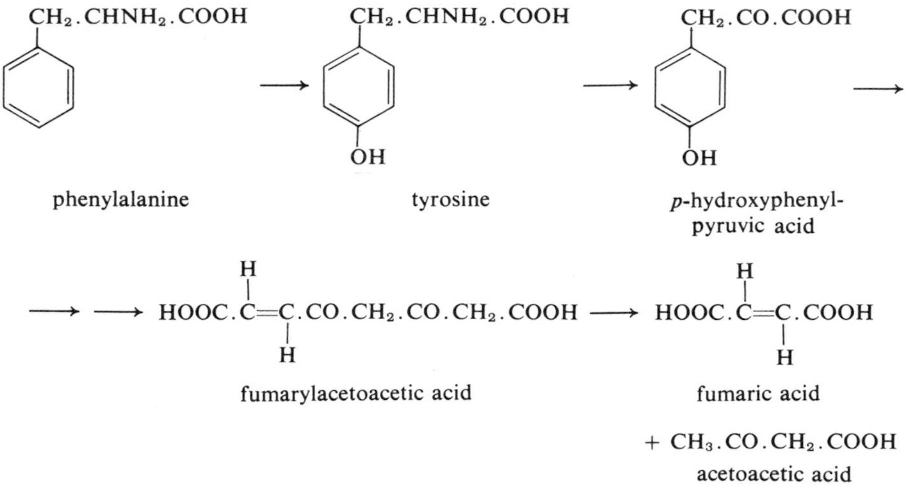

phenylalanine      tyrosine      p-hydroxyphenyl-pyruvic acid

$$\longrightarrow \quad \longrightarrow \quad HOOC.\overset{\displaystyle H}{\underset{\displaystyle H}{C}}{=}C.CO.CH_2.CO.CH_2.COOH \longrightarrow HOOC.\overset{\displaystyle H}{\underset{\displaystyle H}{C}}{=}C.COOH$$

fumarylacetoacetic acid           fumaric acid

$+ CH_3.CO.CH_2.COOH$

acetoacetic acid

Thus these amino acids are ketogenic and also, since fumarate is a constituent of the tricarboxylic acid cycle, glucogenic.

A rare 'inborn error of metabolism' (see p. 426) leads to the absence of the enzyme (known as phenylalanine hydroxylase) that converts phenylalanine to tyrosine. In the individuals affected, phenylalanine is transaminated directly, without hydroxylation, forming phenylpyruvate. This compound is rather toxic, and its accumulation in the blood leads to damage to the central nervous system. This condition is known as phenylketonuria. If detected early in life (as is easily possible by testing the urine of all babies for the presence of phenylpyruvate), phenylketonuria can be managed by omitting phenylalanine from the diet. The nervous degeneration does not then occur.

The degradation of tryptophan is even more complicated. The first step is oxidation to form formyl kynurenine, which is eventually converted to 3-hydroxyanthranilate. This compound is important as a precursor of nicotinic acid – the substance that gives rise to NAD (see p. 79). 3-Hydroxyanthranilate is degraded by a long pathway to α-oxoadipate, which (as we saw under lysine, p. 367) yields acetoacetyl coenzyme A.

tryptophan $\longrightarrow$ formyl kynurenine $\longrightarrow$

3-hydroxyanthranilic acid $\longrightarrow$ α-oxoadipic acid $\longrightarrow$ acetoacetyl coenzyme A

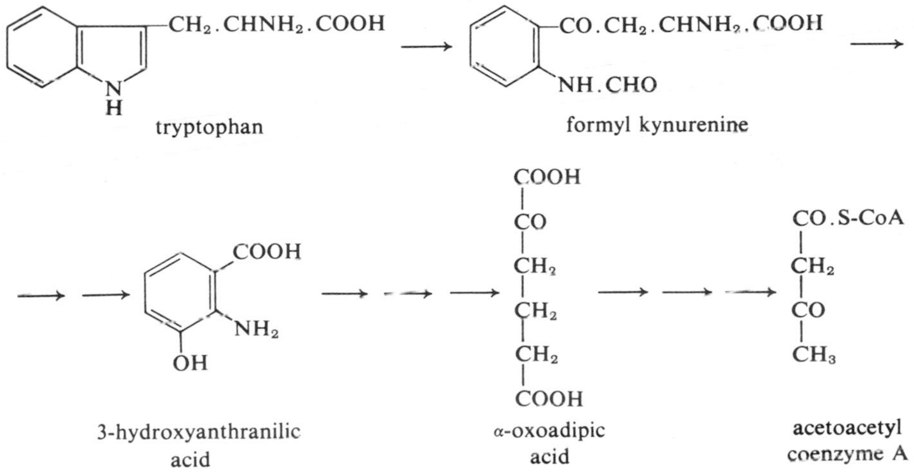

Tryptophan is therefore ketogenic.

## 5. The glycerate-3-phosphate family

The carbon chains of *serine*, glycine and cysteine are derived from glycerate-3-phosphate, which is an intermediate in the Embden–Meyerhof pathway (see p. 263). Oxidation and transamination yield phosphoserine, which is then cleaved by phosphoserine phosphatase.

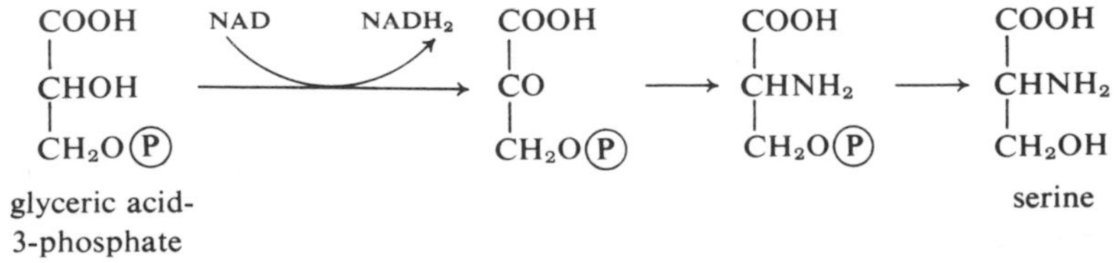

glyceric acid-3-phosphate $\xrightarrow{\text{NAD} \quad \text{NADH}_2}$ $\longrightarrow$ $\longrightarrow$ serine

*Glycine* is formed by loss of the hydroxymethyl group of serine. The reaction involves a compound called tetrahydrofolate, which we shall discuss in detail later (p. 383).

$$\begin{array}{l}\text{COOH} \\ | \\ \text{CHNH}_2 \\ | \\ \text{CH}_2\text{OH}\end{array} + \text{tetrahydrofolate} \longrightarrow \begin{array}{l}\text{COOH} \\ | \\ \text{CH}_2\text{NH}_2\end{array} + \text{methylenetetrahydrofolate} + \text{H}_2\text{O}$$

serine                    glycine

These reactions can readily be carried out in most organisms, so that neither serine nor glycine is an essential amino acid for mammals.

*Cysteine* is synthesized from serine and homocysteine, which is a product of degradation of methionine (p. 368). These two compounds react together in the presence of cystathionine synthetase to yield cystathionine (p. 367). Water is eliminated.

$$\begin{array}{l}\text{COOH} \\ | \\ \text{CHNH}_2 \\ | \\ \text{CH}_2\text{OH}\end{array} + \begin{array}{l}\text{COOH} \\ | \\ \text{CHNH}_2 \\ | \\ \text{CH}_2 \\ | \\ \text{CH}_2\text{SH}\end{array} \longrightarrow \begin{array}{ll}\text{COOH} & \begin{array}{l}\text{COOH} \\ | \\ \text{CHNH}_2 \end{array}\\ | & | \\ \text{CHNH}_2 & \text{CH}_2 \\ | & | \\ \text{CH}_2\!-\!\text{S}\!-\!\text{CH}_2 \end{array} + \text{H}_2\text{O}$$

serine      homocysteine        cystathionine

Now we saw that in the synthesis of methionine (p. 367) cystathionine is split on one side of the sulphur atom to yield homocysteine, liberating pyruvate and ammonia. In the synthesis of cysteine, cystathionine undergoes the opposite cleavage, forming cysteine and liberating α-oxobutyrate and ammonia (see p. 368).

$$\begin{array}{ll}\text{COOH} & \begin{array}{l}\text{COOH} \\ | \\ \text{CHNH}_2 \end{array}\\ | & | \\ \text{CHNH}_2 & \text{CH}_2 \\ | & | \\ \text{CH}_2\!-\!\text{S}\!-\!\text{CH}_2 \end{array} + \text{H}_2\text{O} \xrightarrow[\quad]{\text{NH}_3 \;\; \begin{array}{l}\text{COOH}\\|\\\text{CO}\\|\\\text{CH}_2\\|\\\text{CH}_3\end{array}} \begin{array}{l}\text{COOH} \\ | \\ \text{CHNH}_2 \\ | \\ \text{CH}_2\text{SH}\end{array}$$

cystathionine                                     cysteine

Thus the three-carbon skeleton of cysteine arises from glycerate-3-phosphate and the sulphur atom from methionine.

We cannot, however, leave matters quite like that, because in discussing the synthesis of methionine we said that the sulphur atom comes from cysteine, and we have now claimed that the sulphur atom of cysteine comes from methionine. We can clarify things in the following way: Methionine is an essential amino acid, which in mammals is used to supply sulphur in the synthesis of cysteine. The synthesis of methionine occurs in plants and micro-organisms, and these organisms use cysteine to supply sulphur in that synthesis. But how do they make cysteine in the first place? The answer, at any rate as far as many micro-organisms are concerned, is that it is made by the direct fixation of sulphide. The acceptor for sulphide is O-acetyl serine, which is made by acetylating serine with acetyl coenzyme A.

$$
\underset{\text{serine}}{\overset{\displaystyle \text{COOH}}{\underset{\displaystyle \text{CH}_2\text{OH}}{\overset{\displaystyle |}{\underset{\displaystyle |}{\text{CHNH}_2}}}}}
\xrightarrow[\text{CoA.SH}]{\text{CH}_3\text{CO.S.CoA}}
\underset{\text{O-acetyl serine}}{\overset{\displaystyle \text{COOH}}{\underset{\displaystyle \text{CH}_2\text{O.CO.CH}_3}{\overset{\displaystyle |}{\underset{\displaystyle |}{\text{CHNH}_2}}}}}
\xrightarrow{\text{H}_2\text{S}}
\underset{\text{cysteine}}{\overset{\displaystyle \text{COOH}}{\underset{\displaystyle \text{CH}_2\text{SH}}{\overset{\displaystyle |}{\underset{\displaystyle |}{\text{CHNH}_2}}}}} + \text{CH}_3.\text{COOH}
$$

Notice that in this pathway, too, the carbon atoms of cysteine are derived from serine, and hence from glycerate-3-phosphate.

The first step in the degradation of glycine is conversion to serine by transfer of a one-carbon unit (p. 385):

$$
\underset{\text{glycine}}{\overset{\displaystyle \text{COOH}}{\underset{\displaystyle \text{CH}_2\text{NH}_2}{\overset{\displaystyle |}{\underset{\displaystyle |}{}}}}} + \text{methylenetetrahydrofolate} + \text{H}_2\text{O} \longrightarrow
\underset{\text{serine}}{\overset{\displaystyle \text{COOH}}{\underset{\displaystyle \text{CH}_2\text{OH}}{\overset{\displaystyle |}{\underset{\displaystyle |}{\text{CHNH}_2}}}}} + \text{tetrahydrofolate}
$$

Serine undergoes dehydration deamination (see p. 369) to yield pyruvate.

$$
\underset{\text{serine}}{\overset{\displaystyle \text{COOH}}{\underset{\displaystyle \text{CH}_2\text{OH}}{\overset{\displaystyle |}{\underset{\displaystyle |}{\text{CHNH}_2}}}}}
\xrightarrow{\text{H}_2\text{O}}
\overset{\displaystyle \text{COOH}}{\underset{\displaystyle \text{CH}_2}{\overset{\displaystyle |}{\underset{\displaystyle \|}{\text{CNH}_2}}}}
\xrightarrow[\quad]{\text{H}_2\text{O} \quad \text{NH}_3}
\underset{\substack{\text{pyruvic}\\\text{acid}}}{\overset{\displaystyle \text{COOH}}{\underset{\displaystyle \text{CH}_3}{\overset{\displaystyle |}{\underset{\displaystyle |}{\text{CO}}}}}}
$$

Cysteine can be degraded in several ways. The simplest is desulphydration deamination, a reaction analogous to dehydration deamination.

$$
\underset{\text{cysteine}}{\overset{\displaystyle \text{COOH}}{\underset{\displaystyle \text{CH}_2\text{SH}}{\overset{\displaystyle |}{\underset{\displaystyle |}{\text{CHNH}_2}}}}}
\xrightarrow{\text{H}_2\text{S}}
\overset{\displaystyle \text{COOH}}{\underset{\displaystyle \text{CH}_2}{\overset{\displaystyle |}{\underset{\displaystyle \|}{\text{CNH}_2}}}}
\xrightarrow[\quad]{\text{H}_2\text{O} \quad \text{NH}_3}
\underset{\substack{\text{pyruvic}\\\text{acid}}}{\overset{\displaystyle \text{COOH}}{\underset{\displaystyle \text{CH}_3}{\overset{\displaystyle |}{\underset{\displaystyle |}{\text{CO}}}}}}
$$

Clearly glycine, serine and cysteine are all glucogenic.

# 6. Histidine

The pathway of histidine synthesis is long and complicated, and it contains some curious reactions. It is of general interest for two reasons: first because it is linked to the pathway of purine synthesis (see pp. 394 ff.), and secondly because it has been extensively used in studies of genetic and enzymic regulation (see pp. 483 and 499).

In the first reaction, phosphoribosyl pyrophosphate (p. 375) reacts with ATP to form N'-phosphoribosyl-ATP. This loses pyrophosphate, the adenine ring is split, and then the ribosyl moiety isomerizes to the ribulosyl form.

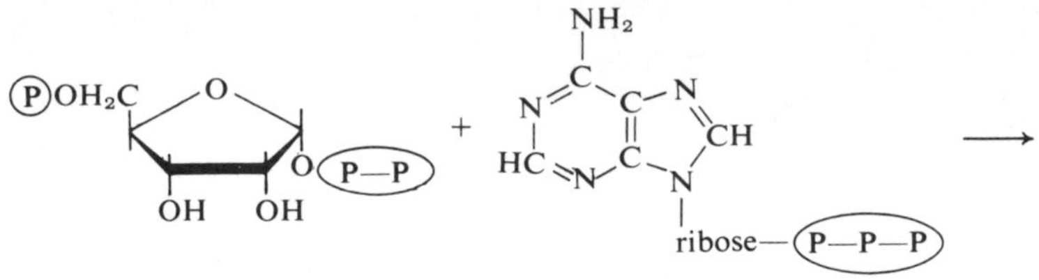

phosphoriboyl pyrophosphate      ATP

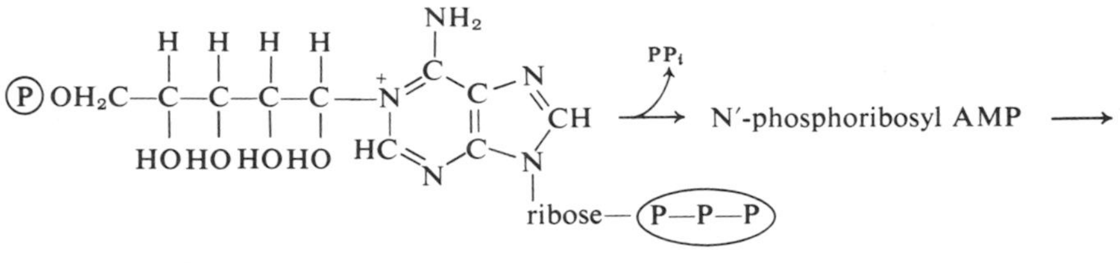

N'-phosphoribosyl ATP

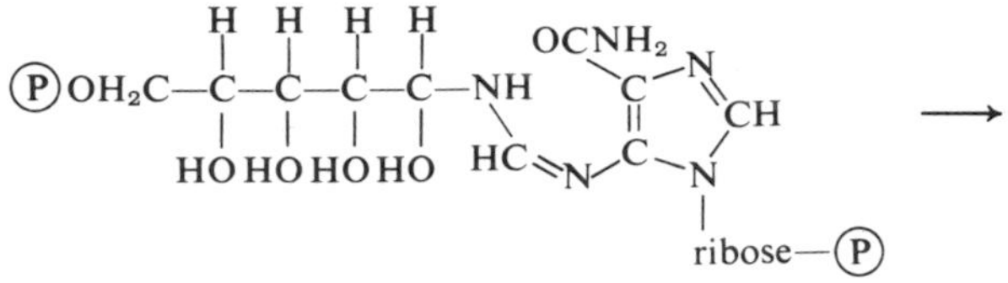

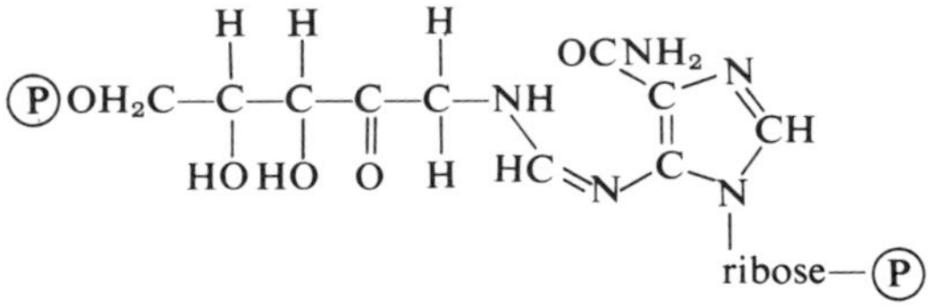

An amino group is now introduced by reaction with glutamine (see p. 363), and the products are imidazole glycerophosphate and phosphoribosyl-5-amino-4-imidazole-carboxamide (known as AICAR for short).

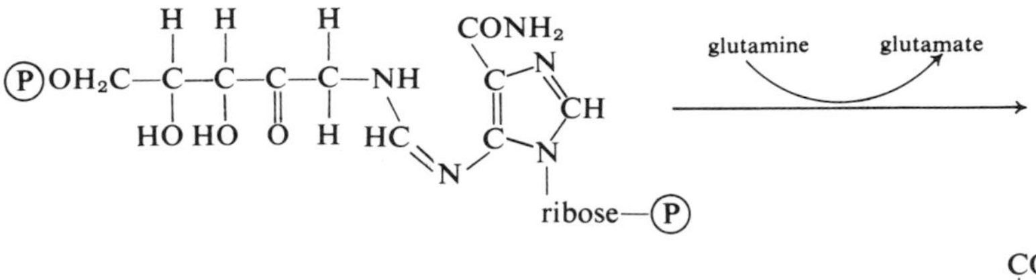

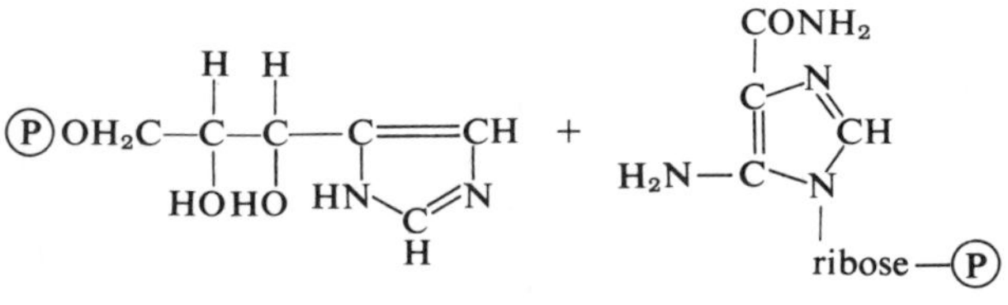

imidazole glycerophosphate      AICAR

We shall meet AICAR again in discussing the synthesis of purines (p. 396).

Imidazole glycerophosphate loses water, and the resulting keto group is transaminated to form histidinol phosphate. After removal of the phosphate, oxidation yields histidine.

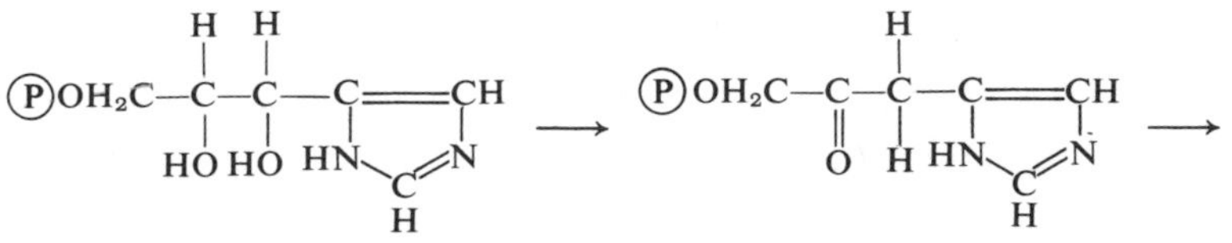

$$\text{(P)OH}_2\text{C--C--C--C==CH} \quad \longrightarrow \quad \text{(P)OH}_2\text{C--C--C--C==CH} \quad \longrightarrow$$

imidazole glycerophosphate

$$\text{(P)OH}_2\text{C--C--C--C==CH} \quad \longrightarrow \quad \text{HOH}_2\text{C--C--C--C==CH} \quad \longrightarrow \quad \text{HOOC--C--C--C==CH}$$

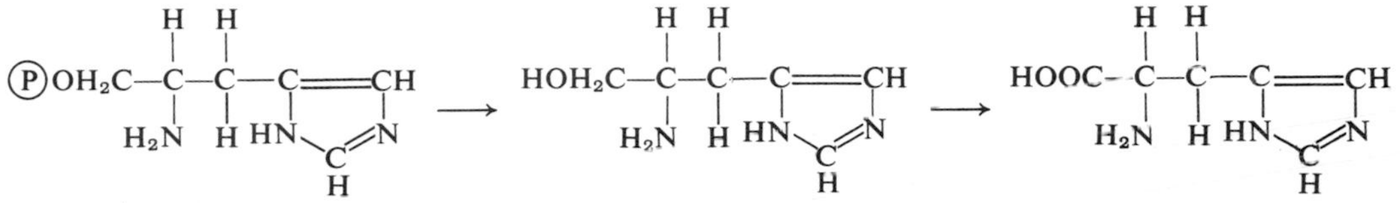

histidinol phosphate                histidine

The degradation of histidine is a good deal less complicated. Histidine is first deaminated directly to urocanate, and this then accepts two successive molecules of water to yield formiminoglutamate.

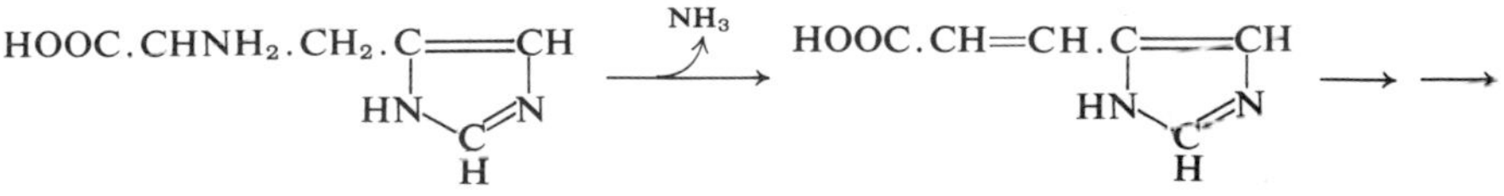

$$\text{HOOC.CHNH}_2.\text{CH}_2.\text{C==CH} \quad \xrightarrow{\text{NH}_3} \quad \text{HOOC.CH==CH.C==CH} \quad \longrightarrow \longrightarrow$$

histidine               urocanic acid

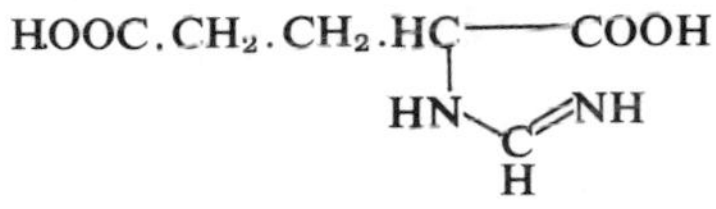

$$\text{HOOC.CH}_2.\text{CH}_2.\text{HC}\text{---COOH}$$

formiminoglutamic acid

The formimino group is removed by reaction with tetrahydrofolate (see pp. 383 ff.) to leave glutamate.

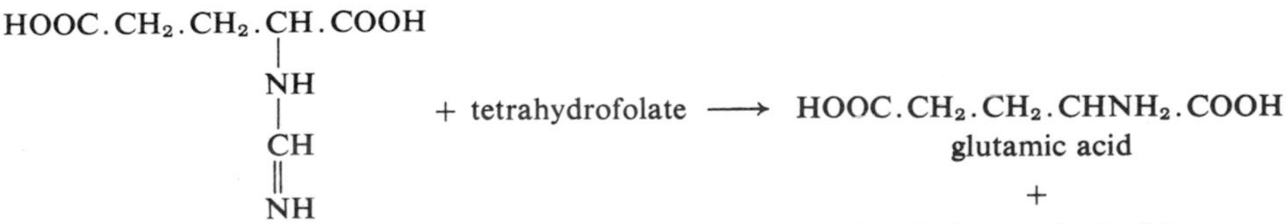

$$\text{HOOC.CH}_2.\text{CH}_2.\text{CH.COOH}$$
$$\overset{|}{\text{NH}}$$
$$\overset{|}{\text{CH}}$$
$$\overset{\|}{\text{NH}}$$

formiminoglutamic acid   + tetrahydrofolate $\longrightarrow$   $\text{HOOC.CH}_2.\text{CH}_2.\text{CHNH}_2.\text{COOH}$

glutamic acid

+

formiminotetrahydrofolate

Histidine is therefore glucogenic.

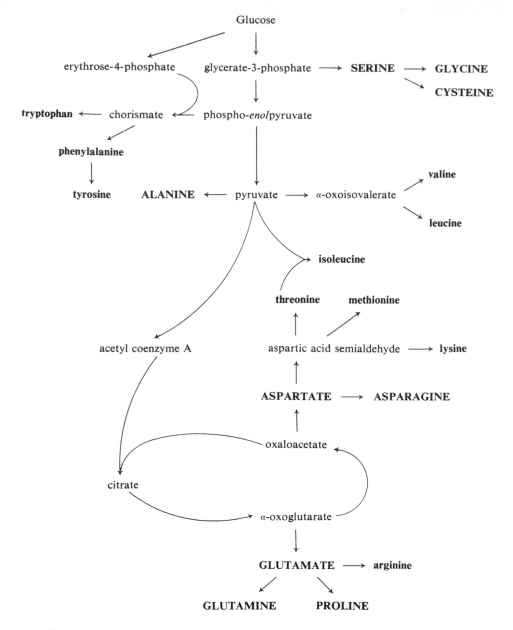

**Fig. 22.1** Synthesis of the amino acids. Amino acids are shown in **bold type**; those that can be synthesized in most species including mammals are in **CAPITALS**, those that are essential in most mammalian species in **lower case**. Histidine is not shown

## SUMMARY OF REACTIONS IN AMINO-ACID SYNTHESIS

It will be convenient at this point if we summarize the pathways that we have described for the synthesis of amino acids. The *amino* groups of amino acids are generally formed by transamination from glutamate, which in turn receives its amino group either by transamination from another amino acid (reaction (22.3),

p. 359), or by fixation of ammonia (reaction (22.1), p. 357). The origins of the *carbon* chains of the amino acids are more diverse; it will be helpful while reading the following discussion to refer to Fig. 22.1, in which the origins of the amino acids are related to key intermediates in the degradation of glucose.

The carbon chains of *glutamine* and *proline* are both formed from that of *glutamate*, and hence from α-oxoglutarate; these conversions take place in mammalian species as well as in plants and micro-organisms. Glutamate also gives rise to *arginine*, but this pathway operates to so slight an extent in mammals that arginine is an essential amino acid, at any rate in young mammals.

*Aspartate* is formed by transamination of oxaloacetate, and can give rise to *asparagine*; neither aspartate nor asparagine is essential in animals. Aspartate is also converted, via aspartic acid semialdehyde, to *lysine*, to *methionine* and to *threonine*, but these pathways do not operate in mammals.

Threonine contributes part of the carbon skeleton of *isoleucine*, the remainder being derived from pyruvate. Pyruvate also supplies the carbon atoms of *valine* and *leucine*. All three of these amino acids are essential in mammals. In addition, pyruvate can be transaminated to yield *alanine*; this reaction occurs in almost all organisms.

The aromatic amino acids *phenylalanine*, *tyrosine* and *tryptophan* arise from chorismate, which is synthesized from a condensation product of phospho-*enol*-pyruvate and erythrose-4-phosphate. This pathway does not operate in mammals.

Glycerate-3-phosphate is converted to *serine*, from which *glycine* and *cysteine* can be formed. These reactions can be carried out by mammals.

*Histidine* arises from phosphoribosyl pyrophosphate and part of the purine ring of ATP. Histidine is an essential amino acid in mammalian species.

We have referred several times in earlier chapters (see for example p. 209) to the fact that intermediates of the tricarboxylic acid cycle are important precursors in synthetic reactions. The pathways that we have described in this chapter include the chief routes by which Krebs cycle intermediates give rise to other compounds. Thus the glutamate family of amino acids (glutamate, glutamine, proline and arginine) draws carbon from the cycle via α-oxoglutarate, and the aspartate family (aspartate, asparagine, lysine, methionine and threonine – as well as two-thirds of isoleucine) draws carbon from the cycle via oxaloacetate. We have already stressed (p. 281) that this kind of leakage of carbon from the Krebs cycle makes its replenishment via carboxylation of pyruvate indispensable for the proper functioning of metabolism.

## TRANSFER OF ONE-CARBON FRAGMENTS

In the chapters in which we discussed the metabolism of carbohydrates and fats, we described many reactions in which two-carbon or three-carbon units were transferred from one molecule to another, but none (except those involving carbon dioxide) in which a one-carbon unit was transferred. Such reactions do, however, occur – though rarely. They are involved chiefly in the metabolism of amino acids (see pp. 368, 379 and 381) and in the synthesis of purines (see pp. 395 and 396), and we must now discuss their mechanism.

The most important feature of one-carbon transfer is that it needs a special coenzyme, called tetrahydrofolate, to which the one-carbon unit becomes attached. We give the formula of tetrahydrofolate on p. 386; we shall here write it as:

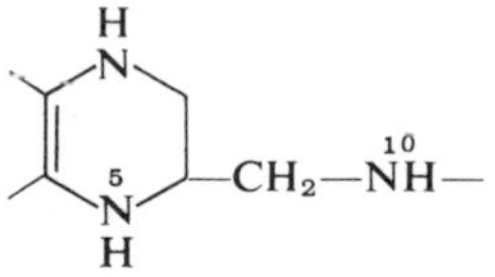

$$\text{—CH}_2\text{—}\overset{10}{\text{N}}\text{H—}$$

to make clear how the one-carbon unit is attached. (Note the nitrogen atoms labelled 5 and 10.)

Tetrahydrofolate can accept and donate one-carbon units of various kinds. In the degradation of histidine, for example, it accepts a formimino group from formiminoglutamate (see p. 381).

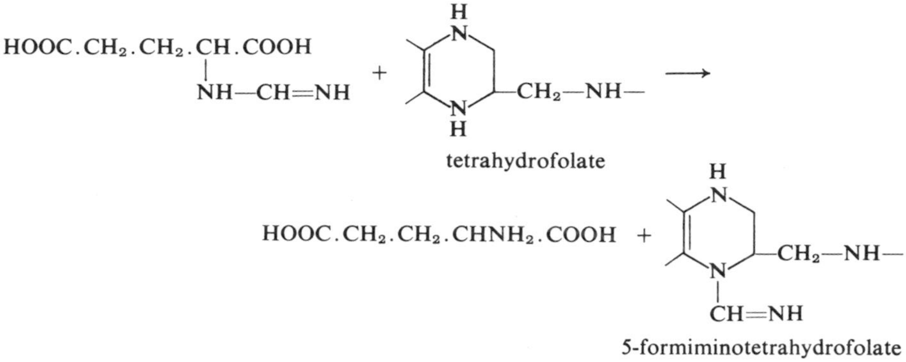

$$\text{HOOC.CH}_2\text{.CH}_2\text{.CH.COOH}$$
$$\underset{\text{NH—CH=NH}}{|} \quad + \quad \text{—CH}_2\text{—NH—} \quad \longrightarrow$$

tetrahydrofolate

$$\text{HOOC.CH}_2\text{.CH}_2\text{.CHNH}_2\text{.COOH} \quad + \quad \text{—CH}_2\text{—NH—}$$
$$\underset{\text{CH=NH}}{}$$

5-formiminotetrahydrofolate

Now the one-carbon unit can undergo metabolic conversions while still attached to the tetrahydrofolate. For instance the formimino group can lose ammonia to form 5,10-methenyltetrahydrofolate.

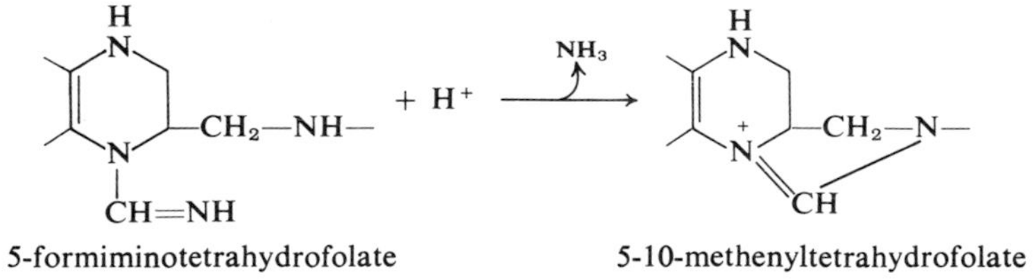

$$\text{—CH}_2\text{—NH—} \quad + \text{ H}^+ \quad \overset{\text{NH}_3}{\longrightarrow} \quad \text{—CH}_2\text{—N—}$$
$$\underset{\text{CH=NH}}{} \qquad\qquad\qquad\qquad \underset{\text{CH}}{}$$

5-formiminotetrahydrofolate          5-10-methenyltetrahydrofolate

5,10-Methenyltetrahydrofolate can be converted to 10-formyltetrahydrofolate:

$$\text{—CH}_2\text{—N—} \quad + \text{ H}_2\text{O} \rightleftharpoons \quad \text{—CH}_2\text{—N—} \quad + \text{ H}^+$$
$$\underset{\text{CH}}{} \qquad\qquad\qquad\qquad\qquad \underset{\text{CHO}}{}$$

5-10-methenyltetrahydrofolate          10-formyltetrahydrofolate

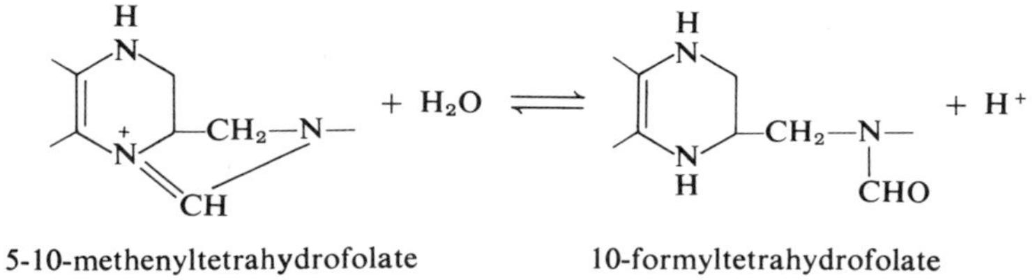

or can be reduced to 5,10-methylenetetrahydrofolate.

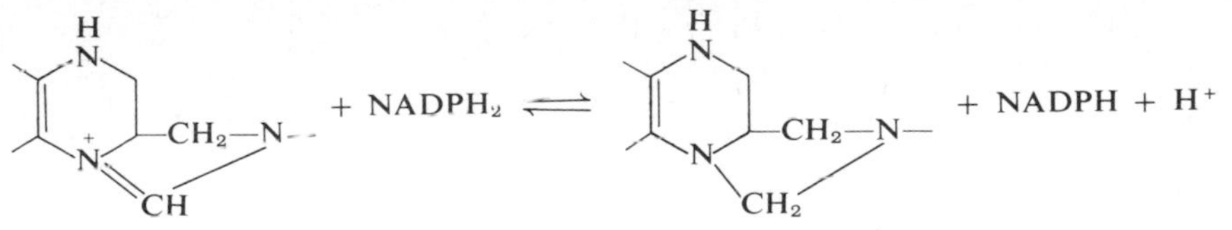

$+ \text{NADPH}_2 \rightleftharpoons$ 　　　　　$+ \text{NADPH} + \text{H}^+$

5,10-methenyltetrahydrofolate 　　　　　 5,10-methylenetetrahydrofolate

The latter can be reduced still further to yield 5-methyltetrahydrofolate.

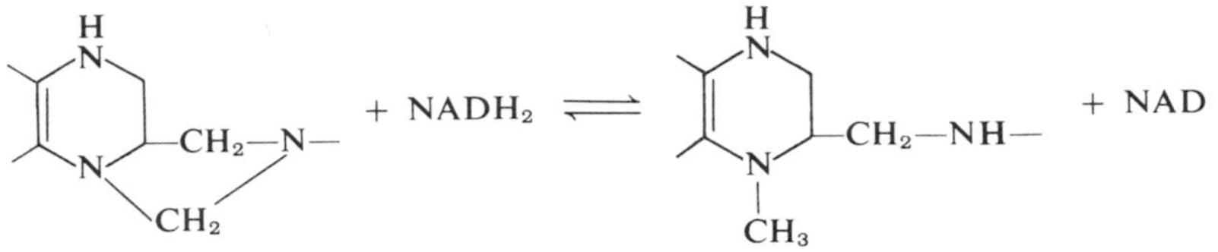

$+ \text{NADH}_2 \rightleftharpoons$ 　　　　　$+ \text{NAD}$

5,10-methylenetetrahydrofolate 　　　　　 5-methyltetrahydrofolate

All these forms of folate function in reactions involving the transfer of one-carbon units. The 5,10-methenyl derivative donates a one-carbon fragment in a reaction in purine biosynthesis (p. 395) as does the 10-formyl derivative in another reaction (p. 396). The 5,10-methylene derivative is used in the synthesis of serine from glycine (p. 379) and is formed in the reverse reaction (p. 378). Finally, the 5-methyl derivative is used in the synthesis of methionine from homocysteine (p. 368). In this way the substituted folate molecules act as a pool of metabolically active one-carbon units (see Fig. 22.2).

We must also consider one other kind of one-carbon transfer, namely that of methyl groups. We have just seen how methionine is synthesized from homocysteine at the expense of 5-methyltetrahydrofolate. Methionine can, in turn, serve as a methylating agent. In order to do so it must first be activated. This is accomplished in an unusual reaction with ATP; all three phosphate groups are split out, and the product is S-adenosylmethionine.

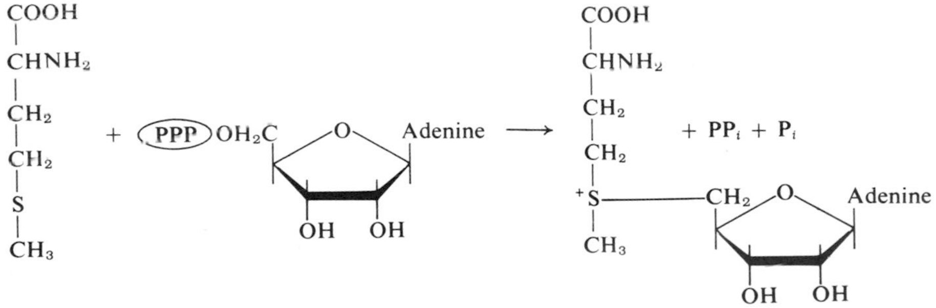

COOH
|
CHNH$_2$
|
CH$_2$
|
CH$_2$      $+$ (PPP)OH$_2$C —O— Adenine $\longrightarrow$
|
S
|
CH$_3$

OH  OH

methionine 　　　　　 ATP

COOH
|
CHNH$_2$
|
CH$_2$
|
CH$_2$      $+ \text{PP}_i + \text{P}_i$
|
$^+$S——CH$_2$—O— Adenine
|
CH$_3$

OH  OH

S-adenosylmethionine

S-Adenosylmethionine has a very high potential for donating methyl groups. We have previously described (p. 349) how it can be used to methylate phosphatidyl ethanolamine. It also acts in the methylation of other compounds,

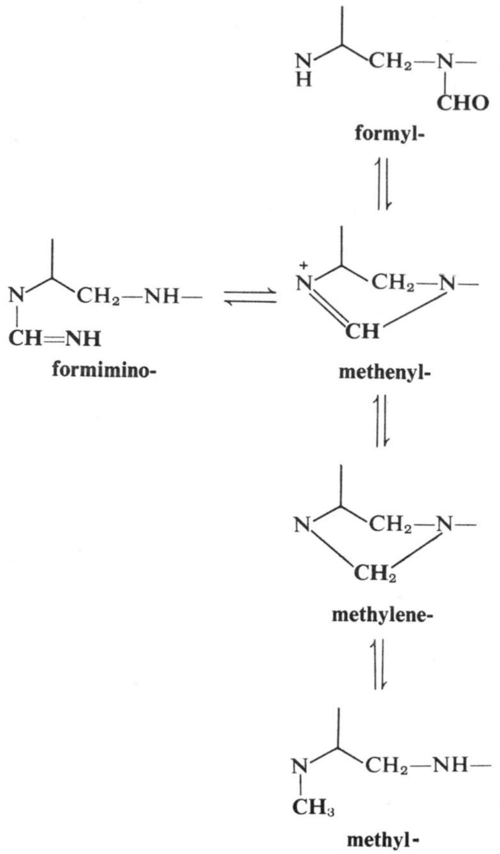

**Fig. 22.2** Interconversions of the folate derivatives. The one-carbon unit can be modified while attached to the folate molecule

thereby forming, for example, creatine and adrenaline. The S-adenosylhomocysteine that is the other product of such methylations is split to adenosine (see p. 368) and homocysteine. Homocysteine can either be converted back to methionine (p. 368) or be degraded via cystathionine (p. 378).

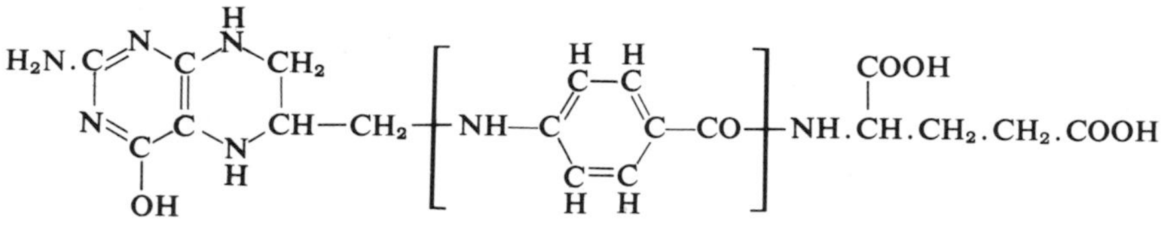

**Fig. 22.3** Tetrahydrofolic acid. Note the p-aminobenzoic acid moiety (enclosed in square brackets) and refer to p. 80 where the action of the mimic compound sulphonamide is discussed

# Chapter 23  Synthesis of other nitrogenous compounds

In the last chapter we discussed the synthesis of the amino acids, starting from inorganic nitrogen in the form of ammonia and from the carbon skeletons of the intermediates in carbohydrate breakdown. We still have to consider the synthesis of two other important classes of nitrogen-containing molecules, namely the purine and pyrimidine bases that are constituents of DNA and RNA (see Ch. 7). The synthetic pathways leading to these bases are the same in almost all species of organism. In addition we shall describe the synthesis of urea, the form in which many vertebrates, including ourselves, excrete surplus nitrogen.

We have already seen that there are two reactions, both of great importance, by means of which ammonia can be fixed into organic molecules – the reductive amination of $\alpha$-oxoglutarate to yield glutamate (p. 357) and the amination of glutamate to yield glutamine (p. 363) – and we have shown how these molecules function as donors of amino groups. There is a third reaction for the fixation of ammonia which we must now consider. This is the formation of a compound called carbamoyl phosphate, in an unusual reaction involving carbon dioxide, ammonia and ATP. N-Acetylglutamic acid (p. 364) has a catalytic role in the reaction, the nature of which is not fully understood.

$$CO_2 + NH_3 + 2ATP \rightleftharpoons NH_2.COO\textcircled{P} + 2ADP + P_i$$

Carbamoyl phosphate is a high-energy compound which can readily transfer its carbamoyl group in syntheses, as we shall see below.

This synthesis completes our repertoire of ammonia-fixing reactions. We recall that in animals the ammonia used in such reactions generally comes from the deamination (see p. 359) of amino acids that are surplus to requirements because they are present, in ingested proteins, in proportions different from those needed for protein synthesis (see p. 361).

## EXCRETION OF NITROGEN – SYNTHESIS OF UREA

The first problem we shall consider is how such surplus nitrogen is excreted.* Many animals that have a ready supply of water, such as fish, use the simplest possible means: they excrete the ammonia as such, having brought it to the kidney in the form of glutamine in the blood (p. 363). To excrete ammonia,

---

* The question of *excreting* nitrogen only arises, of course, after the requirements for nitrogen in *synthetic* reactions (see, for example, pp. 357 and 363) have been met

however, requires the formation of an extremely dilute urine, since ammonia is toxic in concentrated solution, and terrestrial animals cannot afford the waste of water that this involves. Most land-dwelling vertebrates, therefore, turn the ammonia first into another compound which can be excreted in more concentrated solution – namely urea.*

Urea is synthesized in the liver by a cyclical series of reactions which were elucidated by Krebs and Henseleit. We shall now describe the cycle, which involves the formation of arginine from ornithine (p. 364) and the hydrolysis of arginine to yield urea and regenerate ornithine.

In the first reaction of the urea cycle, ornithine reacts with carbamoyl phosphate, the synthesis of which we have just described. The enzyme is ornithine transcarbamylase.

$$
\begin{array}{l}
\text{COOH} \\
|\\
\text{CHNH}_2 \\
|\\
\text{CH}_2 \quad + \text{NH}_2.\text{COO}\,\text{(P)} \rightleftharpoons \\
|\\
\text{CH}_2 \\
|\\
\text{CH}_2\text{NH}_2
\end{array}
\qquad
\begin{array}{l}
\text{COOH} \\
|\\
\text{CHNH}_2 \\
|\\
\text{CH}_2 \qquad\qquad + \text{P}_i \\
|\\
\text{CH}_2 \\
|\\
\text{CH}_2.\text{NH}.\text{CO}.\text{NH}_2
\end{array}
$$

ornithine    carbamoyl phosphate            citrulline

The product of this reaction is citrulline, and this now condenses with aspartate in a reaction driven by the hydrolysis of ATP to AMP.

$$
\begin{array}{l}
\text{COOH} \\
|\\
\text{CHNH}_2 \\
|\\
\text{CH}_2 \\
|\\
\text{CH}_2 \\
|\\
\text{CH}_2.\text{NH}.\text{CO} \\
\qquad | \\
\qquad \text{NH}_2
\end{array}
+
\begin{array}{l}
\text{COOH} \\
|\\
\text{CH}_2 \\
|\\
\text{NH}_2.\text{CH} \\
|\\
\text{COOH}
\end{array}
+ \text{ATP} \rightleftharpoons
\begin{array}{l}
\text{COOH} \\
|\\
\text{CHNH}_2 \\
|\\
\text{CH}_2 \\
|\\
\text{CH}_2 \\
|\\
\text{CH}_2.\text{NH}.\text{C}{=}\text{N}.\text{CH} \\
\qquad\quad | \qquad\quad | \\
\qquad\quad \text{NH}_2 \quad \text{COOH}
\end{array}
\begin{array}{l}
\text{COOH} \\
|\\
\text{CH}_2 \\
|\\
\text{CH}_2 \\
\end{array}
+ \text{AMP} + \text{PP}_i
$$

citrulline              aspartic acid                argininosuccinic acid

The argininosuccinate thus produced is split to arginine and fumarate.

$$
\begin{array}{l}
\text{COOH} \\
|\\
\text{CHNH}_2 \\
|\\
\text{CH}_2 \\
|\\
\text{CH}_2 \\
|\\
\text{CH}_2.\text{NH}.\text{C}{=}\text{N}.\text{CH} \\
\qquad\quad | \qquad\quad | \\
\qquad\quad \text{NH}_2 \quad \text{COOH}
\end{array}
\rightleftharpoons
\begin{array}{l}
\text{COOH} \\
|\\
\text{CHNH}_2 \\
|\\
\text{CH}_2 \\
|\\
\text{CH}_2 \\
|\\
\text{CH}_2.\text{NH}.\text{C}{=}\text{NH} \\
\qquad\qquad\; | \\
\qquad\qquad\; \text{NH}_2
\end{array}
+
\begin{array}{l}
\text{COOH} \\
|\\
\text{CH} \\
\|\\
\text{HC} \\
|\\
\text{COOH}
\end{array}
$$

argininosuccinic acid                    arginine          fumaric acid

* Birds excrete uric acid, a compound which requires still less water for its disposal. We shall not consider the synthesis of uric acid here.

Note that this is the normal pathway of synthesis of arginine from ornithine (p. 364).

We shall pursue the fate of the fumarate in a moment, but first we must see how arginine is converted to urea. This reaction is brought about by the hydrolytic action of arginase (p. 364).

$$
\begin{array}{c}
\text{COOH} \\
|\\
\text{CHNH}_2 \\
|\\
\text{CH}_2 \\
|\\
\text{CH}_2 \\
|\\
\text{CH}_2.\text{NH}.\text{C}=\text{NH} \\
\qquad |\\
\qquad \text{NH}_2 \\
\text{arginine}
\end{array}
\; + \text{H}_2\text{O} \rightleftharpoons \;
\begin{array}{c}
\text{COOH} \\
|\\
\text{CHNH}_2 \\
|\\
\text{CH}_2 \\
|\\
\text{CH}_2 \\
|\\
\text{CH}_2\text{NH}_2 \\
\text{ornithine}
\end{array}
\; + \text{NH}_2\text{CONH}_2 \\
\qquad\qquad \text{urea}
$$

Ornithine can now react with another molecule of carbamoyl phosphate and so the cycle begins again.

Of the two —NH$_2$ groups in urea, one evidently arises from carbamoyl phosphate and hence directly from ammonia. The origin of the other is the α-amino group of aspartate; we shall see that reactions in which aspartate is used as a donor of the —NH$_2$ group occur also in the synthesis of purines (pp. 395 and 396). But what is the origin of the α-amino group of aspartate? The answer is that it arises by transamination of oxaloacetate (p. 365) with glutamate, and the glutamate in turn may have acquired it either by transamination (p. 359) or by direct fixation of ammonia (p. 357). In this way we can see how both of the - -NH$_2$ groups of urea arise ultimately from surplus nitrogen derived from amino acids.

Now after donating its α-amino group, aspartate becomes converted in the urea cycle to fumarate. Fumarate can continue around the tricarboxylic acid cycle to form oxaloacetate (p. 278), and the latter can readily be converted again to aspartate by transamination. We can thus depict the reactions as interlocking cycles:

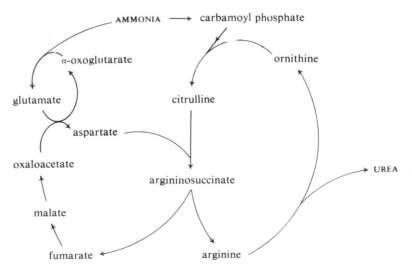

in which ammonia enters in the form of carbamoyl phosphate and glutamate and leaves in the form of urea.

## SYNTHESIS OF PYRIMIDINE NUCLEOTIDES

Carbamoyl phosphate, the synthesis of which we discussed above, is an intermediate in the formation not only of urea but also of pyrimidines. The first reaction leading to pyrimidines is the formation of carbamoyl aspartate from aspartate and carbamoyl phosphate.

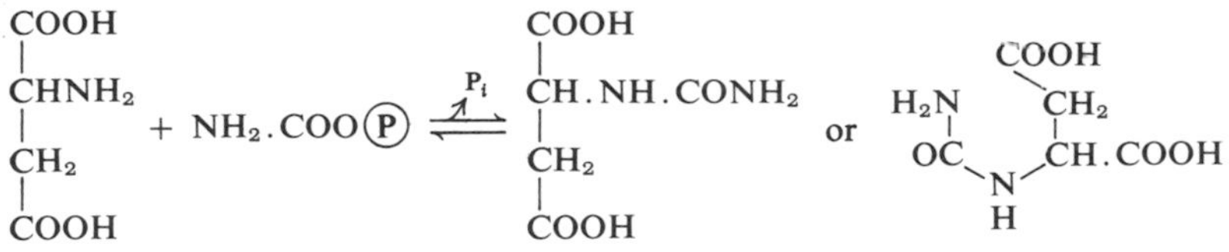

| aspartic acid | carbamoyl phosphate | carbamoyl aspartic acid |

The enzyme that catalyses this reaction is called aspartate transcarbamylase (compare ornithine transcarbamylase, p. 388). It has been studied in great detail as an example of enzymes that are subject to feedback inhibition (see p. 131), and we shall discuss its properties and its ability to regulate the rate of pyrimidine synthesis in Ch. 29.

Carbamoyl aspartate is dehydrated to yield dihydro-orotate, and oxidation with NAD produces orotate.

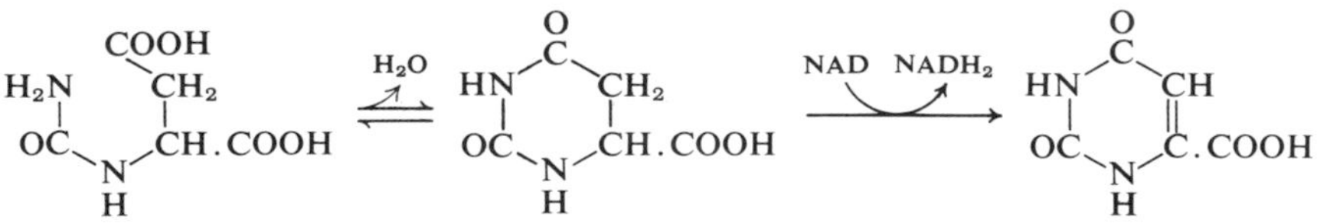

| carbamoyl aspartic acid | dihydro-orotic acid | orotic acid |

At this stage ribose-5-phosphate becomes attached to the molecule, with the result that a ribonucleotide (p. 135) is formed. This first nucleotide of the pathway is called orotidine-5'-phosphate; 5-phosphoribosyl pyrophosphate is involved in its formation, as it is in the synthesis of tryptophan (p. 375). We shall see shortly that 5-phosphoribosyl pyrophosphate also supplies ribose-5-phosphate in the synthesis of purine nucleotides.

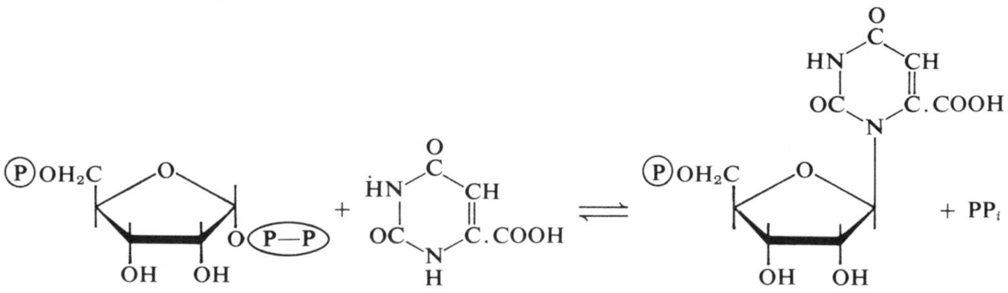

| 5-phosphoribosyl pyrophosphate | orotic acid | orotidine-5'-phosphate |

Note that inorganic pyrophosphate is liberated, so that this reaction may be regarded as going virtually to completion, for the reason that we explained on p. 284. Note also that the *first* reaction of the pathway of pyrimidine synthesis, that catalysed by aspartate transcarbamylase, involves the transfer of phosphate from a high-energy compound. We remarked on p. 285 that the same is generally true also of degradative pathways.

Decarboxylation of orotidine-5'-phosphate yields uridine-5'-phosphate, otherwise called uridylic acid or UMP.

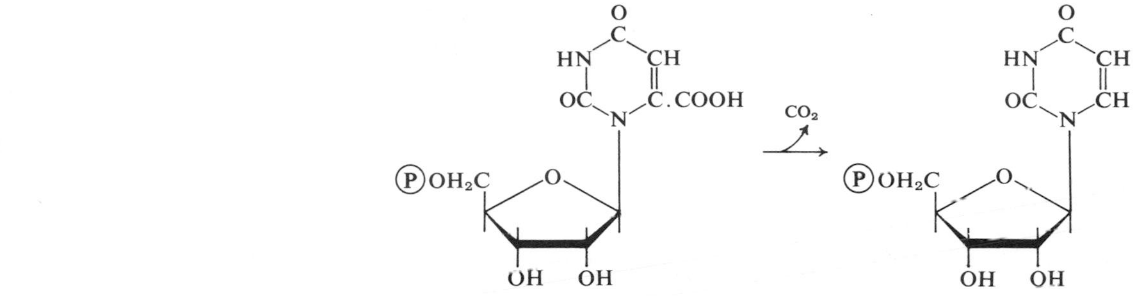

orotidine-5'-phosphate            uridine-5'-phosphate

We have thus reached one of the constituents of RNA. However, we shall see in Ch. 25 that the precursors for RNA synthesis are not the nucleoside monophosphates (UMP, etc.) but the nucleoside triphosphates (UTP, etc.). In fact UMP can be converted to UTP by two successive reactions with ATP catalysed by nucleotide phosphotransferase.

$$UMP + ATP \rightleftharpoons UDP + ADP$$
$$UDP + ATP \rightleftharpoons UTP + ADP$$

This completes the synthesis of UTP, the use of which in the synthesis of RNA we shall describe in Ch. 25. The other pyrimidine precursor of RNA is CTP; this can be formed from UTP by amination.

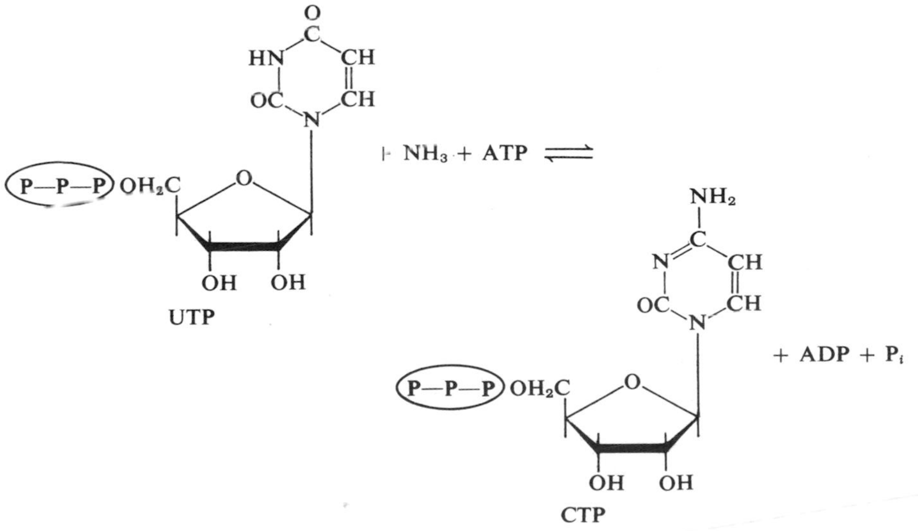

UTP      $+ NH_3 + ATP \rightleftharpoons$      $+ ADP + P_i$      CTP

## Pyrimidine deoxyribonucleotides

We must now discuss the synthesis of the pyrimidine precursors of DNA. We shall see in Ch. 25 that the synthesis of DNA, like that of RNA, requires nucleoside triphosphates. One of these is deoxycytidine triphosphate, which differs from cytidine triphosphate (CTP) only in that the 2'-carbon of the sugar carries —$H_2$— instead of —HOH—. However, the other pyrimidine precursor of DNA is not deoxyUTP, since uracil does not occur in DNA, but its methylated derivative deoxythymidine triphosphate (see pp. 136 ff.). Both deoxyCTP and deoxyTTP are formed from CTP via deoxyCDP.

The first reaction in the synthesis of both deoxyribonucleoside triphosphates is the conversion of CTP to CDP.

$$CTP + ADP \rightleftharpoons CDP + ATP$$

CDP can now be reduced directly to deoxyCDP. The reducing agent is a protein called thioredoxin, and it exists in a reduced form that has two —SH groups and an oxidized form that has an —S—S— bridge (see p. 70).

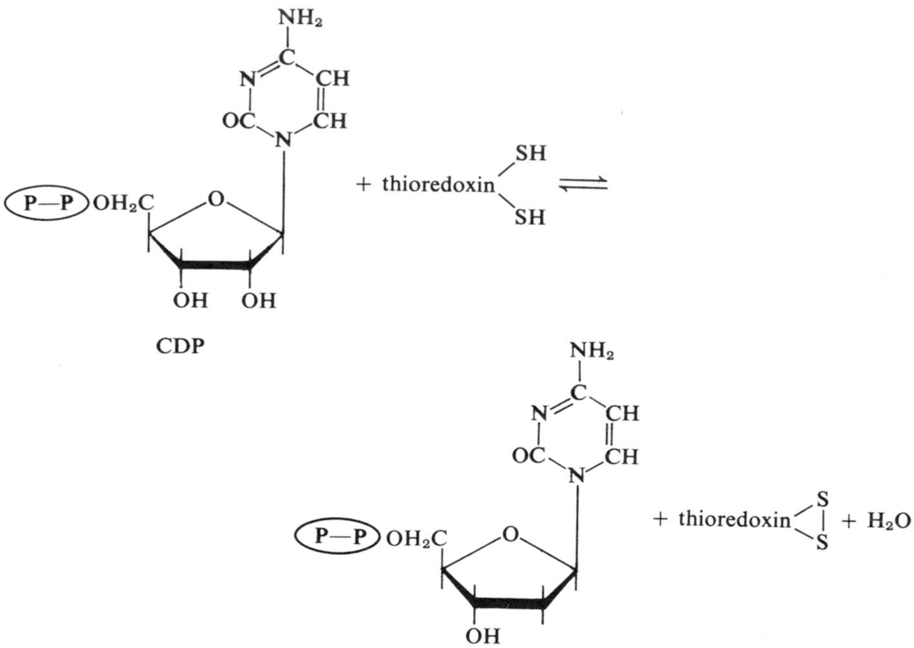

CDP

deoxyCDP

The oxidized form of thioredoxin can be reduced again by $NADPH_2$, so that thioredoxin functions catalytically in the formation of the deoxyribonucleoside diphosphate. DeoxyCDP finally reacts with ATP to form deoxyCTP and ADP.

The synthesis of deoxyTTP from deoxyCDP is rather more complicated. First deoxyCDP donates a phosphate group to ADP and thus becomes deoxyCMP, and this then loses ammonia to form deoxyUMP.

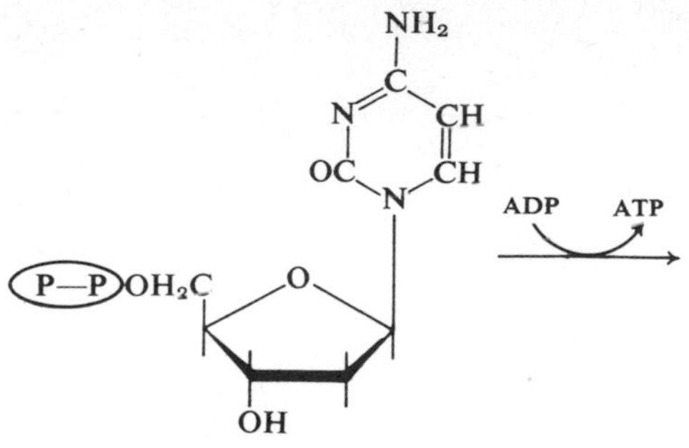

NH$_2$

N

OC

CH

CH

N

ADP    ATP

P—P OH$_2$C    O

OH

**deoxyCDP**

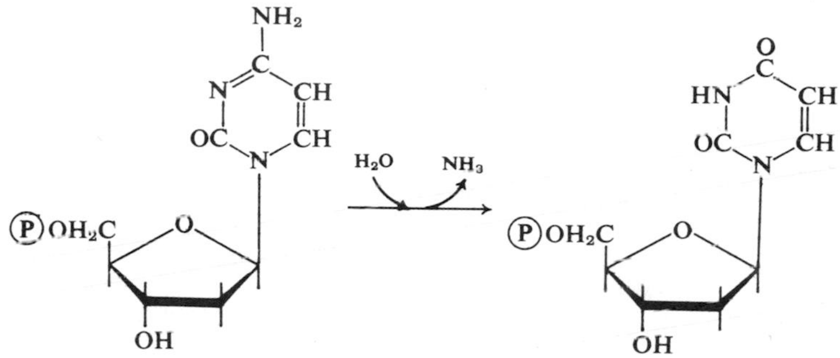

NH$_2$

C

N

OC

CH

CH

N

H$_2$O    NH$_3$

P OH$_2$C    O

OH

**deoxyCMP**

O

C

HN

OC

CH

CH

N

P OH$_2$C    O

OH

**deoxyUMP**

DeoxyUMP is now converted to deoxyTMP in a very unusual reaction with 5,10-methylenetetrahydrofolate (p. 385). The co-factor donates not only the methylene group but also one hydrogen atom from the tetrahydrofolate nucleus

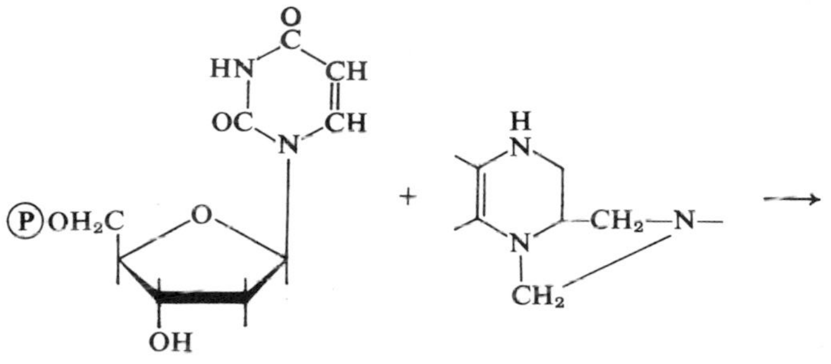

O

C

HN    CH

OC    CH

N

P OH$_2$C    O

OH

**deoxyUMP**

+

H

N

N

CH$_2$

CH$_2$—N—

→

**5,10-methylenetetrahydrofolate**

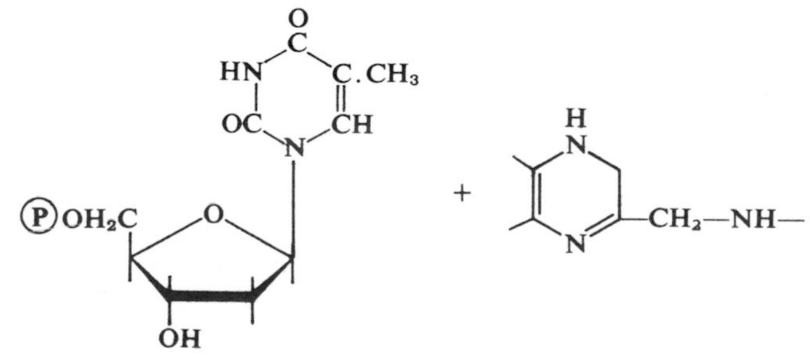

O

C

HN    C.CH$_3$

OC    CH

N

P OH$_2$C    O

OH

**deoxyTMP**

+

H

N

N

CH$_2$—NH—

**dihydrofolate**

itself, leaving dihydrofolate – the latter can be converted back to tetrahydrofolate by reduction with $NADPH_2$.

Finally deoxyTMP is converted to deoxyTTP by two reactions with ATP.

$$\text{deoxyTMP} + \text{ATP} \rightleftharpoons \text{deoxyTDP} + \text{ADP}$$

$$\text{deoxyTDP} + \text{ATP} \rightleftharpoons \text{deoxyTTP} + \text{ADP}$$

Notice that CTP is an important compound in this pathway, being not only one of the final products of pyrimidine biosynthesis, but also an intermediate in the synthesis of both the pyrimidine precursors of DNA. It would be reasonable, then, if the rate at which the pathway as a whole operated, was made to depend on the concentration of CTP; in this way the concentrations of the precursors of both RNA and DNA would be maintained fairly constant. We shall see in Ch. 29 that this is, in fact, exactly the way in which the rate of synthesis of pyrimidines is controlled.

Since three of the four carbon atoms in the pyrimidine ring are derived from aspartate and hence from oxaloacetate (p. 365), the synthesis of pyrimidines is another of the major pathways that draw carbon from intermediates in the tricarboxylic acid cycle (see p. 210).

## SYNTHESIS OF PURINE NUCLEOTIDES

The pathway of synthesis of the purine nucleotides is lengthy and rather complicated. In some ways it resembles the pathway of histidine synthesis, and in fact one of the byproducts of histidine synthesis can be used as an intermediate in the formation of purines (see p. 381). One feature of the purine pathway, and a point of difference from the pathway of pyrimidine synthesis, is that the purine ring is built on a ribose-5-phosphate base rather than being synthesized separately and then attached to ribose-5-phosphate. The effect of this is that when the purine ring system is finished the molecule is already in the form of a nucleotide.

In the first reaction 5-phosphoribosyl pyrophosphate (p. 375) reacts with glutamine. The product is the $\beta$-anomer of 5-phosphoribosylamine; this next reacts with glycine.

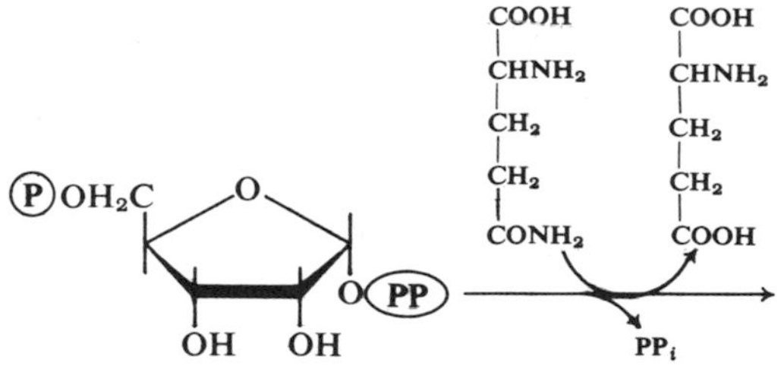

5-phosphoribosyl
pyrophosphate

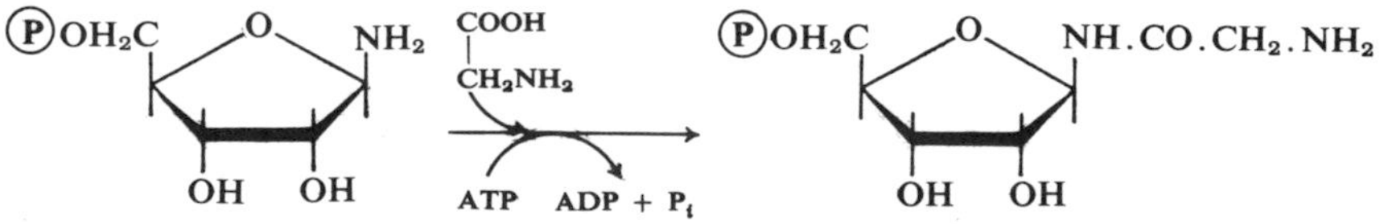

5-phosphoribosylamine

5-phosphoribosyl-glycineamide

The 5-phosphoribosyl moiety remains in place as the molecule receives first a one-carbon unit from 5,10-methenyltetrahydrofolate (p. 384) and then another amino group from glutamine.

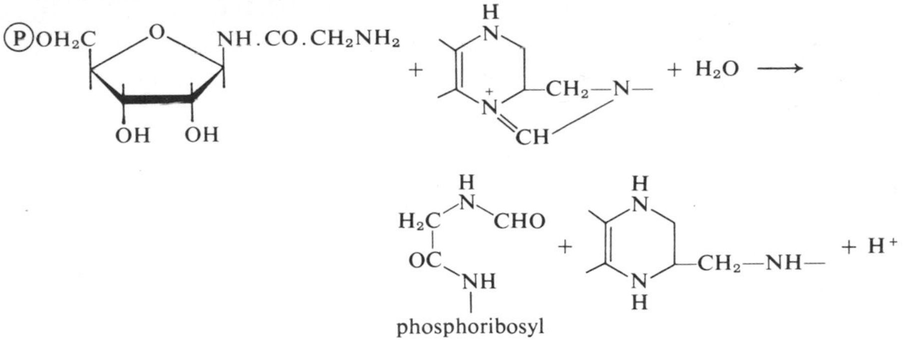

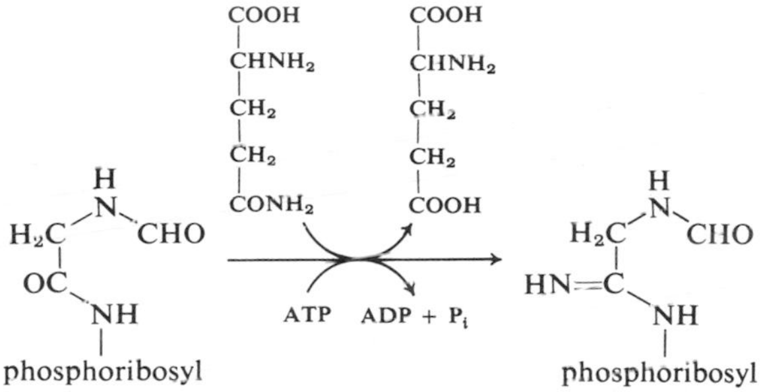

$$\text{(P)OH}_2\text{C} - \text{(ribose ring, OH, OH)} - \text{NH.CO.CH}_2\text{NH}_2 \quad + \quad \text{(methenyltetrahydrofolate)} \quad + \text{H}_2\text{O} \longrightarrow$$

$$\underset{\text{phosphoribosyl}}{\text{H}_2\text{C}-\text{N(H)}-\text{CHO, OC, NH}} \quad + \quad \text{(tetrahydrofolate)} \quad + \text{H}^+$$

$$\underset{\text{phosphoribosyl}}{\text{H}_2\text{C}-\text{N(H)}-\text{CHO, OC, NH}} \quad \xrightarrow[\text{ATP} \quad \text{ADP} + \text{P}_i]{\text{glutamine} \to \text{glutamate}} \quad \underset{\text{phosphoribosyl}}{\text{H}_2\text{C}-\text{N(H)}-\text{CHO, HN}=\text{C, NH}}$$

Now the ring closes, and another carbon atom is added, this time from carbon dioxide.

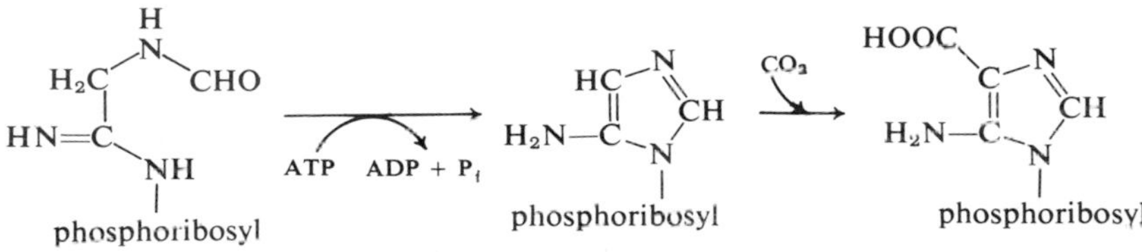

$$\underset{\text{phosphoribosyl}}{\text{H}_2\text{C}-\text{N(H)}-\text{CHO, HN}=\text{C, NH}} \xrightarrow[\text{ATP} \quad \text{ADP} + \text{P}_i]{} \underset{\text{phosphoribosyl}}{\text{HC}-\text{N}, \text{CH, H}_2\text{N}-\text{C}-\text{N}} \xrightarrow{CO_2} \underset{\text{phosphoribosyl}}{\text{HOOC}-\text{C}-\text{N}, \text{CH, H}_2\text{N}-\text{C}-\text{N}}$$

At this stage another nitrogen atom is needed. This is supplied from aspartate, in a reaction exactly analogous to the synthesis of arginine from citrulline (p. 388). Aspartate condenses with the molecule to form a succinyl derivative, and then fumarate is lost. The result is simply to replace a —COOH with —CONH$_2$.

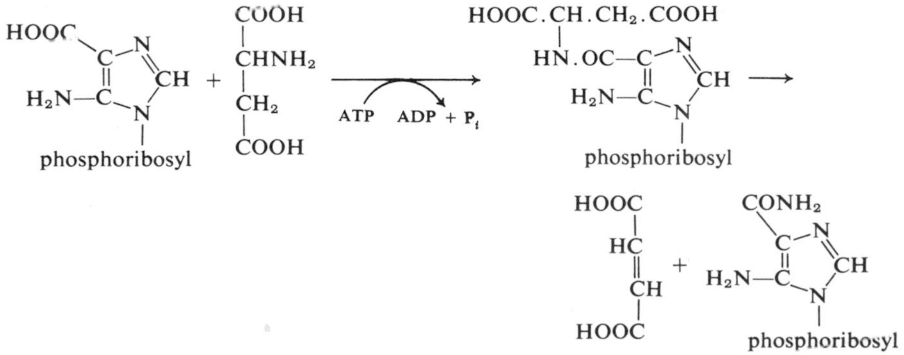

$$\underset{\text{phosphoribosyl}}{\text{HOOC}-\text{C}-\text{N}, \text{CH, H}_2\text{N}-\text{C}-\text{N}} + \underset{\text{COOH, CHNH}_2, \text{CH}_2, \text{COOH}}{} \xrightarrow[\text{ATP} \quad \text{ADP} + \text{P}_i]{} \underset{\text{phosphoribosyl}}{\overset{\text{HOOC.CH.CH}_2\text{.COOH}}{\text{HN.OC}-\text{C}-\text{N}, \text{CH, H}_2\text{N}-\text{C}-\text{N}}} \longrightarrow$$

$$\underset{\text{HOOC}}{\text{HC}=\text{CH}} \quad + \quad \underset{\text{phosphoribosyl}}{\overset{\text{CONH}_2}{\text{C}-\text{N}, \text{CH, H}_2\text{N}-\text{C}-\text{N}}}$$

The product of this reaction is AICAR, which was formed (you will recall) in the synthesis of histidine (p. 380). AICAR that is left over from histidine synthesis can join the sequence at this point.

AICAR is converted to a purine nucleotide by a sequence of two reactions. First another one-carbon unit is introduced from 10-formyltetrahydrofolate (p. 384), and then the purine ring cyclizes with loss of water.

The product is inosinic acid (IMP), which occurs in the cytoplasm along with its derivatives IDP and ITP, and which is also a constituent of transfer RNA (see Table 7.2).

IMP is a precursor both of AMP and of GMP. The conversion to AMP is interesting because it makes use once again of the ability of aspartate to condense with a compound and then lose fumarate.

In the formation of GMP, inosinic acid is first oxidized to xanthylic acid, and this is then aminated by reaction with glutamine.

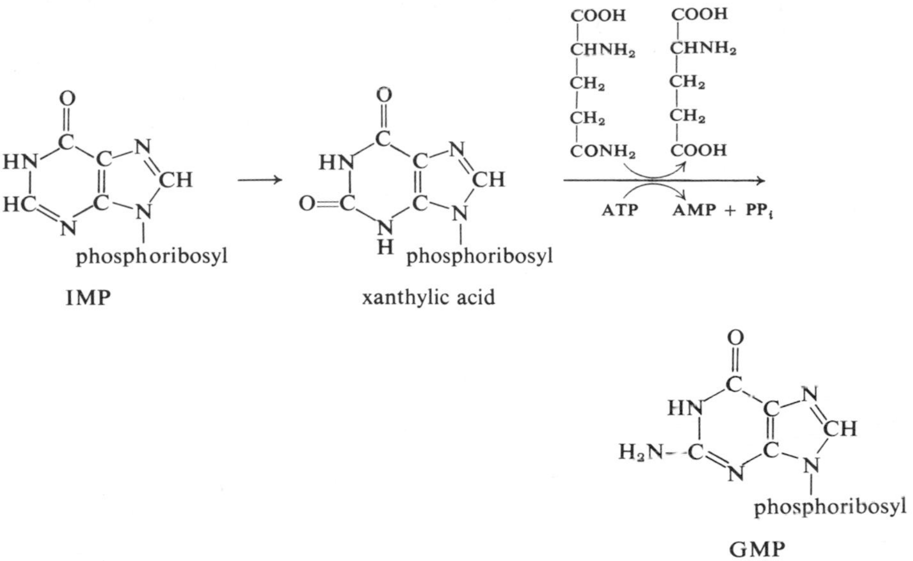

phosphoribosyl

IMP

xanthylic acid

GMP

AMP and GMP can be phosphorylated by ATP to give ADP and GDP, and GDP can be phosphorylated again.

$$\text{GDP} + \text{ATP} \rightleftharpoons \text{GTP} + \text{ADP}$$

Notice that the synthesis of GMP from IMP requires ATP, and the synthesis of AMP from IMP requires GTP. In this way an excess of either of the purine ribonucleoside triphosphates catalyses the formation of the other, so that the balance of the two is maintained.

DeoxyATP is formed by a pathway similar to that for the formation of deoxyCTP. First ADP is reduced to deoxyADP by thioredoxin (p. 392) and this is then phosphorylated by ATP. In the same way GDP is reduced to deoxyGDP and this is phosphorylated by ATP to form deoxyGTP. Thus the conversion of ribonucleotides to deoxyribonucleotides by the thioredoxin reaction always takes place at the level of the nucleoside diphosphate.

We can summarize the synthesis of the purine ring system by tracing the origin of the atoms that make it up.

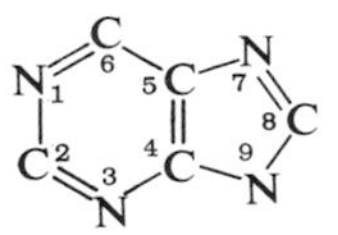

N-9 arises from glutamine, C-4, C-5 and N-7 from glycine, and C-8 from a one-carbon unit carried by tetrahydrofolate. Then N-3 is added from glutamine, C-6 from carbon dioxide, N-1 from aspartate, and finally C-2 from a tetrahydrofolate derivative.

# Part 3 Informational macromolecules

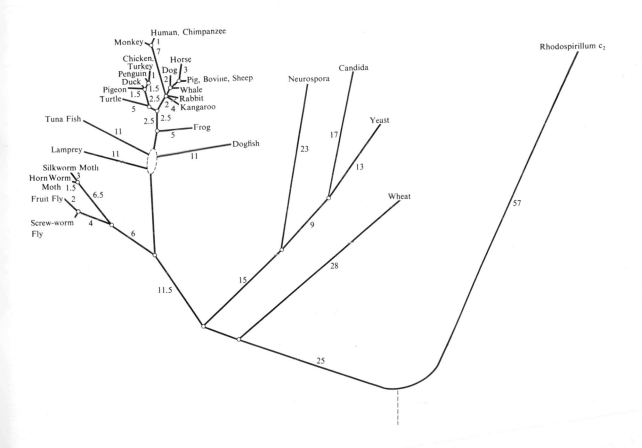

# Chapter 24 **Principles of molecular genetics**

Our attention in the previous Parts of this book has been focused on energy: the coupling of endergonic and exergonic reactions, the liberation of free energy by the degradation of foodstuffs and the use of free energy in the synthesis of small molecules. We now wish to return to the macromolecules, discussion of which dominated Part I of the book. Whereas our concern there was with the structure of macromolecules and the relation between structure and function, we shall approach macromolecules in the next few chapters from a different point of view. We shall be considering the means by which they are synthesized, and in particular the means by which the exact ordering of their constituent residues (Ch. 3) is determined during synthesis.

In the chapters on intermediary metabolism we have mentioned well over a hundred enzymes, and this number represents only a small fraction of the enzymes that are known. Now we have already seen (Ch. 6) that the structure of each one of these enzymes must be dictated by the synthetic machinery of the cell with absolute precision, and that an error in inserting just one amino-acid residue in the protein chain may be sufficient to inactivate the enzyme. What is true of enzymes is true also of other proteins: we described in some detail the consequences (sometimes very grave) of changing a single amino acid in haemoglobin, and we could have multiplied the examples with other proteins. So we must now consider this question: what is it that is responsible for determining the sequence of amino acids in proteins with such precision? (We shall see that analogous questions can be asked about nucleic acids.)

When we spoke of the changes that have been discovered in the structure of haemoglobin in some individuals (p. 107), we remarked that these represented deviations from the normal amino-acid sequence of human haemoglobin. When we discussed the structure of insulin (p. 101) we remarked that the molecule we were discussing was that of pig insulin. The fact that we can speak of 'human haemoglobin' or of 'pig insulin' implies that a given protein has the same amino-acid sequence throughout most (at any rate) of the individuals of a species. This implication is borne out by the fact that if one examines the amino-acid sequence of haemoglobin from several normal individuals, or the amino-acid sequence of insulin from several normal pigs, one observes there is no variation from one individual to the next. We can make the same point in a different way by saying that the amino-acid sequence of these proteins is constant from one generation to the next of the animal concerned. Each human being (ignoring the rare exceptions) makes haemoglobin that has the same amino-acid sequence as that of his parents' haemoglobin, and this sequence has remained invariant for (at least) dozens of generations.

This argument leads inevitably to the conclusion that the amino-acid sequence of proteins is determined *genetically* – that is to say that the hereditary material that is passed on from one generation to the next is responsible for ensuring the precision of structure that is so striking a feature of proteins.

In describing the way in which the specific structure of proteins is determined during their synthesis, we shall therefore have occasion to refer to work in genetics as much as in biochemistry. We shall see that a combination of the two subjects has been immensely powerful in establishing the principles that we shall discuss. In this chapter we shall concentrate on an outline of the relevant aspects of genetics, and in Chs 25 and 26 we shall return to looking at matters from a more biochemical standpoint.

When subjects like heredity and variation are mentioned, many people tend to think of how children resemble their parents, or perhaps, remembering Mendel, of the height of pea plants or the colour of peas. Now although questions of this kind are extremely important, and a study of them has made great contributions to classical genetics, they are not questions that will concern us here. The reason for our neglecting them is implied in the introduction to this chapter. We wish to focus on those aspects of genetics that might be of direct use in helping us to understand how the structure of macromolecules is determined, and although there is no doubt that in principle such matters as the colour of peas is related to the structure of macromolecules, we are far from understanding such a relationship in detail.

Instead we shall be concentrating on the study that is known as *molecular genetics*, which is concerned with an account of the relationship between the hereditary material and the structure of macromolecules. In molecular genetics we are interested, for example, in the genetic events that lead to deviations from the normal structure of haemoglobin (p. 107) or to loss of the enzyme phenylalanine hydroxylase (p. 376). The ultimate aim of the study would be to show in detail how the genetic material of an organism functions to determine the structure of its total complement of macromolecules. Although we have by no means achieved this aim, there is a great deal of useful information that has been derived from molecular genetics, and this we shall now outline.

Before we can go any further, however, we have to ask what is the chemical nature of the genetic material. The answer to this question turned out to be so surprising that for many years it was regarded with caution or even scepticism. In all organisms that are capable of autonomous existence (plants, animals and most micro-organisms), and in most viruses, the genetic material is DNA. The reason why this fact was regarded as so surprising was that until comparatively recently DNA was thought to be a long, boring molecule, with a monotonous repetition of nucleotides, and it was hard to see how such a molecule could determine anything at all.

It is worth mentioning a few of the lines of evidence that led to the conclusion that DNA is indeed the genetic material. Two of the most convincing came from studies of microbial systems. The first is due to experiments of Avery and his collaborators, in 1944, with a bacterium called *Pneumococcus*. This organism exists in two varieties, one virulent (capable of causing disease in animals), the other not, and the virulence or non-virulence is a permanent genetic characteristic. Avery found that if DNA is extracted from virulent organisms, purified and then added to cultures of the non-virulent strain, some cells of the latter become transformed into the virulent variety. The change is permanent and is passed on to the progeny of the transformed cells. If, however, the DNA is treated with deoxyribonuclease, a highly specific hydrolytic enzyme for DNA, before being added to the culture of the non-virulent strain, no transformed cells appear. Subsequent work has shown that many other genetic characteristics can be transferred from one variety of *Pneumococcus* to another with pure DNA.

The second kind of experiment that we shall cite was done by Hershey and Chase in 1953 using a virulent bacteriophage. A bacteriophage (phage for short) is a virus that attacks a bacterium. Infection by the phage leads to the subversion of the metabolic machinery of the bacterial cell, which turns to the synthesis of phage materials and to the assembly of many more phage particles. Finally the cell lyses, releasing phages that are progeny of the original infecting phage (see p. 223). The phage consists of DNA enclosed in a protein coat; Hershey and Chase were able to show by labelling with isotopes (pp. 230 f.) that, in infection, only the DNA of the phage and not its protein entered the bacterium and therefore that the phage genetic material is DNA.

The evidence that DNA is the genetic material of higher organisms, although less conclusive, leaves no real doubt that what is true of bacteria and phages is also true of plants and animals. It is found, for example, that all types of cell in an organism have the same content of DNA, with one suggestive exception – the germ cells* have half as much DNA as the other cells. Moreover, the ratio of the bases in DNA (p. 144) is constant for all the tissues of a given species. This invariability in quantity and composition is to be expected of the genetic material, which must be passed on unchanged from one generation to the next.

By what means might we be able to investigate how the genetic material of an organism determines the structure of its macromolecules? In our discussion of some of the methods that are used in intermediary metabolism (pp. 228 ff.) we saw that a powerful technique was to perturb the system in some way, for example by the use of an enzyme inhibitor, and study the resulting changes. Similarly we might expect that if we could make a change in the structure of the genetic material – such a change is known as a *mutation* – we might be able to find a resulting change in the phenotype of the organism (by 'phenotype' we mean all of those characteristics that we can directly observe).

We may suspect, for example, that the divergences from the normal amino-acid sequence of haemoglobin that are to be found in some patients with anaemia are due to mutation of the hereditary material. Our suspicion here is justified by the fact that if we examine the families of such patients, we often find that their close relatives have abnormal haemoglobins of exactly the same type. So we may conclude that a mutation has occurred in some ancestor of the family, the result of which is that an amino acid in haemoglobin has been replaced to give an abnormal sequence. Similarly, if we examine the families of

---

* Germ cells are those cells (the ovum and sperm in higher organisms) that are the carriers of genetic material from one generation to the next.

children suffering from phenylketonuria (p. 377), we frequently find that some of their relatives are affected by the same condition.

In order to study the genetics of such conditions we would need large numbers of people with the relevant mutation, and we could perhaps obtain them by persuading the parents of children with phenylketonuria to produce more children. It goes without saying that any such suggestion would be ethically quite unacceptable; we cannot persuade genetically intriguing individuals to mate merely because we are interested in examining their offspring.

Practical difficulties of a different sort make higher organisms in general rather unsuitable as subjects for study in molecular genetics. For one thing, the life cycle of higher organisms is comparatively long, so that we have to wait for weeks or months to examine the progeny of experimental animals or plants. Secondly, it is often difficult to correlate changes in those characteristics of a higher organism that we can immediately observe – changes in coat colour of mice, for instance, or changes in the pattern of bristles of fruit-flies – with changes in the structure of single macromolecules.

For all these reasons, micro-organisms, and particularly bacteria, have become the organisms most used in molecular genetics. One obvious advantage of micro-organisms is that they *reproduce extremely quickly*, so that we obtain billions of them in a short time: some bacteria, for instance, may divide every twenty minutes in favourable conditions. Another important advantage is that they can be grown in defined media, so that we can readily *control their environment*. A third advantage, connected with the second, is that we can *select* organisms with very rare genetic characteristics.

We can illustrate the last two points in the following way. Cultures of many species of bacteria are capable of synthesizing for themselves all twenty of the amino acids that are needed for protein synthesis, when provided with a simple medium containing only glucose and a few inorganic salts including an ammonium salt (a so-called 'minimal' medium). Suppose now that one of the cells in the culture suffers a mutation, the result of which is that the organism can no longer make the enzyme phosphoserine phosphatase (p. 377). This cell will be unable to grow in minimal medium, as it will lack the ability to synthesize serine. If, however, we added serine to the minimal medium, we would find that the organism would grow perfectly normally, so that we could study it even though it is a mutant. In a similar way we can supplement the medium with any amino acid, purine or pyrimidine, or growth factor (see *vitamins*, p. 80). There is therefore no difficulty in studying organisms that carry mutations which, in the absence of the supplement, would altogether prevent growth. Let us now suppose that we are interested in seeing whether the mutant bacteria to which we have to supply serine are capable of undergoing another mutation that allows them to regain the ability to make serine. All that we need to do is to place a large number (say $10^8$) of the mutant cells on a plate of minimal medium lacking serine.* If any of the cells has regained the ability to make serine, it will multiply

---

* A 'plate' is a dish containing medium that has been made solid by the addition of a non-metabolizable polysaccharide called agar. A cell placed on the surface of such solid medium will divide, and eventually a 'colony' will appear, in which all of the cells are the progeny of a single cell. Thus all the cells in a colony have, barring mutation during the growth of the colony, the same genetic constitution.

and form a colony, even if it is the only one out of $10^8$ cells that has this ability. This means of selecting rare events has been immensely useful in studying the genetics of bacteria and fungi.

We shall devote most of the rest of this chapter to an outline of the genetics of a single bacterial species, *Escherichia coli* (which is found in large numbers in the human intestine). This species has been used far more than any other in establishing the principles of molecular genetics. We shall first describe some of the techniques of *E. coli* genetics and then discuss the most important results that have been obtained. At the end of the chapter we shall mention some of the ways in which these principles are applicable to higher organisms.

## GENETIC TECHNIQUES IN *ESCHERICHIA COLI*

### Recombination between nutritional mutants

*Escherichia coli* is one of those species of bacteria of which wild-type cultures can grow in minimal medium (see above). By a 'wild-type' culture we mean a 'normal' culture as opposed to a mutant; the definition of 'wild-type' is not very precise, but in general the meaning is not really in doubt. From wild-type cultures it is possible by various techniques to isolate mutants, each of which will grow only if provided with a single amino acid or a purine or a pyrimidine or a growth factor. From such a mutant it is not difficult to isolate a doubly mutant strain which requires two such supplements; from double mutants one can isolate triple mutants, and so on. Let us now suppose that we have isolated one strain that will grow only if provided with methionine and threonine, and another that will grow only if provided with leucine, tryptophan and histidine. Now it is found in some circumstances that if these two cultures are mixed together (but not if they are kept separate) progeny will arise that are wild-type, that is that can grow without any amino acid in the medium. These progeny will remain wild-type indefinitely, so that they must have the genetic ability to synthesize all twenty amino acids even though neither of the starting cultures was able to do so. They must therefore be *recombinants* – that is to say they must have acquired some of their genetic material from one of the starting strains and some from the other. Specifically, they must have acquired from the first strain the ability to synthesize leucine, tryptophan and histidine, and from the second the ability to synthesize methionine and threonine. We shall see from now on how the occurrence of such recombination helps us to study the structure of the genetic material.

The recombinants that we have described arise from *mating* between the two strains. It turns out that in *E. coli* some strains can act as donors of genetic material and other as recipients. During mating, DNA is passed from the donor cell into the recipient cell, so that we get a situation like that depicted in Fig. 24.1(a). In this figure we show a bacterium that has a complete complement of genetic material corresponding to that in the original recipient cell, plus a segment of genetic material from the original donor cell; such a cell is called a zygote. (We represent for the moment the genetic material of the recipient as a closed loop of DNA: we shall prove below that it does in fact take this form.) The zygote has only a transitory existence, as it soon loses the DNA that is surplus to its normal complement of genetic material. Before that happens, however, some part or parts of the DNA that was originally contributed by the donor can replace corresponding parts of the DNA of the recipient. This process is known as recombination; its exact mechanism does not concern us here, but

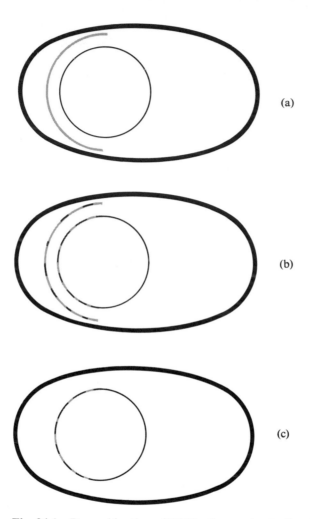

**Fig. 24.1**  Recombination of DNA after mating in *E. coli*. In (a) the recipient cell has its own complete loop of DNA and an additional piece of DNA which has been transferred from the donor. The two pieces of DNA undergo frequent exchange; this process gives rise to the situation shown in (b). The extra DNA is then lost from the cell, which finally contains a single complete loop of *recombinant* DNA as shown in (c)

we can regard it as occurring essentially at random – that is to say, with equal probability at any place along the DNA molecule. In Fig. 24.1(b) we show a cell with recombinant genetic material. You will see that some parts of the donor DNA have exchanged with corresponding parts of the recipient DNA, so that both the complete genetic loop and the additional segment are now recombinant.

In Fig. 24.1(c) the extra segment of DNA has been eliminated, and the recombinant cell has the normal complement of DNA, derived mostly from the recipient parent but in part from the donor parent.

How can the events we have described account for the formation of wild-type recombinant cells? We can answer this question by making the assumption (which seems reasonable) that there is a particular region of the DNA that is responsible, in the wild-type cell, for determining the ability of the organism to synthesize leucine. We shall designate this genetic region $leu^+$. A mutation in this region will cause the organism to lose the ability to make leucine – we shall designate the mutated region $leu^-$. By extension of this argument, we can designate the donor bacteria in the mating we described as having genetic material of the type $leu^+$ $trp^+$ $his^+$ $met^-$ $thr^-$, and the recipients as $leu^-$ $trp^-$ $his^-$ $met^+$ $thr^+$. If we now draw out the donor and recipient DNA as:

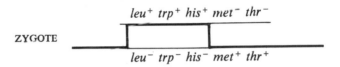

Donor          $leu^+$ $trp^+$ $his^+$ $met^-$ $thr^-$

Recipient      $leu^-$ $trp^-$ $his^-$ $met^+$ $thr^+$

we can imagine that in the course of the random recombination events in the zygotes a few cells happen to have the $leu^-$, $trp^-$ and $his^-$ regions replaced by $leu^+$, $trp^+$ and $his^+$ regions from the donor.

                         $leu^+$ $trp^+$ $his^+$ $met^-$ $thr^-$

ZYGOTE

                         $leu^-$ $trp^-$ $his^-$ $met^+$ $thr^+$

(The vertical lines here symbolize points of exchange or *crossing-over* between the two pieces of DNA in the zygote. Note that the top *horizontal* line, representing the segment of DNA from the donor, is shorter than the bottom horizontal line which represents the complete loop of DNA from the recipient.) The recombinant will now have a complete loop of DNA which is $leu^+$ $trp^+$ $his^+$ $met^+$ $thr^+$ (the thick lines in the above diagram) and a small segment which is $leu^-$ $trp^-$ $his^-$ $met^-$ $thr^-$. This latter will soon be lost from the cell, leaving a loop of DNA which will render the recombinant wild-type.

In the diagram above we have tacitly assumed, first that the order of the genetic regions *leu trp his met thr* is as we have shown it, and secondly that the $leu^+$, $trp^+$ and $his^+$ genetic regions from the donor will replace $leu^-$, $trp^-$ and $his^-$ genetic material from the recipient all together, as a single piece of DNA. Neither of these assumptions is necessarily justified, and we have made them only for the sake of convenience in drawing the diagrams. We shall consider them further below.

You may object that the likelihood of the recombinants' happening to acquire just those regions of DNA from the donor that are $leu^+$, $trp^+$ and $his^+$ and at the same time retaining the recipient's $thr^+$ and $met^+$ is very small. So indeed it may be, but we must remember that we can *select* for very rare events such as an unlikely recombination, in this case by plating, say, $10^8$ bacteria from the mating mixture on minimal medium without any amino-acid supplement. If there are any wild-type recombinants at all resulting from such a mating, we shall see them as giving rise to colonies on minimal medium.

## Mapping the genetic material

*Mapping by interrupted mating*

How can this kind of experiment help us in relating a change in the DNA of *E. coli* to a change in the phenotype? It is not easy at first sight to imagine how a study of nutritional requirements can have any bearing at all on our understanding of the structure and function of DNA. But we shall now see that the great value of mating experiments is that they allow us to make *genetic maps*. A genetic map is a sketch that shows how the regions of DNA that are responsible for particular aspects of the cell's phenotype are arranged in relation to one another. We can illustrate the way in which maps are constructed by looking more closely at the mating we described before. We have a donor strain with genetic material that is $leu^+$ $trp^+$ $his^+$ $met^-$ $thr^-$, and a recipient with genetic material that is $leu^-$ $trp^-$ $his^-$ $met^+$ $thr^+$. We can now do an experiment in which we mix these strains together and at intervals take out a sample of the mixture and subject it to vigorous mixing in a machine something like a food blender. This blending interrupts the mating, and prevents the donor from passing any more of its DNA into the recipient. The DNA that has already been passed into the recipient before blending can, however, recombine with the recipient's own DNA, and we can now see whether the recipient has acquired any or all of the three donor regions $leu^+$, $trp^+$ or $his^+$. We do this by plating the mating mixture blended at various times after the beginning of mating on minimal medium with amino-acid supplements. We exclude threonine and methionine from all plates, and in this way we prevent the donor bacteria themselves from growing. We then use three different types of plate to test the cells at each time point. To one type we add tryptophan and histidine; progeny from the mating will be able to grow on this plate provided that they have received $leu^+$ from the donor. To another type we add leucine and histidine; progeny will grow on this plate if they have received $trp^+$ from the donor. To the third type we add leucine and tryptophan; progeny will grow on this plate if they have received $his^+$ from the donor.

Typical results that might be obtained from an experiment of this sort would be as follows. If the mating is interrupted five minutes after it is begun, none of the three regions from the donor will appear in the recipient. If it is interrupted after ten minutes, $leu^+$ recombinants will be found but not $trp^+$ or $his^+$; if it is interrupted after thirty-five minutes, $leu^+$ and $trp^+$ recombinants will be found but not $his^+$; if it is interrupted after fifty minutes, $leu^+$, $trp^+$ and $his^+$ recombinants will be found.

Repetitions of this experiment with the same two strains will always give much the same results. What can we conclude from them?

We have said that in the zygote there can be crossing-over between the donor's DNA and the recipient's so that, for example, the recipient's $his^-$ region is replaced by the donor's $his^+$. If we interrupt mating at a given time and find that we have *no his⁺ recombinants whatever* we may reasonably assume that $his^+$ from the donor has not yet been transferred into the recipient. Thus the total absence of $his^+$ recombinants when mating is interrupted before fifty minutes suggests that $his^+$ is not transferred before that time; similarly the total absence of $trp^+$ recombinants when mating is interrupted before thirty-five minutes suggests that $trp^+$ is not transferred before that time. These results therefore allow us to draw three tentative conclusions. First, they suggest that the donor strain always transfers its DNA into the recipient in an ordered way, over a period of at least many minutes. Secondly, they suggest that the three regions that we are considering are arranged along the DNA in the order $leu^+ \ldots trp^+ \ldots his^+$. Thirdly, since the time interval between the appearance of $leu^+$ recombinants and of $trp^+$ recombinants is greater than that between the appearance of $trp^+$ recombinants and of $his^+$ recombinants, they suggest that the distance from $leu^+$ to $trp^+$ is greater than from $trp^+$ to $his^+$. You can see that we already have the beginnings of a genetic map which looks like this:

$$leu\ldots\ldots\ldots trp\ldots\ldots his.$$

Now although we shall get much the same results every time we use the donor strain that we have already described, we may find that we get rather different results if we use another donor. It is found that several different strains of *E. coli* can act as donors – the difference between these strains does not lie in their requirement for particular growth supplements, but rather in the *order* in which they transfer their DNA in mating. We could isolate, for instance, another strain which exactly resembles our first donor in being $leu^+ \ trp^+ \ his^+ \ met^- \ thr^-$; but if we mate this with our recipient strain ($leu^- \ trp^- \ his^- \ met^+ \ thr^+$) and do precisely the same interrupted mating experiment as that we have just described, we might get the following results. After twenty minutes $trp^+$ recombinants would be found but no $leu^+$ or $his^+$ recombinants; after thirty-five minutes $trp^+$ and $his^+$ recombinants would be found but no $leu^+$ recombinants; after ninety minutes $trp^+$, $his^+$ and $leu^+$ recombinants will all be found. Notice that $leu^+$, which appeared very early in our first mating, now appears very late; notice also that the time interval between the appearance of $trp^+$ and of $his^+$ recombinants is much the same in this mating as in the other. The genetic map we might construct from this mating would therefore look like this:

$$trp\ldots\ldots his\ldots\ldots\ldots\ldots\ldots leu.$$

The difference between these two results, at first sight very curious, can be easily explained by assuming that the genetic material of *E. coli* is arranged in a *closed loop*. During mating the loop in the donor breaks at one point, and one end of the broken loop is passed first into the recipient followed by the rest of the DNA. However, the point at which the loop breaks is different for each strain of donor. The genetic map can thus be drawn like this:

where $D_1$ and $D_2$ represent the points at which the DNA loop opens in the two donors that we have described. By a repetition of this kind of experiment, using donors and recipients with other mutations, we can enter more regions on the map and in this way build up a fairly complete picture of the arrangement of genetic regions in relation to each other.

*Mapping by recombination frequency*

The technique that we have described, though immensely useful for mapping regions that are fairly far apart, is inapplicable to any attempt to determine the order of genetic regions that are very close, because the time of interrupting mating by blending cannot be determined precisely enough. We therefore need another technique, and this is provided by a study of *recombination frequency*. We can illustrate the idea in the following way. Suppose that we do the mating experiment that we described at first, namely mate our donor which transfers from the point $D_1$ on the above map and is *leu*$^+$ *trp*$^+$ *his*$^+$ *met*$^-$ *thr*$^-$ with our recipient which is *leu*$^-$ *trp*$^-$ *his*$^-$ *met*$^+$ *thr*$^+$. We allow the mating to continue for long enough to get *trp*$^+$ recombinants (say thirty-five minutes) and then plate out the bacteria on minimal medium containing threonine, histidine and leucine but no tryptophan and no methionine. Donor bacteria will not grow because of the absence of methionine. Recombinant bacteria will grow provided they are *trp*$^+$, and because threonine is present they will grow *whether or not* they are *thr*$^+$. We now take the *trp*$^+$ recombinants and see whether they are, in fact, *thr*$^+$ or *thr*$^-$ by determining whether they grow on a minimal medium plate containing, this time, histidine and leucine but no threonine. Provided we test a large enough number, the result we get will be that 50 per cent of the *trp*$^+$ recombinants will be *thr*$^+$ and 50 per cent will be *thr*$^-$.

We can explain this result easily if we draw out the donor and recipient:

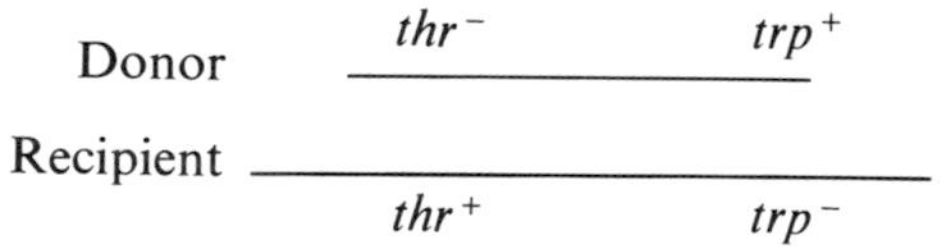

Donor    *thr*$^-$        *trp*$^+$

Recipient

   *thr*$^+$        *trp*$^-$

and then see how recombinants can arise by crossing over. If we get one crossing-over between *thr* and *trp* and one outside:

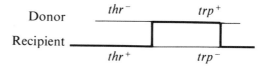

Donor    *thr*$^-$        *trp*$^+$

Recipient

   *thr*$^+$        *trp*$^-$

the recombinant chromosome will be *thr*$^+$ *trp*$^+$. If we get two crossings over between *thr* and *trp*:

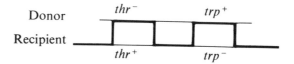

Donor    *thr*$^-$        *trp*$^+$

Recipient

   *thr*$^+$        *trp*$^-$

the recombinant chromosome will be *thr⁻ trp⁺*. If we get three crossings over between *thr* and *trp*:

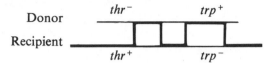

the recombinant chromosome will be *thr⁺ trp⁺* again. In fact an *odd* number of crossings-over between *thr* and *trp* will separate the donor *thr⁻* from *trp⁺*, whereas an *even* number (or *no* crossing-over between *thr* and *trp*) will ensure that recombinants that are *trp⁺* will also be *thr⁻*. The fact that 50 per cent of the total recombinants are *thr⁺* and 50 per cent *thr⁻* – in other words the fact that the recombination frequency is 50 per cent – can be explained by assuming that the distance from *thr* and *trp* is *very great*. The reason is that if the two regions are very far apart, crossing-over between them will be so extremely common that the chance of having an odd number of crossings-over is the same as the chance of having an even number.

By contrast, consider a mating between the same two bacterial strains in which we select *leu⁺* recombinants by plating the mating mixture on minimal medium with histidine, tryptophan and threonine, but no leucine and no methionine. Once again, donor bacteria will not grow because of the absence of methionine. Recombinant bacteria will grow provided they are *leu⁺*, and because threonine is present they will grow whether or not they are *thr⁺*. We now take the *leu⁺* recombinants and see whether they are *thr⁺* or *thr⁻* by determining whether they grow on a minimal medium plate containing histidine and tryptophan, but no threonine. The result we get will be that most of the *leu⁺* recombinants will be *thr⁻* and only a few will be *thr⁺*.

This result can be explained by assuming that *thr* and *leu* are close together – so close that crossing-over between them is rare. The common recombination will be:

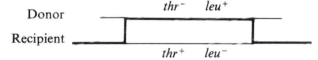

giving a recombinant chromosome which is *thr⁻ leu⁺*. A recombination of the type:

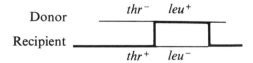

will be uncommon, and its result, *thr⁺ leu⁺*, will be rarer than *thr⁻ leu⁺*. The *recombination frequency* between *thr* and *leu* is given by:

$$\frac{\text{total number of } thr^+ \ leu^+ \text{ recombinants}}{\text{total number of } leu^+ \text{ recombinants}}$$

which will turn out to be about 20 per cent.

It can be shown that the distance between *thr* and *leu*, corresponding to a recombination frequency of some 20 per cent, represents about 1 per cent of the genetic material of *E. coli*. This distance is by no means the limit of resolution of genetic mapping. You will see that the closer two mutations are together, the smaller will be the chance of crossing-over between them and therefore the smaller will be the recombination frequency. It is not difficult to determine a recombination frequency of 0.1 per cent or less by techniques similar to those we have outlined.

Our final illustration will not involve so small a recombination frequency as that, but will indicate instead how one can find the order of three regions that are comparatively close together and also get some idea of the distances between them. Close to *thr* and *leu* is a region called *ara*, which determines the ability of *E. coli* to ferment arabinose. Wild-type strains ferment arabinose and are called *ara*$^+$; mutants unable to ferment arabinose are called *ara*$^-$. Suppose that we mate a donor that transfers from the point $D_1$ on our map (p. 409) and is *met*$^-$ *ara*$^-$ *leu*$^+$ with a recipient that is *met*$^+$ *ara*$^+$ *leu*$^-$. We allow mating to continue for, say, ten minutes and then spread the mixture on a minimal plate with *no* amino acids, containing (as usual) glucose as a source of carbon. Donor bacteria will not grow because of the absence of methionine. Recombinant bacteria will grow provided they are *leu*$^+$, and because glucose is present they will grow whether or not they are *ara*$^+$. We now see whether the recombinants are in fact *ara*$^+$ or *ara*$^-$, and thus we can find the recombination frequency between *ara* and *leu* from the fraction:

$$\frac{\text{total number of } ara^+ \ leu^+ \text{ recombinants}}{\text{total number of } leu^+ \text{ recombinants}}$$

which will turn out to be about 4 per cent.

Clearly *leu* is very close to *ara*, closer than *leu* is to *thr*. But so far we do not know if the map looks like this:

$$thr \ldots \ldots ara.leu \quad \text{(A)}$$

or like this:

$$thr \ldots \ldots leu.ara \quad \text{(B)}$$

We must therefore do another mating to determine the recombination frequency between *ara* and *thr*. This time we might have a donor that is *met*$^-$ *ara*$^-$ *thr*$^+$ and a recipient that is *met*$^+$ *ara*$^+$ *thr*$^-$. After ten minutes' mating we plate the mixture on minimal medium with glucose and no amino acids, so that on this occasion we select for *thr*$^+$ recombinants. We count how many of these are *ara*$^+$ and how many *ara*$^-$, and the fraction:

$$\frac{\text{total number of } ara^+ \ thr^+ \text{ recombinants}}{\text{total number of } thr^+ \text{ recombinants}}$$

gives the recombination frequency between *ara* and *thr*. This will turn out to be about 15 per cent.

The fact that the recombination frequency between *ara* and *thr* is less than that between *leu* and *thr* suggests that (A) above is correct. Had (B) represented the position correctly, one would have expected the recombination frequency between *ara* and *thr* to be more than 20 per cent. We may tentatively conclude from all these results that the *ara* region is between *thr* and *leu*, and much closer to *leu* than to *thr* (see Fig. 24.2).

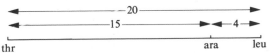

**Fig. 24.2** Mapping of the *thr–ara–leu* region in *E. coli*. The figures between the arrows represent the recombination frequencies between pairs of mutations

This method of determining the recombination frequency allows us to construct very fine genetic maps. We shall see that it is even possible to measure a recombination frequency between two mutants with the *same* phenotype, for instance two mutants both of which need added tryptophan to grow. The ability to do mapping at this extremely fine level has been of enormous importance in the development of our ideas in molecular genetics.

In concluding this account of genetic techniques in *E. coli* we must stress that of the two methods we have described for mapping, the first – mapping by interrupted mating – is applicable *only* to *E. coli* and a few other species of bacteria. By contrast, mapping by recombination frequency is a very general technique indeed, applicable to numerous systems including higher organisms. In most organisms the mechanisms by which a zygote (p. 405) is formed are quite different from that in *E. coli*. But the details of these mechanisms do not affect the principles (a) that recombinants are formed and (b) that the closer together two mutations are the smaller will be the recombination frequency between them. Consequently the measurement of recombination frequencies has enabled geneticists to make genetic maps of a large number of higher organisms as well as of bacteria.

# RESULTS OF GENETIC EXPERIMENTS IN MICRO-ORGANISMS

## One gene – one polypeptide

We shall now discuss some of the most important results that have been derived from genetic experiments with techniques of the kind that we have discussed. For our first illustration we shall return to an example that we used earlier. Let us suppose that from a wild-type culture of *E. coli* we have isolated several independent mutants all of which are unable to grow in minimal medium but are able to grow if supplied with serine. When we previously discussed a serine-dependent mutant we used as an example one in which the inability to grow in the absence of serine was due to inability to make phosphoserine phosphatase. However, if you look at the pathway of serine synthesis (p. 377), you will see that this requires three enzymes – glycerate-3-phosphate dehydrogenase, phosphoserine transaminase and phosphoserine phosphatase – which might become non-functional by mutation; it seems likely that loss of *any* of these would lead to dependence on a supply of serine in the medium. In fact if we study many serine-dependent mutants we shall find that they fall into three biochemical classes; each class will have lost one of the three enzymes. We can now ask whether we can find any *genetic* difference between the three classes that corresponds to this biochemical distinction.

When describing the bacteria involved in the mating experiments above, we spoke of genetic 'regions' responsible for determining the ability of the organisms to synthesize leucine, tryptophan and so on. Similarly we might expect that there will be a genetic 'region' that determines the ability to synthesize serine. In fact, however, if we map the position of the mutations that lead to a requirement for serine, we find that there is not one such region but three. Mutations that lead to loss of the ability to make any one of the enzymes map very far away from mutations that lead to loss of the ability to make either of the other two. On the other hand, *all* mutations that lead to loss of any *one* enzyme map very close to one another.

This latter result is found almost universally, for any given enzyme. If we isolate a set of mutants that are all deficient in one specific enzyme, then the map positions of all the mutations will be found to be very close. They will be confined to a small region of the genetic material – in *E. coli* perhaps one five-thousandth of the total. This kind of consideration leads to one of the most crucial ideas that link genetics and biochemistry. We can define the small region of genetic material within which mutations lead to the loss of a single enzyme as a *gene*, and we can now suggest that this gene *determines the structure* of one enzyme and one enzyme only.

This suggestion, first made by Beadle and Tatum, is known as the 'one gene–one enzyme hypothesis', and there is now no doubt that it is essentially correct. It requires, however, a slight but important modification for the following reason. We have seen in Ch. 6 that some enzymes consist of more than one polypeptide chain. In enzymes in which the constituent polypeptide chains are not identical, changes in any of the separate constituents can lead to loss of enzymic activity. For example, tryptophan synthetase, which catalyses the conversion of indole-glycerophosphate to tryptophan (p. 376), consists of two different subunits, one called $\alpha$ and the other $\beta$. A change in the structure of either the $\alpha$ or the $\beta$ subunit can cause loss of enzymic activity. All the mutations that lead to loss of the $\alpha$ subunit are found to be clustered in one small genetic region in *E. coli*, and all the mutations that lead to loss of the $\beta$ subunit are clustered in another small genetic region. We thus have a gene for tryptophan synthetase $\alpha$ and a separate gene for tryptophan synthetase $\beta$. The modern statement of Beadle and Tatum's hypothesis is therefore 'one gene–one polypeptide chain'.

## Clustering of genes

As we mentioned above, the genes for glycerate-3-phosphate dehydrogenase, phosphoserine transaminase and phosphoserine phosphatase, all of which are specifically involved in the synthesis of serine, are far distant from one another in the genetic material of *E. coli*. Similarly the three genes that are known to be responsible for the enzymes needed for proline synthesis are separated. In many systems, however, it is found that the genes that determine the structure of several enzymes in a single pathway are *clustered*. For instance all of the genes that determine the structure of the tryptophan-synthesizing enzymes are contiguous: there is a block of genetic material, mutations in which lead to dependence on tryptophan, but this block is divisible into several genes corresponding to the several polypeptide chains involved. Similarly there is a block of genetic material, mutations in which lead to dependence on histidine, but this block too is divisible into several genes.

Hitherto we have spoken of genetic capabilities of cells mostly in terms of their capacity to synthesize amino acids. We can, however, consider equally well in genetic terms the capacity to synthesize purines, pyrimidines, growth factors and so on, and also the capacity to utilize unusual carbon sources. *E. coli*, for example, can use sugars other than glucose as sole source of carbon, and the ability to do so is genetically determined in precisely the same way as the ability to synthesize amino acids. The clustering of genes that we have just mentioned is particularly noticeable for those genes that are concerned with the metabolism of unusual sugars. We have seen (p. 412) that there is a block of genetic material, divisible into several genes, involved in the metabolism of arabinose; similarly there is another block of several genes involved in the metabolism of lactose, and so on.

The reason why genes for the enzymes involved in a single biochemical pathway are clustered is not known for certain. One plausible reason that has been suggested is that contiguity of such genes makes the control of their activity much easier than it would be if the genes were scattered. We shall develop this point in Ch. 28. We must, however, stress again that this kind of clustering, though a general rule is, as we saw with mutations in serine synthesis, not invariable in *E. coli*. An interesting intermediate case is that of the nine genes of arginine synthesis. Four of these are clustered in one region, and the other five are in four separate locations.

The clustering of genes seems to be much more characteristic of bacteria than it is of eucaryotic organisms. Genes are known which are far apart in unicellular eucaryotes such as yeast, but which correspond to genes that are clustered in *E. coli*. And in man it is known that the genes for the $\alpha$ chain and for the $\beta$ chain of haemoglobin are not close together.

## The molecular nature of mutations

The results we have described so far can be summarized in this way. The genetic material of *E. coli* includes a number of small regions, or genes, each of which determines the structure of a single polypeptide chain. A mutation in a given region leads to a change in the structure of the corresponding polypeptide, and its effects can often be seen in loss of the enzymic activity that is normally associated with that polypeptide. The question that now obviously presents itself is, what is the detailed relationship between the structure of the gene and the structure of the polypeptide?

Before we can attempt an answer to that question, we must first consider the nature of mutations, since the only way that we have so far been able to explore the structure of genetic material is by a study of the *effect* of mutations (see p. 403). In what we have said previously we have tacitly assumed that mutations just happen, and that although we have techniques for isolating and characterizing mutants we have to work with whatever mutants chance to arise in our cultures of bacteria. In fact, although mutations do arise by chance in cultures of bacteria, we do have means of increasing the rate of their occurrence. Irradiating organisms with X-rays or with ultraviolet light, or treating them with certain chemicals, causes mutations to occur, and these treatments are commonly used in genetics whenever we wish to produce mutants. What these methods have in common is that they increase the frequency of mutation randomly in the genetic material; we have no means of directing mutation to occur in particular genes. Nonetheless some of the techniques are so powerful that they greatly increase our chances of finding a mutation in any gene in which we care to look for it.

It is not known in detail how all of these treatments cause mutation, but it is known for some, and by a study of the effects of these agents as well as by other means it is possible to build up a picture of what happens when a mutation occurs. We recall that the genetic material is DNA (pp. 402 f.), and that DNA is composed of two hydrogen-bonded strands of polynucleotide chains, the constituents of the polynucleotides being deoxyAMP, deoxyGMP, deoxyCMP and deoxyTMP (p. 136). Now one of the kinds of mutation that can occur in DNA is a change from one of the four bases to any of the other three – for example A can change to G, C or T, etc. This kind of mutation is called *base-substitution*. A second kind of mutation involves the loss of a stretch of nucleotides from the DNA; the number lost may vary from one to several thousand. This kind of mutation is called *deletion*. The third kind involves the insertion of a stretch of nucleotides (presumably from another part of the cell's DNA) into a gene. This is called *insertion*. As might be expected, the most readily identifiable changes in the structure of polypeptides are those due to base-substitution, and we shall shortly see how this kind of mutation has been used to correlate mutational events in the DNA with observed changes in the polypeptide.

## Recombination within a gene

We have said already that mutation at any point within a gene can lead to loss of the activity of the corresponding polypeptide, and in fact it was this consideration that enabled us to define a gene (p. 414). If a gene is a stretch of DNA within which we can isolate several mutations leading to loss of the same enzymic activity, we can now ask how long such a stretch of DNA is. There are at least three kinds of possible answer to that question. One answer might be in terms of recombination frequency (see p. 411) between mutations located near the two ends of a gene, and this would be the kind of answer we might expect a geneticist to give. Another kind of answer, perhaps more interesting to a biochemist, would be in terms of the number of nucleotide pairs that the gene comprises. A third answer would be the number of separate sites at which mutation within the gene can occur, and this answer would have implications of great interest to both geneticists and biochemists.

The system in which the most painstaking studies have been done in an attempt to determine the length of a gene is as follows. In a certain strain of bacteriophage, called T4, it is easy to isolate mutations that occur within a particular gene (known as rIIA), since the integrity of this gene is necessary for growth on one strain of *E. coli* but not necessary for growth on another. (The same is true of the neighbouring gene rIIB; see p. 419.) It is easy, too, to select wild-type recombinants resulting from a *cross* between such mutants* by plating the progeny of the cross on the bacterial strain that will support growth only of the wild-type phage. We can thus estimate the maximum recombination frequency (see p. 411) between mutations in the rIIA gene, bearing in mind that the largest recombination frequency will be between mutations at the extreme ends of the gene. In addition we know the recombination frequency between mutations in many *different* genes located at various places in the same phage, and from these various recombination frequencies we can calculate what fraction of the DNA of the phage is represented by the rIIA gene. Moreover from chemical studies (Ch. 7) we know the total number of nucleotide pairs in the DNA of the phage. We can consequently estimate the length of the rIIA gene in terms of nucleotide pairs. Finally, by isolating very many mutants, it is possible to see how many different sites of mutation can be found within the same gene.

The results of experiments with the rIIA gene of phage T4 are due to extremely detailed work by Benzer, and they can be easily summarized. The recombination frequency between mutations at the two ends of the gene is about 5 per cent. The gene is composed of approximately 2500 nucleotide pairs, and the number of sites within it at which mutations have been located is about 300.

At first sight the discrepancy between 2500 and 300 looks rather striking, and we might be tempted to conclude that the 'unit of mutation' is several nucleotides long. On the other hand we must bear in mind two facts. First, the isolation and mapping of many hundreds of mutants is immensely laborious, and we can by no means conclude that because Benzer, after several years' work, succeeded in isolating mutants at some 300 sites in the rIIA gene there are *only* 300 possible sites of mutation. Secondly, it is possible that mutations at some sites in rIIA will be 'silent', that is to say, will change the polypeptide product of the gene so slightly that it will still be functional. (We shall see in Ch. 26 that examples of such 'silent' changes are known.) In view of these two facts, it seems probable that a mutation can occur at *any* point in a gene, by a change from one nucleotide pair to another. Subsequent work has confirmed that this supposition is correct. Subsequent to Benzer's study, too, estimates have been made by similar and by different methods of the sizes of several other genes. We now know that, generally speaking, a gene consists of something between a few hundred and a few thousand nucleotide pairs.

## Colinearity of gene and polypeptide chain

Work with the rIIA gene and, as we shall see, with the rIIB gene of phage T4 has made important contributions to molecular genetics. There is one major drawback, however, to studying this system: the polypeptides corresponding to the rII genes have not been isolated, and we are therefore in no position to correlate the structure of the genes with the structure of their polypeptide products. To make this correlation we must turn to another system.

We have already seen that tryptophan synthetase of *E. coli* is composed of

* Crosses between different strains of phage are easily performed by infecting a bacterial culture (one that allows the growth of the two strains) with both simultaneously. In the infected cell, the genetic material of the two strains of phage undergoes frequent recombination.

polypeptides whose structure is determined by two genes in the tryptophan cluster (p. 414). The α subunit of tryptophan synthetase has been studied in great detail by Yanofsky and his collaborators, and as a result of these studies the entire amino-acid sequence of the α-polypeptide is known. In addition Yanofsky has isolated a number of mutants in which changes have occurred in the gene specifying this polypeptide. By purifying the mutated α-polypeptide from each

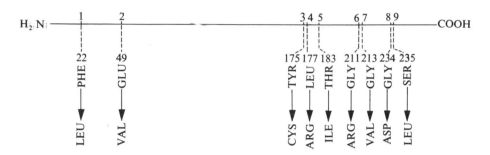

**Fig. 24.3** Colinearity between gene and polypeptide I. The horizontal line represents the tryptophan synthetase α-polypeptide, with the amino terminus on the left and the carboxyl terminus on the right. The vertical lines mark the positions of the amino acids that are replaced in each of nine mutants. Each of the numbers below the horizontal line is the number of the residue affected. counting from the amino terminus. 'A→B' means that amino acid A in the wild type is replaced by amino acid B in the mutant.

**Fig. 24.4** Colinearity between gene and polypeptide II. The horizontal line represents the gene that specifies the structure of the tryptophan synthetase α-polypeptide. The vertical lines mark the map positions (drawn to scale) of the nine mutations, the results of which are shown in Fig. 24.3

one of these mutants and determining which amino acid had been changed he was able to make a *polypeptide* map of the mutations (see Fig. 24.3). By genetic techniques he was able to make a *DNA* map of the mutations (see Fig. 24.4). As you see these two maps correspond remarkably well. Yanofsky was therefore able to draw a very significant conclusion, namely that the *linear sequence of nucleotides* in the DNA corresponds to the *linear sequence of amino acids* in the polypeptide. This conclusion, which is known as the principle of 'colinearity', allows us to suppose that the sequence of amino acids in a polypeptide is determined by the sequence of nucleotides in the corresponding gene. (The term that is generally used for this concept is 'coding': we speak of a sequence of nucleotides 'coding for' a sequence of amino acids.) We shall spend a good part of the remainder of this chapter, and some of Ch. 26, in seeing how this specification of the sequence of amino acids actually occurs.

The results that we have just described allow us to draw a further conclusion about the relation of nucleotide sequence to amino-acid sequence. This conclusion is implicit in the results of the experiment with tryptophan synthetase $\alpha$, but it is worth our making it explicit. In each mutant shown in Fig. 24.4 it was possible to find a polypeptide that differed from the wild-type $\alpha$ subunit by only *one* amino acid. We can therefore conclude that at least some kinds of mutation effect only subtle changes in the amino-acid sequence of a polypeptide – it is reasonable to guess that these might be base-substitutions, since that would appear to be the least drastic of all types of mutation. Now since each mutation caused an alteration in only one amino acid, we may conclude that a single nucleotide pair in the DNA does not contribute to specifying more than one amino acid. A shorthand way of making this point is to say that 'the code is non-overlapping'.

## The triplet code

If a change in a single nucleotide pair results in a change in only one amino acid, can we then conclude that each nucleotide pair determines, or 'codes for', a single amino acid? Plainly this conclusion does not logically follow, since we have no evidence that a single amino acid could not be changed by alterations in any one of a few neighbouring nucleotide pairs. Moreover, if we think about how a nucleotide code might be constructed to specify amino-acid sequences, we have to bear in mind that we must provide combinations of four kinds of nucleotide (deoxyadenylic, deoxyguanylic, deoxycytidylic and deoxythymidylic acids) for each of twenty amino acids. If a single nucleotide represented an amino acid, we could specify only four amino acids, which is of course far too few. If each amino acid were represented by two adjacent nucleotides, we would have sixteen combinations, and that would mean that some combinations would be ambiguous – that is, they would have to specify more than one amino acid. Since the sequence of amino acids is determined extremely rigorously (see p. 401), ambiguity of this sort is obviously not permissible. The smallest number of nucleotides in each 'codeword' that would give twenty unique combinations is three, and that would provide $4^3$, or sixty-four, combinations. The shorthand for this state of affairs is to say that the code is 'triplet'. But the argument is only a theoretical one, and a quadruplet, quintuplet, etc. code would serve the purpose just as well. An important experiment by Crick and his collaborators settled the question of the number of nucleotides actually used for each codeword.

The experiment involved the rIIB gene of bacteriophage T4, which, together with the rIIA gene, is essential for the growth of the phage on a particular strain of *E. coli* (see p. 416). In order to describe the work we shall have to assume one of its conclusions, namely that the product of the rIIB gene is unusually tolerant of changes in its amino-acid sequence. An rIIB polypeptide that contains several amino acids altered from the wild type can often function reasonably well, though not perfectly; the phages that contain such mutations can be distinguished experimentally from the wild type and are called 'pseudo-wild'.

The experiment relied on the properties of a particular chemical called acridine, which causes mutations by introducing *frameshifts* into the DNA. A frameshift is the addition or deletion of a *single* nucleotide pair: acridine is capable of causing both additions and deletions. Crick and his colleagues isolated a mutant phage resulting from acridine treatment and found that the rIIB gene product was completely non-functional, that is, that growth of the phage on the *E. coli* strain that required the product was completely abolished.

How can we explain this result? Let us assume that a certain small segment of the rIIB gene has the sequence:

A–A–C–T–G–G–G–G–T–T–C–C–A–T–G–C–T–A
T–T–G–A–C–C–C–C–A–A–G–G–T–A–C–G–A–T.

Just for the sake of argument, we shall assume that the code is in fact triplet, so that the machinery of the cell that is responsible for translating nucleotide sequence into amino-acid sequence reads off the nucleotides in groups of three.

A–A–C–T–G–G–G–G–T–T–C–C–A–T–G–C–T–A
T–T–G–A–C–C–C–C–A–A–G–G–T–A–C–G–A–T

Let us now suppose that treatment with acridine *inserts* a nucleotide pair (say A–T) after the fourth position of this sequence. Reading off the new sequence in groups of three will give:

A–A–C–T–A–G–G–G–G–T–T–C–C–A–T–G–C–T–A
T–T–G–A–T–C–C–C–C–A–A–G–G–T–A–C–G–A–T.

It is easy to see that after the point of insertion of a nucleotide pair, the triplet sequence is completely disrupted. This means that from this point to the end of the polypeptide chain *every* amino acid that is incorporated is likely to be different from that in the wild-type sequence. It is not surprising, therefore, that the polypeptide produced is completely non-functional.

By further acridine treatment of the mutant phage, Crick and his colleagues were able to isolate wild-type phage, which could be shown to result from a deletion of the newly inserted nucleotide pair. In addition they isolated several strains of pseudo-wild phage. We can explain the production of these by assuming that at some point near to the original insertion, a deletion of a single nucleotide pair has occurred – say the T–A pair at the fifth position from the right-hand end. Reading in groups of three once more we get:

A–A–C–T–A–G–G–G–G–T–T–C–C–A–G–C–T–A
T–T–G–A–T–C–C–C–C–A–A–G–G–T–C–G–A–T.

In this sequence the first group of three (before the point of the original insertion) is as it was in the wild type. The next four groups are not as they were in the wild type, and the four amino acids corresponding to these groups will be different from the corresponding four in the wild-type polypeptide. But the deletion of a nucleotide pair has restored to its original sequence the last of those groups that we have written out, and from this point on to the end of the polypeptide the amino-acid sequence will be exactly as in the wild type. We can speak of the first mutation as altering the 'reading-frame' and the second as restoring the reading frame. (You will now see why these deletions and additions of single nucleotide pairs are called frameshift mutations.) Only four amino acids in the whole polypeptide will therefore have been changed, and if (as we suggested above) this polypeptide is unusually tolerant of amino-acid changes it will be partly functional.

Thus the second mutation induced by acridine partially cancels or *suppresses* the effect of the first. In order for a frameshift mutation to suppress another frameshift mutation, one of them must be an insertion of a nucleotide pair and the other a deletion. There is no way of knowing which is actually an insertion and which a deletion, but by testing for suppression we can assign all frameshift mutations to one of two classes, which we can arbitrarily call (+) and (−).

By crossing two strains of phage each of which carried a frameshift Crick and his colleagues were able to construct doubly mutant strains in which two (+) frameshifts were close to one another in the rIIB gene. All these double mutants showed no rIIB polypeptide activity; the same was true of doubly (−) mutants. By further crosses triply mutant phages were constructed, some having three (+) mutations and one having three (−). Test of these triple mutants gave a remarkable result: they were all pseudo-wild.

In order to account for this result, let us go back to our original short DNA sequence and insert three nucleotide pairs at the point indicated by the arrows.

$$\downarrow \qquad\qquad \downarrow \qquad\qquad \downarrow$$

A–A–C–T–G–G G–G–T–T–C–C–A–T–G–C–T–A
T–T–G–A–C–C–C–A–A–G–G–T–A–C–G–A–T

$$\downarrow$$

A–A–C–T–A–G–G–G–A–T–T–C–G–C–A–T–G–C–T–A
T–T–G–A–T–C–C–C–C–T–A–A–G–C–G–T–A–C–G–A–T

If we read these off in groups of three we get:

A–A–C–T–A–G–G–G–G–A–T–T–C–G–C–A–T–G–C–T–A
T–T–G–A–T–C–C–C–C–T–A–A–G–C–G–T–A–C–G–A–T

Although this procedure will cause the insertion of an extra amino acid into the polypeptide, we shall, after the third addition, have restored the reading frame. Thus the last two of the triplets in our fragmentary sequence, and all of the subsequent triplets in the gene, will code for the *same* amino acids as in the wild-type polypeptide. The resulting polypeptide will therefore be similar to the wild-type protein save for the presence of one extra amino acid, and may well be functional. If on the other hand you try to read off the nucleotides in group of four or five, you will see that you get a completely different sequence after the insertion of three nucleotide pairs from that in the wild type. Similarly the deletion of three nucleotide pairs will restore the reading frame to normal, even though the resulting polypeptide will have one amino acid fewer than the wild-type polypeptide, *provided again that the nucleotides are read in groups of three*.

Apart from proving that the code is triplet, these experiments suggest another conclusion. In some of the phages that carry both a (+) and a (−) mutation, the two are separated by a sequence of about sixty nucleotide pairs. Nonetheless these phages are pseudo-wild, i.e. the second of the mutations suppresses the first. This result must mean that each of the twenty intervening triplets codes for some amino acid or other and thus allows the synthesis of the polypeptide chain to proceed. Now each of these triplets is out of the normal reading-frame. If of the sixty-four possible triplets only twenty were used to specify amino acids – on the basis of one triplet to one amino acid – it is extremely probable that one or more of the triplets intervening between the (+) and the (–) mutations would belong to the group of forty-four that code for no amino acid; if it did, the synthesis of the polypeptide might be expected to come to a halt. We can therefore conclude tentatively that most, if not all, of the possible triplets code for amino acids – certainly more than twenty must do so. Another way of putting this point is to say that some or all amino acids are specified by more than one triplet. In shorthand, the code is *degenerate*.

## Summary of results of genetic studies

In all organisms (except for a few viruses which have RNA instead – see p. 486) the genetic material is DNA. Alterations in the DNA are called mutations, and they often have consequences that can be observed in the biochemical behaviour of the mutants. Examples of altered biochemical behaviour are that bacteriophages become unable to grow in particular strains of bacteria, or that bacteria become unable to synthesize particular amino acids, purines, pyrimidines or growth factors (so that they cannot grow in a medium lacking the appropriate substance), or become unable to utilize particular carbon sources.

Frequently the occurrence of such a mutation can be correlated with the loss of a certain polypeptide that has or contributes to enzymic activity. If many mutants, all of which lack the same polypeptide, are examined, the mutations are almost always found to lie in a small region of the organism's DNA. This small region is called a gene, and each such gene directs the synthesis of one polypeptide.

Several kinds of mutational event are possible. The least drastic is the change of a single nucleotide in the DNA. If several mutants of this kind are collected, each deficient in the same polypeptide, it can be shown that the positions of the mutations in the DNA correspond to the positions of the amino-acid changes in the polypeptide. This result proves that the linear sequence of nucleotides in the DNA determines the linear sequence of amino acids in the polypeptide.

Experiments in which single nucleotide pairs are either inserted into or deleted from the DNA prove that such 'frameshifts' completely inactivate a gene. However, a combination of three nearby insertions or three nearby deletions may restore partial activity to the polypeptide product of the gene. It can be concluded that the nucleotides are read in groups of three – that is, that a sequence of three nucleotides determines which amino acid is to be incorporated into a polypeptide.

It follows that the number of nucleotide pairs in a gene is three times greater than the number of amino acids in the polypeptide for which the gene codes. Estimates derived from genetic studies are in good agreement with this conclusion. They suggest that most genes are from several hundred to several thousand nucleotide pairs in length, corresponding to the fact that most polypeptides are from a hundred to a thousand amino-acid residues in length. There is good reason to think that a mutation can occur in any one of the nucleotide pairs in a gene.

If many bacterial mutants are collected, each lacking a *different* one of the enzymes necessary for the synthesis of a particular amino acid or one of the enzymes necessary for the utilization of a particular carbon source, then these can be used in mapping experiments to determine the location of the *several* genes involved in that biochemical pathway. The results show that the genes for a single pathway are often, but not always, clustered. Thus all of the genes for tryptophan synthesis or histidine synthesis are contiguous in the DNA of *E. coli*, as are the genes for arabinose utilization or lactose utilization.

## PLOIDY: DOMINANT AND RECESSIVE GENES

The genetic material of *E. coli* is, we have seen, arranged in a single closed loop; each gene is (with a very few exceptions) represented only once along this loop – that is to say *E. coli* is a *haploid* organism. The result is that single mutations can have an almost immediate effect on the phenotype of the cell. If, for example, a mutation occurs that inactivates the gene determining the $\alpha$ subunit of tryptophan synthetase (p. 376) the mutant cell will soon stop growing unless supplied with tryptophan.

Occasionally it is possible to find (or to prepare) strains of *E. coli* that have, in addition to their normal complement of genetic material, an additional small loop of DNA. This additional DNA is only a few per cent of the size of the normal genetic loop, and it contains duplicate copies of all the genes that are located within a small region of the latter. In distinction to the large loop of DNA, which is sometimes (by analogy with higher organisms) called the chromosome of *E. coli*, the small extra piece of DNA is called an *episome*. You will see that its presence makes the organism that contains it partially *diploid*; since the genes that it carries are also represented on the chromosome, each of them is present in two copies in the cell.

Let us now suppose that in such a partial diploid in which the episome carries the tryptophan region (p. 414), a mutation occurs in the chromosomal gene that determines the structure of the $\alpha$ subunit of tryptophan synthetase. At first sight you might expect this mutation to lead to dependence on a supply of tryptophan, as it would in the haploid organism. But in fact the organism still contains, in addition to a mutant gene for the $\alpha$ subunit, a wild-type gene. In consequence the wild-type $\alpha$ subunit will continue to be made under the direction of the episomal gene, and the cell will be able to grow without any external supply of tryptophan. When the cell divides, each daughter cell will receive a copy of the chromosome carrying the mutated gene and a copy of the episome carrying the wild-type gene, and will be able to grow without an external supply of tryptophan. Thus the occurrence of the mutation will not be apparent in the phenotype of the cell. You will see that diploidy considerably complicates the genetic analysis of an organism, and the fact that *E. coli* is normally haploid is another important advantage that this organism has over higher organisms as a tool in the study of molecular genetics.

We are now in a position to give an explanation of a genetic phenomenon that often appears puzzling – that of dominance. In the example that we have just given, the partially diploid *E. coli* cell contains two alleles of the gene for tryptophan synthetase $\alpha$ (an allele is one of the possible wild-type or mutant forms in which a gene can exist) – namely the wild-type allele and a mutant allele. The wild-type allele will ensure the synthesis of wild-type $\alpha$ subunit, irrespective of the presence of any other allele: it will therefore *determine the phenotype* of the cell, in this case ability to grow without a supplement of tryptophan. We can say that the wild-type allele is *dominant* over the mutant allele, using as a criterion of dominance this ability to determine the phenotype.

This example suggests that dominance of an allele is due, in molecular terms, to that allele's determining the synthesis of a *biochemically active* polypeptide. The other, or *recessive*, allele will be responsible for the synthesis of either an inactive polypeptide or of no polypeptide at all. We shall see in Ch. 28 that important information can sometimes be obtained, even in the study of poorly understood biochemical systems, by determining which of two or more alleles is dominant.

In *E. coli*, and in bacteria in general, diploidy is an exceptional situation. In higher organisms, of course, diploidy is the rule. We can now discuss how far the principles of molecular genetics that we have derived in this chapter from a consideration of a haploid organism must be modified for diploid organisms. The answer turns out to be that, provided we bear in mind the explanation of dominance that we have just mentioned, little modification is required.

We shall illustrate the way in which an outline of molecular genetics may be sketched for higher organisms by referring once again to phenylketonuria, which we mentioned on pp. 376 and 404. In phenylketonuria all of the symptoms of the disease are attributable to the absence in an active form of a single enzyme, phenylalanine hydroxylase. How can we interpret the absence of this enzyme in genetic terms?

Just as a single gene determines the structure of a single polypeptide chain in *E. coli*, so too in higher organisms, including man, the principle one gene–one polypeptide chain holds good. We may therefore think of each of the species of polypeptides in human beings as being coded for by a single gene, and this principle applies both to polypeptides with enzymic activity, such as phenylalanine hydroxylase, and to those without, such as insulin or haemoglobin. The genes in higher organisms, like their counterparts in *E. coli*, are composed of DNA and are subject to mutation.

Let us suppose that in either the ovum or the sperm that formed a human individual, a mutation occurred in the gene that determines the amino-acid sequence of phenylalanine hydroxylase. (We might call the wild-type gene *phl*$^+$ and the mutant gene *phl*$^-$: see *leu*$^+$, *leu*$^-$ etc. on p. 407.) In principle this mutation is exactly similar to a mutation that might occur in the gene that determines the amino-acid sequence of tryptophan synthetase $\alpha$ in *E. coli*. But the phenotypic consequences are different. In the fertilized ovum the mutant allele of the gene will be accompanied by a wild-type allele, contributed by the germ cell that did not undergo mutation. The result will be that the cells of the individual concerned will each contain both of these allelic forms of the gene, *phl*$^+$ and *phl*$^-$.* As we saw above in a similar situation, the wild-type allele will be

* The term 'heterozygous' is often used to refer to individuals with two different allelic forms of a single gene.

dominant to the mutant, and consequently the individual will be phenotypically more or less normal.

(You may ask how it is that an individual that has only one copy in each cell of the wild-type allele will be able to synthesize enough enzyme to hydroxylate phenylalanine, when the normal situation is that two copies of the gene are present per cell. There are two answers. First, most enzymes are present in the cell in a moderate excess, so that a reduction by 50 per cent will in many cases have no effect on the cell's activities. Secondly there are mechanisms to control the rate at which proteins are synthesized under the direction of particular genes (see Ch. 28), and these may be sufficient to allow a doubling of the rate of synthesis of any given protein. Actually very careful study of heterozygous individuals with only one copy of the wild-type allele shows, for example, that after ingesting phenylalanine they retain the amino acid at a high level in the blood for an unusually long time.)

We have just seen that an individual may carry a *dominant* gene for phenylalanine hydroxylase which will render his metabolism of phenylalanine nearly normal even though he also carries a mutant allele of the same gene. Another way of expressing the same fact is to say that he carries a *recessive* gene for phenylketonuria (which is in reality the mutant allele of the gene that determines the enzyme). Now in the meiotic division that gives rise to the haploid germ cells, there is an equal chance of any one of the germ cells' receiving either the wild-type ($phl^+$) or the mutant ($phl^-$) allele of the gene that determines phenylalanine hydroxylase. (If you are not clear about these events, consult any textbook of classical genetics or any modern textbook of general biology.) So long as this individual mates with a wild-type individual, that is one that is *diploid* or 'homozygous' for the wild-type allele of this gene ($phl^+$ $phl^+$), the progeny will all be phenotypically wild-type. Half of them will be diploid for the wild-type allele ($phl^+$ $phl^+$) and half of them will have one wild-type allele and one mutant allele ($phl^+$ $phl^-$). These latter, once again, we can regard as carrying a recessive gene for phenylketonuria, and, as we have just seen, half of the germ cells produced by such individuals will have the mutant gene.

In this way the mutation can persist for many generations after its original occurrence without being expressed in any obvious phenotype. Finally, however, it may happen that two individuals, both descended from the person with the original mutation and both having the genetic constitution $phl^+$ $phl^-$, will mate. Since half of the germ cells from each of these individuals will carry $phl^+$ and half will carry $phl^-$, one-quarter of their offspring (on average) will have $phl^+$ $phl^+$, one-half $phl^+$ $phl^-$, and one-quarter $phl^-$ $phl^-$ (see Fig. 24.5). The first of these classes will be diploid wild-type, the second will be precisely like the parents in carrying a recessive gene, but the third will have a new phenotype that has not hitherto appeared. People in this class will be completely incapable of hydroxylating phenylalanine to form tyrosine, and they will therefore exhibit clear symptoms of the disease phenylketonuria. This result, which is of course that predicted by classical Mendelian genetics, can thus be understood equally well in the context of molecular genetics.

People that lack, through some genetic defect of the kind that we have outlined, a particular enzyme are sometimes said to have an 'inborn error of metabolism'. This picturesque phrase, due to Garrod, can be used to describe many conditions of enzyme deficiency. A large number of inborn errors of

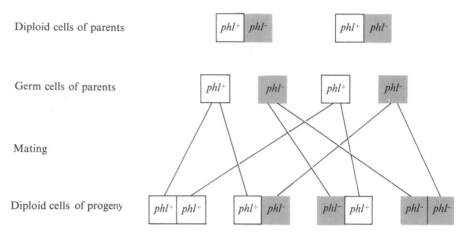

**Fig. 24.5** Inheritance of phenylketonuria. *phl*$^+$ represents the wild-type gene for active phenylalanine hydroxylase and *phl*$^-$ the mutant gene which codes for an inactive polypeptide. Each of the parents shown on the top line is heterozygous, possessing one wild-type and one mutant gene. Half of the germ cells produced by these individuals will have *phl*$^+$ and half *phl*$^-$. At fertilization there is an equal chance of the four crosses *phl*$^+$ × *phl*$^+$, *phl*$^+$ × *phl*$^-$, *phl*$^-$ × *phl*$^+$, and *phl*$^-$ × *phl*$^-$. Therefore on average one-quarter of the progeny will be wild-type, one-half will be heterozygous like their parents, and one-quarter will have no active enzyme and will suffer from phenylketonuria.

metabolism are known, some (like phenylketonuria) having very grave consequences, others with relatively trivial consequences. Many of the inborn errors that have been identified clinically result in lesions in amino-acid metabolism. One of the most striking of the conditions that can arise is albinism, which is due to a defect in the metabolism of tyrosine.

Mutations such as that resulting in phenylketonuria are obviously harmful and in time would be eliminated by natural selection. There are other mutations, however, that are beneficial in effect, and these are likely to be conserved in evolution. The study of *genetics* at the molecular level is beginning to move towards a consideration of *evolution* at the molecular level; Ch. 27 is devoted to a few examples of the results of this intriguing subject. But before we can introduce an account of molecular evolution we must first describe in some detail the mechanisms by which the structure of DNA is expressed in the structure of protein. Part of Ch. 25 and all of Ch. 26 are concerned with these mechanisms.

# Chapter 25 Replication of DNA and synthesis of RNA

The last chapter will have made it clear that DNA has two functions. The first is to provide for its own exact replication. It is the faithful copying of the DNA, and the distribution of the copies to progeny, that permits organisms to show the genetic continuity of which we have given some examples. The second function is to dictate the structure of proteins according to the principle that we derived from considering the work of Yanofsky – namely that the linear sequence of nucleotides in the DNA determines the linear sequence of amino acids in the protein. In this chapter we shall outline what is known of how DNA replication takes place, and shall begin to describe how the sequence of amino acids is dictated. We shall complete the discussion of protein synthesis in the next chapter.

## REPLICATION OF DNA

Studies of DNA replication have been pursued with great vigour for some fifteen years, and you may gain some idea of the complexity of the system if we say that there are, despite all this work, still very many gaps in our knowledge. We can, however, say a good deal about the principles of DNA replication, even though the details are obscure.

What do we mean by the 'principles' of DNA replication? We can best answer this question by referring once again to the structure of the DNA molecule, which we discussed on pp. 144 ff. You will recall that the DNA molecule comprises two strands of nucleotides. Each strand is composed of deoxyribonucleotide residues joined by $3' \rightarrow 5'$-phosphodiester bonds (see p. 46, Fig. 3.2). Hydrogen bonds occur between the bases of the nucleotides in the two strands: adenine in one strand is always bonded to thymine in the other, and guanine in one strand is always bonded to cytosine in the other. The result is that, if we are given the sequence of bases of one strand, we can immediately write down the sequence of bases of the other.

Now when we say that we know a fair amount about the principles of DNA replication, what we mean is that we can describe in outline the way in which we can get two identical DNA molecules from one, so far as the *sequence of bases* is concerned. However, as we shall see, other questions remain to be answered. Because of the point that we just made – namely that if we are given the sequence of bases of one strand we can immediately write down the sequence of bases of the other – you might well guess that the two identical DNA molecules could be formed from one molecule in the following way. The strands of the original molecule would separate, and each strand would then act as a *template* on which a new strand would be built. The sequence of bases of the old strand would determine the sequence of bases of the new one by the specificity of formation of hydrogen bonds that we have mentioned. We would thus have two DNA molecules where we originally had one.

We can make this mechanism clear by means of an example. Suppose we have a small piece of double-stranded DNA that has this sequence.

A–A–C–T–G–G–G–T–T–C–C–A–T–G–C–T–A
T–T–G–A–C–C–C–C–A–A–G–G–T–A–C–G–A–T

The separated strands would be:

A–A–C–T–G–G–G–G–T–T–C–C–A–T–G–C–T–A

and

T–T–G–A–C–C–C–C–A–A–G–G–T–A–C–G–A–T

and these would determine the order of *new* deoxyribonucleotides by the base-pairing rules. By using bold type to indicate the new nucleotides we can write the following double-stranded molecule as the product that is derived from the upper of the two separated strands of the original molecule:

A–A–C–T–G–G–G–G–T–T–C–C–A–T–G–C–T–A
**T–T–G–A–C–C–C–C–A–A–G–G–T–A–C–G–A–T**

and this double-stranded molecule as the product derived from the lower of the two separated strands:

**A–A–C–T–G–G–G–G–T–T–C–C–A–T–G–C–T–A**
T–T–G–A–C–C–C–C–A–A–G–G–T–A–C–G–A–T.

These two molecules are identical to each other, and they are also identical to the original (parental) molecule. Thus a mechanism of this sort could very well satisfy the genetic function of DNA, which is to ensure that on replication each of the two progeny molecules precisely resembles the parental molecule.

To say that a mechanism is attractive, however, is not the same as to prove that it occurs. In fact it *has* been proved that DNA replication does occur by this mechanism, at least in *E. coli* and very probably in higher organisms as well. Before describing the relevant experiment, we must point out an additional feature of the two molecules that were formed by the mechanism that we described above. Each of them contains one old strand, derived from the parental molecule, and one newly synthesized strand. If we now imagine each of these molecules being itself replicated, our next generation of DNA molecules will number four. Two of these four will each contain one strand from the molecule that we started with (the one printed wholly in normal type), and two of them will contain no part of the original molecule at all. So if we have some way of identifying a strand of the original molecule of DNA, we can predict that, if the mechanism we have outlined is correct, we will find *one* of these original strands in *each* of the DNA molecules after one generation, and *one* of these original strands in *half* of the DNA molecules after two generations.

The relevant experiment, by Meselson and Stahl, relied on the fact the DNA molecules of different densities can be distinguished by centrifugation in solutions of heavy-metal salts such as caesium chloride (see p. 526). When *E. coli* is grown in medium in which the common isotope of nitrogen, $^{14}N$, is replaced by $^{15}N$, the DNA is synthesized with $^{15}N$ in it; this DNA can be separated by centrifugation from that synthesized in medium containing $^{14}N$. Meselson and Stahl grew *E. coli* cells for many generations in $^{15}N$ so that all of the DNA was 'heavy' – that is, had a greater density than normal – and then transferred them to medium with $^{14}N$. They took samples at intervals, extracted the DNA, and centrifuged it in caesium chloride.

DNA that was extracted from the culture one generation* after the transfer from $^{15}N$ to $^{14}N$ medium had a 'hybrid' density, exactly halfway between that of DNA containing only $^{15}N$ and that of DNA containing only $^{14}N$; each molecule must therefore have contained one old and one new strand. DNA extracted from the culture after another generation was composed in equal proportions of this 'hybrid' DNA and of completely 'light' ($^{14}N$) DNA. As time went on, more and more 'light' DNA accumulated (see Fig. 25.1).

These results are entirely consistent with the mechanism of DNA replication that we suggested earlier, and are not consistent with other possible mechanisms that one might conceive of. The mechanism that we outlined, which is proved by these experiments to be that actually used in *E. coli*, is called *semi-conservative*, because in each replication half of the parental DNA molecule is passed intact to each daughter molecule. In addition it has been shown that at the point along the double-stranded DNA where replication is taking place *both* strands are replicated at the same time.

Now that we have established the general outline of the system that replicates DNA, you will expect us to go on to describe the details of the enzymic reactions by which replication is accomplished. Unfortunately we are in no position to do so since, as we mentioned earlier, no definitive picture of these reactions is yet available. We must bear in mind that, although a semi-conservative mode of replication provides a neat way of ensuring the correct pairing of bases for the synthesis of two daughter molecules of DNA from one parental molecule, the enzyme complex that is responsible for replication has other tasks to accomplish as well. It has to achieve the polymerization of deoxyribonucleotide residues joined by $3' \rightarrow 5'$ linkage, starting with the deoxyribonucleoside triphosphates whose synthesis we discussed in Ch. 23. Also, in order for replication to take place the enzyme complex has first to ensure that the two strands of DNA are separated. This separation will depend on the strands being unwound from their normal helical configuration; when one considers that the DNA of *E. coli* is more than one millimetre in length, and that a turn of the double helix occurs every 34 Å, it is clear that the geometrical problem of unwinding is very formidable. We saw on p. 147 that in addition a large amount of energy is required to separate the strands.

We may reasonably guess that no one enzyme can fulfill all of these tasks. Most of the effort that has gone into studying the enzymology of replication has been

---

* A 'generation' can be taken to be the time that it takes for a culture of bacteria to double the number of cells it contains.

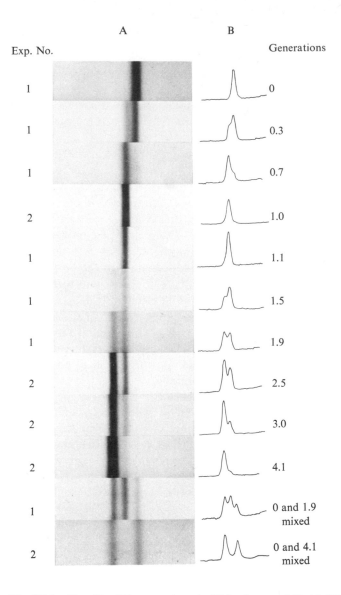

**Fig. 25.1** Results of the experiment of Meselson and Stahl. DNA was extracted from samples taken at intervals from a bacterial culture that had been grown in $^{15}N$-medium for many generations and then transferred to $^{14}N$-medium. For each sample the number of generations that elapsed since the transfer is shown in the right-hand column. The DNA was centrifuged in a solution of caesium chloride, in which DNA samples of different densities can be distinguished by their position (see bottom two lines), the densest band being furthest to the right. Column A shows ultraviolet absorption photographs of the centrifuge cell. Column B shows densitometer tracings of these photographs (a densitometer is a machine that produces a tracing as it scans a photograph, the height of the curve at each position from left to right being proportional to the intensity of the photograph at the corresponding position). Notice in particular the 'hybrid' DNA at 1.0 generation.

directed towards identifying the enzyme responsible for the polymerization itself. A great deal of information has accumulated, largely through the work of Kornberg and his associates, about a polymerizing enzyme that can be isolated from *E. coli* and other cells. This enzyme is called DNA polymerase I, and for many years it was believed that it was responsible for replicating DNA. It is still possible that it may do so, but at present it seems more probable that DNA polymerase I has the function of *repairing* DNA that has become damaged by ionizing radiation (e.g. by cosmic rays). In repair it is necessary to replace a faulty strand of DNA by copying the opposite intact strand, and polymerase I is certainly capable of faithfully copying a strand of DNA. What is in doubt is its ability to replicate the genetic material by semi-conservative replication.

Two other DNA polymerases (II and III) have been isolated from *E. coli*. Although less is known about them than about polymerase I, there is good evidence that polymerase III is involved in some way in semi-conservative replication – which is not to preclude the possibility that polymerase I or II may be involved as well.

Even though we are unclear about the identity of the enzyme responsible for DNA polymerization, we can be reasonably confident about some features of the way in which polymerization must occur. The substrates for the reaction are the deoxyribonucleoside triphosphates (pp. 392–397), and the residues that actually occur in DNA are deoxyribonucleoside monophosphates, so we may presume that in each polymerization step a triphosphate must react with the elimination of inorganic pyrophosphate. We may also presume that each new deoxyribonucleotide must be added to a pre-existing chain or 'primer' (compare the synthesis of amylose, p. 335). We may thus write a polymerization step as:

$$(XMP)_n + XTP \rightleftharpoons (XMP)_{n+1} + PP_i$$

where $(XMP)_n$ represents a DNA chain consisting of the four deoxyribonucleotides in various proportions, XTP an incoming deoxyribonucleoside triphosphate, and $(XMP)_{n+1}$ the DNA chain lengthened by one nucleotide.

Notice that inorganic pyrophosphate is liberated in the reaction; we may therefore regard the polymerization as going essentially to completion, for the reason we gave on p. 284.

This outline does not account for the fact that there must be some means of specifying which nucleotide is to be inserted at each position. We can reasonably suppose that the incoming nucleotide must be able to form hydrogen bonds with its partner in the opposite DNA strand in order to be inserted into the new chain, and that only if the proper pattern of hydrogen bonds is formed will the enzyme accept it.

We must now mention an unsolved problem in DNA replication which has for long exercised biochemists working in this field. You will recall that each strand of DNA has a distinct 'polarity', that is, that one end of the strand has a free 3'-hydroxyl group, and the other end a free 5'-hydroxyl group, in its deoxyribose residue (p. 46). You will recall too that the two strands of the double-helical DNA have *opposite* polarities; if you move along one of the strands from 5' end to 3' end you will be moving along the other strand from 3' end to 5' end (p. 148).

Now the precursors of DNA synthesis are the nucleoside-5′-triphosphates (pp. 392–397), and each polymerization step is presumably of this sort:

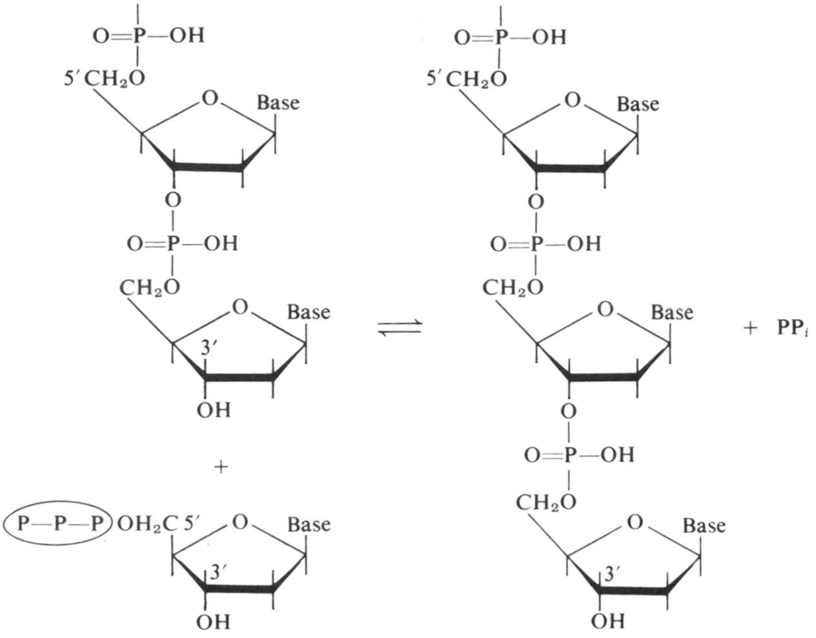

in which the free 3′-hydroxyl end of the growing DNA chain reacts with the 5′-triphosphate of the incoming nucleotide, leaving a new 3′-hydroxyl end to which the next nucleotide can be attached. Consequently the new deoxyribonucleotide strand grows in the 5′ → 3′ direction, and therefore the *old* strand, with which the new nucleotides are forming hydrogen bonds, is that which runs in the 3′ → 5′ direction. The partner of this old strand is the old 5′ → 3′ strand, and we now have a serious problem in asking how this latter strand can be replicated, since for *its* replication (assuming that the replication of the two strands proceeds simultaneously), synthesis in the 3′ → 5′ direction would be required (see Fig. 25.2). Such synthesis would require a polymerizing enzyme with a quite different specificity from that that synthesizes in the other direction.

A hint of a possible solution to this problem comes from the discovery a few years ago that if one extracts broken cells of *E. coli* one recovers newly synthesized DNA in small pieces. One possible interpretation is that the polymerizing enzyme synthesizes DNA in fragments, and that these are later put together by a 'joining enzyme' – such an enzyme (DNA ligase) is known. If this mechanism does, in fact, occur, we could speculate that the polymerizing enzyme might synthesize a fragment from 5′ to 3′ by copying one DNA strand, then cross over to the other strand and copy that one in the *reverse* direction for a short distance, then move back to the point at which it left off copying the first one, and so on. This mode of synthesis would leave pieces like this:

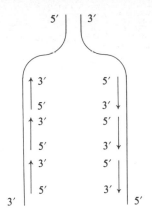

(where the head of each arrow represents a 3'-hydroxyl end of a newly synthesized fragment), which would then be joined together by DNA ligase.

A problem that we have not yet considered results from the fact that none of the known DNA polymerases will synthesize polynucleotide chains without a primer. This fact raises the question of how the synthesis of each piece of DNA is initiated. There is some evidence that each newly synthesized section of DNA is initiated with an *RNA primer*, made by the DNA-directed RNA polymerase (see below). This enzyme does not need a primer to initiate RNA synthesis, and DNA polymerases can use RNA molecules as primers, adding deoxyribonucleotides covalently to a preexisting RNA chain. If in fact RNA polymerase is used to make an RNA primer for DNA synthesis, we must suppose that this RNA is later removed, leaving a short gap which is then filled.

The final problem in DNA replication that we must mention is that of un-

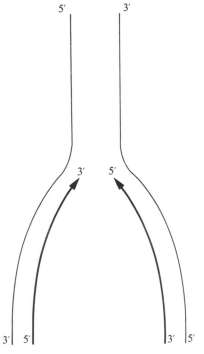

Fig. 25.2 The problem of replicating both strands simultaneously. To copy the 3'→5' strand requires synthesis of DNA in the 5'→3' direction, which is known to occur as shown in the reaction on p. 432. The simultaneous replication of the other strand, however, would require 3'→5' synthesis, and this would require an enzyme with a specificity that is not known to exist.

twisting the strands so that they can be separated for replication. We saw in Ch. 7 that the $\Delta G$ required to separate the strands completely would be enormously high. It seems in fact that it would be impossible to unwind one from another so long as they both remained intact. If, however, one of the strands were broken by nicks near the point of replication, the difficulty of unwinding the strands over a small region would be very much diminished. Once again we do not know for certain whether nicking of a single strand does occur during (or preceding) replication, but it seems likely that it does; enzymes are known (endonucleases) that can introduce single-strand nicks into DNA.

If the speculations that we have made about various aspects of the replication of DNA turn out to be correct, at least three enzymes must be involved. The DNA polymerase would be responsible for making phosphodiester bonds and thus synthesizing small pieces of both strands in the $5' \rightarrow 3'$ direction, presumably inserting only those nucleotides that can form the proper pattern of hydrogen bonds with their partners in the old strands. The DNA ligase would be responsible for joining together the fragments thus formed. During replication the endonuclease would be responsible for breaking one of the old strands of DNA so that the two strands could unwind; presumably the breaks thus made would also be sealed by DNA ligase.

## SYNTHESIS OF RNA

We described in a good deal of detail in Chs 5 and 6 the many cellular functions of proteins, and in Ch. 24 we showed how DNA is the repository of genetic information that is ultimately needed for determining the structure of these proteins. How does RNA fit into this scheme?

The answer is that RNA acts as an *intermediate* carrier of information, being synthesized directly upon a DNA template and in turn serving as a template for protein synthesis. That DNA is not directly involved in protein synthesis is suggested by the following points. In the first place, although the amount of DNA that a given type of cell contains is normally very constant, the amount of RNA per cell varies a great deal. In bacteria, which can easily be made (by incubation in different media) to vary their rate of growth, the concentration of RNA is closely related to the rate of protein synthesis; similar results in higher cells can be obtained even though experimental manipulation is less straightforward. Again, in nucleated cells the nucleus, which contains nearly all of the DNA (p. 218), synthesizes almost no protein, whereas the cytoplasm, which has no DNA except for the small quantity present in mitochondria and chloroplasts, synthesizes protein very actively.

Now if, as we have seen, the linear sequence of nucleotides in DNA determines the linear sequence of amino acids in proteins, and if RNA is the intermediate between DNA and protein, it follows that the sequence of nucleotides in DNA must in some way determine the structure of RNA. The obvious way is for the linear sequence of ribonucleotides in RNA to be determined directly by the linear sequence of deoxyribonucleotides in DNA – that is, for each nucleotide in RNA to correspond exactly to a nucleotide in DNA.

The synthesis of RNA corresponding in sequence to DNA is carried out by an enzyme called DNA-directed *RNA polymerase*, which forms polyribonucleotide strands in a way that is exactly analogous to that we suggested for polydeoxyribonucleotide synthesis (p. 432). Each polymerization step involves the insertion of a ribonucleotide by addition to the 3'-hydroxyl end of the growing RNA chain; the precursor of the nucleotide is one of the four nucleoside triphosphates whose synthesis we described in Ch. 23. The usual base-pairing rules apply.

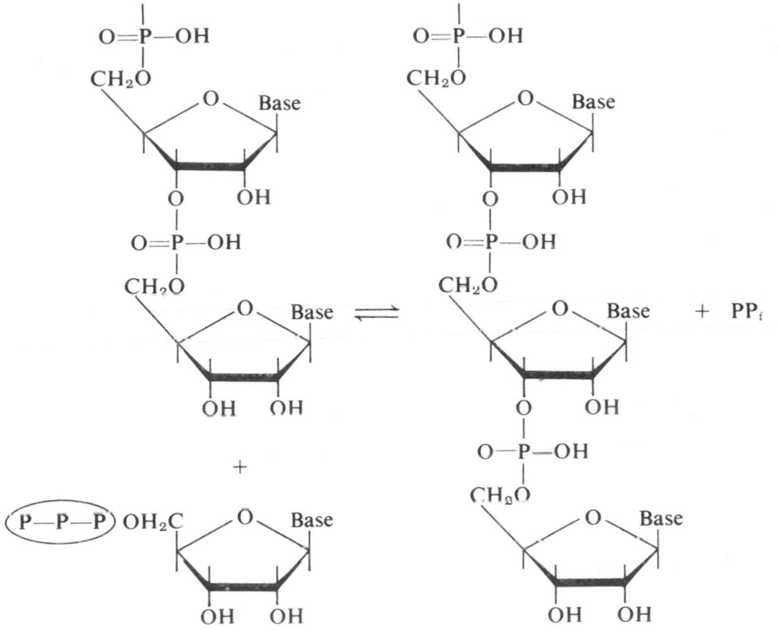

Once again each step creates a new 3'-hydroxyl end to which another nucleotide can be added. Once again, too, inorganic pyrophosphate is split out, so that the polymerization goes virtually to completion (p. 284).

The fact that this method of synthesis causes the RNA chain to grow only in the 5' → 3' direction does not present the problem that it did with DNA synthesis, because the RNA polymerase copies ('transcribes' is the usual jargon) only *one* of the two strands of DNA. (We shall describe below how the correct strand is chosen.) The reaction is therefore much more straightforward than that involved in DNA synthesis, and much more is known about it. The enzyme requires as a template double-stranded DNA which may be from any source, and it transcribes its template faithfully, producing a RNA that is exactly complementary (see p. 148) to *one* of the two DNA strands. (The other strand is not copied.) Deoxythymidylic acid residues are transcribed into adenylic acid residues, deoxycytidylic acid residues into guanylic acid residues, deoxyguanylic residues into cytidylic acid residues, and deoxyadenylic residues into uridylic acid residues (not thymidylic acid, which does not normally occur in RNA). The enzyme will therefore work only if provided with all four kinds of ribonucleoside triphosphate.

We can illustrate the formation of a RNA strand complementary to one of the strands of the DNA template by showing how the RNA polymerase might transcribe the small fragment of DNA whose replication we discussed above

(p. 428). Once again we use bold type to indicate the new nucleotides, this time those *ribo*nucleotides that are inserted into RNA.

$$\text{U–U–G–A–C– C–C–C A–A–G–G–U–A–C–G–A–U}$$
$$\text{A–A–C–T–G–G–G–G–T–T–C–C–A–T–G–C–T–A}$$
$$\text{T–T–G–A–C–C–C–C–A–A–G–G–T–A–C–G–A–T}$$

The strand of RNA thus produced is similar in sequence to the strand of DNA geometrically *opposite* to that which is transcribed (except that uridylic acid is substituted for deoxythymidylic acid).

The fact that the RNA synthesized is exactly complementary to the DNA strand from which it was transcribed is often exploited experimentally. If a particular sample of DNA is heated, the hydrogen bonds holding the two strands together are broken and the strands separate. If RNA is added to the hot solution and the mixture is then cooled, the RNA will form hydrogen bonds with one of the strands of DNA, *provided* that the sequences are exactly complementary. This technique, which is known as *DNA–RNA hybridization*, allows one to estimate how much of the RNA in a sample is complementary to a specific sample of DNA. We shall see in Ch. 26 and 28 that this technique has proved of great value.

Studies of purified RNA polymerase show that the enzyme has an extremely high molecular weight and is composed of several different subunits. One of these subunits, called 'sigma', is required only for *initiation* of RNA synthesis. When we mentioned above that the polymerase produces a RNA that is exactly complementary to one of the two DNA strands, you may have wondered how the enzyme selected the strand that was to be copied. The answer appears to be that the polymerase will start to transcribe wherever a particular initiating sequence of nucleotides occurs in *one* of the DNA strands; however, initiation occurs only if the enzyme includes the subunit sigma. After transcription has been initiated, sigma is released from the enzyme (which can continue to make RNA in the absence of sigma), and it then binds to another enzyme molecule, thus promoting initiation by the latter. Although the sequence of deoxyribonucleotides required for initiation is not known, there is genetic evidence that slightly different sequences have different affinities for binding of the polymerase, which has interesting consequences (see p. 481).

Even less is known about the termination of transcription. A protein factor called 'rho' has been identified which causes synthesis of RNA by the polymerase to stop at certain (unknown) nucleotide sequences in the DNA, but rho is not itself a part of the polymerase enzyme.

By these means, the RNA polymerase makes available an exact RNA copy, or transcript, of any desired length of DNA. The three types of RNA that we described in Ch. 7 are all specified by genes (that is sequences of DNA) and are synthesized by the RNA polymerase in the way that we have just outlined. (You will recall that the unusual bases in transfer RNA (see p. 142) are formed by modification of the normal bases after the RNA molecule itself has been completed). All are needed for protein synthesis, and in the next chapter we shall discuss in detail what their functions are.

# Chapter 26 Synthesis of protein

We mentioned on p. 434 some of the lines of evidence that lead to the conclusion that RNA is intimately involved in the synthesis of protein. We shall therefore begin this chapter by reviewing the different classes of RNA that occur in cells, this time emphasizing not so much their structure (see Ch. 7) as their role in protein synthesis. Later in the chapter we shall discuss what is known of the mechanism of synthesis, in terms both of the manufacture of the peptide bond and of the specificity in ordering amino acids into the primary sequence of the polypeptide.

## THE THREE SPECIES OF RNA

### The ribosome

Early studies of protein synthesis, especially those carried out by Zamecnik and his collaborators during the 1950s, established several important facts. Zamecnik's first results were obtained from experiments with intact animals that had received injections of radioactive amino acids. When the livers of such animals were homogenized and fractionated only a short time after the administration of the amino acids, radioactive protein, formed from the injected amino acids, was found to be associated with the microsomal fraction. Now the microsomes are artefacts of the fractionation procedure, representing fragments from the endoplasmic reticulum with attached ribosomes (see Ch. 11). When the microsomal fraction of livers from animals that had recently received an injection of radioactive amino acids was further fractionated, it was found that the radioactive protein was predominantly associated with the ribosomes. Later experiments showed that ribosomes, if isolated sufficiently gently from the endoplasmic reticulum, will incorporate radioactivity into proteins when incubated together with radioactive amino acids, ATP and a fraction derived from the soluble cytoplasm. All these results suggest that ribosomes are the sites of protein synthesis in the intact cell. We shall now consider their structure and something of their function.

Although ribosomes were first discovered in and isolated from animal cells, the most detailed studies of their structure have been done with ribosomes from bacteria – particularly those from *Escherichia coli*. *E. coli* ribosomes have a particle weight of about 2.6 million and a sedimentation constant in a centrifugal field of about 70 Svedberg units (see p. 525). These '70S particles' are composed of two unequal subunits, which have sedimentation constants of about 50$S$ and 30$S$. Each of these subunits consists of roughly 60 per cent RNA and 40 per cent protein.

The '50$S$' subunit of an *E. coli* ribosome contains two molecules of RNA, which have sedimentation constants of about 23$S$ and 5$S$, and about thirty-five molecules of protein, each molecule probably being different from each of the others. The total molecular weight of the particle is about 1.8 million. The '30$S$' subunit contains a single molecule of RNA, with a sedimentation constant of about 16$S$, and about twenty molecules of protein – again each molecule is probably different from each of the others. The total molecular weight of the particle is about 800 000. None of the fifty-five species of ribosomal proteins is found in *both* the 50$S$ and the 30$S$ subunit.

Despite their great complexity, both the 50$S$ and the 30$S$ particles appear, remarkably enough, to be self-assembling. That is, if each of the twenty species of protein from the 30$S$ subunit is purified and all of the pure proteins are mixed with 16$S$ RNA, active 30$S$ subunits are formed by spontaneous assembly; the same sort of experiment can (although with more difficulty) be done with the 50$S$ subunits. Experimental systems of this kind can be used to derive information about the function of the individual proteins of the ribosomes: one can omit a given protein from the reconstitution mixture and see whether the resulting particle is deficient in any of the functions of protein synthesis. These studies are still in their infancy, but they are bound in the end to tell us a great deal about the relation between structure and function in ribosomes.

The ribosomes of higher organisms are much bigger than those of bacteria, with particle weights of up to 5 million, but they are more variable in size. Animal and plant ribosomes generally have sedimentation constants of about 80$S$, and are composed of subunits of roughly 60$S$ (having RNA with a sedimentation constant of about 28$S$) and 40$S$ (having RNA with a sedimentation constant of about 18$S$). Much less is known about the proteins of eucaryotic ribosomes than about those of bacterial ribosomes.

We saw in the last chapter that RNA is synthesized by the copying ('transcribing') activity of DNA-directed RNA polymerase, and it follows that the information encoded in DNA (p. 419) is transcribed into an RNA code preparatory to being translated into protein. Since so much of the ribosome is RNA, it would be reasonable to conclude that it is this ribosomal RNA that specifies the sequence of amino acids in proteins. A closer look will show us that this expectation, however reasonable, is wrong.

## Messenger RNA

You will by now be clear that proteins come in a wide range of sizes and amino-acid compositions; we may infer that the RNA that codes for proteins must also be very varied in size and nucleotide sequence. By contrast the three species of ribosomal RNA that occur in any given organism (for instance the 23$S$, 16$S$ and 5$S$ RNA of *E. coli*) are all extremely homogeneous in size and, so far as is known, in nucleotide sequence. These facts speak against the suggestion that ribosomal RNA can code for all the proteins that a cell makes.

A further argument against the view that it is the ribosomal RNA that is the template for protein synthesis comes from studies of situations in which bacteria start to make proteins that they have not previously made, *without* synthesizing ribosomal RNA. We shall see (p. 474) that by altering the conditions in which bacteria are cultured one can provoke the synthesis of proteins other than those usually found in the cell. Again, when bacteria are infected with some strains of phage they will start to synthesize proteins that are specifically associated with phage infection and not found in uninfected bacteria. In both situations it can be shown that the synthesis of ribosomal RNA need not necessarily occur before the new proteins appear. These observations suggest that ribosomal RNA does not dictate the amino-acid sequence of proteins.

What, then, does dictate the amino-acid sequence? A closer look at the phage-infected bacteria we have just mentioned provides a hint. These infected cells are found to contain a species of RNA which can easily be distinguished from the bulk of bacterial RNA (which is mostly ribosomal RNA). First, it is synthesized and degraded very rapidly, unlike ribosomal RNA which is generally much more stable. Secondly, it hybridizes specifically with DNA prepared from the phage and not with DNA prepared from the bacteria (see p. 436). These characteristics suggest that this RNA has the function of acting as *intermediate carrier of information* between the phage DNA and the bacterial ribosomes, causing the ribosomes to synthesize the special proteins that are required by the phage.

You might think, and indeed it has been argued, that the situation in phage-infected bacteria cannot be compared to that in uninfected bacteria and that there is therefore no reason to suppose that bacteria normally synthesize a special RNA to carry information from their DNA to their ribosomes. Two points suggest that that view is unsound. First, bacteriophage can only be a successful parasite by subverting the normal protein-synthesizing machinery of its host, and it is reasonable to expect that the means by which the phage directs the bacterium to synthesize proteins is much the same as the system that the bacterium itself uses. Secondly, a rapidly synthesized, rapidly degraded fraction of RNA (similar to that found in phage-infected bacteria) *can* be detected in uninfected bacteria, most easily when growth conditions are chosen in which ribosomal RNA is not being made.

We may conclude that the ribosome itself contains the machinery for assembling proteins (that is, for making the peptide bonds and so on), but not the instructions for determining the sequence of amino acids. These instructions are provided to the ribosome by the species of RNA that we have just described as being present both in phage-infected and in uninfected bacteria. This type of RNA, transcribed in the former case from phage DNA and in the latter case from bacterial DNA, directly dictates the sequence of amino acids. Because it acts as a carrier of information it is called *messenger RNA*.

Although the experiments that led to the discovery of messenger RNA were done in bacteria, there are many experimental results that suggest that higher cells also have messenger RNA. In addition, support for the concept of a messenger RNA comes from experiments designed to decipher the code (pp. 444 ff.).

Now since the RNA polymerase is capable of making a transcript of any desired length of DNA (p. 436), and since a single gene determines the amino-acid sequence of one polypeptide (p. 414), we can think of a messenger RNA molecule as being the transcript of one gene.* When a ribosome binds to this messenger RNA, it will be instructed to synthesize the polypeptide corresponding to the transcribed gene. So long as the messenger RNA remains intact, the ribosome will synthesize that polypeptide. After some time, however, the messenger RNA may be degraded (see p. 485), and the ribosome will then be free to accept another molecule of messenger RNA and synthesize a new protein. Thus the ribosome is the slave of whatever messenger RNA molecule happens to be bound to it at a particular time. With this in mind we can readily understand how it is that the ribosomal RNA can be homogeneous in size and composition, since it does not carry the specificity needed for determining the amino-acid sequences of different proteins.

## Amino-acyl transfer RNA

So far we have spoken loosely of the messenger RNA's 'instructing' the ribosome how amino acids should be arranged to make a protein, but we have said nothing of how the amino acids themselves are brought to the ribosome for polymerization nor of what form the messenger RNA's 'instructions' take. Now the only specificity that a molecule of messenger RNA can possess must lie in its sequence of nucleotides (in just the same way as is true of DNA), since the linkage between adjacent nucleotides is always the $3' \rightarrow 5'$-phosphodiester bond (p. 46). We must therefore consider how the sequence of nucleotides in messenger RNA can determine the sequence of amino acids in protein. We might suspect that the specificity of hydrogen bonding that we have referred to more than once already (pp. 141, 146 and 428) will be used, since the nucleotide bases in messenger RNA are just as capable of making hydrogen bonds as those in DNA. Plainly, though, the messenger RNA cannot form suitable hydrogen bonds directly with each of the amino acids, so the problem is to modify the amino acids by attaching to them some molecule with which the nucleotide bases of messenger RNA *can* form hydrogen bonds.

The required molecule is a third species of RNA called *transfer RNA*, to which amino acids are bound preparatory to their polymerization into proteins. The binding is a two-stage reaction, both stages being catalysed by the same enzyme, which is called an amino-acyl transfer RNA synthetase.

$$R.CHNH_2.COOH + ATP \rightleftharpoons R.CHNH_2.CO-AMP + PP_i$$

$$R.CHNH_2.CO-AMP + transfer\ RNA \rightleftharpoons$$

$$R.CHNH_2.CO\text{-transfer RNA} + AMP$$

This reaction is closely analogous to that in which fatty-acyl coenzyme A is formed from fatty acid (p. 284). That reaction too consists of two stages, the first of which is the formation of an acyl AMP with the liberation of inorganic pyrophosphate; for the usual reason (p. 284) we can regard it as proceeding

* We shall see (p. 478) that occasionally a single molecule of messenger RNA may contain the transcripts of a few genes provided that they are contiguous.

virtually to completion. The fatty-acyl AMP and amino-acyl AMP are both mixed anhydrides, with a linkage between the carboxyl group of the acid and the phosphate group of AMP (see Fig. 26.1). They are thus activated forms of the fatty acid and amino acid from which a fatty-acyl or amino-acyl residue can be readily transferred in synthetic reactions. (The fatty acid and amino acid are still in an activated form when they have been transferred to form fatty-acyl coenzyme A and amino-acyl transfer RNA, which have high-energy bonds.) But whereas in synthetic reactions involving fatty acids (p. 346) very little specificity is needed in selection of the fatty acid, the synthesis of protein requires complete specificity in selection of the amino acid. This specificity is provided by the transfer RNA, any one molecule of which is absolutely specific for a particular amino acid, and by the synthetase enzyme, which is also absolutely specific in that it will transfer only one particular amino acid to its RNA. Consequently every amino acid that enters the above reaction will find itself attached to *its own specific transfer RNA*.

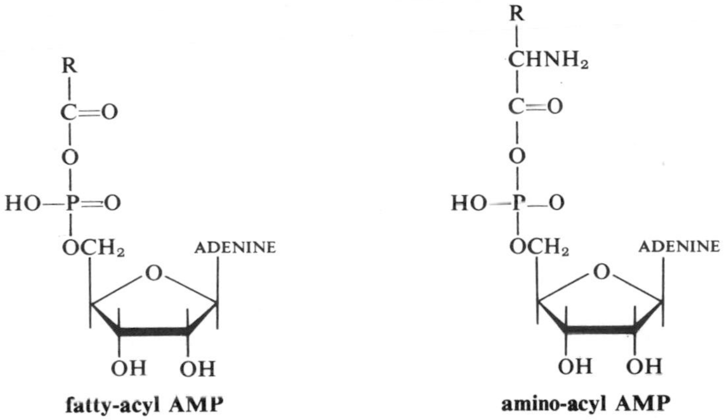

**fatty-acyl AMP**          **amino-acyl AMP**

**Fig. 26.1**   A comparison of fatty-acyl AMP and amino-acyl AMP

In what features of the transfer RNA molecule does the specificity lie? We have seen (p. 159) that all the transfer RNA molecules are similar in many respects. They all contain about seventy to eighty nucleotides, some of which have bases other than the usual adenine, guanine, cytosine and uracil. They are all arranged in a highly hydrogen-bonded structure (see Fig. 7.13a), and they all terminate at the 3'-hydroxyl end with the trinucleotide cytidylic acid–cytidylic acid–adenosine: it is to the 3'-hydroxyl group of the terminal adenosine that the carboxyl group of the amino acid is esterified. What distinguish the transfer RNA molecules one from another are two features. First, as we said on p. 158, there is a sequence of three nucleotides about half-way along the molecule (see Fig. 26.2) which is (for reasons that will become clear) called the *anticodon*, and this sequence is different for every transfer RNA molecule. Secondly, there is (we must presume) a site on each transfer RNA molecule that enables it to be recognized by the synthetase enzyme responsible for esterifying it to its specific amino acid.

Thus because of the specificity of the transfer RNA and of the amino-acyl transfer RNA synthetase, any given transfer RNA molecule can accept only one species of amino acid. Consequently, once the amino acid has been esterified to its transfer RNA, any system that is capable of recognizing the specific transfer RNA will inevitably be recognizing at the same time a specific amino acid. It is rather as if amino acids were coded – if in a laboratory the bottle in which alanine is kept were always cylindrical, that in which glycine is kept were always square in section and so on, then one could always pick out the right bottle even if the label were illegible. We can compare the messenger RNA to someone that can recognize shapes but cannot read labels: by using part of its nucleotide sequence to form hydrogen bonds with part of the nucleotide sequence of the transfer RNA, messenger RNA can select amino-acyl transfer RNA molecules and thereby select the amino acids that are to be inserted into a polypeptide chain.

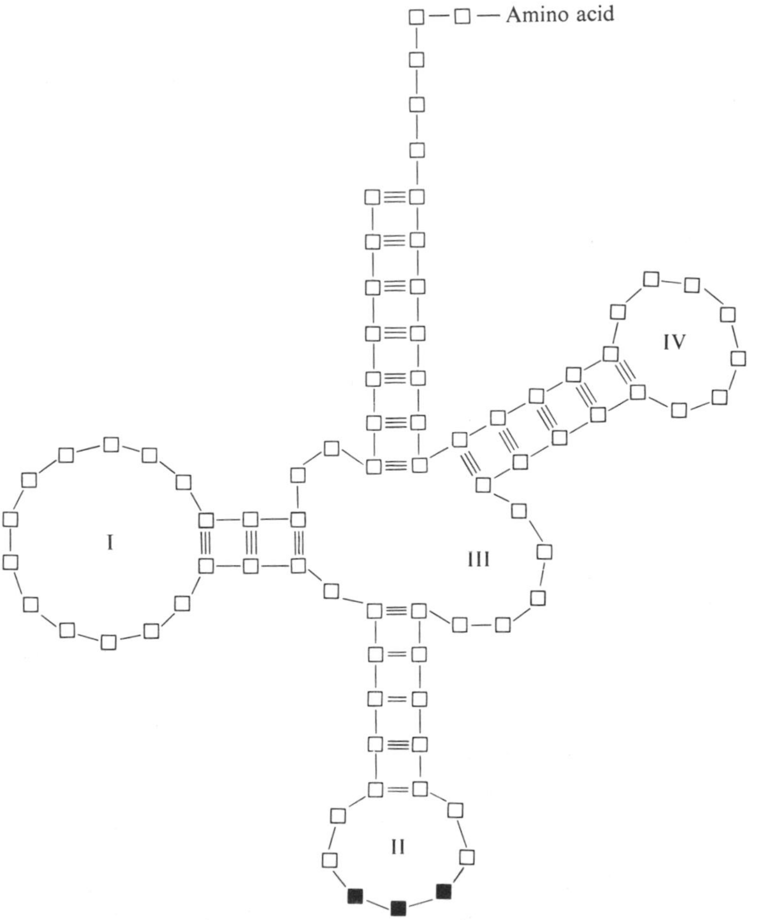

**Fig. 26.2** An amino-acyl transfer RNA. The amino acid is attached to the 3′ end, which has the sequence pCpCpA–OH (see p. 159). The molecule is extensively double-stranded; one of the single-stranded regions contains the anticodon (represented in the figure by the three black squares) which is complementary to the codon of the messenger RNA (see text). Compare Figure 7.13a

We saw in Ch. 24 that a sequence of three nucleotide pairs in DNA corresponds to ('codes for') each amino acid, and in Ch. 25 that each nucleotide pair of DNA is transcribed into a single nucleotide in RNA. It follows that a sequence of three nucleotides in messenger RNA codes for each amino acid; such a sequence is called a *codon*. One might expect that each codon would recognize a transfer RNA molecule by forming hydrogen bonds with three contiguous nucleotides in the molecule, and this expectation is borne out by the finding that we mentioned above, that each transfer RNA has a *specific* sequence of three nucleotides in the second single-stranded region counting from the 5' end; this sequence is therefore the anticodon (see Fig. 26.3). We shall see that hydrogen bonding between the codon and the anticodon occurs as each amino acid is incorporated into the polypeptide chain.

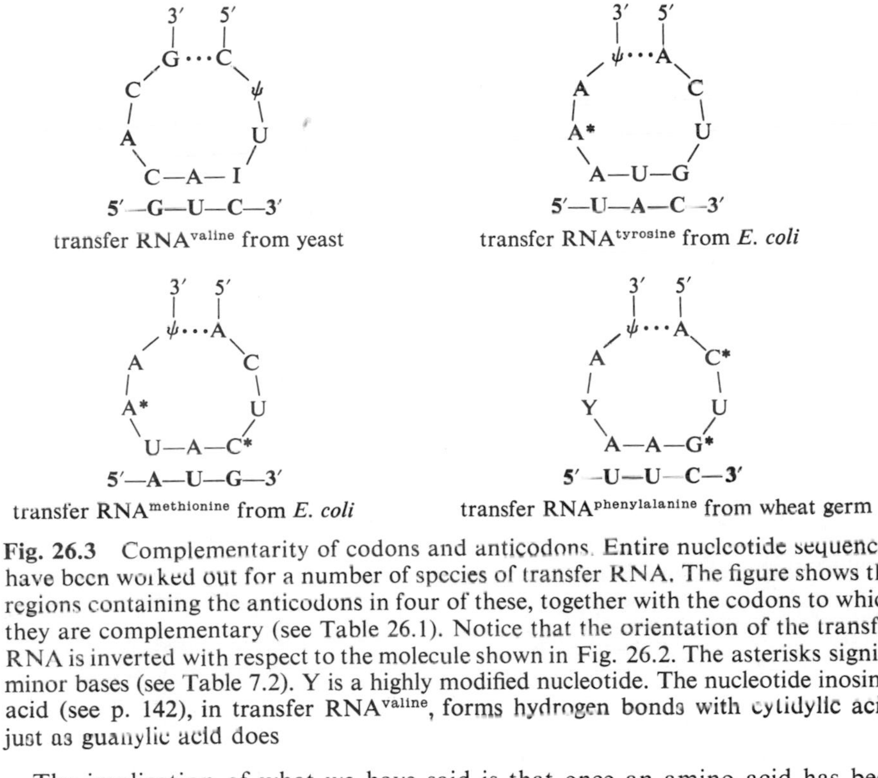

transfer RNA$^{valine}$ from yeast

transfer RNA$^{tyrosine}$ from *E. coli*

transfer RNA$^{methionine}$ from *E. coli*

transfer RNA$^{phenylalanine}$ from wheat germ

**Fig. 26.3** Complementarity of codons and anticodons. Entire nucleotide sequences have been worked out for a number of species of transfer RNA. The figure shows the regions containing the anticodons in four of these, together with the codons to which they are complementary (see Table 26.1). Notice that the orientation of the transfer RNA is inverted with respect to the molecule shown in Fig. 26.2. The asterisks signify minor bases (see Table 7.2). Y is a highly modified nucleotide. The nucleotide inosinic acid (see p. 142), in transfer RNA$^{valine}$, forms hydrogen bonds with cytidylic acid, just as guanylic acid does

The implication of what we have said is that once an amino acid has been bound to its specific transfer RNA, it no longer plays any part in the recognition process that determines its insertion into the polypeptide; it is the transfer RNA alone that the messenger RNA recognizes. That this idea is correct has been proved experimentally: if an amino acid is esterified to its specific transfer RNA and then changed to a different amino acid by a chemical process that does not affect the transfer RNA, the specificity of insertion into a polypeptide is found to be determined by the transfer RNA rather than by the changed amino acid.*

* To give a specific example: cysteine after esterification to its (cysteine-specific) transfer RNA can be reduced to alanine. The protein-synthesizing machinery of the cell, that is the ribosome-messenger RNA complex, will incorporate this alanine into the position reserved for cysteine since it is still esterified to cysteine-specific transfer RNA.

To return to our earlier analogy – if the manufacturers had erroneously put a glycine label on the cylindrical alanine bottle, the bottle would be recognized as if it contained alanine, thus proving that the protein-synthesizing machinery recognizes shapes (transfer RNA molecules) rather than labels (amino acids).

Careful study of the transfer RNA fraction of both bacterial and higher cells has shown that some amino acids have several different transfer RNA molecules that will accept them in the esterification reaction. There are, for example, three different transfer RNAs in *E. coli* that will accept serine. Notice that this fact does not contradict what we said earlier about the specificity of the transfer RNA for its particular amino acid: an amino acid may have more than one transfer RNA that corresponds to it, but no transfer RNA can accept more than one amino acid.

## THE AMINO-ACID CODE

We are now in a position to look at the details of the code itself, that is to ask which amino acids correspond to each of the sixty-four possible arrangements of three nucleotides in the messenger RNA. Several methods have been used to determine this correspondence, but we shall here cite only two. The first was invented by Nirenberg and Leder. If a synthetic trinucleotide is added to a preparation of ribosomes, a complex is formed which is found to be capable of binding one variety of amino-acyl transfer RNA, but not any of the other varieties. For example the trinucleotide pUpUpU (which is composed of three uridylic acid residues joined by $3' \rightarrow 5'$-phosphodiester linkages, with a phosphate group on the 5' end and a free hydroxyl group on the 3' end) will direct the binding to ribosomes of phenylalanyl transfer RNA, but no other amino-acyl transfer RNA. Similarly the trinucleotide pApApA will cause lysyl transfer RNA to bind to ribosomes, and pCpCpC will cause prolyl transfer RNA to bind. The experiments can be extended to trinucleotides that are composed of more than one kind of residue: for instance pUpCpG directs the binding of seryl transfer RNA. It is found that permuting the order of nucleotides in a triplet changes the specific binding: pGpCpU directs the binding of alanyl transfer RNA, pCpGpU of arginyl transfer RNA and pUpGpC of cysteinyl transfer RNA.

These results suggest that the trinucleotides that have been added to ribosomes, and thereby promote the binding of specific amino-acyl transfer RNAs, might be those that actually code for the particular amino acids during protein synthesis. Many lines of evidence support this conclusion. For example it is possible to synthesize RNA-like polymers consisting of repeating short sequences of nucleotides, such as UGUGUGU... or UACUACUAC... Khorana and his colleagues used such polynucleotides as *artificial messenger RNA* in a system in which ribosomes are made to incorporate amino acids into peptide linkage. They then sequenced the polypeptides thus formed, and were able to assign amino acids to particular codons on the basis of their results. The assignments are also consistent with results of studies of the mutant proteins corresponding to the α subunit of tryptophan synthetase in *E. coli* (see p. 418): the observed changes in the amino-acid sequence can be explained by assuming that each mutation is due to a change in a single nucleotide pair. Interestingly enough, alterations in the chains of human haemoglobin can be explained by making the same assumption, which suggests that the same codons are used universally. We can draw this conclusion, too, from studying the amino-acid sequences of proteins from a wide variety of species (see Ch. 27). Moreover, it has been shown that if synthetic trinucleotides are bound to ribosomes from *E. coli*, the complexes thus formed will direct the binding not only of particular amino-acyl transfer RNA molecules from *E. coli*, but also of the *same* kinds of amino-acyl transfer RNA molecules from animal cells.

Sixty-one of the sixty-four possible trinucleotides code for amino acids (see Table 26.1). The three missing codons, UAG, UAA and UGA, are not used for any amino acid, but instead to signal the termination of the polypeptide chain, as we shall discuss on pp. 458 f. Inspection of the table will show that we can confirm in biochemical terms some of the conclusions that we drew from a consideration of genetic results in Ch. 24. Perhaps the most striking feature is the *degeneracy* of the code (p. 422). Nearly all the amino acids are represented by more than one codon – some (serine, leucine and arginine) by as many as six codons. This fact helps to explain a point that we made earlier (p. 417). In discussing Benzer's results with the rIIA gene of phage T4 we suggested that there may be sites in the DNA at which mutation leads to no observable change in phenotype. We can now see from Table 26.1 that (for example) a mutation that leads to replacement of the codon CUU in the messenger RNA by CUC, CUA or CUG will not change the amino-acid sequence of the corresponding protein, since all these codons correspond to leucine. Again, by reference to the table, we can interpret a similar but rather subtler finding, which also bears on the question whether one can expect to find observable mutations at all sites in the DNA. In his studies of tryptophan synthetase α, Yanofsky found that the replacement of the glycine residue at position 211 by glutamate or by arginine inactivated the protein. From the mutant bacteria with glutamate at position 211 he recovered bacteria that had mutated further and now contained active tryptophan synthetase, and examination of their α subunits showed that in many strains the glutamate residue had been replaced by other amino acids, all of which permitted enzyme activity. From a study of all these events it was possible

**Table 26.1**   The amino-acid code

| | | | | | | | |
|---|---|---|---|---|---|---|---|
| UUU | Phenylalanine | UCU | | UAU | Tyrosine | UGU | Cysteine |
| UUC | | UCC | Serine | UAC | | UGC | |
| UUA | Leucine | UCA | | UAA | * | UGA | * |
| UUG | | UCG | | UAG | * | UGG | Tryptophan |
| CUU | | CCU | | CAU | Histidine | CGU | |
| CUC | Leucine | CCC | Proline | CAC | | CGC | Arginine |
| CUA | | CCA | | CAA | Glutamine | CGA | |
| CUG | | CCG | | CAG | | CGG | |
| AUU | | ACU | | AAU | Asparagine | AGU | Serine |
| AUC | Isoleucine | ACC | Threonine | AAC | | AGC | |
| AUA | | ACA | | AAA | Lysine | AGA | Arginine |
| AUG | Methionine | ACG | | AAG | | AGG | |
| GUU | | GCU | | GAU | Aspartic acid | GGU | |
| GUC | Valine | GCC | Alanine | GAC | | GGC | Glycine |
| GUA | | GCA | | GAA | Glutamic acid | GGA | |
| GUG | | GCG | | GAG | | GGG | |

* These codons signify termination of the polypeptide.

to infer that the original codon had been GGA (glycine) which had mutated to give GAA (glutamate). When this in turn had mutated to give (for example) GCA (alanine) the enzymic activity was restored. We can infer that if the original GGA had mutated directly to GCA the change would never have been observed because the organism would have remained capable of synthesizing tryptophan: the mutation would have been 'silent'.

These factors act to diminish the immediate effect of mutations, but they permit slight modifications to occur during evolution that may be beneficial in the long run. For example, it is possible that with glycine present at position 211 in tryptophan synthetase $\alpha$, a particular change elsewhere in the polypeptide may inactivate the protein completely; on the other hand this same change occurring in tryptophan synthetase $\alpha$ that contains alanine at position 211 might conceivably make the protein *more* active than that of wild-type *E. coli*. Note, too, that a single base-change in a codon frequently has only a slight effect on coding, in the sense that it leads to the replacement of an amino acid with a chemically similar amino acid. We have seen that if CUU changes to CUC, CUG or CUA, leucine will still be inserted; even if CUU changes to UUU, GUU or AUU, the resulting amino acid will still be hydrophobic. This fact helps organisms to resist drastic alterations in their proteins while still permitting subtle changes, which can be subject to natural selection, to occur (see also Ch. 27).

## MECHANISM OF PROTEIN SYNTHESIS

We have now described the three principal components of the complex reaction that leads to the formation of polypeptides. The ribosome is the site of protein synthesis; it no doubt has many functions which are not yet understood, but for the present we can concentrate on two functions. It serves to bring the messenger RNA into the correct spatial juxtaposition with a succession of amino-acyl transfer RNA molecules, and it also holds the growing polypeptide chain in place during synthesis. The *messenger RNA* encodes in its nucleotide sequence the desired arrangement of amino acids. The *transfer RNA* brings the amino acid to the ribosome in an activated state and in a form in which it can recognize, by hydrogen bonding, its codon in the messenger RNA. We are now in a position to discuss the way in which these three components take part in protein synthesis.

### Direction of synthesis and of translation

The first point that we have to consider is whether the polypeptide chain grows from its amino terminus or from its carboxyl terminus. An experiment first done by Dintzis settled this question. Dintzis incubated a haemoglobin-synthesizing system from reticulocytes with radioactive leucine, isolated the *completed* haemoglobin at various times of incubation, and digested it with trypsin (see p. 82). The amount of radioactivity in the tryptic peptides turned out to be a function of the position of these peptides in the haemoglobin molecule: after a short incubation only peptides from near the carboxyl terminus contained radio-active leucine, and as time went on radioactivity was found progressively nearer to the amino terminus (see Fig. 26.4). Dintzis reasoned that the molecules that showed labelling at early times were those that had been nearly completed at the moment the radioactive leucine was added; to complete these molecules required the addition of only a few amino-acid residues, and since these molecules included radioactive leucine near the carboxyl terminus, it follows that the carboxyl terminus is synthesized *last*. Note that this experiment depends on isolating *completed* haemoglobin molecules and seeing where the radioactivity lies; plainly if *fragments* of haemoglobin were isolated, the pattern of radioactivity would be different.

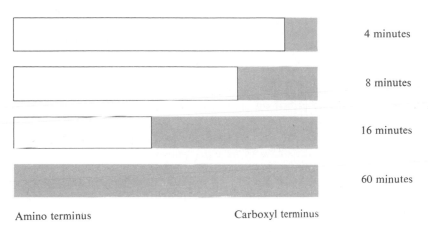

Amino terminus    Carboxyl terminus

**Fig. 26.4**  Schematic representation of the results of the experiment of Dintzis. A haemoglobin-synthesizing system from reticulocytes was labelled with radioactive leucine for the periods of time shown on the right. *Completed* haemoglobin was extracted and digested with trypsin, and the radioactivity of the isolated peptides was measured. When the peptides were arranged according to their position in the polypeptide chain, it was found that at early times the radioactivity appeared only near the carboxyl terminus. The grey areas represent the radioactive regions of the protein

The conclusion must be that polypeptides are synthesized sequentially from the amino terminus to the carboxyl terminus. Similar experiments with protein-synthesizing systems from bacteria have led to the same conclusion.

The fact that protein synthesis proceeds from amino terminus to carboxyl terminus has been used to determine the direction in which the messenger RNA is translated. It turns out that a synthetic messenger RNA of the type $(AAA)_nAAC$ codes for $(lysyl)_n$–asparagine; similarly the hexanucleotide AAAUUU codes for lysyl–phenylalanine and not for phenylalanyl–lysine. It follows that the 5′ end of the messenger RNA (which is conventionally written at the left) codes for the amino terminus of the protein (also conventionally written on the left). Hence the 5′ end of the messenger RNA is translated first.

We saw in Ch. 25 that the direction in which messenger RNA is synthesized is from 5′ end to 3′ end. The fact that it is translated in the same direction raises the possibility that (in procaryotic cells which have no nuclear membrane) a messenger RNA molecule can begin to be translated before its synthesis is completed – in other words that as soon as the 5' end is made on the DNA template it may be engaged by ribosomes and used in protein synthesis (see also p. 486).

## Elongation of the polypeptide chain

Elucidation of the reactions involved in protein synthesis has proved an extremely difficult and lengthy task, and has been due to the efforts of many laboratories including those of Lipmann, Moldave, Lengyel, Watson and several others. The picture that has emerged most clearly, and that we shall describe, is of events in the two bacterial species *E. coli* and *Bacillus stearothermophilus*. The mechanism of mammalian protein synthesis seems, so far as it has been worked out, to be largely similar to that in bacteria.

In describing protein synthesis in bacteria, we shall take as an example the translation into polypeptide of the fragment of messenger RNA whose synthesis we described on p. 436. This has the sequence ...UUGACCCCAAGGUAC GAU... Since three nucleotides form a codon, corresponding to a single amino acid, we can write the sequence as ...UUG.ACC.CCA.AGG.UAC.GAU..., but when we do so we must bear in mind that we have artificially introduced full stops into the representation; the messenger RNA itself contains no puncutation, but consists of an uninterrupted sequence of nucleotides linked by phosphodiester bonds. By reference to Table 26.1 we can see that the sequence will code for the peptide fragment ....leucyl–threonyl–prolyl–arginyl–tyrosyl–aspartyl...

To begin with, we shall discuss the process by which the polypeptide is elongated, and shall defer for the moment a description of how it is initiated. We shall therefore imagine that the first codon (UUG) of our messenger RNA fragment has just been translated to give leucine, and that this leucine residue is not the first (N-terminal) residue of the peptide, but is preceded by other residues. We therefore have a messenger RNA of which ...UUG has already been translated and ACC.CCA.AGG.UAC.GAU... is waiting to be translated, and an incomplete polypeptide of the form N-terminus......leucine. Now during protein synthesis the messenger RNA is bound to the 30$S$ subunit of the ribosome; at any one time at least two codons (six nucleotide residues) of the messenger RNA are bound to that part of the ribosome at which polypeptides are being synthesized (what we might call the 'active centre' of the ribosome). In a corresponding place on the 50$S$ subunit there exist two sites at which transfer RNA can be bound. We know very little of the configuration of the binding sites for transfer RNA, but we can imagine the transfer RNA as being bound in a ribosomal cleft in such a way that its anticodon is close enough to the binding site for messenger RNA that hydrogen bonds can be formed between the codon and the anticodon.

For reasons that will very soon become clear, the incomplete polypeptide terminates in a transfer RNA molecule, so that the peptide we have imagined as being translated from our notional messenger is actually in the form amino terminus ... leucyl transfer RNA. This peptidyl transfer RNA is held on the ribosome in one of the two transfer RNA sites that we have just mentioned; because it is a peptide that is held there this site is sometimes called the P site

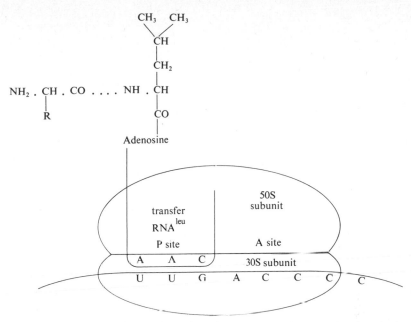

**Fig. 26.5** Protein synthesis I. The growing peptide chain, terminating in leucyl transfer RNA, is held in the P site of the ribosome. Its anticodon is bound through hydrogen bonding with the leucine codon of the messenger RNA. The 5′ end of the messenger RNA, which has already been translated, is towards the left of the figure; the part of the messenger RNA that is waiting to be translated is towards the right

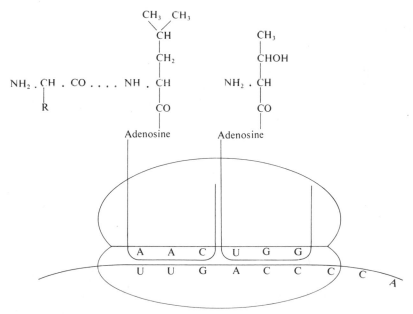

**Fig. 26.6** Protein synthesis II. As a result of *step 1*, threonyl transfer RNA is bound to the A site of the ribosome. Its anticodon is held through hydrogen bonding to the threonine codon of the messenger RNA

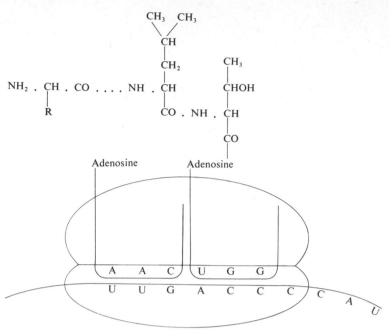

**Fig. 26.7** Protein synthesis III. As a result of *step 2*, a peptide bond has been made between leucine and threonine. The unesterified transfer RNA^leu remains in the P site, and the peptide, lengthened by one amino-acid residue and terminating in threonyl transfer RNA, is held in the A site

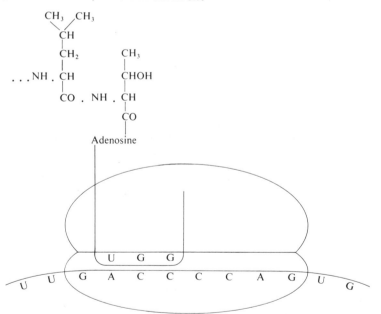

**Fig. 26.8** Protein synthesis IV. As a result of *step 3*, the peptidyl transfer RNA is shifted from the A site to the P site. The messenger RNA is also shifted by the length of one codon with respect to the ribosome, and consequently the hydrogen bonding between the transfer RNA^thr and the threonine codon is maintained. The position is now exactly analogous to that shown in Fig. 26.5, and the ribosome is ready to acept a new amino-acyl transfer RNA in the A site

(on the left of the ribosome as drawn in our Fig. 26.5). In this figure the other site, which is sometimes called A for amino acid, is for the moment empty. Opposite the P site at which N terminus ... leucyl transfer RNA is held is the messenger RNA's leucine codon (UUG), hydrogen-bonded to the anticodon of the leucyl transfer RNA. Further to the left is that part of the messenger RNA that has already been translated; to the right is the remainder of the messenger RNA, waiting to be translated. Notice that the next codon to be translated, ACC, is opposite the A site.

We can summarize the process of elongation of the peptide chain quite briefly before considering it in detail (see Figs 26.5–26.8). Step 1: the amino-acyl transfer RNA (threonyl transfer RNA) specified by the next codon binds at the A site. Step 2: a peptide bond is formed between the amino group of the incoming threonine and the carboxyl group of the leucine at the carboxyl terminus of the peptide; the leucine-specific transfer RNA (no longer esterified) remains bound to the P site, and the A site now contains a lengthened peptide that terminates in threonyl transfer RNA. Step 3: the non-esterified transfer RNA molecule is released, the peptidyl transfer RNA moves from the A site into the P site and simultaneously the messenger RNA moves one codon to the left so that the ACC codon is opposite the P site and the next codon, CCA, opposite the A site.

There are three protein 'elongation factors', called $EFT_u$, $EFT_s$ and $EFG$, which can be isolated from the soluble fraction of broken bacterial cells, that participate in this sequence of events, and in the detailed description that we shall now give we shall discuss the role of these factors. We shall also discuss how energy is provided for the synthesis of polypeptides – interestingly enough by the hydrolysis of GTP to GDP and inorganic phosphate rather than by hydrolysis of ATP.

In *step 1*, amino-acyl transfer RNA becomes bound to the A site of the ribosome. This process has been dissected into various component reactions. First the factor $EFT_u$ forms a complex with GTP and amino-acyl transfer RNA:

$$EFT_u + GTP + \text{amino-acyl transfer RNA} \longrightarrow$$
$$[EFT_u\text{-GTP-amino-acyl transfer RNA}]$$

and this then binds to the ribosome at the A site.

$$[EFT_u\text{-GTP-amino-acyl transfer RNA}] + \text{ribosome} \longrightarrow$$
$$[EFT_u\text{-GTP-amino-acyl transfer RNA-ribosome}]$$

Bear in mind that the only amino-acyl transfer RNA that will form a stable complex with the ribosome and messenger RNA is the one that is specified by the messenger RNA codon opposite the A site. (We can picture $[EFT_u$–GTP–amino-acyl transfer RNA] complexes being formed from all the types of transfer RNA molecules in the soluble fraction of the cell and being ready to participate in the formation of a complex with the ribosome as soon as the appropriate codon appears opposite the A site.) Next the GTP in the quarternary complex is split, and this leaves amino-acyl transfer RNA bound to the ribosome in a form in which it can become involved in the synthesis of a peptide bond; inorganic phosphate is released, and so is a complex of $EFT_u$ and GDP.

[EFT$_u$-GTP-amino-acyl transfer RNA-ribosome] $\longrightarrow$

[amino-acyl transfer RNA-ribosome] + [EFT$_u$-GDP] + P$_i$

It is now necessary for the [EFT$_u$–GDP] complex to be converted again to a [EFT$_u$–GTP–amino-acyl transfer RNA] complex. This reaction is catalysed by another of the protein factors, namely EFT$_s$; the necessary energy is provided at the expense of another molecule of GTP.

[EFT$_u$-GDP] + amino-acyl transfer RNA + GTP $\longrightarrow$

[EFT$_u$-GTP-amino-acyl transfer RNA] + GDP

The result of step 1, therefore, is to place an amino-acyl transfer RNA at the A site of the ribosome ready to take part in peptide synthesis, and this is accompanied by the breakdown of GTP to GDP and inorganic phosphate; the other components are recycled directly. Figure 26.6 shows the position we have now reached.

In *step 2*, a peptide bond is made between the free amino group of the amino-acyl transfer RNA that has just been bound to the ribosome and the carboxyl group of the amino acid that was last inserted into peptide linkage. In effect the amino group of the incoming amino acid displaces the transfer RNA that was covalently bound to the carboxyl group of the last amino acid. In our example, the incoming amino acid is threonine and the preceding one is leucine, so we have:

H$_2$N...leucyl transfer RNA + threonyl transfer RNA $\longrightarrow$

H$_2$N...leucyl–threonyl transfer RNA + transfer RNA$^{leu}$

where by 'transfer RNA$^{leu}$' we mean the leucine-specific transfer RNA which was until just now esterified to leucine, but is now unesterified. This position is shown in Fig. 26.7.

The formation of the peptide bond itself is catalysed by an enzyme called peptidyl transferase, which is not one of the supernatant factors but a component of the 50$S$ ribosome. It requires no splitting of a high-energy phosphate bond at this stage, since the necessary activation of the amino acid has already occurred.

Step 2 leaves the newly lengthened peptidyl transfer RNA bound to the A site of the ribosome, but in order for another amino-acyl transfer RNA to be bound, the peptidyl transfer RNA must be moved to the P site. This movement is called translocation, and is part of *step 3*. The details of step 3 are not very clearly understood, but what is involved is not only the shifting of peptidyl transfer RNA from the A site to the P site, but also a corresponding shifting of the messenger RNA by one codon (so that the hydrogen-bonding between the codon and the anticodon is maintained) and in addition a discharge of the unesterified transfer RNA that was previously in the P site. This complex reaction is dependent on factor EFG and involves the splitting of another molecule of GTP to GDP and inorganic phosphate.

The simultaneous translocation of the peptidyl transfer RNA and of the messenger RNA by the length of one codon is thought to occur by a conformational change in the ribosome itself. One possibility is that this conformational change is catalysed by EFG and driven by the hydrolysis of GTP, and results in the unesterified transfer RNA's being pushed out of the P site. Another possibility (perhaps more likely) is that the chief function of EFG is to empty the P site, and that the ribosome itself is then capable of translocating the peptidyl transfer RNA and the messenger RNA by undergoing the necessary conformational change.

Whatever the mechanism, the position at the end of step 3 is that shown in Fig. 26.8. This is exactly analogous to that shown in Fig. 26.5, and we have now set the scene for a repetition of steps 1, 2 and 3. In this way the insertion of proline, arginine, tyrosine and aspartate into the peptide chain will be catalysed in just the same way as that of threonine.

The transfer RNA molecules that are liberated during step 3 of this cycle are reused by becoming esterified once again to their appropriate amino acids, by the reaction that we described on p. 440. Thus all the intermediate participants in the formation of proteins act catalytically, and the overall process involves amino acids being fed in and peptides being fed out. We may note that the synthesis is energetically very costly: apart from the two high-energy phosphate bonds that are used in the formation of each amino-acyl transfer RNA, a further two are hydrolysed in each cycle of the reactions by which the peptide chain is elongated (one at step 1 and one at step 3). The insertion of each amino acid into the chain therefore needs the hydrolysis of four high-energy phosphate bonds.

## Initiation of the polypeptide chain

It is clear that the mechanism that we have described for elongation of the polypeptide chain will need modification if it is to be used for its initiation. In the first place, we have assumed in the discussion that the P site of the ribosome will already be carrying a peptidyl transfer RNA before an amino-acyl transfer RNA binds to the A site; obviously when the chain is initiated the P site will be empty. Secondly, the enzyme (peptidyl transferase) that synthesizes the peptide bond (see step 2 above) normally has as one of its substrates a peptidyl transfer RNA, which can be regarded as a N-blocked amino-acyl transfer RNA, rather than an amino-acyl transfer RNA with a free $—NH_2$ group. These considerations suggest that a special mechanism is needed to start the synthesis of the peptide chain.

A seemingly unrelated discovery also suggests that there is something special about initiation of the synthesis. If the amino termini of $E.$ $coli$ proteins are examined, it is found that some 40 per cent of all proteins have N-terminal methionine, whereas the average abundance of methionine in $E.$ $coli$ proteins is only about 2.5 per cent. A clue about the reason for this predominance of methionine at the N-terminus, and an important contribution to elucidating the mechanism of initiation, was the discovery by Marcker in 1965 of $N$-formyl-methionyl transfer RNA, the formula of which is

$$CH_3.S.CH_2.CH_2.CHNH.CO\text{-transfer RNA.}$$
$$|$$
$$CHO$$

Marcker found that $E.$ $coli$ contains two kinds of methionyl transfer RNA, of which one (called methionyl transfer $RNA_F^{met}$) can be formylated but the other (called methionyl transfer $RNA_M^{met}$) cannot. Formylation is catalysed by an enzyme at the expense of $N^{10}$-formyltetrahydrofolate (p. 384); the enzyme does not act on free methionine. N-formylmethionyl transfer RNA has a N-blocked amino group, and it is therefore a promising candidate for the role of an initiating amino-acyl transfer RNA.

Another important finding was the discovery that ribosomes *exchange subunits* during growth both of *E. coli* and of eucaryotic cells. By the use of density label (see pp. 527 f.) Kaempfer has shown that the subunits of 70$S$ (and 80$S$) ribosomes are dissociated from one another after completing the synthesis of a polypeptide, with the result that in initiating the synthesis of a new poly-peptide it is free 50$S$ and 30$S$ (or 60$S$ and 40$S$) subunits that take part.

We may now describe the picture of initiation of polypeptide synthesis in *E. coli* that has emerged from these and subsequent studies. The first event is the binding of a 30$S$ ribosomal subunit to the messenger RNA, close to a codon that specifies methionyl transfer RNA (namely AUG). This binding is facilitated by a protein factor called initiation factor 3 (IF$_3$), which also has the function of preventing the free 30$S$ subunit from associating with a 50$S$ subunit until the time is ripe for such an association (see below). Meanwhile N-formylmethionyl transfer RNA binds to GTP and to the initiation factor IF$_2$, to form a complex analogous to the [EFT$_u$-GTP-amino-acyl transfer RNA] complex formed in elongation (p. 451). The [IF$_2$-GTP-N-formylmethionyl transfer RNA] complex now binds to the 30$S$ subunit in response to the AUG codon. Next IF$_3$ is detached, and a 50$S$ subunit, too, is bound to the complex in such a way that the N-formylmethionyl transfer RNA is positioned at the *P site* of the resulting 70$S$ ribosome.

It appears that, in order for IF$_2$ to be released so that it can be recycled, the bound GTP must now be split. It appears, too, that another initiation factor, IF$_1$, is involved in some way in the recycling of IF$_2$, but its role is not fully understood. At all events, the result of the sequence of events (shown in Fig. 26.9) is the formation of a 70$S$ ribosome with N-formylmethionyl transfer RNA in the P site; this situation is analogous to that shown in Fig. 26.5 and the next stage in protein synthesis will be binding, to the A site, of the amino-acyl transfer RNA specified by the next codon in the messenger RNA.

As we have said above, AUG is the codon for formylmethionyl transfer RNA. But it is also the codon for methionyl transfer RNA$_M^{met}$, which has no function in initiation, but is used to insert methionine into *internal* positions in the poly-peptide. Plainly, then, the protein-synthesizing machinery must be able to distinguish between AUG as an initiation codon for formylmethionyl transfer RNA and AUG as an internal codon for methionyl transfer RNA$_M^{met}$, and we must now consider how this distinction is made.

One possibility would be that AUG acts as an initiating codon if it is at the 5′ end of the messenger RNA, and as an internal codon if it is elsewhere. If this idea were correct, translation of messenger RNA would begin right at the 5′ end, and one would expect to find AUG at or very close to this end. There is evidence that this is, at any rate, not always true, and in fact no evidence that it is ever true. A few naturally occurring RNA molecules are known which act as messen-ger RNAs, the best studied being the RNA of the class of bacteriophages that contain RNA instead of DNA as their genetic material (see pp. 159 and 487). In one of these molecules, the nucleotide sequence of some of the RNA has been worked out (see Fig. 7.13), and it is found that the beginning of the RNA

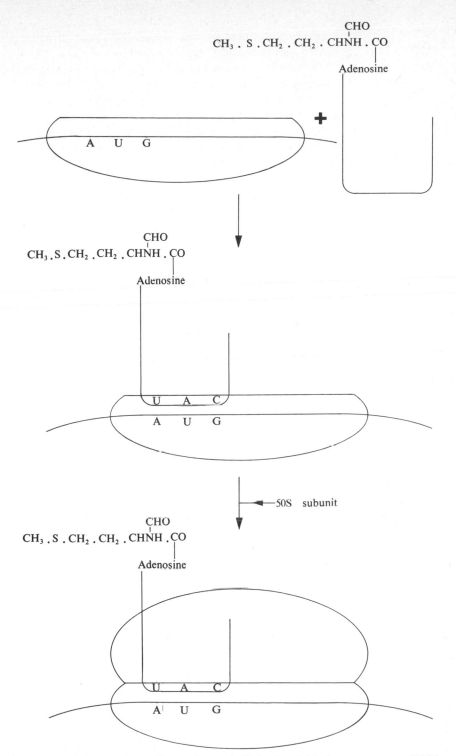

**Fig. 26.9** Initiation of protein synthesis. A 30$S$ subunit binds to messenger RNA at an AUG codon. N-Formylmethionyl transfer RNA and a 50$S$ subunit successively join the complex. The initiation factors are not shown in the figure

sequence that codes for the first of the phage proteins occurs more than 100 nucleotides from the 5′ end.

The fact that the ribosome does not begin to translate the messenger RNA at its physical end now raises another problem. We saw in Ch. 24 that the code has no 'punctuation' but is read continuously in groups of three nucleotides. A property of a code of this sort is that it is essential to set the 'reading frame' correctly; we have seen that if the reading frame is disrupted by the insertion or deletion of a single nucleotide, complete missense results (see p. 420). Thus, if one reads . . . UUG . ACC . CCA . AGG . UAC . GAU . . . one gets . . . leucyl–threonyl–prolyl–arginyl–tyrosyl–aspartate . . . , but if one read for example ( . . . UU)GAC . CCC . AAG . GUA . CGA . (U . . . ) one would get . . . aspartyl–prolyl–lysyl–valyl–arginine. How then does the ribosome know how to set the reading frame so as to read the correct triplets, granted that it does not begin to translate at the 5′ end of the messenger RNA?

The answer appears to be that the occurrence of an initiating AUG itself sets the frame for the ribosome, so that on encountering the AUG the ribosome inserts N-formylmethionyl transfer RNA into the P site of the ribosome by the mechanism that we discussed above. For example, if presented with the synthetic oligonucleotide UUAUGUUUUU as a messenger RNA, the ribosome synthesizes N-formylmethionyl–phenylalanyl–phenylalanine, thus proving that it has read (UU)AUG . UUU . UUU, and not (for example) UUA . UGU . UUU . (UU).

We are still left with the problem of how the protein-synthesizing machinery distinguishes between an initiating AUG and an internal AUG. We can divide this question into two. First, how is the ribosome prevented from inserting N-formylmethionyl transfer RNA into the middle of the polypeptide in response to internal AUG codons? Secondly, how does the $30S$ ribosome recognize an initiating AUG codon as a site to which it should bind, thus promoting the initiation sequence that we have described above?

The probable answer to the first question is this. In order for an amino-acyl transfer RNA to be bound to the A site of the $70S$ ribosome it must first make a complex with $EFT_u$ and GTP (see p. 451). $EFT_u$ appears to be capable of binding to all kinds of amino-acyl transfer RNAs *except* N-formylmethionyl transfer RNA, which instead binds specifically to $IF_2$. This discrimination by $EFT_u$ against N-formylmethionyl transfer RNA ensures that it is used only in initiation.

The answer to the second question is not known. There is reason to believe, however, that messenger RNA has some tertiary structure (see also p. 159 and pp. 486 f.), and it may be that the $30S$ ribosome will bind to the messenger RNA only at a place along its length at which an AUG codon is particularly exposed. The fact that ribosomal RNA is not translated may be due to its lacking AUG codons in the necessary exposed configuration.

The last point about initiation in bacteria that we must consider is this. If N-formylmethionyl transfer RNA is involved in initiating polypeptides, how is it that all completed proteins do not carry N-terminal N-formylmethionine? The answer is that a deformylating enzyme is known that cleaves off the N-formyl function from newly synthesized proteins, or sometimes from incomplete proteins during their synthesis. Similarly there is an enzyme that subsequently removes methionine from the N-terminus of some (but not all) proteins.

The mechanism of initiation in eucaryotic cells is much less clear than that in bacteria. There is some evidence that a special kind of methionyl transfer RNA is used to initiate protein synthesis in reticulocytes and in liver cells. Most work, however, suggests that the methionine esterified to this transfer RNA is not in fact formylated in mammalian cells (although it can be formylated in the test-tube by the enzyme from *E. coli*). It appears, moreover, that methionine is cleaved from the growing peptide quite early in the synthesis when the polypeptide chain is still fairly small. Initiation factors analogous to $IF_1$, $IF_2$ and $IF_3$ have been identified in some mammalian cells.

## Polyribosomes

Our description of the mechanism of initiation of the polypeptide chain will have made it clear that the 30$S$ ribosomal subunit binds to specific initiation sites on the messenger RNA. After N-formylmethionyl transfer RNA and the 50$S$ subunit have joined the complex, the resulting 70$S$ ribosome begins to synthesize peptide bonds. As it does so, it moves along with respect to the messenger RNA (from left to right as we have drawn the diagrams). After the ribosome has moved some distance, the initiation site will again be exposed, and another 30$S$ ribosome will be able to bind in the same place. This in turn will acquire a molecule of N-formylmethionyl transfer RNA and a 50$S$ partner, and will begin to synthesize a polypeptide. Thus, unless the messenger RNA is so short that there is space on it for only one 70$S$ ribosome, it will have several ribosomes bound to it. Those on the right (in our diagram) will have translated more of the messenger RNA than those in the middle, while those on the left will have only just started the translation process. Such a structure is called a *polyribosome* (see Fig. 26.10).

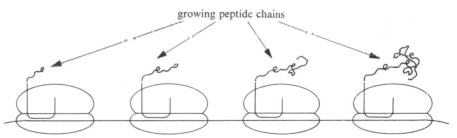

growing peptide chains

**Fig. 26.10**  A polyribosome. Ribosomes move along with respect to the messenger RNA from left to right; as they move the polypeptide chain that they hold is gradually lengthened

Polyribosomes can be seen in the electron microscope, and they are found both in bacterial and in eucaryotic cells. Treatment of them with extremely small quantities of ribonuclease breaks the messenger RNA strand between the ribosomes and liberates single ribosomes attached to small pieces of messenger RNA and to incomplete peptidyl transfer RNA. It can be shown that, both in eucaryotic cells and in bacteria, ribosomes are packed closely together along the strand of messenger RNA with very little space between them.

## Termination of the polypeptide chain

So long as every codon arriving opposite the A site of the ribosome corresponds to a molecule of amino-acyl transfer RNA, and so long as all the required molecules of amino-acyl transfer RNA are available, the ribosome will continue to elongate the polypeptide chain by the mechanism we have described above. How, then, can the synthesis be terminated?

The studies that led to the elucidation of the genetic code (see pp. 444 ff.) showed that three of the possible sixty-four trinucleotides corresponded to no amino-acyl transfer RNA. These codons are UAG, UAA and UGA. Although no amino-acyl transfer RNA is inserted in response to these three, it was soon found that these codons are not simply non-functional. When most synthetic polynucleotides are used as artificial messenger RNAs (pp. 444 f.), the ribosome synthesizes in response to them a polypeptidyl transfer RNA containing amino acids corresponding to the codons in the polynucleotide. However, when the polynucleotide contains, somewhere along its length, a UAG, UAA or UGA codon, the ribosome synthesizes not a polypeptidyl transfer RNA but a free polypeptide. This result suggests that the three codons are used to signify the *termination* of the polypeptide and its *release* from the last transfer RNA molecule to be bound to the A site.

Subsequent work has borne out this conclusion and has given some understanding of the mechanism of termination. When a chain-terminating codon appears opposite the A site of the ribosome, it is recognized not by an amino-acyl transfer RNA but by a specific protein *release factor*. Two such release factors have been isolated from *E. coli*, one recognizing either UAG or UAA, the other recognizing UAA or UGA. The polypeptide held in the P site is thereupon hydrolysed from the transfer RNA to which it has previously been attached (this transfer RNA is that that was specific for the last amino acid to be inserted into the polypeptide chain). It is not known whether the release factors themselves catalyse this hydrolysis, or whether a separate enzyme is needed to achieve it. It is also not known whether all three of the chain-terminating codons are used in natural messenger RNAs as signals for release. In one or two instances it has been proved that UAA is the natural signal.

After the hydrolysis of the completed polypeptidyl transfer RNA and release of the resulting polypeptide, the 70S ribosome remains with a transfer RNA bound to it. It is possible that a separate factor may be required to release the last transfer RNA. The 70S ribosome is detached from the messenger RNA by an unknown mechanism and either dissociates spontaneously into 50S and 30S subunits or, more likely, is dissociated by the action of one of the initiation factors (p. 454).

Since any one of the codons UAG, UAA or UGA acts to cause termination and release of the polypeptide chain, it follows that a mutation in another codon that causes a chain-terminating codon to appear in the messenger RNA will lead to *premature* termination. For example, a mutation that results in a change from AAA to UAA will cause the polypeptide to be terminated at a position at which in the wild-tye protein, a lyrine residue appears. Such chain terminating mutations (often, but rather misleadingly, called *nonsense* mutations) are known and have been extensively studied in some systems, particularly by Brenner and his collaborators in bacteriophage T4). As one might expect, a chain-terminating mutation almost always abolishes activity of the polypeptide in whose gene it occurs; a protein in which a single amino acid is changed may often show some biological activity (see p. 421), but a prematurely truncated protein scarcely ever does (unless the chain-termination mutation occurs close to the end of the gene that codes for the carboxyl terminus, so that only a few amino-acid residues are missing).

## Role of the ribosome

Our description of the mechanism of polypeptide synthesis has paid less attention to the functions of the ribosome than the subject deserves. This neglect of the ribosome is due to our comparative ignorance of the way in which it acts during protein synthesis; descriptions of the process inevitably give the impression that the ribosome is little more than a passive work bench on which the polypeptide is assembled according to the instructions brought by the template messenger RNA.

Although very little is known about the functions of the ribosome, we can do something to correct this impression by referring to *ribosomal mutants*. Many strains of bacteria are known in which one or other of the ribosomal proteins has been altered by mutation. In some of these the mutant protein has been identified, so that it has been possible to correlate the altered behaviour of the bacterial strain with a change in a particular protein. Such studies are still in a preliminary stage, but they make it clear, at any rate, that the ribosome plays an important part in ensuring the specificity of the bonding between codons and their corresponding amino-acyl transfer RNAs. Mutations are known which, by altering the structure of ribosomal proteins, cause occasional mistakes to occur in pairing between codons and anticodons; consequently we may infer that the normal ribosome is responsible for ensuring that this pairing is properly made. However, it is clear that a great deal remains to be discovered about the functions of ribosomal proteins and, even more, of ribosomal RNA, the function of which is almost completely unknown.

# Chapter 27 **The evolution of amino-acid sequences**

The preceding chapters in Part 3 have shown how the sequence of insertion of amino acids is controlled during the biosynthesis of proteins. They have described something of the molecular basis of the genetic process, and especially of mutation and the genetic code. We shall now give an account of a series of experiments which provides both a further instance of the working of the principles described in these chapters and an extremely promising means of studying the events that accompanied the evolution of the species.

These experiments have involved the determination of the amino-acid sequences of proteins that are *genetically related*, and the use of the results in tracing the mutations that accompanied the evolutionary process. We have already described an example of sequence work on genetically related proteins – that on the α protein of tryptophan synthetase (p. 418) – but that was a study of mutations that arose in the laboratory over a time scale of months. If we wish to study evolution we must still look for mutations in amino-acid sequences, but we are now interested in *naturally occurring* mutations with a time scale of hundreds of millions of years.

We shall find it helpful to divide the observations that we are going to describe into three classes: comparisons between the sequences of the same protein drawn from *different* species; comparisons between the sequences of different, but clearly related, proteins from the *same* species; and studies of individual proteins that carry within their own structure an imprint of an evolutionary event that is so clearly discernible that no comparisons with another sequence are needed.

## SEQUENCE COMPARISONS IN THE SAME PROTEIN DRAWN FROM DIFFERENT SPECIES

We shall discuss two examples here: the first is cytochrome *c* (p. 236) and the second fibrinogen (p. 98).

*Cytochrome c* is a member of the electron-transport pathway and is therefore absolutely essential for the survival of a wide range of aerobic organisms. It is found to have an extremely tight tertiary structure in which the polypeptide chain is folded back on itself a great deal. We thus have a protein the structure of which may well be disrupted by even the smallest changes in the nature of its amino-acid side chains and yet which is essential to many organisms. The result is that comparatively few mutations are ever found in cytochrome *c*. No doubt they *occur* as frequently as in any other protein, but with the great sensitivity of the protein to small changes in structure, and the great sensitivity of the organism to deleterious changes in the vital oxidative pathway that would occur as a consequence, few organisms bearing mutations in the gene that specifies cytochrome *c* would survive natural selection.

We find in fact that cytochrome *c* has changed very little in perhaps 2000 million years of evolution. For instance, although the anastral line that gave rise to man diverged from that that gave rise to yeast at least that long ago, the cytochrome *c* of yeast closely resembles that of man. The two proteins are almost identical in size (about 100 amino-acid residues), tertiary structure and, most important, biological activity. If the two cytochromes *c* are used to replace one another in the normal biological tests for electron-transport activity, the results are indistinguishable. This very close similarity is equally true for any pair of cytochromes *c* from the forty or so eucaryotic species that have been studied so far. Even some bacterial cytochromes do not depart too widely from this pattern.

The only amino-acid sequence changes that are permitted are those that do not upset tightly defined requirements. We can see from Table 27.1 that these changes are relatively few. This table shows the amino-acid sequence in eight consecutive positions in the cytochromes *c* obtained from a number of species. We have selected this stretch of sequence because it includes the two cysteine residues that hold the haem ring on to the protein (p. 236). (Although cytochrome *c* is the protein that has been studied in this way the most intensively, there are comparable quantities of information available for other proteins, notably haemoglobin and insulin. If you recall that we said in Ch. 4 that the sequence of

**Table 27.1**  A region of cytochrome *c*

|  | 13 | 14 | 15 | 16 | 17 | 18 | 19 | 20 |
|---|---|---|---|---|---|---|---|---|
| Man | Lys | Cys | Ser | Gln | Cys | His | Thr | Val |
| Rhesus monkey | Lys | Cys | Ser | Gln | Cys | His | Thr | Val |
| Horse | Lys | Cys | Ala | Gln | Cys | His | Thr | Val |
| Pig, cow, sheep | Lys | Cys | Ala | Gln | Cys | His | Thr | Val |
| Dog | Lys | Cys | Ala | Gln | Cys | His | Thr | Val |
| Grey whale | Lys | Cys | Ala | Gln | Cys | His | Thr | Val |
| Rabbit | Lys | Cys | Ala | Gln | Cys | His | Thr | Val |
| Great Austr. kangaroo | Lys | Cys | Ala | Gln | Cys | His | Thr | Val |
| Chicken, turkey | Lys | Cys | Ser | Gln | Cys | His | Thr | Val |
| Penguin | Lys | Cys | Ser | Gln | Cys | His | Thr | Val |
| Pekin duck | Lys | Cys | Ser | Gln | Cys | His | Thr | Val |
| Pigeon | Lys | Cys | Ser | Gln | Cys | His | Thr | Val |
| Snapping turtle | Lys | Cys | Ala | Gln | Cys | His | Thr | Val |
| Rattlesnake | Lys | Cys | Ser | Gln | Cys | His | Thr | Val |
| Bullfrog | Lys | Cys | Ala | Gln | Cys | His | Thr | Cys |
| Tunafish | Lys | Cys | Ala | Gln | Cys | His | Thr | Val |
| Dogfish | Lys | Cys | Ala | Gln | Cys | His | Thr | Val |
| Lamprey | Lys | Cys | Ser | Gln | Cys | His | Thr | Val |
| Fruit fly | Arg | Cys | Ala | Gln | Cys | His | Thr | Val |
| Screw-worm fly | Arg | Cys | Ala | Gln | Cys | His | Thr | Val |
| Silk moth | Arg | Cys | Ala | Gln | Cys | His | Thr | Val |
| Tobacco hornworm moth | Arg | Cys | Ala | Gln | Cys | His | Thr | Val |
| Wheat | Arg | Cys | Ala | Gln | Cys | His | Thr | Val |
| *Neurospora crassa* | Arg | Cys | Ala | Glu | Cys | His | Gly | Glu |
| Bakers' yeast | Arg | Cys | Glu | Leu | Cys | His | Thr | Val |
| *Candida krusei* | Arg | Cys | Ala | Glu | Cys | His | Thr | Ile |

one protein took a matter of man-years to determine, you will gain some idea of the magnitude of the effort that is being put into this kind of work.)

Three types of amino-acid site can be distinguished in Table 27.1. First, there are those in which *no* change has occurred. These are called *totally conservative sites*. These sites will include those most crucially involved in structure and function, but also those for which, by chance, we have yet to come across a species that carries a mutation. In the length of sequence shown in the Table there is, for cytochrome *c*, an unusually large percentage of sites showing no change. The two cysteines are clearly absolutely necessary since they hold on the

**Table 27.2** The amino-acid sequence of wheat-germ cytochrome *c* showing the substitutions that occur in the cytochromes *c* of other organisms

```
                                      Ala
                Lys                   Ile Ala                  Val
           Thr Pro Phe Glu Gln   Ser Val Glu Lys         Thr Thr
           Pro Ala Gly Val Ser Ala  Asp· Ser Lys Asn      Lys Asn Leu
Acetyl-Ala-Ser-Phe-Ser-Glu-Ala-Pro-Pro-Gly-Asn-Pro-Asp-Ala-Gly-Ala-Lys-Ile-Phe-
      −8                           1                              10

Thr                            Ile       Gly            Thr          Thr
Ile Met         Ser Leu        Glu Gly Asn      Leu Pro          Ile
Val Gln Arg     Glu Glu        Gly Cys Glu Lys Asn Gly Lys Gln   Val
Lys-Thr-Lys-Cys-Ala-Gln-Cys-His-Thr-Val-Asp-Ala-Gly-Ala-Gly-His-Lys-Gln-Gly-Pro-
                          20                                         30
      |___HAEM___|

                                     Glu
    Asn                              Val
    Tyr                              Gln                        Glu
    Trp   Ile Ile        His     Ser Val Asp          Thr       Asp
Ala Ser   Phe Tyr Ser    Lys Thr Gln Ala Pro      Phe Ala    Thr Asn
Asn-Leu-His-Gly-Leu-Phe-Gly-Arg-Gln-Ser-Gly-Thr-Thr-Ala-Gly-Tyr-Ser-Tyr-Ser-Ala-
                            40                                     50

                   Asn
    Lys            Lys
    Arg       Val  Gln               Arg
    Gln       Ile  Ala Gln Glu       Met
    Ser   Asn Leu  Asp Asn Pro       Ser Ile          Thr
    Ile Ala Ala Gly Ile Thr   Gly Asp Asp Asn Met Phe Glu      Glu
Ala-Asn-Lys-Asn-Lys-Ala-Val-Glu-Trp-Glu-Glu-Asn-Thr-Leu-Tyr-Asp-Tyr-Leu-Leu-Asn-
                        60                                         70

                                                         Ser
                                                         Ala
                                                         Asp
                                                     Thr Gly
                                        Thr          Lys Thr
                                        Val          Glu Glu
                            Ile    Gly              Asp Lys
                            Ala    Ala     Ile Ser   Ala Asn Glu
Pro-Lys-Lys-Tyr-Ile-Pro-Gly-Thr-Lys-Met-Val-Phe-Pro-Gly-Leu-Lys-Lys-Pro-Gln-Asp-
                        80                                         90

Gln
Lys
Thr
Val
Glu                 Asp       Lys
Gly                 Glu Ser   Asn Glu
Asp                 Gln Thr Ala Cys Ala
Asn Asn Ile    Val Thr Phe Met Leu Ser Lys Cys Ala Lys
Arg-Ala-Asp-Leu-Ile-Ala-Tyr-Leu-Lys-Lys-Ala-Thr-Ser-Ser-
                        100
```

haem ring, on which the function of the protein depends. Histidine 18 is also absolutely necessary, since it provides one of the two non-haem co-ordinating valencies for the iron atom, much as in haemoglobin (Fig. 5.13, p. 105).

The second class is the *functionally conservative* type. We saw on p. 446 that the genetic code is so arranged that changes often leave one with an amino acid that is still within the same functional class. Even if this were not so, we would still expect a large number of *successful* mutations to be functionally conservative because there will then be a lower chance of disrupting the balance of forces than if the change led to an amino acid with radically different properties. There is one clear example of functionally conservative change, that being position 13 (in which the basic amino acids, lysine and arginine, are the only ones to be allowed). Clearly in sites of this type only the general character of the amino acid is important.

The third type of site is *freely variable*. Position 20 is an example in which we see amino acids of several functional classes. These sites presumably have little influence on the structure and function of the molecule.

Table 27.2 summarizes the sequence data for the whole length of a large number of cytochromes *c*. Certain conclusions relating to the influence of individual sites on function can be drawn. One example is that glycine is so often conserved because it has *no* side chain and therefore allows the polypeptide chain to fold back on itself very tightly when crossing over. In these cases any side chain at all would be too large, and we can now see the *absence* of a side chain in glycine to be as much of a positive characteristic for this amino acid as any of the features of the other amino acids that we discussed in Ch. 4.

Another point which concerns the relationship between structure and function can be made about the remarkable length of eleven residues, 70–80, which is the only totally conserved sequence longer than two residues. This must be crucially involved in function to have been protected against evolution for so long and, in fact, a good part of it is in direct contact with the haem ring.

Perhaps the most interesting conclusions that we can draw are those concerning evolutionary change. First of all, it can be said that all the changes in Table 27.1 (and most in the wider summary, Table 27.2) are attributable to single base changes in the genetic code, Table 26.1. Secondly, and perhaps more surprisingly, *the data can be used to draw up an evolutionary tree of the species.* This is done as follows. First, the number of differences between cytochrome *c* of different species can be tabulated (Table 27.3 is an abstract of such a tabulation).

**Table 27.3** Differences in amino-acid sequence between the cytochromes *c* of different species (see text)

|       | yeast | horse | pig |
|-------|-------|-------|-----|
| yeast | —     |       |     |
| horse | 46    | —     |     |
| pig   | 45    | 3     | —   |
| man   | 45    | 12    | 10  |

Thus pig cytochrome *c* differs from human cytochrome *c* at ten sites, yeast cytochrome *c* differs from horse cytochrome *c* in forty-five sites and so on. If we assume that non-lethal mutations occur at more or less constant intervals of time (of the order of millions of years, as we shall see), we can then start to draw up an evolutionary tree in the conventional form:

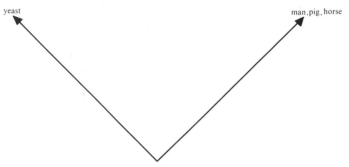

**Fig. 27.1** The first attempt at a phylogenetic tree

where the combined length of the two arrows (about forty-five mutations) is also a measure of time.

We can, further, deal with the group consisting of man, pig and horse.

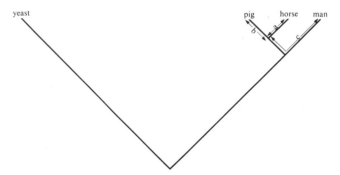

**Fig. 27.2** The second attempt at a phylogenetic tree

We have placed horse and pig on a branch of their own because the table shows them to be closer together than either is to man. We can calculate the lengths of the branches of this part of the tree by using Fig. 27.2 and Table 27.3. Note that $a$ is the distance in mutation (and therefore, we hope, in time) between horse and the common ancestor of horse and pig, $b$ is the distance between this common ancestor and pig, while $c$ is the distance between it and present-day man.

$$a + b = 3 \text{ mutations}$$
$$a + c = 12$$
$$b + c = 10$$

solving for $a$, $b$, $c$ we obtain

$$a = 2\tfrac{1}{2}$$
$$b = \tfrac{1}{2}$$
$$c = 9\tfrac{1}{2}$$

and we can re-draw the tree to scale (Fig. 27.3).

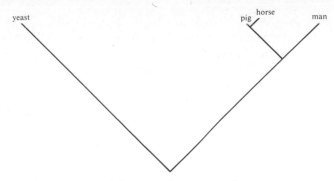

**Fig. 27.3**  The final version of the tree, drawn to scale

By repeating this procedure a complete evolutionary tree can be drawn up for all the sequences for which data are available (Fig. 27.4).

It is found that the shape of this tree is in good agreement with that drawn up on normal taxonomic grounds and that the average interval between successful mutations is about 25 million years. It is worth emphasizing that this tree was obtained almost completely independently of all normal zoological criteria.

This method clearly has promise as a means of solving problems of the inter-relationships between species and is now beginning to be used for this purpose.

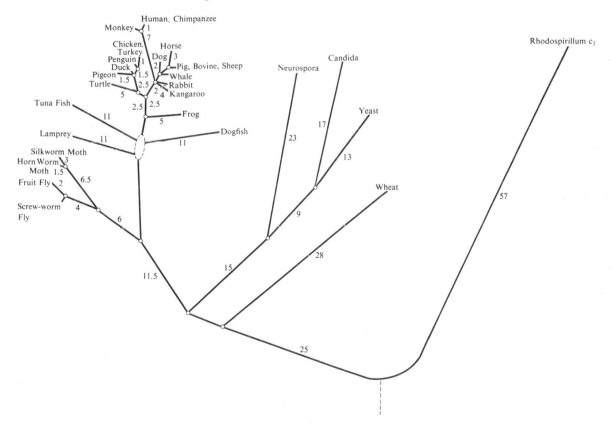

**Fig. 27.4**  The phylogenetic tree, incorporating most of the species for which data exist

(As a curiosity, you might like to use Fig. 27.4 and Table 27.1 to deduce what the amino-acid sequence was of the cytochromes *c* of some of the long-extinct common ancestors. You could even, by using Table 26.1, the genetic code, on the result, deduce something of the nucleotide sequences that specified them.)

A possible criticism of the work that has been done so far on this and other proteins is that it has concentrated too much on the vertebrates at the expense of other groups. (Of the million or so described species of animals, only 50 000 are vertebrates and yet 287 of the 300 known protein sequences are of vertebrate proteins. The entire vegetable kingdom is represented by 17 sequences.) However, there are signs that more attention is now being given to proteins from sources other than the vertebrates.

*Fibrinogen*, as we saw on p. 98, is activated to the blood-clotting protein fibrin as a result of the release of two small peptides by the action of the hydro-

**Table 27.4**  The amino-acid sequences of some fibrinopeptides.

| | Fibrinopeptide A | | | | | | | | | | | | | | | | | | |
|---|---|---|---|---|---|---|---|---|---|---|---|---|---|---|---|---|---|---|---|
| | 1 | 2 | 3 | 4 | 5 | 6 | 7 | 8 | 9 | 10 | 11 | 12 | 13 | 14 | 15 | 16 | 17 | 18 | 19 |
| Human | | | | Ala | Asp | Ser | Gly | Glu | Gly | Asp | Phe | Leu | Ala | Glu | Gly | Gly | Gly | Val | Arg |
| Rhesus monkey* | | | | Ala | Asp | Thr | Gly | Glu | Gly | Asp | Phe | Leu | Ala | Glu | Gly | Gly | Gly | Val | Arg |
| Mandrill | | | | Ala | Asp | Thr | Gly | Asp | Gly | Asp | Phe | Ile | Thr | Glu | Gly | Gly | Gly | Val | Arg |
| Rat | | | Ala | Asp | Thr | Gly | Thr | Thr | Ser | Glu | Phe | Ile | Asp | Glu | Gly | Ala | Gly | Ile | Arg |
| Rabbit | | | | Val | Asp | Pro | Gly | Glu | Ser | Thr | Phe | Ile | Asp | Glu | Gly | Ala | Thr | Gly | Arg |
| Guinea pig | | | | | | | Thr | Asp | Thr | Glu | Phe | Glu | Ala | Ala | Gly | Gly | Gly | Val | Arg |
| Mink | | | | Thr | Asn | Val | Lys | Glu | Ser | Glu | Phe | Ile | Ala | Glu | Gly | Ala | Ala | Gly | Arg |
| Badger | | | | Thr | Asp | Val | Lys | Glu | Ser | Glu | Phe | Ile | Ala | Glu | Gly | Ala | Val | Gly | Arg |
| Dog, fox | | | | Thr | Asn | Ser | Lys | Glu | Gly | Glu | Phe | Ile | Ala | Glu | Gly | Gly | Gly | Gly | Arg |
| Cat | | | | Gly | Asp | Val | Gln | Glu | Gly | Glu | Phe | Ile | Ala | Glu | Gly | Gly | Gly | Val | Arg |
| Grey seal | | | | Thr | Asp | Thr | Lys | Glu | Ser | Asp | Phe | Leu | Ala | Glu | Gly | Gly | Gly | Val | Arg |
| Brown bear | | | | Thr | Asp | Gly | Lys | Glu | Gly | Glu | Phe | Ile | Ala | Glu | Gly | Gly | Gly | Val | Arg |
| Horse, mule variety 1 | | | | | | Thr | Glu | Glu | Gly | Glu | Phe | Leu | His | Glu | Gly | Gly | Gly | Val | Arg |
| Donkey, mule variety 2 | | | | Thr | Lys | Thr | Glu | Glu | Gly | Glu | Phe | Ile | Ser | Glu | Gly | Gly | Gly | Val | Arg |
| Zebra variety 1 | | | | Thr | Lys | Thr | Glu | Glu | Gly | Glu | Phe | Ile | Gly | Glu | Gly | Gly | Gly | Val | Arg |
| Zebra variety 2 | | | | Thr | Lys | Thr | Glu | Glu | Gly | Glu | Phe | Ile | Ser | Glu | Gly | Gly | Ala | Val | Arg |
| Pig, boar | | | Ala | Glu | Val | Gln | Asp | Lys | Gly | Glu | Phe | Leu | Ala | Glu | Gly | Gly | Gly | Val | Arg |
| Llama | | Thr | Asp | Pro | Asp | Ala | Asp | Lys | Gly | Glu | Phe | Leu | Ala | Glu | Gly | Gly | Gly | Val | Arg |
| Vicuna | | Thr | Asp | Pro | Asp | Ala | Asp | Lys | Gly | Glu | Phe | Leu | Ala | Glu | Gly | Gly | Gly | Val | Arg |
| Camel | | Thr | Asp | Pro | Asp | Ala | Asp | Glu | Gly | Glu | Phe | Leu | Ala | Glu | Gly | Gly | Gly | Val | Arg |
| Cow | Glu | Asp | Gly | Ser | Asp | Pro | Pro | Ser | Gly | Asp | Phe | Leu | Thr | Glu | Gly | Gly | Gly | Val | Arg |
| European bison | Glu | Asp | Gly | Ser | Asp | Pro | Ala | Ser | Gly | Asp | Phe | Leu | Ala | Glu | Gly | Gly | Gly | Val | Arg |
| Cape buffalo | | | | | Glu | Asp | Gly | Ser | Gly | Glu | Phe | Leu | Ala | Glu | Gly | Gly | Gly | Val | Arg |
| Water buffalo | Glu | Asp | Gly | Ser | Asp | Ala | Val | Gly | Gly | Glu | Phe | Leu | Ala | Glu | Gly | Gly | Gly | Val | Arg |
| Sheep, goat | Ala | Asp | Asp | Ser | Asp | Pro | Val | Gly | Gly | Glu | Phe | Leu | Ala | Glu | Gly | Gly | Gly | Val | Arg |
| Pronghorn | Ala | Asp | Gly | Ser | Asp | Pro | Val | Gly | Gly | Glu | Ser | Leu | Pro | Asp | Gly | Ala | Thr | Gly | Arg |
| Reindeer* | Ala | Asp | Gly | Ser | Asp | Pro | Ala | Gly | Gly | Glu | Phe | Leu | Ala | Glu | Gly | Gly | Gly | Val | Arg |
| Mule deer | | | | Ser | Asp | Pro | Ala | Gly | Gly | Glu | Phe | Leu | Ala | Glu | Gly | Gly | Gly | Val | Arg |
| Muntjak | Ala | Asp | Gly | Ser | Asp | Pro | Ala | Ser | Gly | Glu | Phe | Leu | Thr | Glu | Gly | Gly | Gly | Val | Arg |
| Sika deer | Ala | Asp | Gly | Ser | Asp | Pro | Ala | Ser | Ser | Glu | Phe | Leu | Ala | Glu | Gly | Gly | Gly | Val | Arg |
| Red deer | Ala | Asp | Gly | Ser | Asp | Pro | Ala | Ser | Ser | Asp | Phe | Leu | Ala | Glu | Gly | Gly | Gly | Val | Arg |
| American elk | Ala | Asp | Gly | Ser | Asp | Pro | Ala | Ser | Ser | Asp | Phe | Leu | Ala | Glu | Gly | Gly | Gly | Val | Arg |
| Kangaroo | | | | | Thr | Lys | Asp | Glu | Gly | Thr | Phe | Ile | Ala | Glu | Gly | Gly | Gly | Val | Arg |
| Wombat | | | | | Thr | Lys | Thr | Glu | Gly | Ser | Phe | Leu | Ala | Glu | Gly | Gly | Gly | Val | Arg |
| Lizard | | | | | | Glu | Asp | Thr | Gly | Thr | Phe | Glu | Glu | Gly | Gly | Gly | His | Val | Arg |

* *Cynomolgus macaque*, green monkey and baboon are identical to Rhesus monkey. European elk is identical to reindeer.

lytic enzyme thrombin. Thus the function of these peptides is, very largely, to be thrown away – once they are released they are of no further interest. Here, then, we have an example of a system on which there are very few structural constraints. We mentioned on p. 99 that the fibrinopeptides are acidic, but all that has to remain absolutely constant is the residue that provides the cleavage point (arginine) – the rest may vary very widely. We may therefore expect to observe a much greater number of successful mutations here, and Table 27.4 shows that this is indeed what happens. In contrast to cytochrome *c* there is almost total variability at all sites. (Other arrangements of this table are possible if we assume that amino acids are occasionally *deleted* instead of being changed to another residue.)

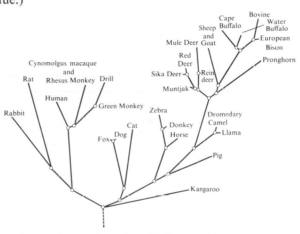

**Fig. 27.5**  A phylogenetic tree based on fibrinopeptide sequences

We cannot expect to be able to use fibrinopeptides to investigate evolution over very great stretches of evolutionary time. The interval between successful mutations can be calculated to be about $6 \times 10^5$ years in comparison with the 25 million years required in the case of cytochrome *c*. But, if the time scale over which we may expect to use the fibrinopeptides has lessened, there is a compensating advantage that the *resolving power* is greater. We can construct a tree from the fibrinopeptide data in a similar way to that for cytochrome *c*. Compare the much clearer picture of the evolution of the mammals that we obtain from Fig. 27.5 (the tree obtained from the fibrinopeptide data) with the one that we obtain from the crowded corner of Fig. 27.4 (the cytochrome *c* tree) devoted to the same group of animals.

## SEQUENCE COMPARISONS BETWEEN DIFFERENT PROTEINS FROM ONE SPECIES

Comparisons of this type would at first seem not to have much to do with evolution, since we are dealing with only one species. However, if we consider a typical example, the relevance to evolution will become clearer.

We saw on p. 112 that the proteolytic enzymes trypsin and chymotrypsin catalyse the same general reaction, but with different specificities. We saw also that the same mechanism appears to hold for the catalysis. When the amino-acid sequences of the two proteases from the pig were determined, it was found that

50 per cent of the residues were strictly conserved in the sense of Table 27.1 and a further 15 per cent were functionally conserved.

This was a much higher degree of homology than chance would allow. Convergent evolution, which could produce such a pattern, is for various reasons highly unlikely.* One can therefore conclude that these two sequences, like those that we considered above, had diverged from a common ancestral sequence. If we accept the rule one gene–one polypeptide chain, how could this ancestral sequence diverge in two directions at once? The answer is that the gene must have *doubled* during a reproductive event somewhere in the early history of the mammals (gene doubling is an event that is well known from other results and is, for example, easily demonstrated by genetic mapping (p. 410)). The two genes that resulted from this doubling were then free to diverge (see Fig. 27.6). Doubling can even be repeated many times, giving a large number of genes, initially identical, free to diverge in many different directions.

This mechanism is an even more powerful means of improving the genetic stock of the species than it might at first seem. If a vital function (proteolysis in this case) is carried by a single gene, any mutation that seriously impairs the function will kill the organism that carries it. This constraint naturally limits the changes that it is possible to make. When gene duplication occurs, however, we have a situation much as if we had changed from a single- to a multi-engined aeroplane. We can now tinker with one engine while the others keep the plane in the air. That is, a gene could now mutate even so as to inactivate the protein completely without killing the organism carrying the modified gene. The other gene (or genes if the doubling has been repeated) now carries the load while the mutated gene, removed entirely from the pressures of natural selection (the protein is totally useless while in this state so it cannot get any more useless), can mutate very rapidly. If, as has been observed in the laboratory for tryptophan synthetase α, a mutation occurs later on that restores activity, it may well be that the large number of others that occurred in the period intervening between the inactivating mutation and the re-activating mutation will be found to have conferred a great advantage on the protein (e.g. a new range of specificity). (There are, however, theoretical reasons for supposing that the reactivating mutation would have to follow the inactivating mutation fairly quickly, or the gene might be lost by the ancestral line.)

Gene doubling therefore allows improvement and innovation to occur at a much greater rate than would be possible without it. This is an advantage that is additional to and quite distinct from those that are conferred simply by the possession of two genes where we had one before – an example of this other type of benefit is given by vertebrate haemoglobin, in which the remarkably subtle responses to oxygenation and deoxygenation depend on the molecule's having two types of subunit, which are specified by duplicated genes.

* In *convergent evolution* the structures start off very different, but because there is only one sequence that will produce a completely satisfactory protein, they mutate toward that structure and thus toward each other. The reason that is easiest to explain for believing that it is unlikely to occur is as follows: There is clear evidence that the versatility of protein structure is such that there is not one sequence that alone will produce the desired properties but many. For example, the proteolytic enzymes from bacteria behave in very similar ways to those that we are discussing, but many of them have totally different sequences from those of the mammalian enzymes. Thus there would be no selective pressure that could cause sequences to converge to the extent that is observed.

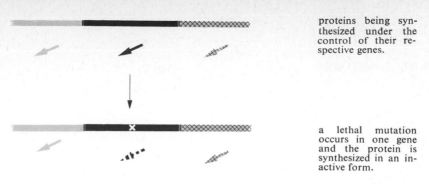

proteins being synthesized under the control of their respective genes.

a lethal mutation occurs in one gene and the protein is synthesized in an inactive form.

**Fig. 27.6** (a) A mutation that is lethal in the absence of gene duplication

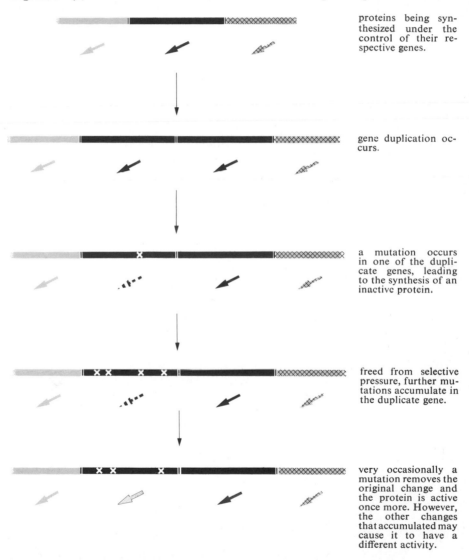

proteins being synthesized under the control of their respective genes.

gene duplication occurs.

a mutation occurs in one of the duplicate genes, leading to the synthesis of an inactive protein.

freed from selective pressure, further mutations accumulate in the duplicate gene.

very occasionally a mutation removes the original change and the protein is active once more. However, the other changes that accumulated may cause it to have a different activity.

**Fig. 27.6** (b) Gene duplication followed by the same mutation in one of the genes that occurred in (a). As a result of the duplication the mutation is no longer lethal. The figure shows how the mutated gene may subsequently regain activity, although possibly of a different sort. The broad arrows symbolize the action of each gene in directing the synthesis of a protein

We have selected one example of the many single sequences that of themselves give a clue to an evolutionary event. The protein is haptoglobin, which is found in the blood stream and has the task of complexing with any haemoglobin that becomes detached from the erythrocytes and finds itself in free solution. (It has a very strong binding constant for this molecule, about $10^{-11}$ M.) Vertebrate haptoglobin has, usually, a molecular weight of 9 000 and this is a useful addition to the molecular weight of the haemoglobin and helps keep it from passing through the filtration system of the kidney and being lost (with its vital complement of iron) from the body.

There are two allelic forms of this protein in human beings and one is almost

Val–Asn–Asp–Ser–Gly–Asn–Asp–Val–Thr–Asp–Ile–**Ala–Asp–Asp–Gly–**
10

**–Gln–**Pro–Pro–Pro–Lys–Cys–Ile–Ala–His–Gly–Tyr–Val–Glu–His–Ser–
20 30

–Val–Arg–Tyr–Gln–Cys–Lys–Asn–Tyr–Tyr–Lys–Leu–Arg–Thr–Gln–Gly–
40

–Asp–Gly–Val–Tyr–Thr–Leu–Asn–Asn–Lys–Lys–Gln–Trp–Ile–Asn–Lys–
50 60

–Ala–Val–Gly–Asp–Lys–Leu–Pro–Glu–Cys–Glu–**Ala–Asp–Asp–Gly–Gln–**
70

–Pro–Pro–Pro–Lys–Cys–Ile–Ala–His–Gly–Tyr–Val–Glu–His–Ser–Val–
80 90

–Arg–Tyr–Gln–Cys–Lys–Asn–Tyr–Tyr–Lys–Leu–Arg–Thr–Gln–Gly–Asp–
100

–Gly–Val–Tyr–Thr–Leu–Asn–Asn–Glu–Lys–Gln–Trp–Ile–Asn–Lys–Ala–
110 120

–Val–Gly–Asp–Lys–Leu–Pro–Glu–Cys–Glu–Ala–Val–Cys–Gly–Lys–Pro–
130

–Lys–Asn–Pro–Ala–Asn–Pro–Val–Gln
140

**Fig. 27.7** The partially duplicated sequence of haptoglobin. The residues in **bold** type are intended to help you find the start of the duplicated sequences.

twice the normal size (the molecular weight is about 15 000 as opposed to 9000 for the normal form). This doubling in size does not appear to have affected its binding properties for haemoglobin, and it is of course even better at preventing loss from the kidney. Examination of the amino-acid sequence of the large form shows that it mainly consists of the same sequence repeated twice (Fig. 27.7). It is concluded that the gene for the small protein doubled, but that the 'terminate' and 'initiate' signals that should have separated the new pair were deleted either at the time of the doubling or subsequently. So few differences exist between the two halves of the enlarged molecule that it is concluded that the doubling has taken place extremely recently. Anthropological data suggest that the frequency of this doubled gene is on the increase in the human population, and it is possible that this new gene is conferring some advantage in natural selection on those that have it over those that do not.

Many other examples could have been quoted of these three types of sequence study. We hope, however, that the ones that we have selected will be sufficient to illustrate both some of the features of molecular genetics and also the powerful nature of this approach to the study of evolution.

# Part 4  **Regulation**

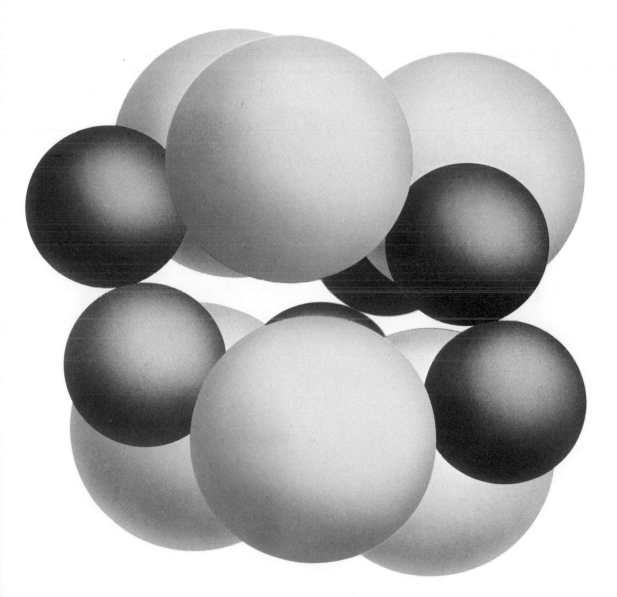

# Chapter 28  **Control of protein synthesis**

## INTRODUCTION. THE NEED FOR CONTROL MECHANISMS

In this book we have described a large number of biochemical pathways. We have outlined the degradation and the synthesis of the major classes of metabolites and have examined the chief biochemical capabilities of living cells. A second's reflection will convince you that not all of these capabilities can be expressed simultaneously. For example, no cell will be at the same time degrading and synthesizing all twenty amino acids at the maximum rate of which it is potentially capable. What has obviously been missing from earlier chapters is any consideration of the *regulation* of the pathways we have discussed. By what means can organisms ensure that the flow of metabolites along each of the biochemical pathways is exactly that amount that is consistent with the most economical organization?

The answer is that there are two means of control. Organisms can regulate both the *rate of synthesis* of enzymes (and, no doubt, of other proteins, although we have much less knowledge of their regulation) and their *catalytic activity*. The use of the two mechanisms in concert enables organisms to achieve great flexibility in their adaptation to changes in conditions, and also great economy of energy and of materials.

We can best illustrate the importance of these mechanisms by referring to a culture of bacteria – *Escherichia coli* is once again the best studied organism in this regard – growing in minimal medium. In this medium the bacteria make all their amino acids, purines, pyrimidines, co-factors, etc. from inorganic salts and a single carbon source such as glucose. Let us concentrate for a moment on the synthesis of tryptophan. Tryptophan is made from chorismate by the pathway that we described on p. 375, a pathway for the operation of which five polypeptides are essential. These five polypeptides are synthesized under the direction of the five genes of the tryptophan cluster (p. 414); plainly when the organisms are growing in minimal medium all five of these genes must be expressed. Suppose now that we add tryptophan to the culture. With an external source of tryptophan the bacteria have no need to make tryptophan for themselves, since they are capable of taking it up from the medium. If, in the presence of tryptophan, they continue to synthesize the five polypeptides required for the pathway, they will be making superfluous protein, and we saw in Ch. 26 that the synthesis of protein is energetically a very costly process. It would therefore be advantageous to the economy of the organisms to have a mechanism by which the synthesis of these polypeptides in particular (but not the synthesis of others) can be switched off, or (to use the technical term) *repressed*. What is true of tryptophan is true of amino acids generally and also of pyrimidines and of purines and of co-factors. All of these are compounds that the organisms are capable of making for themselves, but which might be encountered in the medium; when they *are* available in the medium it is advantageous for the organisms to be able to repress the synthesis of the relevant enzymes.

We can use the same example to illustrate the second regulatory mechanism. If a culture of *E. coli* has been growing for many generations in the absence of tryptophan, so that it has been obliged to synthesize tryptophan at a high rate to ensure rapid growth, it will contain the relevant synthetic enzymes at a concentration that provides for this high rate of synthesis. When tryptophan is added to the medium, even if the manufacture of further quantities of the enzymes is repressed instantaneously, the enzymes that are already present would be expected to continue to be active. Unless they are rapidly denatured or degraded (and there is evidence that most enzymes are stable for reasonable periods of time) they will continue to synthesize tryptophan despite its presence in the medium. It is true that as the bacteria grow the enzymes will become diluted out by the accumulation of other proteins, but it may take several generations of growth before the dilution is sufficient to prevent further synthesis of tryptophan. This synthesis, however, is costly in terms of ATP and raw materials. It would therefore be advantageous for organisms to possess a second mechanism for ensuring that the amino acid actually *inhibited the activity* of the existing biosynthetic enzymes. Again, what is true of tryptophan is true of many other amino acids, pyrimidines, purines and co-factors.

Both such mechanisms exist, and both are of great importance to the economy of living systems. The first mechanism – alteration of the rate of synthesis of particular enzymes – is most conspicuous in bacteria and is of less importance in higher organisms except during development. We shall spend most of the remainder of this chapter in discussing it. The second mechanism – alteration of the activity of particular enzymes – appears to occur universally. We shall defer an account of it to the next chapter.

## INDUCTION AND REPRESSION

We have already referred to the way in which, in bacteria, the presence of a compound that is the end-product of a biosynthetic pathway represses the synthesis of the biosynthetic enzymes. There are other compounds that have exactly the opposite effect: their presence in the medium provokes (or *induces*, to use the technical term) the synthesis of enzymes that are otherwise present only at negligible concentration. Most inducing compounds are carbon sources other than the sugars (glucose and fructose) of the common glycolytic pathway; the enzymes whose synthesis they induce are those responsible for their degradation. Thus lactose, for example, induces the synthesis in *E. coli* of the enzyme β-galactosidase, which hydrolyses lactose to glucose and galactose (see p. 338); similarly glycerol induces the synthesis of the enzyme glycerokinase, which phosphorylates glycerol to α-glycerophosphate, and also of the enzyme α-glycerophosphate dehydrogenase, which produces dihydroxyacetone phosphate (see p. 268).

Induction and repression have certain features in common, even though their effects are opposite. As the last example will have shown, a single compound can induce the synthesis of more than one enzyme: the two enzymes glycerokinase and α-glycerophosphate dehydrogenase are both needed for the degradation of glycerol, and the synthesis of both is induced by glycerol. Similarly the synthesis of all five of the polypeptides that are specific to the synthesis of tryptophan is repressed by tryptophan. (Notice that many other enzymes are needed for the degradation of glycerol to carbon dioxide; these are, however, not *specific* for glycerol, but are needed equally for the degradation of glucose and many other compounds, and they are not induced by glycerol. Similarly many other enzymes are needed for the synthesis of tryptophan, but these are not specific for tryptophan but are needed for the synthesis of phenylalanine and tyrosine, and they are not repressed by tryptophan.) We shall see, moreover, that the mechanism of induction has something in common with that of repression.

The reason why more is known about induction and repression in *E. coli* than in any other organism is this. The study of these phenomena has involved not only biochemical but also genetic techniques, and the very complete understanding that we have of some aspects of induction and repression would never have been achieved without these techniques. What we said in Ch. 24 will have made it clear that our knowledge of genetics, and our ability to perform genetic manipulations, is more advanced for *E. coli* than for any other organism. In what follows, therefore, we shall be referring largely to work with *E. coli*; towards the end of the chapter we shall see how far the concepts that have been derived from this work are applicable to other organisms. Even with *E. coli* we shall refer more to one particular system – that needed for the degradation of lactose – than to any other, since that has been investigated in the greatest detail.

## The *lac* system

When cultures of *E. coli* are growing in the absence of lactose or other β-galactosides, they synthesize very small (barely measurable) quantities of β-galactosidase, an enzyme that hydrolyses β-galactosides. When a β-galactoside is added to a culture, the rate of synthesis of the enzyme rises enormously: in the presence of a high concentration of lactose the rate of synthesis is several hundred-fold greater than in the absence of lactose. Many β-galactosides have this effect of *inducing* the synthesis of β-galactosidase, including some chemically synthesized compounds that cannot themselves be hydrolysed by β-galactosidase. It is possible to show that the β-galactosidase made on induction is synthesized *de novo* from amino acids, rather than formed by modification of a pre-existing macromolecule. Consequently this induction of β-galactosidase synthesis in the presence of β-galactosides, and the virtual absence of the enzyme from cultures grown in the absence of β-galactosides, may serve as an example of specific regulation of the *synthesis* of a protein.

Since we mentioned earlier that genetic studies had been of great importance in elucidating the mechanism of this regulation, you will not be surprised to learn that many strains of *E. coli* are known that carry *mutations* in the genetic apparatus concerned with lactose metabolism. When mutants are isolated that cannot degrade lactose, they are found to fall predominantly into two classes. One class lack β-galactosidase. The other possess β-galactosidase, but lack a protein that is essential for the transport of β-galactosides into the cell (cf. p. 200 f.) – *galactoside permease*. Strains of the first class carry mutations that map in a small region of the genetic material that we may consider as the gene for β-galactosidase (see p. 414); this gene is known as *lac Z*. The mutations in strains of the second class map extremely close to but distinct from those in strains of the first class, in a region that we may consider as the gene for galactoside permease; this gene is known as *lac Y*. Careful mapping shows that some *Z* mutations are so close to some *Y* mutations that the genes must be contiguous.

A third kind of mutant is known, which is (unlike the other two) capable of degrading lactose, but incapable of *controlling* the rate of synthesis of β-galactosidase. Such mutants behave as if they were permanently induced. Mutations of this class, too, can be mapped, and they are found to lie close to *lac Z* on the opposite side from *lac Y*, in a region which, because it seems to be involved with the induction mechanism, is called *lac I*. Such $I^-$ mutants, which synthesize β-galactosidase at a rapid rate all the time irrespective of the presence of an inducer, are known as *constitutive*.

Further study of constitutive mutants shows that they synthesize, even in the absence of inducer, not only β-galactosidase but also galactoside permease. Study of wild-type cells also shows that in the absence of inducer both β-galactosidase and galactoside permease are synthesized at a very low rate; in the presence of a high concentration of inducer both are synthesized at a very high rate; in the presence of a small concentration of inducer both are synthesized at an intermediate rate. It is plain, then, that the synthesis of the two proteins is regulated together, or, to use the technical term, is *co-ordinate*. How is this regulation achieved?

An important experiment that helps to answer this question depends on the fact (which we mentioned on p. 423) that it is possible to prepare partially diploid strains of *E. coli*, which carry in addition to the normal complement of chromosomal genes some additional genes on an *episome*. We can therefore take a strain that has the genotype $I^- Z^- Y^+$ (one that synthesizes permease constitutively because it is $I^-$, but no β-galactosidase because it is $Z^-$) and insert into it an episome having the genotype $I^+ Z^+ Y^+$. In such a strain we can ask whether the β-galactosidase synthesized under the direction of the $Z^+$ gene on the episome is made constitutively or inducibly. The answer is that the synthesis is inducible. Similarly we can put into a strain that has the genotype $I^- Z^+$ an episome having the genotype $I^+ Z^-$. Again the diploid synthesizes β-galactosidase inducibly. These experiments prove that $I^+$ *is dominant to* $I^-$, i.e. in a cell that has both $I^+$ and $I^-$ it is the $I^+$ allele that is expressed.

We saw on p. 424 that the appropriate conclusion from such a result is that the dominant allele is synthesizing a biochemically active macromolecule, and the recessive allele an inactive one. We may conclude here, therefore, that the *lac I⁺* allele is synthesizing an active compound which is responsible for *maintaining inducibility*. Consequently the $I^-$ mutant is constitutive for want of this active compound. Evidently in the $I^+$ (wild-type) strain, the active compound prevents synthesis of β-galactosidase and permease unless an inducer is present. We may therefore think of this compound as being an active *repressor* of the synthesis of the two *lac* proteins. Moreover, since an episomal $I^+$ gene can cause repression of synthesis of *lac* proteins that is directed by chromosomal genes, the repressor must be able to diffuse through the cytoplasm to act at a distance.

Before we discuss the way in which the repressor, the product of the *lac I* gene, acts, we shall pause to examine the possible means by which bacteria might be able to effect specific control of protein synthesis. We saw in Ch. 26 that the ribosome can be regarded as the 'slave' of whatever messenger RNA molecule it finds itself attached to; we might conclude, therefore, that the rate of synthesis of any given protein is proportional to the concentration of the relevant messenger RNA, which would in turn depend on the rate of its synthesis and the rate of its degradation. There is also a further possibility for control that we ought to consider – that the ribosome might discriminate among different messenger RNA molecules and translate some at a greater rate than others.

We have, then, three possible points at which control mechanisms might operate in bacteria. First, the rate of synthesis of a particular species of messenger RNA will affect the rate of synthesis of the corresponding protein. Secondly, there is a possibility that the rate of degradation might be different for different species of messenger RNA. Thirdly, there is a possibility that the rate of translation might be different for different species of messenger RNA. In practice there is good evidence in bacteria for control of the first stage, and not much

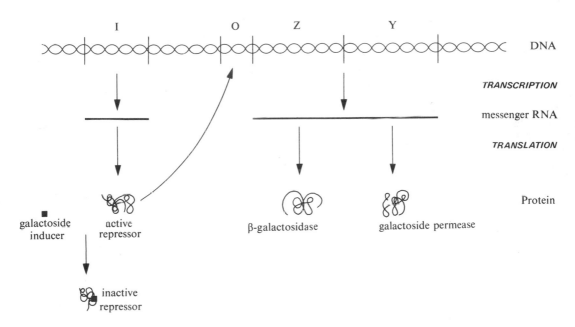

**Fig. 28.1**   Control of the expression of the *lac* operon according to Jacob and Monod. The *I* gene is transcribed into messenger RNA, and this is translated to give an active repressor which, by binding to the operator, prevents the transcription of the *Z* and *Y* genes. β-Galactosides bind to the repressor and reversibly inhibit its binding to the operator; the *Z* and *Y* genes are then transcribed into polycistronic messenger RNA and this is translated to give β-galactosidase and galactoside permease

evidence for control at either of the other two. We shall see, in particular, that in the *lac* system in *E. coli* (which, as we have mentioned, has been studied more than any other) control is largely if not entirely effected by altering the rate of synthesis of messenger RNA.

We must now return to the question: How does the *lac* repressor act to prevent the synthesis of β-galactosidase and galactoside permease? And how does the addition of an inducing galactoside permit synthesis of these proteins? Our understanding of this system derives from a brilliantly ingenious hypothesis, proposed by Jacob and Monod in 1961 and supported by experimental work both in their laboratory and elsewhere.

Jacob and Monod suggested that the repressor's action in preventing synthesis of the two proteins might be due to its binding to the DNA close to, and on one side of, the Z and Y genes. To this hypothetical DNA binding site they gave the name *operator* (O). They proposed that the binding of the repressor to the operator prevented the transcription of the Z and Y genes into messenger RNA, and consequently the synthesis of β-galactosidase and galactoside permease. They suggested that when galactosides that could induce were added to the cells, they bound to the repressor in such a way as to prevent it from binding to the operator; in these circumstances the operator would be free, transcription of the Z and Y genes would proceed, and consequently β-galactosidase and galactoside permease would be synthesized. Finally they explained the fact that regulation of the two proteins is co-ordinate by suggesting that transcription of the Z and Y genes produces a *single* molecule of messenger RNA, corresponding in size to the *two* genes. Such a molecule of messenger RNA, the single product of transcription of more than one gene, is called *polygenic* or *polycistronic*.* The set of contiguous genes that are transcribed into a polycistronic messenger RNA, together with their operator gene, is known as an *operon* (see Fig. 28.1).

These suggestions of Jacob and Monod allow one to make several predictions, all of which have been confirmed experimentally. In the first place, if the operator – the site to which the repressor can bind – is DNA, it must, like all DNA, be subject to mutation. A mutant operator might be unable to bind to the repressor, and in a mutant cell carrying such an operator repression would no longer occur, or would at any rate be less complete than in a wild-type cell. Such mutants, called *operator-constitutive* mutants, have been found; they have the properties expected of them by the hypothesis of Jacob and Monod, and the site of mutation is found to lie between *lac I* and Z. Again, the *lac I* gene, which directs the synthesis of the repressor, is also subject to mutation. Since the repressor has, in the hypothesis of Jacob and Monod, to have two binding sites, one for the operator and one for the inducing galactoside, one can predict mutations in the I gene of two kinds. One kind of mutant will have a repressor that cannot bind to the operator, so that this mutant too will be constitutive: it will, in fact, be the $I^-$ constitutive mutant to which we have already referred. The other kind of mutant will have a repressor that continues to bind to the operator, but has lost its binding site for galactosides: one would predict that such an organism would be incapable of being induced. Mutants of this kind, too, are

---

* The definition of *cistron* is rather technical. You can regard it as equivalent to 'gene' in the sense that we introduced on p. 414 and have used since then.

known; they cannot be induced to synthesize β-galactosidase or galactoside permease, and the site of mutation is found to lie in the *I* gene. They are called *super-repressed* mutants.

In addition to these genetic confirmations of the hypothesis of Jacob and Monod, there is also biochemical evidence in support of the hypothesis. It is possible by DNA–RNA hybridization (p. 436) to show that constitutive mutants and induced wild-type cells contain much more *lac* messenger RNA than uninduced wild-type cells. Moreover this messenger RNA is larger in molecular weight than would be expected if it were the transcript of just a single gene, but of the size to be expected if it were polycistronic.

Thus Jacob and Monod's model pictures a *lac* operon consisting of two* genes,

**Fig. 28.2**  A modern view of the *lac* region. The lengths of the genes are drawn approximately to scale but the promoter and operator regions are in reality even smaller than shown here

*Z* and *Y*, which code for β-galactosidase and galactoside permease, plus an operator region. Close to but outside the operon is the *I* gene; the order of the genes is *I . . . O–Z–Y*. The *I* gene codes for a repressor which, in the absence of inducing galactosides, binds to the operator and prevents transcription of *Z* and *Y*. In the presence of an inducer the repressor is prevented from binding to the operator, so that transcription of *Z* and *Y* proceeds unimpeded and a polycistronic messenger RNA is synthesized; translation of this messenger RNA yields β-galactosidase and galactoside permease.


This summary of the hypothesis of Jacob and Monod allows us to focus attention on some further questions. First, at what point in the *lac* region of DNA does transcription by the RNA polymerase begin? Secondly, what is the chemical nature of the repressor? Thirdly, by what mechanism does the binding of the repressor prevent transcription?

The answers to the first two of these questions have come from experiments that have been carried out since the formulation of the original hypothesis. Transcription of the *Z* and *Y* genes is initiated at a region that lies between the *I* gene and the operator; this region is called the promoter (*P*). Mutations in *P* result in a reduction of the rate of synthesis of both β-galactosidase and galactoside permease, but this slow synthesis is still subject to induction and repression in the normal way. We may think of the promoter as being the site on the DNA at which RNA polymerase binds to initiate transcription (see p. 436).

The repressor turns out to be a protein, which can be isolated (though in

---

* A third gene, *lac A*, which lies contiguous to *Y* on the side away from *Z*, codes for an enzyme called thiogalactoside transacetylase. The synthesis of this enzyme is co-ordinate with that of β-galactosidase and galactoside permease, and *A* evidently belongs to the *lac* operon. However, the physiological function of thiogalactoside transacetylase is unknown. Mutants lacking the enzyme metabolize lactose normally.

extremely low yield, corresponding to perhaps twenty molecules per cell) from wild-type cells. In the test tube this protein binds with exceptionally high affinity (see p. 14) to *lac* DNA provided that it has an intact operator, and with much lower affinity to *lac* DNA isolated from cells with operator-constitutive mutations. Binding of the wild-type repressor protein to wild-type *lac* DNA is prevented by the presence of inducing galactosides, but if repressor is isolated from super-repressed mutants the binding is scarcely affected by inducers.

The answer to the third question – How does the binding of the repressor to the operator prevent transcription of the operon ? – is not yet clear. One possibility is this. After the RNA polymerase has bound to the promoter, it has to traverse the operator before beginning transcription of the *Z* and *Y* genes. The binding of the repressor protein to the operator may well impede the progress of RNA polymerase into the *Z* gene. Alternatively, the binding of the repressor protein to the operator may prevent the binding of RNA polymerase at the neighbouring promoter site. There is some experimental evidence that favours each of these ideas, and it remains to be determined which is correct.

At all events, the important feature of the *lac* system that was originally proposed by Jacob and Monod and abundantly confirmed by later work is that it is subject to *negative* control. Whatever the precise mechanism of its action, the repressor *prevents* synthesis of β-galactosidase and galactoside permease until the repression is relieved by an inducer. This negative control was so well understood and so widely accepted as accounting for regulation of the *lac* operon, that it was most unexpected when an additional *positive* control mechanism for the operon was subsequently discovered. Before we can describe this latter mechanism, we must first refer to a phenomenon that occurs quite widely among bacteria but which we have not yet mentioned.


If a culture of *E. coli* is grown in minimal medium, with glycerol as carbon source, the addition of a high concentration of a non-metabolizable β-galactoside inducer provokes the synthesis of β-galactosidase and galactoside permease at a very high rate. If the bacteria are grown in glucose-minimal instead of in glycerol-minimal medium, then the rate at which the two proteins are synthesized is much lower, even when the bacteria are fully induced. If the sodium salt of gluconic acid (p. 293) is provided as a carbon source in addition to glucose, the rate of synthesis falls still further. This repression cannot be overcome by increasing the concentration of inducer. Thus the rate of synthesis of the *lac* proteins depends on the carbon source. This effect of the carbon source is observed not only with the *lac* proteins but with many other systems, in particular those concerned with the degradation of carbohydrates other than glucose or fructose (e.g. maltose, galactose and arabinose). It was once believed that the synthesis of enzymes necessary for degrading these compounds was repressed by a catabolic product of glucose, and the term *catabolite repression* was introduced to describe the phenomenon.

The molecular basis of this effect of carbon source was partly clarified when Sutherland discovered that *E. coli* grown in glucose contained much lower concentrations of the unusual adenine nucleotide 3′,5′-cyclic AMP (see p. 144) than those grown in glycerol. We shall have occasion to refer to 3′,5′-cyclic AMP at some length in the next chapter; here we shall mention merely that the nucleotide has an essential role in mediating the effect of some hormones. We do not know by what means the concentration of 3',5'-cyclic AMP is regulated by the carbon source on which the bacteria are growing. We do know, however, that it is by changes in the concentration of the nucleotide that the rate of synthesis of the *lac* proteins is altered in accordance with the carbon source.

The way in which 3',5'-cyclic AMP affects the synthesis of the *lac* proteins is as follows. *E. coli* contains a specific protein that is capable of binding to 3',5'-cyclic AMP. This protein, with its attached 3',5'-cyclic AMP, has a high affinity for the *lac* promoter region. In some way that is not understood, the binding of the protein-cyclic AMP complex to the promoter is essential for the binding of RNA polymerase. Consequently a high concentration of 3',5'-cyclic AMP in the cell favours the binding of RNA polymerase to the *lac* promoter and hence the transcription of the *lac* operon. If the concentration of 3',5'-cyclic AMP in the cell is insufficient, the binding of RNA polymerase to the promoter is prevented and the *lac* genes are not transcribed. We can thus speak of cyclic AMP and its specific binding protein as effecting a *positive* control of expression of the *lac* genes. (By this we mean that 3',5'-cyclic AMP is a compound *without* which expression is impossible.) Such positive control is independent of, and complementary to, the negative control exerted by the *lac* repressor.

As 3',5'-cyclic AMP is not essential for the expression of every bacterial gene, we must presume that it is the particular structure (that is, base sequence) of the *lac* promoter region that necessitates the presence of 3',5'-cyclic AMP for the expression of the *lac* genes. There is reason to believe that other systems that are subject to catabolite repression (see above) also have promoters that require 3',5'-cyclic AMP and its binding protein if they are to allow binding of RNA polymerase. Promoters of those systems that are unaffected by catabolite repression are presumably indifferent to 3',5'-cyclic AMP.

Apart from their sensitivity or insensitivity to 3',5'-cyclic AMP, different promoters must, we may reasonably guess, differ in their affinity for RNA polymerase. We may contrast in this respect the promoter for the *lac* proteins, β-galactosidase and galactoside permease, with the promoter for the repressor synthesized under the direction of the *lac I* gene. As we remarked above, the repressor is present at a concentration of around twenty molecules per wild-type cell. β-Galactosidase, on the other hand, may be found at a concentration of $10^5$ molecules per fully induced cell. So far as we know, the synthesis of the *lac* repressor is not regulated by any system analogous to that which regulates the synthesis of β-galactosidase and galactoside permease. The cell needs to make the repressor at a constant but extremely low rate, so that there is a sufficient concentration to maintain repression. Conversely the cell needs to have the *potentiality* of making β-galactosidase and galactoside permease at a very high rate, so that it can degrade lactose when necessary, but needs also to keep this potentiality under control. These ends are apparently achieved by ensuring that the promoter for the *I* gene is very weak and that for the *Z* and *Y* genes very strong. That is to say, the promoter of the *I* gene appears to have a low affinity for RNA polymerase, so that the gene is transcribed extremely seldom and the concentration of the corresponding messenger RNA is extremely low, while the opposite is true of the promoter of the *lac* operon.

## Other bacterial systems

*Inducible systems*

Our account of the *lac* system in *E. coli* will have made it clear that a great deal is known about the regulation of synthesis of β-galactosidase and galactoside permease. There is a temptation to extrapolate the results of studies with *lac* to other bacterial systems and even to other types of organism. We must try to resist this temptation. Although what is true of *lac* is also true, at any rate in part, of some other bacterial systems, in no case is our information as clear as it is for *lac*. Moreover, it has been shown that generalizations based on *lac* are inapplicable to several other systems.

The principle of negative control, based on a repressor that prevents expression of a set of genes under its control, is established in *E. coli* for a number of enzymes involved in the degradation of carbon sources other than glucose. Two examples that we might cite are the enzymes needed for the utilization of galactose and those needed for the utilization of glycerol. Both systems have regulatory genes that direct the synthesis of specific repressors, and in this and several other respects they are similar to *lac*.

On the other hand some sets of genes are subject not to negative but to positive control. We have already referred to positive control in the *lac* operon, by which the presence of 3′,5′-cyclic AMP is essential for expression of the operon. This type of positive control is apparently common to those systems that are liable to catabolite repression. For some other systems there is a positive control that is *specific* in the same way that repression in *lac* by the product of the *I* gene is specific. The two best studied examples are the synthesis of the enzymes involved in the degradation of arabinose (see p. 412) and of maltose. Each system has a regulatory gene which directs the synthesis not of a *repressor* (which *prevents* expression of the relevant genes *unless* an inducer is present) but of an *activator* (which *permits* expression of the relevant genes *only* when an inducer is present). In the *ara* system, for instance, three contiguous genes direct the synthesis of three enzymes needed for the degradation of arabinose. Close by is a regulatory gene, and this directs the synthesis of a protein activator. In the absence of arabinose the three contiguous genes are not expressed; when arabinose is added it becomes bound to the activator, and thereupon synthesis of the three enzymes begins.

*Repressible systems*

So far we have confined our attention to inducible systems, and we must now turn to a discussion of the *repression* of enzyme synthesis in bacteria. In general, rather less is known about repression than about induction. The original hypothesis of Jacob and Monod suggested that the two phenomena were strictly parallel: whereas in inducible enzyme synthesis a macromolecular repressor prevented synthesis until an inducer was added, in repressible enzyme synthesis a macromolecular repressor was made which became active only when a small molecule 'co-repressor' was added. Thus according to Jacob and Monod, a macromolecular repressor for (say) the *trp* genes would be present in the cell in an inactive form. When tryptophan was added to the cell it would be bound to the repressor, and this would then be converted to a form in which it actually repressed expression of the *trp* genes. Similar inactive repressors would exist for all the genes involved in the synthesis of biosynthetic enzymes, one for the histidine-synthesizing enzymes, one for the purine-synthesizing enzymes, etc.

There is no repressible system that is sufficiently clearly understood to allow us to be certain that this hypothesis is correct in respect of even one set of bio-synthetic enzymes. The two most intensively studied systems are those for tryptophan synthesis and for histidine synthesis. It seems at present as if the tryptophan-synthesizing enzymes are probably regulated in the way that the hypothesis of Jacob and Monod demands; the histidine-synthesizing enzymes are almost certainly not.

There are, as we mentioned in Ch. 24, five *trp* genes which are contiguous in *E. coli*; synthesis of their enzyme products is repressed many-fold when tryptophan is added to cultures of wild-type bacteria that have previously been growing in the absence of the amino acid. As with *lac*, constitutive mutants are known in which enzyme synthesis continues at a high rate despite the presence of tryptophan. There are two major classes of these constitutive mutants. The mutation in one class maps very close to the five *trp* genes to which we have referred, and these mutants appear analogous to operator-constitutive mutants in *lac*. The mutation in the other class maps in a gene (called *trp R*) that is very distant from *trp*, and these mutants appear analogous to repressor-deficient ($I^-$) mutants in *lac*. If an episome (p. 423) carrying a *trp* $R^+$ gene is inserted into a *trp* $R^-$ mutant, repression is restored; if the same episome is inserted into an operator-constitutive mutant, constitutive synthesis continues.

These results are consistent with the view that *trp* is regulated in exactly the way suggested by Jacob and Monod. *trp R* would make a repressor which would prevent transcription of the five genes of the *trp* cluster only when tryptophan were present.* The binding site for the repressor would be a tryptophan operator contiguous to the five genes, and mutation in either the gene for the repressor or the operator would lead to constitutivity (see Fig. 28.3). It is known that the product of *trp R* is a protein, and this has been partially purified; in a test-tube system in which DNA of the *trp* cluster is transcribed by RNA polymerase, this protein product of *trp R* represses transcription when tryptophan is present but not when it is absent. If the *trp* DNA is isolated from an operator-constitutive mutant, the *trp R* protein does not repress transcription even in the presence of tryptophan, presumably because it cannot bind to the mutant operator.

Control of the histidine-biosynthetic system has been studied intensively in *Salmonella typhimurium*, an organism that is closely related to *E. coli*. The system presents some curious features, and the details of the means by which it is regulated are by no means clearly understood. There is a group of nine contiguous genes, expression of which is repressed co-ordinately in wild-type cells by adding histidine to the medium. Constitutive mutants are found, but these fall into an alarmingly large number of classes. The most easily understood seem to be ordinary operator-constitutive mutants; their map position is adjacent to the

* Notice that the fact that *trp R* is distant from the *trp* gene cluster is no bar to this hypothesis, since a repressor can diffuse through the cytoplasm to bind to distant DNA. In fact there is no apparent reason why the *lac I* gene should be adjacent to the *lac* operon.

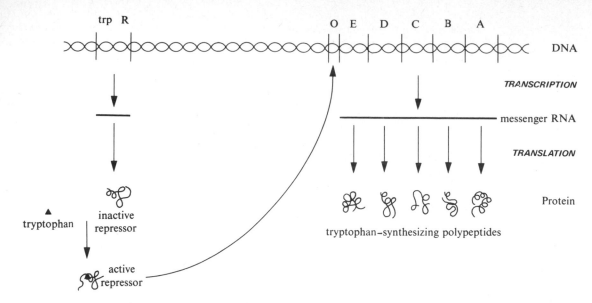

**Fig. 28.3** Control of the expression of the *trp* operon according to Jacob and Monod. The *R* gene is transcribed into messenger RNA, and this is translated to give an inactive repressor. The operator region is therefore free, and the five genes of the *trp* operon are transcribed into polycistronic messenger RNA which is translated to give the polypeptides necessary for tryptophan synthesis. The addition of tryptophan converts the repressor to an active form which binds to the operator and prevents transcription of the five *trp* genes

nine *his* genes. Of the several other classes, at least three seem to be deficient in some respect in the synthesis of histidyl transfer RNA or in the modification of its bases (see p. 142). These results suggest that histidyl transfer RNA is concerned in some way with maintaining repression of *his*; such a situation is not what is to be expected from the hypothesis of Jacob and Monod, and its significance is not clear.

*The operon*

Turning now from details of the mode of regulation of these systems, can we make any generalizations about another aspect of the Jacob–Monod hypothesis, the notion of the operon? The essential feature of an operon is that there be a set of two or more genes, transcribed into polycistronic messenger RNA. Such an arrangement has an obvious advantage in regulation. In both synthetic and degradative pathways, either all of the enzymes of the pathway are needed or none of them is needed. For the synthesis of tryptophan all five of the *trp* genes must be expressed; the expression of only four of them would not enable the cell to make tryptophan. For the degradation of lactose both of the *lac* genes must be expressed; the expression of either one of them alone would not enable the cell to use lactose. Plainly control of the expression of a set of genes is facilitated if they are arranged contiguously and transcribed into a single molecule of messenger RNA. How far is this situation actually found in practice?

Of the systems we have mentioned, some have genes that are arranged in single operons. Apart from *lac*, the five *trp* genes, the three *gal* genes, the nine *his* genes, etc. are clustered, and are transcribed into polycistronic messenger RNA molecules. Similarly there are three *ara* genes which are transcribed into polycistronic messenger RNA, and these code for three enzymes essential for the degradation of arabinose. However, there is a fourth *ara* gene, which codes for arabinose permease, and this is located at a great distance from the other three. The *arg* system is somewhat similar: there are four *arg* genes, coding for enzymes needed for arginine biosynthesis, that are apparently contiguous, but there are five other genes that are scattered. Even such scattered genes, however, are often subject to the same regulation as the clustered genes; for instance the synthesis of arabinose permease is subject to the same positive regulatory system as synthesis of the other three enzymes. This last example suggests that for the expression of several genes to be controlled together it is not essential that they be linked. A more extreme example leading to the same conclusion is provided by the system involved in the utilization of glycerol; here the three relevant genes are all scattered, but the expression of all three is regulated together.

## The level at which control is exerted

Earlier we suggested that there are, in principle, three main possibilities for regulating the rate of synthesis of specific proteins in bacteria. Changes in the rate of synthesis of a protein could result from changes in (a) the rate of synthesis of the corresponding messenger RNA, (b) the rate of degradation of the messenger RNA, (c) the rate of translation of the messenger RNA into protein. How far are these possibilities put into practice in controlling protein synthesis in micro-organisms?

Let us first consider the degradation of messenger RNA. It is possible by a number of means to measure the rate of degradation of bacterial messenger RNA. For example one can add an agent that prevents the synthesis of further RNA and measure the rate of disappearance of messenger RNA synthesized up to that time. Alternatively one can withdraw an inducer (for example of *lac*) and study the decay of the corresponding messenger RNA. Experiments of this kind always lead to the conclusion that most, if not all, messenger RNA in bacteria is very short-lived: it has a half-life of at most a few minutes at 37°. Inducing or repressing the synthesis of particular enzymes makes no very obvious difference to the rate of degradation of messenger RNA. Certainly any slight change that may occur in the rate of degradation of messenger RNA is insufficient to account for the large changes observed in the rate of synthesis of inducible or repressible enzymes in bacteria.

To return to the first of the three possibilities we mentioned: can these large changes be accounted for by changes in the rate of *synthesis* of messenger RNA? In other words, is all control of protein synthesis in bacteria due to changes in the rate of transcription of particular genes or are there also changes in the rate of translation of messenger RNA? There is a voluminous and vexed literature that has debated this point during the past ten years or so, and it is perhaps fool-hardy to try to summarize it. But we think it reasonable to draw this conclusion: that most examples of regulation of enzyme synthesis that have been studied in bacteria can be perfectly well accounted for by assuming that control is exerted primarily on the rate of transcription of genes. In those systems – particularly *lac*, *gal* and *trp* – in which it is possible by DNA–RNA hybridization to measure the quantity of specific messenger RNA with considerable accuracy, it is found that induced (or derepressed) bacteria contain far higher concentrations of the particular messenger RNA than do uninduced (or repressed) bacteria. It is conceivable that, superimposed upon a primary control of transcription, there may be an additional control of translation – in other words, that repression involves not only a great reduction in the rate at which messenger RNA is synthesized, but also a reduction in the rate at which the low concentration of messenger RNA is translated. But there is no conclusive evidence for such an effect.

There is one class of cases in which it appears that all messenger RNA is not translated with the same efficiency, namely when polycistronic messenger RNA is translated to give several proteins. If there were no discrimination by the ribosome one would expect the protein products of a single polycistronic messenger RNA all to be synthesized in equimolar ratios. There is evidence that such a result is, at any rate, not invariably obtained. In the *his* system, for example, it is found that molar ratios of the proteins synthesized from four of the nine genes of the operon are 3:1:1:1; it is the protein corresponding to the gene nearest to the operator that is present in the largest quantity.

This discovery suggests that the number of times that messenger RNA is translated is not the same for the whole length even of a single molecule of messenger RNA (and it is naturally possible that *different* molecules of messenger RNA are translated to quite different extents). There are various conceivable explanations. For instance, ribosomes may be able to bind to the polycistronic messenger RNA only near to its 5′ end (which is equivalent to the end closest to the operator, since that is synthesized first – see pp. 435 and 479) to translate the first section of the messenger RNA; some of them may then become detached before having the opportunity of translating the remainder of the messenger RNA. A second possibility assumes that ribosomes can bind both at the beginning of the messenger RNA, to the transcript of the first gene, and also to internal sites to initiate translation of the transcripts of other genes; if the binding sites on the RNA had different binding constants for the ribosomes, it would be possible for the extent of translation to differ from one section of messenger RNA to another. And other possible explanations can be envisaged.

It is difficult to investigate the translation of bacterial messenger RNA because of its instability. Good progress has, however, been made with the use of viral RNA, particularly that extracted from a class of bacteriophages that have RNA instead of DNA as their genetic material. Such phages, of which the best studied is called R17, contain a long single strand of RNA wrapped in a protein coat. This RNA has two functions. It acts as a polycistronic messenger RNA which, when it is injected into the bacterial host, is translated to yield specific phage proteins, and it also carries the genetic information of the phage. The great experimental value of R17 RNA lies in the facts that it can be readily purified in large quantities from the phage, that it can be used to direct protein synthesis in a cell-free system prepared from bacteria, and that it codes for only three proteins. These three proteins are the coat protein, the RNA synthetase necessary to replicate the phage RNA, and the 'A protein' necessary in some way for maturation of the phage (see also p. 159).

Investigations of R17 and closely related phages have produced the following information. The arrangement of the three genes along the RNA is: 5′ end–A protein–coat protein–RNA synthetase–3′ end. The RNA that codes for the N terminus of the A protein, which we may call the 'initiation site' for the A protein, is more than 100 nucleotides from the 5′ end of the molecule. There are several nucleotides between the A protein gene and the coat protein gene, and again between the coat protein gene and the RNA synthetase gene, that are not translated. In bacteria infected with the phage, the proteins are produced in the rough ratios: 1 A protein:20 coat protein:4 RNA synthetase. In cell-free systems capable of synthesizing protein, ribosomes become bound at first to the initiation sites of the RNAs that code for A protein and coat protein; only after part of the RNA that codes for coat protein has been translated do they become bound to the initiation site for making the RNA synthetase. The great difference in the rates of synthesis of the A protein and the coat protein can be attributed to different affinities of the two initiation sites for ribosomes. The delay before ribosomes are bound to the initiation site for the RNA synthetase gene can be attributed to this initiation site's being buried inside a complex tertiary structure of the intact molecule: we may presume that the translation of the RNA for the coat protein disrupts this tertiary structure and exposes the initiation site for RNA synthetase.

Thus these results suggest that there are at least two mechanisms by which control of the translation of RNA may be effected. The initiation sites at which ribosomes become bound to start translation of different portions of a poly-cistronic RNA may differ in affinity for the ribosomes. In addition, a particular initiation site may be unavailable to ribosomes through being buried in a part of the molecule that they cannot reach. (In fact these two mechanisms may well be more or less the same, since different 'affinities' of initiation sites for ribosomes may be related to the ease or difficulty of access by ribosomes.) These interesting findings allow us to wonder whether control over the extent of translation may be accomplished similarly for non-viral RNA – for example for polycistronic messenger RNA in bacteria. Before we conclude that such mechanisms do exist in bacteria we should pause to remember that phage RNA is peculiar in at least two respects. First, it needs to be arranged in such a way as to be capable of being packed into a phage coat, and the tertiary structure required for this packaging may well not be appropriate for bacterial messenger RNA. Secondly, the RNA of the phage is, of course, complete when it is injected into a bacterium and there translated; bacterial messenger RNA, by contrast, may well be engaged by ribosomes and begin to be translated before its synthesis is finished, since it is both synthesized and also translated from the 5′ end (pp. 435 and 447) We repeat what we said near the beginning of this section: most examples of regulation of enzyme synthesis that have been studied in bacteria can probably be accounted for without invoking control of the rate of translation of messenger RNA.

# REGULATION AND ENZYME SYNTHESIS IN EUCARYOTES

Yeasts and fungi can, like bacteria, be grown either in minimal medium or in media supplemented with amino acids, purines and pyrimidines. We might therefore expect to find that these organisms can repress the synthesis of the enzymes of biosynthetic pathways. Such repression does occur, but it is generally less dramatic than in bacteria. In yeasts and fungi the difference between the rate of enzyme synthesis when a system is repressed and that when it is derepressed is often only some three- to ten-fold. The mechanism of repression is far less well understood in these organisms than it is in bacteria. Genetic techniques – particularly for the fungus *Neurospora* – are well developed, and have shown that there is little or no clustering of genes such as would be necessary for the existence of operons.

The situation in higher animals is different again. Micro-organisms have to adapt to different conditions in which they may find themselves, and it is therefore (as we explained at the beginning of this chapter) advantageous for them to be able to induce or repress the synthesis of particular enzymes. Higher animals, on the other hand, maintain a fixed internal environment, and we would therefore not expect them to need to repress or derepress the synthesis of many enzymes. One exception to this rule is the liver. The liver has to cope with digestion products from the portal circulation, and it is found that the concentration of some liver enzymes can be increased many-fold by changes in the diet. The best-studied examples are enzymes needed for the breakdown of some amino acids, whose concentration in liver rises when an animal's diet is supplemented with large concentrations of the relevant amino acids. (Examples are tryptophan pyrrolase and threonine dehydratase.) The mechanism of this induction of enzyme synthesis is not understood. It is sometimes found that an injection of a hormone may induce the synthesis of one or more enzymes, and it has been suggested that some hormones may act in a way analogous to an inducer in the Jacob–Monod hypothesis. Such a view would require the presence of specific repressors for a number of vertebrate genes, for which there is little good evidence.

Much the most important problem in the regulation of protein synthesis in higher organisms is, however, a quite different one. The fertilized ovum contains the genetic material necessary to code for all the proteins that are present in a mature animal or plant. But during the course of development of the organism specialized tissues emerge, and in the fully differentiated adult specialization has gone so far that many cells are expressing only a small fraction of their genetic material. No cells other than those in the islets of Langerhans in the pancreas make insulin; no cells other than reticulocytes and their precursors make haemoglobin. It has been shown that this specialization is due not to *loss* of some genetic material from differentiated cells but rather to its permanent *repression*. We must therefore ask how it is that this repression is maintained. We must also ask a further question: How is the process of differentiation controlled *temporally*? During the course of development, different sections of the genome are expressed at different times – a striking example is metamorphosis in amphibia, where it is found that the change from tadpole to adult is accompanied by a highly specific and regular change in the pattern of protein synthesis.

These fascinating questions are far more easily asked than answered. Until a short time ago the study of differentiation was more akin to that of natural history than to that of an exact science. Recently, however, various model systems – ranging from sporulation in bacteria to embryonic development in vertebrates – have been developed and are beginning to be dissected by bio-chemical and, in some cases, by genetic techniques. Now that the study of such phenomena has moved from the descriptive to the analytical, we may reasonably hope that within a few years we shall begin to have some idea of the mechanisms of differentiation.

# Chapter 29  Control of enzyme activity

The mechanisms that we considered in the last chapter can, in the long run, achieve large changes in the rate at which a metabolic pathway operates. To recapitulate the same example that we used there, the addition of tryptophan to a culture of bacteria prevents the synthesis of further quantities of the enzymes that synthesize tryptophan, and after pre-existing enzyme has been diluted out the synthesis of tryptophan will cease. But for the greatest economy of the cell it is essential for such an effect to be accompanied by a regulatory system that has quicker effects, so that the synthesis of tryptophan is brought to a halt immediately. To move to a quite different example – the liver of a mammal that has just had a meal may well be busily engaged in synthesizing glycogen from glucose arriving from the portal blood; the same liver several hours later will probably be degrading glycogen so as to maintain the concentration of glucose in the systemic blood.

Superimposed, then, on regulation of enzyme *synthesis* we find an extensive pattern of regulation of enzymic *activity*. The culture of bacteria that has just been supplied with tryptophan quickly cuts off the flow of material along the tryptophan-synthesizing pathway by inhibiting an enzyme of that pathway. The liver of a mammal that has not eaten for many hours diverts its activity from synthesis of polysaccharide to its degradation, by inhibiting glycogen synthetase and activating phosphorylase. In this chapter we shall give some account of this kind of regulation. We shall first describe the general principles of such enzymic inhibition and activation. We shall then give some examples of their use in regulating specific biosynthetic pathways. Finally, we shall try to show how far they are applicable also to the central pathways of degradation and synthesis of carbohydrates and fats.

## THE CONTROL OF MULTI-ENZYME PROCESSES

We hinted from time to time in Part 2 that certain reactions in metabolic pathways were susceptible to control of their activity, and other enzymes in the same pathway were not. We indicated further that it was the control of the activity of these particular enzymes that enabled the cell to control the rate at which material passed down the pathway as a whole.

It is generally found, in fact, that these controllable reactions are those that are able to permit a useful net synthesis of product in one direction but not in the other. We must now explain why such reactions are preferred for the purposes of metabolic control.

## Flux in metabolic pathways

We shall find the concept of metabolic *flux* convenient in considering control. In order to define it let us consider a multi-step pathway in which A produces B and B in its turn produces C and so on.

$$A \rightleftharpoons B \rightleftharpoons C \rightleftharpoons D \rightleftharpoons E \rightleftharpoons F$$

Since this is a living system, it is not at true equilibrium, but at a steady state (p. 18) in which material is put in at one end and withdrawn from the other at the same rate.

$$(\text{input}) \longrightarrow A \rightleftharpoons B \rightleftharpoons C \rightleftharpoons D \rightleftharpoons E \rightleftharpoons F \longrightarrow (\text{output})$$

It is this rate that is the *flux* through the reaction, and we represent it by the symbol $J$. $J$ has units of, say, micromoles of substrate passed down the chain per second.

## The response to changes in catalytic activity – the substrate-saturated step

We shall now explore the effect on $J$, and on the concentrations of metabolites in the system, of variations in the activity of one of the enzymes. Let us imagine that the activity of the enzyme that catalyses the reaction $A \rightleftharpoons B$ is suppressed. The immediate response will be, if material continues to be poured in at one end and drawn off at the other, that $[A]'$ will rise and $[B]'$ will fall (cf. the accumulation of metabolites that occurs when inhibitors are used for the exploration of metabolic pathways, p. 229).*

Will this affect $J$? If it does, then we shall have achieved our aim of regulating the flux through the system and shall have done so by altering the catalytic activity of an enzyme. However, the answer is (perhaps surprisingly at first sight) that $J$ is *not* permanently affected unless certain special conditions are met. We shall explore why this is, and what the necessary conditions are.

The reason that $J$ will not in general be permanently affected is that, as $[A]'$ rises, the rate of the forward reaction (which equals $k_{+1}[A]'$ (see p. 35)) rises also. $[A]'$ will continue to rise until the consequent increase in the velocity of the forward reaction ensures that material is passing through this step at the same rate at which it is arriving at it. That is, $[A]'$ will rise until $J$ through the reaction $A \rightleftharpoons B$ is the same as $J$ through the rest of the system.†

We therefore appear to have failed to influence $J$. This situation is saved, however, if we take a further possibility into account, namely, that this step is, or becomes, saturated with substrate, and we shall now explain why this should be.

We have said that the flux is soon restored to its original value by the rise in

---

* We will use here, as we did in Ch. 2, the convention that $[A]'$ represents a concentration of a metabolite A that is displaced from the equilibrium value $[A]$ (see p. 21).

† $J$ equals $\text{Velocity}_{(\text{forward reaction})} - \text{Velocity}_{(\text{back reaction})}$. At equilibrium these velocities are equal (p. 35) and $J = 0$. This is why living systems that are *producing* anything cannot be at equilibrium.

the rate of the forward reaction that occurs as a consequence of the rise in [A]′. Suppose, however, that [A]′ rose to, or was already at, a value at which the enzyme was *saturated with substrate* (the asymptote of the curve in Fig. 29.1, cf. p. 126). In such cases it is no longer within the power of the forward reaction to increase in rate, since it has reached its maximum velocity. This means that a rise in [A]′ no longer has the ability to compensate for any decrease in enzyme activity, and the value of $J$ through this step will fall as a consequence of such a decrease. In such cases A $\rightleftharpoons$ B can be called the *rate-limiting step.*

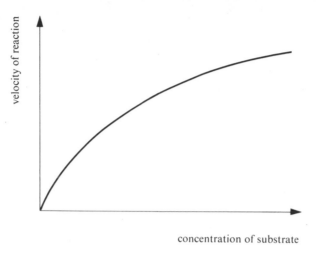

**Fig. 29.1**　The relationship between the velocity of an enzyme-catalysed reaction and the concentration of substrate

We have now achieved what we set out to find, a means of regulating $J$ by regulating catalytic activity. We have found that the enzyme must be one that is, or rapidly becomes, saturated with substrate if there is to be anything other than a *transient* response of $J$ to a change in catalytic activity.

An enzyme that is saturated with substrate spends virtually all its time catalysing the reaction in the direction A $\rightarrow$ B and hardly any time catalysing the reaction in the direction B $\rightarrow$ A. Therefore the step A $\rightleftharpoons$ B is, when in the substrate-saturated condition, one of that class of reaction in which there is a measurable net synthesis of product in one direction only (p. 22). In such a case Velocity$_{\text{(forward reaction)}}$/Velocity$_{\text{(back reaction)}}$ $(=K/\Gamma$, p. 35) is large.

We shall discuss in a moment an actual case in which a substrate-saturated step provides a control point (the role of phosphorylase in the degradation of glycogen), but we must now move on to see, in spite of all we have said, that enzymes that are *not* saturated (or saturatable) with substrate *do* have a part to play in control.

This apparent *volte-face* is explained in a few words. Let us rewrite our metabolic pathway in the following way

$$(\text{input}) \longrightarrow A \rightleftharpoons B \rightleftharpoons C \rightleftharpoons D \rightleftharpoons E \longrightarrow (\text{output})$$

in which the symbol A $\rightleftharpoons$ B indicates that the interconversion of A and B is being operated almost exclusively from A to B. Let this step still be the one that is saturated with substrate, but we shall now consider the effect of depressing the catalytic activity of the reaction $C \rightleftharpoons D$ which, let us assume, is not anywhere near saturation. Once again $[C]'$ will rise until the consequent increase in the velocity of the forward reaction $C \rightleftharpoons D$ restores the flux to its original value. Therefore $J$ (the flux through the pathway as a whole) will not be affected unless one of two things happens. We shall discuss them in turn.

*Use of non-substrate-saturated steps (i): feedback*

It might be thought that there was only one way in which the value of $[C]'$ could influence $A \rightleftharpoons B$, namely simple mass-action: a rise in $[C]'$ would cause a rise in $[B]'$ which would tend to accelerate the velocity of the process $B \rightarrow A$ and thus decrease $J$. However, we have said that, in the substrate-saturated case, the velocity of $B \rightarrow A$ is negligible compared with that of $A \rightarrow B$: a small change in the velocity of $B \rightarrow A$ can therefore make no appreciable difference to $J$, the balance between the forward and the backward reactions.

Since simple mass-action will not help us we must look for an alternative. It is provided by the facts that enzymes frequently have the power to bind substances other than their substrates and that such binding can have an effect on catalytic activity (p. 127). If the enzyme catalysing $A \rightleftharpoons B$ binds C in this way, it will almost certainly be able to do so irrespective of the values of $[A]'$ or $[B]'$, since the configuration and balance of forces of the binding site for C need bear no relation to that designed to accommodate A and B. If, further, the binding of C depresses the activity of the enzyme catalysing the rate-limiting step $A \rightleftharpoons B$, then anything that causes $[C]'$ to rise (such as a depression of the activity of the enzyme catalysing $C \rightleftharpoons D$) will be able to cause the desired variation in $J$. This situation may be symbolized as follows:

$$(\text{input}) \longrightarrow A \xrightleftharpoons{} B \rightleftharpoons C \rightleftharpoons D \rightleftharpoons E \longrightarrow (\text{output}).$$

This is an example of *feedback inhibition*, which is defined more fully below (p. 495).

Lest all of this be thought too far-fetched, let us say at once that many actual examples are known of just the process that we have described. We shall discuss some of them below.

*Use of non-substrate-saturated steps (ii): shunting*

The other way that $[C]'$ can influence $J$ without $C \rightleftharpoons D$ being the substrate-saturated step is by *shunting* the metabolites down another, alternative pathway. We can use the diagram

$$(\text{input}) \longrightarrow A \xrightleftharpoons{} B \rightleftharpoons C \rightleftharpoons D \rightleftharpoons E \longrightarrow (\text{output})_1$$
$$\updownarrow$$
$$F \rightleftharpoons G \rightleftharpoons H \longrightarrow (\text{output})_2$$

to help make this clear.

Let us once again inhibit $C \rightleftharpoons D$, a non-saturatable step. A rise in $[C]'$ will cause a compensating forward acceleration in $C \rightleftharpoons D$, but also in $C \rightleftharpoons F$. So long as the sum total of (output)$_1$ and (output)$_2$ remains constant and equal to $J$ (= input), it would be quite possible for depression of the activity of the enzyme catalysing $C \rightleftharpoons D$ to cause a decrease in flux through $D \rightleftharpoons E$ and the subsequent steps of this branch.

## Displacement from equilibrium as an indication of a regulatory reaction

We have seen that the substrate-saturated type of regulatory reaction ($A \rightleftharpoons B$) is one for which $K/\Gamma$ is large and which is being used for effective synthesis in one direction only.

Reactions that are not substrate-saturated can still be greatly displaced from equilibrium (large $K/\Gamma$), and so it would be possible for our hypothetical non-saturated step to be of this type also and allow useful net synthesis in one direction only.

In the two cases of regulation that we have just discussed need $C \rightleftharpoons D$ have a large $K/\Gamma$? The answer is that they *need* not, but that if they do they are more effective in regulation. This can be demonstrated by substituting some figures into our hypothetical examples.

Let us first assume that $K/\Gamma$ is low for $C \rightleftharpoons D$. (Remember that $K/\Gamma$, the degree of displacement from equilibrium, equals $\text{Velocity}_{C \rightarrow D}/\text{Velocity}_{D \rightarrow C}$, see p. 35. Therefore, for $J$, which equals $\text{Velocity}_{C \rightarrow D}$ minus $\text{Velocity}_{D \rightarrow C}$, to have a high value, both $\text{Velocity}_{C \rightarrow D}$ and $\text{Velocity}_{D \rightarrow C}$ must be even higher. This is the only way in which the quotient of two numbers can be near 1 and yet the difference between them be great.)

If we want to change $J$, what effect will this have on $[C]'$ and $[D]'$ in the present (low $K/\Gamma$) case?

Suppose that $J = 100$ units, that $\text{Velocity}_{C \rightarrow D} = 1000$ units and that $\text{Velocity}_{D \rightarrow C} = 900$ units. (Note that $K/\Gamma = 1000/900 = 1.1$ and we are very near equilibrium.)

If we decrease $J$ to 50 units, we could imagine that we might now have $\text{Velocity}_{C \rightarrow D} = 950$ units while $\text{Velocity}_{D \rightarrow C}$ remains 900 units. Thus a large change in $J$ (a fall of 50 per cent) has been accompanied by a small change (a fall of 5 per cent in one case) in the concentrations of the metabolites.

The fact that, when $K/\Gamma \simeq 1$, the levels of concentration of the metabolites are buffered against large changes in flux could clearly be of advantage to the cell, and this might account for the fact that a large number of steps in metabolic pathways are of this type. However, a difficulty arises if such a reaction (let us continue to call it $C \rightleftharpoons D$) is to be used to control the rate-limiting step, for example by feedback inhibition of $A \rightleftharpoons B$ brought on by a rise in $[C]'$. The enzyme that binds C would have to be extremely sensitive to the small changes in $[C]'$ for control to be at all effective.

This difficulty does not arise if $K/\Gamma$ is large. When $K/\Gamma$ is large,

$$\frac{\text{Velocity}_{C \rightarrow D}}{\text{Velocity}_{D \rightarrow C}}$$

is large, i.e.

$$\text{Velocity}_{C \rightarrow D} \gg \text{Velocity}_{D \rightarrow C}.$$

Now
$$J = \text{Velocity}_{C \rightarrow D} - \text{Velocity}_{D \rightarrow C} \simeq \text{Velocity}_{C \rightarrow D}.$$

$J$ thus depends on Velocity$_{C \rightarrow D}$ alone. A fifty per cent change in $J$ is accompanied by a fifty per cent change in [C]′. In the case of a large $K/\Gamma$, therefore, the enzyme catalysing step $A \rightleftharpoons B$ does not have to be so sensitive to [C]′ for a reasonably effective degree of control to be possible. A similar argument shows that shunting is only really effective when $K/\Gamma$ is large.

The treatment that we have just given is the justification for our statement that reactions that are far displaced from equilibrium, and which are operated for a useful net synthesis in one direction only, are the key sites in metabolic regulation.

## ALLOSTERIC ENZYMES OF BIOSYNTHESIS AND THEIR EFFECTORS

We shall now turn to some specific examples and see how all the principles that we have discussed work out in practice. We pointed out in Ch. 6 that some enzymes have, in addition to a specificity site and an active site, an *allosteric site*. At this site a molecule (very often one that is chemically dissimilar to the substrate) can bind; such binding causes a conformational change in the enzyme which can either enhance or inhibit its activity. We remarked that enzymes of this kind have quaternary structure and can be composed of different kinds of polypeptide subunit.

Now let us imagine a pathway leading to the biosynthesis of a substance T (let us say an amino acid, purine or pyrimidine).

$$P \longrightarrow Q \longrightarrow R \longrightarrow S \longrightarrow T$$

Notice that the pathway is unbranched, so that all molecules of P, on being converted to Q, are as it were 'committed' to being transformed into T. Suppose that the enzyme that converts P into Q has an allosteric site for T, such that *the binding of T inhibits its catalytic activity*. It is plain that such a system would conveniently permit T to act as an *inhibitor of its own synthesis* or *feedback inhibitor*. If, for example, T is tryptophan, P would be chorismate and Q would be anthranilate (pp. 374–376), and we should have achieved the inhibition by tryptophan of its own synthesis that we mentioned at the beginning of this chapter. The fact that the mechanism of enzymic inhibition is allosteric is important: tryptophan does not greatly resemble chorismate or anthranilate, and only if the enzyme possessed an allosteric site specific for tryptophan would it be possible for inhibition to be achieved.

### Aspartate transcarbamylase

Feedback inhibition is extremely prevalent in biosynthetic systems. We can describe its characteristics by discussing in some detail one particular enzyme, aspartate transcarbamylase (p. 390), Aspartate transcarbamylase catalyses the formation of carbamyl aspartate from aspartate and carbamyl phosphate; it is the first enzyme that is specific to the synthesis of pyrimidines. Its kinetic properties have been studied in great detail, and we shall now outline some of the results obtained from investigations of the purified enzyme in the test tube (refer also to p. 127).

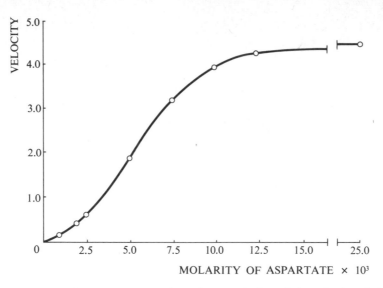

**Fig. 29.2** Aspartate transcarbamylase: variation of the velocity of the reaction with the concentration of aspartate. The velocity is expressed in arbitrary units. Notice the sigmoid shape of the curve

A plot of velocity of the enzyme reaction against substrate concentration (see pp. 520 ff.) gives a sigmoid curve (Fig. 29.2). You will recall that this kind of curve can be interpreted (p. 106) as suggesting co-operativity of binding of substrate molecules: the binding of one molecule of substrate facilitates the binding of the next. We can now ask whether the enzyme might be subject in addition to allosteric regulation. As the enzyme is involved in pyrimidine synthesis, it would be profitable to investigate the effect of pyrimidines on its activity. Of all the pyrimidine derivatives that have been tried, only CTP is found to have a dramatic effect on the enzyme: the addition of CTP to the reaction mixture greatly inhibits activity. Figure 29.3 shows a plot of reaction velocity against substrate concentration in the presence of CTP. We can interpret the shifting of the curve by CTP by saying that the presence of CTP decreases the affinity of the enzyme for its substrate aspartate.

We shall not, in this chapter, go further into the details of enzyme structure that bring about this effect (but see p. 127) or into a rigorous treatment of the kinetics of inhibition (see Appendix 1). Instead we shall consider the biological effects of the inhibition. Notice that CTP affects the affinity of the enzyme for its substrate, and not the maximum velocity. Experimentally, therefore, we can overcome the inhibition by adding a large quantity of aspartate. In the cell, on the other hand, it is likely that the enzyme is working in the range represented in Figs 29.2 and 29.3 towards the left-hand edge of the graph – an educated guess for the concentration of aspartate might be of the order of 1–5 mmolar. In this range inhibition by CTP is very severe: note that to achieve the substantial inhibition shown on the graph the concentration of CTP need be only 0.2 mmolar.

If we think of the function of feedback inhibition, as exemplified by aspartate transcarbamylase, solely in terms of inhibiting the synthesis of compounds that a bacterium might find already present in the medium, we might well ask why

CTP in particular should be the feedback inhibitor. It is extremely improbable that bacteria would ever encounter CTP in their medium, and if they did they would certainly not be able to take it up – like most phosphorylated compounds it does not pass across cell membranes (p. 319). But if we widen a little our way of looking at feedback inhibition we can see that CTP is a very appropriate inhibitor of aspartate transcarbamylase Reference to p. 394 will show you that CTP occupies a central position in the network of reactions that connect the pyrimidine nucleotides. An increase in the concentration of CTP will lead to an increase in the concentrations of other pyrimidine nucleotides, and by monitoring the concentration of CTP the cell will have a good impression of the general availability of pyrimidines.

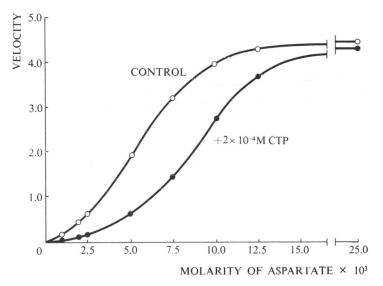

**Fig. 29.3** Aspartate transcarbamylase: the effect of CTP. The velocity is expressed in arbitrary units. Notice that CTP shifts the curve of velocity against substrate concentration, but does not affect the maximum velocity

We are thus led to the conclusion that feedback inhibition has a wider function than merely to prevent the synthesis of compounds that are available to micro-organisms exogenously. It is also used to control rates of synthesis *within* the cell. If pyrimidines have temporarily been over-produced, the concentration of CTP will rise. This rise will signal an excess of pyrimidines, and it will cause, by feedback inhibition, a reduction in the rate of their synthesis. This mechanism applies to all cells, not only to micro-organisms. In micro-organisms, however, CTP can be used *in addition* to signal the availability of exogenous pyrimidine. Both uracil and cytosine are taken up and converted to CTP, so that an external supply of pyrimidine will also lead to feedback inhibition of further synthesis.

Although the mechanism that we have described provides considerable economy in preventing excess synthesis, you may feel that it lacks flexibility. Suppose, for example, that bacteria are growing very fast and there is a tendency for their rate of RNA synthesis to rise. In order to achieve this rise it may be necessary for the concentration of CTP to rise too. But such a rise would have the effect of damping down the synthesis of pyrimidines and thus in fact preventing an increase in the rate of RNA synthesis.

This problem is elegantly solved in the following way. Although aspartate transcarbamylase is inhibited by CTP, this inhibition is relieved by ATP (Fig. 29.4). In circumstances where the increase in concentration of CTP is due not to specific over-production of pyrimidines but to a general increase in the rate of biosynthesis, the concentration of ATP will rise as well. The presence of the extra ATP will prevent feedback inhibition of aspartate transcarbamylase, and consequently a high rate of pyrimidine synthesis will be maintained to meet the need for rapid synthesis of RNA.

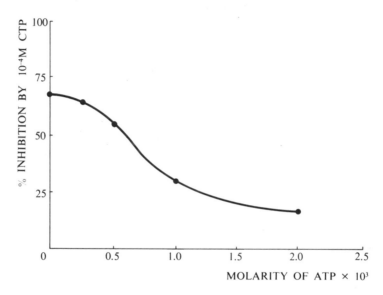

**Fig. 29.4**   Aspartate transcarbamylase: the effect of ATP on inhibition by CTP. The reaction was run with constant concentrations of aspartate (2.5 mmolar) and of CTP (0.1 mmolar), and various concentrations of ATP. Notice that the ordinate is *percentage inhibition* – that is, the percentage by which the reaction is inhibited in comparison with a reaction in the *absence* of CTP

We pointed out earlier (p. 491) that when an enzyme is inhibited, the concentration of its substrate will tend to rise. In the present case we may ask what effect the inhibition of aspartate transcarbamylase will have on the concentration of aspartate. Plainly the concentration will *tend* to increase, but what effect will this tendency have in practice? We should at first sight expect the greater availability of aspartate to lead to an increase in the flux down the other pathways leading from aspartate (p. 365), with the result that the surplus aspartate would be diverted towards threonine, lysine and methionine. However, any tendency for the concentrations of these amino acids to increase would be opposed (as we shall see shortly) by feedback inhibition by the amino acids on the enzymes that lead to their own synthesis. Hence the tendency of the concentration of aspartate to increase cannot in fact be prevented by diverting aspartate towards the synthesis of other amino acids. On the other hand it seems unlikely that there will in reality be any large increase in the concentration of aspartate, since this would vitiate (as we saw earlier) the effect of inhibiting aspartate transcarbamylase. It seems likely, instead, that the synthesis of aspartate must *itself* be inhibited by some mechanism that we do not yet understand.

This example demonstrates that feedback inhibition of a single pathway may have far-reaching consequences. We can follow the effect of an increase in the concentration of CTP back as far as the inhibition of aspartate transcarbamylase, but we do not yet know what the long-range consequences will be. It is conceivable that they may reach back as far as inhibiting (to a marginal extent) the rate at which glucose is degraded to provide energy, or alternatively that they may result in an increase in growth rate of the organism.

There are two further points about aspartate transcarbamylase that we must mention since they are characteristic of enzymes that are subject to feedback inhibition. First, the reaction catalysed by the enzyme is the first reaction that is specific to the synthesis of pyrimidines. Once a molecule of aspartate has reacted to form carbamyl aspartate it is 'committed' (see above) to being converted to a pyrimidine. Evidently, then, this reaction is a particularly suitable one to suffer feedback inhibition by CTP. Secondly, the reaction is one whose equilibrium is greatly displaced in the cell: carbamyl aspartate is never used to generate carbamyl phosphate and aspartate (see p. 390).

These points are important in the general context of feedback inhibition. By inhibiting the first reaction leading to a biosynthetic product the organism prevents the flow of material into unneeded intermediates. Thus in the present example inhibition of aspartate transcarbamylase conserves aspartate; it would have been wasteful to allow the synthetic pathway to proceed as far as (say) orotic acid and inhibit it at that point. (We may contrast here control of enzyme synthesis, which we discussed in the last chapter. There maximum economy is achieved by repressing the synthesis of all the enzymes of a biosynthetic pathway.) Again, by inhibiting a reaction that is used, in physiological conditions, to effect a net synthesis in one direction only, an organism ensures that it can control the flow of material in one direction at a time. We shall see later in this chapter that organisms can separately achieve control of glycolysis and gluconeogenesis by inhibiting the reactions that are not common to the two pathways.

## Other unbranched biosynthetic pathways

Aspartate transcarbamylase is the best studied example from a group of allosteric enzymes that catalyse reactions that stand first in a particular biosynthetic sequence and are inhibited by an end-product of that sequence. Many enzymes with these properties are known. For most of them the kinetic characteristics have not been worked out, so that the details of the mechanism of inhibition are not understood. What is known about these enzymes is that they are regulated by feedback inhibition and that they are important in the economy of the cell in the same way as aspartate transcarbamylase is.

The synthesis of amino acids provides many such examples; it is generally correct to say that the enzyme that is first in the pathway leading to each amino acid is subject to feedback inhibition by the amino acid. Phosphoribosyl-ATP pyrophosphorylase (p. 379) is inhibited by histidine, the end-product of the pathway in which it catalyses the first reaction. Threonine dehydratase (p. 369) is inhibited by isoleucine; anthranilate synthetase (p. 374) is inhibited by tryptophan, and so on. In a similar way, the enzyme catalysing the synthesis of 5-phosphoribosylamine (p. 394) is inhibited by purine nucleotides.

## Branched biosynthetic pathways

So far we have considered systems in which a particular enzyme reaction initiates a sequence leading to just one biosynthetic product. Such systems present no particular difficulties from the point of view of feedback inhibition. By contrast, consider the following sequence.

To arrange for T to inhibit the reaction leading from Q to R, and for Z to inhibit the reaction leading from Q to X, would provide no great problem. But what of the reaction leading from P to Q? If this were inhibited by T alone or by Z alone, an excess of either one of these products would prevent the synthesis of the other. If, on the other hand, the reaction were not subject to any regulatory mechanism, the presence of both T and Z in excess would lead to the wasteful accumulation of the useless intermediate Q.

This problem is particularly severe in amino-acid metabolism, where the same reaction often leads to the synthesis of several different amino acids. Thus, for example, the reaction catalysed by aspartokinase (p. 366) converts aspartate to aspartyl phosphate, which is a precursor of threonine, lysine and methionine. How is such a reaction to be controlled?

Different organisms have adopted different solutions to this kind of problem. One solution is for the reaction to be catalysed by *isoenzymes* – that is, separate species of protein molecule having the same catalytic activity. Suppose that in the theoretical sequence that we have just given, the reaction:

$$P \rightleftharpoons Q$$

were catalysed by two isoenzymes, one sensitive to feedback inhibition by T, the other sensitive to feedback inhibition by Z. In the absence of an exogenous supply of both T and Z, both enzymes would be active, and the sum of their activities would produce just enough Q to satisy the need for synthesis of T and Z. If T were added, the T-sensitive isoenzyme would be inhibited, and the rate of synthesis of Q would be diminished. T would also inhibit the enzyme catalysing the reaction Q → R, so that the diminished quantity of Q would be directed towards Z. If now Z were added as well, the other isoenzyme would also be inhibited, and the conversion of P to Q would cease. Such a system would thus provide the virtues of both economy and flexibility – though at the expense of needing the synthesis of two enzymes to catalyse a single reaction.

A mechanism of this kind is used to control the synthesis of the aspartate family of amino acids – threonine, lysine and methionine – in *E. coli*. Figure 29.5 summarizes the pathways of synthesis of these three amino acids, and will

remind you that the first step, common to the three pathways, is the phosphorylation of aspartate in a reaction catalysed by aspartokinase. It turns out that there
are three aspartokinase isoenzymes. One suffers feedback inhibition by threonine,
one of the end-products of the synthetic pathway, and the second suffers
inhibition by lysine. It would complete a beautiful picture if we could report that

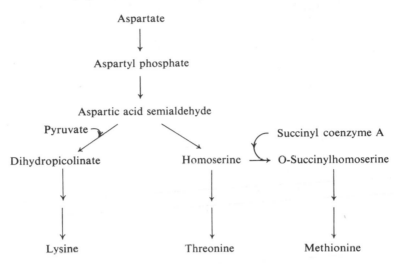

**Fig. 29.5** Pathways of synthesis of amino acids from aspartate

the third is inhibited by methionine. Unfortunately, though there is a third
aspartokinase it appears not to be subject to feedback inhibition; biochemistry is
not always quite as elegant as one would wish.

Further down the metabolic pathways we find that dihydropicolinate synthetase, the first enzyme in the sequence leading specifically to lysine, is allosterically inhibited by lysine. Similarly the enzyme catalysing the succinylation
of homoserine, which is first in the specific pathway of methionine synthesis, is
allosterically inhibited by methionine.

If we go back now to our sequence:

$$P \longrightarrow Q \begin{array}{c} \nearrow R \longrightarrow S \longrightarrow T \\ \searrow X \longrightarrow Y \longrightarrow Z \end{array}$$

we can postulate another possible mechanism for its control. As an alternative
to having isoenzymes catalysing the reaction $P \to Q$, we could imagine that there
might be a single enzyme which would suffer feedback inhibition if *both* T and Z
were present but not be inhibited by either alone. Such a mechanism is known.
In the bacterium *Bacillus polymyxa* there is, unlike in *E. coli*, a single aspartokinase. This enzyme is insensitive to inhibition by any one of the three amino
acids threonine, lysine and methionine. However, a mixture of lysine and
threonine causes considerable inhibition of its activity. This situation is known
as *concerted feedback inhibition*. Even in the presence of both lysine and threonine
inhibition is not complete: we can rationalize this fact by noting that complete

inhibition in such circumstances would cause the organism to be starved of methionine.

Neither of the two arrangements that we have mentioned for the control of aspartokinase produces an absolutely perfect regulatory system. In *E. coli* an excess of methionine will leave the rate of synthesis of aspartyl phosphate completely unaffected, with a consequent danger of overproduction of lysine and threonine (although admittedly an increase in the concentration of either of these will inhibit the synthesis of aspartyl phosphate). In *B. polymyxa* an excess of any one of the amino acids will have no effect at all on regulation of the aspartokinase reaction.

Finally we might cite another imperfection in a regulatory system which has a curious consequence. You will recall that the enzymes that catalyse the synthesis of isoleucine from α-oxobutyrate also catalyse the synthesis of valine from pyruvate (p. 373). The first of these, α-aceto-α-hydroxyacid synthetase, is first in the biosynthetic sequence that leads specifically to valine, and is as we might expect allosterically inhibited by valine. The addition of valine to cultures of many species of bacteria therefore prevents the synthetase from acting, and since it is shared by the isoleucine-synthesizing pathway the bacteria become starved of isoleucine. Valine is therefore growth-inhibitory for many bacteria; the inhibition is relieved by adding isoleucine.

## CONTROL OF THE DEGRADATION AND SYNTHESIS OF CARBOHYDRATE AND FAT

Despite the blemishes that we have uncovered in the last few pages, the control of specific biosynthetic pathways by feedback inhibition is of vital importance to the economy of living organisms. How far can the principles that we have drawn from studying the control of these pathways be extended? In particular, are they applicable to the regulation of central pathways such as glycolysis or gluconeo-genesis?

Even if we were tempted to try to conclude that these central pathways are regulated by allosteric control, we should obviously expect modifications in the general principles that we have so far derived. First, in a biosynthetic scheme the required regulation can be achieved simply by ensuring that the end-product inhibits its own synthesis. With the central metabolic pathways more sophisticated controls are desirable – for example, the organism might want to ensure that in the presence of excess carbohydrate it converted some carbohydrate to fat. Secondly, for a biosynthetic pathway leading to the production of (say) an amino acid, it is easy to name the end-product, or set of end-products. For a pathway like glycolysis it is by no means easy to define an end-product.

What is the effect of these complications? As far as the first point is concerned, we find that substantial extra flexibility in control is achieved by arranging that allosteric *inhibition* is supplemented by allosteric *activation*. If you refer once again to pp. 127 ff., you will see that examples are known of enzymes whose activity is enhanced by the binding of allosteric effectors. We shall not treat in this chapter the kinetics of these enzymes – you will find them mentioned in Appendix 1 – but will merely remark that activation, no less than inhibition, of enzymic activity is of great importance in the regulation of carbohydrate and fat metabolism.

What can we say of the second point – identification of the end-products of such pathways as glycolysis? In Ch. 9 we pointed out that the central degradative pathways of metabolism had two functions. The first was to provide energy in the form of ATP, the second to provide carbon skeletons for other pathways. We might tentatively suggest, therefore, that the 'end-products' of glycolysis could be regarded as ATP, and one or more of the following carbon compounds resulting from the degradation of carbohydrate: lactate, acetyl coenzyme A, citrate, etc. Similarly the 'end-products' of fatty-acid oxidation might be taken to be ATP, and acetyl coenzyme A, citrate, etc. Conversely the 'end-product' of gluconeogenesis and fatty-acid synthesis might be ADP or AMP; we probably cannot regard glycogen and fats as 'end-products' that are likely to act allosterically since they tend to be sequestered into granules or droplets rather than to accumulate in free solution.

Earlier (p. 499) we remarked that in order for an organism to be able to achieve control of glycolysis and gluconeogenesis independently, the enzymes involved in regulation could not be those that catalyse reactions that are used to achieve net synthesis in both directions. For example the ability to inhibit or activate triose phosphate isomerase would not help to promote glycolysis and inhibit gluconeogenesis or *vice versa*. With this point in mind, we can write the scheme of glycolysis and gluconeogenesis in the following abbreviated form:

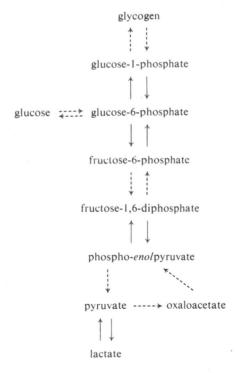

where ordinary double arrows signify reactions that, in the cell, are found to proceed in either direction, and dashed arrows those reactions where different enzymes are found to catalyse the forward and reverse directions.

This scheme allows us to speculate that the following enzymes are candidates for allosteric control. In glycolysis: hexokinase, glycogen phosphorylase, phosphofructokinase and pyruvate kinase; in gluconeogenesis: pyruvate carboxylase, phospho-*enol*pyruvate carboxykinase, fructose diphosphatase, glucose-6-phosphatase and glycogen synthetase. Which of these are important in regulation, and by what effectors are they controlled?

## Glycolysis

In the common glycolytic pathway, i.e. those reactions that are involved in the degradation of all carbohydrate sources – the most important regulatory enzyme appears to be *phosphofructokinase*. Phosphofructokinase is subject to allosteric control by several compounds. It is inhibited by *excess* ATP (note that ATP is a substrate of the reaction), and this inhibition is counteracted by AMP. In the presence of ATP, the enzyme is further inhibited by citrate.

The inhibition by citrate corresponds with what we suggested above – that citrate can be regarded as an 'end-product' of glycolysis. An increase in the concentration of citrate signals excessive flux through the glycolytic pathway and it is reasonable that it should lead to inhibition at the level of phosphofructokinase. But what are we to make of an inhibition by ATP together with relief of this inhibition by AMP?

AMP, ADP and ATP are interconvertible in the cell, the reaction being catalysed by the enzyme adenylate kinase.

$$AMP + ATP \rightleftharpoons 2ADP$$

Consequently an increase in the concentration of ATP is accompanied by a decrease in the concentration of AMP, and *vice versa*. When the glycolytic pathway is active ATP is produced; if ATP is not being used up, it will accumulate, and this will signal (like an increase in the concentration of citrate) excessive flux through the pathway. It is, therefore, easy to see why it is useful for phosphofructokinase to be inhibited by ATP, but it seems at first sight an unnecessary complication for this inhibition to be reversed by AMP. We can explain this dual control by considering the ratio of the concentration of AMP:ADP:ATP, which is (in insect muscle, where it has been extensively studied) about 1:10:60. Hence the equilibrium constant of the above reaction

$$K = \frac{10^2}{1 \times 60} = 1.67$$

and the concentration of AMP will be kept at

$$\frac{[ADP]^2}{1.67 \times [ATP]}.$$

Let us now consider what will happen if out of sixty moles of ATP two break down. They will form ADP, thus increasing the number of moles of ADP from ten to twelve. The number of moles of AMP will rise from one to

$$\frac{12^2}{1.67 \times 58}.$$

Thus a 3 per cent decrease in the concentration of ATP leads to almost a 50 per cent increase in the concentration of AMP. In this way changes in the concentration of AMP will very sensitively reflect the accumulation or the disappearance of ATP and hence the need to inhibit or activate glycolysis. We can thus regard changes in the concentration of AMP as 'amplifying' changes in the concentration of ATP.

Just as we did with aspartate transcarbamylase, we can ask what effect the inhibition of phosphofructokinase will have on the concentration of fructose-6-phosphate. The initial tendency will be for the concentration to rise; if this rise continued, it would overcome the inhibition of phosphofructokinase. However, an increase in the concentration of fructose-6-phosphate will be reflected (owing to the presence of hexose phosphate isomerase) in an increase in the concentration of glucose-6-phosphate. Hexokinase is subject to *product inhibition* (cf. p. 520) by glucose-6-phosphate, so that such an increase will tend to inhibit hexokinase. (There may also be more remote consequences, which we do not yet fully understand, on the entry of glucose across the cell membrane.) In addition glucose-6-phosphate can be converted to glucose-1-phosphate and can thence reach UDP-glucose and glycogen. Consequently it seems reasonable to believe that inhibition of phosphofructokinase (resulting as we have seen from an increase in the concentration of citrate or a decrease in the concentration of AMP) will cause both a decrease in the entry of glucose into the glycolysis pathway and an increase in the rate of synthesis of glycogen.

The other major site of regulation of glycolysis is the degradation of glycogen catalysed by *phosphorylase*. Phosphorylase exists in the liver cell in two forms, phosphorylase *a*, which is enzymically active, and phosphorylase *b*, which is largely inactive under physiological conditions. Phosphorylase *a* is a phosphorylated protein (the phosphate being attached to a serine residue), and its conversion to phosphorylase *b* is catalysed by a phosphatase. The reverse reaction (phosphorylase *b* → phosphorylase *a*) uses ATP and is catalysed by an enzyme called phosphorylase *b* kinase. This kinase too exists in two forms, an inactive form which is not phosphorylated, and an active phosphorylated form. The conversion:

$$\text{Inactive phosphorylase } b \text{ kinase} \xrightarrow{\text{ATP ADP}} \text{Active phosphorylase } b \text{ kinase}$$

is catalysed by yet another enzyme, which is called *protein kinase*.

So if protein kinase is active, it activates phosphorylase *b* kinase, and this in turn activates phosphorylase by converting it to phosphorylase *a*, and this in turn catalyses the breakdown of glycogen to glucose-1-phosphate. But what determines the activity of protein kinase? The answer is that protein kinase requires for its activity the nucleotide 3′,5′-cyclic AMP, the formula of which is (see Table 7.3):

3′,5′-Cyclic AMP is produced from ATP by an enzyme called adenylate cyclase, which is located at the cell membrane of many cells. This enzyme is well situated to act as a receptor of signals conveyed from *outside* the cell in the blood, and it is found to be extremely sensitive to the hormone adrenalin. (We are thus beginning to understand the action of this hormone in molecular terms – we should emphasize that the biochemical basis of the action of many other hormones, especially those that are slower acting, is still far from clear.)

The regulation of the breakdown of glycogen can thus be portrayed in this way:

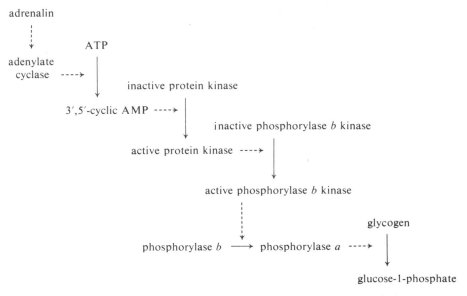

Fig. 29.6   The cascade by which adrenalin activates the breakdown of glycogen

where dashed arrows signify an activation, by the substance at the tail of the arrow, of the reaction at its head. What purpose, other than to torment the student of biochemistry, is served by an arrangement of such complexity? The answer seems to be that at each stage of such a *cascade* an amplification is introduced: a very few molecules of adrenalin can activate adenylate cyclase to produce a larger number of molecules of 3′,5′-cyclic AMP, which can activate protein kinase to produce a still larger number of molecules of active phosphorylase *b* kinase, etc. Thus a very small quantity of adrenalin can give rise to the degradation of a large quantity of glycogen.

## Gluconeogenesis

A key regulatory enzyme involved in the synthesis of carbohydrate from lactate and from alanine is *pyruvate carboxylase*. The activity of this enzyme is found to be dependent on the presence of acetyl coenzyme A. An accumulation of acetyl coenzyme A signals a large flux of material from the degradation of carbohydrate, of fat or of amino acids. In any of these cases, an appropriate response of the cell is to synthesize oxaloacetate from pyruvate. The oxaloacetate can be used either as a substrate for gluconeogenesis or alternatively to couple with the excess acetyl coenzyme A for the synthesis of citrate.

The other enzyme in the common pathway of gluconeogenesis that appears to be of importance in regulation is *fructose diphosphatase*. This enzyme is inhibited by AMP. We have discussed above how changes in the concentration of AMP give a sensitive indication of small changes in the concentration of ATP. It is reasonable for the cell to promote gluconeogenesis when ATP is abundant, and to inhibit it when the concentration of ATP falls and that of AMP correspondingly rises.

The most complex regulatory system in gluconeogenesis is that for the conversion of uridine diphosphate glucose to glycogen, just as in glycolysis the most complex is that for the degradation of glycogen. Like phosphorylase, glycogen synthetase exists in two forms; one (glycogen synthetase D) is far less active under physiological conditions than the other (glycogen synthetase I). The D form is converted to the I form by a phosphatase, and the reverse reaction (I → D) is catalysed by protein kinase, which, as we saw above, is activated by 3′,5′-cyclic AMP.

The regulation of the synthesis of glycogen can be portrayed in this way.

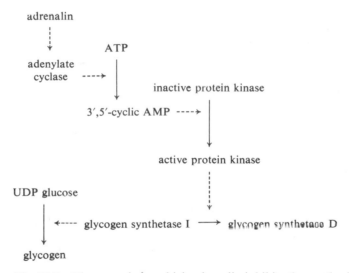

Fig. 29.7   The cascade by which adrenalin inhibits the synthesis of glycogen

Comparison of this system with that shown in Fig. 29.6 shows that the *results* of the cascades are opposite. In Fig. 29.6 it is the *activation* of phosphorylase that is finally produced by adrenalin; in Fig. 29.7 it is the *de-activation* of glycogen synthetase that is finally produced by adrenalin. The hormone thus has pleasingly opposite effects on the two reactions.

## Degradation and synthesis of fat

A good deal less is known, in molecular detail, about the regulation of fat metabolism than about the regulation of carbohydrate metabolism. The rate of oxidation of fatty acids in animals appears to be affected largely by the concentration of the fatty acids in the blood, which in turn is affected largely by the rate at which lipase is active in adipose tissue. Several hormones, including adrenalin (possibly via its effect on increasing the concentration of 3′,5′-cyclic AMP) activate the lipase.

The crucial enzyme in control of fatty-acid synthesis is acetyl coenzyme A carboxylase (p. 341). The activity of this enzyme is dependent on the presence of citrate or isocitrate. An increase in the concentration of these tricarboxylic acids, resulting from an increased rate of glycolysis, will stimulate the production of fatty acids. On the other hand long chain fatty-acyl coenzyme A inhibits acetyl coenzyme A carboxylase. Hence if long-chain acyl coenzyme A accumulates (for instance because of lack of α-glycerophosphate for the synthesis of triglycerides), the synthesis of further fatty acid will be cut off.

## Conclusion

It will be obvious that we have done little more than highlight some aspects of the regulation of carbohydrate and fat metabolism. A great deal remains to be discovered. Research in these important areas is hampered by several difficulties. In the first place, we can do no more than make intelligent guesses about which enzyme reactions are likely to be regulatory in the cell. Secondly, when the putative regulatory enzymes are purified their properties may well be changed, especially since regulatory enzymes are often proteins of complicated structure and particularly hard to purify. Thirdly, after discovering effectors for the isolated enzymes by studies in the test-tube we can by no means necessarily conclude that these control their activities in the cell – particularly because of the difficulty in discovering the intracellular concentration of putative effectors.

Nonetheless such information as we do have draws attention to a different aspect of ATP from that that we have been accustomed to stress. In biosynthesis of specific products it is generally the accumulation of end-products that inhibits reactions that stand first in their pathways. But in the central pathways of carbohydrate metabolism it is ATP that allosterically inhibits glycolysis and activates gluconeogenesis, and AMP that has the reverse effects. ATP, whose importance as a high-energy compound we have so often emphasized, here appears in a new role – a striking example of the versatility of biological molecules.

# Appendix 1  Enzyme kinetics

## How to use this appendix

There are three sections. These are:

1. The derivation of a relationship between the concentration of substrate and the velocity of reaction. This brings in the concept of the Michaelis constant, which under some circumstances can be a measure of the binding constant of the enzyme for its substrate.
2. The kinetics of non-competitive inhibition.
3. The kinetics of competitive inhibition.

The mathematical treatment is quite straightforward, but even so some people may have difficulty with it. However, we shall see that, to take the derivation in section 1 first, only the simplest principles of physical chemistry are involved. The main points, which appear in Ch. 2, or follow readily from it, are as follows:

(a) In a reaction

$$A + B \rightleftharpoons AB$$

if the kinetic constant of the forward reaction is $k_{+1}$, then the forward reaction will proceed at the rate $[A][B]k_{+1}$. Similarly the back reaction will proceed at the rate $[AB]k_{-1}$, where $k_{-1}$ is the kinetic constant of the backward reaction.

(b) At equilibrium, when the rate of the forward reaction = rate of the backward reaction

$$[AB]k_{-1} = [A][B]k_{+1}$$

i.e.

$$\frac{k_{-1}}{k_{+1}} = \frac{[A][B]}{[AB]}.$$

This, by definition, is the equilibrium constant of the reaction for the *dissociation* of AB.

You should therefore have no difficulty in learning to follow and to reproduce the treatment in section 1.

In sections 2 and 3 things appear more complicated, although in fact no new logical steps are introduced. Here, there may be a few more people that cannot handle the equations. Therefore, although it is desirable to be able to follow and reproduce these sections, if this is not possible you should at least note the existence of equations (A1.15) and (A1.19) and compare them with each other. You should compare their form with that of equation (A1.6), which is the chief equation derived in section 1. Perhaps most important of all, you should take note of the assumptions used in each derivation. You *must* be familiar with the graphs corresponding to these equations (Figs A1.3, A1.4 and A1.5) and with the qualitative explanations of the differences between the three cases – not inhibited, competitively inhibited and non-competitively inhibited, which occur at the beginning of sections 2 and 3.

## A note on symbols and terminology

The symbols used in enzyme kinetics have been standardized and the ones used in this section are collected here for reference. They are also defined when first used in the text.

| | | |
|---|---|---|
| E | Enzyme | concentration written as $e$ |
| S | substrate | concentration written as $s$ |
| ES | enzyme–substrate complex | concentration written as $p$ |
| I | inhibitor (competitive or non-competitive) | concentration written as $i$ |
| EI | enzyme–inhibitor complex | concentration written as $q$ |
| EIS | enzyme–substrate–inhibitor complex | concentration written as $p'$ |

$v$ = velocity (usually, but not necessarily, initial velocity)

$V$ = maximum velocity

$k_{+1}$ kinetic constant in the forward reaction in the first of a series of reactions

$k_{-1}$ kinetic constant in the reverse reaction in the first of a series of reactions

$k_{+2}$ kinetic constant in the forward reaction in the second of a series of reactions

$K_s$ the equilibrium constant (dissociation*) for the reaction

$$E + S \rightleftharpoons ES$$

$K_i$ the equilibrium constant (dissociation*) for the reaction

$$E + I \rightleftharpoons EI$$

$K_m$ Michaelis constant (see section 1).

*Example:*

$$E + S \underset{k_{-1}}{\overset{k_{+1}}{\rightleftharpoons}} ES \xrightarrow{k_{+2}} products + E$$

Equilibrium constant, of the first step $K_s = (k_{-1})/(k_{+1})$.

* In enzyme kinetics the equilibrium constants of binding are always written for the dissociation reaction. They are therefore the reciprocals of the association constants.

## Following an enzymic reaction

The quantity most frequently measured in the study of enzyme behaviour is the velocity of reaction, $v$. This velocity is best measured by the *appearance of products* rather than by the *disappearance of substrate*. This is because in the latter case a relatively small decrease (say from 100 per cent of the original value to 99 per cent) might be missed, while in the former case even a small increase (from 0 to a concentration equivalent to 1 per cent of the starting material) can be determined with accuracy.

If we plot the amount of product formed against time, we obtain a so-called 'progress curve' for the reaction. This usually looks like Fig. A1.1. As soon as

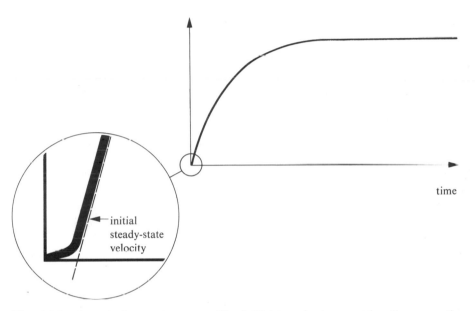

time

**Fig. A1.1** A typical progress curve. The initial transient cannot be shown on the main graph: it is shown in the inset, with a much expanded scale

the 'steady state' (see below) is established (usually a matter of milliseconds only) such a curve takes the form of a straight line. The straight-line period is called the *steady state* (cf. Ch. 29). The changes that occur before the steady state is achieved are called the *initial transient*. If enzymes were perfect catalysts, working under ideal conditions, the progress curve would then continue as a straight line indefinitely. However, in time, the enzyme is denatured, the substrate or co-factor is depleted, or the pH may change as a result of the reaction. These changes, together with many others that may arise, tend to slow the reaction down and in time to stop it completely. A progress curve soon departs from the steady state and is thus very difficult to use as a precise index of enzyme activity under defined conditions. You will recall that we only used a progress curve once in Ch. 6, when we were looking at the effect of temperature on activity.

Only at the start is there an approach to ideal conditions, when none of the factors tending to diminish reaction rate have had time to become significant.

The initial steady-state velocity, given by the tangent to the curve near the origin (Fig. A1.1), is a true and useful measure of enzyme activity under the conditions originally chosen for the experiment. Throughout this treatment, therefore, $v$ will refer to the *initial velocity* of reaction and it is, in fact, essentially the same quantity as the 'efficiency of catalysis' given on the $y$-axis of many of the graphs given in the latter part of Ch. 6. We saw in Ch. 6 that a number of factors affect $v$, among them enzyme concentration, temperature and pH. However, perhaps of most general concern is the response of $v$ to changes in substrate concentration and inhibitor concentration. Analyses of the responses of enzymes to changes in these variables form the majority of the quantitative information available on enzyme action, and it is frequently possible to draw conclusions of considerable theoretical interest from them. What is required is a rigid mathematical treatment of the relation of $v$, $s$ and (where applicable ) $i$ which can be applied in any system.

## Kinetic equations, section 1 – effect of substrate concentration

*The fundamental equation*

Let us assume that an enzyme E combines with its substrate to form an intermediate complex ES which then decomposes to form the products of the reaction and liberate the enzyme again. If we assume, for the moment, that the formation of the complex is readily reversible but that the reversal of the breakdown of the complex to form the products is insignificant,* we can write

$$E + S \rightleftharpoons ES \rightleftharpoons E + products.$$

Let $e$ and $s$ be the concentration of enzyme and substrate, respectively. Let $p$ be the concentration of ES at any instant. Let $k_{+1}$ be the rate constant for the forward reaction in the formation of ES and $k_{-1}$ the rate constant for the back reaction. Let $k_{+2}$ be the rate constant for the decomposition of ES to products and E.

Enzyme reactions are extremely rapid and it will normally only be a very short time before the steady state is achieved and $p$ builds up to its maximum value. It will stay constant at that value until the processes that lead to falling off of $v$ (mentioned above) start to become significant. Until this time, however, $p$ can be said to be constant. That is

$$\text{rate of formation of ES} = \text{rate of breakdown.}$$

If we remember the way in which reaction rates, concentrations and kinetic constants are related we can convert this expression to:

$$k_{+1}(e - p)s = k_{-1}p + k_{+2}p.$$

(the one way in which ES is formed)     (the two ways in which it can break down).

---

* The assumption that the rate of reversal is insignificant is a reasonable one, since the concentration of products at $t = 0$ (we are dealing with the initial steady state) is effectively zero. Therefore the velocity of the reverse reaction which is a function of this concentration will also be effectively zero.

Why has $(e - p)$ been given as the effective concentration of enzyme on the left hand? This is because an amount has been taken from the original concentration, $e$, to make ES. This amount is, of course, the same as $p$. It might be asked, why is not $s$ given as $(s - p)$? The reason is that, since enzymes are present only in very low catalytic amounts, $e$ will be low; therefore $p$, which cannot exceed $e$, will always be very much smaller than $s$. Therefore $(s - p)$ may be taken for all purposes to be the same as $s$.

Re-arranging (A1.1) we have

$$p = \frac{k_{+1}es}{(k_{-1} + k_{+2}) + k_{+1}s} \tag{A1.2}$$

We have already said that $v$, the observed (initial) velocity, is given by the rate of appearance of products. This is, of course, the initial rate of breakdown of ES by the forward reaction governed by $k_{+2}$, i.e.

$$v = k_{+2}p. \tag{A1.3}$$

From (A1.2) and (A1.3)

$$v = \frac{k_{+2}es}{s + \left(\dfrac{k_{-1} + k_{+2}}{k_{+1}}\right)} \tag{A1.4}$$

Now consider the maximum velocity, $V$, for the particular conditions of pH, temperature, etc. This maximum velocity will be attained when all the molecules of the enzyme are engaged in catalysis all the time. When this is true the enzyme will be all in the form ES and therefore $p$ will be equal to $e$. Therefore from (A1.3)

$$V = k_{+2}e \tag{A1.5}$$

and thus

$$v = \frac{Vs}{s + \left(\dfrac{k_{-1} + k_{+2}}{k_{+1}}\right)}. \tag{A1.6}$$

## The Michaelis constant

Equation (A1.6) has a claim to be called the fundamental equation of enzyme kinetics. Let us, for the moment, replace $(k_{+2} + k_{-1})/k_{+1}$ by the symbol $K_m$. Equation (A1.6) then becomes

$$v = \frac{Vs}{s + K_m}. \tag{A1.7}$$

What does $K_m$ mean? If we are able to make one assumption, it becomes a very useful quantity indeed. It is sometimes the case that the back reaction in the step $E + S \rightleftharpoons ES$ is very rapid – much more rapid, even, than the breakdown into products controlled by $k_{+2}$. In these cases we may write

$$k_{+2} \ll k_{-1}$$

$K_m$ is then $k_{-1}/k_{+1}$ which, it will be observed, is the equilibrium (dissociation) constant of the reaction $E + S \rightleftharpoons ES$, that is the binding constant (p. 14) of the substrate for the enzyme. This constant is written $K_s$. In physical terms what matters is that it is only if the forward rate of the second step is small compared with the backward rate of the first step that $[ES]$ will not be disturbed from its equilibrium (as opposed to its steady-state) value. In fact Michaelis and Menten originally assumed that ES was in equilibrium with E and S in deriving equation (A1.7), and that the $K_m$ was the equilibrium constant. We have chosen not to follow Michaelis and Menten, but to derive the expression for the most general case when

$$K_m = \frac{k_{-1} + k_{+2}}{k_{+1}}, \quad \text{i.e.} \quad = K_s + \left( \frac{k_{+2}}{k_{+1}} \right)$$

This expression was originated by Briggs and Haldane. You may already have met the terms 'Michaelis–Menten kinetics' and 'Briggs–Haldane kinetics'. The general form of the derivation was given here because the assumption needed to make $K_m = K_s$ is often untrue and because the logical progression in the general derivation as we have given it is actually easier to follow than in the special 'simplified' case.

In those cases when $k_{+2} \ll k_{+1}$ equation (A1.6) becomes

$$v = \frac{Vs}{s + K_s} \tag{A1.8}$$

and kinetic measurements could be used to study $\Delta G^\circ$ for the formation of the enzyme–substrate complex from equation (2.16, p. 18). Even when this is not completely true, it will be seen that $K_m$ contains a measure of the affinity of enzyme for substrate. This is of some importance in understanding the behaviour of some regulatory enzymes (Ch. 29). It is in fact often found that a remarkable amount can be deduced about enzyme behaviour from the study of such responses of initial rate to changes in other conditions.

There is a further unlooked-for consequence of equation (A1.7). This is that, when $K_m = s$, $v = V/2$. That is, the Michaelis constant is given by the substrate concentration when half the maximum velocity has been reached. (This statement is now the internationally accepted definition of the Michaelis constant, $K_m$, in order to avoid any confusion in going from the Michaelis–Menten case to the Briggs–Haldane case. This is in spite of the fact that, as we have seen, Michaelis himself defined it differently, as $K_s$, the equilibrium constant of the dissociation of ES to E + S.)

*The Lineweaver–Burk plot*

Note that (A1.7) can be written

$$\frac{1}{v} = \frac{1}{s} \cdot \frac{K_m}{V} + \frac{1}{V} \tag{A1.9}$$

simply by turning both sides upside down.

The similarity to $y = mx + c$, the equation representing a straight line, is obvious. A plot of $1/v$ against $1/s$, proposed by Lineweaver and Burk, gives a straight line (Fig. A1.3) rather than the rectangular hyperbola given by equation (A1.7) (Fig. A1.2). $1/V$ is given by the $1/v$ intercept (since, when $1/s = 0$, i.e. substrate concentration is infinite, velocity will naturally be a maximum). $1/K_m$ is given by the negative intercept on the $1/s$ axis.

It is obviously much easier to get values of $V$ and $K_m$ from intercepts on a straight-line plot than to try to deduce the position of the asymptote on the curve in Fig. A1.2. Particular difficulty is caused by the fact that a curve governed by equation A1.7 approaches the asymptote very slowly. For example $v$ is still 10 per cent below $V$ when $s = 9K_m$.

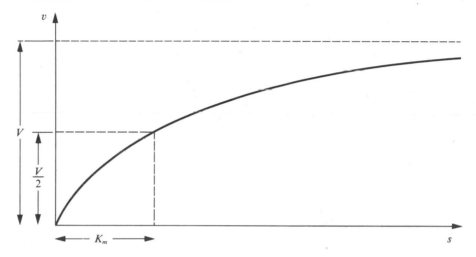

**Fig. A1.2**  The hyperbolic relationship between initial velocity and substrate concentration of a simple enzyme – catalysed reaction

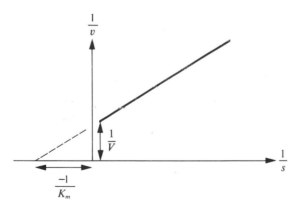

**Fig. A1.3**  Lineweaver–Burk plot (see text)

## INHIBITORS

We described in Ch. 6 the main way by which enzyme activity could be inhibited. Re-read the sections of Chapter 6 dealing with competitive and non-competitive reversible inhibitors.

$K_m$, which is an intrinsic property of the enzyme and substrate, is unaffected by a non-competitive process and is unchanged. The value of $V$, however, is obviously diminished in proportion to the amount of enzyme inactivated. A Lineweaver–Burk plot of enzyme with and without some inhibitor would in this case resemble Fig. A1.4. $K_m$, as we have said, is unchanged and so the two

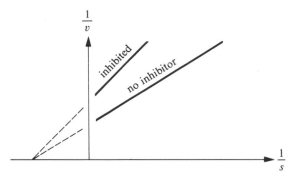

**Fig. A1.4** Lineweaver–Burk plot in the case of non-competitive inhibition, with the plot for the enzyme without inhibitor for comparison

lines have the same intercept on the $1/s$ axis. $V$, however, will fall and so the line for the inhibited sample will be *above* that for the enzyme alone on the $1/v$ *versus* $1/s$ plot. The slope of line in Fig. A1.4 for the enzyme with inhibitor equals, from equation (A1.15, below), $K_m/V(1 + i/K_i)$.

It will be seen later that the Lineweaver–Burk plot in the competitive case is affected differently. This provides a useful means of distinguishing between the two types of inhibition experimentally.

The reactions involved in this type of inhibition are (if I = inhibitor)

$$E + S \rightleftharpoons ES \tag{A1.10}$$

and

$$E + I \rightleftharpoons EI \tag{A1.11}$$

The equilibrium (dissociation) constant of reaction (A1.10) is $K_s$, that of (A1.11) is $K_i$. Also, since the inhibition shows no competition between S and I, a single enzyme molecule can interact with both S and I at once. This leads to the reactions

$$EI + S \rightleftharpoons EIS \tag{A1.12}$$

$$ES + I \rightleftharpoons EIS. \tag{A1.13}$$

Finally,

$$ES \rightleftharpoons E + \text{products} \tag{A1.14}$$

Assuming that I makes no difference to the binding of S to E and therefore that S makes no difference to the binding of I to E, the equilibrium constant for reaction (A1.12) is still $K_s$ and that for (A1.13) is still $K_i$.

As before, $v = k_{+2}p$ and to find $p$ we use the equation that, from the law of mass-action, represents the equilibrium of reaction (A1.10):

$$(e - p - p' - q)s = K_s \cdot p$$

but before it is used to find $p$ we eliminate from it $p'$ and $q$ by using $p' = p \cdot i/K_i$ (from A1.13) and $q = K_s \cdot p'/s$ from (A1.12),

$$q = \frac{K_s \cdot i \cdot p}{K_i \cdot s} \cdot$$

Hence

$$es - ps - \frac{pis}{K_i} - \frac{pi}{K_i} K_s = pK_s$$

so that

$$v = \frac{k_{+2}es}{s + \dfrac{i}{K_i} s + K_s + \dfrac{K_s i}{K_i}} \quad \text{or} \quad v = \frac{Vs}{\left(1 + \dfrac{i}{K_i}\right)(s + K_s)} \quad \text{(A1.15)}$$

By comparison with equation (A1.8) it will be seen that at a fixed concentration of inhibitor the velocity depends on substrate concentration in the same way, but that the effect of non-competitive inhibition is to diminish $V$ by a factor of $(1 + i/K_i)$. Note that we have taken $K_s = K_m$ and that this value is unchanged. Experimentally, therefore, all that is needed is to determine $V$ for a fixed value of $i$ and we will obtain $K_i$, the binding constant of the inhibitor to the enzyme – which is related to $\Delta G^{\circ\prime}{}_{(\text{binding})}$ in the usual way.

The effect of non-competitive reversible inhibitors is mimicked by an irreversible inhibitor that inactivates completely. A quantity of inhibitor that is less than sufficient to inactivate all the enzyme molecules will simply lower the effective concentration of the enzyme while not interfering at all with the catalytic activity of those molecules that remain. Thus, once again $K_m$ will be unchanged as it is a property of the enzyme that is independent of concentration, while $V$ is decreased in proportion to the fall in the effective concentration of the enzyme.

## Kinetic equations, section 3 – competitive inhibition

You will recall from Ch. 6 that in this case the inhibitor competes with the substrate for its binding site, often because it resembles it in structure (Fig. 6.21). Therefore, when the amount of substrate is very large the competitive inhibitor will be swamped out and its effect will not be felt. That is to say $V$, the velocity at infinite substrate concentrations, will be unchanged. At low levels of substrate, however, the inhibitor is able to act. By competing with the substrate for the enzyme it effectively lowers the affinity of enzyme for substrate, i.e. *increasing* the dissociation constant of ES back to E + S and thus increasing $K_m$.

The Lineweaver–Burk plot will look like the one in Fig. A1.5. $V$ is unchanged but the intercept on the $1/s$-axis changes because $K_m$ has changed. You can now see how the two types of inhibition can be distinguished.

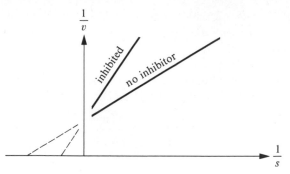

**Fig. A1.5** Lineweaver–Burk plot in the case of competitive inhibition, with the plot for the enzyme without inhibitor for comparison

In the case of competitive inhibition the slope of the line for the enzyme with inhibitor (Fig. A1.5) $= K_m/V(1 + i/K_i)$, from equation (A1.19) below. The equations now are

$$E + S \rightleftharpoons ES \qquad\qquad (A1.16)$$

$$E + I \rightleftharpoons EI \qquad\qquad (A1.17)$$

$$ES \rightleftharpoons E + products \qquad\qquad (A1.18)$$

(Question: Why are there no analogues of equations (A1.12) and (A1.13)?)

As before, applying the steady-state condition to equations (A1.16) and (A1.18)

$$k_{+1}(e - p - q)s = k_{+2}p + k_{-1}p$$

or, writing

$$K_m = \frac{k_{+2} + k_{-1}}{k_{+1}},$$

$$(e - p - q)s = K_m p.$$

Before this can be used to give $p$, $q$ is eliminated from it by using the relation $(e-p-q)i = K_i q$, from equation (A1.17), i.e.

$$q = \frac{(e - p)i}{K_i + i}.$$

Hence

$$es - ps - \frac{(e - p)i.s}{K_i + i} = K_m p$$

so that

$$p = \frac{es\left(1 - \dfrac{i}{K_i + i}\right)}{K_m + s\left(1 - \dfrac{i}{K_i + i}\right)}.$$

Since as usual $v = k_{+2}p$ and $V = k_{+2}e$

$$v = \frac{Vs}{s + K_m(1 + i/K_i)}. \qquad\qquad (A1.19)$$

Again, at a constant concentration of inhibitor, the velocity obeys a relation of the type $v = Vs/(s + K)$ but whereas $V$ has its original value, the apparent $K_m$ is now $K_m(1 + i/K_i)$.

$K_i$ is, as before, a measure of the efficiency of the inhibitor and of its $\Delta G^{\circ\prime}$ of binding and it is instructive to compare variations of $K_i$ when inhibitors of different structure are used. Correlations between $K_i$ and the structure of an inhibitor are another tool for probing the molecular nature of the active site.

Figures A1.6 and A1.7 show that we could have derived the same type of information from the simple plots of $v$ against $s$ of the type shown in Fig. A1.2.

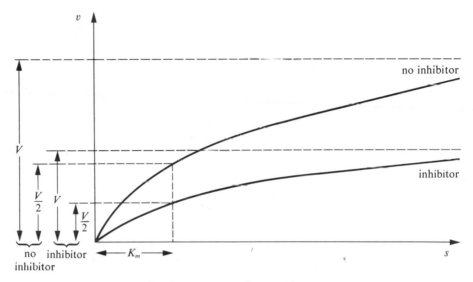

**Fig. A1.6**  The hyperbolic plot corresponding to Fig. A1.4

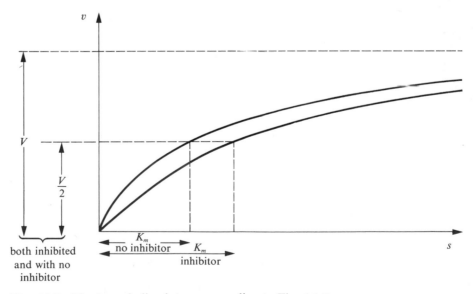

**Fig. A1.7**  The hyperbolic plot corresponding to Fig. A1.5

$V$ is the asymptote of the hyperbola in such a figure and so it is changed in the non-competitive case (Fig. A1.6), but unchanged in the competitive case (Fig. A1.7). $K_m$ is the value of $s$ corresponding to a velocity of $V/2$ (p. 514) and this is unchanged in Fig. A1.6 but changed in Fig. A1.7. It is clear, however, that the straight-line plots are much easier to use.

## Other types of inhibition

*Allosteric inhibition*    Allosteric inhibition (p. 131) is really a type of competitive inhibition. We saw in Ch. 6 that, even though the inhibitor does not bind to the active site, the effect of its binding is transmitted there through the tertiary structure of the protein. However $1/v$ *versus* $1/s$ plots and $s$ *versus* $v$ plots will often distinguish the allosteric from the ordinary, competitive inhibition (Fig. A1.8).

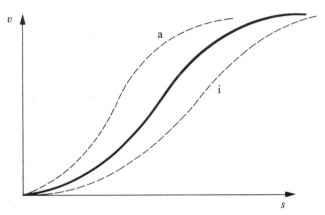

**Fig. A1.8**    Plot showing the sigmoidal relationship between $v$ and $s$, typical of many allosterically inhibited enzymes. Curve $a$ shows the effect of an allosteric *activator* ($K_m$ decreases) while curve $i$ shows the effect of an allosteric inhibitor ($K_m$ increases). (Compare Fig. 29.3)

*Inhibition by excess substrate*

This will give rise to plots of the type shown in Fig. A1.9. They approximate to the normal, inhibited case at low substrate levels (large values of $1/s$), but the velocity progressively falls below what is expected as $s$ increases. In other words, $1/v$ rises above what is expected as $1/s$ decreases.

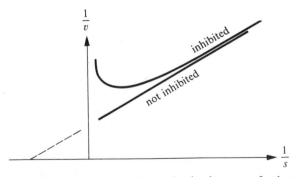

**Fig. A1.9**    Lineweaver–Burk plot in the case of substrate inhibition (see text)

# Appendix 2 **Separation of biochemical compounds**

Very little of what has been described so far could have been discovered had biochemists not been able to separate the compounds found in living systems from each other in sufficient quantity and purity for further study. If we imagine a mixture of just those compounds mentioned in this book, we begin to appreciate something of the problem. We should further reflect that many macromolecules in particular have several variant forms which differ in only a few atoms out of hundreds or thousands (e.g. nucleic acids of slightly different sequence and variant forms of proteins) and that this poses particularly severe problems of separation.

All of the advances in knowledge that we have described have therefore had to await the development of the appropriate separation methods. There now exists a very wide range of subtle and discriminating techniques and the number is constantly growing.

## A general approach to separation and purification

All separation schemes are based on one fundamental principle: taking the original mixture and causing certain of the components to travel to a different location than the others. This difference in location might be one side or the other of a filter after some of the components of the mixture had been rendered insoluble. It might on the other hand be the result of the migration of molecules in solution under the influence of electrical or other fields produced in highly complex and expensive equipment. Whatever system we employ, we are causing different substances to behave in different ways and are exploiting that difference to achieve a separation. We hope that you will have been convinced by now that the behaviour of materials depends on recognizable features of molecular structure. Thus substances of different molecular structure will behave differently – which is what we need for separation. On the basis of the knowledge that we have acquired of the relation between structure and behaviour, we can appreciate the way in which separation methods may be designed on a rational basis to achieve any particular result that is desired.

We will now describe in turn the separating systems that are contributing most notably to the development of biochemistry, emphasizing the underlying physical principles rather than the details of the construction of particular pieces of apparatus, except where these are essential to understanding.

Table A2.1 sets out those separating systems with which we shall deal.

**Table A2.1**

I. Centrifugation – the use of artificial gravitational fields:
  (a) sedimentation to a pellet
  (b) gradient centrifugation
      (i) sucrose type (differential gradient)
      (ii) caesium-chloride type (isopycnic)

II. Methods that depend on differences in solubility:
  (a) the use of changes in pH
  (b) the use of changes in concentration of salts
      (i) salting-in
      (ii) salting-out
  (c) the use of water-soluble organic solvents
  (d) the partition between an aqueous and an organic phase
      (i) countercurrent distribution
      (ii) partition chromatography: column, paper and gas–liquid chromatography

III. Methods that depend on the state of ionization of the molecules:
  (a) electrophoresis
      (i) in free solution
      (ii) on paper
      (iii) in gels
      (iv) isoelectrophoresis
  (b) ion-exchange methods
      (pH-precipitations could have been included in this section also)

IV. Methods that depend on other surface properties of the molecules:
  (a) adsorption chromatography
      (i) on columns
      (ii) in thin-layers
  (partition methods could have been included in this section also)

V. Methods that depend on molecular size and shape:
  (a) dialysis and membrane ultrafiltration
  (b) gel filtration
  (certain of the centrifugation and electrophoretic methods might have been included here)

## Centrifugation

Objects move at a rate that depends on their mass and the forces acting on them. If an object is suspended in a liquid, and is more dense than that liquid, it will tend to sink at a rate controlled by the balance between gravitational forces seeking to accelerate it and frictional forces (viscosity) trying to retard it. Do molecules in solution show this kind of behaviour? That is, if we leave a test tube of a solution of a protein on the bench, will the protein molecules eventually be found to have migrated to the bottom? Proteins are more dense than water, but even so, most people would answer intuitively that the protein would never collect at the bottom of the tube, and they would be correct. The reason is that the average energy of random thermal motion of a macromolecule is considerably greater than the change in gravitational potential energy involved in moving a few centimetres nearer the centre of the earth. As a result of this the molecules *diffuse* through the solution and there is, for any object small enough to be called a molecule as opposed to a macroscopic particle, only an imperceptible gradient of concentration down the tube.

We can show that this is so by means of a short calculation. Consider an aqueous solution of a protein, molecular weight $2 \times 10^4$, density 1.4 gm/cm³.

One gram molecule of this substance weighs 20 kg, and this effective weight is reduced by buoyancy to $(1 - 1/1.4) \times 20$ kg, i.e. approximately 6 kg. The downward force on 1 gm molecule is thus $6 \times 10$ newtons (acceleration due to gravity, $g \simeq 10$ m sec$^{-2}$) and the work done (=energy gained by the system) in moving this quantity of protein downwards by 1 metre is 60 joules. If we wish the protein to form a sharp band, of the order of 1 mm thick, the energy gained by moving downward through it is only 0.06 J/mole.

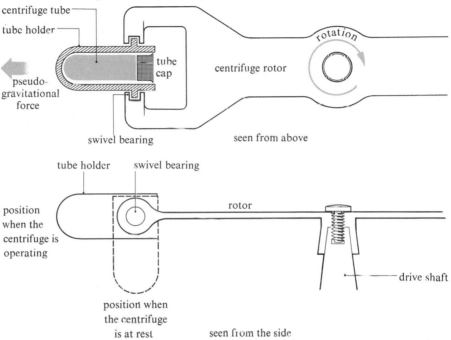

**Fig. A2.1**  A centrifuge rotor of the 'swing-out' type. The tube holder drops to the vertical position as the rotor stops

This value of 0.06 J/mole can be compared with the average energy of random thermal motion which, as we saw on p. 53, is of the order of 1 kcal/mole = $4.2 \times 10^3$ J/mole. Therefore, under normal gravitation, the energy change in falling 1 mm is about $10^{-5}$ of that of random thermal motion, which will tend all the time to redistribute the material in the tube.

Thus, if we could increase gravity by a factor of $10^5$, even protein molecules might fall to the bottom of the tube. Although we are unable to control true gravitational forces, it is well known that artificial gravitational fields can be produced in rotating objects. This fact is exploited by the *centrifuge*, a device for whirling tubes of liquid about a fixed axis (Fig. A2.1). An artificial field is formed perpendicular to the axis of the centrifuge rotor; at a distance $r$ metres from the axis it will have a magnitude of $r\omega^2$ m sec$^{-2}$ where $\omega$ is the angular velocity of rotation in radian sec$^{-1}$ ($2\pi$ radian = 360°).

Thus at 0.1m from the axis of an object rotating at 1000 rev/min ($= 1000.2\pi/60$ radian $\sec^{-1}$) the artificial field can be shown to have a magnitude of $\simeq 10^3$ m $\sec^{-2}$, i.e. $10^2$ times the force of gravity. Doubling $\omega$ increases the strength of the field by a factor of 4. It is mechanically possible to produce rotors that will withstand the tendency to fly apart produced by an angular velocity of tens of thousands of rev/min. It is therefore possible to produce artificial fields which are between $10^4$ and $10^5$ that of gravity, which is the order of magnitude of force that we calculated that we would require for the sedimentation of small proteins. As a result of our discussion it will be clear that the usefulness of the increase in the apparent gravitational field is not just to speed up the velocity of sedimentation, although that is important as well, but to overcome this tendency of the molecules to redistribute themselves.

The gravitational potential energy is thus increased to a figure above that of random thermal motion and it is possible to bring down all but the smallest protein molecules in a reasonable time as a discrete pellet at the bottom of the centrifuge tube. The mass of really small molecules (say below a molecular weight of $6 \times 10^3$, the size of the smallest proteins) is insufficient to produce a change in gravitational potential energy that is sufficient for sedimentation even in the higher artificial fields. However the sedimentation of larger molecules,

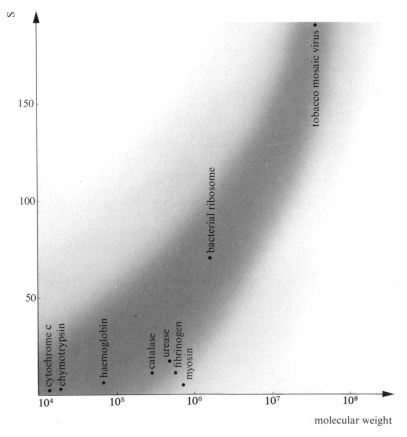

**Fig. A2.2**   The relationship between molecular size and the rate of sedimentation

viruses, parts of cells and cells themselves need lower angular velocities in proportion to their increased mass (Fig. A2.2).

A particle in a centrifuge will accelerate until the friction due to its passage builds up and exactly balances the pseudo-gravitational force. The velocity would then remain constant if it were not for the fact that the pseudo-gravitational field, as we have seen, increases with distance from the axis. It is therefore convenient to define the rate of sedimentation by means of the *sedimentation coefficient*, $S$, where

$$S = \frac{dr/dt}{r\omega^2}.\tag{A2.1}$$

(You will recall that we have already used the $S$ value, when describing the nucleic acids (pp. 155 and 438).)

$S$ depends on a number of factors, notably the mass of the molecule (corrected for buoyancy) and the frictional resistance to its passage. This resistance depends on the shape (i.e. degree of streamlining) of the molecule and the viscosity of the solution.

*Zonal methods*

We have so far restricted ourselves to considering how molecules can be made to pellet themselves at the bottom of the tube. Although this method is suitable for separating molecules of very different $S$ values (Fig. A2.3a, b) there is no point at which molecules that are closer in size are clearly resolved (Fig. A2.3c). The pellet will always contain a considerable amount of the lighter molecule, even if centrifugation is stopped before all the heavier molecule is sedimented. This problem can be avoided if we start the separation with all the material in a zone at the top of the tube (Fig. A2.4).

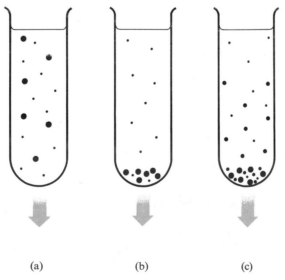

(a)           (b)           (c)

**Fig. A2.3** Centrifugation (see text). The arrow indicates the direction of the applied field

How may this be done? One way is to fill the tube almost completely with a solution somewhat more dense than water, but less dense than that of the molecules that we wish to separate (a solution of sucrose for example) and gently place the sample solution on top. We could even fill the tube with a solution that steadily increases in concentration (and thus density) as we move down the tube. This is better because this concentration gradient will stabilize the state of affairs shown in Fig. A2.4b against mixing, even if we stop the rotor and remove the tube from the machine at the point shown in the figure – before a pellet has formed. The reason is that any portion of the solution that is transferred up the tube by turbulence will find itself in a region of lower density and will tend to sink back. Similarly, any portion transferred downward will tend to rise. Thus the bands shown in Fig. A2.4b may be collected by moderately careful removal of the liquid from the tube.

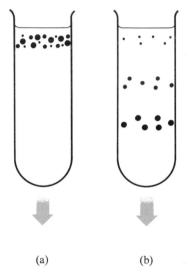

(a)                      (b)

**Fig. A2.4** Zonal centrifugation (see text). The arrow indicates the direction of the applied field

This sucrose-gradient type of separation (*differential gradient*) must be distinguished from centrifugations that employ a gradient of caesium chloride (*isopycnic* separations). This second type makes use of the fact that concentrated solutions of CsCl have a high density (about 3 gm/cm³ at saturation) which can easily exceed that of the macromolecules. A concentration gradient of CsCl can be formed by prolonged centrifugation – diffusion is sufficient to prevent the formation of a pellet even at the highest rev/min, but a marked concentration difference can be achieved down the tube after 24 hours at the highest possible artificial gravitational field. If we then produce a sufficiently strong gravitational field in the presence of our sample, molecules will move to the part of the gradient that has the same density as they do themselves. There they will stop, since buoyancy will exactly cancel the gravitational force. At $10^5$ $g$, the change in gravitational potential energy with distance (p. 523) is such that bands will be extremely sharp and, as we saw on p. 429, normal DNA (density 1.703) can be separated from otherwise similar molecules in which one-half the nitrogen atoms are $^{15}$N rather than $^{14}$N (density 1.723) and from those where all the nitrogen atoms are $^{15}$N (density 1.745). It is possible, in fact, to separate substances that differ in density by as little as 0.001 gm/cm$^3$. Note that in this method, as opposed to the sucrose method, the density gradient has a more positive role to play than simply stabilizing the bands against mechanical disturbance.

## Methods that depend on differences in solubility

In order to appreciate the principles that underlie the various precipitation and extraction methods, we must look at the forces experienced by molecules in solution. Consider the situation shown in Fig. A2.5, in which molecules of solute interact with each other and with the molecules of solvent. If the forces between

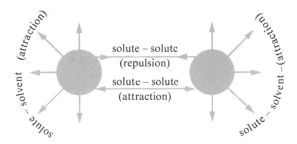

**Fig. A2.5** The balance between solute–solute forces and solvent–solute forces (see text)

solute and solvent are *stronger* than those between solute and solute, the substance will be *soluble*; if the forces between solute and solvent are *weaker* than the forces between solute and solute, the substance will be *insoluble*. The forces here will be the ones we discussed in Ch. 3. They depend on the properties of the groups of atoms in the solute molecules (particularly those on the surface). They will therefore vary from one type of molecule to another and they can constitute the basis of a separation.

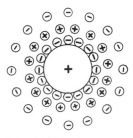

**Fig. A2.6** The ionic atmosphere about a charged ion

The solute–solute forces are a balance between attraction and repulsion (Fig. A2.5). The attraction is due to dipole and dispersion forces, the repulsion is that occurring between molecules of like charge. The forces between solute and solvent are very largely attractive. The attraction results partly from the polarization of molecules of solvent by charged groups on the surface of the solute and, more importantly, if there are inorganic ions present, from the acquisition by each charged group of an *ionic atmosphere*, that is shells of ions as in Fig. A2.6.

These then are the interactions that we must manipulate if we are to precipitate a substance from solution, or extract it from one solvent into another. We shall consider in turn the separation methods that depend on them (given in Section II of Table A2.1) and shall show how the general principles that we have described in the last few paragraphs operate in practice.

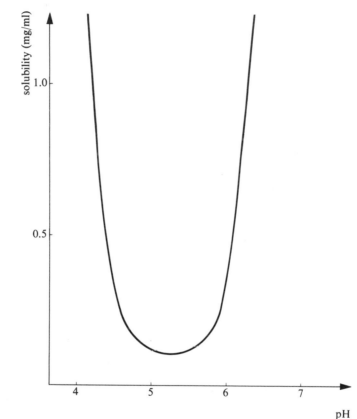

**Fig. A2.7**   The variation of solubility of insulin with pH

*Changes in pH* affect principally the repulsive component of the solute–solute forces. The net charge on a macromolecule varies with pH much as the net charge of an amino acid does (p. 68), but the titration curve of the macromolecule is of a more complex shape, since it is the sum of a large number of titration curves of the various charged groups that it possesses. However, the important feature for our purposes is that the curve passes through a point at which the molecule is electrically neutral (compare p. 65). At this pH (*the isoelectric point*, pI) the electrostatic repulsion that exists between like molecules at all other pH values is neutralized. With the repulsion gone, the attractive forces between one molecule of solute and another will be at a maximum, and will tend to diminish solubility. Figure A2.7 shows the variation of solubility of insulin with pH. Many other proteins, with different isoionic points, will still be soluble at the isoionic point of insulin, and so a purification can be achieved by centrifuging down the precipitate that is formed when the pH is adjusted to the isoelectric point.

*Changes in ionic strength* influence principally the solute–solvent forces.

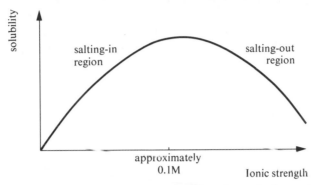

**Fig. A2.8**  The variation in solubility of a typical protein with ionic strength

Figure A2.8 shows the type of relationship that exists between solubility and ionic strength. (Ionic strength $I$ is defined by the relationship

$$I = \tfrac{1}{2}\sum_i c_i z_i^2. \tag{A2.2}$$

That is, ionic strength is one-half of the sum of the products of the concentration of each ion and the square of its valency. Thus a 0.1 M solution of NaCl has $I = 0.1$, but a 0.1 M solution of $(NH_4)_2SO_4$ has $I = 0.3$. Check this latter figure for yourself, noting that the concentration of $NH_4^+$ is *not* 0.1 M.)

The rise in solubility that occurs as we leave very low ionic strengths is known as *salting-in*. It is usually over, and solubility is at its maximum, by the time that $I$ reaches 0.1. Salting-in occurs because at very low ionic strengths there are insufficient ions to form the ionic atmosphere around the charged groups on the molecule. The ionic atmospheres contribute greatly to solute–solvent forces, and thus until sufficient ions are present solubility will not reach a maximum. Both proteins and nucleic acids can be salted-in – for example DNA forms a sticky solution at the best of times but if $I < 10^{-3}$ it is so sticky as to be quite unmanageable.

If $I$ continues to rise, solubility falls again and *salting-out* is said to occur. (Salting-out is a much favoured method in protein purification, probably more than any other solubility method.) The fall in solubility is largely a result of the fact that at very high concentrations of inorganic salts there is a noticeable decrease in the concentration of water and thus of the solute–solvent forces. This last point may be illustrated by considering a saturated solution of $(NH_4)_2SO_4$. (This compound is the one most frequently used for salting-out because it is easy to obtain in large amounts, because of the high ionic strengths that are obtainable even in quite dilute solutions (see above), and because of its great solubility.) A saturated solution of $(NH_4)_2SO_4$ contains over 500 gm of salt/litre. Its density is 1.25, and it therefore contains less than 750 gm $H_2O$/litre and the molarity of water (see p. 13) falls to about 0.7 of its normal value.

The use of this method in practice involves first bringing the concentration of

$(NH_4)_2SO_4$ up to some starting value by the addition of the solid or a saturated solution. The starting value is normally chosen by trial experiments as the one that causes the precipitation of the maximum possible amount of unwanted protein with the retention in solution of the maximum possible amount of the desired protein (point 'A' in Fig. A2.9). The unwanted protein is removed by centrifugation, and the ionic strength is then further increased until the maximum possible amount of the desired protein has been thrown down with the retention in solution of the maximum possible amount of all other protein (point 'B' in Fig. A2.9). Because the procedure is suitable for large-scale operation, it is frequently the first step in the extraction of a protein from the crude homogenate of a piece of tissue.

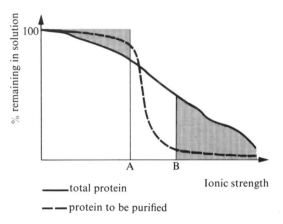

**Fig. A2.9** Fractionation with ammonium sulphate. The areas of tone indicate the amount of unwanted protein that is removed

*Water-soluble organic solvents* affect solubility in two ways. First, they decrease the concentration of water. This reduces the strength of solute–solvent forces for macromolecules (such as most proteins) that have primarily hydrophilic surfaces and intensifies the forces for those (such as most lipids) that are primarily hydrophobic. Secondly, they influence the various interatomic forces. We saw in Ch. 3 that the electrostatic forces were of the form

$$\text{Force} \propto \frac{1}{\varepsilon r^n}$$

where $n$ varies between 1 and 7, depending on the type of force involved. Thus, if we could reduce the dielectric constant, $\varepsilon$, from the exceptionally high value for water of 81, the forces will become more intense. This can be done by adding acetone ($\varepsilon = 21$) or ethanol ($\varepsilon = 24$) to the aqueous solution. Both attractive and repulsive forces are intensified, but, for most macromolecules, the attractive forces eventually predominate and solute–solute forces become stronger. The net effect of all these changes is that proteins and nucleic acids are precipitated, while lipids are made more soluble. The method is suitable for large-scale work, but does not provide particularly fine resolution.

Because of the fact that interatomic forces are altered so profoundly, care must be taken to avoid denaturation. One measure that is usually employed is to cool the solvents to $-20°$ before use.

*Partition between an aqueous and an organic phase* relies on the fact that the balance of the forces in Fig. A2.5 is better satisfied for some substances in an organic phase than in an aqueous one. (In other words, some materials are more hydrophobic than others.) The tendency of a substance to prefer one or other of two phases can be expressed by a constant, specific to the substance and to the solvent system, called the *partition constant*:

$$K_{(partition)} = \frac{(\text{concentration of substance in one phase})}{(\text{concentration of substance in the other phase})}. \quad (A2.3)$$

That is, the substance will distribute itself between the two phases in a constant ratio. You will probably already have recognized $K_{(partition)}$ as the equilibrium constant of the partition process and we can add it to our collection of constants to which the usual free-energy relationship applies.

$$\Delta G°_{(\text{transfer from one phase to another})} = RT \log_e K_{(partition)} =$$
$$1364 \log_{10} K_{(partition)} \quad (A2.4)$$

We shall see in a moment that this expression is useful in predicting the behaviour of substances in separating systems that depend on partition.

By using organic solvents that are partially miscible with water, a wide range of types of phase can be produced with different degrees of hydrophobicity. Solvent systems can be selected that bring about separations between members of all classes of the macromolecules and of their fragments. This includes the lipids which, because of their very poor solubility in water, are not readily handled by the methods that we have mentioned so far.

We saw in Ch. 4 that proteins have some hydrophobic residues at their surfaces and thus that different proteins have different degrees of hydrophobicity. Similarly, different peptides produced by the degradation of proteins during sequence analysis (p. 82) have a different balance between hydrophobic and hydrophilic residues. These differences (and analogous ones in the other classes of macromolecule) give rise to variations in $K_{(partition)}$ which can be exploited to bring about separations. Whole macromolecules are usually handled in *counter-current distribution machines*, which consist of a complicated, interconnected system of containers for shaking the two layers, organic and aqueous, together and then separating them. The reason for having so complex an arrangement, rather than simply shaking in a test tube, is to make it possible to repeat the partition again and again automatically so that even substances with $K_{(partition)}$ values that are very close will eventually be separated.

Fragments of macromolecules can be handled by partition chromatography on paper (the term is usually shortened to '*paper chromatography*').

This method depends on a peculiar property of cellulose, of which paper is almost entirely composed. We saw in Ch. 8 that cellulose fibres have a highly organized sub-structure. It appears that there exist on the surface of the fibres arrays of —OH groups (which of course abound in cellulose) that are capable of hydrogen-bonding a large number of water molecules. If one hangs a sheet of paper in a closed volume in which there is a vapour with an aqueous component, quite large quantities of water are taken up in preference to any organic vapours present. Fortunately, the paper merely becomes damp, and is not saturated, and if the end of the sheet is dipped in an organic solvent, there will still be sufficient open spaces between the fibres to draw up a large volume of solvent by capillarity. The result is a moving front of the organic mixture (Fig. A2.10a). The —OH groups bind the water molecules sufficiently strongly to ensure that the water is

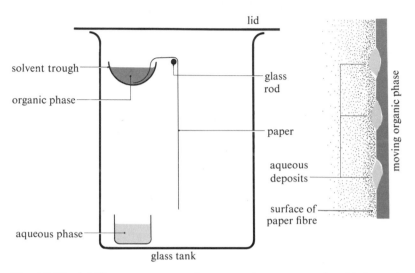

**Fig. A2.10**   (a) The arrangements for paper chromatography. (b) A much-magnified view of part of (a) showing interaction between the moving organic phase and the stationary aqueous pockets

not moved by the flow of solvent. We have now achieved what we require for separation by partition: the organic solvent is drawn across the aqueous deposits (Fig. A2.10b) and a very efficient partition takes place between them, owing to the large surface-to-volume ratio of such small deposits of water.

Typical chromatograms are shown in Fig. A2.11. The original paper-chromatography systems used two phases in the way that we have described. The organic

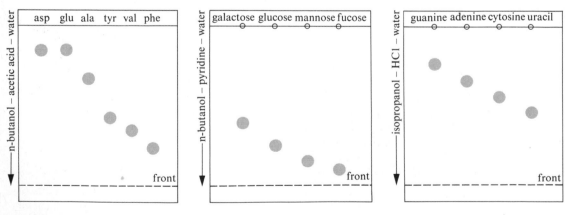

**Fig. A2.11**   Some typical paper chromatograms

layer was used as the mobile phase and the aqueous layer was placed in the bottom of the enclosure or tank to saturate the atmosphere with water and provide the stationary phase. A highly hydrophobic substance will partition in favour of the mobile (organic) phase. In molecular terms this means that the substance spends more of its time in the mobile phase and will be carried along with it in proportion to the time spent in it. Conversely, a hydrophilic substance will spend more of its time in the stationary (aqueous) phase and will be carried along much more slowly.

It was later found that one-phase systems (i.e. an aqueous solution of miscible organic compounds) could also be used. Both one- and two-phase systems are now in use. The apparent paradox – partition chromatography without a second phase to partition in – is resolved when we recall what we mentioned above, that the water is *preferentially* drawn by the paper out of the vapour. The bound phase will thus be more aqueous than the mobile phase or its vapour.

We shall now show how the chromatographic behaviour of a molecule can be predicted from its structure.

By applying the laws for partition (equation A2.3, p. 53) and diffusion, it can be shown that, as a first approximation, the chromatographic mobility is given by the relationship

$$R_f = \frac{A_1}{A_1 + K_{(partition)}A_s} \tag{A2.5}$$

where $R_f$ is the ratio between the distance moved by the substance and that moved by the front (Fig. A2.11) – this ratio should be constant for the substance and solvent system concerned, no matter how far the front travels. $A_1$ is the effective cross-section (taken perpendicular to the direction of travel of the solvent) of the mobile phase, and $A_s$ is that of the stationary phase. $A_1$ and $A_s$ can be determined by measuring the paper and weighing it in the vapour-saturated and dry states.

Combining equations (A2.5) and (A2.4) we obtain

$$\Delta G^\circ_{(transfer)} = RT \log_e \frac{A_1}{A_s}\left(\frac{1}{R_f} - 1\right). \tag{A2.6}$$

Now $\Delta G^\circ_{(transfer)}$ for a substance is, as a first approximation, the sum of the $\Delta G^\circ_{(transfer)}$ terms for each of the individual parts of the molecule. Thus, for a peptide of $n$ residues we may consider separately the transfer of the $\alpha$-$NH_3^+$ group, the $\alpha$-$COO^-$ group, $(n - 1)$ peptide bonds and $n$ side chains (including the $\alpha$ carbon atom).

$$\Delta G^\circ_{(transfer\ of\ peptide)} = (n - 1)\,\Delta G^\circ_{(transfer\ of\ peptide\ bonds)} + \Delta G^\circ_{(transfer\ of\ -NH_3^+)}$$
$$+ \Delta G^\circ_{(transfer\ of\ -COO^-)} + \sum \Delta G^\circ_{(transfer\ of\ each\ side\ chain)} \tag{A2.7}$$

If we were simply transferring the free amino acids

$$\Delta G^\circ_{(transfer\ of\ amino\ acids)} = n\,\Delta G^\circ_{(transfer\ of\ -COO^-)} + n\,\Delta G^\circ_{(transfer\ of\ -NH_3^+)}$$
$$+ \sum \Delta G^\circ_{(transfer\ of\ each\ side\ chain)}. \tag{A2.8}$$

But, from equation (A2.6)

$$\Delta G^{\circ}_{\text{(transfer of amino acids)}} = \sum RT \log_e \frac{A_1}{A_s} \left( \frac{1}{R_{\text{f (each amino acid)}}} - 1 \right). \quad \text{(A2.9)}$$

$\Delta G^{\circ}_{\text{(transfer of side chains)}}$ is the same in both equations (A2.7) and (A2.8); it is therefore possible to equate the rest of these two expressions. This, taken together with equation (A2.9), leads to the relationship

$$RT \log_e \left( \frac{1}{R_{\text{f (peptide)}}} - 1 \right) = (n-1)A + B + \sum RT \log_e \left( \frac{1}{R_{\text{f (amino acids)}}} - 1 \right)$$

$$\text{(A2.10)}$$

where A and B are constants. (You can verify this for yourself if you wish.)

If the $R_f$ values are determined experimentally for all the amino acids and A and B determined from the measured $R_f$ values for peptides of known structure, the $R_f$ values of all others can be predicted. Alternatively, the proposed composition of a peptide can be checked by comparing the predicted $R_f$ with the experimentally obtained value. It is even sometimes possible to make tentative identifications of new substances (e.g. a minor nucleotide base or a modified amino acid) from their $R_f$ values since, for example, an extra methyl group would give a characteristic change in $R_f$ relative to the unmodified material.

Once again you can see that what appears at first sight to be a wholly arbitrary event – in this case the separation of closely similar compounds into a defined order on a piece of paper – can in fact be explained quite well in terms of the basic physical principles set out in the early chapters of this book.

*Gas–liquid chromatography* depends on much the same principles as the previous partition methods. The stationary phase is a liquid, trapped in some porous solid, while the mobile phase is a stream of gas. Small quantities of volatile substances partition between the liquid and the vapour and can be very rapidly and effectively separated by this means. The requirement for volatility restricts the use of the method to rather small, non-polar molecules. The volatility of molecules that might otherwise be disqualified can sometimes be enhanced by chemical modification.

## Methods that depend on the state of ionization of the molecules

We have described in Part 1 of this book the factors that govern the state of ionization of molecules (see especially pp. 10 ff., 64 ff., and 139). You will recall that the net charge depends on the number of ionizable groups, the p$K$ values of these groups and the pH of the solution in which they find themselves. We therefore have in net charge a property of the molecule that is closely related to structure (and will therefore differ between different structures), but which can be influenced by the outside environment. These are the ideal factors on which to build a separation system: all that is needed is to use the property of net charge to induce movement in space. We have already given, in describing pH precipitation, a crude example of their use. We shall now mention two more. The first is *electrophoresis*, in which the movement of the molecules is migration, at different rates and directions, in an electric field; the second is *ion exchange*, in which the molecules are made to move past a charged surface to which they may or may not bind.

*The rate of electrophoretic mobility* depends on the intensity of the electric field, the net charge, the mass of the molecule and the degree of frictional resistance to its passage. (Compare this with the factors affecting the $S$ value, p. 525).

The intensity of the electric field ($X$) is the voltage applied to the system divided by the distance between the electrodes (Fig. A2.13). In a precise analogy with the centrifugal force, it can be shown that ideally the intensity should be large, not only to provide a reasonably rapid rate of migration, but also to be more effective in overcoming the tendency of random thermal motion to redistribute the molecules throughout the volume of the separating system.

The force $F$ on a molecule of charge $e$ exerted by an electric field of intensity $X$ is given by

$$F = -\frac{e}{\varepsilon} X \qquad (A2.11)$$

where $\varepsilon$ is the dielectric constant of the medium (p. 52).

This force will accelerate the molecule until it is balanced by the frictional forces due to the viscosity of the medium. This frictional force was shown many years ago by Stokes for a spherical macroscopic object at an effectively infinite distance from the walls of the vessel to be given by

$$F = 6\pi r\eta V \qquad (A2.12)$$

where $\eta$ is the viscosity of the solution and $V$ the velocity.

If Stokes' law holds at the molecular level, we can equate the two expressions and obtain

$$V = \frac{-Xe}{6\pi r\eta\varepsilon}. \qquad (A2.13)$$

Since the mass of a sphere is proportional to the cube of the radius we can see that

$$V \propto \frac{-Xe}{M^{1/3}\eta\varepsilon}. \qquad (A2.14)$$

This relationship might govern the migration of molecules in free solution. If, however, our condition that the walls of the vessel should be infinitely distant is not satisfied, the equation will not apply. (We shall meet such systems in a moment.) In extreme cases, in fact, the system will resemble, not a sphere dropping through a large tank of water, but a spherical object sliding over a plane. It can be seen intuitively that in this case the frictional resistance is proportional to the surface area of the molecule, which will be, in its turn, proportional to $r^2$, that is $(\sqrt[3]{M})^2$.

Therefore in such a case

$$V \propto \frac{-Xe}{M^{2/3}\eta\varepsilon}. \qquad (A2.15)$$

This relationship should govern the migration of molecules in restricted environments (such as the interstices of moistened paper, see below).

Using such relationships it is easy to see how we can take advantage not only of differences in the *sign* of the net charge (which brings about migration in different directions), but also in its magnitude (which brings about migration at different rates). This may be done in a variety of types of apparatus. If the migration is to occur in free solution, the sample is usually introduced as a zone (compare zonal centrifugation methods, p. 526). As was true for centrifugation, a

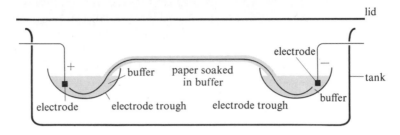

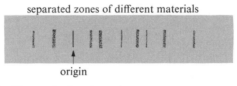

Fig. A2.12   Paper electrophoresis

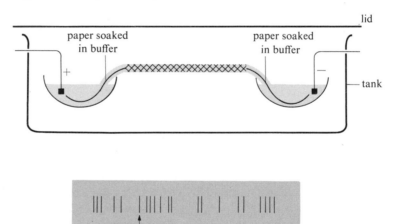

Fig. A2.13   Gel electrophoresis

gradient of concentration of sucrose is often used to stabilize the system against mechanical mixing of all kinds, including that caused by convection due to the heat generated by the passage of electric current. Alternatively, stabilization may be achieved by soaking the separating system up into paper or cellulose acetate (Fig. A2.12) or on to a gel of starch or synthetic polymer (Fig. A2.13). In the latter case the pore size of the gel can be controlled, and this gives further possibilities of separation that are based on the size and shape of the molecules, since a pore size can be chosen that will permit only some of them to pass.

Proteins and nucleic acids, having many charged groups of their own, are very amenable to electrophoretic separation. Lipids, however, are generally not separable because they are usually too insoluble in the polar solvents that the passage of electric current requires. Carbohydrates can be separated by electrophoresis even if they have no charged groups. This is achieved by allowing them

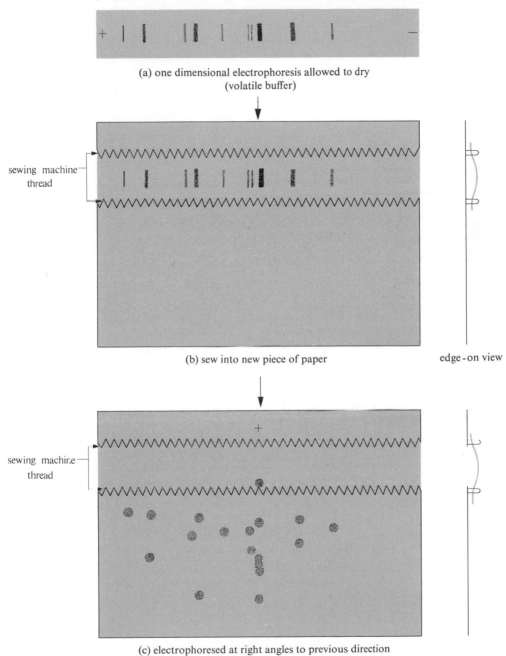

(a) one dimensional electrophoresis allowed to dry
(volatile buffer)

sewing machine thread

(b) sew into new piece of paper

edge-on view

sewing machine thread

(c) electrophoresed at right angles to previous direction

**Fig. A2.14** Two-dimensional paper electrophoresis (the zones are not revealed by staining until the last step – they are drawn in the earlier step for clarity)

to bind (which they do very readily) to borate ions, the charges of which then promote the electrophoretic migration of the carbohydrates.

We said before that *e* can be varied by changing pH. This can be used to bring about the further separation of substances that migrate together at one pH. Thus consider, for example, two proteins of the same molecular weight and shape. Let one have ten positively charged groups and five negatively charged groups and the other five positively charged and no negatively charged groups.

You can use the p$K$ values quoted on p. 68 to satisfy yourself that (if histidine is not present in either protein) both are likely to have the same mobility at

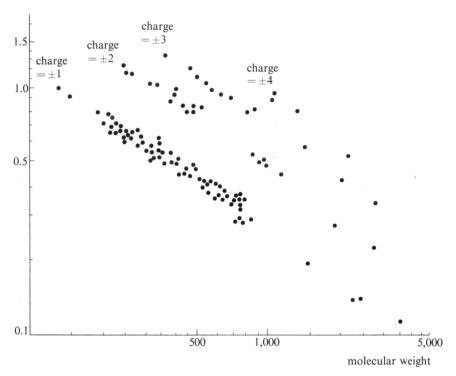

**Fig. A2.15** The relationship between mass, charge and electrophoretic mobility of peptides at pH 6.5 (see text)

pH 6.5, but that they will be easily separable at pH 2. If the support is paper, such a two-dimensional separation is easy to achieve. The mixture is electrophoresed at one pH; the paper is then dried, turned through a right angle, moistened with a buffer at the second pH and the voltage is applied once more (Fig. A2.14).

The fact that electrophoretic mobility is so closely related to structure is of assistance when determining amino-acid sequences. As with paper chromatography, the possible composition of a peptide can be checked against the expected behaviour on paper electrophoresis. This is because different compositions may give rise to significant differences in charge and molecular weight. Figure A2.15 shows the measured mobilities of a large number of peptides of known composition. Show for yourself that the slope and separation of the lines on this logarithmic plot confirms the validity, in this case at least, of equation (A2.15).

You will probably have been struck by the many analogies between centrifugation and electrophoresis. These analogies are not surprising in view of the fact that the two processes differ only in the nature of the force used to move the molecules. We can add a further analogy – the fact that there is an electrophoretic equivalent to equilibrium centrifugation (in CsCl), which we described on p. 526.

It is found that certain buffers, if subjected to an electric field in a tube, will cause the formation of a pH gradient in the tube (low pH at the positive end, high pH at the negative end), much as CsCl will, after centrifugation, form a density gradient.

Any molecule with ionizable, titratable groups will migrate until it passes to the position at which the pH is equal to its pI value. At this point its positive and negative charges will balance and, with a net charge of zero, motion will cease. As with the centrifugation method, the bands will be very sharp and fine separations will be possible, because the effect of random thermal motion (diffusion) is resisted by the following mechanism. As soon as a molecule strays in the direction of the negative electrode, i.e. into a region of pH that is higher than its pI, it will acquire a negative charge because fewer protons are present. It will therefore be repelled from the negative electrode. A similar argument shows that molecules are repelled from the positive electrode.

This method, which is called *isoelectrophoresis*, is used in free solution or on a solid support. It is probably the most discriminating of all the electrophoretic systems (if not of all the systems of any type). It is possible to separate molecules that differ in pI by as little as 0.01 of a pH unit.

*Ion-exchange chromatography* depends on electrostatic interactions between charged molecules and a charged solid substance. If we imagine a tube filled with granules of a polymer that bears many positively charged groups (Fig. A2.16), we can see at once that negatively charged molecules will be abstracted from a mixture that is allowed to flow past. These can be subsequently desorbed from the solid granules in two ways. The first is by raising the concentration of some other negatively charged ion, $Cl^-$ say, until competition for the positively charged sites becomes too severe and the original molecule drops off the granules and can be washed away. The increase in $[Cl^-]$ could be gradual, so that the more weakly bound substances fall away first (Fig. A2.17). The second method is to change the pH. If you worked through the example on p. 537 you will have seen that such changes can drastically alter the net charge of a molecule. A negative charge could easily be neutralized or turned to a positive charge. Again the pH change could be progressive, so that different substances came off at different times. (If the positively charged group on the polymer is itself titratable, e.g. if it is $-NH_3^+$, then this charge can be neutralized also, and the molecules desorbed in that way. This is less desirable than using the effect of pH on the charge of the adsorbed molecules themselves, since if the granules are discharged all the adsorbed molecules will come off together.)

We have now described the operation of a positively charged ion exchanger.

Interchanging 'positive' and 'negative' throughout, our description also covers the action of a negatively charged ion exchanger. (Much of our description may have been familiar to you from earlier chemistry courses since it refers equally

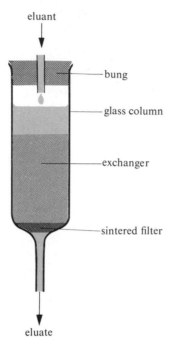

Fig. A2.16   Ion exchange chromatography

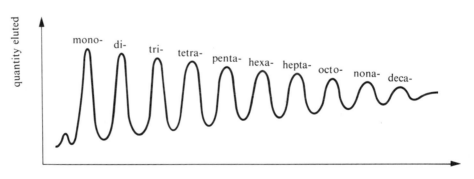

Fig. A2.17   The separation of a mixture of oligonucleotides into fractions that are homogeneous with respect to chain length

well to the commercial water-softening process, where $Ca^{2+}$ is adsorbed on to a negatively charged exchanger and subsequently desorbed by $Na^+$ when the exchanger is regenerated with NaCl.)

There are two major types of granular material in use – one is a polystyrene-based plastic and the other is cellulose. Chemical modifications allow the introduction of positive or negatively charged groups into these materials. Two modified celluloses that are particularly popular for the separation of proteins are carboxymethyl ($-CH_2-COO^-$) cellulose* and diethyl aminoethyl

$$\left(-CH_2-CH_2-N\begin{array}{c} CH_2-CH_3 \\ \\ CH_2-CH_3 \end{array}\right)$$

cellulose.† DEAE cellulose, having a positive charge, is also suitable for separations of oligonucleotides.

These two exchangers are only mildly acidic and basic respectively, since they have charged groups very similar to those of the macromolecules themselves. This is in contrast to the sulphonated polystyrenes, for example, which can have as many acidic groups in a given volume as 0.5 M $H_2SO_4$ and are actually used as hydrolytic agents in carbohydrate chemistry. With the use of pH and salt gradients, quite fine separations are possible on a scale much larger, if desired, than that possible with most electrophoretic methods. Chromatography on

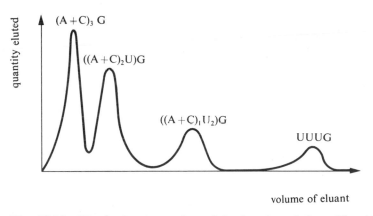

Fig. A2.18   The further separation of the fourth peak from Fig. A2.17

modified cellulose is probably a more common step in the extraction and purification of proteins than any save salting-out with ammonium sulphate.

Once again the behaviour of molecules in these systems is a comprehensible consequence of structure. For example, consider the mixture of oligonucleotides produced by digestion of RNA by ribonuclease $T_1$ (p. 164). We saw that all of them will terminate in G and will have for the rest of their length any number of A, C or U residues. Use the pK values given on pp. 137, 140 to show that, at pH 7, the net charge of each oligonucleotide is determined by its chain length (i.e. number of bases) alone and not by the base compositions. Having done so you will be able to appreciate the molecular basis of the elegant separation shown in Fig. A2.17. (In this experiment urea was added to the solutions to overcome all but the electostatic interactions, see pp. 134 ff.) Each peak contains all the oligonucleotides of a given chain length and can be further subjected to chromatography. Use the pK values again to explain the separation at pH 2.7 (Fig. A2.18) of the fourth peak from the separation shown in Fig A2.17.

* Abbreviated to CM cellulose.
† Abbreviated to DEAE cellulose.

Since paper is a cellulose product and since ion exchange can take place on modified cellulose, it is possible to make *ion-exchange papers*. These can be used for ion-exchange chromatography or for electrophoresis. Figure A2.19a shows a two-dimensional separation of an oligonucleotide digest of RNA. The first

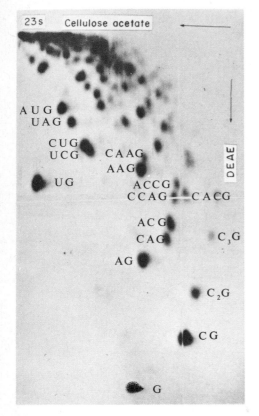 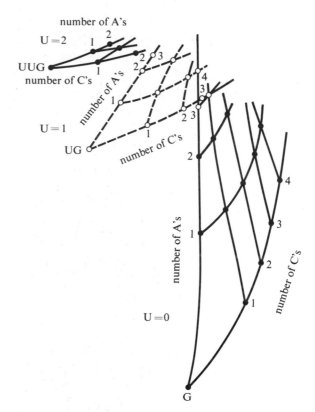

**Fig. A2.19** (a) A two-dimensional separation of oligonucleotides. (b) A drawing that shows the relationship between position in (a) and nucleotide composition

dimension relied on electrophoresis on cellulose acetate, and the second employed electrophoresis on DEAE paper. The complex interactions between the electric field, the charged molecule and charged support give rise to very subtle separations. Figure A2.19b shows that nucleotide structure is obviously controlling the position of the spots. It is this method, developed by Sanger and his colleagues, that has made it possible to separate closely similar oligonucleotides and thus to allow the determination of the nucleotide sequences of RNA molecules (see p. 163 f.).

## Methods that depend on other surface properties

*Adsorption chromatography* is a rather similar process to ion-exchange chromatography. Here, however, the solid support does not retain molecules from the solution primarily by strong electrostatic forces (although these are not excluded). If you have used activated charcoal in preparative organic chemistry, you will probably have learned that some solid substances have an extremely finely-pitted surface, the pits being of molecular size. Some substances will fit into these pits quite well (although with nothing like the specificity and tenacity of the interactions involved in the binding sites of proteins). Such substances will therefore be retained until desorbed by some competing molecule. Alternatively, if they are even more weakly bound, they will spend an appreciable fraction of the time in the desorbed condition. While desorbed they can be carried on by the flow of solvent through the system. The more weakly they are bound, the more time they will spend in solution and the higher will be their $R_f$ (p. 533). Partition phenomena (p. 531) may also play a part in many of these separations.

Adsorption chromatography can be carried out in columns packed with fine granules of such compounds as silica, alumina or nylon (these are generally superior to activated charcoal, which is, however, sometimes used for sugars).

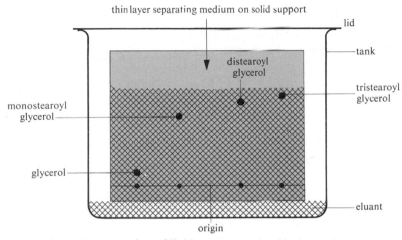

**Fig. A2.20** The separation of lipid monomers by thin-layer chromatography

Alternatively, a thin layer of powdered adsorbent can be deposited on a plate of glass or plastic and handled in much the same way as a piece of paper in partition chromatography. If the powder is extremely fine, capillarity causes the solvent to flow very rapidly giving an exceptionally fast separation. (Thin layer chromatography of physiological fluids is popular for this reason for drug tests at sporting events or to enable the correct antidote to be selected in the case of poisoning by an unknown drug.) The fineness of the powder also cuts down the interstitial volume between the granules and this minimizes diffusion and gives rise to very sharp spots. Adsorption chromatography, usually in its thin-layer form, is extremely useful for another reason – adsorption and desorption take place well in non-polar media. Lipids and other non-polar molecules are therefore better separated in such systems than any other (Fig. A2.20).

## Methods that depend on molecular dimensions

We have seen that considerations of molecular size and shape are often relevant to the behaviour of molecules in separating systems. However there are systems where these properties are *exclusively* important. (You will see in a moment why we distinguish here between molecular *weight*, as in most of our previous discussions, and, in this case, molecular *dimensions*.)

It is possible to obtain thin but tough sheets of certain materials (usually cellulose-based substances) that are pierced by countless tiny pores. The process of manufacture can be adjusted so as to give holes of any desired average size. Sheets and tubes of these materials are available with pore sizes that range from that corresponding to the molecular dimensions of, say, sucrose, to those corresponding to virus particles. Clearly, if a mixture of molecules of various sizes is confined on one side of such a membrane, and pure water placed on the other, diffusion (for once our friend and not our enemy) will distribute those molecules small enough to pass the membrane throughout the total liquid volume on both sides. The larger molecules, in contrast, will remain in the original volume. Separations are therefore possible that are based purely on size and shape. One very simple arrangement has been in use for many years. It is particularly suitable for the crude but necessary purpose of separating salt from protein, after an ammonium sulphate precipitation, for example. In this method, the protein solution is tied in a small sac of the membrane material (in early work it was a pig's bladder; the material now used, which is sold by supply houses under a variety of dignified names, is actually sausage skin) and exposed to a much larger volume of water. The salt is able to diffuse out and its concentration in the sac will therefore fall by a factor equal to the ratio between the volume of the sac and the total liquid volume. This ratio could easily be 1:100 and, therefore, if the external liquid is replaced by fresh water after the equilibrium is reached, the concentration of salt in the sac can be lowered by a further factor of 100 (that is, a factor of $10^4$ altogether) without affecting that of the protein at all. The process as a whole is known as *dialysis*.

More modern (but not always more effective) equipment employs a pre-formed disc of the membrane material with a more closely controlled pore size. The disc is fitted into a holder and liquid is forced through under pressure. You should satisfy yourself this arrangement (called 'pressure dialysis' or 'ultra-filtration') concentrates the large molecule without (unless fresh water is added) diluting the salt.

While membrane methods of this type are extremely useful, *gel filtration* permits us to separate substances that are much closer together in size. For this process we have, as in many of the previous methods, a tube or column filled with hydrated granules and interstitial liquid. This time the separating ability of the column resides not in the surface properties of the granules, but in their overall architecture. Each granule is a three-dimensional network of chain-polymerized molecules. A particularly popular type consists of dextran chains (i.e. glucose units linked $1 \to 6$–$\alpha$, p. 174) cross-linked by chemical modification. If the cross-linking is extensive the average pore size of the network is small (Fig. A2.21a); if the cross-linking is less complete the average pore size is greater (Fig. A2.21b). If a mixture of molecules of various sizes is applied to the top of one of these columns, some molecules might be so large that they would not be able to enter the network and so would be excluded from the granules. Such molecules would be able to enter only the interstitial volume of the column, and if solvent were flowing through it, they would soon be driven through the column. Other molecules, however, might be so small as to be allowed to enter the network quite freely. Such molecules would find that the column had a far greater effective volume than did the larger molecules, and it would take a greater volume of solvent to expel them from it. They would therefore emerge *after* the larger molecules, and a separation would have been achieved based on molecular size. What is just as important for our purposes is that molecules of intermediate size would be allowed into the granules to an intermediate extent. This is because the dimensions of the network are only *averages*, being the result of a random stitching together of the chains. Some meshes of the net will be very much wider than others. Because of this some molecules that would not be excluded at all if the meshes were uniform will, if they are near the borderline size, find parts of the net barred to them. Similarly, some molecules that ought to be excluded completely, but are near the

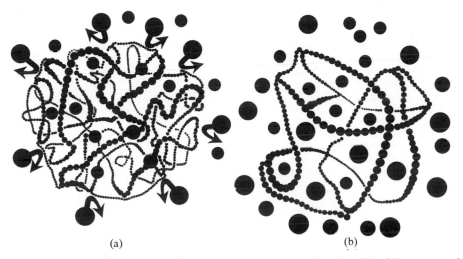

(a)          (b)

**Fig. A2.21** Gel filtration media: (a) excluding large molecules, (b) of large enough mesh to allow both types of molecule to enter

borderline size, will find certain parts of the net open to them and will be retarded relative to those that are excluded completely. This gives rise to graduated separations which depend on molecular size (see Fig. A2.22). Since the degree of cross-linking can be varied during manufacture, different grades of gel-filtration material are available to separate different ranges of sizes of material, from the smallest molecules to the largest viruses.

Gel filtration is, because it takes place under such mild conditions and because of its good resolving power, more popular than all other methods except salting out and ion-exchange chromatography. It is possible to obtain gel filtration materials that will accept penetration by non-polar solvents, and so lipid materials may be handled also.

The survey of separation methods that we have given has touched on only a few of the more widely used methods. However we hope that we have made at least two things clear. First, we wish to stress the extent of capabilities of modern separating methods. (Imagine, if they did not exist, trying to separate a few

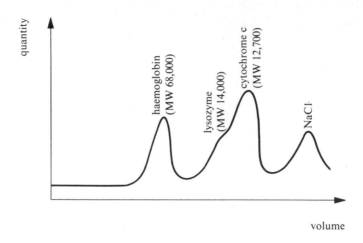

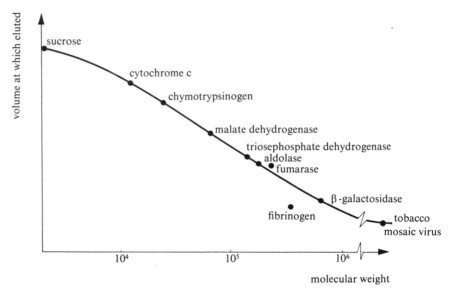

**Fig. A2.22** (a) A typical separation by gel filtration. (b) The relation between elution volume and molecular weight on one mesh size of dextran gel-filtration medium. Other sizes of mesh are available which give separations over different ranges of molecular weight

microgram of a mixture of peptides of different composition by fractional crystallization.) Secondly, we hope that we have made clear that most methods have an underlying logic to their mode of operation, which renders what seems at first sight to be a totally mysterious set of operations easily comprehensible. This is a point that we have sought to make about *all* the aspects of biochemistry that we have discussed in this book.

# Further reading

Advanced reviews on most of the topics that we have covered in this book appear from time to time in serial periodicals. These include *Annual Reviews of Biochemistry* (Annual Reviews Inc); the various 'Advances' volumes, such as *Advances in Enzymology* (Wiley-Interscience) and *Advances in Protein Chemistry* (Academic Press); the various 'Progress' volumes, such as *Progress in Biophysics and Molecular Biology* (Pergamon Press) and *Recent Progress in Hormone Research* (Academic Press); and others. Exercises, and their solutions, can be found in *Biochemical Reasoning*, edited by D. Kerridge and K. Tipton (Benjamin, 1973.

The following list of recommended reading refers to the different chapters of the book.

(2) Moore, W. J. *Physical Chemistry*. Longman (5th edn. 1972).

## Part 1

(3) Setlow, R. B. and Pollard, E. C. *Molecular Biophysics*. Pergamon Press (1962).

(4) and (5) Neurath, H. *The Proteins* (6 vols). Academic Press (2nd edn. 1963).

(4) and (5) Dickerson, R. E. and Geis, I. *The Structure and Action of Proteins*. Harper and Row (1969).

(6) Dixon, M. and Webb, E. C. *Enzymes*. Longman (2nd edn. 1964).

(6) Jencks, W. P. *Catalysis in Chemistry and Enzymology*. McGraw-Hill (1969).

(7) Davidson, J. N. *The Biochemistry of the Nucleic Acids*. Chapman & Hall (7th edn. 1972).

(8) *Advances in Carbohydrate Chemistry*. Academic Press.

(8) Chapman D. *Introduction to Lipids*. McGraw-Hill (1969).

## Part 2

*General reading*
Dagley, S. and Nicholson, D. E. *An Introduction to Metabolic Pathways*. Blackwell (1970).

Greenberg, D. M. (ed). *Metabolic Pathways* (4 vols). Academic Press (3rd edn. 1967–1970).

Gunsalus, I. C. and Stanier, R. Y. (eds). *The Bacteria*. Academic Press (vol. 2, 1961; vol. 3, 1962).

*Reading for individual chapters*

(9) Kornberg, H. L. 'The Co-ordination of Metabolic Routes', *Symposium Soc. General Microbiol.* **15**, 8 (1965).

(9) Krebs, H. A. and Kornberg, H. L. *Energy Transformations in Living Matter*. Springer (1957).

(10) Fawcett, D. W. *The Cell – An Atlas of Fine Structure*. Saunders (1966).

(10) and (11) Brachet, J. and Mirsky, A. E. *The Cell*, vol. 2. Academic Press (1961).

(11) De Duve, C. 'The Separation and Characterization of Subcellular Particles', *Harvey Lectures*, **59**, 49 (1965).

(12) Florkin, M. and Stotz, E. H. (eds). *Biological Oxidations*, vol. 14 of *Comprehensive Biochemistry*. Elsevier (1966).

(12) King, T. E. and Klingenberg, M. *Electron and Coupled Energy Transfer in Biological Systems*. Dekker (1971).

(12) and (13) Greville, G. D. 'Mitchell's Chemiosmotic Hypothesis', *Current Topics in Bioenergetics*, **3**, 1 (1969).

(13) Clayton, R. K. *Light and Living Matter*, vol. **2**, 1. McGraw-Hill (1971).

(13) Avron, M. and Neumann, J. 'Photophosphorylation in Chloroplasts', *Annu. Rev. Plant Physiol.* **19**, 137 (1968).

(14) Axelrod, B. 'Glycolysis'. In *Metabolic Pathways* (see above) **1**, 112 (1967).

(15) Krebs, H. A. 'The History of the Tricarboxylic Acid Cycle', *Perspectives in Biol. and Med.* **14**, 154 (1970).

(15) Lowenstein, J. M. 'The Tricarboxylic Acid Cycle'. In *Metabolic Pathways*, **1**, 146 (1967).

(16) Greville, G. D. and Tubbs, P. K. 'The Catabolism of Long-Chain Fatty Acids in Mammalian Tissues', *Essays in Biochem.* **4**, 155 (1968).

(17) and (19) Axelrod, B. 'Other Pathways of Carbohydrate Metabolism'. In *Metabolic Pathways*, **1**, 271 (1967).

(18) *Cold Spring Harbor Symposium on Quantitative Biology*, vol. 37 (1972).

(19) Ginsburg, V. 'Sugar Nucleotides and the Synthesis of Carbohydrates', *Advances in Enzymol.* **26**, 35 (1964).

(20) Vagelos, P. R. 'Lipid Metabolism', *Annu. Rev. Biochem.* **33**, 139 (1964).

(20) Green, D. E. and Allmann, D. W. 'Biosynthesis of Fatty Acids'. In *Metabolic Pathways*, **2**, 37 (1968).

(21) Chappell, J. B. 'Systems for the Transport of Substrates into Mitochondria', *Brit. Med. Bulletin*, **24**, 150 (1968).

(22) Umbarger, E. and Davis, B. D. 'Pathways of Amino-Acid Biosynthesis'. In *The Bacteria* (see above) **3**, 167 (1962).

(22) Sallach, H. J. and Fahien, L. A. 'Nitrogen Metabolism of Amino Acids'. In *Metabolic Pathways*, **3**, 1 (1969).

(22) Greenberg, D. M. 'Carbon Catabolism of Amino Acids Part 1', *ibid.*, 95 (1969).

(22) Rodwell, V. W. 'Carbon Catabolism of Amino Acids Part 2', *ibid.*, 191 (1969).

(22) Greenberg, D. M. 'Biosynthesis of Amino Acids and Related Compounds Part 1', *ibid.*, 238 (1969).

(22) Rodwell, V. W. 'Biosynthesis of Amino Acids and Related Compounds Part 2', *ibid.*, 317 (1969).

(23) Reichard, P. 'The Enzymic Synthesis of Pyrimidines', *Advances in Enzymol.* **21**, 263 (1959).

(23) Buchanan, J. M. and Hartman, S. C. 'Enzymatic Reactions in the Synthesis of Purines', *ibid.*, 199 (1959).

## Parts 3 and 4

*General reading*
Lewin, B. M. *The Molecular Basis of Gene Expression*. Wiley-Interscience (1970).

Stent, G. S. *Molecular Genetics*. Freeman (1971).

Watson, J. D. *Molecular Biology of the Gene*. Benjamin (2nd edn. 1970).

*Reading for individual chapters*
(24) Hayes, W. *The Genetics of Bacteria and Their Viruses*. Blackwell (2nd edn. 1968).

(24) Benzer, S. 'Genetic Fine Structure', *Harvey Lectures*, **56**, 1 (1961).

(25) Chamberlin, M. 'Transcription 1970: A Summary', *Cold Spring Harbor Symposia on Quantitative Biology*, **35**, 851 (1970).

(25) Sethi, V. S. 'Structure and Function of DNA-dependent RNA Polymerase', *Progress in Biophys. and Molec. Biol.* **23**, 67 (1971).

(26) Lengyel, P. 'The Process of Translation as seen in 1969', *Cold Spring Harbor Symposia on Quantitative Biology*, **34**, 828 (1969).

(26) Lengyel, P. and Soll, D. 'Mechanism of Protein Biosynthesis', *Bacteriol. Rev.* **33**, 264 (1969).

(27) Dayhoff, M. O. (ed). *Atlas of Protein Sequence*. National Biomedical Research Foundation. Published from time to time.

(28) Epstein, W. and Beckwith, J. R. 'Regulation of Gene Expression', *Annu. Rev. Biochem.* **37**, 411 (1968).

(28) Gilbert, W. and Müller-Hill, B. 'The Lactose Repressor' in Beckwith, J. R. and Zipser, D. (eds.). *The Lactose Operon*. Cold Spring Harbor Laboratory (1970).

(29) Cohen, G. N. *The Regulation of Cell Metabolism*. Holt, Rinehart, Winston (1968).

(29) Newsholme, E. A. and Start, C. *Regulation in Metabolism*. Wiley-Interscience (1973).

(29) Scrutton, M. C. and Utter, M. F. 'The Regulation of Glycolysis and Gluconeogenesis in Animal Tissues'. *Annu. Rev. Biochem.* **37**, 249 (1968).

APPENDICES

(1) Dixon, M. and Webb, E. C. *Enzymes*. Longman (2nd edn. 1964).

(2) Morris, C. J. O. R. and Morris, P. *Separation Methods in Biochemistry*. Pitman (1963).

# Index

acetaldehyde, in alcoholic fermentation, 266, 272

acetic acid
  acetate ion as conjugate base of, 12
  pK of, 12, 64
  titration curve for, 11, 12–13

acetoacetate, 290
  from aromatic amino acids, 376; from leucine, 372–3
  interconversion of hydroxybutyrate and, 290
  reaction of, with succinyl CoA, 290

acetoacetyl ACP, in fatty-acid synthesis, 343

acetoacetyl coenzyme A, 289–90
  from amino acids, 361, 377
  cannot give rise to carbohydrate, 361

$\alpha$-aceto-$\alpha$-hydroxyacid synthetase, inhibited by valine, 502

$\alpha$-aceto-$\alpha$-hydroxybutyric acid, in amino-acid synthesis, 370

$\alpha$-acetolactate, in amino-acid synthesis, 370

acetone, fermentation product, 266, and ketone body, 290

acetyl ACP, in fatty-acid synthesis, 342, 343

acetylation
  of serine, 378–9
  of toxic compounds, 271

acetylcholine esterase, DFP as inhibitor of, 128

acetyl coenzyme A (acetyl CoA), high-energy compound, 26, 79, 273
  acetylates serine, 378–9
  activates pyruvate carboxylase, 506
  from amino acids, 361, 371
  cannot give rise to carbohydrate, 339, 361
  carboxylation of, 340, 341–2
  effects of accumulation of, 289–90
  as end-product of glycolysis and fatty-acid oxidation, 503
  from oxidation of fats, 283
  from pyruvate and coenzyme A, 271–3
  reactions of: with acetoacetyl CoA, 290; with ACP, 342; with oxaloacetate, 275, 276, 278, 354
  in syntheses: of amino acids, 364, 372; of fats, 295, 340
  in TCA cycle, 275, 276, 288
  transport of, across mitochondrial membrane, 354

acetyl coenzyme A carboxylase, 341
  control of fatty-acid synthesis through, 508

acetyl fragments, in TCA cycle, 209, 210

N-acetyl-D-galactosamine, 178
  in gangliosides, 193

N-acetyl-D-glucosamine, 178
  in bacterial cell walls, 180, 181, 182
  in chitin, 183
  in hyaluronic acid, 183

N-acetyl glutamate
  in amino-acid metabolism, 364
  catalyst in synthesis of carbamoyl phosphate, 387

N-acetyl glutamate semialdehyde, 364

N-acetyl muramic acid, in bacterial cell walls, 180, 181

N-acetyl neuraminic acid, 180
  in gangliosides, 193

N-acetyl ornithine, in amino-acid metabolism, 364

O-acetyl serine, in fixation of sulphide by micro-organisms, 378–9

acetyl transacylase, 342

acid
  conjugate, of a base, 12, 65
  definition of, 10
  pK as index of strength of, 12
  strong, 13

aconitase, 276

cis-aconitate
  and oxidation of carbohydrate by muscle preparations, 274
  in TCA cycle, 276

acridine, causes additions and deletions in DNA, 420

actin, muscle protein, 98
  fibrous (F) form of (thin filaments of muscle), 312, 313, 314
  globular (G) form of, 312, 315
  relations of, with myosin, 310, 313, 314

activated complexes, temporary, in enzyme action, 37
  bound to enzyme more closely than substrate is, 116, 118, 131
  effects of analogues of, 116, 118, 131

activators
  of enzyme action, 126–8, 502

activators—*contd.*

of synthesis of enzymes, permitting expression of genes only in presence of inducers, 482

actomyosin, 312, 313, 314

N-acyl amino sugars, in structural polysaccharides, 178

acyl carrier protein (ACP), in fatty-acid synthesis, 341–6

acyl phosphates, high-energy compounds, 26, 28

adenine (purine), 135, 136

in DNA from different sources, 144

hydrogen bonds between thymine and, 141, 146, 427

adenosine (nucleoside), 135

adenosine diphosphate (ADP), high-energy compound, 24, 26

exchange of, with ATP, across mitochondrial membrane, 353

interconversions of AMP, ATP, and, 24, 504, 507

phosphorylation of, *see* adenosine triphosphate, production of

reactions of: with creatine phosphate, 300; with CTP, 392

adenosine diphosphate glucose, 333n

adenosine monophosphate (AMP)

conversion of IMP to, 396

derivatives of, in activation of amino acids, 440–1, and of fatty acids, 284, 440–1

inhibits fructose diphosphatase, and hence gluconeogenesis, 507, 508

interconversions of ADP, ATP and, 24, 504, 507

promotes glycolysis, 504–5, 508

adenosine monophosphate, 3′,5′-cyclic (cAMP), 143, 144

in *E. coli* grown on different carbon sources, 480–1

and protein synthesis in *E. coli*, 481

required by protein kinase, 505–6, 507

adenosine triphosphatases (ATPases), 239

adenosine triphosphate (ATP), high-energy compound, 5, 24–6, 300–1

abundance of: inhibits glycolysis, 504, 508; promotes gluconeogenesis, 353, 507, 508

in active transport, 317–19

counteracts inhibition of aspartate transcarbamylase by CTP, 498

cyclic AMP from, 506

as end-product of glycolysis and fatty-acid oxidation, 503

exchange of, with ADP, across mitochondrial membrane, 353

interconversions of ADP, AMP, and, 24, 504, 507

in movement of cilia and flagella, 302

in muscle contraction, 98, 313–16

net gain of: in complete oxidation of glucose, 280, 289; in glycolysis, 269; in oxidation of fatty acids, 288–9; in TCA cycle, 280

adenosine triphosphate (ATP)—*contd.*

phosphorylations by: aspartate, 366; fructose-6-phosphate, 260; glucose, 259; glyceraldehyde, 267; glycerate-3-phosphate, 322, 331; glycerol, 268–9, 346; nucleoside mono- and di- phosphates, 335, 391, 392, 394, 397; phosphorylase *b*, 505; ribulose-5-phosphate, 326, 328

production of, 208; in glycolysis, 27, 263, 264; in oxidative phosphorylation, 29, 233, 239–41, 245; in photo-synthesis, 219, 247, 248, 249, 252, 253, 254, 255; in substrate-level phosphorylations, 27, 264, 278; theories of mechanism of production of, in oxidative phosphorylation, 242–5

required for conversions: of aspartate to asparagine, 365; of citrate and CoA to oxaloacetate and acetyl CoA, 354; of glutamate to glutamine, 363; of pyruvate to oxaloacetate, 330; of UTP to CTP, 391

required for syntheses: of S-adenosyl methionine, 385; of amino acids, 363, 365, 368, 369, 375, 379; of amino-acyl tRNAs, 440; of carbamoyl phosphate, 387; of CoA derivatives, in oxidation of fatty acids, 283, 284; of fatty acids, 341, 343, 345, 355; of glycogen from lactate, 335; of purine nucleotides, 395; of urea, 388

S-adenosyl homocysteine, 349, 386

S-adenosyl methionine, methyl donor, 349, 385–6

adenylate cyclase, 506

adenylate kinase, 504

adenylic acid, 135, 435; *see also* adenosine monophosphate

adipose tissue, 346, 508

adrenaline

activates adenylate cyclase, 506, lipase, 508, and phosphorylase, 506

inhibits glycogen synthetase, 507

synthesis of, 386

affinity between reactants, 7–8

ΔG and order of intrinsic, 17, 18

increased concentration of a reactant as way of reversing unfavourable order of, 8–10

agar plate, selection of bacteria by cultivation on, 404–5, 407

AICAR (phosphoribosyl-5-amino-4-imidazole-carboxamide), 380–1, 396

alanine

in bacterial cell walls, 181, 182

configuration of, 64

interconversion of pyruvate and, 357n, 369

metabolism of, 361, 362, 369, 382, 383

trinucleotide codes for, 444, 445

albinism, 426

alcohols, in lipids, 186, 188, 191

aldehydes, in lipids, 186, 188, 189

aldolase, 261, 264, 267, 324

aldoses, 166, 167

algae, 247

alkaloids, fractional crystallization of enantiomers by means of enantiomers of, 50
alleles, 424, 425
allergy, 103
D-allose, 167
allosteric sites on enzymes
  binding of activators to, 127–8, 495, 502
  binding of inhibitors to, 131, 495, 502
D-altrose, 167
amination, *see* ammonia, *and* amino groups, donors of
D-amino acid oxidase, in liver and kidney, 51, 359n
L-amino acid oxidase, non-specific, 359
amino acids, 60–1, 62–3
  activation of, by reaction with ATP, 440–1
  carbohydrates from, 329
  concentration of, by active transport, 319
  deamination of, 359, 360, 361, 379, 387
  essential and non-essential, 80, 360–1
  'families' of, 231, 362
  glucogenic and ketogenic, 361–2
  ionization of, 64–6
  metabolism of, 356; acetyl CoA in, 271; anabolic pathways in, 210, 212, 214; branched pathways in, 500; central reactions in, 357–60; by families, 362–82; summary and diagram of, 382–3; transfer of one-carbon fragments in, 379, 383–6
  naturally occurring, not in proteins, 60, 66
  in proteins, 4, 45, 47, 60, 149; *see also* proteins
  separation of, on paper chromatograms, 532
  sources of: intermediates of TCA cycle, 280; products of glycolysis, 269
  specific properties of, 66–70
  specific tRNAs for, 441–4
  stereochemistry of, 49, 50–1, 61, 64
  trinucleotide code in DNA for, 419–22, 444–6
amino-acyl AMP, 440–1
amino-acyl transfer RNA synthetases, specific for each amino acid, 158–9, 440, 441
amino-acyl transfer RNAs, *see* ribonucleic acids, amino-acyl transfer
α-aminoadipic acid, in degradation of lysine, 367
p-aminobenzoic acid, 79
  mimicked by sulphonamides, 80–1, 130
  in tetrahydrofolic acid, 386
L-α-aminobutyric acid, naturally occurring non-protein amino acid, 66–7
amino groups, 10n
  of amino acids, 64–6, 382–3
  donors of: asparagine, 365; aspartate, 388, 389, 395, 396, 397; glutamine, 363
  of nucleotide bases, 139–40
  in peptide bonds, 45, 71
amino sugars
  in structural polysaccharides, 178, 221
  synthesis of, 334
amino terminus of proteins, *see under* proteins

ammonia
  from deamination of amino acids, 359, 387
  as excretion product of nitrogen metabolism in aquatic animals, 225, 387–8
  fixation of: into carbamoyl phosphate, 387; into glutamate, 357, 358–9, 363, 387; into CTP, 391
ammonium sulphate, salting out by means of, 529–30
amphibia, change in pattern of protein synthesis at metamorphosis in, 488
amphibolic pathways of metabolism, 208, 356
amylopectin, 174, 175, 267
  synthesis of, 337
amylose, 173, 174, 175
  synthesis of, 335
anabolic pathways of metabolism, 208, 210–12
  enzymes of, often in different cell compartment from those of catabolism, 351
  not reverse of catabolic pathways, 210, 261, 265, 340
anaemia
  caused by deficiency of vitamin $B_{12}$, 289
  caused by mutation in haemoglobin, 107
anaerobic organisms, fermentations by, 256, 257, 266
animal cell, 213–19
anomers, anomerism, 169–70
anthranilate, in synthesis of tryptophan, 374–375
anthranilate synthetase, 374
  inhibited by tryptophan, 500
antibiotics
  competitive inhibitors among, 130–1
  cyclic, conferring increased permeability on membranes, 200
antibodies, 102
anticodons, on tRNAs, 158, 159, 441, 442, 443
antigens, 100
  'memory' for, 102
  mucopolysaccharides as, 184
Antimycin A (antibiotic), blocks respiratory chain between cytochromes $b$ and $c$, 239, 241
D-arabinose, 167
  in gum arabic, 177
  transport of, 202
L-arabinose, transport of, 202
  system for metabolism of, in *E. coli*: control of, 482, 485; mapping of genes of, 413, 415
arabinose permease, in *E. coli*, 485
arachidic acid, 185
arachidonic acid, 185
arginase, 364, 389
arginine
  cleavage of peptides at, by trypsin, 83, 111
  essential amino acid for most mammals, 364–5, 383
  genes for synthesis of, in *E. coli*, 415, 485
  in histones, 157
  hydrophobic side chain of, 70

arginine—*contd.*
  interchange of lysine and, in amino-acid sequences, 463
  metabolism of, 361, 363–4, 383
  positively charged group of, 68, 71
  trinucleotide codes for, 444, 445
  in urea cycle, 388–9
arginine phosphate, as phosphagen, 300n
argininosuccinic acid, in urea cycle, 388, 389
ascorbic acid, as electron donor to cytochrome *c*, 241
asparaginase, 365
asparagine, metabolism of, 361, 362, 365, 382, 383
asparagine synthetase, 365
aspartate
  as donor of amino group: in conversion of IMP to AMP, 396; in synthesis of purine nucleotides, 395, 397; in urea cycle, 388, 389
  interconversion of oxaloacetate and, 357n, 365, 389
  in lysozyme, 118, 119
  metabolism of, 361, 362, 365, 382, 383
  negatively charged group of, 68, 71
  phosphorylation of, 366, 500, 501
  trinucleotide codes for, 445
  use of isotopes in study of metabolism of, 231
aspartate family of amino acids, 365–9, 383
aspartate semialdehyde, 365, 369, 383
aspartate transcarbamylase, 390
  feedback inhibition of, by CTP, 495–9
  subunits of, 127
aspartokinase, 500
  single form of, in *B. polymyxa*, subject to concerted feedback inhibition, 501–2
  three isoenzymes of, in *E. coli*, 501
aspartyl phosphate, 366, 500, 501
ATP, *see* adenosine triphosphate
auxiliary coupling factor, in chemical-coupling hypothesis for oxidative phosphorylation, 242–3
axial ratio, of proteins, 71

*Bacillus polymyxa*, concerted feedback inhibition of aspartokinase in, 501–2
*Bacillus stearothermophilus*, mechanism of protein synthesis studied in, 448
bacteria
  can convert acetyl CoA to carbohydrate, 339
  cells of, 221–3
  cell walls of, 180–2, 221–2
  concentration of RNA in, related to rate of protein synthesis, 434
  DNA of, 157
  flagella of, 302
  generation time of, 429
  of gut: cellulase of, 176; vitamins produced by, 80
  of hot springs, enzymes of, 134
  inhibition of enzymes of, by antibiotics, 131

bacteria—*contd.*
  methods of studying metabolism of, 225–6, 229–30
  phosphorylation of sugars by, 319n
  photosynthetic, 247–8, 251
  producing sulphuric acid, 14
  subunits of ribosomes of, 438
  *see also Escherichia coli and other species*
bacteriophages
  DNA of, 136, 144, 156, 231, 403
  pseudo-wild types of, 420, 421
  R 17, order of translation of RNA of, 486–487
  recombination of genetic material of, 417n
  mRNA and synthesis of proteins in bacteria infected with, 439
  T 4, mutation sites and length of genes in, 417, 445
basal granules, of bacterial flagella, 302
base
  conjugate, of an acid, 12, 65
  definition of, 10
  strong, 13
bases in nucleotides, 46, 135, 136
  constant ratio of, in given species, 403
  in different samples of DNA, 144
  minor, 137, 142–3
  properties of, 139–41
  substitution of one for another, as cause of mutations, 416
  *see also* purines, pyrimidines
benzoic acid, product of metabolism of phenyl-labelled fatty acids with odd number of carbon atoms, 283
beri-beri, 271
binding (association and dissociation) constants, 14–15, 116
binding energies, and enzyme action, 115
biological preparations, different forms of, 225–8
biotin (vitamin H), 79
  in carboxylases: acetyl CoA, 341–2; propionyl, 289; pyruvate, 281
  structure of, 341
blood
  buffering power of, 106
  clotting of, 98–9
  constituents of: fatty acids, 508; glucose, 268; lactate, 258
  pH of, 13
Boltzmann equation, for distribution of molecules in different states of free energy, 39, 53
bond strength, 18, 27
  of covalent bonds, 51
  of dispersion forces, 55
  of hydrogen bonds, 54
  of hydrophobic interactions, 57, 107, 147
  of ionic bonds, 53
  just broken by random thermal agitation, 53
  of resonance stabilization, 71
bonds, *see* covalent bonds, non-covalent bonds

chemical reactions—*contd.*
rates of, 33–41, 114–15; enzymes and, 115–123
chitin, 183
covalent bonds in, 52
synthesis of, 335
chloride ion: free passage of, through membranes, 200
chlorophyll, 79
*a* and *b*, 250–1
P 670, 251
P 700, 251, 253
chloroplast, 23, 249, 250
synthesis of ATP in, 219
cholesterol, 186, 188
choline, phosphoglyceride containing, 191, 348, 349
chondroitin, chondroitin sulphate, 183
chorismate, in synthesis of aromatic amino acids, 374, 375, 495
chromatographic mobility ($R_f$), 533–4
prediction of, 534
chromatography
adsorption, 542–3
gas-liquid, 534
ion-exchange, 539–43
on paper, 84, 531–4
chromosomes, 157, 218
chymotrypsin
amino-acid sequences of trypsin and of, 468
cleavage of peptides by, 82–3, 119
effect of indole on, 123–4
inhibitor for, 128
cilia, 302, 303
*cis* and *trans* forms of unsaturated fatty acids, 188
cisternae, of endoplasmic reticulum, 214, 216
cistron, 478n
citrate
activates acetyl CoA carboxylase, 508
carrier for, in mitochondrial membrane, 354
as end-product of glycolysis, 503
impermeability of many bacterial cells to, 226
inhibits phosphofructokinase (in presence of ATP), 504
and oxidation of carbohydrate in muscle preparations, 274
produced by reaction of acetyl CoA and oxaloacetate, 275, 276, 278, 354
reaction of, with ATP and CoA, to give oxaloacetate and acetyl CoA, 276, 354
in TCA cycle, 276
citrate cleavage enzyme, 354
citrate synthetase, 276
citrulline, in urea cycle, 388, 389
codons, on mRNA, 442, 443
for termination of polypeptide, 458
coenzyme A (CoA), 271
reactions of: with acetoacetyl CoA, 290; with citrate and ATP, 276, 354; with fatty acids and AMP, 284; with fatty-acyl

coenzyme A (CoA)—*contd.*
reactions of—*contd.*
ACP, 346; with $\beta$-ketoacyl CoA, 286; with oxoglutarate, 277; with pyruvate, 271–3
co-factors of proteins, 78–80, 143–4
removal of, in study of metabolism, 229
colinearity, of gene and polypeptide chain, 418–19, 422
collagen, structural protein, 67, 92
superhelix of, 96
colour, determination of amino-acid residues by agents imparting, 85
compartmentation, intracellular, 350–5
concentration
of reactants: and balance in equilibria, 8–10, 14, 15; in enzyme action, 121–2, 125–126; and rate of reaction, 34–5
of solute, free energy required for increase of, 17
of substrates, and rate of transport across membranes, 201–2, 204
of water as a reactant, convention for, 13, 34
condensation (combination of molecules with elimination of water), 45
coniferyl alcohol, in lignin, 177
conjugate acids and bases, 12, 65
constitutive mutants, 476, 480, 483
operator-, 478, 480
control of reactions, *see* regulation
cooperativity, of binding of substrate molecules, 496
copper ion, in donation of electrons to molecular oxygen, 237
CORN law for stereochemistry of amino acids, 61
counter-current distribution, 531
covalent bonds (sharing electrons), 51, 53, 58
macromolecules predominantly constructed with, 52, 71
creatine, synthesis of, 386
creatine phosphate, high-energy compound, 26, 28
in muscle, 300–1
creatine phosphokinase, 301
cristae, of mitochondria, 217, 219
crossbridges, between myofilaments, 305, 310, 312, 314
cross-linking, in macromolecules, 47
in bacterial cell walls, 180–1, 221
between peptide chains, 97–8
in polysaccharides, 173–7
crotonic acid, interconversion of derivatives of butyric acid and of, 29
crotonyl ACP, in fatty acid synthesis, 343
crotonyl coenzyme A, 367
cyanogen bromide, cleavage of peptides at methionine by, 83
cystathionine, 376–8
cysteine
metabolism of, 361, 362, 367, 378–9, 382, 383

deoxyribonucleic acids (DNAs)—*contd.*
strands of: bonding of nucleotides between, 141, 146, 427; separated by heating, 436; untwisting of, 433–4
of viruses, 156, 159, 223; of bacteriophages, 136, 144, 156, 231, 403
deoxyribose, 136, 178
hydroxyl groups of, in phosphodiester linkage, 137
synthesis of, 210
deoxyribonucleoside triphosphates, substrates for DNA synthesis to, 429, 431
deoxythymidine monophosphate (deoxyTMP), 393–4, 435
deoxythymidine triphosphate (deoxyTTP), 392
from deoxyCDP, 392–4
deoxyuridine monophosphate (deoxyUMP) from deoxyCMP, 392–3
deoxyTMP from 393–4
desulphuration deamination, of cysteine, 379
detergents, 58
denaturation of proteins by, 77
dextrans, 171, 174–5
for gel filtration, 544
diabetes, ketone bodies in, 291
dialysis, 200
removal of activators by, 126; removal of co-factors by, 229
separation of compounds by, 544
under pressure (ultra-filtration), 544
diaminopimelic acid, 366
diastereoisomers, 50n
dielectric constant, and strength of ionic bond, 52, 53, 118–19
diesterases: from snake venom and spleen, hydrolysis of nucleic acids at different sites by, 137–8, 139, 164
diethyl aminoethyl (DEAE) cellulose, 541
differentiation of tissues, repressors in regulation of, 488
diffusion across membranes, 103, 200
exchange, 203, 204, 317
dihydrofolate, 393, 394
dihydrolipoyl dehydrogenase, 273
dihydro-orotate, in synthesis of pyrimidines, 390
dihydropicolinate, in synthesis of lysine, 366
dihydropicolinate synthetase, inhibited by lysine, 501
dihydroxyacetone, as parent structure of ketoses, 166, 168
dihydroxyacetone phosphate
carrier out of mitochondria for, 352
interconversions: of fructose-1,6-diphosphate with glyceraldehyde-3-phosphate and, 261, 324, 332; of fructose-1-phosphate with glyceraldehyde and, 267; of glyceraldehyde-3-phosphate and, 19–20, 261, 269, 324, 346, 352, 475; of glycerophosphate and, 261, 269, 346, 352, 475
reaction of, with erythrose-4-phosphate, 325

di-isopropyl fluorophosphate (DFP), inhibitor of enzymes, nerve poison, 128, 129
2,4-dinitrophenol, uncoupler of oxidative phosphorylation, 243, 246
dipoles, 53–5, 73
disaccharides, of glucose, 172
dispersion (London, Heitler) forces, 55, 59, 75
disulphide bridges, 70
in clotting of blood, 99
in cross-linking of peptide chains, 97
denaturation of protein by cleavage of, 77, 78
in protein structure, 71, 75, 76; in thioredoxin, 392
duodenum, pH in, 14

Edman reaction, for removal of single amino-acid residues from proteins, 84
efficiency, thermal: of oxidative phosphorylation, 233, 246
elastin, structural protein, 92, 93, 97
electrode potentials, convertible to $\Delta G$, 23n, 240, 317
electrons
artificial donors of, in locating sites of ATP synthesis, 241
donation of = undergoing oxidation, 23
energy from fall of energized, 22–3, 30, 251–2, 254–5
electrophoresis, separation of compounds by, 534–9
electrostatic interactions, in macromolecules, 52–5
electrostatic repulsion, between negatively charged groups of ATP, 27
elongation factors (EFT$_u$, EFT$_s$, EFG), in synthesis of proteins, 451, 452
EFT$_u$ binds to all tRNAs except N-formylmethionyl tRNA, 456
Embden-Meyerhof pathway, 258, 267
enantiomers, 49
endoplasmic reticulum, smooth and rough, 214–15, 216
triglycerides synthesized in, 350
end-products of synthetic pathways
inhibit first enzyme of pathway, 495, 499–500, 508
repress synthesis of enzymes of pathway, in bacteria, 474
ene-diolate, activated complex in action of triosephosphate isomerase, 116
energy
coupling of processes requiring and releasing, 5, 16, 22–30
from fall of energized electrons, 22–3, 30, 251–2, 254–5
required for breaking bonds, *see* bond strength
energy, Gibbs free ($\Delta G$), 16–18
in denaturation of trypsin, 77
electrode potentials convertible to, 23n, 240, 317
in enzyme action, 115–21

fatty acids—*contd.*
  oxidation of: even-numbered, 282–9, and odd-numbered, 289; in mitochondria, 217, 283, 340–1; regulated by concentration in blood, 508
  synthesis of, 210, 211, 271, 340–6; in cytoplasm, 214, 340–1, 354–5; NADPH$_2$ as reducing agent in, 295–6
  unsaturated: *cis* and *trans* forms of, 188; produced at low temperatures, 199
fatty-acyl AMP, 284
fatty-acyl carnitine, high-energy compound, passes through mitochondrial membrane, 353
fatty-acyl coenzyme A, high-energy compound, 284, 441
  comparison of formation of, with that of amino-acyl tRNA, 440–1
  long-chain, inhibits acetyl CoA carboxylase, 508
  $\beta$-oxidation of, 285–8; in mitochondria, 252–3
  in synthesis of triglycerides, 347
fatty-acyl coenzyme A dehydrogenase, 285
feedback inhibition of enzymes, 390, 493, 495–9
  in branched synthetic pathways, 500–2
  concerted, 501
fermentation
  energetics of, 257, 269
  of glucose to lactate, 256–66; to other compounds, 266, 272
  of other carbohydrates, 267–9
ferredoxin, in photosynthesis, 252, 253
ferredoxin-NADP reductase, 252
fibrin, 98–9
fibrinogen, 98–9, 466
fibrinopeptides, 99
  amino-acid sequences of one of (peptide A), in different vertebrates, 465; phylogenetic tree from, 466–7
fibroin, structural protein, 95–6
filtration, separation of compounds by, 543–4
  gel-, 544–5, 546
  ultra- (pressure dialysis), 544
flagella, 301–2
flagellin, 301
flavin adenine dinucleotide (FAD), 79, 234, 236
  in amino-acid metabolism, 371
  *see also* flavoproteins
flavin mononucleotide (FMN), 234, 236, 252
flavoproteins, hydrogen carriers in respiratory chain, 234, 236, 238, 239
  in dehydrogenases: fatty-acyl CoA, 285; $\alpha$-glycerophosphate, 352n; succinate, 278, 280
fluorescence, determination of amino-acid residues by combination with substances imparting, 85
fluoride, enzyme inhibitor, 259
  removes magnesium, 264
  toxicity of compounds containing, 130
flux in metabolic pathways (J), 491, 504

folate derivatives, interconversions of, 386
formaldehyde, modification of amino group by, 65
formic acid, pK of, 12
formiminoglutamate, 381, 384
5-formiminotetrahydrofolate, 384, 385
formyl kynurenine, in degradation of tryptophan, 377
N-formylmethionyl transfer RNA, in initiation of protein synthesis, 453, 454, 455
  binds to IF$_2$, 456
  N-formyl part of, removed from newly synthesized polypeptide by deformylating enzyme, 456
  sets 'reading frame' for ribosome, 456
10-formyltetrahydrofolate, 384, 385
  formylation of methionyl tRNA by, 453
  in synthesis of purine nucleotides, 395
free radical reactions, cross-linking of peptide chains by, 97
fructokinase, 267
D-fructose, 168
  $\alpha$- and $\beta$-configurations of, 169
  phosphorylation of 285; to 1-phosphate, 267; to 6-phosphate, 267
  from sucrose, 34–5
fructose diphosphatase, 260–1, 324
  control of gluconeogenesis through, 507
fructose-1,-6-diphosphate, 258–9
  hydrolysis of, to fructose-6-phosphate, 25, 260, 325, 332
  interconversion of triose phosphates and, 261, 324, 332
  phosphorylation of fructose-6-phosphate to, 25, 260, 324
fructose-1-phosphate
  from fructose, 267
  interconversion of, with dihydroxyacetone phosphate and glyceraldehyde, 267
fructose-6-phosphate
  from fructose-1,6-diphosphate, 25, 260, 325, 332
  interconversion of glucose-6-phosphate and, 9, 19, 260, 333, 503
  phosphorylation of, 25, 260, 324
  produced in pentose phosphate pathway, 297, 298
  reactions of: with glyceraldehyde-3-phosphate, 325; with UDP-glucose, 338
L-fucose, 179
  transport of, 202
fumarase, 278
fumarate
  from aromatic amino acids, 376
  and oxidation of carbohydrate by muscle preparations, 274, 275
  released: in synthesis of purine nucleotides, 395; in urea cycle, 388, 389
  in TCA cycle, 278, 279
fumaryl acetoacetate, 376
fungi
  mutants of, for studying metabolism, 229
  repression of genes in, 488
furanose ring, 170–1

galactokinase, 267, 337
D-galactosamine, 178
D-galactose, 167
   in gangliosides, 193; in gum arabic, 177; in phospholipids, 189
   from lactose, 337, 474
   phosphorylation of, 267–8, 285, 337
   regulatory system for metabolism of, in *E. coli*, resembling that for lactose, 482
   transport of, 202
L-galactose, transport of, 202
galactose-1-phosphate
   conversions of: to glucose-1-phosphate, 267; to UDP-galactose, 337
β-galactosidase
   induced in *E. coli* by lactose, 474, and by other β-galactosides, 475; mutants of *E. coli* lacking, and permanently induced for, 475–6
   number of molecules of, per cell of fully induced *E. coli*, 481
galactoside permease
   mutants of *E. coli* lacking, and permanently induced for, 476
β-galactosides, and *lac* system of *E. coli*, 475, 477, 478, 480
D-galacturonic acid, 177, 179
gangliosides, 190, 193, 197
gel filtration, 544–5, 546
genes, each determining structure of one polypeptide, 414, 422
   clustering of, in *E. coli*, 414–15, 423; little or none in eucaryotes, 488
   dominant and recessive, 423–6
   doubling of, 468–70
   estimates of numbers of nucleotide pairs in, 417, 422
   for histidine synthesis, in *S. typhimurium*, 483–4
   for lactose metabolism, in *E. coli*, 475–81,
   for tryptophan synthesis, in *E. coli*, 483, 484
genetic mapping
   demonstration of gene doubling by, 468
   by interrupted mating in *E. coli*, 408–10
   by recombination frequency, 410–13
genetics, molecular, 148, 402–5
   knowledge of, for *E. coli*, used in study of enzyme control, 475
germ cells, haploid, 425
   amount of DNA in, 403
globin, 104
gluconate, rate of synthesis of *lac* proteins in *E. coli* grown on, 480
gluconate-6-phosphate, 293
gluconate-6-phosphate dehydrogenase, 293
gluconeogenesis, 329, 353, 354
   control of rate of, 506–7
   end products of, 503
gluconolactone-6-phosphate, 293
D-glucosamine, 178
D-glucose, 167
   disaccharides of, 172
   enantiomers of, 49

D-glucose—*contd.*
   fermentation of, to lactate, 207–9, 214 256–66; *see also* glycolysis
   flow of ¹⁴carbon from fatty acids to, 232
   in gangliosides, 193
   interconversion of α- and β-forms of, 36–7, 169
   phosphorylation of, to 6-phosphate, 202–3, 259, 285, 319
   polysaccharides of, 174, 176
   production of: from glucose-6-phosphate, 259, 268, 319, 333; from lactose, 337, 474; from sucrose, 34–5
   required by brain, 291
   synthesis of, 210, 211
   transport of, 202
   yield of ATP from complete oxidation of, 280, 289; from glycolysis of, 269
L-glucose, transport of, 202
glucose-1,6-diphosphate, cofactor for phosphoglucomutase, 268
glucose-6-phosphatase, 259, 333
glucose-1-phosphate
   from galactose-1-phosphate, 267, 337, 338
   from glycogen and starch, 267
   interconversion of glucose-6-phosphate and, 267, 268, 334, 338, 503
glucose-6-phosphate
   dehydrogenation of, 292–3
   glucose from, 259, 268, 319, 333
   from glucose, 202–3, 259, 285, 319
   inhibits hexokinase, 505
   interconversions of: with fructose-6-phosphate, 9, 19, 260, 333, 503; with glucose-1-phosphate, 267, 268, 334, 338, 503
   synthesis of, from trioses, 332–3
   used in syntheses, 333
glucose-6-phosphate dehydrogenase, 293
D-glucuronic acid, 179, 183
glutamate
   in bacterial cell walls, 181
   as donor of amino group in urea cycle, 389, 390
   from histidine, 381
   interconversion of α-oxoglutarate and, 357, 387
   metabolism of, 361, 362, 382, 383
   negatively charged group in, 68, 71
   pK of, in lysozyme, 118–19, 122
   transamination of α-keto acids by couple of α-oxoglutarate and, 357–9, 365, 369, 389
   trinucleotide codes for, 445
glutamate dehydrogenase, 357, 359, 362
glutamate family of amino acids, 362–5, 383
glutamate semi-aldehyde, 362, 364
glutaminase, 363
glutamine
   as carrier of ammonia, 363, 387
   metabolism of, 361, 362, 363, 382, 383
   in syntheses: of histidine, 380; of purine nucleotides, 394, 395, 397; of tryptophan, 374–5
   trinucleotide codes for, 445

glutamine synthetase, 363
glutaryl coenzyme A, 367
D-glyceraldehyde
  interconversion of fructose-1-phosphate with dihydroxyacetone phosphate and, 267
  as parent structure of aldoses, 166, 167
  phosphorylation of, 267
glyceraldehyde-3-phosphate
  eliminated in synthesis of tryptophan, 376
  glycerate-1,3-diphosphate from, 262
  glycerate-1,3-diphosphate to, 322, 331
  interconversions: of dihydroxyacetone phosphate and, 19–20, 261, 269, 324, 346, 352, 475; of fructose-1,6-diphosphate with dihydroxyacetone phosphate and, 261, 324, 332
  in photosynthesis, 322, 323–8
  in pentose pathway, 296, 297, 298
  reactions of: with fructose-6-phosphate, 325; with sedoheptulose-7-phosphate, 297, 326
glycerate-1,3-diphosphate, high-energy compound, 26, 28
  glyceraldehyde-3-phosphate from, 322, 331
  glyceraldehyde-3-phosphate to, 262
  glycerate-3-phosphate to, 322, 331
  phosphorylation of ADP by, 263
glycerate-2,3-diphosphate, cofactor for phosphoglucomutase, 263
glycerate-2-phosphate
  in glycolysis, 259
  interconversions of: with glycerate-3-phosphate, 263–4, 330–1; with phospho-enolpyruvate, 264, 330
glycerate-3-phosphate
  in glycolysis, 259
  interconversion of glycerate-2-phosphate and, 263–4, 330–1
  phosphorylation of, 322, 331
  produced in photosynthesis, 322
  serine from, 269
glycerate-3-phosphate dehydrogenase, 377, 413, 414
glycerate-3-phosphate family of amino acids, 377–9
glycerokinase, 269, 282, 346
  synthesis of, induced by glycerol, 474, 475
glycerol, 178
  enzyme syntheses induced by, in E. coli, 474–5
  in lipids, 185, 187, 189, 191
  permeability of membranes to, 200
  phosphorylation of, 268–9, 282, 285, 346
  rate of synthesis of lac proteins in E. coli grown on, 480
  regulatory system for metabolism of, in E. coli, 481, 485
  in teichoic acids, 182
α-glycerophosphate
  dehydrogenation of, 268, 269
  from glycerol, 269–9, 282, 285, 346
  interconversion of dihydroxyacetone phosphate and, 261, 269, 346, 352, 475

α-glycerophosphate—contd.
  reducing equivalents of $NADH_2$ conveyed into mitochondria as, 352
glycerophosphate dehydrogenase, 269
  induced by glycerol in E. coli, 474–5
  of mitochondria, 352n
glycine
  in bacterial cell walls, 181
  in collagen, 96
  conserved in evolution of amino-acid sequence of cytochrome c, 463
  ionization of, 64–6
  metabolism of, 361, 362, 378, 379, 382, 383
  in synthesis of purine nucleotides, 394, 397
  trinucleotide codes for, 445
glycogen, 173, 174, 256
  cascade of activation of breakdown of, 506
  phosphorolysis of, to glucose-1-phosphate, 267
  synthesis of, 337; rate of, increased by inhibition of phosphofructokinase, 505
glycogen synthetase, active (I) and inactive (D, phosphorylated) forms of, 507
glycolipids, 189, 196
glycolysis, of glucose to lactate, 207–8, 214, 256–66 (to other compounds, 266)
  control of rate of, through phosphofructokinase, 504–5, and phosphorylase, 505–6
  energetics of, 257, 269
  impairment of, prevents functioning of TCA cycle, 281
  in muscle, 300
  provides intermediates for syntheses, 269
glycoproteins, 183–4
glycosidic bond
  in nucleic acids, 139
  in polysaccharides, 46, 47, 71, 171
Golgi complex, 216–17
grana of chloroplasts, 250
guanidine phosphates, high-energy compounds, 26, 28
guanidinium ion, resonance of, 28
guanido group, 68
guanine (purine), 135, 136
  in DNA from different sources, 144
  hydrogen bonds between cytosine and, 141, 146, 427
guanosine (nucleoside), 135
guanosine diphosphate (GDP), 277, 397
guanosine diphosphate glucose, 333n
guanosine monophosphate (GMP), 135, 397, 435
guanosine triphosphate (GTP)
  in conversions: of IMP to AMP, 396, 397; of oxaloacetate to phospho-enolpyruvate, 330
  energy for synthesis of proteins from, 451–452, 453, 454
  phosphorylates ADP, 277–8
guanylic acid (guanosine monophosphate), 135, 397, 435
D-gulose, 167
gums, of plants, 177, 180

hydrophobic regions, in macromolecules, 56, 57, 58, 74
hydrophobic substances
'apolar' amino acids as, 66–7
lipids as, 185
in paper chromatography, 533
β-hydroxyacyl coenzyme A, 285–6
β-hydroxyacyl coenzyme A dehydrogenase, 286
3-hydroxyanthranilate, precursor of nicotinic acid, 377
β-hydroxybutyrate, 290
β-hydroxybutyryl ACP, in fatty acid synthesis, 343
β-hydroxybutyryl coenzyme A, 371
hydroxyethyl groups
carried by TPP, 370
of serine, 378
hydroxyl ions
in chemi-osmotic theory of oxidative phosphorylation, 245
in photosynthesis, 255
5-hydroxymethyl cytosine (pyrimidine), in DNA of bacteriophages, 136, 144
β-hydroxy-β-methylglutaryl coenzyme A, 372, 373
precursor of steroids, 290
p-hydroxyphenylpyruvate, in degradation of aromatic amino acids, 376
3-hydroxyproline, 61
in collagen, 67, 96

D-idose, 167
imidazole glycerophosphate, in synthesis of histidine, 380, 381
imidazole group, pK of, 69
imino groups, 61
imino-peptidases, 67
immunization, 102
immunoproteins, immunoglobulins, 99, 100
indole, and action of chymotrypsin, 123–4
indole glycerophosphate, in synthesis of tryptophan, 375–6
induction of synthesis of enzymes, in E. coli, 474–5
inhibitors of enzyme action, 119
allosteric, 131, 495, 502, 520
binding at active sites, 128–9
blocking metabolic pathways, 229–30
causing denaturation, 128
competitive, 129–31, 517–20
irreversible, inactivating incompletely, 517
non-competitive, 131, 516–17
substrates in excess as, 520
use of, in studies of: glycolysis, 258–9, 263; sites of ATP synthesis, 241; steps in respiratory chain, 238–9
initial transient, in progress curve, 511
initiation factors in protein synthesis (IF$_1$, IF$_2$, IF$_3$), 454
N-formylmethionyl tRNA binds to IF$_2$, 456
in mammalian cells, 457
inosinic acid (IMP), 395–6
hydrogen bonds between CMP and, 443

inositol, phosphoglyceride containing, 191
insects
exo-skeleton of, 92, 97–8
ratio AMP:ADP:ATP in, 504
structure of flight muscle of, 305, 309
wings of, 183
insulin, 100, 101
amino-acid sequence in, 81–2, 401
pH and solubility of, 528
inulin, 174
iodine, in test for starch, 174
iodoacetate, inhibits enzymes by alkylating SH group, 128, 258, 263
ion-exchange chromatography, 539–43
ionic atmosphere, of charged groups in solution, 528
ionic bonds, 52–3, 71–2
ionic product, of water, 13
ionic strength, and solubility, 529–30
ionization, 10–13
of amino acids, 64–6
free energy of, 18
methods of separation depending on, 534–543
of nucleotide bases, 140–1
of water, 13–14
iron
in catalase, 109n
in cytochromes, 236
in ferredoxin, 252
in haemoglobin, 104, 105
isoalloxazine ring, 237
isocitrate
activates fatty-acyl CoA carboxylase, 508
and oxidation of carbohydrate by muscle preparations, 274
in TCA cycle, 276–7, 279
isocitrate dehydrogenase, 276
isoelectric point, 528
isoelectrophoresis, 539
isoenzymes, in control of branched synthetic pathways, 500
isoionic point, 66
isoleucine
essential amino acid for mammals, 373
inhibits threonine deaminase, 500
metabolism of, 361, 362, 370–1, 382, 383
synthesis of, inhibited by valine in some bacteria, 502
trinucleotide codes for, 445
two asymmetric carbon atoms in, 64
isomerases, classification of, 110
isopropanol, fermentation product, 266
isotopes, in studies of
active-site residues in enzymes, 129
metabolism, 230–1
photosynthesis, 321

keratin, structural protein, 92, 93–5
α-keto acids
from deamination of amino acids, 359, 361
oxidative decarboxylation of, 361
transamination of, by glutamate, 357–9, 365, 369, 389

β-ketoacyl ACP reductase, 343
β-ketoacyl coenzyme A, 286, 287
keto-aldo isomerism, 166
keto-enol tautomerism, 28
  in nucleotide bases, 140
ketone bodies, 289–91
ketoses, 166, 168
ketosis, 291
β-keto thiolase, 286, 287, 371
kidney, metabolizes acetoacetate, 290
kinetic constant
  and energy barrier, 38–9
  in enzyme-catalysed reactions, 39
kinetic stability, 28
kinetics, chemical, 7, 33-41
  of denaturation of trypsin, 77
  of enzymes, 509–10; allosteric inhibition
    in, 520; competitive inhibition in, 517–
    520; non-competitive inhibition in, 516–
    517; substrate concentration in, 512–15
  progress curves in, 511–12
  symbols and terminology used in, 510
kinetosomes, of cilia and flagella, 302

lactate
  carbohydrate from, 329
  fermentation of glucose to, 207–9, 214,
    257–66; see also glycolysis
  interconversion of pyruvate and, 265, 329,
    503
lactate dehydrogenase, 329
lactonase, 293
lactose, 337
  synthesis of, 338
  system of genes in E. coli concerned with
    metabolism of (lac system), 474, 475–81
lauric acid, 185
layer line spacing, in X-ray crystallography,
    88
leucine
  essential amino acid for mammals, 373
  in haemoglobin molecules: effect of change
    of, to valine, 107
  labelled, for following protein synthesis,
    446–7
  metabolism of, 361, 362, 372, 383
  trinucleotide codes for, 445
Leuconostoc mesenteroides, dextran-producer,
    174–5
ligases, classification of, 110
light
  absorption of energy of, by chloroplast, 23,
    251–2
  wave-lengths of, absorbed in photosynthe-
    sis, 253
lignin, 177, 219
  covalent bonds in, 52
Lineweaver-Burk plots, for relation between
    initial velocity and substrate concentra-
    tion of enzyme-catalysed reaction, 514–
    515, 516, 517–18, 520
linoleic acid, 185
linolenic acid, 185

lipases
  in adipose tissue, 508
  intracellular, 282
lipids, 184–5
  food storage in, 192–3
  interaction between protein and, 199
  in membranes, 198–9, 214, 217, 222
  in mitochondria, 233
  separation of: by absorption chromato-
    graphy, 543; by gel filtration, 545; not
    separable by electrophoresis, 537
  structure of, 51, 58, 185–92, 194–7
  synthesis of, 340–9
  see also fatty acids, etc.
lipoic acid, 272–3, 277
lipopolysaccharides, 222
lipoproteins, 222
liver
  D-amino acids destroyed in, 51
  cell from, 214–19
  fatty acids degraded in, 289
  fructose phosphorylated in, 267
  glucose-6-phosphate hydrolysed in, 259,
    268
  glycogen in, 256, 260
  induction of enzymes in, 488
  triose phosphates produced in, 329
  urea cycle in, 388
lubricants, 183, 184
lyases, classification of, 110
lysine
  in bacterial cell walls, 181
  cleavage of peptide by trypsin at, 83, 111
  essential amino acid for mammals, 369
  in histones, 157; in trypsin, 132
  hydrophobic side-chain of, 70
  inhibition of aspartokinases by, 501
  interchange of arginine and, in amino-acid
    sequences, 463
  metabolism of, 361, 362, 366, 382, 383, 501
  positively charged group of, 68, 71
  trinucleotide codes for, 444, 445
lysosomes, 217
  in differential centrifugation, 228
lysozyme
  active and specificity sites on, 124
  binding of substrate to, 116–18, 121, 122
  cleavage of mucopeptide by, 114, 182
  conformation of, 75, 118–19; changes on
    binding of substrate, 127
D-lyxose, 167

macromolecules, 3, 4
  binding of smaller molecules to, 15, 33
  bonds in formation of, 51–9
  common features of, 45–7; differences be-
    tween, 47–8
  control of order of constituents in, 48
  molecular asymmetry and, 49–51
magnesium
  in complexes with ATP and ADP, 24
  in formation of acetyl CoA from pyruvate,
    271
  in phosphorylation of glucose, 259

magnesium—*contd.*
  removed by fluoride, 264
L-malate
  interconversion of oxaloacetate and, 275, 279, 353
  and oxidation of carbohydrate by muscle preparations, 274, 275
  from phospho-*enol*pyruvate, 281
  in TCA cycle, 278–9
malate dehydrogenase, 279
  mitochondrial and cytoplasmic, 353
malonate, inhibits succinate dehydrogenase, 81, 129–30, 273, 275
malonyl ACP, in fatty acid synthesis, 342, 343, 344, 345
malonyl coenzyme A, in fatty acid synthesis, 340–1, 341–2
malonyl transacelase, 342
maltose: control of system for metabolism of, in *E. coli.*, 482
mannans, 177
D-mannose, 167
  transport of, 202
marine invertebrates, $Q_{10}$ of enzymes of, 133
mass action, law of, 9–10
mass action ratio, 22
mating in *E. coli*
  donor and recipient of genetic material in, 405–7
  genetic mapping by interruption of, 408–10
mechanical energy
  conversion of chemical energy into, 300–1; structures involved, 301–16
  coupling of chemical energy with, 16, 25n
  equivalence of heat energy and, 30, 31
media for bacteria
  agar plates, 404–5, 407
  minimal, 404; colonies of wild-type recombinants on, 408
melting points, of saturated and unsaturated fatty acids, 188
membranes
  adenylate cyclase in, 506
  of bacteria, 222; of photosynthetic bacteria, 215
  cell or plasma, 214; in plants, 219
  of eucaryote cilia and flagella, 302
  hormones acting on, 100
  of mitochondria, 217, 240, 244
  no intracellular, in prokaryotes, 213
  of nucleus, 218
  passage of substances across: by diffusion, 103, 200; by mediated transport, *see* transport
  structure of, 198–9
mesosomes, of bacterial cells, 222
metabolism
  inborn errors of, 426
  intermediary, 3, 4, 5
  methods of studying: biological preparations used, 225–8; blocking of pathways, 229–30; labelling of precursors of intermediates, 230–2
  rate-limiting steps in, 492

metals
  heavy, poisonous, 31, 128
  in X-ray crystallography, 87
5, 10-methenyl tetrahydrofolate, 384, 385
methionine
  at amino terminus of 40% of *E. coli* proteins, 453; enzyme removing, from amino terminus of some proteins, 456, 457
  cleavage of peptides at, by cyanogen bromide, 83
  essential amino acid for mammals, 369
  inhibits succinylation of homoserine, 501
  in lysozyme, 75
  may be replaced by nor-valine in some micro-organisms, 81
  metabolism of, 361, 362, 368, 382, 383, 501
  as methylating agent, 385
  supplies SH group to cysteine, 369, 378
  trinucleotide code for, 445
methionyl tRNA, same code for N-formyl-methionyl tRNA and for, 454
methylamine
  methylamine ion as conjugate acid of, 12
  pK of, 12, 64
  titration curves for, 11, 12–13
methylation, by S-adenosylmethionine, 349, 385–6
β-methylcrotonyl coenzyme A, 372, 373
5-methyl cytosine, 136
  in DNA from different sources, 144
methylene blue, as hydrogen acceptor, 273
5,10-methylene tetrahydrofolate, 378, 379, 384, 385, 386
  in syntheses: of deoxyTMP, 393–4; of purine nucleotides, 395
α-methyl-β-hydroxybutyryl coenzyme A, 371
methylmalonyl coenzyme A, 289
  from valine, 371
methylmalonyl mutase, 289
5-methyltetrahydrofolate, 385, 386
Michaelis constant (measure of affinity of enzyme for substrate), 513–14
microsomes (fragments of endoplasmic reticulum with attached ribosomes), 220, 437
mimicry, molecular, 80–1
mitochondria, 23
  in animal cell, 217–18; in plant cell, 219
  in differential centrifugation, 228
  fatty-acid oxidation in, 217, 283, 340–1
  membranes of, 217, 240, 244; conveyance of metabolites across, 350, 351–5
  oxidative phosphorylation in, 23, 217, 233, 239
  TCA cycle in, 217, 278, 330n
monosaccharides, 165–9
  modifications of, in structural polysaccharides, 177–84
  ring structure of, 169–73
mucopeptide of bacterial cell walls, 181, 182
  cleavage of, by lysozyme, 114, 182
mucopolysaccharides, 183–4
muramic acid, 180
muscle
  ATP and creatine phosphate in, 300–1

muscle—*contd.*

fermentation of glucose by extracts of, 258, lactate in, 257–8

oxidation of carbohydrate in, 273–5

proteins of, 98, 312–14; during contraction, 314–16

smooth and striated, 303–4; ultrastructure of striated, 304–12

mutarotation, 170

mutations, alterations in DNA, 78, 403, 415–416, 420–1, 422

average time interval between successful, 465

in doubled genes, 468–9

in genes of *E. coli*: for lactose metabolism, 476, 478–9, 480; for tryptophan metabolism, 417–19, 444, 445–6, 483

means of producing, 81, 416, 420

nutritional, in *E. coli*: recombination between, 405–8

polypeptide-chain-terminating ('nonsense'), 458

in ribosomes, 459

'silent' (polypeptide remains functional), 417, 446

myofibrils of muscle, 304, 308

myofilaments of muscle, 308, 309, 310

thick (myosin) and thin (actin), 309, 312, 313–14

myosin, muscle protein, 98, 310, 312, 313

ATPase activity of, 313, 314, 315

myristic acid, 185

myristoyl coenzyme A, 287

neuraminic acid, 180

*Neurospora*, little clustering of genes in, 488

nicotinamide, 234

nicotinamide adenine dinucleotide (NAD, $NADH_2$), hydrogen carrier, 29, 210, 234, 238, 270

in alcoholic and other fermentations, 266

in amino-acid metabolism, 362, 365, 371–2, 377

in fatty-acid oxidation, 286, 287, 288

in glycolysis, 262, 265–6

mitochondrial membrane and, 240, 351, 352, 354

oxidations by: dihydro-orotic acid, 390; $\alpha$-glycerophosphate, 268–9; reduced lipoic acid, 273

in reversible reactions between: acetoacetate and hydroxybutyrate, 290; glycerate-1,3-diphosphate and glyceraldehyde-3-phosphate, 262, 331; oxoglutarate and glutamate, 357; pyruvate and lactate, 265–6, 329

structure of, 79

in TCA cycle, 210, 276–7, 279, 280

nicotinamide adenine dinucleotide phosphate (NADP, $NADPH_2$), reducing agent in synthetic reactions, 29–30, 295

production of reduced form of: in light reaction of photosynthesis, 248, 249, 251, 252, 253; in pentose phosphate pathway, 293, 298, 355

nicotinamide adenine dinucleotide phosphate (NADP, $NADPH_2$)—*contd.*

reduction of glycerate-1,3-diphosphate by reduced form of, in dark reaction of photosynthesis, 322, 323, 328, 345

in reversible reaction between oxoglutarate and glutamate, 357

in syntheses: of fatty acids, 295–6, 343, 344, 345; of leucine and valine, 370; of pyrimidine nucleotides, 394

nicotinic acid, 377

nitrogen

excretion products of metabolism of, 225, 387–90

use of heavy isotope of, 231, 429, 430

nonactin, cyclic antibiotic conferring increased permeability on membranes, 200, 201

non-covalent bonds, 14–15, 51–2

types of, 52–9

*see also* hydrogen bonds, ionic bonds, *etc.*

nor-leucine, may replace methionine in some micro-organisms, 67

nucleases, endo- and exo-, 139

*see also* deoxyribonucleases, ribonucleases

nucleic acids, 2, 45, 58

bases in, 139–43

control over nucleotide sequence in, 48

ends of (3' and 5'), 46, 138; in replication, 431–3

hydrolysis of, at different sites, 137–8, 139, 163–4

hydrophobic interactions in structure of, 57

separable electrophoretically, 537

structure of, 135–9, 140n; methods of determining, 140, 161, 163–4

*see also* deoxyribo- and ribonucleic acids

nucleolus, 218–19

nucleoside diphosphate sugars, 334–5

nucleosides, 135

nucleotide phosphototransferase, 391

nucleotides

bases in, 140–2; minor, 142, 143

compounds mimicking, 81

in nucleic acids, 2, 46, 47–8, 135; not in nucleic acids, 143–4

ribose of, 295

synthesis of: purine, 394–7; pyrimidine, 390–4

nucleus, 218

oleic acid, 185

oligomycin, inhibitor of respiratory chain, 243

oligonucleotides, 135

separation of, by ion-exchange chromatography, 540, 541, 542

operator gene, site of DNA to which repressor of adjacent genes is bound, 476, 479, 480

mutations in, 478, 480, 483

operon, set of contiguous genes, together with operator gene, transcribed into one molecule of mRNA, 478, 484–5 operon—*contd.*

*lac*, of *E. coli*: negative control of, by repressor, 477–80; positive control of, by carbon source, through cAMP, 480–1

*trp*, of *E. coli*, 484

optical isomers, 49

order: imposition of, on random situation, requires energy, 31–2

organelles of cell, 214, 350

in differential centrifugation, 228

organs, perfusion of, 226, 227

orientation of reacting molecules, enzyme and, 120–1

ornithine, 364

arginine from, in three reactions, 388–9

from arginine, in one reaction, 389

orotate, in synthesis of pyrimidines, 390

orotidine-5′-phosphate, 390

ouabain, specific inhibitor of transport of sodium ions, 318–19

oxaloacetate

carboxylation of pyruvate to, 281, 330, 339, 383, 506–7

formed by reaction of citrate with ATP and CoA, 354

interconversions: of aspartate and, 357n, 365, 389; of malate and, 275, 279, 353

and oxidation of carbohydrate by muscle preparations, 274, 275

phosphorylative decarboxylation of, to phospho-*enol*pyruvate, 330

reacts with acetyl CoA to form citrate, 275, 276, 278, 354

from some amino acids, 361

supply of, necessary for TCA cycle, 276

oxalosuccinate, in TCA cycle, 276–7, 279

oxidation

coupling of reduction and, 28–30

denaturation of proteins by, 77

donation of electrons as, 23

energy from, 233

oxidative phosphorylation, 23, 24, 29, 233–4, 270

enzymes of: in bacterial membrane, 222; in mitochondria, 23, 217, 233, 239

hydrogen and electron transport in, 234–8

mechanism of: on chemical coupling hypothesis, 242–3; on chemi-osmotic hypothesis, 244–6

methods for studying, 238–9

in plants, occurs only in the dark, 219

sites of ATP synthesis in, 239–41

oxidoreductases, classification of, 110

α-oxoadipate, in amino-acid metabolism, 366, 367, 377

α-oxobutyrate, in amino-acid metabolism, 368, 369, 370

α-oxoglutarate

from amino acids, 361

interconversion of glutamate and, 357, 387

and oxidation of carbohydrate by muscle preparations, 274

in TCA cycle, 276–7, 279

transamination of α-keto acids by couple of glutamate and, 357–9, 365, 369, 389

α-oxoisovalerate, 370, 372

oxygen

evolved in photosynthesis, 247, 248

as electron acceptor from cytochromes, 23, 237, 238

as final hydrogen acceptor in oxidation of glucose, 270

haemoglobin and, 104–6, 127

number of molecules of ATP produced, per atom of, reduced (P/O ratio), 239, 240, 245

use of heavy isotope of, 231

palmitic acid, 185

palmitoleic acid, 185

palmitoyl coenzyme A, 287

pantothenic acid, in coenzyme A, 271

partition constant of a substance, between phases, 531

pectic acid, 177

pectins, 177, 219

penicillin, 81, 131

pentose phosphate: conversion of, to hexose phosphate, 296–9

pentose phosphate isomerase, 294

pentose phosphate pathway (hexosemonophosphate shunt, phosphogluconate pathway), 292–4

enzymes of, in cytoplasm, 355

functions of, 294–9

pentoses, 167, 168

from glucose, 209, 210

in nucleic acids, 46, 47

pepsin, pH optimun of, 132

peptidases, 67, 84

peptide bonds, 45, 71

as backbone of protein, 124

in clotting of blood, 99

formation of, in synthesis of proteins, 451, 452

peptides

in determination of primary structure of proteins: by overlap method, 81–3; by stepwise method, 83–5

relations between mass, charge, and electrophoretic mobility of, 537

peptidyl transferase, component of ribosome, 452, 453

pesticides, in wax on fruit, 191

pH, 11, 13–14

denaturation of protein by extremes of, 77, 128

in electrophoresis, 538, 539

and enzyme action, 131–2

variations in solubility with, 528–9

phenotype, 403

phenylacetic acid, product of metabolism of phenyl-labelled amino acids with even number of carbon atoms, 283

phenyl alanine

esential amino acid for mammals, 376

metabolism of, 361, 362, 373–4, 376, 382, 383

trinucleotide codes for, 444, 445

phenylalanine hydroxylase
    gene for, 424–6
    lacking in phenylketonuria, 376–7, 402, 424
phenylisothiocyanate, in Edman reaction, 84
phenylketonuria, 377, 404, 424–5
phenyl pyruvate, in phenylketonuria, 377
phosphagens, 300n
phosphate
    in casein, 92; in nucleic acids, 46
    in phosphorolysis of glycogen and starch,
        268, 269
phosphate esters, do not pass membranes
    easily, 259
phosphatidic acids, 189, 190
    substituents on, 191
    in syntheses: of phospholipids, 347–8; of
        triglycerides, 346–7
phosphatidyl choline, 191, 348, 349
phosphatidyl ethanoline, 348, 349
    methylation of, 349, 385
phosphatidyl serine, 349
phosphodiester bond, in nucleotides, 46, 71,
    136, 137–9
phospho-*enol*pyruvate, high-energy com-
    pound, 26, 27, 28
    carboxylation of, to malate, 281
    conversion of pyruvate to, 329–30
    in formation of aromatic amino acids, 269
    interconversion of glycerate-2-phosphate
        and, 264, 330
    from oxaloacetate, 330
    phosphorylation of ADP by, 27, 263, 264
    phosphorylation of sugars by, in bacteria,
        319n
    pyruvate from, 24, 264
    reaction of, with erythrose-4-phosphate,
        373
phospho-*enol*pyruvate carboxykinase, 330,
    353–4
phospho-*enol*pyruvate family of amino acids,
    375–7
phosphofructokinase, 260
    control of glycolysis through, 208, 504–5
phosphoglucomutase, 267, 268, 334
phosphoglycerate kinase, 263, 331
phosphoglycerides, 189, 195
phosphoglycollate, analogue of activated
    complex in action of triosephosphate
    isomerase, 116
phospholipids, synthesis of, 347–9
4'-phosphopantetheine, in acyl carrier pro-
    tein, 341
5-phosphoribosylamine, 394
    synthesis of, inhibited by purine nucleo-
        tides, 500
phosphoribosyl - 5 - amino - 4 - imidazole -
    carboxamide (AICAR) in syntheses: of
    histidine, 380–1; of purine nucleotides,
    396
5-phosphoribosylanthranilate, 375
N'-phosphoribosyl ATP, 379, 380
phosphoribosyl-ATP pyrophosphorylase, in-
    hibited by histidine, 499
5-phosphoribosyl-glycineamide, 394

5-phosphoribosyl pyrophosphate, 375
    in syntheses: of histidine, 379–80; of
        purine nucleotides, 394; of pyrimidine
        nucleotides, 390; of tryptophan, 375
phosphoric-anhydride bond, free energy of
    hydrolysis of, 24
phosphorolysis, of glycogen and starch, 267
phosphorylase, catalysing breakdown of
    glycogen to glucose-l-phosphate, 267,
    505
    catalytic power of, 109
    interconversion of *a* (active, phosphory-
        lated) and *b* (inactive) forms of, 505–6
phosphorylase *b* kinase, active (phosphory-
    lated) and inactive forms of, 505–6
phosphorylase *a* phosphatase, 505
phosphorylation
    oxidative, 23, 24, 29; *see also* oxidative
        phosphorylation
    photosynthetic, 24, 247–55; *see also*
        photosynthesis
    substrate-level, 27, 242, 262, 264, 278
phosphorylcholine, 347
phosphorylethanolamine, 347
phosphoryl-glucoside bond, high-energy, 335
phosphorylglyceromutase, 263–4, 330
phosphoserine, 377
phosphoserine phosphatase, 377, 413
    mapping of gene for, in *E. coli*, 414
phosphoserine transaminase, 377, 413
    mapping of gene for, in *E. coli*, 414
photosynthesis, primary source of free energy,
    23
    apparatus of, 250–1
    dark and light reactions in, 247–50
    details of light reaction of, 251–5
    synthesis of carbohydrate in dark reaction
        of (Calvin cycle), 321–9
    use of isotopes in study of, 231
phycobilins, 251
phylogenetic trees, from differences between
    amino-acid sequences for pairs of
    species
    in cytochrome *c*, 463–5
    in fibrinopeptide A, 465, 466–7
pigments, photosynthetic, 247
    systems of: I, 251–2; II, 253–5
pipecolic acid, 366, 367
pK, 12, 14
    for ionizing groups: of ATP, 27; of gluta-
        mate in lysozyme, 119, 122; of glycine,
        64–6; in proteins, 68–9
plants
    can convert acetyl CoA to carbohydrate,
        339
    cells of, 219–20
    cytochrome *f* of, 254
    proteins of, in diets, 360n
    synthesis of methionine in, 378
    *see also* chloroplasts, photosynthesis
plasmalogens, 186, 189, 196
plastoquinone, 254
β-pleated sheet structure in proteins, 72, 74,
    92, 95

β-pleated sheet structure in proteins—*contd.*
  disrupted by proline, 67
  interlocking of side-chains in, 95–6
ploidy, dominant and recessive genes and, 423–6
*Pneumococcus:* transfer of virulence to non-virulent strain of, by means of DNA, 402–3
poisons, as inhibitors of enzymes, 131
polarized light; rotation of plane of, by enantiomers, 49, 64n
polyamines, associated with nucleic acids, 148
polypeptides
  one gene for each, 414, 422
  sequence of amino acids in, corresponds to sequence of nucleotides in gene, 418, 422
polyribosomes, 457
polysaccharides, 45, 46–7, 48
  food storage in, 173–6
  reducing and non-reducing ends of, 47
  structural, 176–7
  structure of, 171
polystyrene, for ion-exchange chromatography, 540
  sulphonated, 541
porphyrin ring, 104
  compounds containing, 79, 80
  in cytochromes, 236
potassium ion
  active transport of, 318–19
  free passage of, through membranes, 200
potato ring necrosis virus, without protein, 159
prephenate, in synthesis of phenylalanine, 374
procaryotes
  cells of, 213, 221–3
  subunits of ribosomes of, 438
product inhibition of enzymes, 505
progress curves, of chemical reactions, 511–512
proinsulin, 100, 101
proline
  in collagen, 96
  imino group in, 61, 67
  mapping of genes for enzymes for synthesis of, in *E. coli,* 414
  metabolism of, 361, 362–3, 382, 383
  stereochemistry of, 61
  trinucleotide codes for, 444, 445
promoters, sites on DNA at which RNA polymerase binds to initiate transcription, 479
  binding of RNA polymerase to, favoured by binding of protein-cAMP complex, 481
  with different affinities for RNA polymerase, 481
propionic acid, pK of, 12
propionyl coenzyme A
  from amino acids, 368, 371
  carboxylation of, 289
  from oxidation of odd-numbered fatty acids, 289
prosthetic groups of proteins, 78–80

proteases
  and imino peptide bond, 67
  in lysosomes, 217
  relative specificity of, 82–3, 83–4, 111
protein kinase
  activated by cAMP, 505–6, 507
  phosphorylations by: glycogen synthetase I, 507; phosphorylase *b* kinase, 505
proteins, 2, 45, 47, 48, 58
  associated with nucleic acids, 157, 218
  binding other molecules, 99–107; binding cAMP, 481; binding cytochromes, 236
  carrier, 100, 103–4, 320
  co-factors and prosthetic groups of, 78–80
  in cytoplasm, 214
  ends of, amino and carboxyl, 45; amino end of, coded for by 5′ end of mRNA, 447; carboxyl end of, synthesized last, 446–7
  evolution of amino-acid sequences in, 460–467; gene doubling in, 468–70
  food storage in, 91–2
  globular and fibrous, 71, 99
  interactions between lipids and, 199
  in membranes, 198, 199, 202, 214, 217, 222
  metabolic turnover of, 231
  in mitochondria, 233
  in nucleus, 218
  phosphorylated (HPr), used by bacteria for phosphorylating sugars, 319n
  of plants and animals, essential amino acids in, 360n
  in ribosomes, 438
  separable by electrophoresis, 537
  structural, 92–9
  structure of, 60; interactions controlling, 57, 70–6; primary, 70, 81–5; proline and, 67; quaternary, 73, 75–6, 159; secondary, 72; tertiary, 70–1, 76–8, 85–90
  synthesis of: control of amino-acid sequence in, 48, 149, 401–2, 427; control of rate of, 425; elongation factors in, 451; initiation of, 453–7; mechanism of, 446–453; ribosomes as site of, 437, 440, 446, 457, 459; termination of, 445, 458–9
  of virus coats, 159, 163, 223; of R 17 bacteriophage, 486–7
proton magnetic resonance spectroscopy, of lipids, 188
protons, 10
  concentration of, in cell, 122
  sites on enzymes binding, 119, 122
D-psicose, 168
purine nucleotides
  inhibit synthesis of 5-phosphoribosyl-amine, 500
  synthesis of, 394–7
purine ring, 135
purines
  on paper chromatograms, 532
  synthesis of, 210, 379, 383, 385; origins of carbon atoms in, 397
  *see also* adenine, guanine
pyranose-furanose isomerism, 170–1

pyranose ring
  'chair' and 'boat' forms of, 119, 171
  in transport of sugars, 202
pyridoxal phosphate, 79
  coenzyme: in threonine metabolism, 369;
    of transaminases, 358
pyridoxamine phosphate, in transmination,
  358
pyridoxine (vitamin), 358
pyrimidine nucleotides
  CTP in control of metabolism of, 497
  synthesis of, 390–4
pyrimidine ring, 135, 140
pyrimidines
  on paper chromatograms, 532
  synthesis of, 210, 280; aspartate carbamy-
    lase as first enzyme specific to, 495, 499
pyrophosphatase, ubiquitous, 284, 285
pyrophosphate
  from ATP in syntheses: of fatty-acyl AMP,
    284; of fatty-acyl tRNAs, 440
  from nucleoside triphosphates in synthesis
    of nucleic acids, 431, 435
  from 5-phosphoribosyl pyrophosphate in
    formation of pyrimidine nucleotides,
    390–1
  reactions producing, go to completion, 285
  from UTP in formation of UDP-glucose,
    334
pyrrole derivatives, synthesis of, 280
δ-pyrroline 5-carboxylic acid, 363
pyruvate
  to acetaldehyde in alcoholic fermentation,
    266, 272
  from amino acids, 361, 379; in metabolism
    of amino acids, 366, 370
  carbohydrate from, 329
  carboxylation of, to oxaloacetate, 281, 330,
    339, 383, 506–7
  conversion of, to phospho-enolpyruvate,
    329–30
  interconversions: of alanine and, 357n, 369;
    of lactate and, 265, 329, 503
  from phospho-enolpyruvate, 24, 264
  produced in cytoplasm, conveyed into
    mitochondria, 351–2; oxidized in mito-
    chondria, 270, 352
  reaction of, with coenzyme A, 271–3
pyruvate carboxylase, 281, 330
  control of gluconeogenesis through, 506–7
pyruvate dehydrogenase system, 271–3
pyruvate family of amino acids, 369–73


$Q_{10}$
  for denaturation of proteins, 133
  for enzyme-catalysed reactions, 133
  for non-enzymic reactions, 40

rate-limiting step in a metabolic pathway, 492
rate constant of a chemical reaction, 34, 35
  energy barrier and, 37, 39
reaction intermediates, unstable compounds,
    35–9, 207, 242

reactivity, abnormal, of amino-acid residues
    in enzymes, 118–20
recombination
  genetic mapping by frequencies of, 410–
    413
  of nutritional mutants of E. coli, 405–8
  within a gene, 416–17
reducing agents, required for
  fatty acid synthesis, 295
  fixation of carbon dioxide as carbohydrate,
    247, 248
reduction
  denaturation of proteins by, 77
  involved in many synthetic processes, 28–9
regulation, systems of, 5
  of activity of enzymes, 473–4, 490; in
    branched synthetic pathways, 500–2; in
    carbohydrate and fat metabolism, 502–4,
    507–8; in gluconeogenesis, 506–7; in
    glycolysis, 504–6; in multi-enzyme pro-
    cesses, 490–4; in unbranched synthetic
    pathways, 495–500
  displacement from equilibrium as sign of,
    494–5
  facilitated by splitting metabolism into
    small steps, 208
  heterotropic, of enzymes by molecules
    other than substrates, and homotropic,
    of enzymes by substrates, 127
  intracellular compartmentation in, 351
  potential points of, 285, 295, 490
  of rate of synthesis of enzymes, 473–4; by
    induction and repression, 474–5, (in
    bacteria) 475–87, (in eucaryotes) 487–9
  of synthesis: of malonyl CoA, 342; of
    pyrimidines, 394
release factors, terminating protein synthe-
    sis, 458
rennin, 91
repressors of enzyme synthesis
  inducible systems
    for lac enzymes in E. coli, 477, 479–80;
      mutants of, 478–9; number of molecules
      of, per cell of wild-type E. coli, 480, 481;
    rate of synthesis of, 481
    for systems other than lac, 482
  repressible systems
    for maintaining specialization of cells, 488
    postulated to be inactive until inducer is
      added, 482–3
resilin, structural protein, 92, 93, 97
resonance
  in high-energy compounds, 27–8, 68
  stabilization of peptide bond by, 71
respiratory chain, 233–4
  hydrogen and electron transport in, 234–8
  methods of studying, 238–9
  reoxidation of $NADH_2$ through, 270, 277
  sites of ATP synthesis in, 239–41
revesible chemical reactions, 9, 10
  approach to equilibrium by, 17; free energy
    in, 19–22
L–rhamnose, 177, 179
  transport of, 202

sedoheptulose-1, 7-diphosphate, 325
  hydrolysis of, 326
sedoheptulose-7-phosphate
  in pentose pathway, 296, 297
  reaction of, with glyceraldehyde-3-phosphate, 326
  from sedoheptulose-1,7-diphosphate, 326
selection, of bacterial strains on agar plates, 404–5, 407
semi-permeability, 200
separation of biochemical compounds, 521–2, 545–6
  by centrifugation, 522–5; zonal methods, 525–7
  by differences in molecular dimensions, 543–4; dialysis, 544; gel filtration, 544–5, 546
  by differences in solubility, 527–31; counter-current distribution, 531; gas-liquid chromatography, 534; paper chromatography, 531–4
  by differences in state of ionization: electrophoresis, 534–9; ion-exchange chromatography, 539–43
serine
  in casein, esterified with phosphate, 92
  configuration of, 64
  in enzymes: acylation of, 119, 128; inactivation of enzyme by change of OH to SH in, 69
  hydroxyl group of, in protein structure, 67–8
  metabolism of, 269, 361, 362, 377, 378, 379, 382, 383
  phosphoglyceride containing, 191
  in synthesis of tryptophan, 376
  three enzymes required for synthesis of, 377, 412; mapping of genes for, in E. coli, 414; three mutant types of E. coli corresponding to, 413–14
  three tRNAs for, in E. coli, 444
  trinucleotide codes for, 444, 445
serine proteases, proposed mechanism for action of, 120, 121
shikimate, in synthesis of aromatic compounds, 374
shunting of metabolites, to alternative pathway, 493
shuttles, across mitochondrial membrane, 352, 354
sigma unit of RNA polymerase, for initiation of RNA synthesis, 436
sigmoid relationship
  for allosteric enzymes, 128
  between partial pressure of oxygen and saturation of haemoglobin molecule with oxygen, 105–6
  suggests co-operativity of binding of substrate molecules, 496
silk, structural protein, 92, 95–6
soaps, 193
sodium ion
  active transport of: alone, 319; coupled to potassium ion, 318–19
  free passage of, through membranes, 200

solubility
  as limit to concentration of reactants, 9
  separation of substances by differences in, 527–34
solvents, organic, in separations, 530–4
D-sorbose, 168
spectroscopy, of intermediates in respiratory chain, 239
spermidine, 148
spermine, 148
sphingamine, dihydro-, 187, 188
sphingolipids, 190, 197
sphingosines, 187, 188, 189
Staphylococcus aureus, cell walls of, 181
starch, 171, 173, 256
  glucose-1-phosphate from, 267
starvation, ketone bodies in, 291
steady state, of a chemical reaction, 22, 491, 511, 512
stearic acid, 185
  pathway of oxidation of, 283–9
stearoyl ACP, 344
stearoyl coenzyme A, 284
stereochemistry, 49–51
  of amino acids, 61, 64
  of phosphoglycerides, 189
  of sugars, 166–8
sterols, 188–9, 194
  acetyl CoA in synthesis of, 271
  $NADPH_2$ in synthesis of, 295
stomach, pH in, 14
stroma of chloroplasts, 250
substrate
  effect of concentration of, in enzyme reactions, 512–15
  inhibition by excess of, 520
succinate
  and oxidation of carbohydrate by muscle preparations, 274, 275
  in TCA cycle, 277–8, 279
succinate dehydrogenase, 278, 280
  inhibited by malonate, 81, 129–30, 273, 275
succinyl coenzyme A
  from amino acids, 361, 368, 369, 371
  from methylmalonyl CoA, 289
  reaction of, with acetoacetate, 290
  in TCA cycle, 277, 279, 281
O-succinyl homoserine, 367, 501
succinyl thiokinase, 277
sucrose
  kinetics of hydrolysis of, 34–5
  synthesis of, 338
sucrose-6-phosphate, 338
sugars
  clustering of genes for enzymes metabolizing, in E. coli, 415
  ring structure in, 169–71
  separation of, on paper chromatograms, 532
  stereochemistry of, 50, 166–8
  transport of, across membranes, 202–3, 319
sulphamic acid group, 179

sulphation
  of galactose in cerebrosides, 190
  of monosaccharides, 179
sulphide: fixation of, by micro-organisms,
    378–9
sulphonamides, 80–1, 130
sulphydryl group
  in coenzyme A, 271
  in triosephosphate dehydrogenase, 262
  see also disulphide bridges
superhelix, in collagen, 96
syntheses
  favoured by reducing conditions, 332
  intermediates for, produced in catabolism,
    3, 208, 257, 269; in TCA cycle, 280–1, 394
  NADPH$_2$ as reducing agent in, 29–30, 295
  repression of enzymes involved in, by end-
    products of, 474
syringaldehyde, in lignin, 177

D-tagatose, 168
D-talose, 167
tanning, of insect exo-skeleton by cross-link-
    ing, 92, 97–8
teichoic acids, 182
temperature
  and degree of saturation of body fats, 199
  and rate of enzyme action, 132–4
tendons, 92
tetrahydrofolate, 378, 381
  as coenzyme in transfer of one-carbon
    fragments, 384, 397
  structure of, 386
tetrahydropicolinic acid, 366
thermodynamic instability, of high-energy
    compounds, 28
thermodynamics, 7
  of affinity, 7–8
  of concentration of reactants, 8–10
  of equilibria, 10–15; of relations between
    energy and, 16–18
  of free energy, 19–22; in denaturation of
    protein, 31, 77; sources and coupling of,
    22–30; thermodynamic significance of,
    30–3
thiamine, as vitamin, 271
thiamine pyrophosphate (TPP), 79
  in amino acid metabolism, 370
  coenzyme: in oxidative decarboxylations of
    oxoglutarate, 277, and pyruvate, 266, 271,
    272; for transketolase in conversion of
    pentose phosphate to hexose phosphate,
    297, 323
thiazole ring, 271
thioesters, high-energy compounds, 26, 28,
    242, 262, 273
thioether bonds, in attachment of haem to
    cytochrome, 69
thiogalactoside transacetylase, enzyme with
    unknown function, 479
thiohemiacetal, oxidized to high-energy thio-
    ester, 242
thiokinases, 284
thiolases, 286, 288, 290

thiolysis, 286
thioredoxin (protein), 392
  catalyst in conversions of: ADP to
    deoxyADP, 397, CDP to deoxyCDP,
    392, and GDP to deoxyGDP, 397
threonine
  essential amino acid for mammals, 369
  inhibits aspartokinase, 501
  metabolism of, 361, 362, 368–9, 382, 383,
    501
  trinucleotide codes for, 445
  two asymmetric carbon atoms in, 64
threonine deaminase, inhibited by isoleucine,
    500
threonine dehydratase, 369
  induction of, in liver, 488
D-threose, 167
thrombin, 119
thymidine (nucleoside), 135
thymidine monophosphate (TMP), 135, 435
thymidylic acid (thymidine monophosphate),
    135, 435
thymine (pyrimidine), 135, 136
  can be replaced by uracil in base-pairs, 149
  in DNA from different sources, 144
  hydrogen bonds between adenine and, 141,
    146, 427
tissues
  homogenates of, 226–7; differential centri-
    fugation of, 228
  slices of, 226
titration curves, for acetic acid and methyl-
    amine, 11, 12–13
tobacco mosaic virus, 159, 162
trace elements, 80
transaldolase, 296, 297, 324
transaminases, 357–8
transamination, 357–9, 382
trans and cis forms of unsaturated fatty acids,
    188
transferases, classification of, 110
transketolases, 296–7, 323, 325
translocation, of peptidyl tRNA and mRNA,
    in protein synthesis, 452–3
transport across membranes
  active, 16, 103, 203, 204, 214, 222, 316–17;
    of ions, 317–19; of neutral substances,
    319–20
  exchange diffusion or coupled, 203, 204,
    317
  in mitochondria, 350, 351–5
  passive non-coupled, 201–3
tricarboxylic acid (Krebs) cycle (TCA cycle),
    208–9, 273–80
  intermediates of, in syntheses, 280–1, 394
  links of amino-acid metabolism with, 357,
    383
  in mitochondria, 217, 278, 330n
  resemblances of stages in fatty acid oxida-
    tion to last three reactions of, 286
triglycerides, 189
  hydrolysis of, 282; in cytoplasm, 352
  synthesis of, 346–7; in endoplasmic reticu-
    lum, 350

trinucleotides in mRNA, binding of specific amino-acyl tRNAs to, 419–22, 444–6
triose phosphate
    synthesis of: in Calvin cycle, 321–9; from gluconeogenic precursors, 329–32
    synthesis of carbohydrate from, 332–8
triosephosphate dehydrogenase, 119, 262, 278, 331
    of chloroplast, 322; specific for NADP, 323
triosephosphate isomerase, 261, 324, 503
    binding of substrate to, 116, 120, 122
    inhibitor for, 128–9
tropomyosin, muscle protein, 313, 314
    in muscle contraction, 315–16
troponin, muscle protein, 313, 314
    binds calcium, 314
trypsin
    amino-acid sequences of chymotrypsin and of, 468
    denaturation of, 77
    inhibitor for, 128
    specificity of cleavage of peptides by, 82–3, 111, 112, 119
tryptophan
    cleavage of peptides at, by chymotrypsin, 123
    clustering of genes for five enzymes synthesizing, 414–15, 473
    essential amino acid for mammals, 376
    indole nitrogen of, 67
    inhibits anthranilate synthetase, 500
    in lysozyme, 75
    metabolism of, 361, 362, 373–4, 374–6, 383
    represses synthesis of enzymes synthesizing, 474, 475
    synthesis of, hypothesis for mechanism of control of, 483–4
    trinucleotide code for, 445
tryptophan pyrrolase: induction of, in liver, 488
tryptophan synthetase, 376
    $\alpha$- and $\beta$-subunits of, 414
    amino-acid sequence of $\alpha$-subunit of, and corresponding DNA map of mutations in, 417–19, 444, 445–6, 468
tubules, in cytoplasm, 214
tyrosine
    hydroxyl group of, 67; cross-linking through, 97
    metabolism of, 361, 362, 373–4, 376, 382, 383; albinism from defect in, 426
    trinucleotide codes for, 445

ubiquinone, 80
    between flavoproteins and cytochromes in respiratory chain, 237, 238, 239
ultraviolet light
    absorption photographs by, of DNA of different densities, 433
    production of mutations by, 416
uncouplers, of oxidative phosphorylation, 243
uracil (pyrimidine), 135, 136, 497
    can replace thymine in base pairs, 149

urea, excretion product of nitrogen metabolism, 225
    added to solutions for ion-exchange chromatography, overcomes all but electrostatic interactions, 541
    cycle involved in synthesis of, 387–90
urease, specificity of, 111
uric acid, excretion product of nitrogen metabolism, 225
    in birds, 388n
uridine (nucleoside), 135
uridine diphosphate (UDP), phosphorylated by ATP, 335
uridine diphosphate galactose, 337
    interconversion of UDP glucose and, 337, 338
    phosphorylation of, 391
    reaction of, with glucose, to give lactose, 338
uridine diphosphate glucose, 143
    interconversion of UDP galactose and, 337, 338
    reactions of: with fructose-6-phosphate, 338; with galactose-l-phosphate, 337
    synthesis of glycogen from, 337, 507
uridine diphosphateglucose glycosyl transferase, 335
uridine diphosphateglucose pyrophosphorylase, 334
uridine monophosphate (UMP), 135, 391
    in RNA, taking place of TMP in DNA, 435, 436
uridine triphosphate (UTP), 334, 337
    amination of, to CTP, 391
    from UDP, 391
uridylic acid (uridine monophosphate), 135, 391, 435, 436
urocanate, in breakdown of histidine, 381
uronic acids, 179

vaccenic acid, 185, 188
vacuoles
    of Golgi complex, 216, 217
    of plant cell, 219
valine
    essential amino acid for mammals, 373
    inhibits $\alpha$-aceto-$\alpha$-hydroxyacid synthetase, 502
    metabolism of, 361, 362, 370–1, 383
    trinucleotide codes for, 445
Van der Waals forces, 54–5, 115, 118
vanillin, in lignin, 177
viruses, 223–4
    DNA of, 156, 159, 223
    encapsulation of nucleic acid of, 99
    RNA of, 159, 223, 422, 486–7
    *see also* bacteriophages
vitamin $B_{12}$, 289
vitamin H (biotin), 79
vitamin K, 80
vitamins, 79, 80, 234, 271, 281, 289
    concentration of, by active transport, 319
    distinction between essential amino acids and, 80, 360

waste products, produced in fermentations, 257, 266
water
  active transport of, 319
  convention for concentration of, in equilibria involving, 13, 34
  dielectric constant of, 52, 53, 54
  displacement of, from enzyme binding sites, 33, 119
  elimination of, in synthesis of macromolecules, 45
  hydrogen bonding of: to cellulose, 531; to nucleotide bases, 41
  interactions: of hydrophilic groups with, 57; of ionized groups of proteins with, 72, 77
  ionization of, 13–14
  labelled, for study of photosynthesis, 248
  lattice structure of, 56
  in living systems, 59
  manipulation of concentration of, in chemiosmotic theory of oxidative phosphorylation, 245
  in photosynthesis, 247, 253, 254
  stabilization of interface between membranes and, 199
waxes, 186, 191, 194
'wild-type' (normal), as distinct from mutant organisms, 405
  from recombination of mutants, 405
wood, structural carbohydrates in, 177

xanthylic acid, 397
X-rays
  diffraction patterns obtained with: for DNA, 145–6; for flagellin, 301; for keratin, 93–4; for muscle, 312; for β-pleated sheet structure, 96; for proteins, 75, 78, 85–90; for viruses, 159
  production of mutations by, 416
xylans, 177, 219, 334
D-xylose, 167
  transport of, 202
L-xylose, transport of, 202
D-xylulose, 168
xylulose-5-phosphate
  interconversion of ribulose-5-phosphate and, 294, 325, 326
  reactions of: with erythrose-4-phosphate, 297–8; with ribose-5-phosphate, 296

yeast
  action of, on pyruvate, 266, 272
  cytochrome c of, 461
  fermentation of glucose by cell-free extract from, 258
  genes of, 415; repression of, 488
  tRNAs of, 443

zwitterion structure, 66
zygotes
  of E. coli, 405
  of other organisms, 413